Generalized Concavity
in
Optimization and Economics

Papers Presented at the Proceedings of the NATO Advance Study Institute Held at the University of British Columbia, Vancouver, Canada, August 4-15, 1980.

Generalized Concavity in Optimization and Economics

Edited by

Siegfried Schaible

Faculty of Business Administration and Commerce
University of Alberta
Edmonton, Alberta, Canada

William T. Ziemba

Faculty of Commerce and Business Administration
University of British Columbia
Vancouver, B.C., Canada

ACADEMIC PRESS
A Subsidiary of Harcourt Brace Jovanovich, Publishers
New York London Toronto Sydney San Francisco 1981

ACADEMIC PRESS, INC.
111 Fifth Avenue, New York, New York 10003

United Kingdom Edition published by
ACADEMIC PRESS, INC. (LONDON) LTD.
24/28 Oval Road, London NW1 7DX

Library of Congress Cataloging in Publication Data
Main entry under title:

Generalized concavity in optimization and economics.

Proceedings of the NATO Advanced Study Institute held at the University of British Columbia in Vancouver, Canada, Aug. 4-15, 1980.

1. Mathematical optimization--Congresses. 2. Concave functions--Congresses. 3. Economics, Mathematical--Congresses. I. Schaible, Siegfried. II. Ziemba, W. T. III. NATO Advanced Study Institute (1980 : University of British Columbia)
QA402.5.G45 519.4 81-17644
ISBN 0-12-621120-5 AACR2

PRINTED IN THE UNITED STATES OF AMERICA

81 82 83 84 9 8 7 6 5 4 3 2 1

Dedicated to our wives
Ingrid and Sandra

Contents

V. FRACTIONAL PROGRAMMING

VI. APPLICATIONS OF GENERALIZED CONCAVITY IN MANAGEMENT SCIENCE AND ECONOMICS

VII. APPLICATION TO STOCHASTIC SYSTEMS

Contributors

Numbers in parentheses indicate the pages on which the authors' contributions begin.

M. AVRIEL (21), Faculty of Industrial Engineering and Management, Technion–Israel Institute of Technology, Technion City, Haifa 32000, Israel

A. BEN-ISRAEL (301), Department of Mathematics, University of Delaware, Newark, Delaware 19711

A. BEN-TAL (301), Faculty of Industrial Engineering and Management, Technion–Israel Institute of Technology, Technion City, Haifa 32000, Israel

J.M. BORWEIN (335, 379), Department of Mathematics, Carnegie-Mellon University, Pittsburgh, Pennsylvania 15213

S.L. BRUMELLE (399), Faculty of Commerce and Business Administration, University of British Columbia, Vancouver, B.C., Canada V6T 1W5.

A. CAMBINI (491), Department of Operations Research, University of Pisa, Piazza Dei Cavalieri, 2, Pisa, Italy

B.D. CRAVEN (473, 661), Department of Mathematics, University of Melbourne, Parkville, Victoria 3032, Australia

J.-P. CROUZEIX (109, 199, 207), Département de Mathématiques Appliquèes, Université de Clermont, B.P. 45, F-63170 Aubière, France

W.E. DIEWERT (21, 51, 511), Department of Economics, University of British Columbia, Vancouver, B.C., Canada V6T 1W5

F. DI GUGLIELMO (281), Faculté des Sciences d'Avignon, 33 Rue Louis Pasteur, F-84000 Avignon, France

W. EICHHORN (627, 637), Institut für Wirtschaftstheorie und Operations Research, Universität Karlsruhe, Kaiserstrasse 12, D-7500 Karlsruhe 1, West Germany

J.A. FERLAND (169, 199), Département d'informatique et de recherche opérationelle, Université de Montréal, Case Postale 6128, Montréal, P. Q., Canada H3C 3J7

W. GEHRIG (637), Institut für Wirtschaftstheorie und Operations Research, Universität Karlsruhe, Kaiserstrasse 12, D-75 Karlsruhe 1, West Germany

T. IBARAKI (441), Department of Applied Mathematics and Physics, Faculty of Engineering, Kyoto University, Kyoto 606, Japan

R. G. JEROSLOW (689), College of Management, Georgia Institute of Technology, Atlanta, Georgia 30332

J.G. KALLBERG (719), Graduate School of Business, New York University, 100 Trinity Place, New York, New York 10006

Y. KANNAI (543), Department of Theoretical Mathematics, Weizmann Institute of Science, Rehovot, Israel

S. KARLIN (703), Department of Mathematics, Stanford University, Stanford, California 94305

P.O. LINDBERG (153), Department of Optimization, Royal Institute of Technology, S-1044 Stockholm, Sweden

D.H. MARTIN (95), NRIMS–CSIR (National Research Institute, Mathematical Sciences), P.O. Box 395, Pretoria 0001, South Africa

B. MOND (263), Department of Pure Mathematics, La Trobe University, Bundoora, Victoria 3083, Australia

W. OETTLI (227), Lehrstuhl für Mathematik VII, Universität Mannheim, D-6800 Mannheim–Schloss, West Germany

U. PASSY (239), Faculty of Industrial Engineering and Management, Technion–Israel Institute of Technology, Technion City, Haifa 32000, Israel

M.L. PUTERMAN (399), Faculty of Commerce and Business Administration, University of British Columbia, Vancouver, B.C., Canada V6T 1W5

Y. RINOTT (703), Department of Mathematics, Stanford University, Stanford, California 94305

S. SCHAIBLE (21, 183, 417), Faculty of Business Administration and Commerce, University of Alberta, Edmonton, Alberta, Canada T6G 2G1

B. VON HOHENBALKEN (643), Department of Economics, University of Alberta, Edmonton, Alberta Canada T6G 2G1

R.J. WEBER (613), Kellogg Graduate School of Management, Northwestern University, Evanston, Illinois 60201

T. WEIR (263), Department of Pure Mathematics, La Trobe University, Bundoora, 3083, Melbourne, Australia

H. WOLKOWICZ (379), Department of Mathematics, The University of Alberta, Edmonton, Alberta, Canada T6G 2G1

I. ZANG (131) Faculty of Commerce and Business Administration, University of British Columbia, Vancouver, B.C., Canada V6T 1W5

W.T. ZIEMBA (21, 719), Faculty of Commerce and Business Administration, University of British Columbia, Vancouver, B.C., Canada V6T 1W5

Preface

There is a large and growing interest in the theory and applications of generalized concave functions. These functions relax mathematical properties of concave functions and hence are not concave, yet they retain key qualitative characteristics of concave functions. An important property of concave functions is that their upper level sets are convex sets: in fact, the family of generalized concave functions, called quasiconcave, is actually defined by this property. A well-known property of differentiable concave functions is that every stationary point is a global maximum. However, this property holds for more general types of functions, such as pseudoconcave functions. There are many examples of important problems where these (and other analogous) properties are desirable yet the functions involved are not concave.

Since 1949, when quasiconcave functions were mentioned by de Finetti, numerous authors have defined and studied many classes of generalized concave functions whose properties are of use in particular applications. Indeed, to date over 20 separate classes of generalized concave functions can be found in the literature. These functions are used in a wide variety of fields including business administration, economics, engineering, mathematics, statistics, operations research, and the sciences.

To assess the impact and current status of the knowledge concerning generalized concave functions, a NATO Advanced Study Institute was organized by Professors M. Avriel of the Technion–Israel Institute of Technology, S. Schaible of the University of Alberta, and W.T. Ziemba of the University of British Columbia. This Institute was held on the campus of the University of British Columbia in Vancouver, Canada, August 4–15, 1980. It brought together students as well as leaders in their fields, who presented surveys of current knowledge and described new research results concerned with various aspects of the study of generalized concave functions. The 61 lecturers and students represented a truly international gathering with fifteen countries represented. The lectures and discussions were organized in the broad categories of characterizations of generalized concavity, quadratic and C^2-functions, concavification, duality in generalized concave programs, multiobjective optimization problems, fractional programming, algorithms for the solution of generalized concave programs, and applications of generalized concave functions in economics, management science, and statistics.

This book constitutes the proceedings of the Advanced Study Institute. The editors, with the assistance of Professors M. Avriel and W.E. Diewert, have written an introduction to the main concepts, definitions, and representative uses of the nine principal classes of concave and generalized concave functions. This introduction, which is the first paper of the volume, standardizes the notation and terminology concerning these functions and illustrates through examples the differences between the various function classes. All of the survey and research papers appearing in this volume were presented at the Advance Study Institute. Each of these papers has also been refereed by an anonymous referee and one of the editors. The final versions of these papers reflect the comments of these referees and our editing to achieve consistency in style throughout the volume. Thanks are due to all the individuals who served as referees. We would also like to thank the authors for their prompt and efficient help in the timely completion of this volume and their patience with our red penciling of their manuscripts.

We wish to express our sincere gratitude to Professor M. Avriel who assisted us greatly in planning the Advance Study Institute and editing this volume. In particular his valuable knowledge of the subject aided us in the development of the Institute's program, the refereeing procedure and in the preparation of the introduction to the volume.

We are deeply indebted to the North Atlantic Treaty Organization for generous financial support of the Advanced Study Institute. Thanks are also due to the Natural Sciences and Engineering Research Council, Dean Peter Lusztig of the Faculty of Commerce and Business Administration of the University of British Columbia and Dean Roger Smith of the Faculty of Business Administration and Commerce of the University of Alberta for additional financial support. Ulrike Hilborn, Peggy Kallberg, Marlys Rudiak, Monika Spafford, and Suzanna Jackson aided us greatly in the smooth organization and running of the Advanced Study Institute. We would like to thank Colleen Colclough, Evelyn Fong, Monika Spafford, and Barbara Strouts for their excellent typing of a most difficult manuscript, and Ulrike Hilborn for her outstanding artwork. Special thanks go to Debbie Janson and Barbara Joyce, whose pleasant and efficient work in proofreading, editorial assistance, and correspondence with authors has aided us immensely in producing this volume. Thanks are also due to the staff of Academic Press for their usual outstanding help in the preparation and production of this volume.

INTRODUCTION

The papers appearing in this volume encompass a wide range of subjects in optimization and economics. This introduction summarizes the various papers and indicates the linkages between them. The unifying theme of all the papers is the use of generalized concavity. No prior knowledge of this subject is required from the reader, but familiarity with the basic concepts of optimization and economics could be beneficial.

Part I of the book deals with characterizations of generalized concave functions. In order to acquaint the reader of this volume with generalized concavity, an introductory paper was written by Avriel, Diewert, Schaible, and Ziemba in which the main types of generalized concave functions are defined and characterized. The difference between the various types of (generalized) concave and (generalized) convex functions is only a matter of sign -- that is, the negative of a (generalized) convex function is (generalized) concave. For the sake of completeness, definitions and characterizations are presented in this paper for both the convex and concave cases. Subsequent papers in this volume usually deal with either generalized convex or generalized concave functions. Readers can make the transition from one case to the other with the help of this introductory paper.

Since the terminology used in the generalized concavity literature is not unified, an attempt has been made in this paper to present a unified terminology for the different kinds of concavity and generalized concavity. Nine classes of concave and generalized concave functions are introduced. Each class is formally defined for the general, differentiable, and twice differentiable cases. Several characterizations and properties are provided for each class, as well as additional necessary or sufficient conditions that can be helpful in identifying these function classes. Each function class is also illustrated by applications drawn from optimization and economics as well as numerical and graphical examples. An illustration of hierarchical relationships between the various function classes and a series of examples that illustrate their differences conclude the paper.

Characterizations of generalized concavity for differentiable functions in terms of the gradient were presented in the introductory paper by Avriel, Diewert, Schaible, and Ziemba. These characterizations can be easily generalized if only one-sided derivatives of the function exist. This

ISBN 0-12-621120-5

generalization is especially useful for concave functions for which one-sided derivatives always exist in the interior of their domain. However, one-sided derivatives do not necessarily exist for the various types of generalized concave functions. Diewert provides characterizations for six classes of nondifferentiable generalized concave functions in terms of Dini derivatives. The main tool used by Diewert is a certain generalization of the classical mean value theorem to nondifferentiable functions in terms of inequalities and Dini derivatives. The paper concludes with the development of a set of necessary optimality conditions for a nonlinear program in which the objective and constraint functions are, respectively, nondifferentiable pseudoconcave and quasiconcave. It is shown that these necessary conditions are also sufficient.

Martin addresses the question of uniqueness in optimization as related to the connectedness of level sets. The existence of such a relationship is intuitively clear if one thinks of unimodal versus multimodal functions. In the first case level sets are usually connected, whereas in the case of multimodal functions the level sets are usually disconnected as one approaches the relative maxima. Connectedness of level sets is also a generalization of the convexity of upper level sets which characterizes quasiconcave functions. The motivation for this work was provided by studying the behavior of the optimal value of a linear program as a function of the constraint coefficient matrix. It was found that the optimal value function has path-connected, but not necessarily convex, level sets. By appropriately defining the notion of unique optima the author presents some necessary and sufficient conditions relating topological connectedness of level sets to unique optima.

Continuity and differentiability properties of convex functions are extensively described in the literature. The study of these properties for quasiconvex functions is the subject of Crouzeix's paper in this part. In particular, Crouzeix shows that a quasiconvex function f defined on R^n is continuous at x_0 if and only if the greatest lower semicontinuous function majorized by f is also continuous at x_0. Similarly, one can characterize lower semicontinuous quasiconvex functions f by the lower semicontinuity of certain one dimensional functions formed from f. Crouzeix also shows that positively homogeneous quasiconvex functions defined on R^n whose values are either nonnegative for all $x \in R^n$ or are strictly negative on their domain, are actually convex functions.

Consequently, it is easy to see that every positively homogeneous quasiconvex function can be expressed as the minimum of two convex functions. Concerning differentiability properties, it can be shown that quasiconvex functions on R^n are almost everywhere differentiable and for these functions the concepts of hemi-differentiability, Gateaux-differentiability, and Fréchet-differentiability are equivalent for the first derivative.

A function is said to be concavifiable or concave transformable if it can be transformed into a concave function by a continuous and strictly monotone increasing function. An interesting question concerns the conditions under which there exists such a transforming function. These conditions restricted to the case of twice differentiable functions, are the subject of Zang's paper. He surveys and unifies the necessary and sufficient conditions for the concavifiability of C^2-functions, conditions that were first developed by Fenchel in 1951 and discussed by others in recent years. Zang derives here a hierarchy of four conditions, two of local nature, involving properties of the gradient and an augmented Hessian, and two global conditions. A C^2-function can be transformed into a concave function if and only if the four conditions are satisfied. The paper concludes with a discussion of functions that are concavifiable by an exponential transformation (see e.g. the last section of the Kallberg-Ziemba paper in Part VII for an application).

In the final paper in Part I of the volume Lindberg studies special cases of concavifiable functions, discussed more generally in this volume by Kannai and Zang. The special case concerns functions f such that f^p/p is convex for some real number $p \neq 0$, and are called by Lindberg "power convex" functions. The motivation for the study of these functions arises from the fact that power convex functions are (almost) the only concavifiable functions which are closed under positive scalar multiplication -- that is, if f is concavifiable by a certain continuous monotone transformation h, then also λf, $\lambda > 0$ is concavifiable by the same h. It is also shown by Lindberg that power convex functions are closed under addition if $p \leq 1$, and more generally, under taking r-means of power convex functions with $r \geq p$. Similarly, they are closed under taking pointwise supremum or under infimal convolution.

The papers in Part II focus on properties of twice differentiable generalized convex functions. Ferland reviews criteria for quasi- and pseudoconvexity for C^2-functions and quadratic functions in nonnegative variables. Schaible derives known as well as new criteria for six kinds of generalized convex quadratic functions on solid convex sets using a unified approach. The final article by Crouzeix and Ferland relates several criteria for quasi- and pseudoconvexity of C^2-functions on open convex sets.

Ferland surveys characterizations of quasi- and pseudoconvex C^2-functions on the nonnegative orthant with emphasis on quadratic functions. He presents necessary and sufficient conditions for quasi- and pseudoconvexity of C^2-functions derived earlier by Arrow and Enthoven and extended by Ferland. These conditions are given in terms of the leading principal minors of a bordered Hessian. There is a gap between the necessary and sufficient conditions which even exists for quadratic forms. He then reviews properties of positive subdefinite matrices that were introduced by Martos to characterize quasi- and pseudoconvexity of quadratic forms. Positive subdefinite matrices are related to quasiconvex quadratic forms in nonnegative variables in a similar way that positive semidefinite matrices are to convex quadratic forms. Finally Ferland presents finite criteria for quasi- and pseudoconvex quadratic forms and quadratic functions on the nonnegative orthant which were derived earlier by Cottle and Ferland, Ferland, and Martos.

Schaible characterizes generalized convex quadratic functions on solid convex sets of R^n. He derives criteria for quadratic functions that belong to one of the six different classes of generalized convex functions introduced in the paper by Avriel, Diewert, Schaible and Ziemba in this volume. In the case of quasiconvex, pseudoconvex and strictly pseudoconvex quadratic functions five different criteria are presented. All these are both necessary and sufficient. A particularly useful result for obtaining some of the other criteria is the fact that quasiconvex quadratic functions are convexifiable. This result can also be used to derive the earlier results by Cottle and Ferland, Ferland, and Martos for quadratic functions on the nonnegative orthant which these authors obtained via positive subdefinite matrices. Finally Schaible presents criteria for semistrictly quasiconvex, strictly quasiconvex, and strongly pseudoconvex quadratic functions on open convex sets.

Crouzeix and Ferland characterize twice differentiable quasiconvex and pseudoconvex functions on open convex sets. Two first order conditions are given for quasiconvex functions which are pseudoconvex. The authors then survey several second order characterizations of pseudoconvex and quasiconvex functions. During the last twenty years different types of necessary and/or sufficient conditions were derived. These are given in terms of a bordered Hessian or an augmented Hessian or in terms of the Hessian restricted to a subspace orthogonal to the gradient. The relationship between these criteria is analysed using a matrix-theoretic framework. In their survey the authors also present an extended version of Katzner's criterion. By an example they show that a quasiconvex function cannot be characterized by purely local conditions.

In the preceding parts generalized convex functions are analysed. Part III deals with optimization problems involving generalized convex functions. The emphasis is put on duality. The four papers by Crouzeix, Oettli, Passy, Mond and Weir develop duality concepts for generalized convex programs using different approaches. Finally DiGuglielmo is concerned with estimating the duality gap in the absence of convexity/quasiconvexity in discrete and continuous optimization problems.

Crouzeix develops a framework for duality in quasiconvex programming using so-called dual functions. These involve perturbation functions that can be chosen rather freely. Bidual functions then play the role of the biconjugate function in convex duality. The author obtains two types of dual programs which are related to each other. In both duals the objective function to be maximized is quasiconcave. One of these duals reduces to the surrogate dual(under suitable assumptions)which was considered by Luenberger, Greenberg and Pierskalla earlier. The other dual is applied to generalized fractional programming. The author also derives within this general framework an early duality result in nonconvex programming that von Neumann had obtained in a different way for his model of an expanding economy.

Oettli is concerned with quasiconvex duality based on modified optimality conditions. His results are derived for minimization problems in general topological vector spaces. These programming problems involve a quasiconvex objective function and a quasiconvex multivalued mapping. The classical Kuhn-Tucker optimality conditions are inappropriate for such problems. Extending earlier work of Luenberger, Oettli obtains optimality conditions suitable for quasiconvex programs. These conditions are then used to derive duality relations. The dual is a maximization problem with a quasiconcave objective function. The author also studies biduality. An example illustrates the concepts introduced in the paper.

Duality for nonconvex mathematical programs is also the subject of Passy's paper. In such programs one has to relax well-known optimality conditions of minima and maxima in convex programs and to replace them with those of stationary points, satisfying necessary conditions for minima and maxima. The principal tool used by Passy for establishing "pseudo-" duality results in the nonconvex case is the Legendre transform, that is closely related to Fenchel's conjugate transform in the convex case. The general results are also specialized in this paper to different types of quasiconvex fractional programs.

Mond and Weir suggest new duals for differentiable generalized concave programs that are modifications of Wolfe's dual. As Mangasarian showed, Wolfe's dual is no longer useful for maximization problems when the objective function is pseudoconcave rather than concave and the inequality constraints are quasiconcave rather than concave. The authors modify Wolfe's dual in various ways. One or more terms in Wolfe's dual objective function are dropped and suitably introduced in the constraints. For the new duals weak and strong duality theorems hold under suitable generalized concavity assumptions. A converse duality theorem is proved for one of the new duals. The authors also derive a symmetric duality result for pseudoconcave/pseudoconvex objective functions modifying suitably a symmetric dual suggested by Dantzig, Eisenberg and Cottle in concave programming. In the last part of their paper Mond and Weir show that for skew-symmetric objective functions the nonlinear program is self-dual.

DiGuglielmo determines estimates of the duality gap for discrete and continuous optimization problems. In the first section of his paper he focuses on discrete programs. The estimates of the duality gap are expressed in terms of the lack of γ-convexity of the objective function. They are also applied to Lagrangian relaxations of discrete optimization problems. In the second section the author derives estimates for the duality gap in nonconvex programming. In case of convex constraints such an estimate is given in terms of the lack of quasiconvexity of the objective function using the second quasiconjugate. For nonconvex constraints an estimate of the duality gap is determined which is expressed in terms of the lack of convexity of the objective function and constraints defined via the second conjugate.

The papers in Part IV define and utilize new classes of generalized concave functions. Ben-Tal and Ben-Israel develop a new class of functions termed F-convex, based on nonlinear rather than linear supports as with convex functions, and illustrate the implications and use of these functions. Borwein presents a unified analysis of optimality criteria for multiple criteria optimization problems. Borwein and Wolkowicz develop optimality conditions for abstract convex programs where the constraint functions must lie in a convex cone rather than say the nonnegative orthant of R^n. Brumelle and Puterman develop a class of operators termed W-convex and illustrate their use in the solution of nonlinear equations.

Generalizations of concavity are usually attained by relaxing one or more properties of concave functions. In the paper of Ben-Tal and Ben-Israel the property that the hypograph of a concave function is supported at each point by an affine function is generalized. They replace the supporting affine functions by a family $\mathcal{F}$ of suitably chosen functions. A function f is called $\mathcal{F}$-concave if for every point of its effective domain there exists a function $F \in \mathcal{F}$ such that F is a support of f. Replacing affine functions by a more general family results, as expected, in losing some of the nice properties of concave function but, on the other hand, many important properties of concave functions are retained or generalized.

Ben-Tal and Ben-Israel derive in their paper an "$\mathcal{F}$-convex" analysis along the lines and extend the Fenchel-Rockafellar type convex analysis. For a given family $\mathcal{F}$ a conjugate family $\mathcal{F}^*$ is defined, as well as $\mathcal{F}$-subgradients and $\mathcal{F}$-conjugate functions. Under suitable assumptions on $\mathcal{F}$, the basic results of convex analysis are generalized to the $\mathcal{F}$-convex case.

In general, $\mathcal{F}$-concave functions do not have the local-global maximum property of concave functions. However, it can be shown that a difference of $\mathcal{F}$-concave and $\mathcal{F}$-convex functions is unimodal. This property together with results in $\mathcal{F}$-convex analysis provide a Fenchel-type duality scheme in which the primal problem is to maximize the difference of $\mathcal{F}$-concave and $\mathcal{F}$-convex functions and the dual problem is to minimize the difference of associated $\mathcal{F}$-conjugate functions. An application of $\mathcal{F}$-concavity to the solution of nonlinear equations is also given, where a generalized Newton method is derived using "$\mathcal{F}$-approximation", instead of the linearization in Newton's method.

Borwein develops a comprehensive analysis of vector-valued optimization problems subject to convex inequality and affine equality constraints.

The analysis is presented in a unified way by using the notion of convex relations. First the topological and analytic properties of convex relations (multivalued mappings) are developed. These properties are subsequently applied to the analysis of vector-valued optimization problems. In this context a Lagrange multiplier theorem and a Lagrange-type duality is developed, and it is shown that the approach used in this paper is at least as general as the vector-valued version of the Fenchel-Rockafellar bifunctional approach to duality. Finally, several Karush-Kuhn-Tucker-type conditions are derived with the aid of the author's earlier presented results.

Borwein and Wolkowicz deal with optimality conditions in abstract convex programs, where the constraint functions must lie in a convex cone, rather than in the nonnegative (or nonpositive) orthant of R^n, as in the case of ordinary convex programs. In addition, the special abstract convex program, in which the constraints are represented by linear functionals and the program variables are restricted to a polyhedral set is also studied in detail. Such programs arise in several situations such as in optimal control problems, and in stability problems of differential equations. The main topics discussed in this paper are: optimality conditions and constraint qualifications, Lagrange multiplier theorems, stability and sensitivity of optimal solutions, and a generalization of the Farkas lemma.

Brumelle and Puterman deal with a generalization of convexity useful in numerical analysis. They extend Newton's method to solve nonlinear equations $F(x) = 0$ where the operator F is W-convex rather than convex. W-convexity is a pointwise property defined with regard to a wedge W whereas convexity reflects the behavior of F over line segments. W-convexity is closely related to order-convexity introduced by Ortega and Rheinholdt. It turns out that W-convex operators are essentially those which have a support at each point. A Newton-type method is defined using supports in place of gradients and subgradients. The authors show that the sequence of points generated with this method converges monotonically to a root of the equation. In addition rates of convergence and error bounds are obtained. The rate of convergence is $1+p$ if F has a support at the root which is Lipschitz continuous of order p.

The articles in Part V focus on the special class of generalized concave programs commonly called fractional programs where the objective function is a ratio of two given functions. The papers by Schaible and Ibaraki provide surveys of major developments in this area. Schaible puts more emphasis on theory and applications whereas Ibaraki covers in greater detail algorithms. The two surveys of fractional programming by Schaible and Ibaraki are followed by articles by Craven and Cambini in which the authors present results on special classes of fractional programs.

Schaible reviews applications of linear, quadratic and concave-convex fractional programs as well as generalized fractional programs where the minimum of several ratios is to be maximized. These different types of fractional programs are of interest in a variety of contexts such as production, transportation, portfolio selection, stochastic programming, large scale programming, stochastic processes, information theory and numerical analysis. Schaible describes various generalized concavity properties of the objective function. He then provides a survey of different duality approaches for linear, quadratic and concave-convex fractional programs. He discusses both the differentiable and the general case. Schaible also outlines a primal, a dual and a parametric solution method. Some recommendations are given concerning which of these algorithms is most appropriate for given fractional program classes. The article concludes with an extensive bibliography of fractional programming.

Ibaraki's survey emphasises solution methods in linear and nonlinear fractional programming problems having continuous or discrete variables. He considers three different strategies: variable transformation, a direct nonlinear programming approach and in particular a parametric approach. Various algorithms and their modifications that solve the equivalent parametric program are presented and related. Ibaraki also reports on numerical results testing some of these algorithms using quadratic fractional programs as the setting.

Craven addresses the problem of duality for generalized concave fractional programs. Recall that the articles in Part III deal with duality for nonconcave problems involving generalized concave objective functions. In contrast to other duals in fractional programming the dual objective function in Craven's approach is also a ratio; indeed the primal and dual objective functions are the same. Weak and strong duality is established. These duality relations hold under suitable generalized concavity assumptions on a parameterized linear combination of the numerator and denominator as well as on nonnegative combinations of the constraint functions. In particular the results are true for concave-convex fractional programs considered in most other duality approaches. For linear programs, Craven's dual reduces to the classical linear programming dual. He applies his duality theory to derive duals for fractional programs that arise in portfolio theory and information theory. The paper concludes with some sensitivity results in fractional programming. Based on a recent result by Maurer in general nonlinear programming, the sensitivity of an optimal solution with regard to parameter changes in the objective function and the constraints is measured by a dual optimal solution.

Cambini presents algorithms for the special class of generalized concave fractional programs where the objective function to be maximized is a product of a concave function N(x) and an integer power p of a positive affine-linear function,and the constraints are linear. For negative p this problem is a fractional program which is concave-convex. For any integer p, positive or negative, the objective function is quasiconcave if N(x) is nonnegative. Cambini derives optimality conditions for this model which are based on a theorem of alternatives by Giannessi. He then develops solution methods in which a sequence of feasible solutions is generated that converges to a local optimum. This is done for the special cases of linear fractional programs, of nonlinear fractional programs where N(x) is affine-linear, and of quadratic-linear fractional programs (N(x) strictly concave and quadratic, $p = -1$). All methods exploit the special structure of the objective function and are finite.

The papers in Part VI are concerned with applications of generalized concavity in Economics and Management Science. For the special case of quasiconcave fractional programming such applications are reviewed already in Part V. The first paper in Part VI by Diewert illustrates the use of different types of generalized concave functions in economic theory. Kannai discusses the problem of concavifiability of convex preferences. Weber characterizes attainable sets of markets involving concave and quasiconcave utility functions. Eichhorn describes conditions under which production functions are concave and satisfy the law of eventually diminishing marginal returns. Eichhorn and Gehrig show that inequality measures used in income distribution or industry concentration studies are related to particular generalized convex function classes. Von Hohenbalken describes the simplical decomposition approach for optimization of functions over convex sets and provides a survey of computational experience with applications particularly in portfolio theory. Craven describes optimality criteria for multiobjective minimization problems involving differentiable objective and constraint functions. Finally, Jeroslow focuses on the interface between disjunctive and integer programming and generalized convexity concepts.

Diewert's paper is a survey of the role that generalized concavity plays in economic theory. In particular, he discusses the applications of the various kinds of generalized concave functions introduced in the earlier paper of Avriel, Diewert, Schaible, and Ziemba. First, the problem of minimizing the cost of production, subject to fixed input prices and a fixed technology is studied. Under the very weak assumption of upper semicontinuity on the production function, Diewert shows that the producer's cost function satisfies a series of desired properties, such as nonnegativity, continuity, linear homogeneity, and concavity in prices for fixed output. Next, it is shown that there is a duality between production and cost functions. Under the further assumption of quasiconcavity and nondecreasing monotonicity on the production function, the cost and production functions are mutually completely characterized by each other. Strict quasiconcavity of the production function is a sufficient condition for the cost function to be once differentiable with respect to input prices. Similarly, strong pseudoconcavity (quasiconcavity) of the production function is related to the twice differentiability of the cost function with respect to input prices.

Other types of generalized concave functions arise in problems of consumer theory. Here the consumer maximizes a continuous utility function with respect to a consumption vector, subject to a budget constraint. It can be shown that the maximal value of the utility (the indirect utility function) is a continuous, nonincreasing, quasiconvex function of the (normalized) commodity prices. If the utility function is semistrictly quasiconcave then the indirect utility function has no "thick" indifference surfaces, and if it is strictly quasiconcave, then the optimal consumption is unique.

Strong quasiconcavity, on the other hand, implies that the consumer's demand function is continuously differentiable. Pseudoconcavity and strict pseudoconcavity play a role in finding demand functions consistent with utility maximizing behavior. Finally, strict and strong concavity arise naturally in comparative statics analysis by using a producer's profit maximization model instead of cost minimization, as mentioned earlier.

Kannai discusses the problem of concavifiability of convex preference orderings. In particular he describes conditions that a convex preference ordering defined on a convex set must satisfy for there to exist a concave utility function representing these preferences on this set. For convex preferences the utility function must, of course, be quasiconcave so the thrust concerns what additional conditions are required to obtain concavity. Kannai presents the analysis for functions that are general, once differentiable and twice differentiable. The basis for the development utilizes standard definitions for the concavity of a function of a single variable: a) in general one has a condition relating the values of the function at three collinear points; b) in the once differentiable case one has a condition relating the values of the derivative at two points; and c) in the twice differentiable case one has a pointwise condition on the second derivative. For several variables a) is unchanged, b) is either a monotonicity condition on the gradient or one-sided directional derivative (this case, of course, lends itself to generalizations using subgradients, etc.) and c) is a condition on the definiteness of the matrix of second derivatives. Each of these three types of conditions leads naturally to a construction of a concave utility function. If preferences are concavifiable then certain concave utility functions representing these preferences are *less* concave than other utility functions representing these preferences. By *least* concave we mean that all other concave utility functions can be represented as concave functions of the least concave function. Kannai's constructions yield least concave utility functions and he discusses some of their economic and behavioral properties in particular in bargaining problems.

Weber surveys results on attainable sets of markets. Such a set is defined as the set of all utility outcomes which can be achieved by the traders through a redistribution of goods. Based on the common assumption that preference sets are closed and convex, utility functions have to be upper semicontinuous and quasiconcave. Two classes of attainable sets are therefore of interest; those arising from upper semicontinuous quasiconcave utility functions and the subclass obtained from continuous concave utility functions. For these two classes of attainable sets various characterizations are derived. Furthermore, he attempts to determine lower and upper bounds on the "complexity" of attainable sets in these two classes. Finally, Weber studies market games referring to results on attainable sets. At the end of the paper several open problems are addressed.

A familiar concept in production theory is the law of eventually diminishing marginal returns (LEDMR): for input values no smaller than some given level the output is a strictly concave function of the inputs. In the case of utility theory this phenomenon is referred to as Gossen's law. Eichhorn studies conditions on the production or utility function that imply concavity of this function on the nonnegative orthant. For example a production function is linear homogeneous and satisfies the LEDMR if and only if it is strictly concave. Hence functions that are initially convex cannot satisfy the LEDMR. However, for production functions that are homogeneous of positive degree less than unity (having the property that all inputs are required to obtain positive output) initial convexity means that the LEDMR does not hold and conversely. A production function is called homothetic when input levels that yield the same output have the property that fixed multiples of these inputs also yield the same output. Homogeneous functions of positive degree less than or equal unity are homothetic. Upper semicontinuity, free disposal, strict quasiconcavity and decreasing returns to scale (along rays) are sufficient to yield strictly concave production functions and hence the LEDMR.

Eichhorn and Gehrig show that inequality measures used in income distribution or industry concentration studies are related to particular generalized convex function classes. The analysis involves symmetric strictly quasiconvex and strictly Schur-convex functions. These latter functions, termed strictly S-convex, are such that argument transforms by doubly stochastic matrices that are not permutations yield lower function values. Strictly quasiconvex functions are strictly S-convex but not conversely. The authors' main result is that an inequality measure is symmetric and satisfies the principle of progressive transfers (i.e. individual transfers from richer to poorer individuals lower group inequality) if and only if it is strictly S-convex. They also discuss related inequality measures.

Von Hohenbalken describes the simplicial decomposition approach for optimization of functions over convex sets and provides a survey of computational experience with algorithms and applications. The simplicial approach relies on Caratheodory's Theorem which implies that points in a compact convex set can be described in terms of the extreme points of the set. The algorithms iterate between master and subproblems as in the Dantzig-Wolfe decomposition principle for linear programs and utilize the general strategy of inner linearization followed by restriction. The discussion is concerned with differentiable objectives on polyhedrans (bounded polytopes) although generalizations are possible. In this case under mild assumptions the subproblems are generally exact, and the entire procedure is globally convergent and finite for pseudoconcave quadratics and certain homogeneous functions that arise, for example, in portfolio selection. In general, of course, both the master and subproblems are infinite procedures to obtain convergence. The emphasis is on the general development of such algorithms and on methods to streamline and otherwise make efficient the resulting algorithms. The paper concludes with a discussion of applications of simplicial decomposition algorithms to portfolio selection and least distance problems. In the former case problems with as many as 1000 variable securities are feasible and the algorithm is such that in relatively few iterations one can obtain very acceptable near optimal policies.

Craven discusses optimality for multicriteria minimization problems involving differentiable objective and constraint functions. The discussion utilizes the concepts of strong minima, vector minima and weak minima , where strong implies vector implies weak. These minimization concepts are with respect to an underlying convex cone in objective value space. If the objectives are completely contradictory this cone might be, for example, R^n_+. The weak minima correspond to the minima obtained from linear combinations of the various objectives for suitable weights. In contrast strong minima then require the simultaneous minimization of each objective function which is rarely appropriate. Thus in applications strong minima are generally defined on polyhedral cone subsets of R^n_+. Given a constraint qualification a weak minimum implies the Kuhn-Tucker conditions. The Kuhn-Tucker conditions are sufficient for a weak minimum if the inequality constraints are quasiconvex and the vector valued objective function satisfies a pseudoconvexity property. Craven also develops analogous results for strong minimization. In this case the Kuhn-Tucker necessary conditions take on a simplified form and a vector duality theory can be developed using a vector Lagrangian. The Lagrange multipliers form a linear mapping or matrix instead of a vector. If the Lagrangian satisfies a pseudoconvexity property then one has vector valued duality and converse duality theorems, and there is no duality gap.

Jeroslow discusses recent results in disjunctive programming and integer programming as they pertain to the general theme of this volume. Both types of optimization problems generalize linear programs which are a special case of convex programs. In the first part the author focuses on disjunctive programs which are obtained from linear programs by adding (generally nonconvex) logical conditions. It is shown how ideas of infinitary disjunctive programming may be beneficial in nonconvex nonlinear programming. The remaining part of the paper is devoted to integer linear programming. A certain class of nonconvex functions, so-called Chvatal functions, plays the role in integer programming that linear functions play in the dual of two equivalent statements. Indeed among the various equivalence theorems regarding linear inequalities one often obtains true statements when the variables of the "primal" are required to be integers and the linear functions of the "dual" are allowed to be Chvatal functions. The "integer analogues" represent developments that are parallel to some extent to generalized duality schemes which extend Lagrangian duality for convex programs to more general duality results for nonconvex programs.

The papers by Karlin and Rinott, and Kallberg and Ziemba in Part VII are concerned with the study of generalized concave functions that arise in a natural way in the study of stochastic systems. Karlin and Rinott focus on the study of inequalities that arise from concepts of univariate and multivariate total positivity whereas Kallberg and Ziemba deal with problems of stochastic programming and portfolio theory including a review of logconcave functions and measures.

A function $K(x,y)$ defined on a product of linearly ordered sets X and Y is said to be totally positive, abbreviated as TP, if its integral with respect to some well-defined function f in the Y space produces a function $g(x) = \int K(x,y)f(y)dy$ in X space with no more sign changes than the function f where y is a σ-finite measure on Y. Total positivity of various orders is characterized by simple determinental conditions. Examples of totally positive densities are various exponential, binomial and Poisson distributions. Additional TP functions such as the noncentral t and F distributions may be generated by observing that integrals of products of suitably defined TP functions are TP. Total positivity relates to a wide range of areas including stochastic processes, differential equations, approximation theory, and matrix theory in addition to the probabilistic and statistical models and inequalities described in the paper. Totally positive functions have a number of useful properties. For example, they preserve the number of sign changes, monotonicity and certain convexity properties upon integration of given functions. These properties are equivalent to various concepts of stochastic ordering and stochastic dominance. These concepts may be used in the analysis of various problems in statistical decision theory including classical Neyman-Pearson hypothesis testing. Generalized convex functions of various types are related to determinental conditions of TP functions. These conditions are related to the number of sign changes of certain measures and the results constitute a theory of inequality derivation.

Finally, Karlin and Rinott give a brief survey of some recent results on multivariate total positivity, abbreviated as MTP. Multivariate functions defined on a lattice formed as a product of ordered sets are MTP if they are nonnegative and satisfy a determinental condition involving the meet and join operations on the lattice. A number of common multivariate distributions such as the gamma, logistic, F, t, and the normal for certain covariance structures are MTP. Results on MTP functions yield a series of useful inequalities including results on stochastic ordering and bounds on the reliability of systems. Other applications appear in statistics, statistical mechanics and in the derivation of limit theorems for random variables and probability inequalities. The latter results are useful in chance constrained programming and in the determination of bounds for probabilities associated with multivariate confidence sets related to statistical estimation of parameters.

Kallberg and Ziemba discuss generalized concave functions that arise in a natural way in the study of particular stochastic programming problems. They begin with a survey of results concerning logconcave functions and measures. These constructs are useful in a wide variety of applications;

their focus is on applications to chance constrained, aspiration and two stage programming. The main properties of logconcave and logconvex functions, developed largely by Klinger and Mangasarian, and their relations with other ordinary and generalized concave and convex functions are presented in tabular form. Prékopa has pioneered the development and use of logconcave measures as a useful construct particularly in the study of chance constrained programming problems. The main results of his theory and extensions of Borell and Rinott concerned with the relationship between the generalized concavity of measures and that of related functions is reviewed. Applications to various mathematical programming problems that yield generalized concave and convex objective and constraint functions and convex feasible regions follow directly from the general theory. Deterministic equivalents of the stochastic programs may or may not be explicitly defined; several of the latter are developed in particular instances.

The second part of the Kallberg-Ziemba paper is concerned with portfolio selection models. These models are particular stochastic programs and are generally concave programs when the investor is risk averse. However generalized concave functions, usually fractional functions, arise in at least three instances: 1) when one wishes to decompose the portfolio problem into subproblems that are easier to solve and analyse; 2) it is not economically feasible to solve the exact portfolio problem and "good" approximate solutions are desired; and 3) the security market is not perfect and the level of investment alters the distribution of the security returns.

The two stage approach to the portfolio problem when there are joint normal or symmetric stable distributions and a risk free asset illustrates the first situation. In this case stage one is a fractional program to determine the optimal proportions of the risky assets and stage two is a stochastic search problem to determine the optimal ratios of risky and safe assets. Both problems are easy to solve and the procedure is economically feasible even if the number of securities is very large. The case when the investor is faced with securities having joint lognormally distributed assets is an example of the second situation. Exact calculations are economically as well as numerically intractable except when there are very few securities. Using an approximation that the sum of the lognormal assets is lognormal yields an explicit deterministic nonlinear program that is easy to solve numerically even when there are a large number of securities; the ratio of computing times of the exact and approximate solutions is essentially exponential in the ratio of the computing times between the exact and approximate solutions. In addition the approximation is quite accurate both in expected utility and portfolio allocation spaces.

The third situation is illustrated by two parimutuel betting models for win, and place and show respectively. In these cases the bets made by the investor influence the odds on the various horses. According to the capital growth criteria one maximizes the rate of asset growth in such repeated trials as successive horseraces by maximizing the expected logarithm of final wealth race by race. The resulting models are then the expectation of a concave function of fractional functions of the bets on the various horses. Kallberg and Ziemba discuss the concavity and generalized concavity properties of these models. The models are easy to solve and modifications yield a practical system for betting that exploits market inefficiencies.

PART I

CHARACTERIZATIONS OF GENERALIZED CONCAVE FUNCTIONS

INTRODUCTION TO CONCAVE AND GENERALIZED CONCAVE FUNCTIONS†

M. Avriel W. E. Diewert S. Schaible W. T. Ziemba

This paper provides an introduction to concave and generalized concave functions for readers of this volume. Definitions of the nine main function classes are given for the general, differentiable and twice continuously differentiable cases. Illustrative examples and situations where such functions arise naturally in optimization and economics are provided for each function class. A diagram and a table with examples are used to illustrate the relationships between the various function classes.

1. PRELIMINARIES

We will be concerned with vectors which may be thought of as points in the n-dimensional Euclidean space R^n. R^1 will be written as R. Subscripts will refer to components of vectors and superscripts to particular vectors. For example for n=3, points 1 and 2 are $x^1 \equiv (x_1^1, x_2^1, x_3^1)^T$ and $x^2 \equiv (x_1^2, x_2^2, x_3^2)^T$. All vectors are considered to be columns and T denotes transposition. If $x, y \in R^n$ then $x^T y = \sum_{i=1}^{n} x_i y_i$ is their inner product. If f is a real differentiable function of n variables then $\nabla f(x)$, termed the gradient, is the n-dimensional vector of first partial derivatives of f at x, $\partial f(x)/\partial x_i$, $i = 1,\ldots,n$. If f is twice differentiable then $\nabla^2 f(x)$, termed the Hessian, is the nxn symmetric matrix of second partial derivaties of f at x, $\partial^2 f(x)/\partial x_i \partial x_j$, $i,j = 1,\ldots,n$. If the first (second) order partial derivates of f exist and are continuous functions, we say that f is once (twice) continuously differentiable. It will sometimes be convenient to let primes denote differentiation, i.e. $f'(x) = df(x)/dx$ and $f''(x) = d^2 f(x)/dx^2$, etc. A direction $v \equiv (v_1,\ldots,v_n)$ in R^n is an n dimensional nonzero vector whose squared components, for convenience, are

†This research was supported by the Fund for the Advancement of Research at the Technion (Avriel), the Social Sciences and Humanities Research Council (Diewert), and the Natural Sciences and Engineering Research Council (Schaible and Ziemba). Without implicating them we would like to thank A. Ben-Tal and I. Zang for helpful comments on an earlier version of this paper.

ISBN 0-12-621120-5

assumed to sum up to one, i.e., $v^T v = \sum_{i=1}^{n} v_i^2 = 1$. The directional derivative of a function f in the direction v evaluated at a point x belonging to the interior of the domain of definition of f is defined as $D_v f(x) \equiv \lim_{t\to 0}[f(x+tv) - f(x)]/t$ if the limit exists. If the directional derivative of f exists for a point x in the interior of the domain of definition of f and for all directions v, then we say f is directionally differentiable at x. If $D_v f(x) = v^T \nabla f(x)$ for all directions v, then f is weakly or Gateaux differentiable at x. If, in addition, $D_v f(x) = \lim_{v'\to v, t\to 0}[f(x+tv') - f(x)]/t$, then f is Fréchet differentiable at x. Crouzeix (1981a) uses these three kinds of differentiability in studying quasiconcave functions. Similarly, Diewert (1981b) uses the one sided directional derivative concept, defined as $D_v^+ f(x) \equiv \lim_{t\to 0^+} [f(x+tv) - f(x)]/t$. In this definition, the scalar t must be positive. In this introductory paper, however, we shall only deal with Gateaux differentiable functions. For results dealing with the other types of differentiable functions the reader is referred to Crouzeix (1981a) and Diewert (1981b). Finally, we note that a function f is upper semicontinuous over its domain of definition if and only if the upper level sets $U(\alpha) \equiv \{x: f(x) \geqslant \alpha\}$ are closed for every real number α.

In the definitions and propositions to follow we provide at least one reference to the literature where the definition appears or the proposition is proven. As far as can be determined we also attempt to make reference concerning the original source of the definition/proposition. The coverage in this paper is limited to the nine kinds of concavity and generalized concavity discussed by Diewert, Avriel and Zang (1981). A number of additional types of generalized concave functions that arise in special circumstances appear in the literature and some of the papers of this volume. The intent here is not to provide an exhaustive survey but to provide an introduction to major classes of functions encountered in practice and studied in this volume.

Further readings on properties of generalized concave functions can be found in textbooks of nonlinear optimization such as Mangasarian (1969), Martos (1975), and Avriel (1976), or in a book by Avriel, Diewert, Schaible, and Zang (forthcoming). For a review of some classes of generalized concave functions see Greenberg and Pierskalla (1973).

2. CONCAVE AND STRICTLY CONCAVE FUNCTIONS

The origins of convex sets and functions can be traced back to the turn of the century, see Hölder (1889), Jensen (1906), and Minkowski (1910, 1911).

DEFINITION 1: A subset C *of* R^n *is called a convex set if*

$$x^1 \in C,\ x^2 \in C,\ 0 \leqslant \lambda \leqslant 1 \Rightarrow (\lambda x^1 + (1-\lambda)x^2) \in C \qquad (1)$$

In other words, for any two points of a convex set C the line segment joining them also belongs to C.

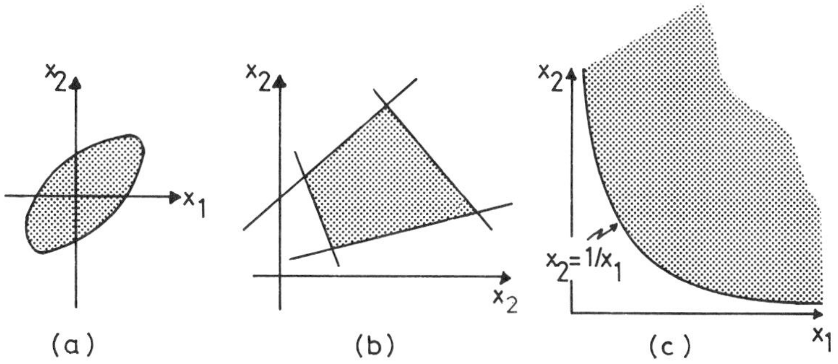

FIGURE 1. Convex sets

In Figure 1 we show examples of convex sets in R^2: (a) the area enclosed by an ellipse, (b) the polyhedron formed by the intersection of four half-spaces, and (c) the subset of R^2 lying above the curve $x_2 = 1/x_1$. Convex sets need not be closed, open or compact. A single point and the whole space R^n are trivial examples of convex sets. Convex sets are important for the study of concave or convex functions and some of their generalizations since the domain of these functions is always a convex subset of R^n. For further discussion of convex sets consult Eggleston (1958), Valentine (1964) or Rockafellar (1970).

DEFINITION 2: a) A real function f *defined on a convex set* $C \subset R^n$ *is called concave if*

$$x^1, x^2 \in C,\ 0 \leqslant \lambda \leqslant 1 \Rightarrow f(\lambda x^1 + (1-\lambda)x^2) \geqslant \lambda f(x^1) + (1-\lambda) f(x^2). \tag{2}$$

b) It is called strictly concave if

$$x^1, x^2 \in C,\ x^1 \neq x^2,\ 0 < \lambda < 1 \Rightarrow f(\lambda x^1 + (1-\lambda)x^2) > \lambda f(x^1) + (1-\lambda) f(x^2). \tag{3}$$

c) If f *is (strictly) concave, then* $g \equiv -f$ *is (strictly) convex.*

It is easy to see that all strictly concave (convex) functions are concave (convex) but not conversely. Furthermore we find that convexity of C is required so that f is defined at $\lambda x^1 + (1-\lambda)x^2$.

Figure 2 depicts (a) the strictly concave function $f(x) = \log(x+1)$ defined on the open interval $C = (-1,\infty)$, (b) the strictly convex function $g(x) = x^2$ defined on the whole real line R, and (c) the linear function $h(x) = 2 - \frac{1}{2}x$ defined on the interval $C = (-\infty, 2]$; this function is both concave and convex but neither strictly concave nor strictly convex.

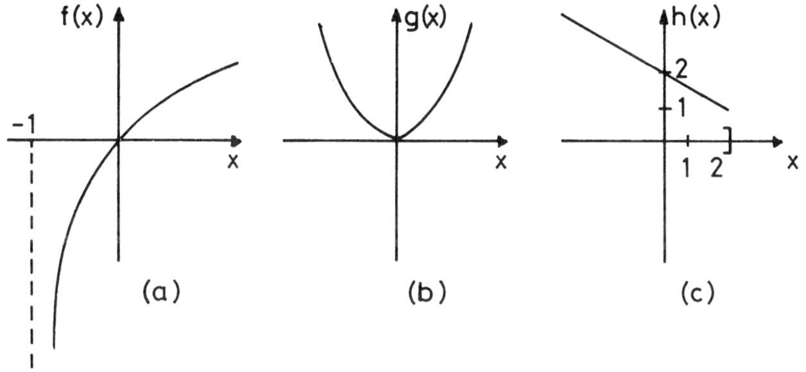

FIGURE 2. Concave and convex functions

Concave, convex, strictly concave and strictly convex functions arise in a natural way in a wide variety of circumstances. In management science, convex cost functions are met in models for production planning for example. This is the case if marginal cost is increasing with an increased output. In some applications one faces decreasing rather than increasing marginal cost. Then the cost function is concave. Functions describing the total revenue are often concave as seen from the example $R(x) = \sum_{i=1}^{n} p_i(x_i)x_i$ where $p_i(x_i)$ is the price-demand function which is assumed to be decreasing and linear. The concave function $R(x)$ is strictly concave for $x_i \geqslant 0$ unless for any good the price p_i does not depend on the demand x_i. Combining a concave revenue function and a convex cost function, one obtains a concave profit function. In risk minimization models in portfolio theory or stochastic programming, the total risk is often a strictly convex function of the decision variables.

In utility theory, the definition of concavity is precisely equivalent to the notion that an individual would never prefer an actuarily fair gamble and hence is risk averse. That is: if the status quo is is $w^o = \lambda w^1 + (1-\lambda)w^2$ and gambles 1 and 2 result in final wealth of w^1 and w^2 with probabilities λ and $(1-\lambda)$ respectively then $u(w^o) \geqslant \lambda u(w^1) + (1-\lambda)u(w^2)$. If the status quo is actually preferred to any fair gamble then f is strictly concave as in (3). In economics, production functions are often assumed to be concave, particularly when the producer is maximizing profits. An example is the Cobb-Douglas functional form $f(x_1,\ldots,x_n) \equiv \alpha_0 \prod_{i=1}^{n} x_i^{\alpha_i}$, where $x_i \geqq 0$ is the amount of input i used and the $\alpha_i > 0$ are technological parameters. The function f is concave on the non-negative orthant if $\sum_{i=1}^{n} \alpha_i \leqslant 1$. Concave functions also occur in the producer's cost minimization problem, in the consumer's expenditure minimization problem, in the theory of the cost of living index and in cost benefit analysis (see Diewert (1981a)).

Strictly concave functions also occur frequently in economics. To obtain a unique solution to the producer's profit maximization problem, it is usually assumed that the production function is strictly concave. The Cobb-Douglas functional form is strictly concave over the positive orthant if $\sum_{i=1}^{n} \alpha_i < 1$. Further applications of concavity in production theory appear in Eichhorn (1981). A typical (strictly) convex function is $x^T x$ which is used in regression analysis and econometrics when one wishes to estimate relationships that minimize a sum of squared residuals.

Concave functions are important in optimization, since every local maximum of such a function is global and if the function is strictly concave, then its maximum can be attained at no more than one point. Optimality conditions and duality relations for mathematical programs involving concave or convex functions were the first to be derived and have motivated research in generalized concave (convex) functions to be introduced below. For reading on the role of concave and convex functions in optimization , see Mangasarian (1969), Rockafellar (1970), Stoer and Witzgall (1970), Martos (1975), Avriel (1976), and Bazaraa and

Shetty (1976).

The functions shown in Figure 2 are continuous on their respective domains and are also differentiable on any open set contained in their domains. To see that these properties are not necessary, consider the function in Figure 2(c). Suppose as above that $C = (-\infty, 2]$, but now define $\bar{h}$ by

$$\bar{h}(x) \equiv \begin{cases} 2 - \tfrac{1}{2}x & \text{if } -\infty < x < 2 \\ 0 & \text{if } x = 2. \end{cases} \tag{4}$$

This function is discontinuous on the boundary of its domain, but satisfies (2), thus it is a concave function; it is no longer linear and, therefore, it cannot be both concave and convex. In general, concave and convex functions are continuous in the interior of their domain, but may have discontinuities on boundary points. It is easy to construct non-differentiable concave or convex functions. For example

$$f(x) = \begin{cases} x & \text{if } 0 < x \leqslant 1 \\ \tfrac{1}{2}x + 1/2 & \text{if } 1 \leqslant x < 2 \end{cases} \tag{5}$$

is concave although it is not differentiable at $x = 1$. Without assuming some kind of differentiability, concavity or convexity of a function can be only established by the conditions of Definition 2 -- that is, an inequality involving function values at three collinear points. For a thorough account of the differentiability properties of convex functions consult Rockafellar (1970).

An important property of a concave function is that every local maximum is a global one over its domain of definition (see Fenchel (1953)). For a strictly concave function such a maximum, if attained, is always a strict (unique) maximum (see Roberts and Varberg (1973)).

We now consider differentiable functions (unless mentioned specifically we refer to Gateaux differentiable functions simply as "differentiable").

PROPOSITION 1: (Mangasarian (1969); Avriel (1976)): a) Let f *be a differentiable function on an open convex set* $C \subset R^n$. *Then* f *is concave if and only if*

$$x^1 \in C,\ x^2 \in C \Rightarrow f(x^2) \leqslant f(x^1) + (x^2 - x^1)^T \nabla f(x^1). \qquad (6)$$

b) It is strictly concave if and only if the inequality in (6) is strict for $x^1 \neq x^2$. *For a (strictly) convex function the sense of the (strict) inequality in (6) is reversed.*

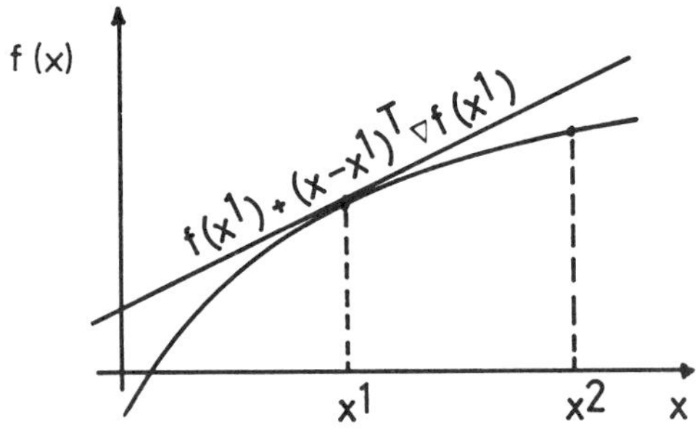

FIGURE 3. A concave function and its linear approximation

Figure 3 illustrates the idea behind Proposition 1: Taking any point x^1 in the domain of f and constructing a linear (first order Taylor) approximation of f at x^1 we obtain a linear function that is tangent to f at x^1 and at every other point in the domain of f it lies above f or coincides with it.

Another characterization of differentiable concave functions, generalizing the monotonicity property of the first derivative of a single variable concave function, is given next.

PROPOSITION 2 (Katzner (1970), Avriel (1976), Bazaraa and Shetty (1976)):
a) Let f *be a real differentiable function on an open convex set* $C \subset R^n$. *Then* f *is concave if and only if*

$$x^1 \in C,\ x^2 \in C \Rightarrow (x^2 - x^1)^T [\nabla f(x^2) - \nabla f(x^1)] \leq 0. \qquad (7)$$

b) It is strictly concave if and only if the inequality in (7) is strict for $x^1 \neq x^2$.

c) For a (strictly) convex function the sense of the (strict) inequality in (7) is reversed.

Note that both characterizations of differentiable concave functions involve conditions on two points. If the function under consideration is twice differentiable, there is a characterization by second derivatives at only one point:

PROPOSITION 3 (Fenchel (1953)): a) Let f *be a real twice continuously differentiable function on an open convex set* $C \subset R^n$. *Then* f *is concave if and only if*

$$x \in C \Rightarrow \nabla^2 f(x) \text{ is negative semidefinite.} \qquad (8)$$

b) Similarly, f *is convex if and only if* $\nabla^2 f(x)$ *is positive semidefinite for every* $x \in C$.

In other words, the eigenvalues of the Hessian matrix of a concave (convex) function are nonpositive (nonnegative). The characterization in the above Proposition cannot be fully extended to strictly concave functions by requiring the Hessian matrix $\nabla^2 f$ to be negative definite. To illustrate this argument consider the strictly concave function $f(x) = -x^4$ on R. Clearly, the second derivative of f vanishes at the origin, thus it is not negative at every point of its domain. However, we have

PROPOSITION 4 (Fenchel (1953)): a) Let f *be a real twice continuously differentiable function on an open convex set* $C \subset R^n$. *If*

$$x \in C \Rightarrow \nabla^2 f(x) \quad \textit{is negative definite} \tag{9}$$

then f *is strictly concave.*

b) Similarly, if $\nabla^2 f(x)$ *is positive definite for every* $x \in C$, *then* f *is strictly convex.*

For example, the function $f(x) = \log x$, defined on $C = \{x: x \in R, x > 0\}$ is strictly concave since its second derivative is $f''(x) = -1/x^2 < 0$ for $x > 0$. Similarly, $f(x) = -\frac{1}{2}x^T x$, defined on R^n is strictly concave, since its Hessian matrix consisting of the elements $\partial^2 f/\partial x_j \partial x_j = -1$, $j = 1,\ldots,n$, $\partial^2 f/\partial x_i \partial x_j = 0$, $i = 1,\ldots,n$, $j = 1,\ldots,n$, $i \neq j$ is negative definite.

An important property of (strictly) concave and (strictly) convex functions is that they are closed under addition and positive scalar multiplication. Moreover, the sum of a strictly concave and a concave function is also strictly concave, and similarly in the convex case. This last property has probably motivated the definitions of strongly concave and strongly convex functions to be introduced next.

3. STRONGLY CONCAVE AND STRONGLY CONVEX FUNCTIONS

DEFINITION 3 (Poljak (1966), Rockafellar (1976), Diewert, Avriel and Zang (1981)): a) A real function f *defined on a convex set* $C \subset R^n$ *is called strongly concave if* $f(x) = \phi(x) - \frac{1}{2}\alpha x^T x$, *where* ϕ *is a concave function defined on* C, *and* α *is a positive number.*

b) Similarly, g *is called strongly convex if* $g(x) = \psi(x) + \frac{1}{2}\alpha x^T x$, *where* ψ *is convex and* $\alpha > 0$.

Since f is the sum of a concave function and a strictly concave quadratic function, every strongly concave function is also strictly concave. A strongly convex function is similarly defined as the sum of a convex function and a stictly convex quadratic function. An example of a strictly concave function which is not strongly concave is again the function $f(x) = -x^4$ on R. To show this let $f(x) = \phi(x) - \frac{1}{2}\alpha x^2$, where ϕ is some concave function defined on R, and $\alpha > 0$. We can see that at the origin $f''(0) = 0 = \phi''(0) - \alpha$, or $\phi''(0) = \alpha > 0$, contradicting that ϕ is concave.

We can characterize strongly concave functions as follows:

PROPOSITION 5 (Poljak (1966), Rockafellar (1976), Diewert, Avriel and Zang (1981)):

a) Let f *be a real function defined on a convex set* $C \subset R^n$. *Then* f *is strongly concave if and only if there exists* $\alpha > 0$ *such that*

$$x^1 \in C,\ x^2 \in C,\ 0 \leqslant \lambda \leqslant 1 \Rightarrow f(\lambda x^1 + (1-\lambda)x^2) \geqslant \lambda f(x^1) + (1-\lambda) f(x^2) + \tfrac{1}{2}\lambda(1-\lambda)\alpha \| x^2 - x^1 \|^2. \tag{10}$$

b) If f *is differentiable on an open convex set* $C \subset R^n$, *then it is strongly concave if and only if there exists* $\alpha > 0$ *such that*

$$x^1 \in C,\ x^2 \in C \Rightarrow f(x^2) \leqslant f(x^1) + (x^2 - x^1)^T \nabla f(x^1) - \tfrac{1}{2}\alpha \| x^2 - x^1 \|^2. \tag{11}$$

c) If f *is twice continuously differentiable on an open convex set* $C \subset R^n$, *then it is strongly concave if and only if there exists* $\alpha > 0$ *such that*

$$x \in C \Rightarrow \nabla^2 f(x) + \alpha I \text{ is negative semidefinite} \tag{12}$$

where I *is the identity matrix.*

d) For a strongly convex function g, *conditions (10) to (12) become*

$$x^1 \in C,\ x^2 \in C,\ 0 \leqslant \lambda \leqslant 1 \Rightarrow g(\lambda x^1 + (1-\lambda)x^2) \leqslant \lambda g(x^1) + (1-\lambda) g(x^2) - \tfrac{1}{2}\lambda(1-\lambda)\alpha \| x^2 - x^1 \|^2, \tag{13}$$

$$x^1 \in C,\ x^2 \in C \Rightarrow g(x^2) \geq g(x^1) + (x^2-x^1)\nabla g(x^1) + \tfrac{1}{2}\alpha \|x^2 - x^1\|^2, \tag{14}$$

and

$$x \in C \Rightarrow \nabla^2 g(x) - \alpha I \text{ is positive semidefinite,} \tag{15}$$

respectively.

It follows then that the eigenvalues of the Hessian matrix of a twice differentiable strongly concave function are bounded away from zero by the negative number $-\alpha$.

Strong concavity is used in the economics literature, see e.g. Ginsberg (1973), Jorgenson and Lau (1974), and Lau (1978). If the producer's production function is strongly concave and twice continuously differentiable, then we can deduce the once continuous differentiability of the associated system of input demand and output supply functions that are generated by solving the producer's profit maximization problem. Moreover, under these conditions the supply function slopes upward and the input demand functions slope downward (with respect to their own prices); see Lau (1978) and Diewert (1981a).

The Cobb-Douglas production function $f(x_1,\ldots,x_n) \equiv \alpha_0 \prod_{i=1}^{n} x_i^{\alpha_i}$, where the positive parameters α_i satisfy $\sum_{i=1}^{n} \alpha_i < 1$ is a strongly concave function over any compact convex subset of the positive orthant.

4. QUASICONCAVE AND QUASICONVEX FUNCTIONS

DEFINITION 4 (Fenchel (1953), Avriel (1976)): a) A real function f *defined on a convex set* $C \subset R^n$ *is quasiconcave if*

$$x^1 \in C,\ x^2 \in C,\ f(x^2) \geq f(x^1),\ 0 \leq \lambda \leq 1 \Rightarrow f(\lambda x^1 + (1-\lambda)x^2) \geq f(x^1). \tag{16}$$

b) If f *is quasiconcave, then* $g \equiv -f$ *is quasiconvex.*

Comparing Definitions 1 and 4 one can immediately conclude that a concave function is also quasiconcave and a convex function is also quasiconvex. Examples of quasiconcave functions which are not concave are shown in Figure 4.

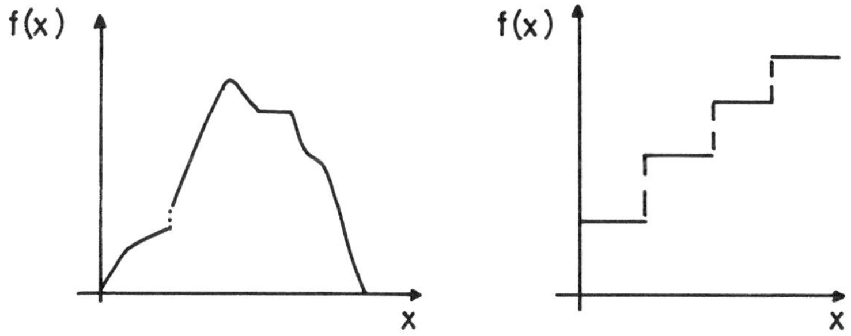

FIGURE 4. Quasiconcave functions

Note again that without making differentiability assumptions, the definition of quasiconcavity or quasiconvexity is given in terms of three collinear points. Contrary to concave functions, quasiconcave functions can be discontinuous in the interior of their domain, and they can have local maxima that are not global.

An alternative characterization of quasiconcave functions can be given in terms of their upper level sets; indeed this was the way they were originally defined by de Finetti.

PROPOSITION 6 (deFinetti (1949), Fenchel (1953)): a) Let f *be a real function defined on the convex set* $C \subset R^n$. *It is quasiconcave if and only if*

$$x \in C,\ \alpha \in R \Rightarrow U(\alpha) \equiv \{x : x \in C,\ f(x) \geqslant \alpha\} \text{ is a convex set.} \qquad (17)$$

b) It is quasiconvex if and only if

$$x \in C,\ \alpha \in R \Rightarrow L(\alpha) \equiv \{x : x \in C,\ f(x) \leqslant \alpha\} \text{ is a convex set.} \qquad (18)$$

Property (17) is a justification for the assumption that a consumer's utility function must be at least quasiconcave since it is a minimal assumption that the commodity purchases must have convex level sets, see Arrow and Enthoven (1961), and Diewert (1981a) for more discussion. Quasi-concave functions also arise in the producer's cost minimization problems, see Diewert (1981a).

In some management science applications, ratios such as cost/time, profit/time, profit/capital, benefit/cost, return/risk are to be optimized, see Schaible (1978) and Kallberg and Ziemba (1981). A function $f(x) = f_1(x)/f_2(x)$ is quasiconcave if $f_1(x)$ is concave, $f_2(x)$ is convex and both functions are positive. This can be seen from (17) by crossmultiplication. A ratio $f(x) = f_1(x)/f_2(x)$ is quasiconvex if $f_1(x)$ is convex, $f_2(x)$ is concave and f_1, f_2 are positive as seen from (18). Hence linear fractional functions $f(x) = a^T x/b^T x$ are both quasiconcave and quasiconvex for $b^T x > 0$ but $f(x)$ is neither concave nor convex in general. For a survey on ratio optimization see Schaible (1981a). Another quasiconcave function that is not concave is the Cobb-Douglas production function $f(x) = \alpha_0 \prod_{i=1}^{n} x_i^{\alpha_i}$ defined on the non-negative orthant when $\sum_{i=1}^{n} \alpha_i > 1$, and $\alpha_i > 0$. In addition to products and ratios, quadratic functions $f(x) = x^T Ax + b^T x$ have been investigated with respect to quasiconcavity and quasiconvexity properties; see Schaible (1981b). For example the function $f(x) = x_1^2 - x_2^2 - x_3^2$ is quasiconcave on the cones $\{x \in R^3 | f(x) \geqslant 0, x_1 \geqslant 0\}$, $\{x \in R^3 | f(x) \geqslant 0, x_1 \leqslant 0\}$. For a survey of criteria for quasiconcave quadratic functions see Schaible (1981c).

Turning to differentiable functions we have

PROPOSITION 7 (Arrow and Enthoven (1961), Avriel (1976)): a) Let f *be a real differentiable function defined on an open convex set* $C \subset R^n$. *Then* f *is quasiconcave if and only if*

$$x^1 \in C,\ x^2 \in C,\ f(x^2) \geq f(x^1) \Rightarrow (x^2-x^1)^T \nabla f(x^1) \geq 0. \qquad (19).$$

b) For a quasiconvex function the senses of the two inequalities in (19) are reversed.

As in the concave case, for differentiable functions the characterization involves conditions on two points. Comparing Propositions 1 and 7 it is also easy to see that concavity implies quasiconcavity.

Consider now a twice continuously differentiable function. The characterization of quasiconcavity in this case is more complicated than in the various types of concave functions. First we need

DEFINITION 5 (Diewert, Avriel and Zang (1981)): A real function f *defined on an open interval* $J \subset R$ *is said to attain a semistrict local minimum at a point* $t^0 \in J$ *if there exist two points* $t^1 \in J$, $t^2 \in J$ *such that*

$$t^1 < t^0 < t^2,\quad 0 \leq \lambda \leq 1 \Rightarrow (i)\ f(t^0) \leq f(\lambda t^1 + (1-\lambda)t^2) \text{ and } (ii)\ f(t^0) < \min\{f(t^1), f(t^2)\}. \qquad (20)$$

This definition means that a semistrict local minimum is a special type of local minimum in the sense that even if f is constant around x^0, the function eventually increases on both sides of x^0. Strict local minima are, of course, also semistrict local minima. Figure 5 illustrates this concept. A semistrict local maximum is defined analogously.

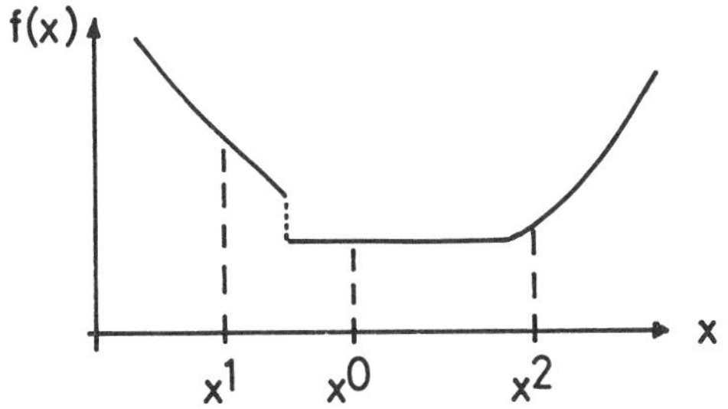

FIGURE 5. Semistrict local minimum

Now we can state another characterization of quasiconcave functions:

PROPOSITION 8 (Diewert, Avriel and Zang (1981)): a) Let f *be a real twice continuously differentiable function defined on an open convex set* $C \subset R^n$. *Then* f *is quasiconcave if and only if*

$$x \in C,\ v^T v = 1,\ v^T \nabla f(x) = 0 \Rightarrow (i)\ v^T \nabla^2 f(x) v < 0 \ \text{or} \tag{21}$$
$$(ii)\ v^T \nabla^2 f(x) v = 0 \ \text{and}\ F(t) = f(x+tv)$$

does not attain a semistrict local minimum at $t = 0$.

b) For a quasiconvex function the sense of the inequality in (i) is reversed and in (ii) $F(t)$ *does not attain a semistrict local maximum at* $t = 0$.

An equivalent characterization of quasiconcave functions is given next.

PROPOSITION 9 (Crouzeix (1977), Crouzeix and Ferland (1981)): a) Let f *be a real twice continuously differentiable function defined on an open convex set* $C \subset R^n$. *Then* f *is quasiconcave if and only if*

$$x \in C,\ v^T \nabla f(x) = 0 \Rightarrow (i)\ v^T \nabla^2 f(x) v < 0 \ \text{or} \tag{22}$$
$$(ii)\ v^T \nabla^2 f(x) v = 0 \ \text{and}\ F(t) \equiv f(x+tv)$$

is quasiconcave on the interval $J(x) = \{t : t \in R,\ x+tv \in C\}$

b) For a quasiconvex function the sense of the inequality in (i) is reversed and in (ii) $F(t)$ *is quasiconvex on* $J(x)$.

5. STRICTLY QUASICONCAVE AND STRICTLY QUASICONVEX FUNCTIONS

We now consider a special class of quasiconcave and quasiconvex functions.

DEFINITION 6 (Ponstein (1967), Katzner (1970)): a) A real function f *defined on a convex set* $C \subset R^n$ *is called strictly quasiconcave if*

$$x^1 \in C,\ x^2 \in C,\ x^1 \neq x^2,\ f(x^2) \geq f(x^1),\ 0 < \lambda < 1 \Rightarrow \quad f(\lambda x^1 + (1-\lambda)x^2) > f(x^1). \tag{23}$$

b) If f *is strictly quasiconcave, then* $g \equiv -f$ *is strictly quasiconvex.*

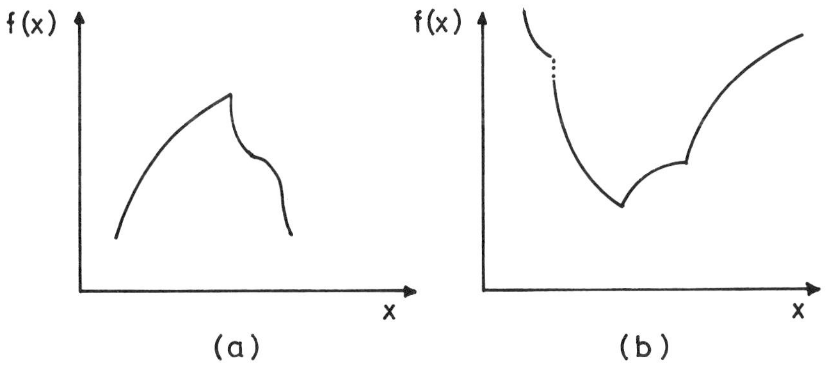

FIGURE 6. Strictly quasiconcave and strictly quasiconvex functions

Figures 6a and 6b depict strictly quasiconcave and strictly quasiconvex functions, respectively.

Strictly quasiconcave functions are also called "strongly quasiconcave" (Avriel (1976)) "unnamed concave" (Ponstein (1967)) and "X-concave" (Thompson and Parke (1973)).

Every strictly quasiconcave function is quasiconcave and every strictly concave function is strictly quasiconcave. However, not every concave function is strictly quasiconcave; for example a concave function which has a "flat" region cannot be strictly quasiconcave. A property shared by strictly quasiconcave and strictly concave functions is that they attain global maxima over their domain at no more than one point, and every local maximum is global. For more discussion see e.g. Diewert, Avriel and Zang (1981).

Strictly quasiconcave functions are also used frequently in economics. The assumption of strict quasiconcavity is made in both production and consumer theory to ensure that the solution to the producer's cost minimization problem or to the consumer's utility maximization problem is unique (see Diewert (1981a)). The assumption of strict quasiconcavity ensures the existence of cost minimizing input demand functions or utility maximizing consumer demand functions.

The Cobb-Douglas utility or production function $f(x) = \alpha_0 \prod_{i=1}^{n} x_i^{\alpha_i}$ for $\alpha_i > 0$ and $\sum_{i=1}^{n} \alpha_i \leqslant 1$ is strictly quasiconcave over the positive orthant. The ratio $f(x) = f_1(x)/f_2(x)$ is strictly quasiconcave if $f_1(x)$ is strictly concave, $f_2(x)$ is convex and f_1, f_2 are positive (see Schaible (1981a)).

For differentiable functions we can characterize strict quasiconcavity as follows:

PROPOSITION 10 (Diewert, Avriel and Zang (1981)): a) Let f *be a real differentiable function defined on an open convex set* $C \subset R^n$. *Then* f *is strictly quasiconcave if and only if*

$$x \in C,\ v^T v = 1,\ v^T \nabla f(x) = 0 \Rightarrow F(t) \equiv f(x+tv) \text{ does not attain a local minimum at } t = 0. \tag{24}$$

b) For a strictly quasiconvex function F(t) *does not attain a local maximum at* t = 0.

Note that whereas strict quasiconcavity is given in Definition 6 above is a simple strengthening of the definition of quasiconcavity, no such simple strengthening of the characterization given in Proposition 7 is known for differentiable strictly quasiconcave functions.

PROPOSITION 11 (Diewert, Avriel and Zang (1981)): a) Let f *be a real twice continuously differentiable function defined on an open convex set* $C \subset R^n$. *Then* f *is strictly quasiconcave if and only if*

$$x \in C,\ v^T v = 1,\ v^T \nabla f(x) = 0 \Rightarrow \textit{(i)}\ v^T \nabla^2 f(x) v < 0 \textit{ or} \qquad (25)$$
$$\textit{(ii)}\ v^T \nabla^2 f(x) v = 0 \textit{ and } F(t) \equiv f(x+tv)$$
does not attain a local minimum at $t = 0$.

b) For a strictly quasiconvex function the sense of the inequality in (i) is reversed and in (ii) $F(t)$ *does not attain a local maximum at* $t = 0$.

6. SEMISTRICTLY QUASICONCAVE AND SEMISTRICTLY QUASICONVEX FUNCTIONS

Let us focus our attention on a new class of functions.

DEFINITION 7 (Elkin (1968), Ginsberg (1973)): a) A real function f *defined on a convex set* $C \subset R^n$ *is called semistrictly quasiconcave if*

$$x^1 \in C,\ x^2 \in C,\ 0 < \lambda < 1,\ f(x^2) > f(x^1) \Rightarrow f(\lambda x^1 + (1-\lambda)x^2) > f(x^1). \quad (26)$$

b) if f *is semistrictly quasiconcave, then* $g \equiv -f$ *is semistrictly quasiconvex.*

Semistrictly quasiconcave functions are also called "strictly quasiconcave" (Mangasarian (1965), Ponstein (1967), Thompson and Parke (1973) and Avriel (1976)), "functionally concave" (Hanson (1964), and "explicitly quasiconcave" (Martos (1965)). Our terminology here (following Elkin (1968) and Ginsberg (1973)) is motivated by the fact that semistrictly quasiconcave functions, having certain continuity properties, have an intermediate position between strictly quasiconcave and quasiconcave functions. Lacking these continuity properties, semistrictly quasiconcave functions need not be quasiconcave. For example, let $C = \{x : x \in R,\ -1 \leqslant x \leqslant 1\}$ and let

$$f(x) = \begin{cases} 0 & \text{if } x = 0 \\ 1 & \text{if } x \in C,\ x \neq 0. \end{cases} \qquad (27)$$

This function satisfies (26), thus it is semistrictly quasiconcave, but it is not quasiconcave. However, it can be shown (see Karamardian (1967) or Mangasarian (1969)) that an upper semicontinuous semistrictly quasiconcave function is quasiconcave. Comparing Definitions 6 and 7 we can immediately conclude that a strictly quasiconcave function is also semistrictly quasiconcave. Similarly, a concave function is semistrictly quasiconcave. Every local maximum of a continuous semistrictly quasiconcave function is also global but, contrary to strictly quasiconcave functions, the maximum can be attained at more than one point (see Martos (1965) or Mangasarian (1965)).

In the differentiable case one can clearly see the intermediate position of semistrictly quasiconcave functions between strictly quasiconcave and quasiconcave functions. First we define a special type of local minimum.

DEFINITION 8 (Diewert, Avriel and Zang (1981)): A real function f *defined on an open interval* $J \subset R$ *is said to attain a one-sided semistrict local minimum at a point* $t^0 \in J$ *if there exist two points* $t^1 \in J$, $t^2 \in J$ *such that*

$$t^1 < t^0 < t^2,\ 0 \leqslant \lambda \leqslant 1 \Rightarrow \textit{(i)}\ f(t^0) \leqslant f(\lambda t^1 + (1-\lambda)t^2)\ \textit{and}$$
$$\textit{(ii)}\ f(t^0) < f(t^1)\ \textit{or}\ f(t^0) < f(t^2). \tag{28}$$

A one-sided semistrict local maximum is defined analogously. Clearly, a semistrict local minimum is also a one-sided semistrict local minimum which, in turn, is a local minimum. At a one-sided semistrict local minimum point the function must eventually increase at least on the right or the left, as can be seen in Figure 7.

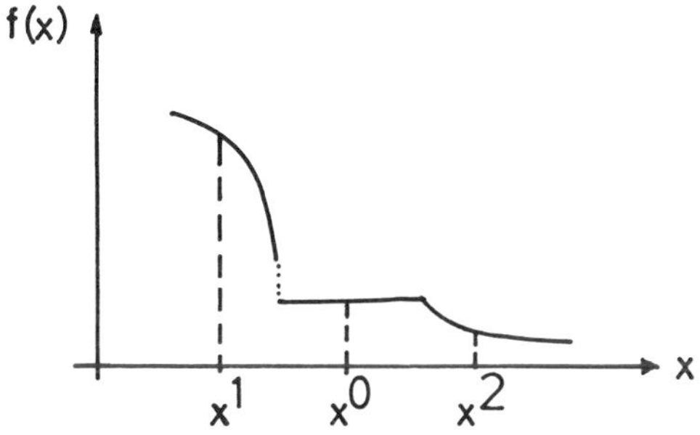

FIGURE 7. One-sided semistrict local minimum at x^0

PROPOSITION 12 (Diewert, Avriel and Zang (1981)): a) Let f *be a real differentiable function defined on an open convex set* $C \subset R^n$. *Then* f *is semistrictly quasiconcave if and only if*

$$x \in C,\ v^T v = 1,\ v^T \nabla f(x) = 0 \Rightarrow F(t) \equiv f(x+tv) \text{ does not attain a one-sided semistrict local minimum at } t = 0. \tag{29}$$

b) For a semistrictly quasiconvex function F(t) *does not attain a one-sided semistrict local maximum at* t = 0.

PROPOSITION 13 (Diewert, Avriel and Zang (1981)): a) Let f *be a real twice continuously differentiable function defined on an open convex set* $C \subset R^n$. *Then* f *is semistrictly quasiconcave if and only if*

$$x \in C,\ v^T v = 1,\ v^T \nabla f(x) = 0 \Rightarrow (i)\ v^T \nabla^2 f(x) v < 0 \text{ or } (ii)\ v^T \nabla^2 f(x) v = 0 \text{ and } F(t) \equiv f(x+tv) \text{ does not attain a one-sided semistrict local minimum at } t = 0. \tag{30}$$

b) For a semistrictly quasiconvex function the sense of the inequality in (i) is reversed and in (ii) F(t) *does not attain a one-sided semi-strict local maximum at* t = 0.

Semistrict quasiconcave functions have not been used in the economics literature a great deal.

7. PSEUDOCONCAVE, STRICTLY PSEUDOCONCAVE, PSEUDOCONVEX, AND STRICTLY PSEUDOCONVEX FUNCTIONS

An important property of a differentiable concave function is that a stationary point implies a global maximum. This property is, however, not restricted to concave functions only. The family of pseudoconcave functions includes all concave functions and also has this property. We introduce this family without assuming differentiability.

DEFININITION 9 (Ortega and Rheinboldt (1970), Thompson and Parke (1973)):

a) A real function defined on a convex set $C \subset R^n$ *is called pseudoconcave if*

$$x^1 \in C,\ x^2 \in C,\ 0 \leqslant \lambda \leqslant 1,\ f(x^2) > f(x^1) \Rightarrow f(\lambda x^1+(1-\lambda)x^2) \geqslant \geqslant f(x^1) + \lambda\beta(x^1,x^2) \tag{31}$$

where $\beta(x^1,x^2)$ *is a positive number, depending on* x^1 *and* x^2.

b) It is called strictly pseudoconcave if $x^1 \neq x^2$ *and*

$$f(x^2) \geqslant f(x^1),\ 0 < \lambda < 1 \Rightarrow f(\lambda x^1+(1-\lambda)x^2) \geqslant f(x^1) + \lambda\beta(x^1,x^2). \tag{32}$$

c) If f *is (strictly) pseudoconcave then* $g \equiv -f$ *is (strictly) pseudoconvex.*

A slightly more general definition of pseudoconcavity in the non-differentiable case can be found in Diewert (1981b). Pseudoconcave functions may also be defined on nonconvex sets.

The following characterization of a differentiable pseudoconcave function often serves as a definition of pseudoconcavity. This was the definition used when it was introduced by Mangasarian.

PROPOSITION 14 (Mangasarian (1965, 1969)): a) Let f *be a real differentiable function defined on an open convex set* $C \subset R^n$. *Then* f *is pseudoconcave if and only if*

$$x^1 \in C,\ x^2 \in C,\ f(x^2) > f(x^1) \Rightarrow (x^2 - x^1)^T \nabla f(x^1) > 0. \quad (33)$$

b) (Ponstein (1967)): The function f *is strictly pseudoconcave if and only if*

$$x^1 \in C,\ x^2 \in C,\ x^1 \neq x^2,\ f(x^2) \geqslant f(x^1) \Rightarrow (x^2 - x^1)^T \nabla f(x^1) > 0. \quad (34)$$

c) For a (strictly) pseudoconvex function the senses of the inequalities in (33) and (34) are reversed.

The property of (strictly) pseudoconcave functions that every stationary point is a (strict) global maximum over C can be easily seen by taking the contrapositives of (33) and (34) and setting $\nabla f(x^1) = 0$ (Mangasarian (1969)). It is also easy to see that (strict) concavity implies (strict) pseudoconcavity and, in turn, (strict) pseudoconcavity implies (strict) quasiconcavity (see Ponstein (1967)). Pseudoconcave functions are also semistrictly quasiconcave (Mangasarian (1969)).

Pseudoconcavity is used in many economic, statistical and optimization applications that are formulated as unconstrained maximization problems (e.g., the producer's profit maximization problem or econometric problems involving maximum likelihood estimation). If a differentiable objective function can be shown or assumed to be pseudoconcave, then the usual first order conditions for a maximum will serve to characterize the solution.

If strict pseudoconcavity can be assumed or shown, then the solution to the first order conditions is unique. Strictly pseudoconcave fractional functions also arise in portfolio theory. They have the form $f(x) = f_1(x)/f_2(x)$ where f_1 is positive and affine-linear and f_2 is positive and strictly convex. The function f_2 can take many forms such as $f_2(x) = (x^T A x)^{1/2}$ with A positive definite or $f_2(x) = [\sum_{i=1}^{n} \beta_i x_i^{\alpha}]^{1/\alpha}$ for

$\beta_i > 0$ and $1 < \alpha \leqslant 2$. The ratio $f(x) = f_1(x)/f_2(x)$ is pseudoconcave but not strictly so in the case $f_2(x) = [\sum_{j=1}^{n} (\beta_j)^{\alpha} |\sum_{i=1}^{n} a_{ij} x_i|^{\alpha}]^{1/\alpha}$ for $\beta_j > 0$ and $1 < \alpha \leqslant 2$. Then the maximum may not be unique. See Kallberg and Ziemba (1981) for more details.

An equivalent characterization to Proposition 14 is:

PROPOSITION 15 (Diewert, Avriel and Zang (1981)): a) Let f *be a real differentiable function defined on an open convex set* $C \subset R^n$. *Then* f *is (strictly) pseudoconcave if and only if*

$$x \in C,\ v^T v = 1,\ v^T \nabla f(x) = 0 \Rightarrow F(t) \equiv f(x+tv) \text{ attains a (strict) local maximum at } t = 0. \tag{35}$$

b) For a (strictly) pseudoconvex function F(t) *attains a (strict) local minimum at* t = 0.

In the twice differentiable case we have

PROPOSITION 16 (Diewert, Avriel and Zang (1981)): a) Let f *be a twice continuously differentiable function defined on an open convex set* $C \subset R^n$. *Then* f *is (strictly) pseudoconcave if and only if*

$$x \in C,\ v^T v = 1,\ v^T \nabla f(x) = 0 \Rightarrow \text{(i) } v^T \nabla^2 f(x) v < 0 \text{ or (ii) } v^T \nabla^2 f(x) v = 0 \text{ and } F(t) \equiv f(x+tv) \text{ attains a (strict) local maximum at } t = 0. \tag{36}$$

b) For a (strictly) pseudoconvex function the sense of the inequality in (i) is reversed and in (ii) F(t) *attains a (strict) local minimum at* t = 0.

8. STRONGLY PSEUDOCONCAVE AND STRONGLY PSEUDOCONVEX FUNCTIONS

The last family of functions introduced in this paper is a subfamily of differentiable strictly pseudoconcave functions with the property that if the directional derivative of the function at x is zero in a direction v then the function decreases at least quadratically in the neighborhood of x along v.

DEFINITION 10 (Diewert, Avriel and Zang (1981)): a) A real differentiable function f *defined on an open convex set* $C \subset R^n$ *is called strongly pseudoconcave if it is strictly pseudoconcave and*

$x \in C,\ v^T v = 1,\ v^T \nabla f(x) = 0 \Rightarrow$ *there exist positive numbers* ε *and* α *such that* $(x \pm \varepsilon v) \in C$ *and*

$$F(t) \equiv f(x+tv) \leqslant F(0) - \tfrac{1}{2}\alpha t^2 \text{ for } |t| \leqslant \varepsilon . \qquad (37)$$

b) Similarly, g *is called strongly pseudoconvex if it is strictly pseudoconvex and*

$x \in C,\ v^T v = 1,\ v^T \nabla f(x) = 0 \Rightarrow$ *there exist positive numbers* ε *and* α *such that* $(x \pm \varepsilon v) \in C$ *and* $G(t) \equiv g(x+tv) \geqslant G(0) + \tfrac{1}{2}\alpha t^2$

$$\text{for } |t| \leqslant \varepsilon \qquad (38)$$

The concept of strong pseudoconcavity is used in the theory of consumer demand in order to ensure the existence and continuous differentiability of the consumer's system of commodity demand functions; see Blackorby and Diewert (1979)or Diewert (1981a).

The Cobb-Douglas function $f(x) \equiv \alpha_0 \prod_{i=1}^{n} x_i^{\alpha_i},\ \alpha_i > 0,\ \sum_{i=1}^{n} \alpha_i \leqslant 1$ is strongly pseudoconcave over the positive orthant.

A definition in the nondifferentiable case appears in Diewert (1981b). Strongly pseudoconcave functions are also called "strongly quasiconcave" (Newman (1969), Ginsberg (1973))and "differentially strictly quasiconcave" (McFadden (1978)).

Next we define the concept of a strong local maximum.

DEFINITION 11 (Diewert, Avriel and Zang (1981)): A real function f *defined on an open interval* $J \subset R$ *attains a strong local maximum at* t^0 *if there exist positive numbers* ε *and* α *such that* $(t^0 \pm \varepsilon) \in J$ *and* $f(t) \leqslant f(t^0) - \tfrac{1}{2}\alpha(t-t^0)^2$ *for* $|t| \leqslant \varepsilon$.

A strong local minimum is defined analogously. We now have

PROPOSITION 17 (Diewert, Avriel and Zang (1981)): a) Let f *be a real differentiable function defined on an open convex set* $C \subset R^n$. *Then* f *is strongly pseudoconcave if and only if*

$$x \in C,\ v^T v = 1,\ v^T \nabla f(x) = 0 \Rightarrow F(t) \equiv f(x+tv) \text{ attains a strong local maximum at } t = 0. \qquad (39)$$

b) For a strongly pseudoconvex function F(t) *attains a strong local minimum at* t = 0.

Finally,

PROPOSITION 18 (Diewert, Avriel and Zang (1981)): a) Let f *be a real twice continuously differentiable function defined on an open convex set* $C \subset R^n$. *Then* f *is strongly pseudoconcave if and only if*

$$x \in C,\ v^T v = 1,\ v^T \nabla f(x) = 0 \Rightarrow v^T \nabla^2 f(x) v < 0. \qquad (40)$$

b) For a strongly pseudoconvex function the sense of the inequality in (40) is reversed.

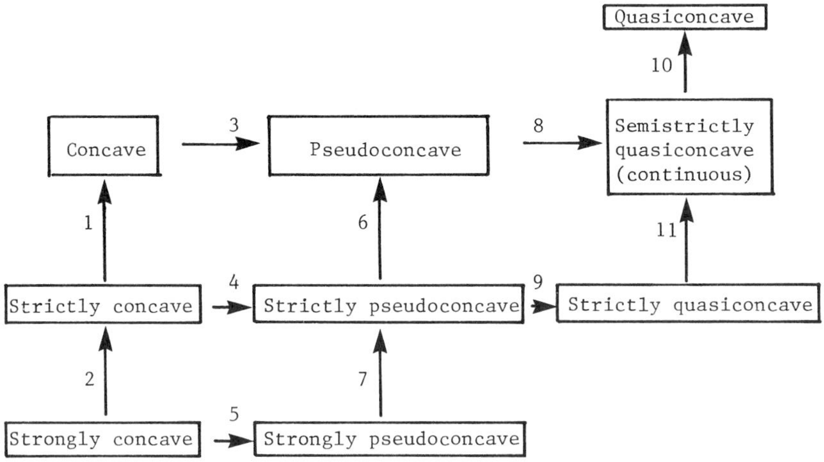

FIGURE 8. Relationships between families of concave and generalized concave functions

9. RELATIONSHIPS BETWEEN THE NINE KINDS OF CONCAVITY AND GENERALIZED CONCAVITY

The relationships between the various function classes is illustrated in Figure 8 and the hierarchy of these classes is exhibited. Table 1 supplements this figure by giving specific examples of functions showing that the inclusions indicated are proper. The numbers appearing next to implications correspond to the numbers of the examples in Table 1.

TABLE 1: Examples of Various Concave and Generalized Concave Functions

	Function	Domain	
1.	$f_1(x) \equiv x$	$-\infty < x < +\infty$	concave, but not strictly concave
2.	$f_2(x) \equiv -(x)^4$	$-\infty < x < \infty$	strictly concave, but not strongly concave
3.	$f_3(x) \equiv x + (x)^3$	$-\infty < x < +\infty$	pseudoconcave, but not concave
4.	$f_4(x) \equiv e^{-(x)^2}$	$-\infty < x < +\infty$	strictly pseudoconcave, but not strictly concave
5.	$f_5(x) \equiv \begin{cases} -(x)^4 \\ 3-4x \end{cases}$	$-1 \leqslant x \leqslant 1$; $x > 1$	strongly pseudoconcave, but not strongly concave
6.	$f_6(x) \equiv \begin{cases} e^{-(x)^2} \\ 1 \end{cases}$	$-\infty < x \leqslant 0$; $x > 0$	pseudoconcave, but not strictly pseudoconcave
7.	$f_7(x) \equiv e^{-(x)^4}$	$-\infty < x < +\infty$	strictly pseudoconcave, but not strongly pseudoconcave
8.	$f_8(x) \equiv (x)^3$	$-\infty < x < +\infty$	semistrictly quasiconcave, but not pseudoconcave
9.	$f_9(x) \equiv \begin{cases} (x)^2 \\ -(x)^2 \end{cases}$	$0 \leqslant x < +\infty$; $-\infty < x \leqslant 0$	strictly quasiconcave, but not strictly pseudoconcave
10.	$f_{10}(x) \equiv \begin{cases} 0 \\ (x)^2 \end{cases}$	$-\infty < x \leqslant 0$; $x > 0$	quasiconcave, but not semistrictly quasiconcave
11.	$f_{11}(x) \equiv \begin{cases} -(x)^2 \\ 0 \\ -(x-1)^2 \end{cases}$	$-\infty < x \leqslant 0$; $0 < x \leqslant 1$; $x > 1$	semistrictly quasiconcave, but not strictly quasiconcave

REFERENCES

ARROW, K.J. and ENTHOVEN, A.C. (1961). "Quasiconcave Programming.", *Econometrica 29*, 779-800.

AVRIEL, M. (1972). "r-Convex Functions." *Mathematical Programming 2*, 309-323.

AVRIEL, M. (1976). *Nonlinear Programming: Analysis and Methods.* Prentice-Hall, Englewood Cliffs, N.J.

AVRIEL, M., DIEWERT, W.E., SCHAIBLE, S. and ZANG, I. *Generalized Concavity*, (forthcoming).

BAZARAA, M.S. and SHETTY, C.M. (1976). *Foundations of Optimization.* Springer-Verlag, Berlin.

BELLMAN, R. (1960). *Introduction of Matrix Analysis.* McGraw-Hill Book Co., New York.

BERNSTEIN, B. and TOUPIN, R.A. (1962). "Some Properties of the Hessian Matrix of a Strictly Convex Function." *Journal für die Reine und Angewandte Mathematik 210*, 65-72.

BLACKORBY, C. and DIEWERT, W.E. (1979). "Expenditure Functions, Local Duality and Second Order Approximations." *Econometrica 47*, 579-601.

CROUZEIX, J.-P. (1977). "Contributions a l'étude des Fonctions Quasi-convexes." Ph.D. Dissertation, Université de Clermont, France.

CROUZEIX, J.-P. (1981a). "Continuity and Differentiability Properties of Quasiconvex Functions on R^n." This volume, 109-130.

CROUZEIX, J.-P. (1981b). "A Duality Framework in Quasiconvex Programming." This volume, 207-225.

CROUZEIX, J.-P. and FERLAND, J.A. (1981). "Criteria for Quasi-convexity and Pseudoconvexity: Relationships and Comparisons."This vol.,199-204.

DEBREU, G. (1952). "Definite and Semidefinite Quadratic Forms." *Econometrica 20*, 295-300.

DIEWERT, W.E. (1981a). "Generalized Concavity and Economics." This volume, 511-541.

DIEWERT, W.E. (1981b). "Alternative Characterizations of Six Kinds of Quasiconcavity in the Nondifferentiable Case with Applications to Nonsmooth Programming." This volume, 51-93.

DIEWERT, W.E. and WOODLAND, A.D. (1977). "Frank Knight's Theorem in Linear Programming Revisited." *Econometrica 45*, 375-398.

DIEWERT, W.E., AVRIEL, M. and ZANG, I. (1981). "Nine Kinds of Quasi-concavity and Concavity." *Journal of Economic Theory*, (forthcoming).

EGGLESTON, H.D. (1958). *Convexity.* Cambridge University Press, Cambridge, England.

EICHHORN, W. (1981). "Concavity and Quasiconcavity in the Theory of Production." This volume, 627-636.

ELKIN, R.M. (1968). "Convergence Theorems for Gauss-Seidel and Other Minimization Algorithms." Ph.D. Dissertation, University of Maryland, College Park, Md.

FENCHEL, W. (1953). "Convex Cones, Sets and Functions." Lecture Notes, Department of Mathematics, Princeton University, Princeton, N.J.

FERLAND, J.A. (1972). "Mathematical Programming Problems with Quasi-Convex Objective Functions." *Mathematical Programming 3*, 296-301.

DE FINETTI,B. (1949). "Sulle Stratificazioni Convesse." *Ann. Math. Pura Appl. 30*, 173-183.

GINSBERG, W. (1973). "Concavity and Quasiconcavity in Economics." *Journal of Economic Theory 6*, 596-605.

GREENBERG, H.J. and PIERSKALLA, W.P. (1973). "A Review of Quasi-Convex Functions." *Operations Research 19,* 1553-1570.

HANSON, M.A. (1964). "Bounds for Functionally Convex Optimal Control Problems." *Journal of Mathematical Analysis and Applications 8,* 84-89.

HÖLDER, O. (1889). "Über einen Mittelwertsatz." *Nachr. Ges. Wiss. Goettingen,* 38-47.

JENSEN, J.L.W.V. (1906). "Sur les Fonctions Convexes et les Inégalities Entre les Valeurs Moyennes." *Acta Math. 30,* 175-193.

JORGENSON, D.W. and LAU, L.J. (1974). "Duality and Differentiability in Production." *Journal of Economic Theory 9,* 23-42.

KALLBERG, J.G. and ZIEMBA, W.T. (1981). "Generalized Concave Functions in Stochastic Programming and Portfolio Theory." This volume, 719-767.

KANNAI, Y. (1977). "Concavifiability and Constructions of Concave Utility Functions." *Journal of Mathematical Economics 4,* 1-56.

KANNAI, Y. (1981). "Concave Utility Functions - Existence, Constructions and Cardinality." This volume, 543-611.

KARAMARDIAN, S. (1967). "Duality in Mathematical Programming." *Journal of Mathematical Analysis and Applications 20,* 344-358.

KATZNER, D.W. (1970). *Static Demand Theory.* MacMillan, New York, N.Y.

LAU, L.J. (1978). "Applications of Profit Functions." In *Production Economics: A Dual Approach to Theory and Applications.* (M. Fuss and D. McFadden, eds.), Vol. 1 133-216. North-Holland, Amsterdam.

MANGASARIAN, O.L. (1965). "Pseudo-Convex Functions." *SIAM Journal Control 3,* 281-290.

MANGASARIAN, O.L. (1969). *Nonlinear Programming.* McGraw-Hill Book Co., New York, N.Y.

MARTOS, B. (1965). "The Direct Power of Adjacent Vertex Programming Methods." *Management Science 12,* 241-252.

MARTOS, B. (1975). *Nonlinear Programming: Theory and Methods.* American Elsevier, New York, N.Y.

MCFADDEN, D. (1978). "Convex Analysis." In *Production Economics: A Dual Approach to Theory and Applications.* (M. Fuss and D. McFadden, eds.), Vol. 1, 383-408. North-Holland, Amsterdam.

MINKOWSKI, H. (1910). *Geometrie der Zahlen.* Teubner, Leipzig.

MINKOWSKI, H. (1911). *Theorie der Konvexen Körper, Insbesondere Begründung ihres Oberflächenbegriffs.* Gesammelte Abhandlungen II, Leipzig.

NEWMAN, P. (1969). "Some Properties of Concave Functions." *Journal of Economic Theory 1,* 291-314.

ORTEGA, J.M. and RHEINBOLDT, W.C. (1970). *Iterative Solution of Nonlinear Equations in Several Variables.* Academic Press, New York, N.Y.

POLJAK, B.T. (1966). "Existence Theorems and Convergence of Minimizing Sequences in Extremum Problems with Restrictions." *Doklady Akademii Nauk C.C.P. 166,* 287-290. (English translation in *Soviet Mathematics 7,* 72-75.)

PONSTEIN, J. (1967). "Seven Kinds of Convexity." *SIAM Review 9,* 115-119.

ROCKAFELLAR, R.T. (1970). *Convex Analysis.* Princeton University Press, Princeton, N.J.

ROCKAFELLAR, R.T. (1976). "Saddle Points of Hamiltonian Systems in Convex Lagrange Problems Having a Nonzero Discount Rate." *Journal of Economic Theory 12,* 71-113.

ROBERTS, A.W. and VARBERG, D.E. (1973). *Convex Functions.* Academic Press, New York, N.Y.

SAMUELSON, P.A. (1947). *Foundations of Economic Analysis.* Harvard University Press, Cambridge, Mass.
SCHAIBLE, S. (1978). *Analyse und Anwendungen von Quotienten programmen.* Hain-Verlag, Meisenheim.
SCHAIBLE, S. (1981a). "A Survey of Fractional Programming." This vol.417-440
SCHAIBLE, S. (1981b). "Quasiconvex, Pseudoconvex and Strictly Pseudoconvex Quadratic Functions." *Journal of Optimization Theory and Applications,* (forthcoming).
SCHAIBLE, S. (1981c). "Generalized Convexity of Quadratic Functions." This volume, 183-197.
STOER, J. and WITZGALL, C. (1970). *Convexity and Optimization in Finite Dimensions I.* Springer-Verlag, New York, N.Y.
THOMPSON JR., W. A. and PARKE, D.W. (1973). "Some Properties of Generalized Concave Functions." *Operations Research 21,* 305-313.
VALENTINE, F.A. (1964). *Convex Sets.* McGraw-Hill Book Co., New York, N.Y.

ALTERNATIVE CHARACTERIZATIONS OF SIX KINDS OF QUASICONCAVITY IN THE NONDIFFERENTIABLE CASE WITH APPLICATIONS TO NONSMOOTH PROGRAMMING*

W.E. Diewert

Very useful characterizations of quasiconcavity and pseudoconcavity in the once differentiable case are: (i) $f(x^2) \geq f(x^1)$ implies $(x^2-x^1)^T \nabla f(x^1) \geq 0$ and (ii) $(x^2-x^1)^T \nabla f(x^1) \leq 0$ implies $f(x^2) \leq f(x^1)$ respectively. These characterizations can readily be generalized to the case where only one sided directional derivatives of the function exist. If a function is concave, then one sided directional derivatives of the function always exist in the interior of the function's domain of definition. However, one sided derivatives do not necessarily exist for quasiconcave functions. The main purpose of the present paper is to provide alternative characterizations of six types of quasiconcavity (quasiconcavity, semistrict quasiconcavity, strict quasiconcavity, pseudoconcavity, strict pseudoconcavity and strong quasiconcavity) similar to the characterizations given above, but without assuming that one sided directional derivatives exist. Instead of directional derivatives or generalized gradients, we use the classical Dini derivates. Our results rest on a generalization of the Mean Value Theorem valid for Dini derivates of upper semicontinuous functions. Our generalization is similar to a generalization developed by Hiriart-Urruty. Finally, we develop some very simple necessary conditions for a nonsmooth programming problem and show that they are also sufficient if the objective function is pseudoconcave and the constraint functions are quasiconcave. We also show how our new characterizations of quasiconcavity and pseudoconcavity can be used in order to prove the sufficiency of some generalized Kuhn-Tucker conditions for a nonsmooth nonlinear programming problem. Our proof of this theorem is a modification of an approach pioneered by Mangasarian.

1. INTRODUCTION

Quasiconcave, pseudoconcave and other generalizations of concave functions have proven to be very useful in economics, operations research and optimization theory. One of the most useful characterizations of

*A preliminary version of this paper was presented at the Tenth International Symposium on Mathematical Programming, Montreal, August 27-31, 1979. This research has been supported by the Social Sciences and Humanities Research Council of Canada. The author thanks M. Avriel, F. Clarke, J.B. Hiriart-Urruty, Y. Kannai and W. Oettli for helpful suggestions and the National Bureau of Economic Research, Stanford, California, for providing office space.

ISBN 0-12-621120-5

quasiconcave functions, due to Arrow and Enthoven (1961; 780), is: $x^1 \in S$, $x^2 \in S$, $f(x^2) \geq f(x^1)$ implies $(x^2-x^1)^T \nabla f(x^1) \geq 0$ where $S \subset R^N$ is the domain of definition of the function f, $\nabla f(x^1)$ is the gradient vector of f evaluated at x^1 and x^Ty denotes the inner product of the vectors x and y. Similarly, Mangasarian's (1965; 281) definition of a pseudoconcave function, $x^1 \in S$, $x^2 \in S$, $(x^2-x^1)^T \nabla f(x^1) \leq 0$ implies $f(x^2) \leq f(x^1)$, also involves the gradient vector of f. Thus both of the above characterizations require the assumption of differentiability.

However, as Rockafellar (1978) points out, in many practical applied problems, the objective and constraint functions are not necessarily differentiable. Thus there is now a growing body of literature which derives optimality conditions (e.g., Clarke (1976); Hiriart-Urruty (1978)) and provides algorithms which allow one to optimize (e.g., Mifflin (1977a;1977b)) without assuming differentiability.

The reader will recall that in Avriel, Diewert, Schaible and Ziemba (1981), characterizations for nine kinds of generalized concave functions were presented for the general case as well as for the cases where the functions were once and twice differentiable. It can be shown (see for example Fenchel (1953)) that concave, strictly concave and strongly concave functions are always one sided directionally differentiable; i.e., the limit $D_v^+ f(x) \equiv \lim_{t \to 0^+} [f(x+tv) - f(x)]/t$ exists for all points x in the interior of the domain of definition of f and for all directions v if f is concave. Moreover, convenient characterizations in terms of the one sided directional derivatives exist for the above three kinds of concave function.

The main purpose of this paper is to provide alternative characterizations in terms of one sided directional derivatives for the six kinds of quasiconcave functions which are not necessarily concave which were discussed in Avriel, Diewert, Schaible and Zang (1981). We want characterizations similar to those of Arrow and Enthoven and Mangasarian noted above, but without necessarily assuming differentiability, since many practical applications of quasiconcave functions involve nondifferentiable functions.

Unfortunately, in contrast to the case of concave functions, quasiconcave functions do *not* necessarily have one sided directional derivatives, as the following example shows.

Example 1. $f(x) \equiv \begin{cases} 0 & \text{for } -1 < x \leq 0 \\ 1/2^{n+1} & \text{for } 1/2^{n+1} \leq x < 1/2^n;\ n=0,1,. \end{cases}$

f defined above is an upper semicontinuous step function defined for $-1 < x < 1$, which for positive x lies between two linear functions passing through the origin; i.e., $x/2 \leq f(x) \leq x$ for $0 \leq x < 1$. It can be verified that f is quasiconcave over its domain of definition and that the upper one sided directional derivative of f does not exist at $x = 0$. Thus we cannot hope to characterize *general* quasiconcave functions in terms of one sided directional derivatives. In place of the one sided directional derivative concept, we utilize the classical Dini (1892) derivate. We use Dini derivates rather than the more recently developed generalized gradients of Clarke (1975), Hiriart-Urruty (1977;1978) and Rockafellar (1978) since the generalized gradient requires that the function under consideration satisfy a Lipschitz condition, and Example 1 above shows that quasiconcave functions need not satisfy Lipschitz conditions.

It should be noted that in virtually all practical applications of quasiconcave functions, one sided directional derivatives will exist. Hence the empirically oriented reader may well wonder why we have not simply assumed one sided directional differentiability. The reason is that the latter assumption does not significantly simplify our proofs. Thus we establish our new characterizations for the six kinds of quasiconcavity without making any a priori restrictive differentiability assumptions.

Our primary tool is a generalization of the classical mean value theorem to nondifferentiable functions. This generalization appears to be different than other generalizations which have appeared in Young and Young (1909), Wegge (1974), Lebourg (1975) and Hiriart-Urruty (1979a) in that we do not require that the function be either convex, continuous or satisfy a Lipschitz condition. However, our mean value theorem is easily deduced as a corollary to a theorem of Hiriart-Urruty (1979b, Proposition 1).

In section 2 below, we follow the example of Dieudonné (1960; 153) by presenting our generalization of the mean value theorem in terms of inequalities rather than as an equality. As mentioned above, in place of the derivative concept, we utilize the classical Dini (1892) derivates.

In section 3, we use the mean value theorem in section 2 in order to obtain characterizations of quasiconcavity in the nondifferentiable case. A similar program is implemented in section 4 for semistrict quasiconcavity, in section 5 for strict quasiconcavity, in section 6 for pseudoconcavity, in section 7 for strict pseudoconcavity, and in section 8 for strong quasiconcavity (pseudoconcavity).

Finally, in section 9, we develop some very simple necessary conditions for a nondifferentiable programming problem and show that they are also sufficient if the objective function is pseudoconcave and the constraint functions are quasiconcave. We also show how our new characterizations of quasiconcavity and pseudoconcavity can be used in order to prove the sufficiency of some generalized Kuhn-Tucker (1951) conditions for a nonsmooth programming problem.

2. A GENERALIZED MEAN VALUE THEOREM

Let f be a real valued function defined over S, a convex subset of R^N. Let $x \in R^N$, $v \in R^N$ with $v^T v \equiv \sum_{i=1}^{N} v_i^2 = 1$ (and thus v is a directional vector). We define the *Dini (1892) derivates* of f in the direction v at x as:

$$D_v^{+u} f(x) \equiv \sup_{\{t_n\}} \lim_{n\to\infty} \{[f(x+t_n v) - f(x)] / t_n : 0 < t_n \leq 1/n\} ,$$

$$D_v^{+\ell} f(x) \equiv \inf_{\{t_n\}} \lim_{n\to\infty} \{[f(x+t_n v) - f(x)] / t_n : 0 < t_n \leq 1/n\} ,$$

$$D_v^{-u} f(x) \equiv \sup_{\{t_n\}} \{\lim_{n\to\infty} [f(x-t_n v) - f(x)] / [-t_n] : 0 < t_n \leq 1/n\} ,$$

$$D_v^{-\ell} f(x) \equiv \inf_{\{t_n\}} \{\lim_{n\to\infty} [f(x-t_n v) - f(x)] / [-t_n] : 0 < t_n \leq 1/n\} .$$

In the first two definitions above, we require the existence of $t>0$ such that $x + tv \in S$ while the last two definitions require a $t>0$ such that $x - tv \in S$.

$D_v^{+u} f(x)$ is the *upper right derivate*, $D_v^{+\ell} f(x)$ is the *lower right derivate*, $D_v^{-u} f(x)$ is the *upper left derivate*, $D_v^{-\ell} f(x)$ is the *lower left derivate* of f evaluated at x. We allow infinite limits in the above definitions. It is easy to see that the following hold: $D_v^{+\ell} f(x) \leq D_v^{+u} f(x)$; $D_v^{-\ell} f(x) \leq D_v^{-u} f(x)$; $D_v^{-u} f(x) = -D_{-v}^{+\ell} f(x)$; $D_v^{-\ell} f(x) = -D_{-v}^{+u} f(x)$. If $D_v^{+u} f(x) = D_v^{+\ell} f(x)$ (or if $D_v^{-u} f(x) = D_v^{-\ell} f(x)$), then we say f has *right (left) derivative* at x in the direction v and we denote the common value as $D_v^{+} f(x)$ ($D_v^{-} f(x)$). Finally, if $D_v^{+} f(x) = D_v^{-} f(x)$, then we say f has a *derivative* at x in the direction v or that f is *differentiable* at x in the direction v, and we denote the common value as the directional derivative $D_v f(x)$. If f is a function of one variable, then usually we take the directional vector v to be the scalar 1 and we define $D^{+u} f(x) \equiv D_1^{+u} f(x)$, $D^{+\ell} f(x) \equiv D_1^{+\ell} f(x)$, $D^{+} f(x) \equiv D_1^{+} f(x)$, $Df(x) \equiv f'(x) \equiv D_1 f(x)$; i.e., when there is no subscript v, we assume $v \equiv 1$. (It should be noted that Scheeffer (1884) uses the notation D^+,

D_+, D^- and D_- to denote our D^{+u}, $D^{+\ell}$, D^{-u} and $D^{-\ell}$ respectively; however, we have used the symbol D^+ to denote the one sided directional derivative.) Note that in Example 1, $D^{+u}f(0) = 1$, $D^{+\ell}f(0) = \frac{1}{2}$ and $D^{-u}f(0) = D^{-\ell}f(0) = D^-f(0) = 0$.

The useful feature about Dini derivates is that they always exist, whereas right derivatives need not. We use Dini derivates in the theorem below.

THEOREM 1 (Generalized Mean Value Theorem): Let g *be an upper semicontinuous function (i.e., the sets* $U(u) \equiv \{x : g(x) \geq u\}$ *are closed for every real number* u*) of one variable defined over the closed interval* $[t^1,t^2]$ *where* $-\infty < t^1 < t^2 < +\infty$. *Define* $\alpha \equiv (g(t^2) - g(t^1)) / (t^2-t^1)$. *Then* ***either***

$$D^{+\ell}g(t^1) \leq D^{+u}g(t^1) \leq \alpha \leq D^{-\ell}g(t^2) \leq D^{-u}g(t^2) \tag{1}$$

or *there exists* $t^0 \varepsilon (t^1,t^2)$, *where* (t^1,t^2) *is the open interval between* t^1 *and* t^2, *such that*

$$D^{+\ell}g(t^0) \leq D^{+u}g(t^0) \leq \alpha \leq D^{-\ell}g(t^0) \leq D^{-u}g(t^0) \tag{2}$$

or *both* (1) *and* (2) *hold.*

PROOF: Define

$$\alpha \equiv [g(t^2) - g(t^1)]/[t^2-t^1]. \tag{3}$$

Since $g(t) - \alpha t$ is upper semicontinuous over the compact set $[t^1,t^2]$, the function assumes a maximum at some point $t^0 \varepsilon [t^1,t^2]$ by Berge's (1963; 76) Maximum Theorem. Thus, for every $t \varepsilon [t^1,t^2]$,

$$g(t) - \alpha t \leq g(t^0) - \alpha t^0 \quad \text{or} \quad g(t) - g(t^0) \leq \alpha(t-t^0). \tag{4}$$

Case (i): $t^0 \varepsilon (t^1,t^2)$. If $t^0 < t \leq t^2$, then (4) becomes $(g(t) - g(t^0)) / (t-t^0) \leq \alpha$ and thus $D^{+\ell}g(t^0) \leq D^{+u}g(t^0) \leq \alpha$. On the other hand, if $t^1 \leq t < t^0$, then (4) becomes $(g(t) - g(t^0)) / (t-t^0) \geq \alpha$ and thus $\alpha \leq D^{-\ell}g(t^0) \leq D^{-u}g(t^0)$. Thus in this case, (2) is true.

Case (ii): $t^0=t^1$. As in the previous case, $D^{+\ell}g(t^1) \leq D^{+u}g(t^1) \leq \alpha$. Now consider (4) with $t^0 \equiv t^1$ and $t \equiv t^2$. We obtain the following inequality:

$$g(t^2) - g(t^1) \leq \alpha(t^2-t^1). \tag{5}$$

However, from the definition of α, (5) must hold as an equality. Thus $(g(t^2),t^2)$ must also be a solution to the maximization problem below (3). As in case (i), we find that $\alpha \leq D^{-\ell}g(t^2) \leq D^{-u}g(t^2)$. Thus in this case, (i) is true.

Case (iii): $t^0=t^2$. This case is analogous to case (ii), and we find that (1) holds. Q.E.D.

When the maximum for the maximization problem in (3) occurs only at an end point of the interval $[t^1,t^2]$ (e.g., consider $g(t) \equiv 0$ for $t^1 \leq t < t^2$, $g(t^2) \equiv 1$), then we obtain only the inequalities in (1). On the other hand, when the maximum occurs at an interior point of the interval $[t^1,t^2]$ so that case (i) occurs, then we obtain the geometrically appealing inequalities in (2), and a generalization of Wegge's (1974) Mean Value Theorem for concave functions. We also note that Theorem 1 is not true without the hypothesis of upper semicontinuity (e.g., consider $g(t) \equiv 0$ for $t^1 \leq t < t^2, g(t^2) \equiv -1$).

COROLLARY 1: Let g *be a lower semicontinuous function of one variable (i.e., the sets* $L(u) \equiv \{x : g(x) \leq u\}$ *are closed for every real number* L*), defined over* $[t^1,t^2]$ *and define* α *as before. Then* *__either__*

$$D^{+u}g(t^1) \geq D^{+\ell}g(t^1) \geq \alpha \geq D^{-u}g(t^2) \geq D^{-\ell}g(t^2) \tag{6}$$

__or__ there exists $t^0 \varepsilon (t^1,t^2)$ *such that*

$$D^{+u}g(t^0) \geq D^{+\ell}g(t^0) \geq \alpha \geq D^{-u}g(t^0) \geq D^{-\ell}g(t^0) \tag{7}$$

__or__ both (6) and (7) hold.

The proof of this corollary is analogous to the proof of Theorem 1 except that various inequalities are reversed.

COROLLARY 2: If g *is continuous over* $[t^1,t^2]$*, then at least one of the inequalities (2) or (7) will hold.*

PROOF: Suppose g is continuous over $[t^1,t^2]$ and both (1) and (6) occurred. Then for every $t\varepsilon[t^1,t^2]$, $g(t)-g(t^1) \leq \alpha(t-t^1)$ and $g(t)-g(t^1) \geq \alpha(t-t^1)$ so that $g(t) = g(t^1) + \alpha(t-t^1)$. Thus $g'(t) = \alpha$ for every $t\varepsilon(t^1,t^2)$ and (2) and (7) will both hold (as inequalities).

COROLLARY 3 (Mean Value Theorem): If g *is continuous over* $[t^1,t^2]$ *and differentiable over* (t^1,t^2)*, then there exists* $t^0\varepsilon(t^1,t^2)$ *such that* $g'(t^0) = (g(t^2) - g(t^1)) / (t^2-t^1) \equiv \alpha$.

PROOF: From the previous corollary, there exists $t^0\varepsilon(t^1,t^2)$ such that at least one of the inequalities (2) or (7) holds. Since g is differentiable at t^0 by hypothesis, (2) and (7) both reduce to the equality $g'(t^0) = \alpha$.

Q.E.D.

COROLLARY 4: Suppose g *is an upper semicontinuous function of one variable defined over the interval* $[b,c]$ *where* $b < c$. *If for every* $t\varepsilon[b,c)$,

$$D^{+u}g(t) \geq 0, \text{ or } D^{+\ell}g(t) \geq 0 \tag{8}$$

where $[b,c) \equiv \{t : b \leq t < c\}$*, then* g *is a nondecreasing function over* $[b,c]$.

PROOF: Let g be upper semicontinuous over $[b,c]$ and *suppose* g is not nondecreasing. Then there exist finite t^1 and t^2 such that $b \leq t^1 < t^2 \leq c$ and $g(t^1) > g(t^2)$. Thus $\alpha \equiv (g(t^2) - g(t^1)) / (t^2-t^1) < 0$. Now apply Theorem 1 and we obtain the existence of $t^*\varepsilon[t^1,t^2)$ such that $D^{+\ell}g(t^*) \leq D^{+u}g(t^*) \leq \alpha < 0$. This contradicts (8) when $t \equiv t^*$. Thus our supposition is false and the corollary follows.

Q.E.D.

COROLLARY 5: Suppose g *is an upper semicontinuous function defined over* [b,c] *and for every* $t\varepsilon(b,c]$,

$$D^{-\ell}g(t) \leq 0 \text{ or } D^{-u}g(t) \leq 0, \tag{9}$$

where $(b,c] \equiv \{t : b < t \leq c\}$. *Then* g *is nonincreasing over* [b,c].

The proof of this corollary is analogous to the proof of the previous corollary.

Results closely related to the last two corollaries were obtained by Scheeffer (1884) and McShane (1944; 200-201). We also note that counterparts to Corollaries 4 and 5 hold for lower semicontinuous functions. Finally, we note that if g is upper semicontinuous over [b,c] and (8) and (9) both hold, then g is constant over [b,c].

3. QUASICONCAVE FUNCTIONS

DEFINITION 1 (Fenchel (1953; 117)): f *is a* <u>*quasiconcave function*</u> *defined over a convex subset* S *of* R^N *iff:*

$$x^1 \varepsilon S,\ x^2 \varepsilon S,\ 0 < \lambda < 1 \text{ implies } f(\lambda x^1+(1-\lambda)x^2) \geq \min\ \{f(x^1),\ f(x^2)\}. \tag{10}$$

Note that the following theorem imposes no continuity requirements on f.

THEOREM 2: Let f *be a quasiconcave function defined over a convex subset* S *of* R^N. *Then the minimum of* f *over any compact line segment contained in* S *is attained on that line segment; i.e., if* $x^0 \varepsilon S$, $v^T v = 1$, $\bar{t} > 0$ *and* $x^0+\bar{t}v \varepsilon S$, *then*

$$\inf_t\{f(x^0+tv) : t\varepsilon[0,\bar{t}]\} = \min_t\{f(x+tv) : t\varepsilon[0,\bar{t}]\}. \tag{11}$$

PROOF: Using (10), for $t \in [0,\bar{t}]$, $f(x^0+tv) \geq \min\{f(x^0),\ f(x^0+\bar{t}v)\}$. Thus

$$\inf_t\{f(x^0+tv) : t \in [0,\bar{t}]\} = \min\{f(x^0),\ f(x^0+\bar{t}v)\}.$$

The proof of the above theorem also follows readily from some monotonicity properties of quasiconcave functions developed by Stoer and Witzgall (1970; 172).

If a function f defined over a convex subset S of R^N satisfies (11), then we say that f satisfies the *line segment minimum property.* Thus quasiconcave functions have this property.

We require some preliminary definitions and results before we can state our new characterizations of quasiconcavity. In the definitions which follow, let g be a function of one variable defined over the closed interval $[b,c]$, with $b < c$.

DEFINITION 2: g *attains a* *__one sided semistrict local minimum from above__* *at* $t_0 \varepsilon [b,c)$ *iff there exists* $\varepsilon > 0$ *such that* $g(t_0) \leq g(t_0+k)$ *for all* k *such that* $0 < k \leq \varepsilon \leq c-t_0$ *and* $g(t_0) < g(t_0+\varepsilon)$.

DEFINITION 3: g *attains a* *__one sided semistrict local minimum from below__* *at* $t_0 \varepsilon (b,c]$ *iff there exists* $\varepsilon > 0$ *such that* $g(t_0) \leq g(t_0)$ *for all* k *such that* $0 < k \leq \varepsilon \leq t_0-b$ *and* $g(t_0) < g(t_0-\varepsilon)$.

DEFINITION 4: g *attains a semistrict local minimum at* $t_0 \varepsilon (b,c)$ *iff* g *attains a one sided semistrict local minimum from above* *__and__* *below at* t_0.

Thus g attains a semistrict local minimum at t_0 iff it attains a local minimum at t_0 and the function eventually increases as we travel away from t_0 in either direction. Note that our Definition 4 is equivalent to Definition 5 in Avriel, Diewert, Schaible and Ziemba (1981).

Semistrict and one sided semistrict local maxima are defined analogously.

THEOREM 3 (Diewert, Avriel and Zang (1981)): A function f *defined over a convex subset* S *of* R^N *is quasiconcave iff* f *satisfies the line segment minimum property (11) over* S *and the following property:*

$$x^0 \varepsilon S, v^T v = 1,\ \bar{t} > 0,\ x^0+\bar{t}v \varepsilon S \text{ implies } g(t) \equiv f(x^0+tv) \text{ for } t \varepsilon [0,\bar{t}] \quad (12)$$

does *__not__* *attain a semistrict local minimum at any* $t \varepsilon (0,\bar{t})$.

THEOREM 4: Let f *be a quasiconcave function defined over* $S \subset R^N$. *Then* f *satisfies the line segment minimum property (11) over* S *and the following property:*

$$x^0 \varepsilon S, v^T v = 1,\ t > 0,\ x^0 + tv \varepsilon S,\ D_v^{+\ell} f(x^0) < 0 \text{ implies}$$
$$f(x^0 + tv) < f(x^0). \tag{13}$$

On the other hand, if f *is an upper semicontinuous function defined over the convex subset* S *of* R^N *and satisfies properties (11) and (13), then* f *is quasiconcave over* S.

PROOF: Quasiconcavity implies (11) and (13). By Theorem 2, quasiconcavity implies (11). Hence we need only show that (11) and not (13) imply that f is not quasiconcave. Not (13) means there exist $x^0 \varepsilon S$, $v^T v = 1$, $\bar{t} > 0$, such that $x^0 + \bar{t}v \varepsilon S$, $D_v^{+\ell} f(x^0) < 0$ and $f(x^0 + \bar{t}v) \geq f(x)$. All of this implies that $g(t) \equiv f(x+tv)$ attains a semistrict local minimum for some $t^* \varepsilon (0,\bar{t})$ and thus f is not quasiconcave by Theorem 3.

Upper Semicontinuity, (11) and (13) imply f *is quasiconcave.* Let $x^1 \varepsilon S$, $x^2 \varepsilon S$, $x^1 \neq x^2$ and $f(x^1) \geq f(x^2)$. Define $\bar{t} \equiv ((x^2 - x^1)^T (x^2 - x^1))^{1/2} > 0$ and $v \equiv (x^2 - x^1) / \bar{t}$. If $\min_t \{f(x+tv) : t \varepsilon [0,\bar{t}]\} = f(x^1)$, then f is quasiconcave. *Suppose*

$$\min_t \{f(x^1 + tv) : t \varepsilon [0,\bar{t}]\} < f(x^1) \leq f(x^2). \tag{14}$$

Define the set of minimizers for the minimization problem in (14) to be M and define $t_0 \varepsilon [0,\bar{t}]$ to be the lower limit of the minimizers; i.e., $t_0 \equiv \inf_t \{t : t \varepsilon M\}$. We consider three cases:

Case (i):

$$t_0 \equiv \inf_t \{t : t \varepsilon M\} = \min_t \{t : t \varepsilon M\} \tag{15}$$

so that $t_0 \varepsilon M$. Note that (14) and (15) imply $t_0 \varepsilon (0,\bar{t})$. Let the positive integer n be large enough so that $t_0 - 1/n \geq 0$. If $D_v^{+\ell} f(x^1 + tv) \geq 0$ for all $t \varepsilon [t_0 - 1/n, t_0)$, then by Corollary 4 above, $f(x^1 + tv)$ is nondecreasing in t over $[t_0 - 1/n,\ t_0]$ which contradicts (15). Thus for every positive integer n such that $t_0 - 1/n \geq 0$, there exists $t_n \varepsilon [t_0 - 1/n, t_0)$ such that

$$D_v^{+\ell} f(x^1 + t_n v) < 0. \tag{16}$$

Using (13), (16) implies that for all $t \geq t_n$ such that $x^1+tv \varepsilon S$,

$$f(x^1+tv) < f(x^1+t_n v). \tag{17}$$

By (15), $f(x^1+t_n v) \geq f(x^1+t_0 v)$. Since f is upper semicontinuous and $\lim t_n = t_0$, taking limits of (17) yields

$$f(x^1+tv) \leq f(x^1+t_0 v) \quad \text{for all } t \geq t_0. \tag{18}$$

Letting $t=\bar{t}$, (18) implies $f(x^1+\bar{t}v) = f(x^2) \leq f(x^1+t_0 v) = \min_t \{f(x^1+tv) : t \varepsilon [0,\bar{t}]\}$ which contradicts (14).

Case (ii): $t \equiv \sup_t \{t : t \varepsilon M\} = \max_t \{t : t \varepsilon M\}$.

Repeat the argument of case (i) but in the opposite direction and conclude that $f(x^1) \leq f(x^1+t_1 v) = \min_t \{f(x+tv) : t \varepsilon [0,\bar{t}]\}$ which again contradicts (14).

Case (iii): $t_0 \equiv \inf_t \{t : t \varepsilon M\}$, $t_0 \notin M$, $t_1 \equiv \sup_t \{t : t \varepsilon M\}$, $t_1 \notin M$.

In this case, we must have $f(x^1+t_0 v) > \min_t \{f(x^1+tv) : t \varepsilon [0,\bar{t}\} = \lim_{n\to\infty} (\inf_t \{f(x^1+tv) : t_0 < t \leq t_0 + 1/n\}$. Thus $D_v^{+\ell} f(x^1+t_0 v) = -\infty < 0$. Hence by (13), $f(x^1+t_1 v) < f(x^1+t_0 v)$ since $t_1 > t_0$. (If $t_1 = t_0$, $M = \{t_0\}$, which contradicts $t_0 \notin M$.) Similarly, we deduce $f(x^1+t_0 v) < f(x^1+t_1 v)$, a contradiction.

Thus we conclude (14) cannot be true and the quasiconcavity of f follows.

Q.E.D.

Thus for upper semicontinuous functions, properties (11) and (13) characterize quasiconcavity. However, if f is not upper semicontinuous, this is not true. Consider the following example:

Example 2.

$$f(x) \equiv \begin{cases} 1 & \text{for } 0 \leq x < 1 \text{ and } 1 < x \leq 2 \\ 0 & \text{for } x=1 \end{cases} \tag{19}$$

This function satisfies (11) and (13) but it is not upper semicontinuous or quasiconcave.

We also note that $D_v^{+\ell} f(x^0) < 0$ in (13) can be replaced by $D_v^{+u} f(x^0) < 0$ and the resulting version of Theorem 4 remains valid (as do various corollaries listed below).

COROLLARY 6: Theorem 4 remains valid if we replace (13) by (20):

$$x^0 \varepsilon S,\ v^T v=1,\ t>0,\ x^0+tv\varepsilon S,\ f(x^0+tv) \geq f(x^0) \text{ implies}$$
$$D_v^{+\ell} f(x^0) \geq 0 \qquad (20)$$

PROOF: Take the contrapositive of (13) to get (19).

Q.E.D.

COROLLARY 7: A continuous function f *defined over a convex subset* S *of* R^N *is quasiconcave iff* f *satisfies*

$$x^0 \varepsilon S,\ v^T v=1,\ t>0,\ x^0+tv\varepsilon S,\ D_v^{+\ell} f(x^0) < 0 \text{ implies}$$
$$f(x^0+tv) \leq f(x^0). \qquad (21)$$

PROOF: Quasiconcavity implies (21): Quasiconcavity implies (13) which implies (21).

(21) implies quasiconcavity: Since f is continuous, (11) holds and the set of minimizers M defined in the proof of Theorem 4 above is compact. Now modify the proof of case (i) in Theorem 4 and again obtain a contradiction to (14). Hence quasiconcavity of f follows.

Q.E.D.

COROLLARY 8: A one sided directionally differentiable function f *defined over a convex subset* S *of* R^N *is quasiconcave iff* f *satisfies either (22) or (23):*

$$x^0 \varepsilon S,\ v^T v=1,\ t>0,\ x^0+tv\varepsilon S,\ D_v^{+} f(x^0) < 0 \text{ implies}$$
$$f(x^0+tv) \leq f(x^0) \qquad (22)$$

$$x^0 \varepsilon S,\ v^T v=1,\ t>0,\ x^0+tv\varepsilon S,\ D_v^{+} f(x^0) < 0 \text{ implies}$$
$$f(x^0+tv) < f(x^0). \qquad (23)$$

COROLLARY 9: A Gateaux differentiable function f *defined over a convex subset* S *of* R^N *with a nonempty interior is quasiconcave iff*

$$x^1 \varepsilon S,\ x^2 \varepsilon S,\ x^1 \neq x^2,\ \nabla^T f(x^1)(x^2 - x^1) < 0 \text{ implies}$$
$$f(x^2) \leq f(x^1). \tag{24}$$

COROLLARY 10 (Mangasarian (1969; 147)): A Gateaux differentiable function f *defined over an open convex subset* S *of* R^N *is quasiconcave iff*

$$x^1 \varepsilon S,\ x^2 \varepsilon S,\ x^1 \neq x^2,\ \nabla^T f(x^1)(x^2 - x^1) < 0 \text{ implies}$$
$$f(x^2) < f(x^1). \tag{25}$$

COROLLARY 11 (Arrow and Enthoven (1961; 780)): A Gateaux differentiable function f *defined over an open convex subset* S *of* R^N *is quasiconcave iff*

$$x^1 \varepsilon S,\ x^2 \varepsilon S,\ x^1 \neq x^2,\ f(x^2) \geq f(x^1) \text{ implies}$$
$$\nabla^T f(x^1)(x^2 - x^1) \geq 0. \tag{26}$$

PROOF: Take the contrapositive of Corollary 10 above.

COROLLARY 12 (Ponstein (1967; 116)): A Gateaux differentiable function f *defined over an open convex subset* S *of* R^N *is quasiconcave iff*

$$x^1 \varepsilon S,\ x^2 \varepsilon S,\ x^1 \neq x^2,\ f(x^2) > f(x^1) \text{ implies}$$
$$\nabla^T f(x^2)(x^2 - x^1) \geq 0. \tag{27}$$

PROOF: Take the contrapositive of (24) above.

In the above corollaries, we follow the terminology used by Rockafellar (1978; 26-27): f is *one sided directionally differentiable* over an open convex subset S of R^N if $D_v^+ f(x)$ exists for every direction v and every $x \varepsilon S$, f is *directionally differentiable* over S if $D_v^+ f(x^0) = D_v^- f(x^0) = D_v f(x^0)$ for $v^T v = 1$ and $x \varepsilon S$, and f is *Gateaux or weakly differentiable* over S if $D_v f(x)$ exists for $v^T v = 1$ and $x \varepsilon S$ and in addition, $D_v f(x) = v^T \nabla f(x)$ where $\nabla f(x)$ is the vector of partial derivatives of f evaluated at x.

THEOREM 5: Let f *be upper semicontinuous over the convex subset* S *of* R^N. *Then* f *is quasiconcave over* S *iff* f *satisfies*

$$x^0 \varepsilon S,\ v^T v=1,\ t_1 > 0,\ x^0+t_1 v \varepsilon S,\ D_v^{+u} f(x^0) < 0 \text{ implies the function of variable } g(t) \equiv f(x^0+tv) \tag{28}$$

is nonincreasing for $t \varepsilon [0,t_1]$.

PROOF: Quasiconcavity implies (28). We show that not (28) implies that f is not quasiconcave. If (28) is not satisfied, then there exist $x^0 \varepsilon S$, $v^T v=1$, $t_1>0$, $0 \leq t_2 < t_3 \leq t_1$ such that $D_v^{+u} f(x^0) < 0$, $x^0+t_1 v \varepsilon S$ and

$$g(t_2) \equiv f(x^0+t_2 v) < f(x^0+t_3 v) \equiv g(t_3). \tag{29}$$

$D_v^{+u} f(x^0) < 0$ implies that there exists $t_4 \varepsilon (0,t_3)$ such that

$$g(t_4) \equiv f(x^0+t_4 v) < f(x^0) \equiv g(0). \tag{30}$$

If $\inf_t \{g(t) : t \varepsilon [0,t_1]\}$ is not attained as a minimum, then f does not have the line segment minimum property (11) and hence f is not quasiconcave by Theorem 2. If the inf is attained as a minimum, then (29) and (30) show that $g(t) \equiv f(x^0+tv)$ attains a semistrict local minimum over $[0,t_3]$ and hence by Theorem 3, f is not quasiconcave.

(28) implies f *is quasiconcave.* It is straightforward to show (using Corollary 4) that (28) and the upper semicontinuity of f imply that f has the line segment property.

Now let $x^0 \varepsilon S$, $v^T v=1$, $t_1>0$, $x^0+t_1 v \varepsilon S$ and $f(x^0) \leq f(x^0+t_1 v)$. Define $g(t) \equiv f(x^0+tv)$ for $t \varepsilon [0,t_1]$. We wish to show that $g(0) \leq f(t)$ for $t \varepsilon [0,t_1]$. *Suppose* there exists $t^* \varepsilon (0,t_1)$ such that $g(0) > g(t^*)$. Then

$$\begin{aligned} \inf_t \{g(t) : t \varepsilon [0,t_1]\} &= \min_t \{g(t) : t \varepsilon [0,t_1]\} \\ &= g(t_2) < g(0) \leq g(t_1). \end{aligned} \tag{31}$$

where $t_2 \varepsilon (0,t_1)$. If $D^{+u} g(t) \equiv D_v^{+u} f(x^0+tv) \geq 0$ for $t \varepsilon [0,t_2)$, then by Corollary 4, $g(t_2) \geq g(0)$ which contradicts (31). Thus there exists $t_3 \varepsilon [0,t_2)$ such that

$$D^{+u}g(t_3) \equiv D_v^{+u}f(x^0+t_3v) < 0. \tag{32}$$

Using (28), (32) implies that g(t) is nonincreasing for $t\varepsilon[t_3,t_1]$. Since $t_3<t_2$, this contradicts (31). Thus our supposition is false and quasiconcavity of f follows.

Q.E.D.

Inspection of the proof of the above theorem shows that the statement $D_v^{+u}f(x^0) < 0$ in (28) can be replaced by $D_v^{+\ell}f(x^0) < 0$ and we still obtain a valid theorem.

4. SEMISTRICT QUASICONCAVITY

DEFINITION 6 (Hanson (1964), Martos (1965), Mangasarian (1965), Ponstein (1967) and Elkin (1968)): f *is a* *__semistrictly quasiconcave__* *function defined over a convex subset* S *of* R^N *iff*

$$x^1\varepsilon S,\ x^2\varepsilon S,\ f(x^2) > f(x^1),\ 0 < \lambda < 1, \text{ implies}$$
$$f(\lambda x^1+(1-\lambda)x^2) > f(x^1). \tag{33}$$

See Avriel, Diewert, Schaible and Ziemba (1981) for alternative terms for property (33). Recall also that Karamardian (1967) showed that a semistrictly quasiconcave function was not necessarily quasiconcave unless the function was also upper semicontinuous. The following theorem characterizes semistrictly quasiconcave functions in the upper semicontinuous case.

THEOREM 6: Let f *be an upper semicontinuous function defined over the convex subset* S *of* R^N. *Then* f *is semistrictly quasiconcave iff* f *satisfies:*

$$x^0\varepsilon S,\ v^Tv=1,\ t_1>0,\ x^0+t_1v\varepsilon S,\ D_v^{+u}f(x^0) < 0 \text{ implies}$$
$$g(t) \equiv f(x^0+tv) \tag{34}$$

is a decreasing function for $t\varepsilon[0,t_1]$.

PROOF: Semistrict quasiconcavity implies (34). Since upper semicontinuity and semistrict quasiconcavity implies quasiconcavity which in turn implies

(11), we need only show that (11) and not (34) implies that f is not semistrictly quasiconcave. This is straightforward since (11) and not (34) imply the existence of a line segment along which f attains a local minimum, and thus by Theorem 7 in Diewert, Avriel and Zang (1981), f is not semistrictly quasiconcave.

(34) implies semistrict quasiconcavity. Note that (34) implies (28). Thus by Theorem 5, (34) implies that f is quasiconcave. Now let $x^1 \varepsilon S$, $x^2 \varepsilon S$, $f(x^2) > f(x^1)$ and *suppose* there exists $0 < \lambda^* < 1$ such that

$$f(\lambda^* x^1 + (1-\lambda^*)x^2) \leq f(x^1). \tag{35}$$

However, since f is quasiconcave, the strict inequality in (35) cannot hold, so we must have

$$f(\lambda^* x^1 + (1-\lambda^*)x^2) = f(x^1) < f(x^2), \; 0 < \lambda^* < 1. \tag{36}$$

Define $\bar{t} \equiv [(x^1-x^2)^T(x^1-x^2)]^{1/2}$, $v \equiv (x^1-x^2) / \bar{t}$ and $g(t) \equiv f(x^2+tv)$ for $t\varepsilon[0,\bar{t}]$. Let $t_1 \varepsilon (0,\bar{t})$ be such that $\lambda^* x^1+(1-\lambda^*)x^2 = x^2+t_1 v$. Then (36) translates into

$$g(t_1) = g(\bar{t}) < g(0). \tag{37}$$

Suppose $D^{+u}g(t) \geq 0$ for $t\varepsilon[0,t_1)$. Then since g is continuous from above, by Corollary 4, we have

$$g(t_1) \geq g(0)$$

which contradicts (37). Thus our second supposition is false, and we must have $t_0 \varepsilon [0,t_1)$ such that

$$D^{+u}g(t_0) < 0. \tag{38}$$

But (38) and (34) imply that g(t) is a decreasing function for $t\varepsilon[t_0,\bar{t}]$ which contradicts (37). Thus our first supposition is also false and semistrict quasiconcavity of f follows.

Q.E.D.

5. STRICT QUASICONCAVITY

Definition 7 (Ponstein (1967; 118)): f *is a* <u>*strictly quasiconcave*</u> *function defined over a convex subset* S *of* R^N *iff*

$$x^1 \in S,\ x^2 \in S,\ x^1 \neq x^2,\ f(x^1) \leq f(x^2),\ 0 < \lambda < 1 \text{ implies}$$
$$f(x^1) < f(\lambda x^1 + (1-\lambda)x^2). \tag{39}$$

See Avriel, Diewert, Schaible and Ziemba (1981) for alternative terms for (39).

THEOREM 7: Let f *be a function defined over a convex subset* S *of* R^N*. Then* f *is strictly quasiconcave iff* f *is quasiconcave and has the following property:*

f is not constant on any line segment belonging to S. (40)

PROOF: Strict quasiconcavity implies quasiconcavity and (40). Obvious.

Quasiconcavity and (40) imply strict quasiconcavity. We need only show that f is quasiconcave but not strictly quasiconcave implies not (40), which is straightforward.

Q.E.D.

COROLLARY 13: Let f *be an upper semicontinuous function defined over a convex subset* S *of* R^N*. Then* f *is strictly quasiconcave over* S *iff* f *satisfies (28) and (40).*

THEOREM 8: Let f *be an upper semicontinuous function defined over a convex subset* S *of* R^N*. Then* f *is strictly quasiconcave iff* f *is semistrictly quasiconcave and satisfies (40).*

The proof is similar to the proof of Theorem 7.

COROLLARY 14: Let f *be an upper semicontinuous function defined over a convex subset* S *of* R^N*. Then* f *is strictly quasiconcave over* S *iff* f *satisfies (34) and (40).*

THEOREM 9: Let f *be an upper semicontinuous function defined over a convex subset* S *of* R^N*. Then* f *is strictly quasiconcave iff*

$x^0 \varepsilon S$, $v^T v=1$, $t_1>0$, $x^0+t_1 v \varepsilon S$, $g(t) \equiv f(x^0+tv)$ for $t\varepsilon[0,t_1]$, $t_2 \varepsilon[0,t_1]$ and $g(t_2) \equiv \max_t\{g(t) : t\varepsilon[0,t_1]\}$ implies $g(t)$ is an increasing function over $[0,t_2]$ and a decreasing function over $[t_2,t_1]$. (41)

PROOF: Use Corollary 14.

Q.E.D.

6. PSEUDOCONCAVITY

DEFINITION 8 (Mangasarian (1965; 281)): f is a <u>*pseudoconcave*</u> *function defined over a convex subset S of* R^N *iff*

$x^0 \varepsilon S$, $v^T v=1$, $t>0$, $x^0+tv\varepsilon S$, $D_v^{+u} f(x^0) \le 0$ implies

$$f(x^0+tv) \le f(x^0). \qquad (42)$$

Actually Mangasarian assumed that S is an open set and that f was Gateaux differentiable over S, in which case $D_v^{+u} f(x^0) \le 0$ can be replaced by $(x^1-x^0)^T \nabla f(x^0)$ where $x^1 \equiv x^0+tv$. We also note that Tuy's (1964; 113) definition of a semiconcave function is equivalent to (42) when f is a one sided directionally differentiable function.

The geometric interpretation of Definition 8 should be clear. If the one sided upper directional derivate of f evaluated at $x^0 \varepsilon S$ in the direction indexed by the vector v is negative or zero, then the function does not increase in that direction.

The following theorem explains the usefulness of the class of pseudoconcave functions.

THEOREM 10 (Generalization of Mangasarian (1969; 140-141)): Let f be a pseudoconcave function defined over the convex subset S of R^N. *(i) If* $x^0 \varepsilon S$ *is such that* $D_v^{+u} f(x^0) \le 0$ *for every feasible direction v (i.e., for every* $x\varepsilon S$ *such that* $x \ne x^0$, *we have* $D_v^{+u} f(x^0) \le 0$ *where* $v \equiv (x-x^0) / \{(x-x^0)^T(x-x^0)\}^{1/2}$), *then* $f(x^0) \ge f(x)$ *for every* $x\varepsilon S$; *i.e.,* f *attains a global maximum over* S *at* x^0. *(ii) If* f *attains a local maximum at* $x^0 \varepsilon S$, *then* $f(x^0) \geq f(x)$ *for every* $x \varepsilon S$, *i.e.,* f *attains a global maximum over* S *at* x^0.

PROOF: (i) follows directly from the definition of pseudoconcavity. (ii) follows from the following straightforward result: if f attains a local maximum at x^0, then we must have $D_v^{+u} f(x^0) \leq 0$ for all feasible directions v. Now apply (i).

Q.E.D.

We note that we can follow the example of Guignard (1969; 233) or Bazaraa and Shetty (1976; 113) and assume that f in Theorem 10 is only pseudoconcave at $x^0 \varepsilon S$ where the set S is x^0 convex (i.e., if $x \varepsilon S$, $0 \leq \lambda \leq 1$, then $\lambda x^0 + (1-\lambda)x \varepsilon S$). The resulting theorem is still valid.

It appears that our definition of pseudoconcavity captures the spirit of Mangasarian's original definition which assumed differentiability of f; however, other definitions are possible. In particular, Thompson and Parke (1973; 310) define f to be pseudoconcave iff the following condition holds (i) $x^0 \varepsilon S$, $v^T v = 1$, $x^0 + \bar{t}v \varepsilon S$, $f(x^0 + \bar{t}v) > f(x^0)$ implies there exists a number $B(x^0, \bar{t}) > 0$ such that for every t such that $0 < t < \bar{t}$, we have $f(x^0 + tv) \geq f(x^0) + tB(x^0, \bar{t})$. In particular, it can be shown that if f is pseudoconcave in the sense of Thompson and Parke, then f satisfies the following two conditions: (ii) $x^0 \varepsilon S$, $v^T v = 1$, $\bar{t} > 0$, $x^0 + \bar{t}v \varepsilon S$, $f(x^0 + \bar{t}v) > f(x^0)$ implies $D_v^{+\ell} f(x^0) > 0$, and (iii) $x^0 \varepsilon S$, $v^T v = 1$, $\bar{t} > 0$, $x^0 + \bar{t}v \varepsilon S$, $D_v^{+\ell} f(x^0) \leq 0$ implies $f(x^0 + tv) \leq f(x^0)$. However, it can be shown that the classes of functions defined by (i), (ii), or (iii) above are strictly smaller than the class of functions defined by our definition of pseudoconcavity, Definition 8. Consider the following function h defined over [0,1/2]:

Example 3:

$$h(x) \equiv \begin{cases} 1/2 & \text{if } x = 1/2 \\ (1/2)^2 & \text{if } (1/2)^2 \le x < 1/2 \\ (1/2)^{2^2} & \text{if } (1/2)^{2^2} \le x < (1/2)^2 \\ \cdot & \\ \cdot & \\ \cdot & \\ (1/2)^{2^{n+1}} & \text{if } (1/2)^{2^{n+1}} \le x < (1/2)^{2^n} \\ \cdot & \\ \cdot & \\ \cdot & \end{cases}$$

Note that $h(x)$ is a monotonic step function which zigzags between the functions $h^1(x) \equiv x$ and $h^2(x) \equiv x^2$ for x between 0 and 1/2. It can be shown that $h(x) + x^2 \equiv f(x)$ is pseudoconcave according to Definition 8 but not according to (i), (ii) or (iii) above.

Mifflin (1977a; 961) and Hiriart-Urruty (1977) provide other definitions of pseudoconcavity. Mifflin's definition assumes the existence of one sided directional derivatives while Hiriart-Urruty's definition assumes the existence of a Lipschitz condition and he phrases his definition in terms of Clarke's (1975;1976) generalized directional derivative. Both Mifflin and Hiriart-Urruty provide versions of our Theorem 10 above using their definitions of pseudoconcavity.

THEOREM 11: If f *is a concave function of* N *variables defined over a convex set* S*, then* f *is pseudoconcave.*

PROOF: Let $x^0 \varepsilon S$, $v^T v=1$, $t>0$, $x^0+tv \varepsilon S$ and let f be concave over the convex set S with $D_v^{+u} f(x^0) \le 0$. Using a result which can be found in Fenchel (1953; 71), since f is concave, $D_v^{+u} f(x^0) = D_v^{+} f(x^0) \le 0$. Using a result which can be found in Roberts and Varberg (1973; 12), we have

$$f(x^0+tv) \le f(x^0) + D_v^{+} f(x^0) t \le f(x^0)$$

since $D_v^+ f(x^0) \leq 0$ and $t>0$ which shows that f is pseudoconcave.

Q.E.D.

Mangasarian (1965; 283) proved the above theorem for the case where f is differentiable and defined over an open set.

Mangasarian (1965; 283) also showed that pseudoconcave functions are not necessarily concave (e.g., consider the function of one variable $f(x) = x+x^3$ defined over R^1 which is pseudoconcave but not concave) and under his differentiability conditions, pseudoconcave functions are semistrictly quasiconcave and continuous and hence quasiconcave. However, this last statement is not true using our definition of pseudoconcavity (Definition 8) which does not place any differentiability or continuity assumptions on f at all: consider the following example:

Example 4:

$$f(x) \equiv \begin{cases} 1 \text{ for } 0 \leq x < 1 \text{ and } 1 < x \leq 2 \\ 0 \text{ for } x=1. \end{cases} \tag{43}$$

The set S is $\{x: 0 \leq x < 2\}$. It can be verified that f defined by (43) is pseudoconcave according to Definition 8, but that f is not quasiconcave (or strictly quasiconcave either).

However, the following theorems show that if we restrict f to be upper semicontinuous, then pseudoconcavity implies both quasiconcavity and semistrict quasiconcavity.

THEOREM 12: Let f *be an upper semicontinuous function defined over a convex subset* S *of* R^N. *Then* f *is pseudoconcave iff the following condition holds:*

$$x^0 \varepsilon S,\ v^T v=1,\ \bar{t}>0,\ x+\bar{t}v\varepsilon S,\ D_v^{+u} f(x^0) \leq 0 \text{ implies}$$

$$g(t) \equiv f(x^0+tv) \tag{44}$$

is a nonincreasing function of t over $[0,\bar{t}]$.

PROOF: (44) implies pseudoconcavity. Trivial.

Pseudoconcavity implies (44). Let $x^0 \varepsilon S$, $v^T v=1$, $x^0+\bar{t}v \varepsilon S$ and $D_v^{+u} f(x^0) \leq 0$. If $D_v^{+u} f(x^0+tv) \leq 0$ for all $t \varepsilon [0,\bar{t})$, then the definition of pseudoconcavity (42) will imply that $g(t) \equiv f(x^0+tv)$ is nonincreasing over $[0,\bar{t}]$, the desired conclusion. Thus we *suppose* that $D_v^{+u} f(x^0+tv) > 0$ for some $t \varepsilon (0,\bar{t})$. Now define t_0 by

$$t_0 \equiv \inf_t \{t: D_v^{+u} f(x^0+tv) > 0,\ t \varepsilon (0,\bar{t})\}. \tag{45}$$

We have to consider two cases.

Case (i):

$$D_v^{+u} f(x^0+t_0 v) > 0. \tag{46}$$

Since $D_v^{+u} f(x^0) \leq 0$, $t_0 > 0$. By definition (45), $f(x^0+tv) \leq 0$ for $t \varepsilon [0,t_0)$. Thus using (42), for $t \varepsilon [t_0,\bar{t}]$ and $0 < \varepsilon < t_0$, we have $f(x^0+(t_0-\varepsilon)v) \geq f(x^0+tv)$. Take the limit of this last inequality as ε tends to 0 and we obtain the following inequality using the fact that f is upper semicontinuous:

$$f(x^0+t_0 v) \geq f(x^0+tv) \text{ for all } t \varepsilon [t_0,\bar{t}]. \tag{47}$$

However, (46) implies that there exists t_1 such that $t_0 < t_1 < \bar{t}$ and $f(x^0+t_0 v) < f(x^0+t_1 v)$ which contradicts (47). Thus case (i) cannot occur.

Case (ii):

$$D_v^{+u} f(x^0+t_0 v) \leq 0. \tag{48}$$

Using (42), (48) implies (47). Our supposition that $D_v^{+u} f(x^0+tv) > 0$ for some $t \varepsilon (0,\bar{t})$, the definition (45), and (48) imply that there exists t_1 arbitrarily close to t_0 such that

$$D_v^{+u} f(x^0+t_1 v) > 0 \text{ where } t_0 < t_1 < \bar{t}. \tag{49}$$

This in turn implies that there exists t_5 such that $t_1 < t_5 \leq \bar{t}$ and

$$f(x^0+t_1 v) < f(x^0+t_5 v) \leq f(x^0+t_0 v) \tag{50}$$

where the last inequality follows from (47). Using (50) and then the upper semicontinuity property of f, there exists t_4 such that $t_1 < t_4 \leq t_5$

and

$$f(x^0+tv) < f(x^0+t_0v) \text{ for all } t\varepsilon[t_1,t_4]. \tag{51}$$

Using (49) and (51), there exists t_3 such that $t_1 < t_3 \leq t_4$ and

$$f(x^0+t_1v) < f(x^0+t_3v) < f(x^0+t_0v). \tag{52}$$

Now define $h(t) \equiv f(x^0+t(-v))$ for $t\varepsilon[-\bar{t},0]$. Note that (52) translates into $h(-t_1) < h(-t_3) < h(-t_0)$. Define $\alpha \equiv [h(-t_3) - h(-t_1)] / [-t_3+t_1] < 0$ since $t_1 < t_3$. Now apply the Generalized Mean Value Theorem and we obtain the existence of t_2 such that $t_1 \leq t_2 \leq t_3$ and

$$D^{+u}h(-t_2) \equiv D_{-v}^{+u}f(x^0-t_2(-v)) = D_{-v}^{+u}f(x^0+t_2v) \leq \alpha < 0. \tag{53}$$

Using (42), (53) implies that

$$f(x^0+tv) \leq f(x^0+t_2v) < f(x^0+t_0v) \quad \text{for all } t\varepsilon[0,t_2] \tag{54}$$

using (57) and $t_1 \leq t_2 \leq t_3 \leq t_4$.

However, since $t_0\varepsilon[0,t_2]$, letting $t=t_0$ in (54) leads to the contradiction $f(x^0+t_0v) < f(x^0+t_0v)$.

Thus case (ii) cannot occur either and our supposition must be false and the desired conclusion (44) follows.

Q.E.D.

COROLLARY 15: If f *is upper semicontinuous and pseudoconcave over the convex subset* S *of* R^N, *then* f *is quasiconcave.*

PROOF: The proof is immediate, comparing (44) with (28), our third characterization of quasiconcavity.

COROLLARY 16: If f *is upper semicontinuous and pseudoconcave over the convex subset* S *of* R^N, *then* f *is semistrictly quasiconcave.*

PROOF: Let $x^0\varepsilon S$, $v^Tv=1$, $\bar{t}>0$, $x^0+\bar{t}v\varepsilon S$ and $D_v^{+u}f(x^0) < 0$. By Theorem 6, we need only show $g(t) \equiv f(x^0+tv)$ is a decreasing function over $[0,\bar{t}]$. Moreover since $D_v^{+u}f(x^0) = D^{+u}g(0) < 0$, we must have

$$g(t) < g(0) \text{ for } t\varepsilon(0,\bar{t}]. \tag{55}$$

Suppose that g is not decreasing over $[0,\bar{t}]$. Then since it is nonincreasing, g must be constant over a subinterval of $[0,\bar{t}]$, say over $[t_1,t_2]$, where $0 < t_1 < t_2 \leq \bar{t}$. But then $D_{-1}^{+u}g(t_2) = 0$, and since g is pseudoconcave, $g(0) \leq g(t_2)$, which contradicts (55). Thus our supposition is false, and f is semistrictly quasiconcave.

Q.E.D.

The following theorem exhibits precisely the extra condition required to ensure that a semistrictly quasiconcave function is pseudoconcave in the continuous from above case.

THEOREM 13: Let f *be an upper semicontinuous function defined over a convex subset* S *of* R^N. *Then* f *is pseudoconcave iff* f *is semistrictly quasiconcave and* f *satisfies:*

$$x^0\varepsilon S,\ \bar{t}>0,\ v^Tv=1,\ x^0+\bar{t}v\varepsilon S,\ D_v^{+u}f(x^0)=0 \textit{ implies}$$
$$g(t) \equiv f(x^0+tv) \tag{56}$$

is a nonincreasing function of t *over* $[0,\bar{t}]$.

PROOF: Pseudoconcavity implies semistrict quasiconcavity plus (56). By Corollary 16, pseudoconcavity implies semistrict quasiconcavity. By Theorem 12, pseudoconcavity implies (44) which in turn implies (56).

Semistrict quasiconcavity plus (56) implies pseudoconcavity. By Theorem 6, semistrict quasiconcavity implies (34). However, (34) and (56) imply (44), which in turn implies pseudoconcavity.

Q.E.D.

COROLLARY 17: Theorem 13 is valid if (56) is replaced by:

$$x^0\varepsilon S,\ v^Tv=1,\ \bar{t}>0,\ x^0+\bar{t}v\varepsilon S, \text{ and } D_v^{+u}f(x^0)=0 \text{ implies}$$
$$g(t) \equiv f(x^0+tv) \tag{57}$$

attains a one sided local maximum from above at t=0.

PROOF: Obviously (56) implies (57). We need only show that semistrict quasiconcavity plus (57) imply (56).

Let $x^0 \varepsilon S$, $v^T v=1$, $\bar{t}>0$, $x^0+\bar{t}v \varepsilon S$, and $D_v^{+u} f(x^0)=0$. We wish to show that $g(t) \equiv f(x^0+tv)$ is nonincreasing over $[0,\bar{t}]$. By (57), there exists t_1 such that $0 < t_1 < \bar{t}$ and

$$g(0) \geq g(t) \text{ for } t \varepsilon [0,t_1]. \tag{58}$$

We need to analyze several cases.

Case (i):

$$D^{+u} g(t) \geq 0 \text{ for all } t \varepsilon [0,\bar{t}).$$

Then by Corollary 4, g is nondecreasing over $[0,\bar{t}]$. However, (58) implies that g is constant over $[0,t_1]$. But then g must also be constant over $[t_1,\bar{t}]$ as well, since if there exists $t_1 \leq t_2 < t_3 \leq \bar{t}$ such that $g(t_2) < g(t_3)$, then for every $t \varepsilon (0,t_1)$, g attains a local minimum at t which is also a one sided semistrict local minimum, which contradicts the semistrict quasiconcavity of f (see Diewert, Avriel and Zang (1981)). Thus g is nonincreasing (in fact g is constant) if Case (i) occurs.

Now suppose $D^{+u} g(t) < 0$ for $t \varepsilon (0,\bar{t})$ and define

$$t_0 \equiv \inf_t \{t : D^{+u} g(t) < 0, \ t \ (0,\bar{t})\}. \tag{59}$$

Case (ii):

$$t_0 = 0.$$

In this case, by semistrict quasiconcavity of f which implies (34), we have g decreasing over $[0,\bar{t}]$.

Case (iii):

$$t_0 > 0 \text{ and } D^{+u} g(t_0) < 0.$$

Thus by (59), $D^{+u} g(t) \geq 0$ for $t \varepsilon [0,t_0]$. Repeating the analysis of Case (i) shows that g is constant on $[0,t_0]$. However, $D^{+u} g(t_0) < 0$ implies that g is decreasing over $[t_0,\bar{t}]$ using the semistrict quasiconcavity of f. Thus g is nonincreasing over $[0,\bar{t}]$.

Case (iv):

$$t_0 > 0 \text{ and } D^{+u}g(t_0) \geq 0.$$

Repeating the analysis of Case (i) shows that g is constant on $[0,t_0]$. Using the semistrict quasiconcavity of f and the definition of t_0, (50), we have g decreasing over $(t_0,\bar{t})$. Define

$$\alpha \equiv \lim_{t\to t_0, t>t_0} g(t)$$
$$\leq g(t_0)$$

where the last inequality follows from the decreasingness of g over $(t_0,\bar{t}]$ and the fact that g is upper semicontinuous. Thus again g is constant over $[0,t_0]$ and decreasing over $[t_0,\bar{t}]$.

Q.E.D.

COROLLARY 18: Let f *be upper semicontinuous over the convex subset* S *of* R^N. *Then is pseudoconcave over* S *iff* f *satisfies the line segment minimum property (11), (57) and the following property:*

$$x^0 \epsilon S,\ v^T v=1,\ t_1>0,\ x^0+t_1 v\epsilon S,\ g(t) \equiv f(x^0+tv), \tag{60}$$

g *attains a local minimum at* 0 *implies* g *does not attain a one sided semistrict local minimum at* 0.

PROOF: Diewert, Avriel and Zang (1981) show that (11) and (60) characterize semistrict quasiconcavity. Now use Corollary 17.

Q.E.D.

Note that the above Corollary characterizes pseudoconcave functions in terms of the local properties of the function along line segments.

7. STRICT PSEUDOCONCAVITY

DEFINITION 9 (Ponstein (1967; 117)): f *is a* *<u>strictly pseudoconcave</u>* *function defined over a convex subset* S *of* R^N *iff*

$$x^0 \varepsilon S,\ v^T v=1,\ t>0,\ x^0+tv\varepsilon S,\ D_v^{+u} f(x^0) \leq 0 \text{ implies}$$
$$f(x^0+tv) < f(x^0). \tag{61}$$

Ponstein's actual definition assumed differentiability and was:

$$x^0 \varepsilon S,\ x^2 \varepsilon S,\ x^1 \neq x^2,\ f(x^2) \geq f(x^1) \text{ implies}$$
$$(x^2-x^1)^T \nabla f(x^1) > 0.$$

The contrapositive of his definition is:

$$x^1 \varepsilon S,\ x^2 \varepsilon S,\ x^1 \neq x^2,\ (x^2-x^1)^T \nabla f(x^1) \leq 0 \text{ implies}$$
$$f(x^2) < f(x^1).$$

Thus (61) is a straightforward generalization of Ponstein's definition.

Thus if f is pseudoconcave and if the one sided upper directional derivate of f evaluated at x^0 in the direction v is negative or zero, then the function *decreases* along the line segment starting at x^0 in the v direction.

The following theorem shows that strict pseudoconcavity is a generalization of strict concavity.

THEOREM 14 (Ponstein (1967; 118)): Let f *be a function defined over a convex subset* S *of* R^N. *Then: (i)* f *strictly pseudoconcave implies* f *is pseudoconcave, (ii)* f *strictly concave implies* f *is strictly pseudoconcave, and (iii)* f *pseudoconcave and strictly quasiconcave implies* f *is strictly pseudoconcave.*

PROOF: (i) It is obvious that Definition 9 implies Definition 8. (ii) Let $x^0 \varepsilon S$, $v^T v=1$, $t>0$, $x^0+tv\varepsilon S$ and $D_v^{+u} f(x^0) \leq 0$. Since f is strictly concave and hence concave, using a result found in Fenchel (1953; 71), we have $D_v^{+u} f(x^0) = D_v^{+} f(x^0)$. By a characterization of strict concavity (e.g., see Roberts and Varberg (1973; 10)), we have

$$f(x^0+tv) < f(x^0) + D_v^{+} f(x^0) t$$
$$\leq f(x^0) \qquad \text{since } D_v^{+} f(x^0) \leq 0 \text{ and } t>0$$

which shows that f is strictly pseudoconcave. (iii) Let $x^0 \varepsilon S$, $v^T v=1$, $\bar{t}>0$, $x^0+\bar{t}v \varepsilon S$ and $D_v^{+u} f(x^0) \leq 0$. Since f is pseudoconcave, by Definition 8, $f(x^0+tv) \leq f(x^0)$ for all $t \varepsilon [0,\bar{t}]$. *Suppose*

$$f(x^0+t^1 v) = f(x^0) \text{ for some } t^1 \varepsilon (0,\bar{t}]. \tag{62}$$

Since f is strictly quasiconcave and hence quasiconcave, we have $f(x^0+tv) \geq f(x^0) = f(x^0+t^1 v)$ for all $t \varepsilon (0,t^1)$. Thus $f(x^0+tv)$ is constant for $t \varepsilon [0,t^1]$ which contradicts (40), an implication of strict quasiconcavity. Thus our supposition (62) is false and $f(x^0+tv) < f(x^0)$ which establishes strict pseudoconcavity.

Q.E.D.

The following theorem shows why the class of strictly pseudoconcave functions is useful.

THEOREM 15: Let f *be a strictly pseudoconcave function defined over a convex subset* S *of* R^N. *(i) If* $x^0 \varepsilon S$ *is such that* $D_v^{+u} f(x^0) \leq 0$ *for every feasible direction* v, *then* $f(x^0) > f(x)$ *for every* $x \varepsilon S$ *such that* $x \neq x^0$; *i.e.,* f *attains a unique global maximum over* S *at* x^0. *(ii) If* f *attains a local maximum at* $x^0 \varepsilon S$, *then* $f(x^0) > f(x)$ *for every* $x \varepsilon S$ *such that* $x \neq x^0$; *i.e.,* f *attains a unique global maximum over* S *at* x^0.

(We note that the hypothesis that f be strictly pseudoconcave over the convex set S can be weakened to the hypothesis that f be strictly pseudoconcave at $x^0 \varepsilon S$ where S is x^0 convex.)

COROLLARY 19: If f *is a Gateaux differentiable strictly pseudoconcave function defined over a convex subset* S *of* R^N *which has a nonempty interior and* $x^0 \varepsilon S$ *with* $\nabla f(x^0) = 0_N$, *then* f *attains a unique global maximum over* S *at* x^0.

Thus if f is strictly pseudoconcave, the necessary conditions for a local maximum are sufficient for a unique global maximum.

Although strict concavity implies strict pseudoconcavity, it is not the case that strict pseudoconcavity implies strict quasiconcavity or even quasiconcavity in general. Consider the following example of a function of one variable which is strictly pseudoconcave in the sense of Definition 8, but yet is not even quasiconcave (the set S is $\{x : 0 \leq x \leq 2\}$).

Example 5:

$$f(x) \equiv \begin{cases} x & \text{for } 0 \leq x < 1 \\ 0 & \text{for } x=1 \\ 1-x & \text{for } 1 < x \leq 2 \end{cases}$$

The following theorems show that if we restrict f to be upper semi-continuous, then strict pseudoconcavity implies strict quasiconcavity (and hence semistrict quasiconcavity and quasiconcavity as well).

THEOREM 16: Let f *be an upper semicontinuous function defined over a convex subset* S *of* R^N*. Then* f *is strictly pseudoconcave iff the following condition holds:*

$$x^0 \varepsilon S,\ v^T v=1,\ \bar{t}>0,\ x^0+\bar{t}v\varepsilon S,\ D_v^{+u} f(x^0) \leq 0 \text{ implies}$$
$$g(t) \equiv f(x^0+tv) \tag{63}$$

is a decreasing function of t *over* $[0,\bar{t}]$.

PROOF: (63) implies strict pseudoconcavity. Trivial.

Strict pseudoconcavity implies (63). Let $x^0 \varepsilon S$, $v^T v=1$, $\bar{t}>0$, $x^0+\bar{t}v\varepsilon S$, $D_v^{+u} f(x^0) \leq 0$ and define $g(t) \equiv f(x^0+tv)$. Since f is upper semicontinuous and strictly pseudoconcave and hence pseudoconcave, Theorem 12 implies that g is nonincreasing over $[0,\bar{t}]$. *If* g were constant on some subinterval of $[0,\bar{t}]$, then $g'(t) = 0$ for t's belonging to the interior of that sub-interval, and this contradicts the strict pseudoconcavity of g over the subinterval. Thus g cannot be constant over any subinterval of $[0,\bar{t}]$ and thus g(t) is a decreasing function of t over $[0,\bar{t}]$. Q.E.D.

COROLLARY 20: Let f *be an upper semicontinuous strictly pseudoconcave function defined over the convex subset* S *of* R^N. *Then* f *is strictly quasiconcave and hence also semistrictly quasiconcave and quasiconcave.*

PROOF: Since f is upper semicontinuous and strictly pseudoconcave, by Theorem 16, f satisfies property (63) which obviously implies (34). From the proof of Theorem 16, it is also true that f satisfies property (40). By Theorems 6 and 8, an upper semicontinuous f is strictly quasiconcave iff f satisfies (34) and (40). Hence f is strictly quasiconcave.

Q.E.D.

COROLLARY 21: Let f *be upper semicontinuous over the convex subset* S *of* R^N. *Then* f *is strictly pseudoconcave iff* f *is pseudoconcave and strictly quasiconcave.*

PROOF: The first part of the present corollary follows from part (i) of Theorem 14 and Corollary 20 above while the second part of the present corollary follows from part (iii) of Theorem 14.

Q.E.D.

COROLLARY 22: Let f *be upper semicontinuous over the convex subset* S *of* R^N. *Then* f *is strictly pseudoconcave over* S *iff* f *is pseudoconcave and has property (40) over* S.

PROOF: Strict pseudoconcavity implies pseudoconcavity plus (40). Obviously strict pseudoconcavity implies pseudoconcavity. Hence we need only show that f is pseudoconcave over S but f does not satisfy (40) implies that f is not strictly pseudoconcave. This is straightforward.

Pseudoconcavity plus (40) imply strict pseudoconcavity. Pseudoconcavity of f implies (44) if f is upper semicontinuous over S. But (44) and (40) imply (63) which implies f is strictly pseudoconcave by Theorem 16.

Q.E.D.

THEOREM 17: Let f *be an upper semicontinuous function defined over the convex subset* S *of* R^N. *Then* f *is strictly pseudoconcave iff* f *has properties (11), (64) and (65) below:*

$$x^0 \varepsilon S,\ v^T v=1,\ \bar{t}>0,\ x^0+\bar{t}v \varepsilon S,\ x^0-\bar{t}v\ S \text{ implies}$$

$$g(t) \equiv f(x^0+tv) \tag{64}$$

does not attain a local minimum at t=0.

$$x^0 \varepsilon S,\ v^T v=1,\ \bar{t}>0,\ x^0+\bar{t}v \varepsilon S,\ D_v^{+u} f(x^0) = 0 \text{ implies}$$

$$g(t) \equiv f(x^0+tv) \tag{65}$$

attains a one sided strict local maximum from above at t=0.

PROOF: Strict pseudoconcavity implies (11), (64) and (65). By Corollary 21, strict pseudoconcavity implies strict quasiconcavity if f is upper semicontinuous. By Theorem 9 in Diewert, Avriel and Zang (1981), strict quasiconcavity implies (11) and (64). By Theorem 16, strict pseudoconcavity implies (63) and (63) implies (65).

(11), (64) and (65) imply strict pseudoconcavity. (11) and (63) imply that f is strictly quasiconcave by Theorem 9 in Diewert, Avriel and Zang (1981). By Theorem 8, f is semistrictly quasiconcave also. We note that (65) implies (57). Thus f is upper semicontinuous, semistrictly quasiconcave and has property (57). Thus by Corollary 17 f is pseudoconcave. Since f is pseudoconcave and strictly quasiconcave, f is also strictly pseudoconcave by Corollary 21.

Q.E.D.

Note that the above Theorem characterizes strict pseudoconcavity in terms of the local properties of the function along line segments.

8. STRONG QUASICONCAVITY OR STRONG PSEUDOCONCAVITY

Before defining strong quasiconcavity, it is convenient to define strong concavity.

DEFINITION 10 (Poljak (1966; 73)): A function f *defined over a convex subset* S *of* R^N *is strongly concave over* S *iff there exists* $\alpha > 0$ *such that for every* $x^1 \varepsilon S$, $x^2 \varepsilon S$, $0 \leq \lambda \leq 1$,

$$f(\lambda x^1+(1-\lambda)x^2) \geq f(x^1) + (1-\lambda)f(x^2) + \lambda(1-\lambda)\alpha(x^1-x^2)^T(x^1-x^2). \quad (66)$$

We note that f is *concave* over S iff $f(\lambda x^1+(1-\lambda)x^2) \geq \lambda f(x^1) + (1-\lambda)f(x^2)$ and f is *strictly concave* over S iff for every $x^1 \varepsilon S$, $x^2 \varepsilon S$, $x^1 \neq x^2$, $0 < \lambda < 1$, we have $f(\lambda x^1+(1-\lambda)x^2) > \lambda f(x^1) + (1-\lambda)f(x^2)$.

The following Theorem provides a convenient characterization of strong concavity.

THEOREM 18 (Rockafellar (1976; 75)): Let S *be a convex subset of* R^N. *Then* f *is strongly concave over* S *iff there exists* $\alpha > 0$ *such that* g *defined by* $g(x) \equiv f(x) + \alpha x^T x$ *is a concave function over* S.

The definition of strong concavity presented here is equivalent to the one presented in Avriel, Diewert, Schaible and Ziemba (1981).

DEFINITION 11: f *is a* <u>*strongly quasiconcave*</u> *function defined over a convex subset* S *of* R^N *iff* f *is strictly quasiconcave over* S *and in addition has the following property:*

$$x^0 \varepsilon S,\ v^T v=1,\ \bar{t}>0,\ x^0+\bar{t}v \varepsilon S,\ D_v^{+u} f(x^0) = 0 \text{ implies there exist } \varepsilon \text{ and } \alpha \text{ such that } 0 < \varepsilon \leq \bar{t},\ 0<\alpha \text{ and}$$
$$f(x^0+tv) \leq f(x^0) - \alpha t^2 \text{ for } t \varepsilon [0,]. \quad (67)$$

The above definition implies that a strongly quasiconcave function is a strictly quasiconcave function that has the additional property that if the upper one sided directional derivative of the function is zero in a certain direction at a certain point, then the function decreases quadratically (at least locally) along that direction.

The following theorem shows that strong quasiconcavity is a generalization of strong concavity.

THEOREM 19: Let f *be a function defined over a convex subset* S *of* R^N. *Then (i)* f *strongly concave implies* f *is strongly quasiconcave, (ii)* f *strongly quasiconcave implies* f *is strictly quasiconcave and hence quasiconcave also, and (iii)* f *strongly quasiconcave implies* f *is strictly pseudoconcave.*

PROOF OF (i): If f is strictly concave, then using Definition 7, it is easy to show that f is also strictly quasiconcave. Thus we need only show that (67) holds. Since f is strongly concave, by Theorem 18 there exists $\alpha > 0$ such that $f(x) + \alpha x^T x$ is concave over S. Now let $x^0 \varepsilon S$, $v^T v=1$, $\bar{t}>0$, $x^0+\bar{t}v \varepsilon S$ and $D_v^{+u} f(x^0) = D_v^+ f(x^0) = 0$. Since $f(x) + \alpha x^T x$ is concave, for $t \varepsilon [0,\bar{t}]$,

$$f(x^0+tv) + \alpha(x^0+tv)^T(x^0+tv) \leq f(x^0) + \alpha x^{0T} x^0 + D_v^+ f(x^0) + 2\alpha x^{0T} vt$$

or

$$f(x^0+tv) \leq f(x^0) - \alpha t^2$$

since $D_v^+ f(x^0) = 0$ and $v^T v=1$. Thus (67) is satisfied with $\varepsilon \equiv \bar{t} > 0$.

PROOF OF (ii): Shown in part (i) above.

PROOF OF (iii): Let f be strongly quasiconcave. Hence f is also strictly quasiconcave and also has the line segment minimum property (11). Let $x^0 \varepsilon S$, $v^T v=1$, $\bar{t}>0$, $D_v^{+u} f(x^0) \leq 0$ and *suppose* $f(x^0+\bar{t}v) \geq f(x^0)$. If $D_v^{+u} f(x^0) < 0$, it can be seen that $g(t) \equiv f(x^0+tv)$ must attain a minimum on the interval $(0,\bar{t})$. Similarly, if $D_v^{+u} f(x^0) = 0$, using (67), the same conclusion about g follows. However, since f and g are strictly quasiconcave, by Theorem 9 in Diewert, Avriel and Zang (1981), g cannot attain a local minimum over $(0,\bar{t})$. Thus our supposition is false and strict pseudoconcavity of f follows.

Q.E.D.

We note that it is possible to characterize strong quasiconcavity using property (67) plus any of our alternative characterizations of strict quasiconcavity. It is also possible to characterize strongly quasiconcave functions using the concept of a strong local maximum, as is done in Diewert, Avriel and Zang (1981).

In Diewert, Avriel and Zang (1981), it is shown that our definition of strong quasiconcavity is equivalent to the definition of strong quasiconcavity used by Newman (1969; 307) and Ginsberg (1973; 600), assuming that f is twice continuously differentiable and is defined over an open convex set. Thus our definition of strong quasiconcavity can be viewed as a generalization of their definition to the nondifferentiable case. In place of the term "strong quasiconcavity", McFadden (1978; 403) uses "differential strict quasiconcavity".

Figure 1 summarizes the relationships between the various types of concavity and quasiconcavity in the case where the functions under consideration are upper semicontinuous.

9. NECESSARY AND SUFFICIENT CONDITIONS FOR NONSMOOTH PROGRAMMING

In this section, we will let g, f^1, f^2,..., f^M be M+1 *proper* real valued functions defined over R^N. We say that function f defined over R^N is *proper* if $f(x) < +\infty$ for all $x \varepsilon R^N$ and $f(x) > -\infty$ for at least one $x \varepsilon R^N$ (our definition is motivated by Rockafellar's (1970) definition of a proper concave function).

Consider the following constrained maximization problem:

$$\max_x \{g(x) : f^1(x) \geq 0,\ f^2(x) \geq 0, \ldots, f^M(x) \geq 0,\ x \varepsilon R^N\}. \quad (68)$$

DEFINITION 12 (e.g., Bazaraa and Shetty (1976; 84)): The *set of feasible directions* to a point $x \varepsilon R^N$ is defined as the set

$$F(x) \equiv \{v : v \varepsilon R^N,\ v^T v = 1, \text{ there exists } \delta > 0 \text{ such that } f^m(x+tv) \geq 0 \text{ for } 0 \leq t \leq \delta \text{ for } m=1, 2, \ldots, M\}. \quad (69)$$

FIGURE 1

Relations Between Types of Quasiconcavity in the Upper Semicontinuous Case.

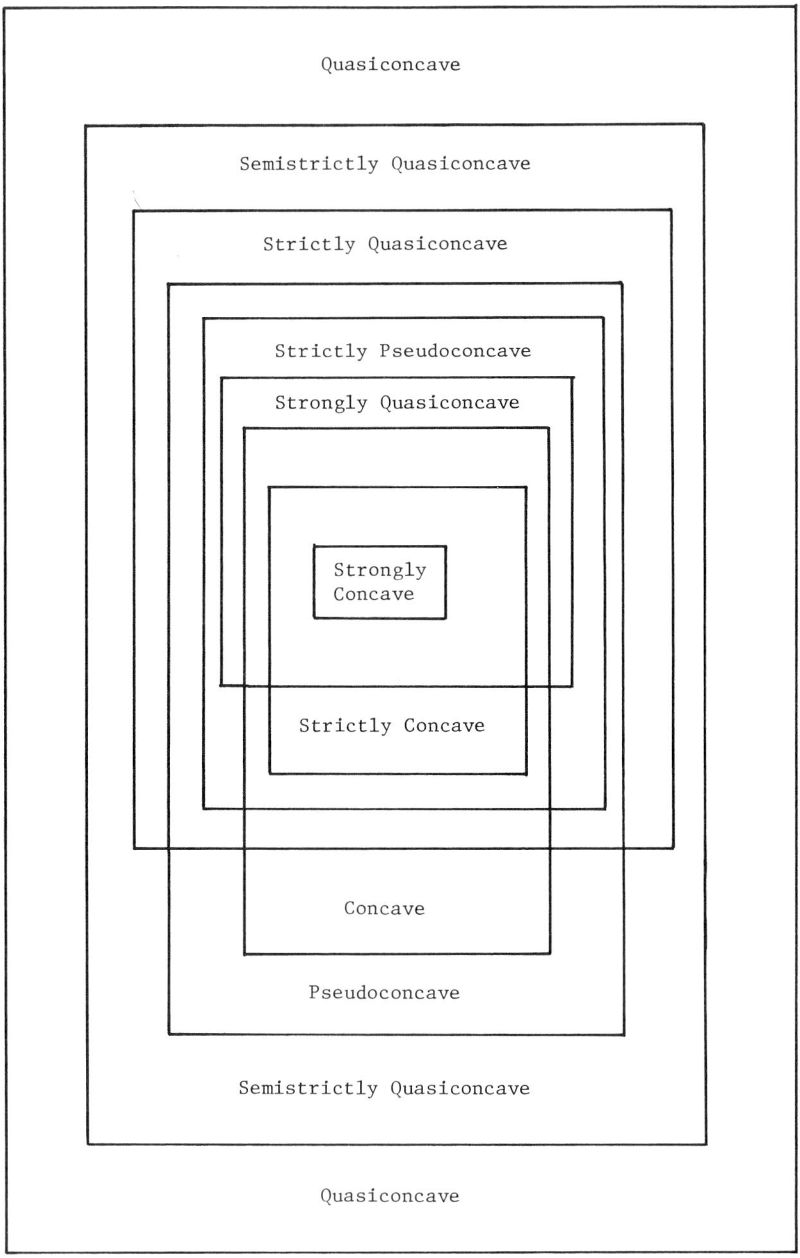

THEOREM 20: Suppose $x^0 \varepsilon R^N$ *solves (68). Then the following* <u>*necessary*</u> *condition holds:*

$$v\varepsilon F(x^0) \text{ implies } D_v^{+u} g(x^0) \leq 0. \tag{70}$$

The proof follows directly from the definitions of $F(x^0)$ and $D_v^{+u} g(x^0)$.

COROLLARY 23: Let x^0 *solve (68) and let* $B(x^0)$ *be the set of binding constraints; i.e.,* $m\varepsilon B(x^0)$ *iff* $f^m(x^0) = 0$. *Then the following* <u>*necessary*</u> *condition holds:*

$$D_v^{+\ell} f^m(x^0) > 0 \text{ for } m\varepsilon B(x^0) \text{ and } D_v^{+\ell} f^m(x^0) > -\infty \text{ for } m \notin B(x^0) \text{ implies } D_v^{+u} g(x^0) \leq 0. \tag{71}$$

PROOF: Let $m\varepsilon B(x^0)$ and $D_v^{+\ell} f^m(x^0) > 0$ for some direction v. Then $f^m(x^0) = 0$ since $m\varepsilon B(x^0)$. *Suppose* for every $\delta>0$, there exists t_δ such that $0 < t_\delta \leq \delta$ and $f^m(x^0+t_\delta v) < 0$. Then $D_v^{+\ell} f^m(x^0) \equiv \lim\inf_{t\to 0^+} [f(x^0+tv)-0] / t \leq 0$ which contradicts $D_v^{+\ell} f^m(x^0) > 0$. Thus our supposition is false and there exists $\delta^m>0$ such that

$$f^m(x^0+tv) \geq 0 \text{ for } 0 < t \leq \delta^m,\ m\varepsilon B(x^0). \tag{72}$$

Now let $m\notin B(x^0)$ and $D_v^{+\ell} f^m(x^0) > -\infty$ for the same direction v. Since $m\notin B(x^0)$, $f^m(x^0) > 0$. *Suppose* for every $\delta>0$, there exists t_δ such that $0 < t_\delta \leq \delta$ and $f^m(x^0+t_\delta v) < 0$. Then $D_v^{+\ell} f^m(x^0) \equiv \lim\inf_{t\to 0^+} [f^m(x^0+tv) - f^m(x^0)] / t = -\infty$ which contradicts $D_v^{+\ell} f^m(x^0) > -\infty$. Thus our supposition is false and there exists $\delta^m>0$ such that

$$f^m(x^0+tv) \geq 0 \text{ for } 0 < t \leq \delta^m,\ m\notin B(x^0). \tag{73}$$

(72) and (73) imply that $v\varepsilon F(x^0)$ and thus by (70), $D_v^{+u} g(x^0) \leq 0$ which verifies (71).

Q.E.D.

We note that although (70) implies (71), the converse is not true; i.e., v could belong to the set of feasible directions $F(x^0)$ but it need not be the case that $D_v^{+\ell} f^m(x^0) > 0$ for $m\varepsilon B(x^0)$ and $D_v^{+\ell} f^m(x^0) > -\infty$ for $m\notin B(x^0)$.

The following theorem shows that the necessary conditions (70) are also sufficient if the objective function g is pseudoconcave and the constraint functions $f^1,\ldots,f^M$ are quasiconcave.

THEOREM 21: Suppose g *is pseudoconcave,* f^1, $f^2,\ldots,f^M$ *are quasiconcave, and* x^0 *is a feasible solution for (68) which satisfies (70). Then* x^0 *solves the nonlinear, nonsmooth programming problem (68).*

PROOF: Let $x \neq x^0$ be any feasible point for (68); i.e., $f^1(x) \geq 0$, $f^2(x) \geq 0,\ldots,f^M(x) \geq 0$. Define $v \equiv (x-x^0) / [(x-x^0)^T(x-x^0)]^{1/2}$, $x=x^0+t_0v$ for $t_0 \equiv [(x-x^0)^T(x-x^0)]^{1/2} > 0$.

By the quasiconcavity of f^m, $f^m(x^0+tv) \geq \min\{f^m(x^0), f^m(x^0+t_0v)\} \geq 0$ since both x^0 and x^0+t_0v are feasible for (68). Thus $v\varepsilon F(x^0)$ and by (70) $D_v^{+u}g(x^0) \leq 0$. Since g is pseudoconcave, by (42) $g(x^0+t_0v) = g(x) \leq g(x^0)$. Thus x^0 solves (68).

Q.E.D.

We say that the proper function g is pseudoconcave at x^0 iff $D_v^{+u}g(x^0) \leq 0$ implies $g(x^0+tv) \leq g(x^0)$ for all $t \geq 0$. Similarly, the proper function f is quasiconcave at x^0 iff $f(x^0)$ and $f(x^0+t_0v)$ finite implies $f(x^0+tv) \geq \min\{f(x^0), f(x^0+t_0v)\}$ for all t such that $0 \leq t \leq t_0$ where v is a direction vector and $t_0>0$. Using the above definitions, it is easy to repeat the proof of Theorem 21 in order to prove the following result.

COROLLARY 24: Suppose g *is pseudoconcave at* x^0, f^1, $f^2,\ldots,f^M$ *are quasiconcave at* x^0*, and* x^0 *is a feasible solution for (68) which satisfies the necessary conditions (70). Then* x^0 *solves (68).*

Theorems 20 and 21 are very similar in spirit to Theorem 1 in Guignard (1969; 234) and to Theorem 2 in Hiriart-Urruty (1978; 78); the main difference is that we do not assume that our objective and constraint functions are either differentiable or locally Lipschitz.

There is another approach for obtaining sufficient conditions for the constrained maximization problem (68) -- namely Kuhn and Tucker's (1951) *saddle point approach.* Kuhn and Tucker assumed that the objective and constraint functions were concave and differentiable and showed that the Kuhn-Tucker conditions were sufficient under these assumptions. Mangasarian (1969; 151-153) weakened the Kuhn-Tucker regularity conditions to a pseudoconcave objective function and quasiconcave constraint functions. Our next theorem is similar to that of Mangasarian except that we do not assume differentiability. Before proving the theorem, we require two auxiliary results.

Let a and b be proper functions defined over R^N with $a(x^0)$ and $b(x^0)$ finite for $x^0 \varepsilon R^N$. Let v be a direction in R^N. Then

$$D_v^{+\ell}[a(x^0)+b(x^0)] \equiv \liminf_{t\to 0^+} [a(x^0+tv)+b(x^0+tv) - a(x^0) - b(x^0)] / t \geq \liminf_{t\to 0^+} [a(x^0+tv) - a(x^0)] / t + \liminf_{s\to 0} [b(x^0+sv) - b(x^0)] / s = D_v^{+\ell}a(x^0) + D_v^{+\ell}b(x^0). \quad (74)$$

Now let $D_v^{+u}a(x^0) = \lim_{n\to\infty}[a(x^0+s_n v) - a(x^0)] / s_n$ for some sequence of positive scalars $\{s_n\}$ converging to 0. Then for some sequence of positive scalars $\{t_n\}$ converging to 0, we have (if the lim does not exist, replace it by the lim inf):

$$D_v^{+u}[a(x^0)+b(x^0)] = \lim_{n\to\infty}[a(x^0+t_n v)+b(x^0+t_n v) - a(x^0) - b(x^0)] / t_n$$

$$\geq \lim_{n\to\infty}[a(x^0+s_n v)+b(x^0+s_n v) - a(x^0) - b(x^0)] / s_n$$

$$= \lim_{n\to\infty}[a(x^0+s_n v)-a(x^0)] / s_n + \lim_{n\to\infty}[b(x^0+s_n v)-b(x^0)] / s_n$$

and by the definition of the sequence $\{s_n\}$,

$$= D_v^{+u}a(x^0) + \lim_{n\to\infty}[b(x^0+s_n v)-b(x^0)] / s_n$$

$$\geq D_v^{+u}a(x^0) + D_v^{+\ell}b(x^0). \quad (75)$$

THEOREM 22 (Local Saddle Point Sufficiency Theorem): Suppose g *is a proper pseudoconcave function and* $f^1,\ldots,f^M$ *are proper quasiconcave functions. Suppose there exists a vector of multipliers* $u^0 \equiv (u_1^0,u_2^0,\ldots,u_M^0)\varepsilon R^M$ *and* $x^0 \varepsilon R^N$ *such that* $g(x^0)$ *is finite and the following conditions are satisfied:*

$$\text{for every direction } v,\ D_v^{+u}\,[g(x^0)+\sum_{m=1}^{M} u_m^0\, f^m(x^0)] \le 0, \tag{76}$$

$$f^1(x^0) \ge 0,\ f^2(x^0) \ge 0,\ldots,f^M(x^0) \ge 0, \tag{77}$$

$$u^0 \ge 0_M, \tag{78}$$

$$\sum_{m=1}^{M} u_m^0\, f^m(x^0) = 0. \tag{79}$$

Then x^0 *solves the constrained maximization problem (68).*

PROOF: Let v be a direction, t>0 and $x \equiv x^0+tv$ be any feasible point; i.e., $f^1(x^0+tv) \ge 0,\ldots,f^M(x^0+tv) \ge 0$. Let $B(x^0)$ be the set of indices m such that $f^m(x^0) = 0$. From (77) - (79), we deduce that $u_m^0 = 0$ if $f^m(x^0) > 0$. Thus from (76),

$$
\begin{aligned}
0 &\ge D_v^{+u}\,[g(x^0) + \sum_{m\varepsilon B(x^0)} u_m^0\, f^m(x^0)] \\
&\ge D_v^{+u}\, g(x^0) + D_v^{+\ell}\,[\sum_{m\varepsilon B(x^0)} u_m^0 f^m(x^0)] \\
&\quad \text{using (75)} \\
&\ge D_v^{+u}\, g(x^0) + \sum_{m\varepsilon B(x^0)} D_v^{+\ell}\,[u_m^0 f^m(x^0)] \\
&\quad \text{by repeated use of (74)} \\
&= D_v^{+u}\, g(x^0) + \sum_{m\varepsilon B(x^0)} u_m^0\, D_v^{+\ell} f^m(x^0)
\end{aligned} \tag{80}
$$

If $B(x^0) = \emptyset$, then (80) and the definition of pseudoconcavity (42) imply that $g(x) \le g(x^0)$ and thus x^0 solves (68). If $B(x^0) \ne \emptyset$, then (80) can be rewritten as

$$
\begin{aligned}
D_v^{+u}\, g(x^0) &\le -\sum_{m\varepsilon B(x^0)} u_m^0\, D_v^{+\ell}\, f^m(x^0) \\
&\le 0
\end{aligned}
$$

since $u_m^0 \geq 0$ and $0 = f^m(x^0) \leq f^m(x^0+tv)$ for $m\varepsilon B(x^0)$ and using (20), $D_v^{+\ell} f^m(x^0) \geq 0$. Hence by pseudoconcavity of g, $g(x) \leq g(x^0)$. Thus x^0 solves (68).

Q.E.D.

COROLLARY 25: Theorem 22 remains valid if (76) is replaced by:

$$v\varepsilon F(x^0) \text{ implies } D_v^{+u} [g(x^0) + \sum_{m=1}^{M} u_m^0 f^m(x^0)] \leq 0. \tag{81}$$

COROLLARY 26: Theorem 22 remains valid if (76) is replaced by:

$$v\varepsilon F(x^0) \text{ implies } D_v^{+u} g(x^0) + \sum_{m\varepsilon B(x^0)} u_m^0 D_v^{+\ell} f^m(x^0) \leq 0. \tag{82}$$

Conditions (77), (78), (79) and (82) are not necessary for x^0 to solve (68) in general, but they are necessary (and sufficient) in the case where the objective function g is pseudoconcave and the constraint functions $f^1,\ldots,f^M$ are quasiconcave. However, conditions (76) to (79) are *not* necessary for x^0 to solve (68) even when g is pseudoconcave and $f^1,\ldots,f^M$ are quasiconcave. In order to make conditions (76) to (79) necessary conditions for x^0 to solve (68), we require differentiability of g and $f^1,f^2,\ldots,f^M$ and a constraint qualification condition (see Kuhn and Tucker (1951) and Guignard (1969) for various constraint qualification conditions).

We conclude this paper by noting that our general necessary conditions for the constrained maximization problem (68) given by (70) and (71) are almost trivial compared to the deeper necessary conditions derived by Clarke (1976; 170) and Hiriart-Urruty (1978; 83); however, these authors assume that the objective and constraint functions are locally Lipschitz whereas we assume no regularity conditions at all, (other than the assumption of properness).

REFERENCES

ARROW, K.J. and ENTHOVEN, A.C. (1961). "Quasiconcave Programming." *Econometrica* *29*, 779-800.

AVRIEL, M., DIEWERT, W.E., SCHAIBLE, S. and ZIEMBA, W.T. (1981). "Introduction to Concave and Generalized Concave Functions." This vol. pp. 21-50.

BAZARAA, M.S. and SHETTY, C.M. (1976). *Foundations of Optimization.* Lecture Notes in Economics and Mathematical Systems 122 Springer-Verlag, Berlin.

BERGE, C. (1963). *Topological Spaces.* Oliver and Boyd, London.

CLARKE, F.C. (1975). "Generalized Gradients and Applications." *Transactions of the American Mathematical Society* *205*, 247-262.

CLARKE, F.C. (1976). "A New Approach to Lagrange Multipliers." *Mathematics of Operations Research* *1*, 165-174.

DIEUDONNÉ, J. (1960). *Foundations of Modern Analysis.* Academic Press, New York.

DIEWERT, W.E., AVRIEL, M. and ZANG, I. (1981). "Nine Kinds of Quasiconcavity and Concavity." *Journal of Economic Theory.* Forthcoming.

DINI, U. (1892). *Grundlagen für eine Theorie der Functionen einer veränderlichen reelen Grösse.* B.E. Teubner.

ELKIN, R.M. (1968). "Convergence Theorems for Gauss-Seidel and Other Minimization Algorithms." Ph.D. Dissertation, University of Maryland, College Park, Maryland.

FENCHEL, W. (1953). "Convex Cones, Sets and Functions." Lecture Notes, Department of Mathematics, Princeton University, Princeton, N.J.

GINSBERG, W. (1973). "Concavity and Quasiconcavity in Economics". *Journal of Economic Theory* *6*, 596-605.

GUIGNARD, M. (1969). "Generalized Kuhn-Tucker Conditions for Mathematical Programming Problems in a Banach Space." *SIAM Journal of Control* *7*, 232-241.

HANSON, M.A. (1964). "Bounds for Functionally Convex Optimal Control Problems." *Journal of Mathematical Analysis and Applications* *8*, 84-89.

HIRIART-URRUTY, J.B. (1977). "New Concepts in Nondifferentiable Programming." *Journées d'Analyse Non Convexe*, Université de Pau, France.

HIRIART-URRUTY, J.B. (1978). "On Optimality Conditions in Nondifferentiable Programming." *Mathematical Programming* *14*, 73-86.

HIRIART-URRUTY, J.B. (1979a). "Mean Value Theorems in Nonsmooth Analysis." Dept. of Mathematics, University of Kentucky, Lexington, Kentucky.

HIRIART-URRUTY, J.B. (1979b). "A Note on the Mean Value Theorem for Convex Functions." Dept. of Mathematics, University of Kentucky, Lexington, Kentucky.

KARAMARDIAN, S. (1967). "Duality in Mathematical Programming." *Journal of Mathematical Analysis and Applications* *20*, 344-358.

KUHN, H.W. and TUCKER, A.W. (1951). "Nonlinear Programming." In *Proceedings of the Second Berkeley Symposium on Mathematical Statistics and Probability* (J. Neyman, ed.). University of California Press, Berkeley, California, 481-492.

LEBOURG, G. (1975). "Valeur moyenne pour gradient généralisé." *Des Comptes Rendus des Séances de l'Académie des Sciences* *Serie A, 281*, 795-797.

MANGASARIAN, O.L. (1965). "Pseudo-Convex Functions." *SIAM Journal on Control* *3*, 281-290.

MANGASARIAN, O.L. (1969). *Nonlinear Programming.* McGraw-Hill, New York.

MARTOS, B. (1965). "The Direct Power of Adjacent Vertex Programming Methods." *Management Science* *12*, 241-252.

MCFADDEN, D. (1978). "Convex Analysis." In *Production Economics: A Dual Approach to Theory and Application, Vol.1* (M. Fuss and D. McFadden, eds.). North-Holland, Amsterdam, 383-408.
MCSHANE, E.J. (1944). *Integration.* Princeton University Press, Princeton, New Jersey.
MIFFLIN, R. (1977a). "Semismooth and Semiconvex Functions in Optimization." *SIAM Journal on Control and Optimization, 15,* 959-972.
MIFFLIN, R. (1977b). "An Algorithm for Constrained Optimization with Semismooth Functions." *Mathematics of Operations Research 2,* 191-207.
NEWMAN, P. (1969). "Some Properties of Concave Functions." *Journal of Economic Theory 1,* 291-314.
POLJAK, B.T. (1966). "Existence Theorems and Convergence of Minimizing Sequences in Extremum Problems with Restrictions." *Doklady Akademii Naul C.C.P., 166,* 287-290.

PONSTEIN, J. (1967). "Seven Kinds of Convexity." *SIAM Review, 9,* 115-119.
ROBERTS, A.W. and VARBERG, D.E. (1973). *Convex Functions.* Academic Press, New York.
ROCKAFELLAR, R.T. (1970). *Convex Analysis.* Princeton University Press, Princeton, N.J.
ROCKAFELLAR, R.T. (1976). "Saddlepoints of Hamiltonian Systems in Convex Lagrange Problems Having a Nonzero Discount Rate." *Journal of Economic Theory 12,* 71-113.
ROCKAFELLAR, R.T. (1978). "The Theory of Subgradients and its Applications to Problems of Optimization." Lecture Notes, University of Montreal, Les Presses de l'Université de Montréal, Montréal, Canada.
SCHEEFER, L. (1884). "Zur Theorie der stetigen Funktionen einer reelen Veränderlichen." *Acta Mathematica 5,* 183-194.
STOER, J. and WITZGALL, C. (1970). *Convexity and Optimization in Finite Dimensions I.* Springer-Verlag, Berlin.
THOMPSON JR., W.A. and PARKE, D.W. (1973). "Some Properties of Generalized Concave Functions." *Operations Research 21,* 305-313.
TUY, H. (1964). "Sur les inégualités linéaires." *Colloquium Mathematicum 13,* 107-123.
WEGGE, L. (1974). "Mean Value Theorems for Convex Functions." *Journal of Mathematical Economics 1,* 207-208.
YOUNG, W.H. and YOUNG, G.C. (1909). "On Derivates and the Theorem of the Mean." *Quarterly Journal of Pure and Applied Mathematics 40,* 1-26.

CONNECTEDNESS OF LEVEL SETS AS A GENERALIZATION OF CONCAVITY

D.H. Martin

The author's purpose in this paper is to show that the property of connectedness of upper level sets has some claim to be regarded as a viable generalization of concavity in optimization theory. After surveying recent work on the natural occurrence of connected (but non-convex) level sets in parametric linear programming and on the relationship between connectedness of level sets and uniqueness, a new result is presented concerning the preservation of connectedness of level sets under a very general form of supremal convolution. This result has direct application to multi-stage decision processes.

1. INTRODUCTION

The theme of this paper is the significance for optimization theory of the property, which a given objective function

$$f : X \to [-\infty,\infty]$$

on a topological space X may or may not have, that all of its GE-level sets

$$GE_\alpha(f) \triangleq \{x \in X | f(x) \geqslant \alpha\}, \quad -\infty < \alpha \leqslant \infty$$

are (topologically) connected, or alternatively, that all of its GT-level sets

$$GT_\alpha(f) \triangleq \{x \in X | f(x) > \alpha\}, \quad -\infty \leqslant \alpha < \infty$$

are connected.

One sees immediately that either of these notions is technically a generalization of the notion of quasiconcavity in the case that X is a linear topological space, but in fact, one can claim much more than a mere technical extension. In three specific ways, the property of connectedness of level sets proves to be a valuable generalization of quasiconcavity. Firstly, as is demonstrated in the next section, functions with connected but non-convex GT-level sets occur naturally when one considers the dependence of the maximum of a linear program upon the constraint coefficient matrix. Secondly, there is an intimate rela-

ISBN 0-12-621120-5

tionship, which is almost an equivalence, between connectedness of level sets and uniqueness of maximizing sets. This theory, full details of which will appear in Martin (1981), is reviewed in Section 3. Thirdly, when one asks what functional operations preserve connectedness properties of level sets, one discovers that the situation in this respect is more favourable than it is for the class of quasiconcave functions. A new general result in this regard is proved in Section 4, with an application to multi-stage decision processes.

2. THE MAXIMUM OF A LINEAR PROGRAM AS A FUNCTION OF THE CONSTRAINT COEFFICIENT MATRIX

Let $b \in \mathbb{R}^m$ and $c \in \mathbb{R}^n$ be fixed vectors, and consider the function $f : \mathbb{R}^{m\times n} \to [-\infty,\infty]$ defined by

$$f(A) = \sup\{c'x \mid x \in \mathbb{R}^n, Ax = b, x \geqslant 0\},$$

in which we adopt the convention that the supremum of the empty set is $-\infty$. If f(A) is finite for some A, then the supremum is achieved at some basic feasible solution x_A, and f(A) can accordingly be expressed, via Cramer's Rule, as a rational expression in the entries of A, where the expression depends upon which components of x_A are basic. Thus $\mathbb{R}^{m\times n}$ can be written as a finite union of regions, on each of which f(A), if not identically $\pm\infty$, is given by some rational expression in the entries of A. Thus f is a piecewise rational function[1)], but as shown by Bereanu (1973) and Martin (1975), need not even be continuous.

It turns out however that there is a further general statement that can be made about the function f - viz. its GT-level sets are all connected, as are its GE-level sets with the sole possible exception of GE_∞. In fact, as the following theorem makes precise, this statement remains true even when the domain of f is restricted in a certain way.

---

1) See Dinkelbach (1969) for a further discussion of this aspect.

THEOREM 1: Let $D_1, D_2, \ldots, D_n$ *be any convex subsets of* $\mathbb{R}^m$ *and let* $D = D_1 \times D_2 \times \ldots \times D_n \subseteq \mathbb{R}^{m \times n}$ *denote the set of coefficient matrices* A *having the i-th column in* D_i, *for each* $i = 1,2,\ldots,n$. *Then all of the* GT-*level sets of* f *on* D *are path-connected (and hence connected), as are all of the* GE-*level sets of* f *on* D *with the sole possible exception of* GE_∞.

This is proved in Martin (1981) by explicit construction of paths joining any two points of any given level set. Also given there is an example showing that GE_∞ may indeed be disconnected, and showing that although the other level sets are connected they may indeed fail to be convex. This theorem thus provides a good affirmative answer to the question as to whether such functions may arise in a natural way.

It is easy to see the importance of the connectedness of the GE-level sets of f when one wishes, as in a Wolfe generalized program (see, for example, Dantzig (1963)), to seek the maximum of f over the domain D. Suppose f is bounded on D and achieves its supremum, which we denote by α. Then the set of global maximizers is precisely $GE_\alpha(f)$, which is connected. Secondly, there can be no *strict* local maximizer which is not global. This statement can be strengthened by the introduction of the notions of GE-maximizing sets and GT-maximizing sets, and this is the purpose of the theory reviewed in the next section.

3. MAXIMIZING SETS AND UNIQUENESS

The above result first stirred the writer's interest in the significance of *connectedness* of upper level sets as a generalization of the convexity of level sets which characterizes quasiconcave functions. Ortega and Rheinboldt (1970) had previously considered a curvilinear generalization of strict quasiconcavity. They called a function $f : X \to \mathbb{R}$ on a path-connected domain X *strictly connected* if for every

$x_o, x_1 \in X$ with $x_o \neq x_1$, there exists a continuous path $p : [0,1] \to X$ with $p(0) = x_o$, $p(1) = x_1$ and

$$f(p(t)) < \max \{f(x_o), f(x_1)\} \quad \text{for all } t \in (0,1).$$

Such functions cannot be characterized in terms of their level sets, but were shown to have at most one local maximizer, which would actually be a strict global maximizer. Avriel and Zang (1981) proved the following partial converse. Let f be a real continuous function on $\mathbb{R}^n$ having compact GE-level sets, and with the additional property that for all $\alpha \in \mathbb{R}$, if GT_α is path-connected then so is $\{\bar{x}\} \cup GT_\alpha$ for every $\bar{x} \in cl(GT_\alpha)$. Then if f has a unique global maximizer and no other local maximizers, f must be strictly connected. The somewhat awkward additional hypothesis here seems unavoidably associated with generalizations of *strict* concavity designed to guarantee uniqueness of maximizers as *points*.

In Martin (1981) it is shown how, by suitably re-defining one's idea of what constitutes a single maximizing entity, one can distil the essence of the idea that connectedness of level sets should imply uniqueness of a maximizer. What is more, a converse theorem shows that the uniqueness is not simply a result of too wide a notion of what constitutes a unity.

Two different theories are involved, the simpler one pertaining to GE-level sets, and the other to GT-level sets. The two theories coincide however for functions having Hausdorff domain X and compact GE-level sets. A brief exposition of these theories follows, the reader being referred however to Martin (1981) for fuller details. It turns out that the topological notion of a *connected component of a set* has precisely the properties which are needed for the analysis of the relationship between uniqueness and connectedness of level sets.

3.1 GE-Maximizing Sets

Let $f : X \to [-\infty,\infty]$ be an extended real function on a topological space X. We shall say that f is GE-connected (GE-compact) on X if every GE-level set is topologically connected (compact).

DEFINITION 1: A non-empty subset $S \subset X$ *is a* GE-*maximizing set* (GEMS) *for* f *on* X *if*

(a) there exists an open neighbourhood $U \supset S$ *in* X *such that*

$$f(y) \leqslant f(x) \quad \textit{for all} \quad y \in U, \quad x \in S,$$

from which it follows that f *is constant on* S, *and*

(b) S *is a connected component of* GE_α, *where* $\{\alpha\} = f(S)$. *If in* (a) *one may take* $U = X$, S *is called a global* GEMS *for* f *on* X.

Note that here we permit the value $\alpha = +\infty$.

Remark. It is easy to see that if f attains its supremum then f must have at least one global GEMS. This observation means that in circumstances where a normal maximizer exists, so does a GEMS. However, it is not true that every local maximizer necessarily belongs to some GEMS. For example, in the function

$$f(x) = \begin{cases} -x & \text{for } 0 \leqslant x \leqslant 1 \\ -1 & \text{for } 1 \leqslant x \leqslant 2, \end{cases}$$

all points $x \in (1,2)$ are in the long accepted, but nevertheless somewhat absurd position of being local maximizers and at the same time global minimizers, and this for a non-constant function on a connected domain. However these points do not lie in any GEMS for this function.

UNIQUENESS THEOREM FOR GEMSs 2: If f *is* GE-*connected on* X, *then either* f *has no* GEMS *or has precisely one, and this one is global.*

PROOF. Let S be a GEMS for f on X with $f(S) = \{\alpha\}$. Since GE_α is connected, and S is a connected component of it, $S = GE_\alpha$. Hence for all $x \in X \setminus S$, $f(x) < \alpha$, which proves both assertions of the theorem. □

In the writer's opinion, the immediacy of this proof, together with the truth of the converse result presented later, strongly justify our regarding the points of a GEMS as *jointly constituting a single maximizing entity.*

3.2 GT-Maximizing Sets

The appropriate definition of a GT-maximizing set for an extended real function f on a topological space X is more involved, but is suggested by the situation for parametric linear programs discussed above.

As a notational convenience, if $Y \supset Z$ are subsets of X with Z contained in a single connected component of Y, that component will be denoted by CC(Y,Z).

DEFINITION 2: A non-empty subset $S \subset X$ *is a* GT-*maximizing set* (GTMS) *for* f *on* X *if*

(a) *there exists an open neighbourhood* $U \supset S$ *in* X *such that*

$$f(y) \leqslant f(x) \quad \text{for all} \quad y \in U, \quad x \in S,$$

from which it follows that f *is constant on* X, *and*

(b) *with* $f(S) = \{\alpha\}$, *we have* $\alpha > -\infty$ *and for all* $\beta < \alpha$, S *is contained in a single connected component of* GT_β *and*

$$\bigcap_{\beta < \alpha} CC(GT_\beta, S) = S.$$

If in (a) one may take U = X, *then* S *is called a global* GTMS *for* f *on* X.

Remarks. It is easy to show that any GTMS is a disjoint union of GEMSs, and that a GTMS is a GEMS iff it is connected. However, in general, a given GEMS need *not* be contained in some GTMS. Consequently one should not regard the GTMS notion simply as an extension or generalization of the GEMS notion - rather the two notions co-exist and each has its realm of application.

We shall say that $f : X \to [-\infty,\infty]$ is GT-connected if every GT-level set is connected.

UNIQUENESS THEOREM FOR GTMSs 3: If f *is* GT-*connected on* X, *then either* f *has no* GTMS, *or has precisely one, and that one is global.*

PROOF. Let S be a GTMS for a GT-connected function $f : X \to [-\infty,\infty]$. Then $f(S) = \{\alpha\}$ with $\alpha > -\infty$, and for all $\beta < \alpha$ we must have $CC(GT_\beta, S) = GT_\beta$, and hence

$$S = \bigcap_{\beta < \alpha} GT_\beta = GE_\alpha .$$

Thus $f(x) < \alpha$ for all $x \in X \setminus S$, which means that S is a global GTMS for f on X, and is unique. □

Once again the immediacy of this proof, together with the converse theorem presented later, justify our regarding the points of a GTMS as a single minimizing entity. It should also be remarked that this theorem does *not* contain the uniqueness theorem for GEMSs as a special case.

3.3 Equivalence When f Is GE-Compact and X Is Hausdorff

Suppose $f : X \to [-\infty,\infty]$ is GE-connected, and consider any non-empty GT-level set, say GT_α. Then for some $\alpha' > \alpha$ the set $GE_{\alpha'}$ is also non-empty, and so the identity

$$GT_\alpha = \bigcup_{\alpha' > \beta > \alpha} GE_\beta$$

expresses GT_α as the union of connected sets with non-empty intersection, from which it follows that GT_α is connected. Thus GE-connectedness implies GT-connectedness, whereas the converse may fail. The following theorem, the proof of which is given in Martin (1981), gives conditions for the converse to hold, and under these conditions the GE-theory and the GT-theory are equivalent.

EQUIVALENCE THEOREM 4: Suppose f *is* GE-*compact and* X *is Hausdorff. Then* f *is* GE-*connected iff* f *is* GT-*connected. A subset* $S \subseteq X$ *is a* GEMS *iff* S *is a* GTMS.

3.4 Connectedness as a Necessary Condition for Uniqueness

Simple examples, such as

$$f(x) = x / (x^2 + 1), \quad x \in \mathbb{R} = X$$

serve to show that a function may have a unique GEMS, which is actually global, without having connected GE-level sets, provided some of the GE-level sets are not compact. This motivates the restriction in the following theorem to GE-compact functions, while Hausdorffness is imposed to ensure that compact sets are closed. Thus the conditions under which the equivalence theorem holds are also required here. Once again, the reader is referred to Martin (1981) for a proof.

CONVERSE THEOREM 5: Let $f : X \to [-\infty, \infty]$ *be a* GE-*compact function on a Hausdorff domain* X. *Then* f *has a unique* GEMS *only if* f *is* GE-*connected.*

Remarks. It is interesting that this converse theorem fails if the notion of topological connectedness is replaced throughout by that of path-connectedness. A counterexample is given in Martin (1981).

This converse theorem, together with the Equivalence Theorem and the Uniqueness Theorems show that the notions of GEMS and GTMS lead to a satisfactory statement of the relationship between connectedness properties of level sets and uniqueness in optimization.

4. OPERATIONS WITH GT-CONNECTED FUNCTIONS

It is well-known that while the sum of concave functions is concave, even the *separable sum*

$$f_1(x_1) + f_2(x_2)$$

of quasiconcave functions is not in general quasiconcave in (x_1, x_2). For the class of GT-connected functions, the situation is better, for it turns out that GT-connectedness is preserved by separable summation. This is a consequence of the following general theorem, which also specialises to the known result (see Martin (1978)) that the operation of *supremal convolution* preserves GT-connectedness.

Consider m given GT-connected functions

$$f_i : X_i \to \mathbb{R}, \qquad i = 1,2,\ldots,m,$$

where each domain X_i is a topological space. We consider combination operations of the form of separable programmes, having the following ingredients. Let Ω be a connected space of parameters (possibly a singleton), and let

$$\psi : X_1 \times X_2 \times \ldots \times X_m \times \Omega \to X$$

be a continuous map into a further topological space X. Finally let

$$\Theta : \mathbb{R}^m \to \mathbb{R}$$

be continuous and non-decreasing in each argument.

THEOREM 6: The function $F : X \to [-\infty,\infty]$ *given by*

$$F(x) = \sup \{\Theta(f_1(x_1), f_2(x_2),\ldots,f_m(x_m)) \,|\, x_i \in X_i, \; i=1,\ldots,m, \; \omega \in \Omega, \; \psi(x_1,x_2,\ldots,x_m,\omega) = x\}$$

is GT-connected.

PROOF. Since $GT_{-\infty}(F) = \bigcup_{\alpha \in \mathbb{R}} GT_\alpha(F)$, it suffices to show that $GT_\alpha(F)$ is connected for all real α. To this end, consider any two points

$$\hat{x}, \tilde{x} \in GT_\alpha(F)$$

for some $\alpha \in \mathbb{R}$.

From the definition of F it follows that there are points

$$\hat{x}_i, \tilde{x}_i \in X_i, \quad i=1,2,\ldots,m \text{ and } \hat{\omega}, \tilde{\omega} \in \Omega$$

such that writing $f_i(\hat{x}_i) = \hat{f}_i$, $f_i(\tilde{x}_i) = \tilde{f}_i$, we have

$$\psi(\hat{x}_1,\ldots,\hat{x}_m, \hat{\omega}) = \hat{x} \quad \text{and} \quad \Theta(\hat{f}_1,\ldots,\hat{f}_m) > \alpha,$$

and likewise

$$\psi(\tilde{x}_1,\ldots,\tilde{x}_m, \tilde{\omega}) = \tilde{x} \quad \text{and} \quad \Theta(\tilde{f}_1,\ldots,\tilde{f}_m) > \alpha.$$

Without loss of generality we may suppose the indices have been relabelled so that for some index n we have

$$\hat{f}_i \leqslant \tilde{f}_i \quad \text{for} \quad i=1,2,\ldots,n$$

while

$$\hat{f}_i \geqslant \tilde{f}_i \quad \text{for} \quad i=n+1,\ldots,m.$$

Of course it may occur that one of these ranges is void. This done, we construct m subsets of X as follows: for k=1,...,m set

$$A_k = \{\psi(\tilde{x}_1,\ldots,\tilde{x}_{k-1}, x_k, x_{k+1},\ldots,\hat{x}_m,\omega) \mid x_k \in X_k, \quad \omega \in \Omega,$$
$$\Theta(\tilde{f}_1,\ldots,\tilde{f}_{k-1}, f_k(x_k), \hat{f}_{k+1},\ldots,\hat{f}_m) > \alpha\}.$$

It follows immediately from the definition of F as a supremum that

$$A_k \subseteq GT_\alpha(F), \quad k=1,\ldots,m.$$

Also, using the continuity and monotonicity of Θ in its k-th argument, we have

$$\Theta(\tilde{f}_1,\ldots,\tilde{f}_{k-1}, f_k(x_k), \hat{f}_{k+1},\ldots,\hat{f}_m) > \alpha$$

if and only if

$$f_k(x_k) > \beta_k,$$

where the interval (β_k,∞) is the inverse image of (α,∞) under $\Theta(\tilde{f}_1,\ldots,\tilde{f}_{k-1}, \cdot, \hat{f}_{k+1},\ldots,\hat{f}_m)$. Hence it follows for k=1,...,m that

$$A_k = \psi(\tilde{x}_1,\ldots,\tilde{x}_{k-1}, GT_{\beta_k}(f_k), \hat{x}_{k+1},\ldots,\hat{x}_m, \Omega).$$

Since f_k is GT-connected, and Ω is connected, it follows that each set A_k is connected.

Consider now points of the form

$$\psi(\tilde{x}_1,\ldots,\tilde{x}_k, \hat{x}_{k+1},\ldots,\hat{x}_m, \omega), \quad \omega \in \Omega, \quad k=1,\ldots,m-1.$$

For k=1,2,...,n we have

$$\Theta(\tilde{f}_1,\ldots,\tilde{f}_k, \hat{f}_{k+1},\ldots,\hat{f}_m) \geqslant \Theta(\hat{f}_1,\ldots,\hat{f}_k, \hat{f}_{k+1},\ldots,\hat{f}_m) > \alpha,$$

while for k=n,...,m-1, we have

$$\theta(\tilde{f}_1,\ldots,\tilde{f}_k,\ \hat{f}_{k+1},\ldots,\hat{f}_m) \geqslant \theta(\tilde{f}_1,\ldots,\tilde{f}_k,\ \tilde{f}_{k+1},\ldots,\tilde{f}_m) > \alpha.$$

It follows that for all k=1,...,m-1 we have

$$\psi(\tilde{x}_1,\ldots,\tilde{x}_k,\ \hat{x}_{k+1},\ldots,\hat{x}_m,\ \Omega) \subseteq A_k \cap A_{k+1}.$$

Thus these intersections are non-empty and an immediate induction shows that the set

$$A = \bigcup_{k=1}^{m} A_k$$

is a connected subset of $GT_\alpha(F)$. Since $\hat{x} \in A_1 \subseteq A$ while $\tilde{x} \in A_m \subseteq A$, this shows that given any pair of points of $GT_\alpha(F)$, there exists a connected subset of $GT_\alpha(F)$ containing both points. It follows that $GT_\alpha(F)$ is itself connected, which completes the proof.

4.1 Remarks and Applications

1. GE-connectedness may replace GT-connectedness in this theorem and proof provided we adjoin the additional condition that the supremum involved in the definition of the function F is always *achieved*.

2. The special case of *supremal convolution* is obtained if we take the parameter space Ω to be a singleton (so that its role falls away), take the spaces X_1, $X_2,\ldots,X_m$, X to be copies of a linear topological space E, and take

$$\psi(x_1,x_2,\ldots,x_m) = \sum_{i=1}^{m} x_i$$

$$\theta(f_1,f_2,\ldots,f_m) = \sum_{i=1}^{m} f_i,$$

so that

$$F(x) = \sup\{\Sigma f_i(x_i) \mid \Sigma x_i = x,\quad x_1,\ldots,x_m \in E\}.$$

As was shown by Mereau (1963) the supremum here is always achieved in the case that the linear space E is Hausdorff and all the functions f_i are GE-compact (i.e. have compact GE-level sets). Thus supremal convolution preserves GT-connectedness, and with these extra provisos, it also preserves GE-connectedness (Martin (1978)).

3. A further class of special cases arises if we again take Ω to be a singleton space, but take

$$X = X_1 \times X_2 \times \ldots \times X_m \text{ with } \psi = \text{id} : X \to X.$$

The supremum operation in the definition of F then becomes trivial, but is (trivially) always achieved, and we have

$$F(x) = F(x_1, x_2, \ldots, x_m) = \Theta(f_1(x_1), f_2(x_2), \ldots, f_m(x_m)).$$

Thus any separable combination of this form preserves both GT-connectedness and GE-connectedness. In particular, choosing Θ again as summation shows that the *separable sum* of GT-connected functions (GE-connected functions) is again GT-connected (GE-connected). This simple result does not seem to have been noted before.

4. Consider a multistage decision process (see, for example, Bellman and Kalaba (1965)) with decision vector $u \in \mathbb{R}^m$ state vector $x \in \mathbb{R}^n$, and dynamics

$$x_{k+1} = \phi_k(x_k, u_k) \quad k=0,1,\ldots,N-1, \quad u_k \in U_k \subseteq \mathbb{R}^m,$$

where we assume invertibility of this equation in the form

$$x_k = \psi_k(x_{k+1}, u_k) \quad k=0,1,\ldots,N-1,$$

where $\psi_k(\cdot,\cdot)$ is continuous. Suppose one wishes to maximize a criterion of the form

$$J = \sum_{k=0}^{N-1} f_k(u_k) + f_N(x(N)).$$

The familiar dynamic programming recursion is then given by

$$J_N(x(N)) = f_N(x(N)),$$

$$J_{k-1}(x(k-1)) = \sup \{J_k(x(k)) + f_{k-1}(u(k-1)) \mid x(k-1) = \psi_{k-1}(x(k), u(k-1))\} \quad k=N, N-1,\ldots,1$$

$$J_{min}(x_o) = J_o(x_o).$$

An immediate induction, making appeal to the above theorem, then shows that *if each of the functions* $f_i(\cdot)$ *is* GT-*connected, then so is each of the optimal return functions* $J_k(\cdot)$.

REFERENCES

AVRIEL, A. and ZANG, I. (1981). "Generalized Arcwise-Connected Functions and Characterization of Local-Global Minimum Properties", *J. of Optimization Theory and Applications*, forthcoming.

BELLMAN, R. and KALABA, R. (1965). *Dynamic Programming and Modern Control Theory*, Academic Press, New York.

BEREANU, B. (1973). "On Stochastic Linear Programming IV", *Proceedings of the Fourth Conference on Probability Theory*, Brasov, Romania.

DANTZIG, G.B. (1963). *Linear Programming and Extensions*, Princeton University Press, Princeton, N.J.

DINKELBACH, W. (1969). *Sensitivitätsanalysen und Parametrische Programmierung*, Springer-Verlag, Berlin-Göttingen-Heidelberg.

MARTIN, D.H. (1975). "On the Continuity of the Maximum in Parametric Linear Programming", *J. of Optimization Theory and Applications*, *17*, 205-210.

MARTIN, D.H. (1978). "Some Function Classes Closed Under Infimal Convolution", *J. of Optimization Theory and Applications*, *25*, 579-584.

MARTIN, D.H. (1981). "Connected Level Sets, Minimizing Sets, and Uniqueness in Optimization", *J. of Optimization Theory and Applications*, forthcoming.

MEREAU, J.-J. (1963). "Inf-Convolution des Fonctions Numériques sur un Espace, Vectoriel", *Comptes Rendus de l'Academie des Sciences de Paris*, *256*, 5047-5049.

ORTEGA, J.H. and RHEINBOLDT, W.C. (1970). *Iterative Solution of Non-linear Equations in Several Variables*, Academic Press, New York.

CONTINUITY AND DIFFERENTIABILITY PROPERTIES OF QUASICONVEX FUNCTIONS ON $\mathbb{R}^n$

Jean-Pierre Crouzeix

This paper is concerned with continuity and differentiability properties of quasiconvex functions on $\mathbb{R}^n$. Some properties of positively homogeneous quasiconvex functions are also developed.

1. DEFINITIONS AND NOTATIONS

By a function on $\mathbb{R}^n$, we mean a function from $\mathbb{R}^n$ to $[-\infty, +\infty]$. Given a function f on $\mathbb{R}^n$, the level sets of f which we denote by $S_\lambda(f)$, $\lambda \in \mathbb{R}$ are defined by

$$S_\lambda(f) = \{ x \in \mathbb{R}^n \ / \ f(x) \leq \lambda \}.$$

A function f on $\mathbb{R}^n$ is said to be quasiconvex if each of its level sets is convex. The two following equivalent characterizations are especially useful.

- The function f is quasiconvex if and only if

  $$f(x + t(y - x)) \leq \text{Max } [f(x), f(y)]$$

  whenever $x, y \in \mathbb{R}^n$ and $t \in [0,1]$.

- The function f is quasiconvex if and only if the subset $\{x/f(x) < \lambda\}$ is convex for all $\lambda \in \mathbb{R}$.

Notice that a convex function is always quasiconvex. A geometric way to see this is to observe that the level set $S_\lambda(f)$ is the projection on $\mathbb{R}^n$ of the epigraph of f and the hyperplane $\mathbb{R}^n \times \{\lambda\}$.

It is an immediate exercise to show that the sets $\{x/f(x) = -\infty\}$ and $\{x/f(x) < +\infty\}$ are convex whenever f is quasiconvex. The last subset is called the domain of f and is denoted by dom(f).

The function f is said to be l.s.c. (lower semi-continuous) at x_o if for each $\lambda < f(x_o)$ there exists a neighbourhood of x_o such that $f(x) > \lambda$ whenever x belongs to this neighbourhood.

ISBN 0-12-621120-5

The function f is said to be u.s.c. (upper semi-continuous) at x_o if for each $\lambda > f(x_o)$, there exists a neighbourhood of x_o such that $f(x) < \lambda$ whenever x belongs to this neighbourhood.

Notice that f is l.s.c. at each x satisfying $f(x) = -\infty$, u.s.c. at each x satisfying $f(x) = +\infty$, continuous at each x belonging to the interiors of the sets $\{x/f(x) = -\infty\}$ and $\{x/f(x) = +\infty\}$.

A function is said to be l.s.c. if it is l.s.c. at each $x \in \mathbb{R}^n$, u.s.c. if it is u.s.c. at each $x \in \mathbb{R}^n$. It is well known that a function is l.s.c. if each of its level sets is closed.

Given any function f on $\mathbb{R}^n$, we denote by $\bar{f}$, f_q and $f_{\bar{q}}$ respectively the greatest lower semi-continuous, the greatest quasiconvex, the greatest lower semi-continuous quasiconvex functions which are majorized by f.

The majority of the results which will be given were already stated in Crouzeix (1977; 1979 a, b; 1980 a, b, c). We shall present them in a unified approach; some of the proofs have been modified for clarity.

2. CONSTRUCTING A FUNCTION FROM A FAMILY SET

Let f be any function on $\mathbb{R}^n$. Then,

$$S_\lambda(f) \subset S_\mu(f) \text{ whenever } \lambda < \mu ,$$

$$S_\lambda(f) = \bigcap_{\mu>\lambda} S_\mu(f) ,$$

$$f(x) = \text{Inf } [\lambda/x \in S_\lambda(f)].$$

Conversely, let us give a family $\{T_\lambda, \lambda \in \mathbb{R}\}$ of subsets of $\mathbb{R}^n$ verifying

$$T_\lambda \subset T_\mu \text{ whenever } \lambda < \mu$$

we can define a function g on $\mathbb{R}^n$ by

$$g(x) = \text{Inf } [\lambda/x \in T_\lambda]$$

with the convention that $g(x) = +\infty$ if x belongs to none of the sets T_λ.

One should not necessarily think of the sets T_λ as being the level sets of g. For instance, consider $T_\lambda = \{x \in \mathbb{R} / x < \lambda\}$ for each $\lambda \in \mathbb{R}$. Then $g(x) = x$ and $S_\lambda(g) = \{x \in \mathbb{R} / x \leq \lambda\}$.

PROPOSITION 1: Let $\{T_\lambda, \lambda \in \mathbb{R}\}$ *be a family of subsets of* $\mathbb{R}^n$ *verifying* $T_\lambda \subset T_\mu$ *whenever* $\lambda < \mu$ *and let* g *be the function defined by* $g(x) = \operatorname{Inf}[\lambda / x \in T_\lambda]$. *Then* $S_\lambda(g) = \bigcap_{\mu > \lambda} T_\mu$. *It follows that if all the subsets are convex the function* g *is quasiconvex and if all the subsets are closed the function* g *is lower semi-continuous.*

PROOF: an immediate exercise. □

Given a subset S of $\mathbb{R}^n$, we denote by co(S) the convex hull of S, by $\overline{\text{co}}(S)$ the closure of co(S), by cl(S) the closure of S.

PROPOSITION 2: Let f *be any function on* $\mathbb{R}^n$. *Then,*

a) $\bar{f}(x) = \operatorname{Inf}[\lambda / x \in \text{cl}(S_\lambda(f))]$ and $S_\lambda(\bar{f}) = \bigcap_{\mu > \lambda} \text{cl}(S_\mu(f))$,

b) $f_q(x) = \operatorname{Inf}[\lambda / x \in \text{co}(S_\lambda(f))]$ and $S_\lambda(f_q) = \bigcap_{\mu > \lambda} \text{co}(S_\mu(f))$,

c) $f_{\bar{q}}(x) = \operatorname{Inf}[\lambda / x \in \overline{\text{co}}(S_\lambda(f))]$ and $S_\lambda(f_{\bar{q}}) = \bigcap_{\mu > \lambda} \overline{\text{co}}(S_\mu(f))$.

PROOF: First notice that $\text{cl}(S_\lambda(f)) \subset \text{cl}(S_\mu(f))$, $\text{co}(S_\lambda(f)) \subset \text{co}(S_\mu(f))$ and $\overline{\text{co}}(S_\lambda(f)) \subset \overline{\text{co}}(S_\mu(f))$ whenever $\lambda < \mu$. Let us consider the functions

$$f_1(x) = \operatorname{Inf}[\lambda / x \in \text{cl}(S_\lambda(f))],$$

$$f_2(x) = \operatorname{Inf}[\lambda / x \in \text{co}(S_\lambda(f))],$$

$$f_3(x) = \operatorname{Inf}[\lambda / x \in \overline{\text{co}}(S_\lambda(f))].$$

According to the last proposition, f_1 is l.s.c., f_2 is quasiconvex and f_3 is l.s.c. and quasiconvex. Let us show that $f_1 = \bar{f}$. Let h be any l.s.c. function which is majorized by f. Then $S_\mu(h) \supset S_\mu(f)$ for all $\mu \in \mathbb{R}$. Hence $S_\mu(h) \supset \text{cl}(S_\mu(f) \supset S_\mu(f)$ because $S_\mu(h)$ is closed. Thus,

$$S_\lambda(h) = \bigcap_{\mu > \lambda} S_\mu(h) \supset S_\lambda(f_1) = \bigcap_{\mu > \lambda} \text{cl}(S_\mu(f)) \supset S_\lambda(f) = \bigcap_{\mu > \lambda} S_\mu(f).$$

It follows that f_1 is the greatest lower semi-continuous function which is majorized by f. By replacing $cl(S_\lambda(f))$ by $co(S_\lambda(f))$ and $\overline{co}(S_\lambda(f))$ one shows in a similar way that $f_2 = f_q$ and $f_3 = f_{\bar{q}}$. □

COROLLARY 3: Let f *be a quasiconvex function on* $\mathbb{R}^n$. *Then* $f_{\bar{q}} = \bar{f}$.

PROOF: an immediate consequence of Proposition 2. □

One should not necessarily think of the level sets of $\bar{f}$ as being the closures of the level sets of the function f. For instance, consider the function on $\mathbb{R}^n$

$$f(x,y) = \begin{cases} y & \text{if } x = 0 \text{ and } 0 < y < 1, \\ 1+x & \text{if } 0 < x < 1, \\ 2 & \text{otherwise.} \end{cases}$$

Note that f is quasiconvex and

$$S_o(f) = \emptyset \qquad \text{and } S_o(\bar{f}) = \{(0,0)\},$$

$$S_1(f) = \{0\} \times (0,1) \quad \text{and } S_1(\bar{f}) = \{0\} \times \mathbb{R}.$$

However, in the specific case where f is a quasiconvex function on $\mathbb{R}^n$, the number of the values λ for which $S_\lambda(\bar{f}) \neq cl(S_\lambda(f))$ is finite. In order to prove this, we must introduce the following definition: α is called a critical value of f if $\dim(S_\lambda(f)) < \dim(S_\mu(f))$ whenever $\lambda < \alpha < \mu$ where by dim(C) we denote the dimension of the affine hull of C (with the convention that dim(C) = -1 when C is empty). Then,

PROPOSITION 4: Let f *be a quasiconvex function on* $\mathbb{R}^n$. *Then* $S_\lambda(\bar{f}) = cl(S_\lambda(f))$ *whenever* λ *is not a critical value of* f.

PROOF: Suppose λ is not a critical value. Then there exists $\mu > \lambda$ such that $\dim(S_\lambda(f)) = \dim(S_\mu(f))$. Hence $ri(S_\lambda(f)) \subset ri(S_\xi(f))$ whenever $\lambda \leq \xi \leq \mu$ (where ri(C) denotes the relative interior of C, i.e., the interior of C which results when C is regarded as a subset of its affine hull). If $S_\lambda(f)$ is empty, then $S_\mu(f)$ is empty and so $S_\lambda(\bar{f})$ is empty. If not $ri(S_\lambda(f))$ is not empty. Let $\bar{x}$ be a fixed point in $ri(S_\lambda(f))$. Let us show that $S_\lambda(\bar{f}) = cl(S_\lambda(f))$. Of course, $cl(S_\lambda(f)) \subset S_\lambda(\bar{f})$ since $S_\lambda(\bar{f}) = \bigcap_{\xi>\lambda} cl(S_\xi(f))$ and $S_\lambda(f) \subset S_\xi(f)$ whenever $\xi > \lambda$. Let $x \in S_\lambda(\bar{f})$, then $x \in cl(S_\xi(f))$ for all $\lambda < \xi \leq \mu$. Hence $x_t = x + t(\bar{x} - x) \in S_\xi(f)$ for all $\xi \in (\lambda,\mu]$ and all $t \in (0,1]$. Since $S_\lambda(f) = \bigcap_{\xi>\lambda} S_\xi(f)$, $x_t \in S_\lambda(f)$ for all $t \in (0,1]$ and thus $x \in cl(S_\lambda(f))$. □

3. CONTINUITY PROPERTIES OF A QUASICONVEX FUNCTION

Let f be a function on $\mathbb{R}^n$ and let $x_o, h \in \mathbb{R}^n$, $h \neq 0$. Consider the function of one real variable $f_{x_o,h}(t) = f(x_o + th)$. If f is quasiconvex, then $f_{x_o,h}$ is quasiconvex. Conversely, it can be easily seen that if the function $f_{x_o,h}$ is quasiconvex for all $x_o, h \in \mathbb{R}^n$ then f is quasiconvex. What can be said about quasiconvex functions of one real variable? It can be easily seen that a function of one real variable is quasiconvex if and only if it belongs to one of the three following classes,

a) f is non-increasing on $\mathbb{R}$,

b) f is non-decreasing on $\mathbb{R}$,

c) there exists an $\bar{x}$ so that f is non-increasing on $(-\infty, x]$ and non-decreasing on $[\bar{x}, +\infty)$.

Due to properties of monotonic functions of one variable, a quasiconvex function of one real variable is almost everywhere continuous and differentiable.

Let us now consider quasiconvex functions of several variables.

PROPOSITION 5: Let f *be a function on* $\mathbb{R}^n$, *then* f *is* l.s.c. *at* x_o *if and only if* $\bar{f}(x_o) = f(x_o)$. *If* f *is quasiconvex, then* f *is continuous at* x_o *if and only if* $\bar{f}$ *is continuous at* x_o.

PROOF: If $f(x_o) = \bar{f}(x_o)$, then f is l.s.c. at x_o because $\bar{f}$ is l.s.c. at x_o and is majorized by f. Conversely, let us assume that f is l.s.c. at

x_o and let us show that $f(x_o) = \bar{f}(x_o)$. Let any $\lambda < f(x_o)$, there exists a neighbourhood V of x_o such that $f(x) > \lambda$ whenever $x \in V$. Hence $x_o \notin cl(S_\lambda(f))$ and so $\bar{f}(x_o) \geq \lambda$. Thus $f(x_o) = \bar{f}(x_o)$. Assume now that f is quasiconvex and continuous at x_o. Then $f(x_o) = \bar{f}(x_o)$. Since f is u.s.c. at x_o and $\bar{f}$ is majorized by f, the function $\bar{f}$ is u.s.c. at x_o. Assume now that f is quasiconvex and $\bar{f}$ is continuous at x_o. Let any λ, μ be such that $\lambda > \mu > \bar{f}(x_o)$, then $S_\mu(\bar{f})$ is a neighbourhood of x_o. Hence $cl(S_\lambda(f))$ and $S_\lambda(f)$ (since $S_\lambda(f)$ is convex) are also neighbourhoods of x_o implying that $f(x_o) \leq \lambda$. It follows that $f(x_o) = \bar{f}(x_o)$ and f is lower semi-continuous at x_o. It remains to be shown that f is u.s.c. at x_o, this is immediate since we have shown that $S_\lambda(f)$ is a neighbourhood of x_o for all $\lambda > f(x_o) = \bar{f}(x_o)$. □

Let f be any function on $\mathbb{R}^n$ and x_o be a fixed point in $\mathbb{R}^n$. We say that

- f is *hemi-l.s.c. at* x_o if for all $h \neq 0$ the function of one real variable $f_{x_o,h}(t) = f(x_o + th)$ is l.s.c. at $t = 0$.
- f is *hemi-u.s.c. at* x_o if for all $h \neq 0$ the function $f_{x_o,h}$ is u.s.c. at $t = 0$.
- f is *hemi-continuous at* x_o if f is hemi l.s.c. and hemi u.s.c. at x_o.

Of course, if f is continuous at x_o f is hemi-continuous at x_o but for general functions on $\mathbb{R}^n$ the converse is not true.

PROPOSITION 6: Let f *be a quasiconvex function on* $\mathbb{R}^n$. *Then* f *is* l.s.c. *at* x_o *if and only if* f *is* hemi-l.s.c. *at* x_o. *Similarly,* f *is* u.s.c. *at* x_o *if and only if* f *is* hemi-u.s.c. *at* x_o.

PROOF:

a) Let us assume that f is not l.s.c. at x_o. Then there exists $\lambda < f(x_o)$ such that $x_o \in cl(S_\lambda(f))$. Since $S_\lambda(f)$ is convex $ri(S_\lambda(f))$ is not empty. Let $\bar{x} \in ri(S_\lambda(f))$ and $h = \bar{x} - x_o$. Then $x_o + th \in S_\lambda(f)$ (implying that

$f(x_o + th) \leq \lambda$) for all $t \in (0,1]$. Hence the function $f_{x_o,h}$ is not l.s.c. at $t = 0$ so that f is not hemi-l.s.c. at x_o.

b) Let us assume now that f is hemi-u.s.c. at x_o. Denote by $e_i (i = 1,2,\ldots,n)$ the vector of $\mathbb{R}^n$ where all of the components are equal to zero except the one of order i which is equal to 1. Let $\lambda > f(x_o)$, since f is hemi-u.s.c. at x_o there exists $t_i > 0$ such that $f(x_o + t_i e_i) \leq \lambda$ and $f(x_o - t_i e_i) \leq \lambda$. Denote by C the convex hull of the finite set $\{x_o + t_i e_i, x_o - t_i e_i / i = 1,2,\ldots,n\}$, C is a neighbourhood of x_o which is contained in $S_\lambda(f)$ because f is quasiconvex. □

PROPOSITION 7: Let f *be a quasiconvex function on* $\mathbb{R}^n$ *and* a *be a fixed point in* $\mathbb{R}^n$. *Let us define* $S = \{x/f(x)<f(a)\}$ *and* $K = \{h/\exists t > 0$ s.t. $a + th \in S\}$ *and assume that the interior of* S *is non-empty. If there exists* $d \in \text{int}(K)$ *such that the function* $f_{a,d}(t) = f(a + td)$ *is continuous at* 0 *then* f *is continuous at* a.

PROOF:

a) First, let us show that f is l.s.c. at a. Due to the assumptions, $\mu < f(a)$ and $t_o > 0$ exist, such that $a + t_o d \in \text{int}(S_\mu(f))$. Let any $\lambda \in (\mu, f(a))$, then $f(a + \bar{t}d) > \lambda$ for some $\bar{t} \in (0,t_o)$. Since $S_\lambda(f)$ is convex there exists $x^* \neq 0$ such that $\langle x, x^*\rangle \leq \langle a + \bar{t}d, x^*\rangle$ whenever $x \in S_\lambda(f)$. Notice that one has necessarily $\langle d,x^*\rangle \neq 0$ because $a + t_o d \in \text{int}(S_\lambda(f))$. It follows that the set $\{x/\langle x, x^*\rangle > \langle a + \bar{t}d, x^*\rangle\}$ is a neighbourhood of a on which $f(x) > \lambda$. q.e.d.

b) Let us now prove that f is u.s.c. at a. Let any $\lambda > f(a)$, then $f(a + td) < \lambda$ for some $t < 0$. Define V to be the convex hull of the set $S \cup \{a + td\}$, V is a neighbourhood of a and $f(x) < \lambda$ whenever $x \in V$. □

Convex functions on $\mathbb{R}^n$ are continuous on the interior of their domain. This property does not remain valid for quasiconvex functions (consider, for instance, functions of one real variable). However, quasiconvex functions on $\mathbb{R}^n$ are almost everywhere (Lebesgue) continuous. See Deak (1962).

4. CHARACTERIZATIONS OF A L.S.C. QUASICONVEX FUNCTION

Let C be a subset of $\mathbb{R}^n$. We shall denote by $\delta(./C)$ and $\delta^*(./C)$ the indicator function and the support function of C, i.e.,

$$\delta(x/C) \;=\; 0 \text{ if } x \in C, +\infty \text{ otherwise}$$

$$\delta^*(x^*/C) \;=\; \text{Sup}\,[\langle x,x^*\rangle/x \in C].$$

Recall that C and $\overline{\text{co}}(C)$ have the same support function.

Let f be any function on $\mathbb{R}^n$, we shall now introduce the following functions on $\mathbb{R}^{n+1}$.

$$F(x^*,\lambda) \;=\; \delta^*(x^*/S_\lambda(f)) = \text{Sup}[\langle x,x^*\rangle/f(x) \leqslant \lambda],$$

$$F^-(x^*,t) = \text{Inf}[f(x)/\langle x,x^*\rangle \geqslant t].$$

where by convention $\text{Sup}[\xi/\xi \in A] = -\infty$ and $\text{Inf}[\xi/\xi \in A] = +\infty$ when A is an empty subset of $\mathbb{R}$.

PROPOSITION 8: Let f *be any function on* $\mathbb{R}^n$. *Then*

$$f_{\bar{q}}(x_o) = \underset{x^*}{\text{Sup}}\ \underset{\lambda}{\text{Inf}}[\lambda/F(x^*,\lambda) \geqslant \langle x_o,x^*\rangle] \leqslant \underset{x^*}{\text{Sup}}\ F^-(x^*,\langle x_o,x^*\rangle) \leqslant$$

$$\leqslant f_q(x_o) \leqslant f(x_o).$$

PROOF:

a) Let any $\lambda > f_{\bar{q}}(x_o)$, then $x_o \in \overline{\text{co}}(S_\lambda(f))$ and so

$$F(x^*,\lambda) = \delta^*(x^*,S_\lambda(f)) \geqslant \langle x_o,x^*\rangle.$$

Hence

$$f_{\bar{q}}(x_o) \geqslant \underset{\lambda}{\text{Inf}}[\lambda/F(x^*,\lambda) \geqslant \langle x_o,x^*\rangle] \qquad \text{for all } x^*.$$

Now, let any $\bar{\lambda} < f_{\bar{q}}(x_o)$. Since $f_{\bar{q}}(x_o) = \text{Inf}[\xi/\xi \in \overline{\text{co}}(S_\xi(f))]$ there exists $\xi > \bar{\lambda}$ such that $x_o \notin \overline{\text{co}}(S_\xi(f))$. Hence, there exists $x^* \in \mathbb{R}^n$ so that

$\text{Sup}[\langle x,x^*\rangle / x \in \overline{\text{co}}(S_\xi(f))] < \langle x_o,x^*\rangle$. But then $\text{Inf}[\lambda / F(x^*,\lambda) \geq \langle x_o,x^*\rangle] \geq \xi > \bar{\lambda}$. It follows that $\text{Sup}_{x^*}[\text{Inf}_\lambda[\lambda/F(x^*,\lambda) \geq \langle x_o,x^*\rangle] = f_{\bar{q}}(x_o)]$.

b) Let $\bar{\lambda} < \text{Inf}[\lambda/F(x^*,\lambda) \geq \langle x_o,x^*\rangle]$. Then $F(x^*,\bar{\lambda}) < \langle x_o,x^*\rangle$ and so

$$f(x) \leq \bar{\lambda} \Longrightarrow \langle x,x^*\rangle < \langle x_o,x^*\rangle,$$

$$\langle x,x^*\rangle \geq \langle x_o,x^*\rangle \Longrightarrow f(x) > \bar{\lambda}.$$

$$F^-(x^*,\langle x_o,x^*\rangle) = \text{Inf}[f(x)/\langle x,x^*\rangle \geq \langle x_o,x^*\rangle] \geq \bar{\lambda}.$$

Hence,

$$F^-(x^*,\langle x_o,x^*\rangle) \geq \text{Inf}[\lambda/F(x^*,\lambda) \geq \langle x_o,x^*\rangle] \quad \text{for all } x^*.$$

It remains to be shown that $F^-(x^*,\langle x_o,x^*\rangle) \leq f_q(x_o)$ for all x^*. Let us consider some $\lambda > f_q(x_o)$. Then $x_o \in \text{co}(S_\lambda(f))$. Hence, x_o is a finite convex combination of elements of $S_\lambda(f)$. That is, there exist $x_1, x_2, \ldots, x_p \in S_\lambda(f)$ and non-negative coefficients $\lambda_1, \lambda_2, \ldots, \lambda_p$ so that $\lambda_1 + \lambda_2 + \ldots + \lambda_p = 1$ and $x_o = \sum_{i=1}^{p} \lambda_i x_i$. Then necessarily $\langle x_i,x^*\rangle \geq \langle x_o,x^*\rangle$ for some $i \in \{1,2, \ldots,p\}$. Since $f(x_i) \leq \lambda$,

$$F^-(x^*,\langle x_o,x^*\rangle) = \text{Inf}[f(x)/\langle x,x^*\rangle \geq \langle x_o,x^*\rangle] \leq f(x_i) \leq \lambda.$$

It follows that $F^-(x^*,\langle x_o,x^*\rangle) \leq f_q(x_o) \leq f(x_o)$. □

COROLLARY 9: Let f *be a quasiconvex function on* $\mathbb{R}^n$. *If* f *is l.s.c. at* x_o, *then* $f(x_o) = \text{Sup}_{x^*} \text{Inf}_\lambda[\lambda/F(x^*,\lambda) \geq \langle x_o,x^*\rangle] = \text{Sup}_{x^*}[F^-(x^*,\langle x_o,x^*\rangle)]$.

PROOF: an immediate consequence of Proposition 8. □

One can easily understand how interesting the above results are. A quasiconvex l.s.c. function f is well known when the function F or the function F^- is given. So, the functions F and F^- can be regarded as being dual functions of f. Properties of the functions F and F^- are extensively studied in Crouzeix (1981) where they are used to construct a duality theory in quasiconvex programming. In the next theorems we show how conditions for convexity of the function f can be stated in terms of its dual functions. Recall that if f is convex then the function $\bar{f}$ is also convex and if $\bar{f}$ is convex, then f coincides with $\bar{f}$ on the relative interior of dom(f).

THEOREM 10: Let f *be a quasiconvex function on* $\mathbb{R}^n$.

i) If f *is convex, then* $F(x^*, .)$ *is concave for all* x^*.

ii) If $F(x^*, .)$ *is concave for all* x^*, *then* $\bar{f}$ *is convex.*

PROOF:

i) Denote by C the epigraph of f, i.e., $C = \{(x,\lambda)/f(x) \leq \lambda\}$.

Since f is convex, then the epigraph of f is a convex set and the indicator function of C is a convex function. Notice that

$$F(x^*,\lambda) = \operatorname*{Sup}_x[\langle x,x^*\rangle - \delta((x,\lambda)/C)].$$

Let x^* be a fixed point in $\mathbb{R}^n$, the function $\phi(x,\lambda) = \langle x,x^*\rangle - \delta((x,\lambda)/C)$ is concave in (x,λ). Hence the function $h(\lambda) = \operatorname*{Sup}_x[\phi(x,\lambda)]$ is concave, q.e.d.

ii) According to Proposition 8,

$$\bar{f}(x_o) = \operatorname*{Sup}_{x^*\neq 0}\ [\operatorname*{Inf}_{\lambda}[\lambda/F(x^*,\lambda) \geq \langle x_o,x^*\rangle].$$

Let x^* be fixed in $\mathbb{R}^n$, consider

$$D_{x^*} = \{(x,\lambda)/\langle x,x^*\rangle - F(x^*,\lambda) \leq 0\}.$$

Since $F(x^*, .)$ is concave, D_{x^*} is a convex subset of $\mathbb{R}^{n+1}$. It follows that the function $\phi_{x^*}(x,\lambda) = \lambda + \delta((x,\lambda)/D_{x^*})$ is convex in (x,λ). Hence $\operatorname{Inf}[\lambda/F(x^*,\lambda) \geq \langle x,x^*\rangle]$ is convex in x for all fixed x^*. It follows that $\bar{f}$ is convex. □

THEOREM 11: Let f *be a quasiconvex funtion on* $\mathbb{R}^n$,

i) If f *is convex, then* $F^-(x^*, .)$ *is convex for all* x^*.

ii) If $F^-(x^*, .)$ *is convex for all* x^*, *then* $\bar{f}$ *is convex.*

PROOF:

i) Let x^* be fixed and consider the function

$$\phi(x,t) = f(x) + \delta((x,t)/C)$$

where $C = \{(x,t)/\langle x,x^*\rangle \geqslant t\}$. The function f is convex and C is a convex set. Hence ϕ is convex. It is then sufficient to note that

$$F^-(x^*,t) = \operatorname*{Inf}_x[\phi(x,t)].$$

ii) Let x* be fixed and consider the function $\phi_{x^*}(x) = F^-(x^*,\langle x^*,x\rangle)$. It is easy to prove the convexity of the function ϕ_{x^*} from the convexity of $F^-(x^*, .)$. Hence the function $g(x) = \operatorname{Sup} \phi_{x^*}(x)$ is convex. It follows that $\bar{f}$ is convex since $\bar{f} \leqslant g \leqslant f$ (due to Proposition 8). □

Theorem 10 generalizes a result of Fenchel (1953) (see also Moulin (1974)). Let us give an economic interpretation of the two theorems. Suppose that the preferences of a consumer are represented by a utility function u on $\mathbb{R}^n$. Assume that u is quasiconcave and upper semi-continuous. The consumer is assumed to determine his optimal consumption by maximizing u(x) subject to the budget constraint $\langle p,x\rangle \leqslant r$ where r is the maximal amount that the consumer can spend and p is the vector of commodity prices.

Consider the functions:

$$U_p(\xi) = \operatorname{Inf}[\langle p,x\rangle/u(x) \geqslant \xi],$$

$$V_p(r) = \operatorname{Sup}[u(x)/\langle p,x\rangle \leqslant r].$$

$U_p(\xi)$ is the minimal price to be paid to obtain a consumption having a utility greater than or equal to ξ while $V_p(r)$ is the maximal value of the utility that the consumer can obtain with the amount r.

Take $f = -u$, $x^* = -p$, $\lambda = -\xi$, $t = -r$. By a direct application of the theorems we obtain:

u is concave if and only if U_p is convex for all p.

u is concave if and only if V_p is concave for all p.

The function of p taking the value $V_p(1)$ is the so-called indirect utility function. Similar results in the economic literature can be found in Diewert (1978).

5. POSITIVELY HOMOGENEOUS QUASICONVEX FUNCTIONS

Positively homogeneous quasiconvex functions will be useful when we study differentiability properties of quasiconvex functions. Recall that a quasiconvex function is said to be positively homogeneous if $f(k\,x) = k\,f(x)$ for all $k > 0$ and $x \in \mathbb{R}^n$. The next proposition improves former results of Newman (1969) and Crouzeix (1977 ; 1980a).

PROPOSITION 12: Let f *be a positively homogeneous quasiconvex function on* $\mathbb{R}^n$.

a) If $f(x) \geq 0$ *for all* $x \in \mathbb{R}^n$, *then* f *is convex.*

b) If $f(x) < 0$ *for all* $x \in \mathrm{ri}(\mathrm{dom}(f))$, *then* $\bar{f}$ *is convex.*

PROOF: Let x^* be a fixed point in $\mathbb{R}^n$. We shall prove that $F(x^*, .)$ is concave. First, note that for all $k > 0$

$$F(x^*,k\lambda) = \mathrm{Sup}[\langle kx,x^*\rangle / f(kx) \leq k\lambda] = k\,\mathrm{Sup}[\langle x,x^*\rangle / f(x) \leq \lambda].$$

Hence $F(x^*,k\lambda) = kF(x^*,\lambda)$ for all $k > 0$. It follows that

$$F(x^*,\lambda) = \begin{cases} (-\lambda)F(x^*, -1) & \text{if } \lambda < 0 \\ \lambda\ F(x^*, 1) & \text{if } \lambda > 0 \end{cases}$$

Since $F(x^*, .)$ is non-decreasing, then $F(x^*, 1)$ is non-negative and $F(x^*, -1)$ is non-positive or equal to $+\infty$.

a) $F(x^*,\lambda) = -\infty$ for all $\lambda < 0$. Hence $F(x^*, .)$ is concave, implying that $\bar{f}$ is convex. Thus

$$\bar{f}(tx_1 + (1-t)x_2) \leq t\bar{f}(x_1) + (1-t)\bar{f}(x_2) \leq tf(x_1) + (1-t)f(x_2)$$

whenever $x_1, x_2 \in \mathrm{dom}(f)$ and $t \in (0,1)$. Recall that $\bar{f}(x) = f(x)$ when x belongs to $\mathrm{ri}(\mathrm{dom}(f))$. It only remains to prove that

$$f(tx_1 + (1-t)x_2) \leq tf(x_1) + (1-t)f(x_2)$$

when $tx_1 + (1-t)x_2$ belongs to the relative boundary of $\mathrm{dom}(f)$. Let us denote by g the restriction of f to the affine hull of the finite set $\{0, x_1, x_2\}$. Applying the above results to g and using the fact that $tx_1 + (1-t)x_2$ belongs to the relative interior of $\mathrm{dom}(g)$, one obtains the result.

b) Assume that $F(x^*, 1) > 0$. Notice that $F(., 1)$ is the support function of $ri(S_1(f))$ and $ri(dom(f)) \subset S_1(f) \subset dom(f)$. Hence $ri(S_1(f)) = ri(dom(f))$. Since $F(x^*, 1) > 0$, there exists $\bar{x} \in ri(dom(f))$ (implying that $f(\bar{x}) < 0$) so that $<\bar{x},x^*> > 0$. Set $\alpha = f(\bar{x})$, then $F(x^*,\alpha) = -\alpha F(x^*, -1) > <\bar{x},x^*> > 0$. Thus $F(x^*, +1) > 0$ implies that $F(x^*, -1) = +\infty$. It follows that $F(x^*, +1) = 0$ whenever $F(x^*, -1) \leq 0$. Thus $F(x^*, .)$ is concave for all x^* and $\bar{f}$ is convex. □

Assumption b) is not sufficient to assure that f is convex. Consider for instance the function on $\mathbb{R}^3$ defined by

$$f(x_1,x_2,x_3) = \begin{cases} -(x_1+x_2+x_3) & \text{if } x_3 > 0,\ x_1 \geq 0 \text{ and } x_2 \geq 0, \\ -(x_1+x_2) & \text{if } x_3 = 0,\ x_1 \geq 0,\ x_2 \geq 0 \text{ and } x_1-x_2 \geq 0, \\ 0 & \text{if } x_3 = 0,\ x_1 \geq 0,\ x_2 \geq 0 \text{ and } x_1-x_2 < 0, \\ +\infty & \text{elsewhere.} \end{cases}$$

However,

COROLLARY 13: Let f *be a positively homogeneous quasiconvex function whose values are strictly negative on its domain. Then* f *is convex.*

PROOF: One must prove that

$$f(tx_1 + (1-t)x_2) \leq tf(x_1) + (1-t)f(x_2)$$

whenever $t \in]0,1[$ and $x_1,x_2 \in dom(f)$. Let us denote by g the restriction of f to the affine hull of the set $\{0,x_1,x_2\}$. Then, due to Proposition 12 b) $\bar{g}$ is convex and it is sufficient to note that $tx_1 + (1-t)x_2$ belongs to $ri(dom(g))$. □

Let f be a positively homogeneous quasiconvex function on $\mathbb{R}^n$. Set $T = \{x/f(x) < 0\}$ and define the functions f_- and f_+ from f by

$$f_-(x) = \begin{cases} f(x) & \text{if } x \in T, \\ +\infty & \text{if not.} \end{cases} \qquad f_+(x) = \begin{cases} 0 & \text{if } x \in T, \\ f(x) & \text{if not.} \end{cases}$$

Clearly, $f(x) = \mathrm{Min}[f_-(x), f_+(x)]$ for all x. According to the above results f_- and f_+ are convex. Thus, a positively homogeneous quasiconvex function on $\mathbb{R}^n$ can be expressed as being the minimum of two convex functions. Assume in addition, that f is a real-valued function, then $\mathrm{dom}(f_+) = \mathbb{R}^n$. It follows that f_+ is continuous on $\mathbb{R}^n$.
Hence f is continuous at each $x \notin \mathrm{cl}(T)$. On the other hand, f is l.s.c. at each $x \in \mathrm{ri}(T)$ and u.s.c. at each $x \in \mathrm{int}(T)$ but not necessarily u.s.c. or l.s.c. on the relative boundary of T. Consider for instance $f : \mathbb{R}^3 \to \mathbb{R}$ defined by

$$f(x_1,x_2,x_3) = \begin{cases} -(x_1+x_2+x_3) & \text{if } x_3 > 0,\ x_1 \geq 0 \text{ and } x_2 \geq 0, \\ -(x_1+x_2) & \text{if } x_3 = 0,\ x_1 \geq 0,\ x_2 \geq 0 \text{ and } x_1-x_2 \geq 0, \\ 0 & \text{elsewhere.} \end{cases}$$

It is easily seen that $\bar{f}(x) = \mathrm{Min}[\bar{f}_-(x), \bar{f}_+(x)]$. The functions $\bar{f}_+$, $\bar{f}_-$ are positively homogeneous l.s.c. convex functions and thus can be considered as being the support functions of two closed convex sets which we denote by C_+ and C_-. Let us now describe properties of C_+ and C_-.

PROPOSITION 14:

(a) $x^* \in C_-$ *and* $\lambda \geq 1$ *implies* $\lambda x^* \in C_-$,

(b) $x^* \in C_+$ *and* $0 \leq \lambda \leq 1$ *implies* $\lambda x^* \in C_+$,

(c) $T^o = \{x^*/\langle x,x^*\rangle \leq 0 \text{ for all } x \in T\}$ *is the closure of the convex cone generated by* C_-,

(d) C_+ *is compact and contained in* T^o.

PROOF:

(a) By definition

$$x^* \in C_- \text{ if and only if } \langle x,x^*\rangle \leq \bar{f}_-(x) \qquad \forall x \in \mathrm{dom}(\bar{f}_-).$$

Recall that $\bar{f}_-(x) \leq 0$ for all $x \in \mathrm{dom}(\bar{f}_-)$. Hence for all $x^* \in C_-$ and $\lambda \geq 1$

$$\langle x,\lambda x^*\rangle = \lambda\langle x,x^*\rangle \leq \lambda\bar{f}_-(x) \leq \bar{f}_-(x) \qquad \forall x \in \mathrm{dom}(\bar{f}_-)$$

Thus $\lambda x^* \in C_-$.

(b) Recall that $\bar{f}_+(x) \geq 0$ for all x and proceed similarly.

(c) The indicator function of T^o is the support function of $\{x/\bar{f}_-(x) \leq 0\}$.

(d) C_+ is compact because $\mathrm{dom}(f_+)$ is the whole space $\mathbb{R}^n$. Note that $\delta(x/T) \geq f_+(x)$ for all x. Hence $\delta^*(x^*/T) = \delta(x^*/T^o) \leq f_+^*(x^*) = \delta(x^*/C_+)$.□

Remember that f is l.s.c. except on the relative boundary of T and $f(x) = \mathrm{Min}[f_-(x), f_+(x)]$. Hence

$f(x) = \delta^*(x/C_-)$ if $x \in \mathrm{ri}(T)$,

$\delta^*(x/C_-) \leq f(x) \leq \delta^*(x/C_+) = 0$ if x belongs to the relative interior of C,

$\delta^*(x/C_+) = f(x)$ elsewhere.

We present now an important property.

THEOREM 15: Let f *be a positively homogeneous quasiconvex function from* $\mathbb{R}^n$ *to* $\mathbb{R}$ *verifying*

$$f(h) + f(-h) = 0 \qquad \forall h \in \mathbb{R}^n.$$

Then f *is linear.*

PROOF: Clearly, f is linear if $f(h) = 0$ for all $h \in \mathbb{R}^n$. If not, T is not empty. Let us show that cl(T) is a half space. If this is not true, an open convex cone C exists in $\mathbb{R}^n$ such that $T \cap C$ and $(-T) \cap C$ are empty. Hence $f(h) \geq 0$ (since $T \cap C = \emptyset$) and $f(h) \leq 0$ (since $(-T) \cap C = \emptyset$) for all $h \in C$. Then C and $-C$ are contained in the convex cone $\{x/f(x) \leq 0\}$. Hence this convex cone is the whole space $\mathbb{R}^n$ and thus a contradiction arises because $f(h) > 0$ for all $h \in -T$. It follows that there exists $c \in \mathbb{R}^n$, $c \neq 0$ such that $T^o = \{\lambda c/\lambda \geq 0\}$. Hence there exists $\lambda_-, \lambda_+ \geq 0$ such that $C_- = \{\lambda c/\lambda \geq \lambda_-\}$ and $C_+ = \{\lambda c/0 \leq \lambda \leq \lambda_+\}$. Then it is easy to see that $\lambda_- = \lambda_+$. Hence $f_-(x) = f_+(x) = \lambda_- \langle c,x \rangle$ and f is linear. □

More results and details concerning quasiconvex positively homogeneous functions can be found in Crouzeix (1979a).

6. DIFFERENTIABILITY PROPERTIES OF QUASICONVEX FUNCTIONS

Since we are concerned with functions whose values are real or $\pm\infty$, we must deal with arithmetic calculations involving $+\infty$ and $-\infty$. Throughout this paper we adopt the following rules

$$\alpha + \infty = \infty + \alpha = \infty \qquad \text{for } -\infty < \alpha \leq +\infty,$$

$$\alpha - \infty = -\infty + \alpha = -\infty \qquad \text{for } -\infty \leq \alpha < +\infty,$$

$$\infty - \infty = -\infty + \infty = 0.$$

Let f be any function on $\mathbb{R}^n$, we define the Dini-directional derivatives of f at a with respect to the direction h as being the limits

$$f'_+(a,h) = \limsup_{t \downarrow 0} \frac{f(a+th) - f(a)}{t},$$

$$f'_-(a,h) = \liminf_{t \downarrow 0} \frac{f(a+th) - f(a)}{t}.$$

If $-\infty < f'_-(a,h) = f'_+(a,h) < +\infty$, we define the one-sided directional derivative of f at a with respect to the direction h to be the (finite) limit

$$f'(a,h) = \lim_{t \downarrow 0} \frac{f(a+th) - f(a)}{t}.$$

If $f'(a,h)$ is defined for all $h \in \mathbb{R}^n$ and verifies

$$f'(a,h) + f'(a,-h) = 0 \qquad \forall h \in \mathbb{R}^n$$

then f is said to be *hemi-differentiable* or *weakly Gâteaux-differentiable* at a.

If f is hemi-differentiable at a and if the function $f'(a, .)$ is linear, then f is said to be *Gâteaux-differentiable* at a. The Gâteaux-derivative of f at a is the vector $f'(a)$ of $\mathbb{R}^n$ such that

$$f'(a,h) = \langle f'(a),h \rangle \qquad \forall h \in \mathbb{R}^n.$$

If f is Gâteaux-differentiable at a and if

$$\lim_{h \to 0} \frac{1}{\|h\|} [f(a+h) - f(a) - \langle f'(a),h \rangle] = 0$$

then f is said to be *Fréchet-differentiable* at a.

Notice that under these conventions a function f is Fréchet-differentiable on the interiors of the sets $\{x/f(x) = +\infty\}$ and $\{x/f(x) = -\infty\}$ (and the derivative is zero).

PROPOSITION 16: Let f *be a quasiconvex function on* $\mathbb{R}^n$ *and let* a *be a fixed point of* $\mathbb{R}^n$. *Then the function* $f'_+(a, .)$ *is quasiconvex and can be expressed as the minimum of two convex functions.*

PROOF: Let $h_1, h_2 \in \mathbb{R}^n$ and let $\lambda \in (0,1)$. Then, due to the quasi-convexity of f

$$f(a + t(\lambda h_1 + (1-\lambda)h_2) \leq \text{Max}[f(a + th_1), f(a + th_2)]$$

for all $t > 0$. It follows easily that

$$\frac{f(a+t(\lambda h_1+(1-\lambda)h_2) - f(a)}{t} \leq \text{Max}\left[\frac{f(a+th_1)-f(a)}{t}, \frac{f(a+th_2)-f(a)}{t}\right]$$

Take the limit when t tends to zero. □

Unfortunately, the second Dini-directional derivative is not necessarily quasiconvex. Consider for instance the function on $\mathbb{R}^2$ defined by:

$$f(x,y) = \begin{cases} -1 & \text{if } x + y \leq -\frac{1}{2} \text{ and } x \leq 0 \text{ and } y \leq 0, \\ -2^{-k} & \text{if } x + y > -2^{-k} \text{ and } x + 2y \leq -2^{-k} \text{ and } x \leq 0 \text{ and } y \leq 0, \\ -2^{-(k+\frac{1}{2})} & \text{if } x + 2y > -2^{-k} \text{ and } 2(x+y) \leq -2^{-k} \text{ and } x \leq 0 \text{ and } y \leq 0, \\ 0 & \text{if } x > 0 \text{ or } y > 0 \text{ or } x = y = 0, \end{cases}$$

where $k = 1, 2, 3, \ldots$

This function is quasiconvex and lower semi-continuous. Let $\alpha \leq 0$, $\beta \leq 0$ such that $\alpha + \beta < 0$, then for all $t > 0$

$$f(t\alpha, t\beta) = \begin{cases} -1 & \text{if } t \geq -[2\alpha + 2\beta]^{-1}, \\ -2^{-k} & \text{if } -2^{-k}(\alpha + 2\beta)^{-1} \leq t < -2^{-k}(\alpha + \beta)^{-1}, \\ -2^{-k+\frac{1}{2}} & \text{if } -2^{-k-1}(\alpha + \beta)^{-1} \leq t < -2^{-k}(\alpha + 2\beta)^{-1}. \end{cases}$$

Thus,

$$f'_+((0,0) \; ; \; (\alpha,\beta)) = \mathrm{Max}[\alpha + \beta, \; \frac{1}{\sqrt{2}} (\alpha + 2\beta)] \; ,$$

$$f'_-((0,0) \; ; \; (\alpha,\beta)) = \mathrm{Min}[\alpha + 2\beta, \; \sqrt{2} \; (\alpha + \beta)] \; .$$

It follows that $f'_-((0,0) \; ; \; (. \; , \; .))$ is not quasiconvex.

An important property of quasiconvex functions on $\mathbb{R}^n$ is that for these functions the three concepts of hemi-differentiability, Gâteaux-differentiability and Fréchet-differentiability coincide for the first derivative. Due to the complexity of the proof of the equivalence between Gâteaux and Fréchet-differentiability, it is not reproduced here; a proof appears in Crouzeix (1979a). To prove the equivalence between hemi-differentiability and Gâteaux-differentiability it is sufficient to note that $f'(a,.)$ is a positively homogeneous quasiconvex function on $\mathbb{R}^n$ (because of Proposition 16) and to apply Proposition 15.

Other important properties of the first derivative of quasiconvex functions on $\mathbb{R}^n$ are listed below

PROPOSITION 17: Let f *be a quasiconvex function on* $\mathbb{R}^n$ *and* a *be a fixed point in* $\mathbb{R}^n$. *Let us define* $S = \{x/f(x) < f(a)\}$ *and* $K = \{h/\exists t > 0 \text{ s.t. } a+th \in S\}$.

(a) Assume that the closure of K *is a half-space and there exists* $d \in \mathrm{int}(K)$ *such that the function* $f_{a,d}(t) = f(a + td)$ *is differentiable at* 0, *then* f *is differentiable at* a.

(b) f *is differentiable at* a *if and only if* $\bar{f}$ *is differentiable at* a.

Proposition 17 a) is analogous to Proposition 7, while Proposition 17 b) is analogous to Proposition 5. The proofs of these two propositions are complex and can be found respectively in Crouzeix (1979a) and (1980b).

PROPOSITION 18: Quasiconvex functions on $\mathbb{R}^n$ *are almost everywhere differentiable.*

This result was proved for quasiconvex continuous functions on R^n in Crouzeix (1979b) and extended to general quasiconvex functions on $\mathbb{R}^n$ in

Crouzeix (1980c). Here again due to the complexity of proofs the reader is referred to these papers. Notice that the result of Deak concerning the almost everywhere continuity of quasiconvex functions can be deduced from Proposition 18. The concepts of subdifferentials for convex functions and generalized gradients for locally Lipschitz functions are extremely useful tools for optimization problems involving these functions. The above two concepts cannot be applied to quasiconvex functions since these are neither convex nor locally Lipschitz functions. We now intend to develop a specific concept for quasiconvex functions. Keeping in mind that the function $f'_+(a,.)$ is quasiconvex while $f'_-(a,.)$ is not so, one only retains the first function. For the sake of simplicity, here we shall limit this study to the two following classes of functions.

$\mathcal{C}$ is the class of the functions f from $\mathbb{R}^n$ to $(-\infty, +\infty]$ which are quasiconvex and such that $f'_+(a,h)$ is finite for all $h \in \mathbb{R}^n$ and $a \in \mathrm{dom}(f)$.

$\mathcal{P}$ is the class of the functions which belong to $\mathcal{C}$ and verify

$$x,y \in \mathrm{dom}(f) \text{ and } f(y) < f(x) \Longrightarrow f'_+(x, y-x) < 0.$$

If f belongs to $\mathcal{C}$, then f is continuous on its domain.
This can be seen by noticing that f is hemi-continuous at each $a \in \mathrm{dom}(f)$ and applying Proposition 6. On the other hand, the added condition to define class $\mathcal{P}$ from class $\mathcal{C}$ is related to pseudoconvexity.

Let $f \in \mathcal{C}$ and a be a fixed point in dom(f). Let us define

$$S = \{x/f(x) < f(a)\},$$

$$K = \{h/\exists t > 0 \text{ s.t. } a + th \in S\},$$

$$T = \{h/f'_+(a,h) < 0\}.$$

Notice that if S is not empty, then int(S) and int(K) are not empty (since f is continuous on its domain). On the other hand, $T \subset K$.

PROPOSITION 19: If T *is not empty, then* $\mathrm{int}(K) \subset T \subset K$.

PROOF: Let $d \in T$ and $h \in \text{int}(K)$. One must show that $f'_+(a,h) < 0$. There exists $\lambda \in (0,1)$ and $k \in K$ such that $h = \lambda d + (1-\lambda)k$. Due to the definition of K, there exists $\bar{t} > 0$ such that $f(a + \bar{t}k) < f(a)$. Set $c = a + \bar{t}k$ and associate with each $t > 0$ the point $m = a + th$ and the point n where the line $\Delta = \{a + \xi d / \xi \in \mathbb{R}\}$ intersects with the line passing through c and m. That is,

$$n = a + \xi d \qquad m = n + \nu(c - n)$$

$$\text{with} \quad \xi = \frac{\lambda t\bar{t}}{\bar{t}-(1-\lambda)t} \text{ and } \nu = (1 - \lambda)\frac{t}{\bar{t}}\,.$$

Note that $0 < \nu < 1$ for small positive values of t. Hence, since f is quasiconvex,

$$f(m) \leq \text{Max}[f(n), f(c)]$$

$$\frac{f(m)-f(a)}{t} \leq \text{Max}\left[\frac{f(n)-f(a)}{t}, \frac{f(c)-f(a)}{t}\right].$$

Since $f(c) = f(a + \bar{t}k) < f(a), \frac{f(c) - f(a)}{t} \to -\infty$ when $t \downarrow 0$.

Hence,

$$\limsup_{t \downarrow 0} \frac{f(a+th)-f(a)}{t} \leq \limsup_{t \downarrow 0} \left[\frac{f(a+\xi d)-f(a)}{\xi} \times \frac{\lambda\bar{t}}{\bar{t}-(1-\lambda)t}\right].$$

Thus $f'_+(a,h) < 0$. □

It follows that int(T) is not empty when T is not empty. As in section 5 , define

$$\mu_-(h) = \begin{cases} f'_+(a,h) & \text{if } h \in T, \\ +\infty & \text{otherwise} \end{cases} \qquad \mu_+(h) = \begin{cases} 0 & \text{if } h \in T, \\ f'_+(a,h) & \text{otherwise.} \end{cases}$$

Then μ_- and μ_+ are convex functions and $f'_+(a,h) = \text{Min}[\mu_-(h), \mu_+(h)]$ for all h. Since $f'_+(a, .)$ takes only finite values it is continuous except perhaps on the boundary of T. The fact that f belongs to $\mathcal{C}$ is not sufficient to assure that $f'_+(a, .)$ is lower semi-continuous on the boundary of T. To see this consider the already given example

$$f(x_1,x_2,x_3) = \begin{cases} -(x_1+x_2+x_3) & \text{if } x_3 > 0,\ x_1 \geqslant 0 \text{ and } x_2 \geqslant 0, \\ -(x_1+x_2) & \text{if } x_3 = 0,\ x_1 \geqslant 0,\ x_2 \geqslant 0 \text{ and } x_1-x_2 \geqslant 0, \\ 0 & \text{elsewhere.} \end{cases}$$

However, we have

PROPOSITION 20: $f'_+(a, .)$ *is upper semi-continuous on* $\mathbb{R}^n$.

PROOF: One must only consider the case where T is not empty. Then T and K have the same boundary. If $f'_+(a, .)$ is not u.s.c. at $h \in \mathbb{R}^n$, then h necessarily belongs to the boundary of T and $f'_+(a,h) < 0$. Then, there exists $t > 0$ such that $f(a + th) < f(a)$. Since f is continuous at $a + th$, h belongs to the interior of K and thus is a contradiction. □

i) Assume that T is not empty. The functions $\bar{\mu}_-$ and $\bar{\mu}_+$ can be regarded as being the support functions of two closed convex sets which we denote by $\partial^- f(a)$ and $\partial^+ f(a)$. Notice that $f'_+(a, .)$ can be deduced from $\partial^- f(a)$ and $\partial^+(a)$. Indeed, due to Proposition 14, one can construct T^o from $\partial^- f(a)$. Then,

$$f'_+(a,h) = \begin{cases} \delta^*(h/\partial^- f(a)) & \text{if } h \in \text{int}(T) = \text{int}(T^{oo}), \\ \delta^*(h/\partial^+ f(a)) & \text{otherwise.} \end{cases}$$

ii) If T is not empty, we define $\partial^- f(a)$ as being equal to T^o and $\partial^+ f(a)$ as being the closed convex set having $f'_+(a, .)$ as support function. Here again $f'_+(a, .)$ is perfectly known from $\partial^- f(a)$ and $\partial^+ f(a)$.

One defines $(\partial^- f(a), \partial^+ f(a))$ as being the quasiconvex gradient of f at a. Optimality conditions for problems involving functions belonging to the class $\mathcal{P}$ can be expressed in terms of quasiconvex gradient. The interested reader is referred to Crouzeix (1980b).

REFERENCES

CROUZEIX, J.P. (1977). "Contributions à l'étude des fonctions quasi-convexes." Thèse, Université de Clermont, (France).

CROUZEIX, J.P. (1979a). "About Differentiability of Quasiconvex Functions." Miméo, Université de Clermont.

CROUZEIX, J.P. (1979b). "Sur l'existence de la dérivée des fonctions quasiconvexes." Miméo, Université de Clermont.

CROUZEIX, J.P. (1980a). "Conditions for Convexity of Quasiconvex Functions." *Math. of Operations Research* Vol.5, 120-125.

CROUZEIX, J.P. (1980b). "Some Differentiability Properties of Quasi-convex Functions on $\mathbb{R}^n$." Miméo, Oberwolfach.

CROUZEIX, J.P. (1980c). "A Review on Continuity and Differentiability Properties of Quasiconvex Functions on $\mathbb{R}^n$." Miméo, Imperial College, London.

CROUZEIX, J.P. (1981). "A Duality Framework in Quasiconvex Programming." This volume, 207-225.

DEAK, E. (1962). "Über Konvexe und Interne Funktionen, Sowie eine Gemeinsame Verallgemeirung Von Beiden." *Ann. Sci. Budapest Sect. Math. 5,* 109-154.

DIEWERT, W.E. (1978). "Hicks' Aggregation Theorem and the Existence of a Real Value Added Function." In *Production Economics: A Dual Approach to Theory and Applications.* (Fuss, M. and McFadden, D., editors), North-Holland Publishing Company, 17-51.

FENCHEL, W. (1953). "Convex Cones, Sets and Functions." Miméo, Princeton University, Princeton, N.J.

MOULIN, H. (1974). "Représentation d'un Préordre par une Fonction d'Utilité Concave ou Differentiable." *Comptes rendus de l'Academie des Sciences Paris 278,* 483-485.

NEWMAN, P. (1969). "Some Properties of Concave Functions." *Journal of Economic Theory 1,* 291-314.

ROCKAFELLAR, R.T. (1970). *Convex Analysis.* Princeton University Press, Princeton, N.J.

CONCAVIFIABILITY OF C^2-FUNCTIONS: A UNIFIED EXPOSITION[1]

Israel Zang[2]

This paper is concerned with necessary and sufficient conditions under which a C^2-function is concave transformable by a strictly monotone increasing C^2-function. The classical approach of Fenchel as well as other recent contributions can be placed within a unified consistent framework. Sufficient conditions for concavifiability over compact sets using exponential transformations are also discussed.

1. INTRODUCTION

The concept of concavifiability is well known throughout the mathematical economics and mathematical programming literature: let f be a real valued continuous function on the convex subset $C \subset R^n$ and let G be a continuous real valued strictly monotone increasing function on $I_f(C)$, where $I_f(C)$ denotes the image of C under f. Then f is said to be *concave transformable, concavifiable, or G-concave* when Gf(x) is concave on C. Properties of these functions can be found in Avriel and Zang (1974). In particular, every G-concave function is semistrictly quasiconcave (in the sense of Diewert, Avriel and Zang (1981); see also Avriel, Diewert, Schaible and Ziemba (1981)), and is pseudoconcave if f is differentiable.

An important problem associated with concave transformable functions concerns the existence of a concavifying function G. One would like to have a set of necessary and sufficient conditions under which a general function (which must be at least semistrictly quasiconcave) is G-concave for some G. Such conditions are the subject of this paper. This question has been treated in the economic and mathematical programming literature. This

---

[1]The author is indebted to the editors of the proceedings and to W. E. Diewert for their valuable remarks on an earlier draft of this paper. This research was supported by the Israel Institute for Business Research at Tel Aviv University.

[2]Faculty of Management, Tel Aviv University, Tel Aviv, Israel. Currently visiting Faculty of Commerce and Business Administration, University of British Columbia, Vancouver, B.C., Canada.

ISBN 0-12-621120-5

paper is only concerned with a portion of this literature namely that dealing with functions which are twice continuously differentiable on open convex subsets of R^n. For these functions, a unified derivation of necessary and sufficient conditions for concavifiability will be presented together with some new conditions. Results referring to functions which are not necessarily in C^2, can be found in de Finetti (1949), Fenchel (1951, 1956), Crouzeix (1977), Debreu (1976) and Kannai (1974, 1977, 1980, 1981). Since Schaible (1971, 1977) (see also Schaible (1981)) has shown that non-concave quasiconcave quadratic functions are actually G-concave, we do not consider the quadratic case.

The first complete set of necessary and sufficient conditions for the concavifiability of C^2-functions was derived by Fenchel (1951, 1956). Kannai (1977, 1981) has dwelt on these results and given some different formulations. Other conditions were derived by Avriel (1972), Avriel and Schaible (1978), Diewert (1978), Gerencsér (1973), Katzner (1970) and Schaible and Zang (1980). In this paper we present all these results within a unified framework based largely on Fenchel's approach. There is an hierarchy of four conditions. The first two are derived in Section 2 and are of a local nature, considering properties of the gradient and of the Hessian of f augmented by a rank-one matrix formed by the self outer product of the gradient. These properties must hold for every point in C. The two other conditions, presented in Section 3, are higher in the hierarchy and are global ones.

A special family of G-concave functions, for which $G(t) = -\exp(-rt)$ for some positive r, is the subject of Section 4. These functions correspond to r-concave functions introduced by Avriel (1972). We call these functions r-concave, although Avriel defined r-concavity differently. In this context we shall consider for example G-concave functions where $G'(t) > 0$ hold for every $t \in I_f(C)$. Functions in this class, once restricted to a compact convex subset of C, are actually r-concave over this subset. For such functions (over compact sets) it is easier to establish G-concavity, since the functional form of G is known and only the value of r must be determined.

2. LOCAL CONDITIONS

The problem is to find necessary and sufficient conditions for G-concavity of a function f which is twice continuously differentiable over the open convex set $C \subset R^n$. Let

$$\alpha \equiv \inf_{x \in C} f(x) \quad , \qquad \beta \equiv \sup_{x \in C} f(x), \tag{1}$$

where $-\infty \leqslant \alpha,\ \beta \leqslant \infty$ hold. We require the concavifying function G(t) to be strictly monotone increasing and twice continuously differentiable over $I_f(C)$. A concave function has no minimum in an open domain. Thus, $I_f(C) = (\alpha,\beta]$ in case f has a maximum in C and $I_f(C) = (\alpha,\beta)$ otherwise. We also rule out the possibility that f is constant over C and thus $\alpha < \beta$ holds.

If f is G-concave and $\nabla f(\bar{x}) = 0$ for some nonmaximal $\bar{x} \in C$, then $\nabla G(f(\bar{x})) = G'(f(\bar{x}))\nabla f(\bar{x}) = 0$, and since $\bar{x}$ is also nonmaximal for G(f(x)) we have a contradiction to the G-concavity assumption. Thus, we obtain the following basic necessary condition for concavifiability expressed by Fenchel (1951):

(A) $\nabla f(x) \neq 0$ *for all* $x \in C$, *except for points in* C *satisfying* $f(x) = \beta$ *if such points exist.*

This condition, as well as some others to be derived later on, is implied by the pseudoconcavity of f which is a necessary condition for the concavifiability of a differentiable function (see Avriel and Zang (1974)). If we now differentiate Gf twice we see that a necessary and sufficient condition for the concavity of Gf is the negative semidefiniteness of the matrix

$$\nabla^2 G(f(x)) = G'(f(x))\nabla^2 f(x) + G''(f(x))\nabla f(x)\nabla f(x)^T \qquad (2)$$

for every $x \in C$. Thus, our problem is to formulate conditions which are equivalent to the negative semidefiniteness of (2) for some G satisfying the hypotheses. In case there exists a point $\tilde{x} \in C$ satisfying $f(\tilde{x}) = \beta$, then the well known necessary optimality criteria (see e.g. Avriel (1976)) imply that $\nabla f(\tilde{x}) = 0$ and that $\nabla^2 f(\tilde{x})$ is negative semidefinite; thus $\nabla^2 G(f(x))$ is negative semidefinite as well for every G satisfying the hypotheses and for all maximal points in C. Moreover, if f is G-concave and $G'(\bar{t}) = 0$ for some $\alpha < \bar{t} < \beta$, then $\nabla G(f(x)) = 0$ for all $x \in C$ satisfying $f(x) = \bar{t}$ contradicting the G-concavity assumption since $\bar{t}$ is nonmaximal. Thus, $G'(t) > 0$ for all $\alpha < t < \beta$ and consequently a necessary and sufficient condition for the G-concavity of f is that the matrix $\nabla^2 f(x) + [G''(f(x))/G'(f(x))]\nabla f(x)\nabla f(x)^T$ be negative semidefinite for all nonmaximal points $x \in C$.

Let

$$H(x;\rho) \equiv \nabla^2 f(x) - \rho\nabla f(x)\nabla f(x)^T \qquad (3)$$

be the augmented Hessian of f. To insure negative semidefiniteness of (2) we must first ensure, for every $x \in C$, the existence of a number ρ such that $H(x;\rho)$ is negative semidefinite. Consequently we shall derive a set of local conditions which are necessary and sufficient for the existence of a function $\rho(x)$ defined on C such that $H(x;\rho(x))$ is negative semidefinite for all $x \in C$. Alternatively, considering the quadratic form associated with (3), we shall establish conditions which guarantee that

$$\rho_0(x) = \sup_{\substack{\|z\| = 1 \\ z^T\nabla f(x) \neq 0}} \{z^T\nabla^2 f(x)z/(z^T\nabla f(x))^2\} \qquad (4)$$

is less than $+\infty$ for every $x \in C$. This implies that $H(x;\rho(x))$ is negative semidefinite for every function $\rho(x)$ satisfying $\rho(x) \geq \rho_0(x)$ and for every $x \in C$. We may consider all points in C, since points where $\nabla f(x) = 0$ hold must be, in view of condition (A), maximal in C and thus have a negative semidefinite Hessian. Therefore, for these points $\rho_0(x)$ can take any value. In particular, from (4) we have that we may let $\rho_0(x) = -\infty$ in these cases.

Following Avriel and Schaible (1978) let

H $\equiv$ family of functions for which a negative semidefinite augmented Hessian exists at every $x \in C$.

Our second necessary (and sufficient together with all the other conditions to come) condition is then

(B) $\quad f \in H$.

We now proceed to characterize functions belonging to H. This is done in terms of conditions which are necessary and sufficient. Since for a given $x \in C$ $\nabla^2 f(x)$ and $\nabla f(x)$ are constants, we can state this characterization problem in a somewhat more general framework: given a symmetric nxn matrix A and a vector $b \in R^n$, find necessary and sufficient conditions for the existence of a number $\bar{r}$ such that the matrix

$$H(r) \equiv A - rbb^T, \tag{5}$$

is negative semidefinite for all $r \geq \bar{r}$.

Recall that a vector $v \in R^n$ is an eigenvector of A restricted to the subspace orthogonal to b, denoted $b^\perp$, and a scalar λ is the corresponding restricted eigenvalue of A, if v is a stationary vector and λ a stationary value to the problem:

$$\begin{aligned} &\text{stat} \quad v^T A v \\ &\text{s.t.} \quad v^T b = 0 \\ &\qquad\quad v^T v = 1. \end{aligned} \tag{6}$$

That is, there exists a scalar β such that v, λ and β satisfy the first order necessary optimality criteria

$$
\begin{aligned}
&Av - \frac{\beta}{2}b - \lambda v = 0 \\
&v^T b = 0 \\
&v^T v = 1.
\end{aligned}
\tag{7}
$$

Since (7) implies that $v^T Av = \lambda$ must hold, it is possible to find n-1 orthogonal vectors satisfying (7). The number of nonzero restricted eigenvalues corresponding to these restricted eigenvectors is the rank of A restricted to $b^\perp$.

Let A and B be defined as above and

$$
B \equiv \begin{pmatrix} 0 & b^T \\ b & A \end{pmatrix} \quad . \tag{8}
$$

We denote by Q_k the set consisting of monotone increasing sequences of k numbers from $\{1,\ldots,n\}$ that is

$$
Q_k = \{\gamma: \ \gamma=(i_1,\ldots,i_k),\ 1 \leqslant i_1 <\ldots< i_k \leqslant n\}. \tag{9}
$$

Let $A_{\gamma,k}$ and $M_{\gamma,k}$, $\gamma \in Q_k$, respectively, denote the principal submatrix and minor of order k of A formed by the $i_1,\ldots,i_k$-th rows and columns of A, and $b_{\gamma,k}$ be the corresponding subvector of b. Also, denote by A_k and M_k, $k=1,\ldots,n$, respectively, the leading principal submatrix and minor of order k x k of A, and let b_k be the corresponding subvector of b. We also associate with $A_{\gamma,k}$ the principal submatrix $B_{\gamma,k}$ and minor

$$
D_{\gamma,k} = \det B_{\gamma,k} = \det \begin{pmatrix} 0 & b^T_{\gamma,k} \\ b_{\gamma,k} & A_{\gamma,k} \end{pmatrix}, \ \gamma \in Q_k \tag{10}
$$

of B. Similarly, D_k, $k=1,\ldots,n$ will denote the leading principal minor of order k + 1 of B. We now have:

PROPOSITION 1. The matrix $H(r)$ *is negative semidefinite for all* $r \geqslant \bar{r}$ *if and only if conditions (i.j) and (ii.l) (j = 1 or 2 or 3, l = 1 or 2 or 3 or 4) below hold:*

(i.1) $v^T b = 0 \Rightarrow v^T A v \leqslant 0$

(i.2) *A is negative semidefinite whenever* $b=0$, *and if* $b \neq 0$ *then for every* $\gamma \in Q_k$, $k=1,\dots,n$ $\quad (-1)^k D_{\gamma,k} \geqslant 0$ *holds.*

(i.3) *A is negative semidefinite whenever* $b = 0$, *and B has exactly one positive eigenvalue whenever* $b \neq 0$.

(ii.1) $v^T b = 0,\ v^T A v = 0 \Rightarrow Av = 0$

(ii.2) *If the rank of A restricted to* $b^{\perp}$ *is* $q-1$ *then the rank of A is at most* q.

(ii.3) *If* $b \neq 0$ *and the rank of the matrix B is* $q + 1$ *then the rank of A is at most* q.

(ii.4) *For every* $\gamma \in Q_k$, $k=1,\dots,n$ $D_{\gamma,k} = 0$ *implies* $(-1)^k M_{\gamma,k} \geqslant 0$.

PROOF. The equivalence of (i.1) and (i.2) is due to Debreu (1952). The equivalence of (i.1) and (i.3) is due to Crouzeix and Ferland (1979). The equivalence of (i.1) and (ii.1) to negative semidefiniteness of H(r) is due to Schaible and Zang (1980). A slightly different but equivalent formulation of (ii.1) due to Kannai (1977) is

(ii.1') $v^T b = 0,\quad v^T A v = 0 \Rightarrow b^T A v = 0.$

Kannai actually showed that (ii.1') (and hence (ii.1)) and (ii.2) are equivalent when (i.1) holds. The quivalence of (i.2) and (ii.4) to the negative semidefiniteness of H(r) was shown by Avriel and Schaible (1978).

To complete the proof we show that (i.1) and (ii.3) are equivalent to negative semidefiniteness of H(r). This is obvious in case $b = 0$ ((ii.3) is empty). Let $b \neq 0$. We show that (ii.3) $\Leftrightarrow$ (ii.2) in this case. This relationship holds since the number of zero eigenvalues of B equals the number of zero eigenvalues of A restricted to $b^{\perp}$. To show this we note that $(v_0, v^T)^T \neq 0$, $v_0 \in R$, $v \in R^n$ is an eigenvector of B with zero eigenvalue iff

$$\begin{aligned} b^T v &= 0 \\ v_0 b + Av &= 0 \end{aligned} \tag{11}$$

hold. Because $b \neq 0$ we must have $v \neq 0$, since otherwise $v_0 = 0$ must hold. Thus, in view of (7) we conclude that (11) holds iff v is an eigenvector of A, restricted to $b^{\perp}$, with zero eigenvalue. Thus, B and A restricted to $b^{\perp}$ have the same number of zero eigenvalues. □

Condition (i.1) of the above proposition is an obvious one: Since the matrix bb^T does not augment A on the subspace orthogonal to b, it is necessary for the negative semidefiniteness of H(r) that A will be negative semidefinite on $b^{\perp}$. However, this condition alone is not sufficient for the negative semidefiniteness of H(r), since it may happen that there exists a sequence $\{v^k\} \to \bar{v}$ such that $(v^k)^T b > 0$, $\{(v^k)^T b\} \to 0$, $(v^k)^T A v^k > 0$ and $\{(v^k)^T A v^k\} \to 0$ such that $\lim (v^k)^T A v^k / ((v^k)^T b)^2 \to \infty$. Along such a sequence it may be necessary to augment A asymptotically infinitely to retain negative semidefiniteness of H(r). Under Condition (ii.1) such a situation can not occur.

Condition (ii.2) is the well known rank condition due to Fenchel (1951). However the equivalence of this condition together with (i.1) to the negative semidefiniteness of H(r) is a new result. The other rank condition (ii.3) is new. The equivalence of conditions (ii.1) and (i.1) to the negative semidefiniteness of H(r), established by Schaible and Zang (1980), is actually an extension to the well known Finsler (1937) Theorem, which states that H(r) is negative definite for all $r > \bar{r}$ iff the matrix A restricted to $b^{\perp}$ is negative definite. Condition (ii.1) was used earlier by Diewert (1978) in the context of sufficient conditions for concavifiability, where it was conjectured that it is equivalent to condition (ii.2). Conditions (i.1) and (ii.1) can also be easily interpreted in terms of solutions of (7). That is,

(i.1)' $\lambda_i \leqslant 0$

(ii.1)" *If* $b \neq 0$ *then* $\lambda_i = 0 \Rightarrow \beta_i = 0$.

See also Crouzeix and Ferland (1979) for some other conditions which are equivalent to (i.1) and (ii.1) and for a numerical analysis of the above criteria. The best way to test negative semidefiniteness of H(r) seems to be via conditions (i.3) and (ii.3). This can be done by computing the inertia (a triplet including the number of negative, the number of positive, and the number of zero eigenvalues) of the matrices A and B. An algorithm which can carry these calculations is suggested by Cottle (1974).

Utilizing conditions (i.1) - (i.3) and (ii.1) - (ii.4) we can easily derive conditions which are necessary and sufficient for f to be in class H. If we compare (3) with (5), then all we should do is to substitute $\nabla^2 f(x)$ for A and $\nabla f(x)$ for b in the above conditions to obtain

THEOREM 2. $f \in H$ *if and only if for every* $x \in C$ *conditions (I.i) and (II.j) (i = 1 or 2 or 3, j = 1 or 2 or 3 or 4) below hold:*

(I.1) $v^T \nabla f(x) = 0 \Rightarrow v^T \nabla^2 f(x) v \leqslant 0.$

(I.2) $\nabla^2 f(x)$ *is negative semidefinite whenever* $\nabla f(x) = 0$, *and if* $\nabla f(x) \neq 0$ *then for every* $\gamma \in Q_k$, $k=1,\dots,n$, $(-1)^k D_{\gamma,k}(x) \geqslant 0$ *holds.*

(I.3) $\nabla^2 f(x)$ *is negative semidefinite whenever* $\nabla f(x) = 0$, *and* $B(x)$ *has exactly one positive eigenvalue whenever* $\nabla f(x) \neq 0$.

(II.1) $v^T \nabla f(x) = 0,\ v^T \nabla^2 f(x) v = 0 \Rightarrow \nabla^2 f(x) v = 0.$

(II.2) *If the rank of* $\nabla^2 f(x)$ *restricted to* $\nabla f(x)^\perp$ *is* $q - 1$ *then the rank of* $\nabla^2 f(x)$ *is at most* q.

(II.3) *If* $\nabla f(x) \neq 0$ *and the rank of* $B(x)$ *is* $q + 1$ *then the rank of* $\nabla^2 f(x)$ *is at most* q.

(II.4) *For all* $\gamma \in Q_k$, $k=1,\dots,n$ *if* $D_{\gamma,k}(x) = 0$ *then* $(-1)^k M_{\gamma,k}(x) \geqslant 0$ *holds.*

In the above conditions $B(x)$, $D_{\gamma,k}(x)$ and $M_{\gamma,k}(x)$ stand for B, $D_{\gamma,k}$ and $M_{\gamma,k}$, as introduced above, where $A = \nabla^2 f(x)$, $b = \nabla f(x)$ is substituted. To establish these conditions it is better, by the preceding discussion, to try to verify conditions (I.3) and (II.3). Thus, it is necessary to compute the inertia of either $\nabla^2 f(x)$ or $B(x)$ for *every* $x \in C$, which is usually an impossible task. Note however, that the same sort of complexity appears when trying to verify concavity of a general C^2-function, by testing whether its Hessian is negative semidefinite everywhere. Some other conditions which are equivalent to (II.1) appear in Kannai (1977,1981). One of these characterizations makes use of the Implicit Function Theorem.

It is quite logical to assume that for many pseudoconcave functions condition (II.1) will not be satisfied for some points $x \in C$. The following example which is due to Avriel and Schaible (1978) is interesting, since the function considered is pseudoconcave but no point $x \in C$ satisfies condition (II.1).

Example 1. Let $f(x) = -x_2/x_1$ on $C = \{x \in R^2 : x_1 > 0\}$. According to Managasarian (1969) f is pseudoconcave, thus (I.1) holds for every $x \in C$. We have

$$\nabla f(x) = \begin{pmatrix} x_2/(x_1)^2 \\ -1/x_1 \end{pmatrix}, \quad \nabla^2 f(x) = \begin{pmatrix} -2x_2/(x_1)^3 & 1/(x_1)^2 \\ 1/(x_1)^2 & 0 \end{pmatrix}, \tag{12}$$

and $v^T \nabla f(x) = 0,\ v \neq 0 \Rightarrow v_2/v_1 = x_2/x_1$. Moreover
$v^T \nabla^2 f(x) v = (-2(v_1)^2 x_2/x_1 + 2v_1 v_2)/(x_1)^2 = 0$ for v_1, v_2 satisfying the above proportion. However

$$\begin{aligned} \nabla^2 f(x) v &= (1/(x_1)^2)(-2x_2 v_1/x_1 + v_2, v_1)^T \\ &= (1/(x_1)^2)(-v_2, v_1)^T \neq 0 . \end{aligned} \tag{13}$$

Thus, condition (II.1) does not hold.

It turns out that for the above function for every $x \in C$, there exists no finite ρ such that $H(x;\rho)$ is negative semidefinite. This example was also considered by Crouzeix (1977), where it was shown that this function can not be concavified on any open neighborhood of C even when non-differentiable transformations are considered. Note however that it suffices to find one point in C, for which the above condition does not hold, to conclude that $f \notin H$.

To conclude this section we state three different mathematical expressions for $\rho_0(x)$, given by (4), which are valid for functions in H and for points in C having a nonzero gradient. First, it is possible to express $\rho_0(x)$ in terms of the coefficients of the characteristic equations of $\nabla^2 f(x)$ and of $\nabla^2 f(x)$ restricted to $\nabla f(x)^\perp$. The resulting expression, due to Fenchel (1951), is

$$\rho_0(x) = \frac{s_{q(x)}(x)}{\nabla f(x)^T \nabla f(x) s^*_{q(x)-1}(x)}, \tag{14}$$

where $q(x)-1$ is the rank of $\nabla^2 f(x)$ restricted to $\nabla f(x)^\perp$ (or equivalently, $q(x) + 1$ is the rank of $B(x)$) and $s_{q(x)}(x)$ and $s^*_{q(x)-1}(x)$ are coefficients of the characteristic equations of $\nabla^2 f(x)$

$$\det(\nabla^2 f(x)-\lambda I) = s_n - s_{n-1}\lambda + \ldots + (-1)^i s_{n-1}\lambda^i + \ldots + (-1)^n s_0 \lambda^n, \tag{15}$$

and of $\nabla^2 f(x)$ restricted to $\nabla f(x)^\perp$

$$\frac{-1}{\nabla f(x)^T \nabla f(x)} \det \begin{pmatrix} 0 & \nabla f(x)^T \\ \nabla f(x) & \nabla^2 f(x)-\lambda I \end{pmatrix} =$$

$$s^*_{n-1} - s^*_{n-2}\lambda + \ldots + (-1)^i s^*_{n-i-1}\lambda^i + \ldots + (-1)^{n-1} s^*_0 \lambda^{n-1} \tag{16}$$

respectively. See Fenchel (1951) for more details.

In view of (14), it is possible to observe that $\rho_0(x)$ is not necessarily a continuous function of x, even though f is a C^2-function and consequently s_j and s^*_j are continuous functions of x. This is because $q(x)$ is discrete and consequently discrete changes in q may cause discrete changes in ρ_0.

The second expression for $\rho_0(x)$ is due to Schaible and Zang (1980) and is stated in terms of principal minors of $\nabla^2 f(x)$ and $B(x)$:

$$\rho_0(x) = \max_{\gamma \in Q_k} \{-M_{\gamma,k}(x)/D_{\gamma,k}(x): \quad (-1)^k D_{\gamma,k}(x) > 0\} \tag{17}$$

$$k = 1,\ldots,n$$

The last expression for $\rho_0(x)$ is due to Diewert (1978). Here $\rho_0(x)$ is represented in terms of the eigenvalues of $\nabla^2 f(x)$ restricted to $\nabla f(x)^{\perp}$:

$$\rho_0(x) = \frac{\nabla f(x)^T \nabla^2 f(x) \nabla f(x)}{(\nabla f(x)^T \nabla f(x))^2} - \sum_{\substack{i=1 \\ \lambda_i(x) < 0}}^{n-1} \beta_i(x)^2 / 4\lambda_i(x), \tag{18}$$

where $\beta_i(x)$ and $\lambda_i(x)$ are the solutions of (7) for $A = \nabla^2 f(x)$ and $b = \nabla f(x)$ and the summation is taken over indices such that $\lambda_i(x) < 0$.

3. GLOBAL CONDITIONS

In this section we consider functions which belong to H as characterized in the last section. Our aim is to establish conditions which are necessary and sufficient for the existence of a concavifying function G for a given $f \in H$. Clearly, the existence of a function $\rho(x)$ such that $H(x;\rho(x))$ is negative semidefinite for every $x \in C$ is already assumed. This however does not guarantee that f is G-concave for some G even if $\rho(x)$ is continuous. The following example, due to Schaible (1971), demonstrates such a situation.

Example 2. The function

$$f(x) = -(x_1)^3 - (x_2)^2 \tag{19}$$

is pseudoconcave on every open convex set contained in

$$Y = \{x \in R^2: x_1 < 0, -[-\tfrac{3}{4}(x_1)^3]^{\frac{1}{2}} < x_2 < [-\tfrac{3}{4}(x_1)^3]^{\frac{1}{2}}\} . \tag{20}$$

Consider an open convex subset $C \subset Y$ such that there exists a point $\bar{x}$ on the boundary of C which is also on the boundary of Y. For example let $C = \{x \in R^2: x_1 > -4/3, x_2 > 0, 3x_1 + 2x_2 < -4/3\}$; then $\bar{x} = (-4/3, 4/3)^T$. Some calculations show that $f \in H$ with

$$\rho_0(x) = -\tfrac{1}{2}[\tfrac{3}{4}(x_1)^3 + (x_2)^2]^{-1}. \tag{21}$$

It is easy to see that $\rho_0(x) > 0$ for $x \in Y$ and that $\rho_0(x) \to \infty$ as x approaches a boundary point of Y. Consider all points $x \in C$ satisfying

$f(x) = \bar{t} = 16/27$. This level curve is shown in Figure 1. Certainly since $\alpha < \bar{t} < \beta$ (see (1)),

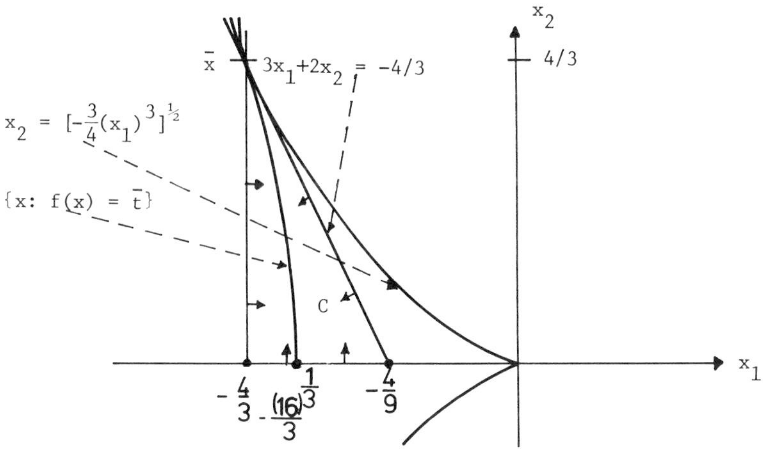

FIGURE 1. A Non-concavifiable function.

the set $\{\rho_0(x): f(x) = \bar{t}\}$ should be bounded from above if f is G-concave for some G. However $\rho_0(x) \to \infty$ as $x \to \bar{x}$ along this level curve. □

In view of the above example the following conditions should be satisfied so that f can be G-concave (Fenchel (1951)):

(C) *The function*

$$g(t) \equiv \sup\{\rho_0(x): x \in C,\ f(x) = t\} \tag{22}$$

is finite for every $t \in (\alpha,\beta)$.

Utilizing (14), (17), or (18) it is possible to obtain the following equivalent expressions for g(t)

$$g(t) = \sup\{\frac{s_{q(x)}(x)}{\nabla f(x)^T \nabla f(x) s^*_{q(x)-1}(x)}: x \in C,\ f(x) = t\}, \tag{23}$$

$$g(t) = \sup_{\substack{x \in C \\ f(x)=t}} \left[\max_{\substack{\gamma \in Q_k \\ k=1,\dots,n}} \{-M_{\gamma,k}(x)/D_{\gamma,k}(x) : (-1)^k D_{\gamma,k}(x) > 0\}\right], \tag{24}$$

and

$$g(t) = \sup_{\substack{x \in C \\ f(x)=t}} \left\{\frac{\nabla f(x)^T \nabla^2 f(x) \nabla f(x)}{(\nabla f(x)^T \nabla f(x))^2} - \sum_{\substack{i=1 \\ \lambda_i(x) < 0}}^{n-1} \beta_i(x)^2/4\lambda_i(x)\right\} . \tag{25}$$

The discussion following (16) implies through (23) that $g(t)$ is not necessarily continuous.

It turns out, however, that condition (C) alone is not sufficient for the G-concavity of f. The following is an example for such a case:

Example 3. (Avriel and Schaible (1978)): Let

$$f(x) = \begin{cases} -\int_0^x (\xi)^4[2+\cos(1/\xi)]d\xi & \text{if } x > 0 \\ 0 & \text{if } x = 0 \\ \int_0^x (\xi)^4[2+\cos(1/\xi)]d\xi & \text{if } x < 0. \end{cases} \tag{26}$$

This function is strictly pseudoconcave on R. However, its second derivative changes sign in every neighborhood of the origin, thus f is not concave. The conditions (A), (B), and (C) hold for this function. To show that f is G-concave we should establish the existence of a function h(t) which is differentiable for all $t \geqslant 0$, such that

$$\frac{d \ln h(t)}{dt} = \frac{h'(t)}{h(t)} \leqslant -g(t). \tag{27}$$

This will imply that f is G-concave with $G'(t) = h(t)$. However, it is possible to show that a function h satisfying (27) does not exist. See Avriel, Diewert, Schaible and Zang (1981) for more details. □

In view of the above example the following condition should also be satisfied so that f can become G-concave (Fenchel (1956)):

(D) *There exists a function* $h(t) > 0$ *for* $\alpha < t < \beta$ *which is differentiable for* $t \in I_f(C)$ *and satisfies*

$$\frac{d}{dt}\ln(h(t)) \leqslant -g(t) \tag{28}$$

for $\alpha < t < \beta$, *where* $g(t)$ *is given by* (22) *or by* (23) - (25).

Kannai obtained different conditions for the G-concavity of f, which are stated in terms of Perron or Lebesgue integrability of g(t). See Kannai (1977, 1981) for further details. The following concluding theorem is due to Fenchel (1951, 1956):

THEOREM 3. Let f *be a* C^2*-function on the open convex set* $C \subset R^n$. *Then* f *is* G-*concave on* C *if and only if conditions (A)-(D) hold.*

4. G-CONCAVE FUNCTIONS WHICH ARE r-CONCAVE OVER COMPACT SETS

In this section we consider a more restricted family of twice continuously differentiable G-concave functions on the open convex subset $C \subset R^n$, which are also r-concave over compact convex subsets $D \subset C$. That is for these G-concave functions, for every compact convex subset $D \subset C$, there exists a nonnegative number r(D) such that f is r(D)-concave or equivalently $-\exp(-r(D)f(x))$ is concave. A typical example is a G-concave function for which $G'(t) > 0$ for every $t \in I_f(C)$. For these functions the following result, which we state without proof, is immediate (see also Avriel, Diewert, Schaible and Zang (1981)):

PROPOSITION 4. Let f *be a twice continuously differentiable* G-*concave function on an open set containing the convex compact subset* $D \subset R^n$, *and let* G *be a twice continuously differentiable function satisfying* $G'(t) > 0$ *on* $I_f(D)$. *Then,* f *is* r-*concave on* D.

Consequently, once it is known that such a function is G-concave over a compact set, then it may be easier to obtain an expression for G in terms of an exponential function, where it is only necessary to determine the value of r(D). However, as demonstrated by Example 3, a function $f \in H$ is not necessarily r(D) concave. Avriel and Schaible (1978) consider a more restricted class of functions:

$H_c \equiv$ The family of functions in H for which $H(x;\rho(x))$ is negative semidefinite for some continuous function $\rho(x)$ on C.

Avriel and Schaible showed that every function in H_c is pseudoconcave. From Example 3 it follows that the inclusion $H_c \subset H$ is strict, i.e. $H_c \neq H$ holds. A different but equivalent characterization of class H_c is given by Schaible and Zang (1980):

H_c = The family of functions in H satisfying the property that for every compact convex subset $D \subset C$, there exists an $\bar{r}(D)$ such that $H(x;\bar{r}(D))$ is negative semidefinite on D.

In view of this equivalence it is possible to characterize functions in H_c utilizing Theorem 2 and the expressions for $\rho_0(x)$ given in Section 2. The following result is due to Schaible and Zang (1980):

PROPOSITION 5. $f \in H_c$ *if and only if* $f \in H$ *and for every compact convex subset* $D \subset C$

$$r(D) = \max\{0, \sup_{x \in D}[\rho_0(x)]\} < \infty \tag{29}$$

holds, where $\rho_0(x)$ *is given either by (14), (17) or (18).*

Thus a pseudoconcave function f which belongs to H_c is also r(D)-concave, and that in case $r(D) > 0$ then it is the lowest possible bound for r. However, it is not necessarily r-concave or even G-concave on the whole of C, as shown in Example 2.

We now turn to consider a more restricted family of r(D)-concave functions. These are functions for which a negative *definite* augmented

Hessian exists at every $x \in C$. Following Avriel and Schaible (1978), define

$H^s \equiv$ The family of functions for which a negative definite augmented Hessian exists at every $x \in C$.

That is, $f \in H^s$ if there exists a function $\rho(x)$ such that $H(x;\rho(x))$ is negative definite for every $x \in C$. We also let

$H^s_c \equiv$ The family of all functions in H^s for which $H(x;\rho(x))$ is negative definite for some continuous function $\rho(x)$ on C.

Following Schaible and Zang (1980), we can equivalently define H^s_c functions by:

$H^s_c \equiv$ The family of functions in H^s satisfying the property that for every compact convex subset $D \subset C$, there exists an $\bar{r}(D)$ such that $H(x;\bar{r}(D))$ is negative definite on D.

Although we have seen that $H_c \neq H$ we have the following result, due to Avriel and Schaible (1978), which will simplify the discussion to follow.

PROPOSITION 6. $H^s_c = H^s$.

Thus, a function which belongs to H^s is actually $r(D)$-concave for $r(D) = \max\{0,\bar{r}(D)\}$. Moreover, $-\exp(-rf(x))$ is strictly concave on D for every $r > r(D)$.

We now turn, in view of the above proposition, to derive necessary and sufficient conditions for a function f to belong to H^s_c. As we did for class H, we consider first conditions which are necessary and sufficient for the negative definiteness of $H(r)$, given by (5), for all $r > \bar{r}$. Using the notations that preceded Proposition 1, we have

PROPOSITION 7. The matrix $H(r)$ *is negative definite for all* $r > \bar{r}$ *if and only if either condition (iii.1) or (iii.2) or (iii.3) below holds:*

(iii.1) $v \neq 0,\ v^T b = 0$ *implies* $v^T A v < 0$.

(iii.2) *For* $k = 1,\dots,n$

$(-1)^k D_k \geqslant 0$ *holds, and if* $D_k = 0$ *then*

$(-1)^k M_k > 0$ *and* $b_k = 0$ *hold.*

(iii.3) A *is negative definite whenever* $b = 0$, *and* B *is nonsingular and has exactly one positive eigenvalue whenever* $b \neq 0$.

PROOF. Condition (iii.1) is the well known Finsler (1937) Theorem. The equivalence of (iii.2) to the negative definiteness of $H(r)$ was shown by Schaible and Zang (1980). It is left to show that condition (iii.3) is equivalent to the negative definiteness of $H(r)$. Clearly if $b = 0$ then by (iii.1) we are done. Let $b \neq 0$ and assume that $H(r)$ is negative definite. Then (iii.1) holds and in view of (i.1) condition (i.3) holds. It is only left to show that B is nonsingular. Let $(v_0, v^T)^T, v_0 \in R$, $v \in R^n$ be an eigenvector of B with zero eigenvalue, then

$$\begin{aligned} v^T b &= 0 \\ v_0 b + Av &= 0 \end{aligned} \tag{30}$$

holds. Clearly $v \neq 0$ must hold since otherwise $v_0 = 0$ holds, since $b \neq 0$. Thus, multiplying the second equation in (30) by v^T yields $v^T A v = 0$ contradicting (iii.1). Conversely, let $b \neq 0$ and assume that (iii.3) holds. Then (i.3) and consequently (i.1) hold. To show that (iii.1) holds we establish that

$$v \neq 0\ ,\ v^T b = 0,\ v^T A v = 0 \tag{31}$$

contradicts the nonsingularity of B and thus (iii.1) holds. We note that (31) and (i.1) imply that v is an eigenvector of A restricted to $b^{\perp}$ with zero eigenvalue. This implies that (7) holds with $\lambda = 0$, and consequently (30) holds which, in turn, implies that B is singular. □

An earlier version of condition (iii.2) can be found in Debreu (1952) or in Avriel and Schaible (1978). Condition (iii.3) is new. It is also the one that seems to be superior from the computational point of view and it can be tested by using Cottle's (1974) algorithm for computing inertias. Also note that to satisfy the above proposition, A cannot have more than one zero eigenvalue.

Conditions which are necessary and sufficient for f to belong to H_c^s, can now be derived as a straighforward applications of Proposition 7. Letting $A = \nabla^2 f(x)$, $b = \nabla f(x)$, $b_k = (\nabla f(x))_k$ and $B(x)$, $D_k(x)$, $M_k(x)$ stand for B, D_k and M_k where $\nabla^2 f(x)$ and $\nabla f(x)$ substitute A and b respectively, we obtain

THEOREM 8'. $f \in H_c^s$ *if and only if for every* $x \in C$ *either Condition (III.1) or (III.2) or (III.3) below holds:*

(III.1) $v \neq 0$, $v^T \nabla f(x) = 0$ *implies* $v^T \nabla^2 f(x) v < 0$

(III.2) *For* $k = 1, \ldots, n, (-1)^k D_k(x) \geqslant 0$ *holds, and if* $D_k(x) = 0$ *then* $(-1)^k M_k(x) > 0$ *and* $(\nabla f(x))_k = 0$ *hold.*

(III.3) $\nabla^2 f(x)$ *is negative definite whenever* $\nabla f(x) = 0$, *and* $B(x)$ *is non-singular and has exactly one positive eigenvalue whenever* $\nabla f(x) \neq ($

Note that condition (III.1) is actually the definition of strongly pseudoconcave functions (see Diewert, Avriel and Zang (1981) or Avriel, Diewert, Schaible and Ziemba (1981)). Therefore

COROLLARY 9. $f \in H_c^s$ *if and only if* f *is strongly pseudoconcave.*

Thus, every strongly pseudoconcave function is r(D)-concave. Moreover, it is possible to show that strictly pseudoconcave functions become also r(D)-concave under some mild assumptions (Schaible and Zang (1980)):

COROLLARY 10. $f \in H_c^s$ *if and only if* f *is strictly pseudoconcave and for all* $x \in C$ *and* $v \in R^n$, $v \neq 0$ *the function* $\bar{f}(\theta) = f(x+\theta v)$ *has a local maximum at* $\theta = 0$ *only if* $\bar{f}''(0) < 0$.

In view of the above corollary we have that every strictly pseudoconcave function for which the second derivative of a one dimensional restriction is negative in a local maximum, is r(D)-concave.

Finally we derive an expression of $\rho_0(x)$ for functions in H_c^s. In this case, $\rho_0(x)$ need not be continuous and $H(x;\rho_0(x))$ is not necessarily negative definite but negative semidefinite. However, every $\rho(x) > \rho_0(x)$ will result with a negative definite $H(x;\rho(x))$, and there exists such a function which is also continuous on C. We refer first to (14) and note that for points $x \in C$ where $\nabla f(x) \neq 0$ we must have in view of (III.3) that $q(x) = n$. Thus

$$\rho_0(x) = \frac{s_n(x)}{\nabla f(x)^T \nabla f(x) s^*_{n-1}(x)} . \tag{32}$$

From (15) we obtain substituting $\lambda = 0$ that $s_n(x) = \det(\nabla^2 f(x))$, while (16) implies that $s^*_{n-1}(x) = -\det(B(x))/\nabla f(x)^T \nabla f(x)$. Therefore

$$\begin{aligned} \rho_0(x) &= -\det(\nabla^2 f(x))/\det(B(x)) \\ &= -M_n(x)/D_n(x) \end{aligned} \tag{33}$$

holds. This expression was also obtained by Gerencsér (1973), using a different approach. Moreover, in case $M_n(x) \neq 0$ then Schur's formula (see Gantmacher (1959)) implies that

$$D_n(x) = M_n(x)\det(-\nabla f(x)^T \nabla^2 f(x)^{-1} \nabla f(x)), \tag{34}$$

and thus

$$\rho_0(x) = \begin{cases} \dfrac{1}{\nabla f(x)^T \nabla^2 f(x)^{-1} \nabla f(x)} & \text{if } \nabla f(x) \neq 0,\ M_n(x) \neq 0, \\ 0 & \text{if } \nabla f(x) \neq 0,\ M_n(x) = 0, \\ -\infty & \text{if } \nabla f(x) = 0,\ M_n(x) \neq 0. \end{cases} \tag{35}$$

Note that $\rho_0(x)$ given by this expression is not continuous, in case $\nabla f(x) = 0$ for some $x \in C$, and that $\nabla f(x) \to 0$ implies that $(-1)^n D_n(x)$ is approaching zero. Thus in view of (III.2), $(-1)^n M_n(x)$ approach positively, and $-M_n(x)/D_n(x) = 1/\nabla f(x)^T \nabla^2 f(x)^{-1} \nabla f(x) \to -\infty$. Therefore, $\rho_0(x)$ as given by (35) is bounded from above on every compact subset of C. This agrees with the fact that $f \in H_c^s$, and it is possible to find a continuous function $\rho(x)$, satisfying $\rho(x) > \rho_0(x)$ for every $x \in C$, such that $H(x;\rho(x))$ is negative definite on C.

Expression (35), here obtained as a specialization Fenchel's (1951) formula (14) to H_c^s functions, was obtained by Schaible and Zang (1980) utilizing a different approach. It is possible also to obtain another expression for $\rho_0(x)$, whenever $f \in H_c^s$ by specializing (18). The resulting expression will differ from (18) only by dropping the constraint $\lambda_i(x) < 0$ in the summation. However, (35) seems to be preferable from the computational point of view. We summarize the above discussion in the following (Schaible and Zang (1980)):

PROPOSITION 11. Let $f \in H_c^s$. *Then* f *is* r(D)-*concave on every compact convex subset* $D \subset C$ *for every* r(D) *satisfying*

$$r(D) > \max\{0, \sup_{x \in D}[\rho_0(x)]\}, \tag{36}$$

where $\rho_0(x)$ *is given by (35).*

Note that $\sup\{\rho_0(x): x \in D\}$ may be negative, and thus expression (36) assures that r(D) is nonnegative.

REFERENCES

AVRIEL, M. (1972). "r-Convex functions." *Mathematical Programming 2*, 309-323.

AVRIEL, M. (1976). *Nonlinear Programming Analysis and Methods.* Prentice Hall, New Jersey.

AVRIEL, M., DIEWERT, W.E., SCHAIBLE, S., and ZANG, I. (1981). *Generalized Concave Functions.* In preparation.

AVRIEL, M., DIEWERT, W.E., SCHAIBLE, S., and ZIEMBA, W.T. (1981). "Introduction to concave and generalized concave functions." This volume, 21-50.

AVRIEL, M., and SCHAIBLE, S. (1978). "Second order characterizations of pseudoconvex functions." *Mathematical Programming 14*, 170-185.
AVRIEL, M., and ZANG, I. (1974). "Generalized convex functions with applications to nonlinear programming." Chapter 2 in *Mathematical Programs for Activity Analysis*, (P. Van Moeseke ed.). North-Holland Publishing Co., Amsterdam.
COTTLE, R.W. (1974). "Manifestation of the Schur complement." *Linear Algebra and Applications 8*, 189-211.
CROUZEIX, J.P. (1977). "Contributions à l'étude des fonctions quasi-convexes." Thèse de Doctorat, U.E.R. des Sciences Exactes et Naturelles, Université de Clermont-Ferrand II, France.
CROUZEIX, J.P. and FERLAND, J.A. (1979). "Criteria for quasi-convexity and pseudo-convexity: Relationships and numerical analysis." Université de Montreal, Faculté des Arts et des Sciences, Publication #346.
DEBREU, G. (1952). "Definite and semidefinite quadratic forms." *Econometrica 20*, 295-300.
DEBREU, G. (1976). "Least concave utility functions." *Journal of Mathematical Economics 3*, 121-129.
DE FINETTI, B. (1949). "Sulle stratificazioni convesse." *Annali di Mathematica Pura ed Applicata 30*, 173-183.
DIEWERT, W.E. (1978). "Notes on transconcavity." Unpublished manuscript.
DIEWERT, W.E., AVRIEL, M. and ZANG, I. (1981). "Nine kinds of quasi-concavity and concavity." *Journal of Economic Theory*. To appear.
FENCHEL, W. (1951). *Convex Cones, Sets and Functions*. Mimeographed lectures notes, Dept. of Mathematics, Princeton University, N.J.
FENCHEL, W. (1956). "Über konvexe Funktionen mit vorgeschriebenen Niveaumannigfaltigkeiten." *Mathematische Zeitschrift 63*, 496-506.
FINSLER, P. (1937). "Über das Vorkommen definiter und semidefiniter Formen in Scharen quadratischer Formen." *Commentarii Mathematici Helvetici 9*, 188-192.
GANTMACHER, F.R. (1959). *The Theory of Matrices, Vol. I.* Chelsea Publishing Company, New York, N.Y.
GERENCSÉR, L. (1973). "On a close relation between quasiconvex and convex functions and related investigations." *Mathematische Operationenforschung und Statistik 4*, 201-211.
KANNAI, Y. (1974). "Approximation of convex preferences." *Journal of Mathematical Economics 1*, 101-106.
KANNAI, Y. (1977). "Concavifiability and the constructions of concave utility functions." *Journal of Mathematical Economics 4*, 1-56.
KANNAI, Y. (1980). "The ALEP definition of complementarity and least concave utility functions." *Journal of Economic Theory 22*, 115-117.
KANNAI, Y. (1981). "Concave utility functions - existence, construction and cardinality." This volume, 543-611.
KATZNER, D.W. (1970). *Static Demand Theory*. Macmillan, New York, N.Y.
MANGASARIAN, O.L. (1969). *Nonlinear Programming*, McGraw-Hill Book Co., New York, N.Y.
SCHAIBLE, S. (1971). "Beiträge zur quasikonvexen Programmierung." Doctoral Dissertation, Universitat Köln, Germany.
SCHAIBLE, S. (1977). "Second-order characterizations of pseudoconvex quadratic functions." *Journal of Optimization Theory & Applications 21*, 15-26.
SCHAIBLE, S. (1981). "Quasiconvex, pseudoconvex and strictly pseudoconvex quaudratic functions." *Journal of Optimization Theory & Applications*. To appear.
SCHAIBLE, S. and ZANG, I. (1980). "On the convexifiability of pseudo-convex C^2-functions." *Mathematical Programming 19*, 289-299.

POWER CONVEX FUNCTIONS

P.O. Lindberg

This paper studies functions f convexifiable in the sense that f^p/p (or log f for p = 0) is convex for some real p. Such functions are called power convex or p-th power convex when p is given. It is shown that this essentially is the only exterior convexification that makes the function class closed under positive scalar multiplication. For $p \leq 1$ the p-th power convex functions are further seen to be closed under addition. For general p they are seen to be closed under q-means for $q \geq p$. Products of power convex functions are also studied.

1. INTRODUCTION AND OVERVIEW

This paper studies some special types of F-convex functions, i.e. functions f such that for a given strictly increasing function F on R, F(f) is convex. (Here F(f) denotes the function that takes the value F(f(x)) at x.) F-convex functions were introduced in Avriel and Zang (1974) and we refer to this paper for further results. See also Fenchel (1953), § III.7-8. Well known instances of F-convex functions include log-convex functions and r-convex functions (see e.g. Klinger and Mangasarian (1968) and Avriel (1972)) where F(x) is log x and e^{rx}/r, respectively.

In this paper we will specialize F to be of the form $F_p(x) = x^p/p$ or $F_o(x) = \log x$ for p = 0. According to the above terminology thus say that f is *p-th power convex* if $F_p(f)$ is convex and further that f is *power convex* if it is p-th power convex for some p.

In this paper, we will often use the term p-convex as a shorthand for p-th power convex. This term is ambiguous, however, since if a numerical value is inserted for p, one does not know whether the function is p-convex in the above sense or in the senses e.g. of Avriel (1972) or of Popoviciu as described by Roberts and Varberg (1973).

ISBN 0-12-621120-5

Power convex functions have been studied previously by Rado (1935). He studies among other things the relations between the class of p-convex functions and the class of functions f which satisfy

$$\frac{1}{2h}\left(\int_{-h}^{h} f^{\alpha}(x+\xi)\,d\xi\right)^{1/\alpha} \leq \frac{1}{2}(f(x-h) + f(x+h)).$$

In studying the r-convex functions Avriel (1972; 1973) notes that a function f is r-convex (in this sense) if and only if e^f is r-th power convex. (In his terminology an r-th power convex function is termed ρ-convex (with $\rho=r$) (Avriel (1972)) respectively r^+-convex (Avriel (1973)).) Hence the results obtained for r-convex functions may be translated to power convex functions (although some care might be needed, since for $p > 0$ a p-th power convex function may take the value 0.)

Since power convex functions are a specialization of the F-convex functions of Avriel and Zang (1974), their results may be specialized to our case (although it might often be easier to make a direct derivation). One usually wants a generalization of convexity to retain as many properties as possible of the convex functions. Power convex functions do this to a large extent, as will be seen below.

The paper is structured as follows. In section 2 we give a short background to the present study, leading to the definition of power convex functions. We further characterize p-convex functions for $p \leq 1$ and show that they are closed under addition. In section 3 we show that if one wants the set of F-convex functions to be closed under positive scalar multiplication, then essentially F must be of the form $F(x) = x^p/p$ or $F(x) = \log x$. In section 4 we investigate when power convex functions may be multiplied. We further show that the set of p-convex functions are closed under q-means for $q \geq p$. Finally we show that p-convexity or more generally F-convexity is retained under a number of other operations.

2. POWER CONVEX FUNCTIONS

This paper was inspired by Bector (1968) where it was shown that $f^2(x)/g(x)$ is convex when f is convex nonnegative and g concave and positive. By straightforward analysis, this result easily is extended to f^α/g^β for $\alpha \geq \beta + 1$. The function f^2/g is similar to x^2/y which is known to be convex (e.g. Stoer and Witzgall (1970); 137) and obviously is increasing in x and decreasing in y for $x \geq 0$ and $y > 0$. This hints that one could generalize and use the well-known results that if f is convex and nondecreasing on R and g is convex, then f(g) is convex:

PROPOSITION 1: Let X, Y, Z be partially ordered convex sets. Suppose $f: X \times Y \to Z$ *is convex (quasiconvex) in the natural definition for the given orderings, nondecreasing on* X *and nonincreasing on* Y. *Further assume that* $g_1: X_1 \to X$ *is convex and* $g_2: X_2 \to Y$ *is concave. Then* $f(g_1, g_2)$ *is convex (quasiconvex) on* $X_1 \times X_2$.

PROOF: Elementary checks of the definitions using the (quasi) convexity (concavity) and the orderings. □

Note that the sets X, Y and Z need not be included in some linear space. It is enough that convex combinations are defined. (X could e.g. be the convex subsets of R^m ordered by inclusion.) Proposition 1 could be applied to $f(x,y) = x^\alpha y^\beta$ on R^2_{++} (where $R_{++} = \{x \in R | x > 0\}$) using

LEMMA 1: $h(x,y) = x^\alpha y^\beta$ *defined on* R^2_{++} *is*

(i) *convex if* $\alpha + \beta \geq 1$ *and* $\alpha \leq 0$ *or* $\beta \leq 0$ *or if* $\alpha \leq 0$ *and* $\beta \leq 0$

(ii) *concave if* $\alpha \geq 0$, $\beta \geq 0$ *and* $\alpha + \beta \leq 1$

(iii) *quasiconvex if* $0 \leq \alpha + \beta \leq 1$ *and* $\alpha \leq 0$ *or* $\beta \leq 0$

(iv) *quasiconcave if* $\alpha + \beta \leq 0$ *and* $\alpha \geq 0$ *or* $\beta \geq 0$ *or if* $\alpha \geq 0$ *and* $\beta \geq 0$ *and* $\alpha + \beta \geq 1$.

When α *or* $\beta \geq 0$ *we may allow* x *or* y *to be zero also.*

PROOF: This result is in fact a specialization of propositions 3.2 - 3.3 in Schaible (1972). It also follows easily by checking positive/negative definiteness of the Hessian respectively the convexity of the upper or lower level sets. □

Combining with Proposition 1 we get

COROLLARY 1: Suppose f *is a nonnegative convex function on some space* X *and that* g_1 *and* g_2 *are positive concave functions on* X *and* Y *respectively,* X *and* Y *being convex. Then, on* X × Y

(a) $(g_1(x))^\alpha (g_2(y))^\beta$ *is*

(i) *convex for* $\alpha, \beta \leq 0$

(ii) *concave for* $\begin{cases} \alpha + \beta \leq 1 \\ \alpha, \beta \geq 0 \end{cases}$

(iii) *quasiconcave for* $\alpha, \beta \geq 0, \quad \alpha + \beta \geq 1$

(b) $(f(x))^\alpha (g_2(y))^\beta$ *is*

(i) *convex for* $\begin{cases} \alpha + \beta \geq 1 \\ \beta \leq 0 \end{cases}$

(ii) *quasiconvex for* $\begin{cases} 0 \leq \alpha + \beta \leq 1 \\ \beta \leq 0 \end{cases}$

(iii) *quasiconcave for* $\begin{cases} \alpha + \beta \leq 0 \\ \alpha \geq 0. \end{cases}$

This result is very similar to the theorem and corollary in Schaible (1972).† Here the functions are defined on different (not necessarily linear) spaces (which is needed in this paper) whereas Schaible allows products of powers of several functions. In the case of linear spaces, our result follows rather easily from that of Schaible. Corollary 1 shows that if for a function f on some set X, $f^{\frac{1}{2}}$ is convex then f(x)/y is convex on X × R_{++}. This condition also is easily seen to be sufficient for $f^{\frac{1}{2}}$ to be convex.

Trying to generalize this result it is natural to introduce

---

†This reference was pointed out to me by Prof. Schaible at the conference.

DEFINITION 1: Let f *be a positive realvalued function on some convex set* X:

f *is said to be p-th power convex if* f^p/p *is convex for* $p \neq 0$ *or if* log f *is convex for* $p = 0$; *and*

f *is called power convex if it is p-th power convex for some* p. *(For* $p > 0$ *we may allow* f *to be zero.)*

Obviously all the power convex functions are quasiconvex. That log-convexity corresponds to 0-convexity may be seen from that $(x^p - 1)/p \to \log x$ as $p \to 0$, and we could of course have used $(x^p - 1)/p$ instead of x^p/p to define p-convexity. For twice differentiable power convex functions on R, differentiating twice in the definition gives

PROPOSITION 2: A twice differentiable positive function on R *is p-th power convex if and only if* $(p - 1)(f')^2 + ff'' \geq 0$.

By the correspondence between Avriel's r-convexity and r-th power convexity we get

PROPOSITION 3: If f *is p-th power convex then it is q-th power convex for* $q \geq p$.

PROOF: Essentially a translation of Lemma 3.1 in Avriel (1972). □

Also in accordance with Avriel (1972) a nonnegative function f on a convex set X is p-th power convex if and only if for $x_1, x_2 \in X$ and $\lambda \in (0,1)$

$$f(\lambda x_1 + (1 - \lambda)x_2) \leq M_p(f(x_1), f(x_2); \lambda) \qquad (1)$$

where

$$M_p(x,y;\lambda) = (\lambda x^p + (1 - \lambda)y^p)^{1/p}$$

is the weighted p-mean of x and y.

From the properties of M_p, Avriel (1972) deduces that the natural extensions of the r-convex functions by (1) to $r = \pm \infty$ are, respectively, the quasiconvex functions and functions f satisfying

$$f(\lambda x + (1 - \lambda)y \leq \min (f(x_1), f(x_2)).$$

The latter functions are easily seen to be convex functions, constant on any relatively open subset of X. For power convex functions we have the corresponding result, only that we get nonnegative quasiconvex functions respectively positive convex functions constant on relatively open subsets of X.

Now to the characterization of p-convex functions for $p \leq 1$.

THEOREM 1: Let f *be a positive function on some convex set* X. *Then, for* $p \leq 1$, f *is p-th power convex if and only if*

$$f(x)y^{1/q} \text{ is convex on } X \times R_{++}$$

where q *is conjugate to* p, *i.e.* $1/p + 1/q = 1$. *(*$y^{1/q}$ *has to be interpreted as* e^y *for* $p = 0$ *and as* 1 *for* $p = 1$.*)*

PROOF: The case with $p = 1$ is the usual convex case. Thus we may assume $p < 1$. The necessity is straightforward.

For $p \neq 0$, $f = g^{1/p}$ with g convex/concave depending on the sign of p. Thus $f(x)y^{1/q} = (g(x))^{1/p}y^{1/q} = (g(x))^{\alpha}y^{\beta}$ with $\alpha + \beta = 1$.

Since $p \leq 1$ we have $\alpha \geq 1$ or $\alpha \leq 0$, depending on sign p, implying that b(i) of corollary 3 is applicable.

For $p = 0$, $f = e^g$ with g convex and $f(x)y^{1/q} = e^{g(x)+y}$ which obviously is convex.

To prove sufficiency for $p \neq 0$, assume f is not p-convex, i.e. that there are $\lambda \in (0,1)$ and x_1, $x_2 \in X$ such that

$$f(\lambda x_1 + (1-\lambda)\, x_2)^p/p > \lambda f^p(x_1)/p + (1-\lambda) f^p(x_2)/p.$$

Then for $f(x)y^{1/q}$ we get (with y_i to be specified)

$$f(\lambda x_1 + (1-\lambda)x_2)(\lambda y_1 + (1-\lambda)y_2)^{1/q} = [p(f^p(\lambda x_1 + (1-\lambda)x_2)/p)]^{1/p}$$
$$\times\ (\lambda y_1 + (1-\lambda)y_2)^{1/q} > (\lambda f^p(x_1) + (1-\lambda)f^p(x_2))^{1/p}(\lambda y_1 + (1-\lambda)y_2)^{1/q}$$
$$= \lambda f^p(x_1) + (1-\lambda)f^p(x_2) = \lambda f(x_1)y_1^{1/q} + (1-\lambda)f(x_2)y_2^{1/q} ,$$

with $y_i = f^p(x_i)$ and $1/q = (p-1)/p$.

For $p = 0$ the sufficiency follows in the following similar way: If f is not log-convex then there are x_1, x_2 and λ such that

$$\log f(\lambda x_1 + (1-\lambda)x_2) > \lambda \log f(x_1) + (1-\lambda)\log f(x_2).$$

Thus for $f(x)y^{1/q} = f(x)e^y$

$$f(\lambda x_1 + (1-\lambda)x_2)e^{\lambda y_1+(1-\lambda)y_2} = e^{\log f(\lambda x_1+(1-\lambda)x_2)+\lambda y_1+(1-\lambda)y_2}$$
$$> e^{\lambda \log f(x_1)+(1-\lambda)\log f(x_2)+\lambda y_1+(1-\lambda)y_2} = e^a = \lambda e^a+(1-\lambda)e^a$$
$$= \lambda f(x_1)e^{y_1} + (1-\lambda)f(x_2)e^{y_2},$$

with $y_i > 0$ such that $\log f(x_1) + y_1 = \log f(x_2) + y_2 = a$.

From the theorem we see that p-convex functions are convex for $p \le 1$. For $p > 1$ they need not be convex since $x^{1/p}$ is concave and p-convex for $x \ge 0$.

COROLLARY 2: The set of p-th power convex functions (on a given space) is closed under addition if $p \le 1$.

For $p = 0$ and 1 we have the classical results that log convex and convex functions are closed under addition. For $p > 1$, the result is not true as may be seen from trivial counter-examples.

3. F-CONVEX FUNCTIONS CLOSED UNDER POSITIVE SCALAR MULTIPLICATION

By the definition of the p-convex functions it is obvious that they are closed under positive scalar multiplication. In this section we show that these essentially are the only possible instances of F-convex functions with this property. If we want the F-convex functions to be closed under positive scalar multiplication, then the domain of F obviously must be a cone. The most demanding case for $F(\lambda f)$ to be convex when $F(f)$ is convex obviously is when f is the inverse of F.

LEMMA 2: Suppose that F *is a strictly increasing function defined on a subcone of* R *and that* f *is the inverse of* F. *If* $F(\lambda f)$ *is convex for all* $\lambda > 0$ *then in fact* $F(\lambda f)$ *is affine.*

PROOF: The general idea is that if $F(\lambda f)$ is strictly convex at some point x_1 for some $\lambda > 0$, then $F(f/\lambda)$ should be strictly concave at the point x_λ where $f(x_\lambda) = \lambda f(x_1)$. We of course suppose that $\lambda f(x_1)$ is in the interior of the domain of F. Let $\delta_1(x) = f(x_1 + x) - f(x_1)$, where δ_1 is increasing and thus has an inverse δ_1^{-1}. Also let $\delta_\lambda(x) = f(x_\lambda + x) - f(x_\lambda)$ with inverse δ_λ^{-1}. We have

$$F(\lambda f(x_1 + x)) = F(\lambda f(x_1) + \lambda\delta_1(x)) = F(f(x_\lambda) + \lambda\delta_1(x)) =$$

$$= F(f(x_\lambda + \delta_\lambda^{-1}(\lambda\delta_1(x)))) = x_\lambda + \delta_\lambda^{-1}(\lambda\delta_1(x)).$$

By hypothesis $\phi(x) = \delta_\lambda^{-1}(\lambda\delta_1(x))$ is convex. Further

$$F(f(x_\lambda + x)/\lambda) = F((f(x_\lambda) + \delta_\lambda(x))/\lambda) = F(f(x_1) + \sigma_\lambda(x)/\lambda) =$$

$$= F(f(x_1 + \delta_1^{-1}(\delta_\lambda(x)/\lambda))) = x_1 + \delta_1^{-1}(\delta_\lambda(x)/\lambda).$$

But $\delta_1^{-1}(\delta_\lambda(x)/\lambda)$ is the inverse ϕ^{-1} of ϕ and thus concave if ϕ is convex. For both to be convex, they must be affine. □

THEOREM 2: If the class of F*-convex functions is closed under positive scalar multiplication then* F *has one of the following forms (modulo affine transformations):*

(i) $F(x) = sign(x)\cdot|x|^p/p$ *for* $x > 0$ *or* $x < 0$; *and*

(ii) $F(x) = sign(x)\cdot\log|x|$ *for* $x > 0$ *or* $x < 0$.

For positive powers in (i) curves with different p's *for* $x > 0$ *and* $x < 0$ *may be patched together.*

PROOF: By lemma 3, $F(\lambda f(x))$ is affine. Thus $F(\lambda f(x)) = c(\lambda)x + d(\lambda)$ for some functions $c(\lambda)$ and $d(\lambda)$, or $\lambda f(x) = f(c(\lambda)x+d(\lambda))$. From this follows that $c(\lambda)$ cannot be 0 for any λ. Further c and d must be continuous, since

$$d(\lambda) = F(\lambda f(x)) - c(\lambda)x \quad \text{and} \quad c(\lambda) = (F(\lambda f(x_1)) - F(\lambda f(x_2)))/(x_1-x_2)$$

for any different x_1 and x_2 in the domain of f. Now

$$\lambda_1\lambda_2 f(x) = \lambda_1 f(c(\lambda_2)x+d(\lambda_2)) = f(c(\lambda_1)c(\lambda_2)x+c(\lambda_1)d(\lambda_2) + d(\lambda_1)).$$

But also

$$\lambda_1\lambda_2 f(x) = f(c(\lambda_1\lambda_2)x + d(\lambda_1\lambda_2)).$$

Thus, by varying x,

(i) $c(\lambda_1\lambda_2) = c(\lambda_1)c(\lambda_2)$ and

(ii) $d(\lambda_1\lambda_2) = c(\lambda_1)d(\lambda_2) + d(\lambda_1)$.

By Aczel (1969), (i) implies that c is identically zero or $c(\lambda) = \lambda^k$ for some real k. Since c cannot be zero $c(\lambda) = \lambda^k$. Then in (ii) we have two different cases:

(a) $k = 0$. Then $d(\lambda_1\lambda_2) = d(\lambda_1) + d(\lambda_2)$. Substituting $\lambda = e^t$ we get the Cauchy equation whose only continuous solution is linear (Aczel (1969)). Thus $d(\lambda) = d_o \log \lambda$.

(b) $k \neq 0$. Then

$$d(\lambda_1\lambda_2) = \lambda_1^k d(\lambda_2) + d(\lambda_1). \quad \text{Also}$$

$$d(\lambda_1\lambda_2) = \lambda_2^k d(\lambda_1) + d(\lambda_2). \quad \text{Thus}$$

$$(\lambda_1^k - 1)d(\lambda_2) = (\lambda_2^k - 1)d(\lambda_1) \quad \text{or}$$

$$\frac{d(\lambda_2)}{\lambda_2^k - 1} = \frac{d(\lambda_1)}{\lambda_1^k - 1} = \text{const.} = d_o. \quad \text{Thus}$$

$d(\lambda) = d_o(\lambda^k - 1)$ which fits into (ii).

Hence, for $k = 0$, $\lambda f(x) = f(x + d_o \log \lambda)$. Now suppose that $f(x_o) = f_o \neq 0$. Then when y is of the same sign as f_o,

$$F(y) = F(\frac{y}{f_o} \cdot f(x_o)) = F(f(x_o + d_o \log \frac{y}{f_o})) = x_o + d_o \log \frac{y}{f_o}.$$

Since F is increasing, we must have $d_o > 0$ for $y > 0$ and $d_o < 0$ for $y < 0$. Thus, up to affine transformations, F is of the form

$$F(y) = \text{sign}(y) \cdot \log |y|.$$

Of course, only one of the "branches", $y > 0$ or $y < 0$, can be valid in each case.

For $k \neq 0$.

$$f(x) = f(\lambda^k x + d_o(\lambda^k - 1)) = f(\lambda^k(x + d_o) - d_o).$$

Again suppose that $f(x_o) = f_o \neq 0$. Then, when y is of the same sign as f_o,

$$F(y) = F(\frac{y}{f_o} f(x_o)) = F(f((\frac{y}{f_o})^k(x_o + d_o))) = (\frac{y}{f_o})^k(x + d_o) - d_o =$$

$$= |y|^k c_o - d_o .$$

Since F is increasing, it is of the form $F(y) = \text{sign}(y)|y|^k/k$ (up to affine transformations).

Here, for $k < 0$, only one of the branches, $y > 0$ or $y < 0$, is valid for each case.

For k's greater than 0, a branch for $y > 0$ may be fit to a branch for $y < 0$. □

The cases with negative arguments are not studied in this paper.

4. OPERATIONS ON POWER CONVEX FUNCTIONS

It is well known that 0-convexity (i.e. log-convexity) is retained under multiplication. What may be said of products of power convex functions? Suppose f and g are p- and q-convex on some spaces X and Y respectively. Now for $p,q \neq 0$

$$(f(x)\ g(y))^r = (f^p(x))^{r/p}(g^q(x))^{r/q} = (f^p(x))^{\alpha}(g^q(x))^{\beta}.$$

Thus, corollary 1 may be applied, and we get

THEOREM 3: Let f *and* g *be p-th and q-th power convex on some spaces* X *and* Y *respectively. If* $p + q < 0$ *then* f g *is r-th power convex on* $X \times Y$ *where* $1/r = 1/p + 1/q$. *(For* p *or* $q = 0$, r *is* 0.*)*

PROOF: The cases with $p,q \neq 0$ follow by elementary but somewhat tedious checking using corollary 1. If p and q are not nonzero we may without loss of generality assume $p = 0$ and hence $q < 0$. Then f and g are both log-convex and so is their product. □

Thus power convex functions can be added and multiplied. This suggests other combinations of power convex functions. From the definition of p-convexity it follows immediately that p-means of p-convex functions are p-convex. How about q-means for $q \geq p$?

LEMMA 3: $M_p(x,y;\lambda)$ *is componentwise increasing and convex (concave) if* $p \geq 1$ $(p \leq 1)$.

PROOF: The monotonicity is direct. Since M_p is positively homogenous, it is enough to study $M_p(x,1;\lambda)$ to prove the convexity/concavity. But for $p \neq 0$

$$M_p(x,1;\lambda) = (1-\lambda)^{1/p} f_p\left(\left(\frac{\lambda}{1-\lambda}\right)^{1/p} x\right) \quad \text{where}$$
$$f_p(x) = (1 + x^p)^{1/p}.$$

Further, a direct computation shows that

$$f_p''(x) = (p-1)(1 + x^p)^{1/p-2} x^{p-2}$$

which is nonnegative or nonpositive according to if $p \geq 1$ or $p \leq 1$. Further, $M_o(x,1;\lambda) = x^\lambda$ which obviously is concave. □

THEOREM 4: If f_1 *and* f_2 *are p-th power convex on* X *and* Y *respectively then* $M_q(f_1,f_2;\lambda)$ *is p-th power convex on* $X \times Y$ *for* $q \geq p$.

PROOF:

(1) Suppose $p > 0$. Then the f_i^p are convex and $q/p \geq 1$. Thus

$$(M_q(f_1,f_2;\lambda))^p = (\lambda f_1^q + (1-\lambda) f_2^q)^{p/q} = (\lambda (f_1^p)^{q/p} + (1-\lambda)(f_2^p)^{q/p})^{p/q} =$$
$$= M_{q/p}(f_1^p, f_2^p;\lambda)$$

which is convex by lemma 3 and proposition 1. Hence $M_q(f_1,f_2;\lambda)$ is p-convex.

(2) Suppose $p = 0$. We may then assume $q > 0$. Take an arbitrary $r \in (0,q)$. By (1), $M_q(f_1,f_2;\lambda)$ is r-convex and by a limit argument it is seen to be 0-convex.

(3) Suppose $p < 0$. First let $q \neq 0$. Then f_i^p are concave and $q/p \leq 1$. Thus

$$(M_q(f_1,f_2;\lambda))^p = M_{q/p}(f_1^p,f_2^p;\lambda)$$

which is concave by lemma 3 and the concave version of proposition 1. The case $q = 0$ now follows by a limit argument since $M_q(x,y;\lambda) \to M_o(x,y;\lambda)$ as $q \to 0$ (Hardy, Littlewood and Polya (1934); 15). □

We will now add some simple results which are valid already for F-convex functions and hence for power convex functions. The proofs are so simple that they will not be given.

THEOREM 5: If $f\colon X \to R$ *is* F-*convex then* $f(Ax)$ *is* F-*convex for any affine transformation* $A\colon Y \to X$.

THEOREM 6: If $f\colon X \to R$ *is* F-*convex then* $\bar{f}(y) = \inf\{f(x) \mid Ax = y\}$ *is* F-*convex for any affine transformation* $A\colon X \to Y$.

COROLLARY 3: For $p \leq 1$ *the set of p-th power convex functions is closed under infimal convolution.*

THEOREM 7:[1] *Let* X *and* Y *be partially ordered convex sets in any spaces. Suppose* $f\colon Y \to R$ *is* F-*convex and nondecreasing and that* $g\colon X \to Y$ *is convex with respect to the given ordering on* Y.
Then $f(g)$ *is* F-*convex.*

---

[1] A similar statement appears in Ben-Tal (1977), but with a typographical error; in Table 2 (result E2) there, the function g should be $(\phi-\phi)$ convex and not H-ϕ convex.

REFERENCES

ACZÉL, J. (1969). *On Applications and Theory of Functional Equations.* Birkhäuser Verlag, Basel.

AVRIEL, M. (1972). "r-Convex Functions." *Math. Programming 2,* 309-323.

AVRIEL, M. (1973). "Solution of Certain Nonlinear Programs Involving r-Convex Functions." *J. Optimization Theory Appl. 11,* 159-174.

AVRIEL, M., and ZANG, I. (1974). "Generalized Convex Functions with Applications to Nonlinear Programming." In *Mathematical Programs for Activity Analysis* (P. van Moeseke, ed.). North-Holland, Amsterdam.

BECTOR, C.R. (1968). "Programming Problems with Convex Fractional Functions." *Operations Research 16,* 629-635.

BEN-TAL, A. (1977). "On Generalized Means and Generalized Convex Functions." *J. Optimization Theory Appl. 21,* 1-13.

FENCHEL, W. (1953). "Convex Cones, Sets, and Functions." Mimeographed lecture notes. Department of Mathematics, Princeton University, Princeton.

HARDY, G.H., LITTLEWOOD, J.E., and POLYA, G. (1934). *Inequalities.* Cambridge University Press, Cambridge.

KLINGER, A., and MANGASARIAN, O.L. (1968). "Logarithmic Convexity and Geometric Programming." *J. Math. Anal. Appl. 24,* 388-408.

RADO, T. (1935). "On Convex Functions." *Trans. Amer. Math. Soc. 37,* 266-285.

ROBERTS, A.W., and VARBERG, D.E. (1973). *Convex Functions.* Academic Press, New York.

SCHAIBLE, S. (1972). "Quasi-Convex Optimization in General Real Linear Spaces." *Zeitschrift für Operations Research 16,* 205-213.

STOER, J., and WITZGALL, C. (1970). *Convexity and Optimization in Finite Dimensions I.* Springer-Verlag, Berlin-Heidelberg.

PART II

GENERALIZED CONCAVE QUADRATIC FUNCTIONS AND C^2-FUNCTIONS

QUASICONVEXITY AND PSEUDOCONVEXITY OF

FUNCTIONS ON THE NONNEGATIVE ORTHANT

Jacques A. Ferland

This paper is a survey of the earlier efforts to give criteria for the quasi-convexity and pseudo-convexity of functions on the non-negative orthant. General twice continuously differentiable functions are first analysed. Then criteria for quadratic forms and quadratic functions are specified in terms of the positive subdefinite matrices.

1. INTRODUCTION

This paper is a survey of the earlier efforts to give criteria for the quasi-convexity and pseudo-convexity of functions on the non-negative orthant. The pioneers in this area where Arrow and Enthoven (1961) who studied twice continuously differentiable functions quasi-convex on E^n_+. Then came the paper by Martos (1969) introducing the concept of positive subdefinite matrices and giving criteria for the quasi-convexity and pseudo-convexity of quadratic forms on E^n_+ in terms of this concept. Martos extended his results to quadratic functions in Martos (1971). Cottle and Ferland (1972) have given more characterizations of positive subdefinite matrices, and they gave further results on quadratic functions in their papers in 1971 and 1972. More recently Schaible (1979) obtained the criteria by Cottle, Ferland, and Martos as well as a number of other criteria for quasi-convex and pseudo-convex quadratic functions by specializing results of Schaible (1971 ; 1972).

The first section of this paper analyses twice continuously differentiable functions. It includes the results of Arrow and Enthoven (1961). The results are given for twice continuously differentiable functions on convex sets with nonempty interior (as in Ferland (1971; 1972)) because they are as easy to prove as on E^n_+.

ISBN 0-12-621120-5

Criteria for the quasi-convexity and pseudo-convexity of quadratic forms on E^n_+ are given in terms of the positive subdefiniteness of the matrices defining these forms. It is essential to study and characterize first this property. Section 3 of the paper is devoted to this task.

In the first part of Section 4 we introduce the criteria for the quasi-convexity and pseudo-convexity of quadratic forms on the non-negative orthant. Then these criteria are extended to quadratic functions.

To be more concise, the proofs are omitted, but the references where they are found are mentioned.

If f, a real-valued function defined on a subset of E^n, is differentiable at x, then $\nabla f(x)$ (the *gradient* of f at x) denotes the column vector of partial derivatives of f at x; i.e.

$$\nabla f(x)^T = \left[\frac{\partial f(x)}{\partial x_1}, \frac{\partial f(x)}{\partial x_2}, \ldots, \frac{\partial f(x)}{\partial x_n}\right]$$

2. TWICE CONTINUOUSLY DIFFERENTIABLE FUNCTIONS

Necessary conditions and sufficient conditions for the quasi-convexity of twice continuously differentiable functions have been given for the first time by Arrow and Enthoven (1961). These results are for functions restricted to the non-negative orthant $E^n_+ = \{x \in E^n : x_j \geqslant 0 \ , \ 1 \leqslant j \leqslant n\}$. Since the proofs are not more difficult for functions having domains with nonempty interior, this case will be analysed in this paper. These results are taken from Ferland (1971 ; 1972).

The conditions are given in terms of properties of the bordered Hessian. Let f be a twice continuously differentiable function at $x \in E^n$. The bordered Hessian of f at $x \in S$ is defined by the matrix

$$D_n(x) = \begin{bmatrix} 0 & \frac{\partial f(x)}{\partial x_1} & \frac{\partial f(x)}{\partial x_2}, \ldots, & \frac{\partial f(x)}{\partial x_n} \\ \frac{\partial f(x)}{\partial x_1} & \frac{\partial^2 f(x)}{\partial x_1^2} & \frac{\partial^2 f(x)}{\partial x_1 \partial x_2}, \ldots, & \frac{\partial^2 f(x)}{\partial x_1 \partial x_n} \\ \frac{\partial f(x)}{\partial x_2} & \frac{\partial^2 f(x)}{\partial x_2 \partial x_1} & \frac{\partial^2 f(x)}{\partial x_2^2}, \ldots, & \frac{\partial^2 f(x)}{\partial x_2 \partial x_n} \\ \vdots & \vdots & \vdots & \vdots \\ \frac{\partial f(x)}{\partial x_n} & \frac{\partial^2 f(x)}{\partial x_n \partial x_1} & \frac{\partial^2 f(x)}{\partial x_n \partial x_2}, \ldots, & \frac{\partial^2 f(x)}{\partial x_n^2} \end{bmatrix}$$

The submatrices $D_r(x)$ of the bordered Hessian are defined for each r, $1 \leqslant r \leqslant n$ by deleting from $D_n(x)$ rows and columns $r+1, r+2, \ldots, n$.

The next result gives sufficient conditions for the pseudo-convexity of f on a convex set $X \subset E^n$, and the proof relies on second order optimality conditions in mathematical programming.

THEOREM 1: Let $X \subset E^n$ *be convex and* f *be twice continuously differentiable on* X. *If* $\det D_r(x) < 0$ *for* $1 \leqslant r \leqslant n$ *and for all* $x \in X$, *then* f *is pseudo-convex on* X.

PROOF : Theorem 5.19 in Ferland (1971).

Since a pseudo-convex function on a convex set is quasi-convex (see Property 2 in Mangasarian (1965)), Theorem 1 gives also sufficient conditions for quasi-convexity.

Necessary conditions are now given. The proof of this theorem requires the introduction of first order optimality conditions for quasi-convex programming problem.

Consider the following mathematical programming problem

$$\begin{aligned} &\text{Min } f(x) \\ &\text{Subject to } g_i(x) \leqslant 0 \quad , \quad 1 \leqslant i \leqslant m \qquad (1) \\ &\qquad x \in C \subset E^n. \end{aligned}$$

The Kuhn-Tucker optimality conditions for this problem are as follows:

$$\begin{aligned} &\nabla f(x) + \sum_{i=1}^{m} \lambda_i \nabla g_i(x) = 0 \\ &\sum_{i=1}^{m} \lambda_i g_i(x) = 0 \\ &g_i(x) \leqslant 0 \quad , \quad 1 \leqslant i \leqslant m \\ &\lambda_i \geqslant 0 \quad , \quad 1 \leqslant i \leqslant m \\ &x \in C \ . \end{aligned} \qquad (2)$$

LEMMA 2: Let f, $g_1,\ldots,g_m$ *be differentiable functions on* E^n. *Suppose that* $g_1, g_2,\ldots,g_m$ *are quasi-convex on the convex set* $C \subset E^n$. *If* $x^o \in C$ *and* $\lambda^o \in E^n$ *satisfy conditions (2), then* $\nabla f(x^o)^T(x-x^o) \geqslant 0$ *for all* x *in the constraint set of problem (1).*

PROOF: Lemma 3 in Ferland (1972).

LEMMA 3: Let f, $g_1,\ldots,g_m$ *be differentiable functions on* E^n. *Suppose that* $g_1, g_2,\ldots,g_m$ *are quasi-convex on the convex set* C *and that* f *is quasi-convex on a convex set* S *having a nonempty interior such that* $C \subset S$. *Let* $x^o \in C$ *and* $\lambda^o \in E^m$ *satisfy (2). If*

(i) there is an $x^1 \in S$ *such that* $\nabla f(x^o)^T(x^1-x^o) > 0$, *or*

(ii) $\nabla f(x^o) \neq 0$ *and* f *is twice continuously differentiable on* S, *then* x^o *is a solution of (1).*

PROOF: Theorem 4 in Ferland (1972).

This result is used to prove the necessity of the following conditions

THEOREM 4: Let $S \in E^n$ *be a convex set with nonempty interior. If* f *is twice continuously differentiable and quasi-convex on* S, *then*
$\det D_r(x) \leqslant 0$ *for* $1 \leqslant r \leqslant n$ *and for all* $x \in S$.

PROOF: Theorem 5.23 in Ferland (1971).

Notice the gap existing between the necessary conditions and the sufficient conditions. In Section 4.1 counter examples (using quadratic forms) will be given to show that the necessary conditions are not sufficient and that the sufficient conditions are not necessary.

This section concludes with a result on the strong connection between quasi-convexity and pseudo-convexity.

THEOREM 5: Let $S \in E^n$ *be a convex set with nonempty interior and* f *be twice continuously differentiable on* S. *If* f *is quasi-convex on* S, *then* f *is pseudo-convex at any point* $x^1 \in S$ *where* $\nabla f(x^1) \neq 0$.

PROOF: Theorem 12 in Ferland (1972).

3. SUBDEFINITE MATRICES

In 1969 Martos introduced the concept of subdefinite matrices to study the quasi-convexity and the pseudo-convexity of quadratic forms on the non-negative orthant. The relation between these concepts is similar to the relation between semi-definite matrices and convex quadratic forms.

To establish the relation between subdefinite matrices and quasi-convexity and pseudo-convexity it is necessary to analyse and characterize first these matrices. This is the purpose of this section. Section 4 will be devoted to the relation with quasi-convexity and pseudo-convexity. In this section only the relevant properties of subdefinite matrices are summarized. Other properties of these are given in Cottle and Ferland (1971; 1972), Ferland (1980), and Martos (1969).

Let D be a real symmetric matrix of order n and $\psi(x)=x^TDx$ be the *quadratic form* associated with it.

DEFINITION 1: (Martos (1969)). The real symmetric matrix D *of order* n *is positive subdefinite if for all* $x \in E^n$

$$x^TDx < 0 \text{ implies } Dx \geqslant 0 \text{ or } Dx \leqslant 0.$$

DEFINITION 2: (Martos (1969)). The real symmetric matrix D *of order* n *is strictly positive subdefinite if for all* $x \in E^n$

$$x^T Dx < 0 \text{ implies } Dx > 0 \text{ or } Dx < 0.$$

It is interesting to notice the parallel with positive semi-definite matrices having the property that for all $x \in E^n$

$$x^T Dx \leq 0 \text{ implies } Dx = 0.$$

It is evident that positive semi-definite matrices are strictly positive subdefinite (by default), and strictly positive subdefinite matrices are positive subdefinite. Thus, in order to exclude the positive semi-definite matrices, Martos inserts the word "merely" before "positive sub-definite".

Some properties of subdefinite matrices are first introduced. These preliminary results will allow us to define criteria for subdefiniteness. The first two properties are a direct consequence of Definition 1 and 2.

LEMMA 6: Let D *be positive subdefinite. Then*

$$x \geqslant 0, \text{ and } x^T Dx < 0 \text{ imply } Dx \leqslant 0.$$

PROOF: Lemma 2 in Martos (1969).

LEMMA 7: Let D *be strictly positive subdefinite. Then*

$$x \geqslant 0, \text{ and } x^T Dx < 0 \text{ imply } Dx < 0.$$

PROOF: Lemma 3 in Martos (1969).

The following lemmas are used to prove a criterion for mere positive subdefiniteness.

LEMMA 8: Let D *be merely positive subdefinite. Then*

(i) D *has at most one negative eigenvalue;*

(ii) any eigenvector y $(y \neq 0)$ *associated with a negative eigenvalue has the property* $y \geqslant 0$ *or* $y \leqslant 0$.

PROOF: Lemma 4 in Martos (1969).

Let $\bar{D}$ be the following real symmetrix matrix of order (n+1)

$$\bar{D} = \begin{bmatrix} \lambda & \lambda y^T \\ \lambda y & D \end{bmatrix} = \begin{bmatrix} y^T D y & y^T D \\ Dy & D \end{bmatrix}$$

where λ is a negative eigenvalue with an associated normalized eigenvector $y \geqslant 0$.

LEMMA 9: If D *is merely positive subdefinite, then so is* $\bar{D}$.

PROOF: Lemma 5 in Martos (1969).

LEMMA 10: Let D *be merely positive subdefinite. If* D *has a negative diagonal entry, then* $D \leqslant 0$ (i.e. $d_{ij} \leqslant 0$, $1 \leqslant i \leqslant n$, $1 \leqslant j \leqslant n$).

PROOF: Lemma 6 in Martos (1969).

Now a first criterion for mere positive subdefiniteness can be introduced. Notice that the necessity of the condition has been shown by Martos in 1969 and that the sufficiency has been established by Cottle and Ferland in 1972.

THEOREM 11: The matrix D *is merely positive subdefinite if and only if*

(i) $D \leqslant 0$, $D \neq 0$;

(ii) D *has exactly one (simple) negative eigenvalue.*

PROOF: Theorem 11 in Martos (1969) and Theorem 4.1 in Cottle and Ferland (1972).

Given the non-negativity property (i), one can replace condition (ii) on the eigenvalues of D by an equivalent condition on the principal minors of D.

THEOREM 12: The matrix D *is merely positive subdefinite if and only if*

(i) $D \leqslant 0$, $D \neq 0$

(ii) D *has nonpositive principal minors.*

PROOF: Theorem 4.2 in Cottle and Ferland (1972).

Now let us characterize the strict positive subdefiniteness.

THEOREM 13: Let D *be merely positive subdefinite.* D *is strictly merely positive subdefinite if and only if it does not contain any row of zeros.*

PROOF: Theorem 2 in Martos (1969).

It is possible to introduce sufficient conditions for strict mere positive subdefiniteness in terms of the leading principal minors of the matrix. This test is more efficient than the results in Theorem 12 and stress an obvious analogy with a test for positive definiteness.

THEOREM 14: If D *is a real symmetric matrix such that* $D \leqslant 0$, $D \neq 0$ *with negative leading principal minors, then* D *is strictly merely positive subdefinite.*

PROOF: Theorem 4.3 in Cottle and Ferland (1972).

4. QUADRATIC FORMS AND QUADRATIC FUNCTIONS

The first part of this section is devoted to Martos' results in 1969 on quasi-convex and pseudo-convex quadratic forms. Then these results were extended to quadratic functions by Martos (1971) and Cottle and Ferland (1971) and (1972).

4.1 Quadratic Forms on E^n_+

The quasi-convexity and pseudo-convexity of quadratic forms on E^n_+ are defined in terms of the positive subdefiniteness of the matrices associated with these forms.

Recall that a quadratic form $\psi(x) = x^T Dx$ (where D is a real symmetric matrix of order n) is *quasi-convex* on a convex set $X \subset E^n$ if and only if for all $x, y \in X$

$$y^T Dy \leqslant x^T Dx \text{ implies } x^T D(y-x) \leqslant 0. \quad (3)$$

The quadratic form $\psi(x) = x^T Dx$ is *pseudo-convex* on a set $X \subset E^n$ if and only if for all x, $y \in X$

$$x^T D(y-x) \geqslant 0 \text{ implies } y^T Dy \geqslant x^T Dx. \quad (4)$$

THEOREM 15: The quadratic form $\psi(x) = x^T Dx$ *is quasi-convex on the non-negative orthant* E^n_+ *if and only if* D *is positive subdefinite.*

PROOF: Theorem 4 in Martos (1969).

A similar result holds for pseudo-convex quadratic forms.

THEOREM 16: The quadratic form $\psi(x) = x^T Dx$ *is pseudo-convex on the semi-positive orthant,* $E^n_+ \setminus 0 = \{x \in E^n_+ : x \neq 0\}$, *if and only if* D *is strictly positive subdefinite.*

PROOF: Theorem 5 in Martos (1969).

Before we move to analyse quadratic functions, it is interesting to use quadratic forms to illustrate the gap existing between the necessary conditions and the sufficient conditions for the quasi-convexity of twice continuously differentiable functions.

It is easy to see that conditions of Theorem 1 are not necessary for quasi-convexity. Indeed, consider the quadratic form $\psi(x) = x^T Dx$ on E^2 where

$$D = \begin{bmatrix} -1 & 0 \\ 0 & 0 \end{bmatrix}.$$

By theorem 15, $\psi(x)$ is quasi-convex on E^2_+, but $\det D_r(x) = 0$ for $r = 1,2$ and for all $x \in E^2$.

Consider the quadratic form $\psi(x) = x^T Dx$ on E^2 where

$$D = \begin{bmatrix} -1 & -1 \\ -1 & -1 \end{bmatrix}.$$

It is easy to verify that

$$D_2(x) = \begin{bmatrix} 0 & -2(x_1 + x_2) & -2(x_1 + x_2) \\ -2(x_1 + x_2) & -1 & -1 \\ -2(x_1 + x_2) & -1 & -1 \end{bmatrix}$$

Hence $\det D_r(x) \leqslant 0$ for $r = 1,2$ and for all $x \in E^2$. Conditions of Theorem 4 are not sufficient since $\psi(x)$ is not quasi-convex on all E^2. Indeed, let $x^T = [1,0]$ and $y^T = [0,-1]$, and a contradiction to definition (3) is exhibited.

4.2 Quadratic Functions on E^n_+

Since the sum of quasi-convex functions is not necessarily quasi-convex, the quasi-convexity of the quadratic function $\phi(x) = \frac{1}{2}x^T Dx + c^T x$ does not follow from the quasi-convexity of the quadratic form $x^T Dx$ (as it is the case for convexity). Hence the criteria become more complex. Recall that a quadratic function $\phi(x) = \frac{1}{2}x^T Dx + c^T x$ (where D is a real symmetric matrix of order n and $c \in E^n$) is *quasi-convex* on a convex set $X \subset E^n$ if and only if for all $x, y \in X$

$$y^T Dy + c^T y \leqslant x^T Dx + c^T \mathbf{x} \text{ implies } (c + Dx)^T (y-x) \leqslant 0.$$

The quadratic function $\phi(x) = x^T Dx + c^T x$ is *pseudo-convex* on a set $X \subset E^n$ if and only if for all $x, y \in X$

$$(c + Dx)^T (y-x) \geqslant 0 \text{ implies } y^T Dy + c^T y \geqslant x^T Dx + c^T x.$$

The most natural extension of the criteria for the quasi-convexity and pseudo-convexity of quadratic forms follows from the observation

$$\phi(x) = \tfrac{1}{2} \begin{bmatrix} 1 \\ x \end{bmatrix}^T \begin{bmatrix} 0 & c^T \\ c & D \end{bmatrix} \begin{bmatrix} 1 \\ x \end{bmatrix}.$$

Thus, if we denote

$$\psi(\xi,x) = \tfrac{1}{2} \begin{bmatrix} \xi \\ x \end{bmatrix}^T \begin{bmatrix} 0 & c^T \\ c & D \end{bmatrix} \begin{bmatrix} \xi \\ x \end{bmatrix}, \quad \begin{bmatrix} \xi \\ x \end{bmatrix} \in E^{n+1},$$

then the following relationship exists

$$\phi(x) = \psi(1,x).$$

One can expect criteria for the quasi-convexity and pseudo-convexity of $\phi(x)$ to be defined in terms of the positive subdefiniteness of the matrix

$$\begin{bmatrix} 0 & c^T \\ c & D \end{bmatrix}.$$

Before doing this, the following results, stating themselves criteria for the quasi-convexity of ϕ, have to be introduced. Notice the natural extension of the concept of positive subdefiniteness.

THEOREM 17: The quadratic function $\phi(x) = \frac{1}{2}x^TDx + c^Tx$ *is quasi-convex on the non-negative orthant* E^n_+ *if and only if for all* $x \in E^n$

$$x^TDx < 0 \text{ implies } \begin{bmatrix} c^Tx \\ Dx \end{bmatrix} \geqslant 0 \text{ or } \begin{bmatrix} c^Tx \\ Dx \end{bmatrix} \leqslant 0 .$$

PROOF: Theorem 1 in Martos (1971).

COROLLARY 18: If the quadratic function $\phi(x) = \frac{1}{2} x^TDx + c^Tx$ *is quasi-convex on* E^n_+, *then so is the quadratic form* $\psi(x) = x^TDx$.

PROOF: Remark 3 in Martos (1971).

The following result characterizes the matrix D and the vector c.

THEOREM 19: The nonconvex quadratic function $\phi(x) = \frac{1}{2}x^TDx + c^Tx$ *is quasi-convex on the non-negative orthant,* E^n_+ *, if and only if*

(i) D *has exactly one (simple) negative eigenvalue,*

(ii) $D \leqslant 0$, $D \neq 0$,

(iii) $c \leqslant 0$,

(iv) there is a vector $q \in E^n$ *such that* $Dq = c$ and $c^Tq \leqslant 0$.

PROOF: Theorem 2 in Martos (1971).

The following criterion is a more natural extension of the one for quadratic forms.

THEOREM 20: The nonconvex quadratic function $\phi(x) = \frac{1}{2}x^TDx + c^Tx$ *is quasi-convex on the non-negative orthant,* E^n_+ *, if and only if the matrix* $\begin{bmatrix} 0 & c^T \\ c & D \end{bmatrix}$ *is merely positive subdefinite.*

PROOF: Theorem 5.1 in Cottle and Ferland (1972).

Finally, to analyse pseudo-convex quadratic functions on E^n_+ we use the strong relation between quasi-convexity and pseudo-convexity shown in Section 2.

THEOREM 21: If the nonconvex quadratic function $\phi(x) = \frac{1}{2}x^TDx + c^Tx$ *is quasi-convex on the non-negative orthant* E^n_+ *, then it is pseudo-convex on* E^n_+ *provided* $c \neq 0$.

PROOF: Cottle and Ferland (1971).

Notice that Theorem 21 gives a criterion for the pseudo-convexity of $\phi(x)$ on E^n_+ while Theorem 16 gives a criterion for pseudo-convexity of quadratic forms but only on $E^n_+ \setminus 0$.

REFERENCES

ARROW, K.J. and ENTHOVEN, A.C. (1961). "Quasi-concave Programming." *Econometrica, 29*, 779-800.

COTTLE, R.W. and FERLAND, J.A. (1972). "Matrix-Theoretic Criteria for the Quasi-convexity and Pseudo-convexity of Quadratic Functions." *Linear Algebra and its Application, 5*, 123-136.

COTTLE, R.W. and FERLAND, J.A. (1971). "On Pseudo-convex Functions of Non-negative Variables." *Mathematical Programming, 1*, 95-101.

FERLAND, J.A. (1971). "Quasi-convex and Pseudo-convex Functions on Solid Convex Sets." Technical report No.71-4, Dept. Operations Research, Stanford University.

FERLAND, J.A. (1972). "Mathematical Programming Problems with Quasi-convex Objective Functions." *Mathematical Programming, 3*, 296-301.

FERLAND, J.A. (1980). "Positive Subdefinite Matrices." *Linear Algebra and its Applications, 31*, 233-244.

GALE, D. (1960). *The Theory of Linear Economic Models*. McGraw-Hill, New York, N.Y.

GANTMACHER, F.R. (1959). *The Theory of Matrices*. Vol. 1, Chelsea Publishing Company, New York, N.Y.

MANGASARIAN, O.L. (1965). "Pseudo-convex Functions." *SIAM J. Control*, *3*, 281-290.

MANGASARIAN, O.L. (1969). *Nonlinear Programming*. McGraw-Hill, New York.

MARTOS, B. (1969). "Subdefinite Matrices and Quadratic Forms." *SIAM J. Appl. Math.*, *17*, 1215-1223.

MARTOS, B. (1971). "Quadratic Programming with a Quasiconvex Objective Function." *Operations Research*, *19*, 87-97.

SAMUELSON, P.A. (1963). *Foundations of Economic Analysis*. Harvard University Press, Cambridge, Mass.

SCHAIBLE, S. (1971). "Beiträge zur quasikonvexen Programmierung." Doctoral Dissertation, Köln.

SCHAIBLE, S. (1972). "Quasiconcave, Strictly Quasiconcave and Pseudoconcave Functions." *Methods of Operations Research*, *17*, 308-316.

SCHAIBLE, S. (1981). "Quasiconvex, Pseudoconvex and Strictly Pseudoconvex Quadratic Functions." *Journal of Optimization Theory and Applications*, to appear.

USPENSKY, J.V. (1948). *Theory of Equations*. McGraw-Hill, New York, N.Y.

GENERALIZED CONVEXITY OF QUADRATIC FUNCTIONS

Siegfried Schaible†

We characterize quadratic functions that are quasiconvex, semistrictly quasiconvex, strictly quasiconvex, pseudoconvex, strictly pseudoconvex or strongly pseudoconvex on solid convex sets of R^n. Known as well as new criteria are derived from the author's earlier characterization of quasiconvex and pseudoconvex quadratic functions.

1. INTRODUCTION

In Diewert, Avriel and Zang (1981) six kinds of generalized concave functions are discussed: quasiconcave, semistrictly quasiconcave, strictly quasiconcave, pseudoconcave, strictly pseudoconcave and strongly pseudoconcave functions; see also Avriel, Diewert, Schaible and Ziemba (1981). Each type of generalized concave function reflects some of the nice properties of concave functions such as convexity of level sets, local-global properties of maxima, maximal properties of stationary points, uniqueness of maxima etc. For details see the above two papers as well as Avriel (1976) and Martos (1975).

In this paper we characterize quadratic functions that belong to one of these six classes of generalized concave (convex) functions.

This question is of practical as well as theoretical interest. Whenever a utility or production function is approximated by a quadratic function there arises the following question: what are the conditions that the quadratic function, restricted to a convex set, belongs to one of the six classes of generalized concave functions? On the other hand, if the true utility or production function rather than a quadratic approximation is used one would like to know whether such a function is one of these

†This research was supported by the NSERC grant A4534. The author wants to thank Professor W.E. Diewert, Department of Economics, University of British Columbia for helpful remarks.

ISBN 0-12-621120-5

six kinds of generalized concave functions. Criteria for quadratic functions may then serve as a guide to find criteria for those nonquadratic functions. See for example, Avriel and Schaible (1978), Ferland (1979, 1981), Schaible (1973b).

As done often before, we characterize generalized convex rather than generalized concave functions.

In the last decade several criteria for generalized convex quadratic functions were derived most of which were concerned with quasi- and pseudoconvexity. These results fall into two groups: characterizations of quadratic functions on solid convex sets of R^n (Avriel and Schaible (1978), Ferland (1971, 1972, 1979), Martos (1967), Schaible (1971a, 1973a, 1977, 1981)) and finite criteria for quadratic functions on the nonnegative orthant of R^n (Cottle and Ferland (1971, 1972), Martos (1969, 1971), Schaible (1981), Schaible and Cottle (1980)). The second group of criteria will not be considered in this paper. Recently it was shown in Schaible (1981) and Schaible and Cottle (1980) how these finite criteria can be obtained by specializing the more general ones for generalized convex quadratic functions on arbitrary solid convex sets. In Schaible (1981) new characterizations of quadratic functions of nonnegative variables were also derived.

In the present paper we shall concentrate on criteria for generalized convex quadratic functions on arbitrary solid convex sets of R^n. First various criteria for quasiconvex, pseudoconvex and strictly pseudoconvex functions in the quadratic case will be reviewed. We then derive a recent result by Ferland (1979) within our own framework. In addition, we obtain a new criterion for quadratic functions that are strictly pseudoconvex. The proof of these criteria again demonstrates that all characterizations of quasiconvex, pseudoconvex and strictly pseudoconvex quadratic functions can be obtained from an earlier characterization due to Schaible (1971a), see also Schaible (1973a, 1977, 1981).

Finally, we supplement the results on quasiconvex, pseudoconvex and strictly pseudoconvex functions by criteria for semistrictly quasiconvex, strictly quasiconvex and strongly pseudoconvex quadratic functions. For quadratic functions on open convex sets these six classes of generalized convex functions are indeed only two different ones. The smaller class is characterized as a subclass of the other one where in the nonconvex case the coefficient matrix is nonsingular.

2. MAIN CHARACTERIZATIONS OF QUASICONVEX, PSEUDOCONVEX AND STRICTLY PSEUDOCONVEX FUNCTIONS Q(x)

We consider quadratic functions $Q(x) = \frac{1}{2}x^TAx + b^Tx$ on convex sets C of R^n. Let A be a real symmetric $n \times n$ matrix and $b \in R^n$. We assume that C is a solid set, i.e., it has a nonempty interior.

Every quadratic function $Q(x) = \frac{1}{2}x^TAx + b^Tx$ can be reduced to the following normalform using an appropriate affine transformation $x = Wy + v$ ($\det W \neq 0$):

$$Q(x) = q(y) + \delta$$

where

$$q(y) = -\frac{1}{2}\sum_{i=1}^{k} y_i^2 + \frac{1}{2}\sum_{i=k+1}^{r} y_i^2 + \gamma\cdot y_{r+1}, \qquad 0 \le k \le r \le n, \quad \gamma = 0 \text{ or } \gamma = 1, \ \delta \in R. \tag{1}$$

Quasiconvexity and (strict) pseudoconvexity is preserved under an affine transformation of variables. Therefore, it suffices to derive conditions under which $q(y)$ is quasiconvex, pseudoconvex or strictly pseudoconvex on the solid convex set

$$D = \{y \in R^n | Wy + v \in C\} .$$

Let $q(y)$ be nonconvex on D, i.e., $k \ge 1$. The normalform $q(y)$ is not quasiconvex on D if $k > 1$ or $\gamma = 1$ since for these functions the lower level sets are not convex. In the remaining case

$$q_o(y) = -\tfrac{1}{2}y_1^2 + \tfrac{1}{2}\sum_{i=2}^{r} y_i^2 \tag{2}$$

we can ensure quasiconvexity on two convex cones

$$D_1 = \{y \in R^n | q_o(y) \leq 0,\ y_1 \geq 0\},\ D_2 = \{y \in R^n | q_o(y) \leq 0,\ y_1 \leq 0\}.$$

This follows from the fact that $h(y) = -(-q_o(y))^{\frac{1}{2}}$ is convex on D_i as proved in Schaible (1971a); see also Schaible (1981). The convex cones D_i are illustrated for the function $q_o(y) = -\tfrac{1}{2}y_1^2 + \tfrac{1}{2}y_2^2$ in Figure 1. This function is quasiconvex on each of the convex cones D_i.

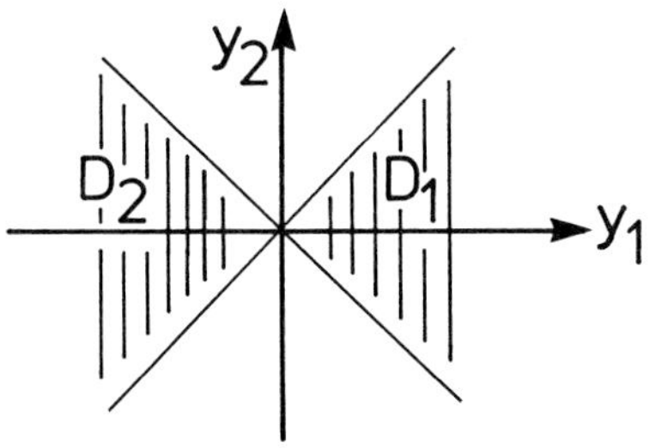

FIGURE 1. $q_o(y) = -\tfrac{1}{2}y_1^2 + \tfrac{1}{2}y_2^2$

Hence we have (Schaible 1971a, 1973a, 1981):

THEOREM 1: A nonconvex function $Q(x)$ *is quasiconvex on a solid convex set* C *of* R^n *if and only if*

(i) *rank* (A,b) = *rank* A,

(ii) A *has exactly one (simple) negative eigenvalue,*

(iii) $C \subset C_1$ *or* $C \subset C_2$ *where*

$$C_1 = \{x \in R^n | Q(x) \leq \delta,\ y_1 \geq 0\},$$

$$C_2 = \{x \in R^n | Q(x) \leq \delta,\ y_1 \leq 0\}.$$

The proof of Theorem 1 reveals that every (nonconvex) quasiconvex quadratic function is convexifiable since $Q(x) = q_o(y) + \delta$ where $-(-q_o(y))^{\frac{1}{2}}$ is convex on D_i. Hence we have

THEOREM 2: A nonconvex function $Q(x)$ *is quasiconvex on a solid convex set* C *of* R^n *if and only if* $Q(x) \leq \delta$ *for some* $\delta \in R$ *on* C *and* $H(x) = -(\delta - Q(x))^{\frac{1}{2}}$ *is convex on* C.

Furthermore the proof of Theorem 1 shows (Schaible (1971a, 1981)):

THEOREM 3: A nonconvex quasiconvex function $Q(x)$ *is pseudoconvex on a solid convex set* C *of* R^n *if and only if* $C \subset C_1^o$ *or* $C \subset C_2^o$ *where*

$$C_1^o = \{x \in R^n | Q(x) \leq \delta,\ y_1 > 0\}$$

$$C_2^o = \{x \in R^n | Q(x) \leq \delta,\ y_1 < 0\}.$$

Thus a quasiconvex quadratic function is even pseudoconvex on C *if* C *is an open convex set.*

The following theorem provides a characterization of strictly pseudoconvex functions (Schaible (1977)):

THEOREM 4: A nonconvex pseudoconvex function $Q(x)$ *is strictly pseudoconvex on* C *if and only if* A *is nonsingular.*

We have seen in Theorem 1 and 3 that for quadratic functions there exist maximal domains of quasiconvexity and pseudoconvexity. These sets C_i (C_i^o) are affine images of convex cones D_i (D_i^o). We can prove the following representation of C_i in terms of eigenvalues $\lambda_1,\ldots,\lambda_n$ and orthonormal eigenvectors $t_1,\ldots,t_n$ of A where λ_1 is the only negative eigenvalue (Schaible (1971a, 1973a, 1981)):

$$\begin{aligned} C_1 &= \{x \in R^n | Q(x) \leq \delta,\ t_1^T x \geq -b^T t_1/\lambda_1\}, \\ C_2 &= \{x \in R^n | Q(x) \leq \delta,\ t_1^T x \leq -b^T t_1/\lambda_1\}, \end{aligned} \tag{3}$$

where

$$\delta = -\tfrac{1}{2} \sum_{i=1}^{r} (b^T t_i)^2/\lambda_i.$$

The formulas for C_1^o, C_2^o are obtained from (3) by replacing the second inequality in the formula for C_1, C_2 by a strict inequality.

Independently of the author and via a different approach Ferland

(1971, 1972) proved another characterization of quasiconvex and pseudoconvex quadratic functions which can be related to Theorems 1 and 3. Condition *(i)* of Theorem 1 is equivalent to Ferland's assumption that $Q(x)$ has a stationary point s in R^n, i.e., $\nabla Q(x) = Ax + b = 0$ has a solution $x = s$ in R^n. Ferland characterizes the maximal domains of quasiconvexity and pseudoconvexity with help of such a stationary point as follows:

$$\begin{aligned} C_1 &= s + \{z \mid z^T A z \leq 0,\ t_1^T z \geq 0\} \\ C_2 &= s + \{z \mid z^T A z \leq 0,\ t_1^T z \leq 0\}. \end{aligned} \tag{4}$$

For C_1^o, C_2^o the second inequality in the formula for C_1, C_2 in (4) is to be replaced by a strict inequality.

Ferland's characterization (4) of C_1, C_2 can be derived within our approach as follows: the affine transformation

$$x = \left(\frac{1}{\sqrt{-\lambda_1}}\, t_1,\ \frac{1}{\sqrt{\lambda_2}}\, t_2,\ \ldots,\ \frac{1}{\sqrt{\lambda_r}}\, t_r,\ t_{r+1},\ \ldots,\ t_n\right) y + s$$

reduces $Q(x)$ to its normalform $Q(x) = q_o(y) + \delta$. Since then $\delta = Q(s)$ and $y_1 = \sqrt{-\lambda_1}\, t_1^T(x - s)$ we have

$$C_1 = \{x \mid Q(x) \leq \delta,\ y_1 \geq 0\} = \{x \mid Q(x) \leq Q(s),\ t_1^T(x - s) \geq 0\}.$$

Introducing $z = x - s$ yields (4).

This shows that the characterizations of quasiconvex and pseudoconvex quadratic functions in Ferland (1971) and Schaible (1971a) are equivalent (Schaible (1971b)).

We see from Theorem 1 that (nonconvex) quasiconvex quadratic functions are bounded from above. Hence in a quadratic program where such a function is to be maximized on a solid convex polyhedron the objective function is bounded from above (Schaible (1973a)). The supremum of the quadratic objective function is actually attained and it is a vertex of the feasible region if there exists one (Ferland (1975)).

3. FURTHER CHARACTERIZATIONS OF PSEUDOCONVEX AND STRICTLY PSEUDOCONVEX FUNCTIONS Q(x)

From Theorems 1 and 3 we see that a quadratic function is quasiconvex on a solid convex set C if and only if it is pseudoconvex on the interior of C. Hence it suffices to characterize quadratic functions that are pseudoconvex on open convex sets. We then simultaneously obtain criteria for quasiconvex functions on solid convex sets.

A main result in our analysis in Section 2 is given in Theorem 2: every quasiconvex (pseudoconvex) quadratic function is convex transformable by a suitable increasing function. This result was reobtained in Crouzeix (1980), using a different approach.

Convexifiability of quadratic functions can be exploited when deriving characterizations of quasiconvex (pseudoconvex) polynomials as shown in Schaible (1973b) in a first attempt.

The fact that every quasiconvex (pseudoconvex) quadratic function can be convexified gives rise to the following result (see Schaible (1977)):

THEOREM 5: A quadratic function $Q(x)$ *is pseudoconvex (strictly pseudoconvex) on an open convex set* C *of* R^n *if and only if there exists a function* $r: C \to R$ *such that the augmented Hessian* $\nabla^2 Q(x) + r(x)\nabla Q(x)\nabla Q(x)^T$ *is positive semidefinite (positive definite) on* C. *For a nonconvex function* $Q(x)$ *take* $r(x) \geq \frac{1}{2}(\delta - Q(x))^{-1}$ $(r(x) > \frac{1}{2}(\delta - Q(x))^{-1})$.

From Theorem 5 another characterization can be derived involving the principal minors of the bordered Hessian

$$M(x) = \begin{pmatrix} 0 & \nabla Q(x)^T \\ \nabla Q(x) & \nabla^2 Q(x) \end{pmatrix}.$$

The following criterion was derived in Avriel and Schaible (1978); for a simpler criterion in case of normalforms q(y) see Schaible (1977).

THEOREM 6: A quadratic function $Q(x)$ *is pseudoconvex (strictly pseudoconvex) on an open convex set of* C *of* R^n *if and only if all principal minors (leading principal minors) of the bordered Hessian* $M(x)$ *are nonpositive on* C*, and if one vanishes then the associated principal minor (leading principal minor) of the Hessian* $\nabla^2 Q(x) = A$ *is nonnegative (positive).*

This criterion was effectively used in Schaible and Cottle (1980) and more fully in Schaible (1981) to relate finite criteria for quasiconvex (pseudoconvex) quadratic functions on the nonnegative orthant of R^n to the general characterizations of quasiconvex (pseudoconvex) quadratic functions on arbitrary convex sets. In Schaible (1981) all known criteria for quasiconvex (pseudoconvex) quadratic functions in nonnegative variables were derived by specializing Theorems 1 and 3. In addition, characterizations of strictly pseudoconvex functions $Q(x)$ as well as new criteria for quasiconvex and pseudoconvex functions in nonnegative variables were obtained.

Recently Ferland (1979) proved a new characterization of quasiconvex quadratic functions on convex sets. He showed that a nonconvex function $Q(x)$ is quasiconvex on a solid convex set if and only if the bordered Hessian $M(x)$ has exactly one negative eigenvalue for all x in C. We now derive this criterion within our own framework. Like the other criteria in Theorems 5 and 6 this result will be proved by using our main characterizations in Section 2. Again we prove the criterion for open convex sets C recalling that $Q(x)$ is quasiconvex on a solid convex set if and only if $Q(x)$ is pseudoconvex on the interior of C. In addition to the result in Ferland (1979) we obtain also a criterion for functions $Q(x)$ that are strictly pseudoconvex.

THEOREM 7: A nonconvex quadratic function Q(x) *is pseudoconvex (strictly pseudoconvex) on an open convex set* C *of* R^n *if and only if the bordered Hessian* M(x) *has exactly one negative eigenvalue for all* x *in* C *(and* M(x) *is nonsingular for all* x *in* C*).*

PROOF: It suffices to prove the assertion for normalforms q(y) since an affine transformation x = Wy + v such that Q(x) = q(y) + δ yields

$$\begin{pmatrix} 0 & \nabla Q(x)^T \\ \nabla Q(x) & \nabla^2 Q(x) \end{pmatrix} = \begin{pmatrix} 1 & 0 \\ 0 & W \end{pmatrix}^T \begin{pmatrix} 0 & \nabla q(y)^T \\ \nabla q(y) & \nabla^2 q(y) \end{pmatrix} \begin{pmatrix} 1 & 0 \\ 0 & W \end{pmatrix}.$$

This shows that

$$M(x) = \begin{pmatrix} 0 & \nabla Q(x)^T \\ \nabla Q(x) & \nabla^2 Q(x) \end{pmatrix} \quad \text{and} \quad N(y) = \begin{pmatrix} 0 & \nabla q(y)^T \\ \nabla q(y) & \nabla^2 q(y) \end{pmatrix}$$

are congruent and thus have the same number of negative (and positive) eigenvalues.

As we saw in Section 2 a nonconvex function Q(x) is pseudoconvex on C if and only if there exists an affine transformation x = Wy + v such that

$$Q(x) = q_o(y) + \delta \text{ and } q_o(y) < 0 \text{ on } D. \tag{5}$$

Hence we have to show that in this case and only in this case N(y) has exactly one negative eigenvalue for all $y \in D$.

According to Jacobi's Theorem (see Gantmacher (1959; 303)) the number of negative eigenvalues of a matrix B with nonvanishing leading principal minors det B_i equals the number of sign changes in the sequence {1, det B_1, ..., det B_n}. The leading principal minors of any normalform

$$q(y) = -\tfrac{1}{2}\sum_{i=1}^{k} y_i^2 + \tfrac{1}{2}\sum_{i=k+1}^{r} y_i^2 + \gamma y_{r+1}$$

are given by Schaible (1977)

$$\det N_i(y) = \begin{cases} (-1)^i (y_1^2 + \ldots + y_i^2) & i = 1, \ldots, k \\ (-1)^k (y_1^2 + \ldots + y_k^2 - y_{k+1}^2 - \ldots - y_i^2) & i = k+1, \ldots, r \\ (-1)^{k+1} \gamma^2 & i = r+1 \\ 0 & i = r+2, \ldots, n. \end{cases}$$

We have to show that (5) holds if and only if there is exactly one sign change in the sequence $\{1, \det N_1(y), \ldots, \det N_n(y)\}$ for all $y \in D$.

Suppose $Q(x)$ is pseudoconvex but not convex on C. Then (5) holds. In this case we have

$$\det N_i(y) = \begin{cases} -y_1^2 & \\ -y_1^2 + y_2^2 + \ldots + y_i^2 & i = 2, \ldots, r \\ 0 & i = r+1, \ldots, n. \end{cases}$$

Since $\det N_i(y) \leq \det N_{i+1}(y)$ for $i = 1, \ldots, r-1$ and $\det N_r(y) = 2q_o(y)$, (5) yields $\det N_i(y) < 0$ for $i = 1, \ldots, r$ for all $y \in D$. In view of Jacobi's Theorem $N(y)$ then has exactly one negative eigenvalue for all $y \in D$.

In addition we see that $\det N(y) = \det N_n(y) \neq 0$ if and only if $r = n$. Hence a (nonconvex) pseudoconvex function $Q(x)$ is strictly pseudoconvex on C if and only if $\det M(x) \neq 0$ for all $x \in C$.

Now suppose $Q(x)$ is not pseudoconvex on C. Thus (5) does not hold. We shall show that we can always find $y \in D$ such that $N(y)$ has more than one negative eigenvalue. We first note that there always exists $\bar{y} \in D$ such that $\det N_i(\bar{y}) \neq 0$ for $i = 1, \ldots, r$ since D is an open set and $\det N_i(y)$ are quadratic functions.

Suppose $k \geq 2$. Then $\det N_1(\bar{y}) = -\bar{y}_1^2$ and $\det N_2(\bar{y}) = \bar{y}_1^2 + \bar{y}_2^2$. Hence in the sequence $\{1, \det N_1(\bar{y}), \ldots, \det N_n(\bar{y})\}$ there are at least two sign changes. Therefore $N(\bar{y})$ has at least two negative eigenvalues.

Assume $k = 1$ and $\gamma = 1$. Then $\det N_1(\bar{y}) = -\bar{y}_1^2$ and $\det N_{r+1}(\bar{y}) = 1$, and therefore in the sequence $\{1, \det N_1(\bar{y}), \ldots, \det N_n(\bar{y})\}$ we have at least two sign changes. Again $N(\bar{y})$ has more than one negative eigenvalue.

Finally let $k = 1$, $\gamma = 0$ and $q_o(y') \geq 0$ for some $y' \in D$. Since D is open there exists $\bar{y}' \in D$ such that $\det N_i(\bar{y}') \neq 0$ for $i = 1, \ldots, r$ and in particular $\det N_r(\bar{y}') = 2q_o(\bar{y}') > 0$. Therefore in the sequence $\{1, \det N_1(\bar{y}'), \ldots, \det N_n(\bar{y}')\}$ we have at least two sign changes. Again $N(\bar{y}')$ has more than one negative eigenvalue. □

4. SEMISTRICTLY QUASICONVEX, STRICTLY QUASICONVEX AND STRONGLY PSEUDOCONVEX FUNCTIONS Q(x)

We now relate three other concepts of generalized convexity to quadratic functions: semistrict quasiconvexity, strict quasiconvexity and strong pseudoconvexity. The definitions are given in Avriel, Diewert, Schaible and Ziemba (1981).

For lower semicontinuous functions semistrict quasiconvexity implies quasiconvexity (Karamardian (1965)). For quadratic functions every quasiconvex function is even semistrictly quasiconvex (Martos (1967)). This can be seen as follows. Let $\bar{x}, \bar{\bar{x}} \in C$ such that

$$Q(\bar{x}) < Q(\bar{\bar{x}}) \tag{6}$$

Now assume there exists $x^1 = \lambda_1\bar{x} + (1 - \lambda_1)\bar{\bar{x}}$, $\lambda_1 \in (0,1)$ such that $Q(x^1) = Q(\bar{\bar{x}})$. Then $Q(x) = Q(\bar{\bar{x}})$ on the line segment between $\bar{\bar{x}}$ and x^1 since otherwise $Q(x^1) > \text{Max}\,(Q(x), Q(\bar{\bar{x}}))$ or $Q(x) > \text{Max}\,(Q(x^1), Q(\bar{\bar{x}}))$ contradicting quasiconvexity of $Q(x)$. A quadratic function being constant on the line segment between x^1 and $\bar{\bar{x}}$ is also constant on the whole line through these points. In particular $Q(\bar{x}) = Q(\bar{\bar{x}})$ contradicting (6). Hence every quasiconvex function $Q(x)$ is even semistrictly quasiconvex. This proof by Martos (1967) shows:

THEOREM 8: A quadratic function $Q(x)$ *is semistrictly quasiconvex on a convex set* C *of* R^n *if and only if* $Q(x)$ *is quasiconvex on* C.

In addition to semistrict quasiconvexity there is the concept of strict quasiconvexity. In general, a strictly quasiconvex function is semistrictly quasiconvex, but the reverse is not true. Even for quadratic functions the gap remains as seen from the example

$$Q(x) = -\tfrac{1}{2}x_1^2 + \tfrac{1}{2}x_2^2, \quad C = \{x \in R^3 \mid Q(x) < 0,\ x_1 > 0\}.$$

This function is semistrictly quasiconvex on C in view of Theorem 1 and Theorem 8. However, Q(x) is not strictly quasiconvex there since Q(x) is constant on lines parallel to the x_3-axis.

Strict quasiconvexity can also be related to strict pseudoconvexity. Diewert, Avriel and Zang (1981) recently proved that every strictly pseudoconvex function on an open convex set C of R^n is also strictly quasiconvex. For quadratic functions we now show that the converse is also true.

THEOREM 9: A quadratic function Q(x) *is strictly quasiconvex on an open convex set* C *of* R^n *if and only if* Q(x) *is strictly pseudoconvex on* C.

PROOF: It remains to show that a strictly quasiconvex function Q(x) is strictly pseudoconvex on C. If Q(x) is convex then the assertion follows easily from the definitions. Let us assume Q(x) is not convex. Certainly a strictly quasiconvex function is quasiconvex. Then from Theorem 1 we know that $Q(x) = q_o(y) + \delta$. Now

$$q_o(y) = -\tfrac{1}{2}y_1^2 + \tfrac{1}{2}\sum_{i=2}^{r} y_i^2$$

is strictly quasiconvex on the open convex set D only if r = n as seen from the definition. Thus A is nonsingular. According to Theorem 4 Q(x) is strictly pseudoconvex on C. □

Finally we turn to strongly pseudoconvex functions. These functions are strictly pseudoconvex (Diewert, Avriel and Zang (1981); for a different proof see Avriel and Schaible (1978)). We now show that for quadratic functions the converse is also true.

THEOREM 10: A quadratic function $Q(x)$ *is strongly pseudoconvex on an open convex set* C *of* R^n *if and only if* $Q(x)$ *is strictly pseudoconvex on* C.

PROOF: It remains to show that every strictly pseudoconvex function $Q(x)$ is strongly pseudoconvex. From the characterization of strictly pseudoconvex quadratic functions in Theorem 5 we know that there exists a function $r: C \rightarrow R$ such that the augmented Hessian $\nabla^2 Q(x) + r(x)\nabla Q(x)\nabla Q(x)^T$ is positive definite for all $x \in C$. Hence $z^T \nabla Q(x) = 0$ implies $z^T \nabla^2 Q(x) z > 0$ for all $x \in C$, $z \in R^n$, $z \neq 0$. Thus $Q(x)$ is strongly pseudoconvex on C. □

Summarizing the results we see that for quadratic functions on open convex sets of R^n the following inclusions hold:

quasiconvex $\Longleftrightarrow$ semistrictly quasiconvex $\Longleftrightarrow$ pseudoconvex

and

strictly quasiconvex $\Longleftrightarrow$ strictly pseudoconvex $\Longleftrightarrow$ strongly pseudoconvex.

Hence considering quadratic functions on open convex sets there are only two different classes of generalized convex functions. The second one is a proper subclass of the first one. It consists of all those quadratic functions where A is nonsingular in the nonconvex case.

For quadratic functions on solid convex sets that are not open not all the inclusions above hold. For instance a quasiconvex function is not necessarily pseudoconvex as we see from Theorems 1 and 3.

REFERENCES

ARROW, K.J., and ENTHOVEN, A.D. (1961). "Quasi-concave Programming." *Econometrica 29*, 779-800.

AVRIEL, M. (1976). *Nonlinear Programming: Analysis and Methods.* Prentice-Hall, Englewood Cliffs, New Jersey.

AVRIEL, M., DIEWERT, W.E., SCHAIBLE, S., and ZIEMBA, W.T. (1981). "Introduction to Concave and Generalized Concave Functions." This vol.,21-50.

AVRIEL, M., and SCHAIBLE, S. (1978). "Second Order Characterizations of Pseudoconvex Functions." *Mathematical Programming 14,* 170-185.

COTTLE, R.W., and FERLAND, J.A. (1971). "On Pseudoconvex Functions of Nonnegative Variables." *Mathematical Programming 1,* 95-101.

COTTLE, R.W., and FERLAND, J.A. (1972). "Matrix-Theoretic Criteria for the Quasiconvexity and Pseudoconvexity of Quadratic Functions." *Linear Algebra and Its Applications 5,* 123-136.

CROUZEIX, J.P. (1980). "Conditions of Convexity of Quasiconvex Functions." *Mathematics of Operations Research 5,* 120-125.

DIEWERT, W.E., AVRIEL, M., and ZANG, I. (1981). "Nine Kinds of Quasiconcavity and Concavity." *Journal of Economic Theory,* forthcoming.

FERLAND, J.A. (1971). "Quasiconvex and Pseudoconvex Functions on Solid Convex Sets." Technical Report 71-4. Operations Research House, Stanford University, Stanford, California.

FERLAND, J.A. (1972). "Maximal Domains of Quasiconvexity and Pseudoconvexity for Quadratic Functions." *Mathematical Programming 3,* 178-192.

FERLAND, J.A. (1975). "On the Maximum of a Quasiconvex Quadratic Function on a Polyhedral Convex Set." Working Paper 129. Département d'Informatique, Université de Montréal.

FERLAND, J.A. (1979). "Necessary and Sufficient Conditions for Quasiconvexity and Pseudoconvexity." Working Paper. Département de Mathématiques, Université de Montpellier.

FERLAND, J.A. (1981). "Quasiconvexity and Pseudoconvexity of Functions on the Nonnegative Orthant." This volume, 169-181.

GANTMACHER, F.R. (1959). *The Theory of Matrices,* Vol. I. Chelsea, New York.

KARAMARDIAN, S. (1965). "Duality in Mathematical Programming." Ph.D. thesis. University of California, Berkeley.

MANGASARIAN, O.L. (1965). "Pseudoconvex Functions." *SIAM J. Control 3,* 281-290.

MARTOS, B. (1965). "The Direct Power of Adjacent Vertex Programming Methods." *Management Science 12,* 241-252.

MARTOS, B. (1967). "Quasiconvexity and Quasimonotonicity in Nonlinear Programming." *Studia Scientarium Mathematicarum Hungarica 2,* 265-273.

MARTOS, B. (1969). "Subdefinite Matrices and Quadratic Forms." *SIAM J. Appl. Math. 17(2),* 1215-1223.

MARTOS, B. (1971). "Quadratic Programming with a Quasiconvex Objective Function." *Operations Research 19,* 82-97.

MARTOS, B. (1975). *Nonlinear Programming: Theory and Methods.* North-Holland, Amsterdam.

PONSTEIN, J. (1967). "Seven Kinds of Convexity." *SIAM Review 9,* 115-119.

SCHAIBLE, S. (1971a). "Beitraege zur Quasikonvexen Programmierung." Doctoral Dissertation, Universitaet Koeln.

SCHAIBLE, S. (1971b). "Letter to J.A. Ferland", September 7.

SCHAIBLE, S. (1973a). "Quasiconcave, Strictly Quasiconcave and Pseudoconcave Functions." *Methods of Operations Research 17,* 308-316.

SCHAIBLE, S. (1973b). "Quasiconvexity and Pseudoconvexity of Cubic Functions." *Mathematical Programming 5,* 243-247.

SCHAIBLE, S. (1977). "Second Order Characterizations of Pseudoconvex Quadratic Functions." *J. of Optimization Theory and Applications 21,* 15-26.

SCHAIBLE, S., and COTTLE, R.W. (1980). "On Pseudoconvex Quadratic Forms." In *General Inequalities II, International Series of Numerical Mathematics*. (E.F. Beckenbach, ed.), Vol.47, 81-88. Birkhaeuser, Basel.

SCHAIBLE, S. (1981). "Quasiconvex, Pseudoconvex and Strictly Pseudoconvex Quadratic Functions - A Unified Approach." *J. of Optimization Theory and Applications*, to appear.

SHEPHARD, R.W. (1970). *Theory of Cost and Production Functions*. Princeton University Press, Princeton, N.J.

CRITERIA FOR QUASICONVEXITY AND PSEUDOCONVEXITY AND THEIR RELATIONSHIPS

Jean-Pierre Crouzeix and Jacques A. Ferland

In this paper, first order conditions for pseudoconvexity are given. Then second order conditions for quasiconvexity and pseudoconvexity introduced by several authors are surveyed. They are stated in a matrix-theoretic framework, and their relationships are studied.

1. INTRODUCTION

Quasiconvexity and pseudoconvexity are two important concepts in optimization theory and economics. One needs defining conditions that are easily verified. This has recently urged the introduction of several second order necessary and/or sufficient conditions for quasiconvexity and pseudoconvexity (Avriel (1976), Avriel and Schaible (1978), Crouzeix (1977), Crouzeix and Ferland (1981), Crouzeix (1980), Diewert et.al (1981), Katzner (1970)). The purposes of this paper are to give first order conditions for quasiconvexity and pseudoconvexity, and to analyse the relationships between several second order conditions.

Given the vectors $x,y \in R^n$, $\langle x,y \rangle = \sum_{i=1}^{n} x_i y_i$ denotes their scalar product. In this paper we consider real valued functions f(x) that are defined on an open convex set C of R^n. f'(x) and f"(x) denote the derivative and the Hessian, respectively. Here we consider Fréchet-differentiability.

2. FIRST ORDER CONDITIONS

Let f(x) be differentiable on C. Even if f(x) is quasiconvex on C, in general f'(x) = 0 is not a sufficient condition for x to be a minimum. This observation leads to the introduction of the concept of pseudoconvexity. A pseudoconvex function is quasiconvex and has a minimum at x whenever f'(x) = 0. Actually this property characterizes pseudoconvexity, as illustrated in the following proposition.

ISBN 0-12-621120-5

PROPOSITION 1 (Crouzeix and Ferland (1981)): f(x) *is pseudoconvex on* C *if and only if* f *is quasiconvex on* C *and* f *has a minimum at* x *whenever* $f'(x) = 0$.

This property can be strengthened by requiring only a local property of the minimum.

THEOREM 2 (Crouzeix and Ferland (1981)): f(x) *is pseudoconvex on* C *if and only if* f *is quasiconvex on* C *and* f *has a local minimum at* x *whenever* $f'(x) = 0$.

3. SECOND ORDER CONDITIONS

Let us assume that f(x) is twice differentiable on C. During the last twenty years several second order necessary and/or sufficient conditions for quasiconvexity and pseudoconvexity have been introduced. Some of these are given in terms of a bordered Hessian (Arrow and Enthoven (1961)), Ferland (1972), Avriel and Schaible(1978)), others in terms of an augmented Hessian (Mereau and Paquet (1974)), and others in terms of the positive semidefiniteness of the Hessian on the orthogonal subspace to f'(x) (Katzner (1970), Diewert et.al (1981), Crouzeix (1977)). The relationships between these conditions are analysed in Section 3.2, but first let us introduce the following extensions of Katzner's result.

3.1 Extensions of Katzner's Result

The first extension is for quasiconvex function.

THEOREM 3: f(x) *is quasiconvex on* C *if and only if*

(i) $< f'(x), h > = 0$ *implies that* $< f''(x) h,h > \geqslant 0$, *and*

(ii) *whenever* $f'(x) = 0$, *then for all* $h \in R^n$, $f_{x,h}(t) = f(x + th)$ *is quasiconvex.*

Katzner's result is Theorem 3 with the additional requirement that f" is continuous and $f'(x) < 0$ for all $x \in C$ (hence condition (ii) vanishes). Diewert et al (1981) extended Katzner's result to the case where f'' is continuous and $f'(x) \neq 0$ for all $x \in C$. Theorem 3 is proved in Crouzeix (1980).

Let us introduce the following Katzner-type condition for pseudo-convexity.

THEOREM 4 (Crouzeix and Ferland (1981)): $f(x)$ *is pseudoconvex on* C *if and only if*

(i) $\langle f'(x), h \rangle = 0$ *implies that* $\langle f''(x)h, h \rangle \geqslant 0$, *and*

(ii)' *whenever* $f'(x) = 0$ *then* f *has a local minimum at* x.

Let us define the following condition:

(ii)" whenever $f'(x) = 0$ then there exists a convex neighborhood V of x, $V \subset C$, such that $\langle f''(y)(y-x), y-x \rangle \geqslant 0$ for all $y \in V$.

Since condition (ii)" implies condition (ii)' and since the converse is not true, it follows that (i) and (ii)" are only sufficient conditions for pseudoconvexity.

The following example illustrates that it is not possible to obtain pure local conditions for quasiconvexity (condition (ii) in Theorem 3 being not local):

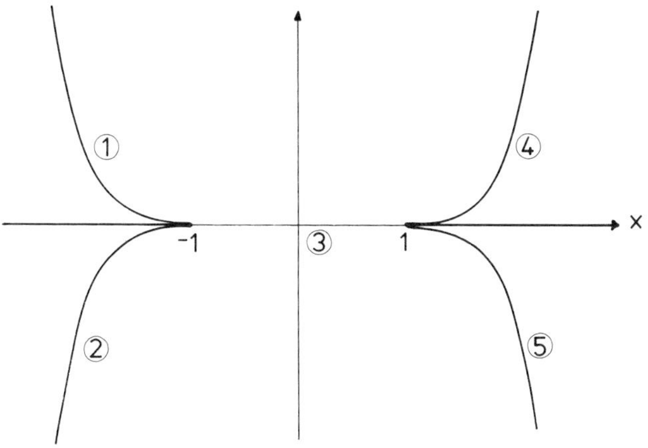

FIGURE 1. No local condition for quasiconvexity.

$$f_{ij}(x) = \begin{cases} (-1)^i \,(x+1)^4 & \text{if } x \leqslant -1 \\ 0 & \text{if } -1 \leqslant x \leqslant 1 \\ (-1)^j \,(x-1)^4 & \text{if } x \geqslant 1 \end{cases} \qquad i,j = 1,2$$

Now, $f_{22}(x)$ (consisting of parts ①, ③, and ④ of the figure) is convex, $f_{21}(x)$ (consisting of parts ①, ③, and ⑤) and $f_{12}(x)$ (consisting of parts ②, ③, and ④) are quasiconvex. But $f_{11}(x)$ (consisting of parts ②, ③, and ⑤) is not quasiconvex. It is easy to verify that condition (ii) in Theorem 3 is violated at $x = -1$ for $f_{11}(x)$.

3.2 Relationships Between Conditions

The relationship between the second order conditions are established in this section. The results are given in a more general matrix-theoretic framework, and the second order conditions are recovered by replacing a by f' and A by f". More details on these are included in Crouzeix and Ferland (1981).

Let $a \in R^n$, $a \neq 0$, and A be a real symmetric matrix of order n. The following conditions are equivalent

(C1) $\langle a, h \rangle = 0$ implies $\langle h, Ah \rangle \geqslant 0$

(C2) Either A is positive semidefinite, or A has one simple negative eigenvalue and there exists a vector $b \in R^n$ such that $Ab = a$ and $\langle a, b \rangle \leqslant 0$.

(C3) For all nonempty subsets $R \subset \{1, 2, \ldots, n\}$,

$$\det \begin{bmatrix} 0 & a_R^T \\ a_R & A_R \end{bmatrix} \leqslant 0$$

where A_R is obtained from A by deleting rows and columns whose indices are not in R, and a_R is obtained analogously from a.

(C4) The bordered matrix $\begin{bmatrix} 0 & a^T \\ a & A \end{bmatrix}$ has one simple negative eigenvalue.

The following conditions (D1), (D2), (D3) are equivalent to each other:

(D1) There is a scalar ν, $0 \leqslant \nu < \infty$, such that $A + \nu\, aa^T$ is positive semidefinite.

(D2) Either A is positive semidefinite, or A has one simple negative eigenvalue and there exists a vector $b \in R^n$ such that $Ab = a$ and $< a, b > < 0$.

(D3) For all nonempty subsets $R \subset \{1, 2, \ldots, n\}$,

$$\det \begin{bmatrix} 0 & a_R^T \\ a_R & A_R \end{bmatrix} \leqslant 0 ,$$

and if equality holds, then $\det A_R \geqslant 0$.

The D-conditions are weaker than the C-conditions since a (Di) condition implies a (Ci) condition, but the converse is not true.

Condition (C3) was discussed earlier by Arrow and Enthoven (1961) and Ferland (1972). Mereau and Paquet (1974) introduced condition (D1), and Avriel and Schaible (1978) discussed conditions (D1) and (D3). For quadratic functions, (D1) fully characterizes pseudoconvexity (see Schaible (1977)).

It is interesting to notice that (C3) requires less operations to be verified than (C1) or (C2). The other conditions require even more operations to be verified.

REFERENCES

ARROW, K.J., and ENTHOVEN, A.C. (1961). "Quasi-Concave Programming". *Econometrica 29*, 779-800.

AVRIEL, M. (1976). *Nonlinear Programming: Analysis and Methods*. Prentice-Hall, New York.

AVRIEL, M., and SCHAIBLE, S. (1978). "Second Order Characterization of Pseudo-Convex Functions". *Mathematical Programming 14*, 170-185.

CROUZEIX, J.P. (1977). "Contributions à l'étude des fonctions quasi-convexes". Thèse de doctorat, U.E.R. des Sciences Exactes et Naturelles, Université de Clermont-Ferrand II, France.

CROUZEIX, J.P. (1980). "A Second Order Condition for Quasiconvexity". *Mathematical Programming 18*, 349-352.

CROUZEIX, J.P., and FERLAND, J.A. (1981). "Criteria for Quasiconvexity and Pseudoconvexity: Relationships and Comparisons". *Mathematical Programming* (forthcoming).

DEBREU, G. (1952). "Definite and Semidefinite Quadratic Forms". *Econometrica 20*, 295-300.

DIEWERT, W.E., AVRIEL, M., and ZANG, I. (1981). "Nine Kinds of Quasi-Concavity and Concavity". *Journal of Economic Theory* (forthcoming).

FERLAND, J.A. (1972). "Mathematical Programming Problems with Quasi-convex Objective Functions". *Mathematical Programming 3*, 296-301.

FERLAND, J.A. (1981). "Matrix-Theoretic Criteria for the Quasiconvexity of Twice Continuously Differentiable Functions". *Linear Algebra and its Applications*, (forthcoming).

KATZNER, D.W. (1970). "*Static Demand Theory*". The MacMillan Company, New York.

MANGASARIAN, O.L. (1969). "*Nonlinear Programming*". McGraw-Hill, New York.

MEREAU, P., and PAQUET, J.G. (1974). "Second Order Conditions for Pseudoconvex Functions". *SIAM J. Appl. Math. 27*, 131-137.

SCHAIBLE, S. (1977). "Second Order Characterizations of Pseudoconvex Quadratic Functions". *Journal of Optimization Theory and Applications 21*, 15-26.

PART III

DUALITY FOR GENERALIZED CONCAVE PROGRAMS

A DUALITY FRAMEWORK IN QUASICONVEX PROGRAMMING

Jean-Pierre Crouzeix

This paper constructs a general framework of duality which is then applied to a quasiconvex mathematical program with inequality constraints and to a generalized fractional program. Finally, the linear von Neumann model of an expanding economy is studied through this general duality framework.

1. INTRODUCTION

Consider the optimization problem

$$\alpha = \inf [f(x)] \qquad \text{(P)}$$

where f is a function on $\mathbb{R}^n$. Problem (P) is called the primal problem.

It is usual to associate with f a perturbation function ϕ on $\mathbb{R}^n \times \mathbb{R}^p$ which verifies

$$\phi(x,0) = f(x) \qquad \forall x \in \mathbb{R}^n.$$

Define a new function h on $\mathbb{R}^p$ by

$$h(u) = \text{Inf } [\phi(x,u)/x \in \mathbb{R}^n].$$

Clearly $\alpha = h(0)$.

A duality framework in convex programming is obtained by considering the conjugate function h* and the biconjugate function h** of h,

$$h^*(u^*) = \sup_u [\langle u,u^*\rangle - h(u)],$$
$$h^{**}(v) = \sup_u [\langle v,u^*\rangle - h^*(u^*)].$$

These two functions are convex and lower semi-continuous and the biconjugate function h** is majorized by h. Let $\beta = h^{**}(0)$.

Then $\beta \leq \alpha$ and

$$\beta = \text{Sup } [-h^*(u^*)] \qquad \text{(Q)}$$

So, we obtain a new problem of optimization which consists in maximizing a concave function. This new problem is called the dual problem

ISBN 0-12-621120-5

of (P). If h is convex, lower semi-continuous at 0 and finite at 0, then $\beta = \alpha$. See, for instance, Rockafellar (1970, 1974) and Joly and Laurent (1971).

We intend to show how the above duality framework can be extended to quasiconvex programming. First, notice that if the function ϕ is quasiconvex then h is also quasiconvex. This can be seen by noticing that

$$\{u/h(u) < \lambda\} = \{u/\exists x \text{ s.t. } \phi(x,u) < \lambda\}.$$

So $\{u/h(u) < \lambda\}$ can be considered as being the projection of the convex set $\{(x,u)/\phi(x,u) < \lambda\}$ on $\mathbb{R}^P$ and thus is convex. Next, dual optimization problems can be obtained by using the dual functions $H(u^*,\lambda)$ and $H^-(u^*,t)$ which we introduced in Crouzeix (1981). Indeed if h is quasiconvex and lower semi-continuous (l.s.c.) at 0 then

$$h(0) = \operatorname*{Sup}_{u^*}[\operatorname*{Inf}_{\lambda}[\lambda/H(u^*,0) \geq 0]] = \operatorname*{Sup}_{u^*}[H^-(u^*,0)] .$$

In the next section, we study properties of the functions H and H^-.

2. PROPERTIES OF THE DUAL FUNCTIONS

Let f be any function on $\mathbb{R}^n$. The dual functions F and F^- are defined by

$$F(x^*,\lambda) = \operatorname*{Sup}_{x}[\langle x,x^*\rangle/f(x) \leq \lambda] ,$$

$$F^-(x^*,t) = \operatorname*{Inf}_{x}[f(x)/\langle x,x^*\rangle \geq t] ,$$

where by convention $\mathrm{Sup}[\xi/\xi \in A] = -\infty$ and $\mathrm{Inf}[\xi/\xi \in A] = +\infty$ if A is empty.

PROPOSITION 1:

a) $F^-(x^*,.)$ *is non-decreasing for all* x^*,

b) $F^-(\mu x^*,\mu t) = F^-(x^*,t)$ *for all* $\mu > 0$,

c) $F^-(.,t)$ *is quasiconcave for all* t.

PROOF: Proofs of a) and b) are immediate. Let us show c). Let $\xi \in (0,1)$, $x_1^*, x_2^* \in \mathbb{R}^n$ and $\mu = \mathrm{Min}[F^-(x_1^*,t),F^-(x_2^*,t)]$.

Then for i = 1, 2

$$\langle x, x_i^* \rangle \geq t \Longrightarrow f(x) \geq \mu ,$$

$$f(x) < \mu \Longrightarrow \langle x, x_i^* \rangle < t .$$

Hence

$$f(x) < \mu \Longrightarrow \langle x, \xi x_1^* + (1-\xi) x_2^* \rangle < t$$

$$\langle x, \xi x_1^* + (1-\xi) x_2^* \rangle \geq t \Longrightarrow f(x) \geq \mu$$

and so $F^-(\xi x_1^* + (1-\xi)x_2^*, t) \geq \mu$. □

PROPOSITION 2:

a) $F(x^*,.)$ *is non-decreasing for all* x^*,

b) $F(kx^*,\lambda) = kF(x^*,\lambda)$ *for all* $k > 0$,

c) $F(.,\lambda)$ *is convex and lower semi-continuous.*

PROOF: An immediate exercise. □

Next we intend to study the relationships between $F(x^*,.)$ and $F^-(x^*,.)$.

PROPOSITION 3: Let x^* *be a fixed non zero vector of* $\mathbb{R}^n$, *then*

a) $F(x^*,\lambda) < t \Longrightarrow F^-(x^*,t) \geq \lambda$,

b) $F^-(x^*,t) > \lambda \Longrightarrow F(x^*,\lambda) \leq t$,

c) $F(x^*,\lambda) > t \Longrightarrow F^-(x^*,t) \leq \lambda$,

d) $F^-(x^*,t) < \lambda \Longrightarrow F(x^*,\lambda) \geq t$.

PROOF:

a) Assume $F(x^*,\lambda) < t$. Then,

$$f(x) \leq \lambda \Longrightarrow \langle x, x^* \rangle < t$$

$$\langle x, x^* \rangle \geq t \Longrightarrow f(x) > \lambda$$

and so $F^-(x^*,t) \geq \lambda$.

b) Assume $F^-(x^*,t) > \lambda$. Then,

$$\langle x, x^* \rangle \geq t \Longrightarrow f(x) > \lambda$$

$$f(x) \leq \lambda \Longrightarrow \langle x, x^* \rangle < t$$

and so $F(x^*,\lambda) \leqslant t$.

c) and d) are immediate consequences of a) and b). □

COROLLARY 4: Let x^* *be a fixed non zero vector of* $\mathbb{R}^n$. *Then,*

i) $\text{Sup}[\lambda/F(x^*,\lambda) < \bar{t}] = \text{Inf}[\lambda/F(x^*,\lambda) \geqslant \bar{t}] \leqslant F^-(x^*,\bar{t}) \leqslant$

$\leqslant \text{Inf}[\lambda/F(x^*,\lambda) > \bar{t}] = \text{Sup}[\lambda/F(x^*,\lambda) \leqslant \bar{t}]$.

ii) $\text{Sup}[t/F^-(x^*,t) < \bar{\lambda}] = \text{Inf}[t/F^-(x^*,t) \geqslant \lambda] \leqslant F(x^*,\lambda) \leqslant$

$\leqslant \text{Inf}[t/F^-(x^*,t) > \bar{\lambda}] = \text{Sup}[t/F^-(x^*,t) \leqslant \bar{\lambda}]$.

PROOF: The equalities are due to the fact that $F^-(x^*,.)$ and $F(x^*,.)$ are non-decreasing functions. Let us show the inequalities:

i) Let $\bar{\lambda} = F^-(x^*,\bar{t})$. From statement a) of Proposition 3, $F(x^*,\lambda) < \bar{t}$ implies that $\bar{\lambda} = F^-(x^*,\bar{t}) \geqslant \lambda$. Hence,

$$\text{Sup}[\lambda/F(x^*,\lambda) < \bar{t}] \leqslant \bar{\lambda} = F^-(x^*,\bar{t}).$$

From statement c), $F(x^*,\lambda) > \bar{t}$ implies that $\bar{\lambda} \leqslant \lambda$. Thus

$$\text{Inf}[\lambda/F(x^*,\lambda) > \bar{t}] \geqslant F^-(x^*,\bar{t}).$$

ii) Let $\bar{t} = F(x^*,\bar{\lambda})$ and proceed similarly as above by using statements b) and d). □

Let us give a geometric interpretation of Proposition 3 and Corollary 4. Let x^* be a fixed point in $\mathbb{R}^n$ and put $\Omega = \mathbb{R}$ and $T = \mathbb{R}$. Consider in $T \times \Omega$ the usual and the strict epigraph of $F^-(x^*,.)$ and in $\Omega \times T$ the usual and the strict hypograph of $F(x^*,.)$. That is,

$$\text{epi}(F^-(x^*,.)) = \{(t,\lambda)/\lambda \leqslant F^-(x^*,t)\} \subset T \times \Omega,$$

$$\widetilde{\text{epi}}(F^-(x^*,.)) = \{(t,\lambda)/\lambda < F^-(x^*,t)\} \subset T \times \Omega,$$

$$\text{hypo}(F(x^*,.)) = \{(\lambda,t)/t \geqslant F(x^*,\lambda)\} \subset \Omega \times T,$$

$$\widetilde{\text{hypo}}(F(x^*,.)) = \{(\lambda,t)/t > F(x^*,\lambda)\} \subset \Omega \times T.$$

Then, from Proposition 3 a):

$$\widetilde{\text{hypo}}(F(x^*,.)) \subset \text{epi}(F^-(x^*,.))$$

and from Proposition 3 b):

$$\widetilde{\text{epi}}(F^-(x^*,.)) \subset \text{hypo}(F(x^*,.)),$$

where the four sets are regarded as being subsets of the same product space T x Ω.

For a general function, the epigraph and the strict epigraph have the same closure; of course, the same thing is true for the hypographs. It follows that the epigraph of $F(x^*,.)$ and the hypograph of $F^-(x^*,.)$ have the same closure when regarded as subsets of the same product space. Hence, the function $F(x^*,.)$ can be considered as the inverse function of $F^-(x^*,.)$ and vice versa. This explains why conditions on the convexity of a quasiconvex function can be expressed in terms of the convexity of the functions $F^-(x^*,.)$ or in terms of the concavity of the function $F^-(x^*,.)$.

Let us give a few examples:

i) $f : \mathbb{R} \to \mathbb{R}$

$$f(x) = \begin{cases} n + 1 & \text{if } n < x < n + 1 \\ n + \frac{1}{2} & \text{if } x = n \end{cases}$$

where n = 0, -1, 1, -2, 2, -3, 3, ...

Let $x^* = 1$, then

$F^-(1,t) = f(t)$ for all t

$F(1,\lambda) = n$ where $n \leq \lambda < n + 1$

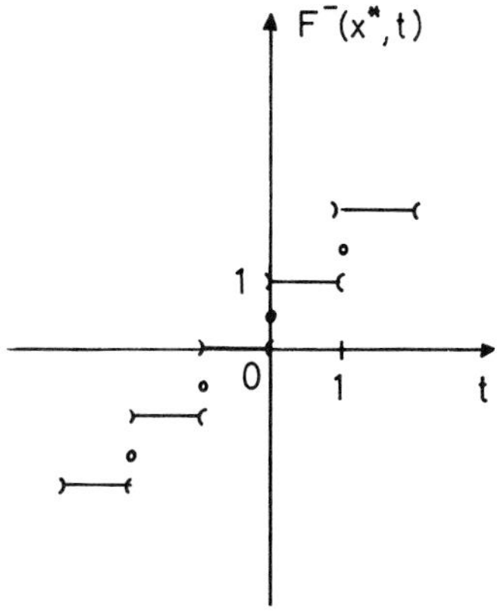

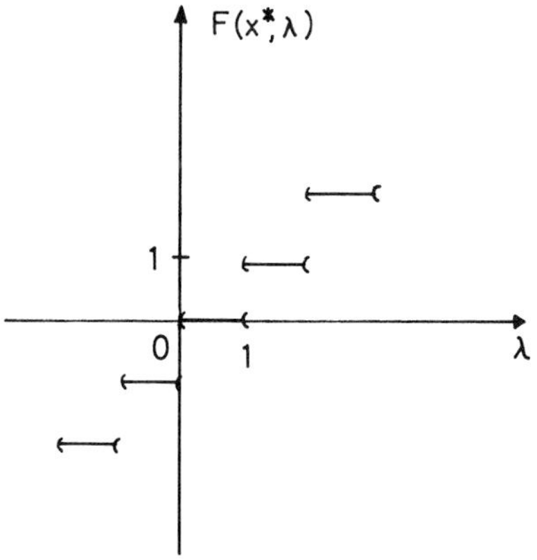

ii) $f : \mathbb{R}^2 \to [-\infty, +\infty]$,

$$f(x_1, x_2) = \begin{cases} -\infty & \text{if } x_1 = 0 \text{ and } x_2 \leq 0, \\ \dfrac{x_2}{x_1} & \text{if } x_1 > 0, \\ +\infty & \text{elsewhere.} \end{cases}$$

Let $x^* = (1,1)$. Then

$$F^-(x^*, t) = \begin{cases} -\infty & \text{if } t \leq 0 \\ -1 & \text{if } t > 0 \end{cases}$$

$$F(x^*, \lambda) = \begin{cases} 0 & \text{if } \lambda \leq -1 \\ +\infty & \text{if } \lambda > -1 \end{cases}$$

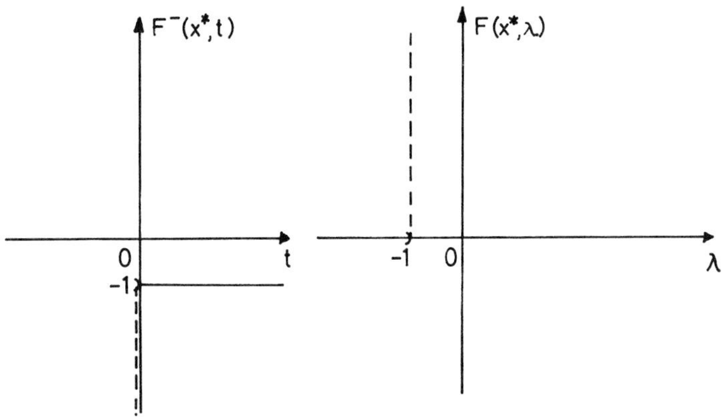

iii) $f : \mathbb{R}^2 \to \mathbb{R}$

$$f(x_1, x_2) = \begin{cases} -x_1 x_2 & \text{if } \max(x_1, x_2) \leq 0, \\ 0 & \text{if } 0 \leq \max(x_1, x_2) \leq 1, \\ \max(x_1, x_2) - 1 & \text{elsewhere.} \end{cases}$$

Let $x^* = (1,0)$. Then

$$F^-(x^*,t) = \begin{cases} -\infty & \text{if } t < 0 \\ 0 & \text{if } 0 \leq t \leq 1 \\ t-1 & \text{if } t \geq 1 \end{cases}$$

$$F(x^*,\lambda) = \begin{cases} 0 & \text{if } \lambda < 0 \\ 1 & \text{if } \lambda = 0 \\ \lambda+1 & \text{if } \lambda > 0 \end{cases}$$

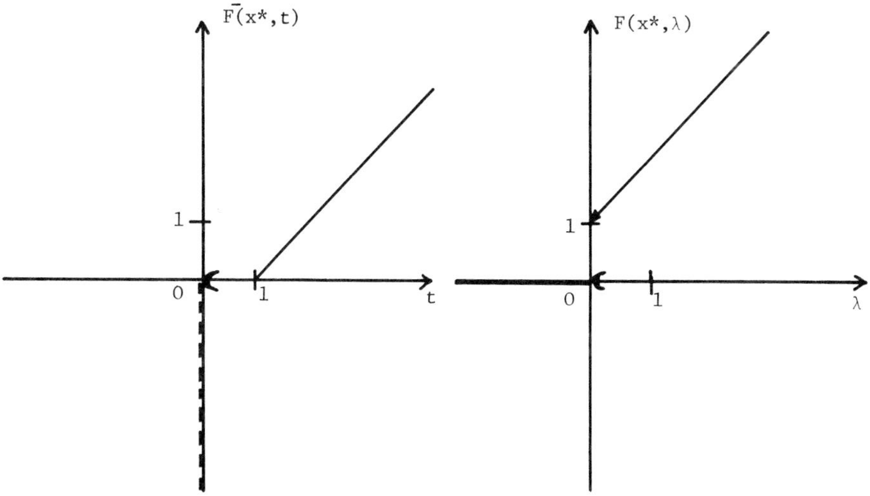

In each of the three examples, the function f is quasiconvex. One notices that in general $F^-(x^*,\bar{t})$ is different from $\text{Inf}[\lambda/F(x^*,\lambda) \geq \bar{t}]$ and $\text{Inf}[\lambda/F(x^*,\lambda) > \bar{t}]$. Indeed in the first example,

$$\text{Inf}[\lambda/F(x^*,\lambda) \geq 0] = 0,\ \text{Inf}[\lambda/F(x^*,\lambda) > 0] = 1,\ F^-(x^*,0) = \frac{1}{2}\ ;$$

whereas in the second example

$$\text{Inf}[\lambda/F(x^*,\lambda) \geq 0] = -\infty,\ \text{Inf}[\lambda/F(x^*,\lambda) > 0] = -1,\ F^-(x^*,0) = -\infty.$$

Notice also that $F(x^*,.)$ is usually neither upper semi-continuous (u.s.c.) (example 2) nor l.s.c. (examples 1 and 3).

For the sake of simplicity, we shall use the following notation:

$$F^-(x^*,\bar{t}_-) = \limsup_{t \uparrow \bar{t}} F^-(x^*,t).$$

Clearly, $F^-(x^*,\bar{t}_-) = \text{Sup}[F^-(x^*,t)/t < \bar{t}]$ because $F^-(x^*,.)$ is non-decreasing.

PROPOSITION 5:

$$\text{Inf}[\lambda/F(x^*,\lambda) \geq \bar{t}] = F^-(x^*,\bar{t}_-)$$

PROOF: Let $\bar{\lambda} = \text{Inf}[\lambda/F(x^*,\lambda) \geq \bar{t}]$ and let $\mu > \bar{\lambda}$ and $t < \bar{t}$. Then $F(x^*,\mu) \geq \bar{t} > t$ and, from Proposition 3 c), $F^-(x^*,t) \leq \mu$. Hence $\bar{\lambda} \geq \text{Sup}[F^-(x^*,t)/t < \bar{t}]$. Now let any $\mu < \bar{\lambda}$. Then $F(x^*,\mu) < \bar{t}$. Hence, there exists $\hat{t}$ such that $F(x^*,\mu) < \hat{t} < \bar{t}$. From Proposition 3 a), $F^-(x^*,\hat{t}) \geq \mu$ and so $\bar{\lambda} \leq \text{Sup}[F^-(x^*,t)/t < \bar{t}]$. □

A function f such that $F^-(x^*,\bar{t}) = F^-(x^*,\bar{t}_-)$ for all x^* and $\bar{t}$ is said to be regular. Regularity conditions can be expressed in terms of lower semi-continuity of the functions $F^-(x^*,.)$ or by saying that the functions $F(x^*,.)$ are increasing. Regular functions are extensively studied in Crouzeix (1977).

Previously, we have noticed that the functions $F^-(.,t)$ are quasi-concave. Now we extend this result to the functions $F^-(.,t_-)$.

PROPOSITION 6: The function $F^-(.,\bar{t}_-)$ *is quasiconcave.*

PROOF: Let $x_1^*, x_2^* \in \mathbb{R}^n$ and $\xi \in]0,1[$. Let $\bar{\mu} = \text{Min}[F^-(x_1^*,\bar{t}_-), F^-(x_2^*,\bar{t}_-)]$. Letting $\mu < \bar{\mu}$, we must prove that $F^-(\xi x_1^* + (1-\xi)x_2^*, \bar{t}_-)$ is greater than μ. Clearly, there exists $t < \bar{t}$ such that $F^-(x_1^*,t) > \mu$ and $F^-(x_2^*,t) > \mu$. Now, it suffices to use the fact that $F^-(.,t)$ is quasi-concave. □

The next property will be useful later.

PROPOSITION 7: Let r *be a function on* $\mathbb{R}^n \times \mathbb{R}^P$ *and let* $h(x) = \text{Inf}[r(x,y)/y \in \mathbb{R}^P]$. *Then*

$$H^-(x^*,t) = R^-((x^*,0),t),$$

$$\textit{and } H^-(x^*,t_-) = R^-((x^*,0),t_-),$$

for all x^* *and* t.

PROOF: an immediate consequence of the definitions. □

3. BIDUAL FUNCTIONS

Let f be any function on $\mathbb{R}^n$. As in Crouzeix (1981), we denote by f_q the greatest function which is quasiconvex and majorized by f, by $f_{\bar{q}}$ the greatest function which is quasiconvex, lower semi-continuous and majorized by f. Then (Proposition 8, Crouzeix (1981))

$$f_{\bar{q}}(x) = \operatorname*{Sup}_{x^*} [\operatorname{Inf}[\lambda / F(x^*,\lambda) \geq \langle x,x^*\rangle]]. \tag{1}$$

Thus by Proposition 5

$$f_{\bar{q}}(x) = \operatorname*{Sup}_{x^*}[F^-(x^*,\langle x,x^*\rangle_-)]. \tag{2}$$

Let us introduce the new function

$$f_G(x) = \operatorname*{Sup}_{x^*}[F^-(x^*, \langle x,x^*\rangle)]. \tag{3}$$

This function is called the "second quasiconjugate function" of f by Greenberg and Pierskalla (1973b) who proved that f_G is quasiconvex and majorized by f (to prove that f_G is quasiconvex, prove that for all x* the function $x \to F^-(x^*, \langle x,x^*\rangle)$ is quasiconvex). On the other hand by Proposition 8 in Crouzeix (1981)

$$f_{\bar{q}}(x) \leq f_G(x) \leq f_q(x) \leq f(x) \quad \text{for all } x.$$

It follows that if f is quasiconvex and lower semi-continuous at x_o then $f(x_o) = f_{\bar{q}}(x_o) = f_G(x_o)$. It should be noticed that if f is only quasiconvex, f_G does not usually coincide with f nor $f_{\bar{q}}$. Consider, for instance, $\alpha \in [0,1]$ and the function f on $\mathbb{R}^n$ defined by

$$f(x_1,x_2) = \begin{cases} 0 \text{ if } x_2 < 0, \\ \alpha \text{ if } x_2 = 0 \text{ and } x_1 < 0, \\ 1 \text{ elsewhere.} \end{cases}$$

Then f is quasiconvex ($f = f_q$), $f(0,0) = 1$, $f_{\bar{q}}(0,0) = 0$, $f_G(0,0) = \alpha$.

The functions $f_{\bar{q}}$ and f_G will play in our quasiconvex duality framework the role of the (convex) biconjugate function f in convex duality.

4. QUASICONVEX DUALITY

Let us turn back to the duality framework of section 1. A first dual program is obtained by considering the bidual function h_G of h,

$$\beta_G = h_G(0) = \operatorname*{Sup}_{u^*} [H^-(u^*,0)] = \operatorname*{Sup}_{u^*} [\phi^-((0,u^*),0)]$$

or more explicitly

$$\beta_G = \operatorname*{Sup}_{u^*} \operatorname*{Inf}_{x,u} [\phi(x,u)/\langle u,u^*\rangle \geqslant 0]. \tag{4}$$

A second dual program is obtained by considering the bidual function $h_{\bar{q}}$ of h

$$\bar{\beta} = h_{\bar{q}}(0) = \operatorname*{Sup}_{u^*} [H^-(u^*,0_-)] = \operatorname*{Sup}_{u^*} [\phi^-((0,u^*),0_-)]$$

thus,

$$\bar{\beta} = \operatorname*{Sup}_{u^*} [\operatorname*{Inf}_{\lambda} (\lambda/\operatorname*{Sup}_{x,u} [\langle u,u^*\rangle/\phi(x,u) \leqslant \lambda] \geqslant 0)]. \tag{5}$$

Let us notice that the function $\phi^-((u^*,0),t)$ is nondecreasing in t. Hence

$$\phi^-((0,u^*),0_-) = \operatorname*{Sup}_{t<0} [\phi^-((0,u^*),t].$$

On the other hand, $\phi^-((0,ku^*),kt) = \phi^-((u^*,0),t)$ for all $k > 0$. Thus

$$\bar{\beta} = \operatorname*{Sup}_{u^*,t<0} [\phi^-((0,u^*),t)] = \operatorname*{Sup}_{u^*} [\phi^-((0,u^*),-\xi)]$$

for all $\xi > 0$.

It easily follows that

$$\bar{\beta} = \operatorname*{Sup}_{u^*} \operatorname*{Inf}_{x,u} [\phi(x,u)/\langle u,u^*\rangle > -1]. \tag{6}$$

Of course, we have

$$\bar{\beta} = h_{\bar{q}}(0) \leqslant \beta_G = h_G(0) \leqslant \alpha = h(0).$$

Let us now define

$$\theta_1(u^*) \equiv \operatorname*{Inf}_{x,u} [\phi(x,u)/\langle u,u^*\rangle \geqslant 0],$$

$$\theta_2(u^*) \equiv \operatorname*{Inf}_{\lambda} [\lambda/ \operatorname*{Sup}_{(x,u)} [\langle u,u^*\rangle/\phi(x,u) \leqslant \lambda] \geqslant 0],$$

$$\theta_3(u^*) \equiv \operatorname*{Inf}_{x,u} [\phi(x,u)/\langle u,u^*\rangle > -1].$$

PROPOSITION 8: θ_1, θ_2 and θ_3 are quasiconcave. Moreover, θ_3 is upper semi-continuous.

PROOF: Propositions 1 and 5 imply that θ_1 and θ_2 are quasiconcave. A slight modification of the proof of Proposition 1 gives the quasiconcavity of θ_3. Let us prove that θ_3 is u.s.c. Let any $\bar{u}^*$ and let us consider any $\lambda > \theta_3(\bar{u}^*)$. Then there exists $\bar{x}$ and $\bar{u}$ such that

$$<\bar{u},\bar{u}^*> > -1 \text{ and } \phi(\bar{x},\bar{u}) < \lambda.$$

In a neighbourhood of $\bar{u}^*$, the inequality $<\bar{u},u^*> > -1$ holds. Hence $\theta_3(u^*) < \lambda$ on this neighbourhood. □

Thus, dual programs (4), (5) and (6) are quasiconcave. Of course, it is possible to replace in (6) the symbol > by the symbol ⩾ but then θ_3 is not necessarily upper semi-continuous. Let us note that in general θ_1 and θ_2 are not necessarily upper semi-continuous.

Let us assume that h is quasiconvex. Then $h(0) = h_{\bar{q}}(0)$ if and only if h is lower semi-continuous at x_o (then $\alpha = \beta_G = \bar{\beta}$). Of course, the equality $\beta_G = \alpha$ may hold while h is not l.s.c. at 0. Generally speaking, if h is upper semi-continuous at each u such that $h(u) < h(0)$, then θ_1 is upper semi-continuous, $\beta_G = \alpha$ and $\theta_1(u^*) = \alpha$ for some u^*. More details can be found in Crouzeix (1977; 60).

5. SURROGATE DUALITY

Let us consider the problem

$$\alpha = \text{Inf } [f_o(x) \;/\; g_i(x) \leqslant 0,\ i = 1, \ldots, p] \tag{7}$$

where $f_o : \mathbb{R}^n \to \bar{\mathbb{R}}$, while $g_i : \mathbb{R}^n \to \mathbb{R}$ $i = 1, \ldots, p$.

Let us take,

$$f(x) = \begin{cases} f_o(x) \text{ if } g_i(x) \leqslant 0 & i = 1, \ldots, p \\ + \infty \quad \text{elsewhere.} \end{cases}$$

$$\phi(x,u) = \begin{cases} f_o(x) \text{ if } g_i(x) + u_i \leq 0 & i = 1, \ldots, p \\ +\infty & \text{elsewhere.} \end{cases}$$

Then $\phi(x,0) = f(x)$ for all x and $\alpha = h(0)$.

Let us calculate the function θ_1

$$\theta_1(u^*) = \operatorname*{Inf}_{x,u} \left[f_o(x) \middle/ \begin{array}{l} g_i(x) + u_i \leq 0 \\ < u, u^* > \geq 0 \end{array} \right]$$

$$\theta_1(u^*) = \operatorname*{Inf}_{x} \operatorname*{Inf}_{u} \left[f_o(x) \middle/ \begin{array}{l} g_i(x) + u_i \leq 0 \\ < u, u^* > \geq 0 \end{array} \right]$$

It is easy to see that

$$\theta_1(u^*) = \begin{cases} \operatorname*{Inf}_x [f_o(x) \ / \ \Sigma u_i^* g_i(x) \leq 0] \text{ if } u^* \geq 0 \\ \operatorname*{Inf}_x [f_o(x)] \quad \text{elsewhere.} \end{cases}$$

Hence,

$$\beta_G = \operatorname*{Sup}_{u^* \geq 0} \operatorname*{Inf}_{x} [f_o(x) \ / \ \Sigma u_i^* g_i(x) \leq 0]. \tag{8}$$

Let us now assume that

(H_1) f_o is quasiconvex, g_i, $i = 1, \ldots, p$, are convex

(H_2) there exists $\bar{x}$ such that $g_i(\bar{x}) < 0 \quad i = 1, \ldots, p$

(H_3) f_o is upper semi-continuous.

Then,

PROPOSITION 9: (Luenberger (1968); see also Greenberg and Pierskalla (1970) ; (1973) and Crouzeix (1977)).

Under the above assumptions, θ_1 *is upper semi-continuous at each non zero* u^*, $\beta_G = \alpha$ *and there exists* u^* *so that* $\theta_1(u^*) = \alpha$.

Remark. Assumption (H_3) is relaxed in Crouzeix (1977).

6. GENERALIZED FRACTIONAL PROGRAMMING

Let us now consider

$$\alpha = \operatorname*{Inf}_{x\in K} \left[\operatorname*{Max}_{i=1,\dots,p} \left[\frac{f_i(x)}{-g_i(x)} \middle/ h_j(x) \leq 0,\ j = 1, \dots, r \right]\right] \tag{9}$$

where i) K is a convex compact set of $\mathbb{R}^n$,

ii) f_i, g_i and h_j are l.s.c. convex functions from K to $\mathbb{R}$,

iii) $f_i(x) > 0$ and $g_i(x) < 0$ $\forall x \in K$, $i = 1, \dots, p$,

iv) there exists $\bar{x} \in K$ such that $h_j(\bar{x}) \leq 0$ for $j = 1, \dots, r$.

These assumptions imply $\alpha > 0$.

If $p = 1$, the above problem can be transformed into a problem of convex programming. This is obtained by using the so-called "transformation of Charnes and Cooper". See for instance Schaible (1976-b), (1976-c). If $p > 1$, the transformation of Charnes and Cooper is not appropriate and it does not seem that (9) can be replaced by an equivalent convex program.

Henceforth, we suppose $p > 1$. Let us note that

$$\alpha = \operatorname*{Inf}_{\substack{x\in K\\ \lambda>0}} \left[\lambda \middle/ \begin{array}{l} f_i(x) + \lambda g_i(x) \leq 0 \\ h_j(x) \leq 0 \end{array} \right] \tag{10}$$

Let us take

$$\phi(x;(u,v)) = \begin{cases} \operatorname*{Inf}_{\lambda} \left[\lambda \middle/ \begin{array}{l} \lambda \geq 0 \\ f_i(x) + u_i + \lambda g_i(x) \leq 0 \end{array} \right] & \text{if } x \in K \text{ and } h_j(x) + v_j \leq 0 \\ +\infty & \text{elsewhere.} \end{cases}$$

where $u \in \mathbb{R}^p$ and $v \in \mathbb{R}^r$,

and according to our general duality framework

$$h(u,v) = \operatorname*{Inf}_{x} [\phi(x;(u,v))]$$

then $\alpha = h(0,0)$.

Let us consider the level sets of ϕ

$$S_\lambda(\phi) = \{(x;(u,v)) \ / \ \phi(x;(u,v)) \leq \lambda\}.$$

Then $S_\lambda(\phi)$ is empty when $\lambda < 0$ and

$$S_\lambda(\phi) = \left\{ (x;(u,v)) \;\middle/\; \begin{array}{l} x \in K \\ h_j(x) + v_j \leqslant 0 \\ f_i(x)+u_i+\lambda g_i(x) \leqslant 0 \end{array} \right\} \quad \text{if } \lambda \geqslant 0.$$

It follows that ϕ is quasiconvex and lower semi-continuous. Hence h is also quasiconvex. In addition,

$$h(u,v) = \operatorname*{Inf}_{x\varepsilon K} [\phi(x;(u,v))].$$

Since K is compact and ϕ lower semi-continuous then h is lower semi-continuous. Hence

$$\bar\beta = \beta_G = \alpha.$$

We intend to consider the dual program obtained from θ_2. Let us define

$$s_\lambda(u^*,v^*) = \operatorname*{Sup}_{x,u,v} [\langle u,u^*\rangle + \langle v,v^*\rangle \;/\; \phi(x;(u,v)) \leqslant \lambda].$$

Clearly, $s_\lambda(u^*,v^*) = -\infty$ for all u^*, v^* if $\lambda < 0$.

Assume that $\lambda \geqslant 0$,

- If $u^*_i < 0$ for some $i \in \{1, \ldots, p\}$ or $v^*_j < 0$ for some $j \in \{1, \ldots, r\}$, then $s_\lambda(u^*,v^*) = +\infty$.
- If $u^* \geqslant 0$ and $v^* \geqslant 0$, then

$$s_\lambda(u^*,v^*) = \operatorname*{Sup}_{x\varepsilon K} [- \sum_i u^*_i(f_i(x) + \lambda g_i(x)) - \sum_j v^*_j h_j(x)].$$

Notice that $s_\lambda(u^*,v^*) \geqslant 0$ if $u^* = 0$ and $v^* \geqslant 0$ (from assumption iv).

Recall that

$$\theta_2(u^*,v^*) = \operatorname{Inf} [\lambda / s_\lambda(u^*,v^*) \geqslant 0] = \operatorname*{Sup}_{\lambda} [\lambda / s_\lambda(u^*,v^*) < 0].$$

Since K is compact

$$\theta_2(u^*,v^*) = \operatorname{Max} \left[0, \operatorname*{Min}_{x\varepsilon K} \left[\frac{\sum u^*_i f_i(x) + \sum v^*_j h_j(x)}{-\sum u^*_i g_i(x)} \right] \right] \quad \text{if } u^* \geqslant 0,\ v^* \geqslant 0 \text{ and } u^* \neq 0,$$

and

$$\theta_2(u^*,v^*) = 0 \text{ elsewhere.}$$

Therefore, since $\alpha > 0$

$$\alpha = \sup_{\substack{u^*\geq 0 \ v^*\geq 0 \\ u^*\neq 0}} \ \min_{x\in K} \left[\frac{\Sigma u_i^* f_i(x) + \Sigma v_j^* h_j(x)}{-\ \Sigma u_i^* g_i(x)} \right]. \tag{11}$$

For the particular case where functions g_i are affine, see Gol'stein (1972).

7. VON NEUMANN MODEL OF AN EXPANDING ECONOMY

In his celebrated paper published in 1937 and in an English translation von Neumann (1945) constructed a model of an economy for which the rate of expansion is equal to the rate of interest. This is a result of duality which cannot be obtained from convex duality. The intention in this section is to show that this result is connected with quasiconvexity.

Let us briefly describe the static von Neumann model which was developed in Kemeny, Morgenstern and Thompson (1956). One considers a finite set of m processes that produce a finite number n of different goods and a pair of *non-negative* matrices, an input matrix A and an output matrix B. If process j operates at a level x_j $(x_j \geq 0)$ it requires $a_{ij}x_j$ units of the goods $i(i=1, \ldots, n)$ and produces $b_{kj}x_j$ units of the goods k $(k = 1, \ldots, n)$. Symbolically, a non-negative activity x $(x \in \mathbb{R}^n)$ requires goods Ax and produces goods Bx. The problem to be solved is:

$$\bar{\theta} = \sup_{\lambda,x} \left[\lambda / \begin{array}{c} x \geq 0 \quad x \neq 0 \\ \lambda Ax \leq Bx \end{array} \right]. \tag{N.P.}$$

Then $\bar{\theta}$ is called the maximal growth rate.

Matrices A and B are assumed to fulfill the following conditions.

(H_1) $x \geq 0$ and $x \neq 0 \Longrightarrow Ax \neq 0$

(every non-zero activity uses some inputs)

(H_2) $\forall b \geq 0$, $\exists x \geq 0$ such that $Bx \geq b$

(any quantity of goods can be produced in the economy).

It is a simple exercise in linear programming to show that condition (H_1) is equivalent to

(H_1') $\forall c \geq 0\ \exists u^* \geq 0$ such that $A^t u^* \geq c$

whereas condition (H_2) is equivalent to

(H_2') $u^* \geq 0$ and $u^* \neq 0 \Longrightarrow B^t u^* \neq 0$.

Let us define,

$$K = \{x \in \mathbb{R}^n \ / \ x_i \geq 0 \quad i = 1, \ldots, n \text{ and } \Sigma x_i = 1\}$$

$$\theta(x) = \text{Sup } [\lambda \ / \ \lambda Ax \leq Bx] \ , \quad x \in K$$

then

$$\bar{\theta} = \text{Sup } [\theta(x) \ / \ x \in K] \ .$$

One can easily see that $\{x/\theta(x) \geq \lambda\} = \{x \in K/\lambda Ax \leq Bx\}$.

Hence, θ is upper semi-continuous on K and $\bar{\theta}$ is attained for some $x \in K$ implying that $\bar{\theta}$ is finite. On the other hand, there exists $\bar{x} \in K$ such that $(B\bar{x})_i > 0$ for all i. It follows that $\bar{\theta}$ is positive.

Let us now consider

$$\alpha = \underset{x,\lambda}{\text{Inf}} \ [\lambda/\lambda \geq 0, \ x \in K \text{ and } Ax \leq \lambda Bx] \tag{12}$$

then $\bar{\theta} = \frac{1}{\alpha}$ and $0 < \alpha < \infty$. Let us define

$$\phi(x,u) = \begin{cases} \text{Inf } [\lambda/\lambda \geq 0 \text{ and } Ax + u \leq \lambda Bx] & \text{if } x \in K, \\ + \infty & \text{otherwise.} \end{cases}$$

$$h(u) = \underset{x}{\text{Inf}} \ \phi(x,u) = \underset{x \in K}{\text{Inf}} \ \phi(x,u)$$

then $\alpha = h(0)$. Put $S_\lambda = \{(x,u)/\phi(x,u) \leq \lambda\}$. Then

$$S_\lambda = \phi \text{ if } \lambda < 0$$

$$S_\lambda = \{(x,u)/x \in K \text{ and } Ax + u \leq \lambda Bx\} \text{ if } \lambda \geqslant 0.$$

Since S_λ is convex and closed for all λ, ϕ is quasiconvex and lower semi-continuous. Hence, h is quasiconvex and lower semi-continuous (because K is compact).

Let us now compute $s_\lambda(u^*) = \underset{u}{\text{Sup}} \ [\langle u,u^*\rangle/(x,u) \in S_\lambda]$.

If $\lambda < 0$, then $s_\lambda(u^*) = -\infty$ (because S_λ is empty)

If $\lambda \geq 0$, then:

i) if u^* has at least one negative component then $s_\lambda(u^*) = +\infty$

ii) if $u^* \geq 0$, then $s_\lambda(u^*) = \operatorname{Sup}_{x \in K} [\langle x, \lambda B^t u^* - A^t u^* \rangle]$. Note that $s_\lambda(u^*)$ is negative if and only if $\lambda B^t u^* - A^t u^* < 0$ (where by $y < 0$ we mean $y_i < 0$ for all i).

Recall that

$$\theta_2(u^*) = \operatorname{Inf}_\lambda [\lambda / s_\lambda(u^*) \geq 0] = \operatorname{Sup}_\lambda [\lambda / s_\lambda(u^*) < 0]$$

Thus, we obtain the dual problem

$$\alpha = \operatorname*{Sup}_{\substack{\lambda, u^* \geq 0 \\ u^* \neq 0}} [\lambda / \lambda B^t u^* < A^t u^*], \tag{13}$$

and by normalizing

$$\alpha = \operatorname*{Sup}_{\lambda, u^* \in K} [\lambda / \lambda B^t u^* < A^t u^*]. \tag{14}$$

Let us turn back to the initial problem. It immediately follows that

$$\bar{\mu} = \operatorname*{Inf}_{\mu, u^*} \left[\mu \middle/ \begin{matrix} u^* \geq 0 \;\; u^* \neq 0 \\ B^t u^* \leq \mu A^t u^* \end{matrix} \right] = \operatorname*{Sup}_{\lambda, x} \left[\lambda \middle/ \begin{matrix} x \geq 0 \quad x \neq 0 \\ \lambda A x < B x \end{matrix} \right] \leq$$

$$\leq \operatorname*{Sup}_{\lambda, x} \left[\lambda \middle/ \begin{matrix} x > 0 \quad x \neq 0 \\ \lambda A x \leq B x \end{matrix} \right] = \operatorname*{Inf}_{\mu, u^*} \left[\mu \middle/ \begin{matrix} u^* \geq 0 \quad u^* \neq 0 \\ B^t u^* < \mu A^t u^* \end{matrix} \right] = \bar{\theta}.$$

In his model, von Neumann made the following additional assumption:

$$a_{ij} + b_{ij} > 0 \quad \text{for all i and j.}$$

Under this assumption $\bar{\mu}$ turns out to be equal to $\bar{\theta}$.

The value $\bar{\mu}$ is the lowest interest rate at which a profitless system is possible.

8. SOME OTHER APPROACHES OF QUASICONVEX DUALITY

Connections with surrogate duality and the related paper of Greenberg and Pierskalla (1973) were investigated above. Concerning surrogate duality see also Greenberg (1970 ; 1973), Glover (1975); this list is not exhaustive.

An approach connected to the Legendre transformation can be found in Passy and Keslassy (1980) for a class of generalized fractional programs.

If one considers the functions $F(x^*,\lambda)$ and $F^-(x^*,t)$, we noticed that $F(kx^*,\lambda) = kF(x^*,\lambda)$ and $F^-(kx^*,kt) = F^-(x^*,t)$ for all $k > 0$, x^*, λ and t. To evaluate, for instance $F^-(x^*,t)$ for all x^* and t it suffices to evaluate $F^-(x^*,-1)$, $F^-(x^*,0)$ and $F^-(x^*,1)$ for all x^*. Given a function f, El Qortobi (1980) associates with f two functions,

$$f^{o}(x^*) = -\operatorname{Inf}\,[f(x) \;/\; \langle x,x^*\rangle > 1] = -F^-(x^*, 1_+)$$

$$f^{\square}(x^*) = -\operatorname{Inf}\,[f(x) \;/\; \langle x,x^*\rangle < 1] = -F^-((-x^*), -1_+)$$

where $F^-(x^*,\bar{t}_+) = \lim_{t \downarrow \bar{t}} F^-(x^*,t)$.

These two functions turn out to be quasiconvex and lower semi-continuous and El Qortobi defines $(f^{o}, f^{\square})$ as being the quasiconjugate of f. Then $f_{\bar{q}}$ can be obtained from f^{o} and $f^{\square}$.

An important class of functions arising in economics is the class of monotonic functions which are defined on the non-negative orthant. For these functions, it is sufficient to evaluate $F^-(x^*,1)$ in order to have $F^-(x^*,t)$ for all t. The function which associates with x^* the value $F^-(x^*,1)$ is connected to the indirect utility function. See, for instance, Lau (1970), Diewert (1974 ; 1978 ; 1981).

Finally, Flachs (1979) investigated the special class of functions $f(x,t)$, $t \in \mathbb{R}$ which are monotonic in t.

REFERENCES

CHARNES, A. and COOPER, V.W. (1962). "Programming with Linear Fractional Functionals." *Nav. Res. Log. Quart.* *9*, 181-186.

CROUZEIX, J.-P. (1977). "Contributions à l'Etude des Fonctions Quasi-convexes." Thèse Université de Clermont.

CROUZEIX, J.-P. (1981). "Continuity and Differentiability Properties of Quasiconvex Functions on $\mathbb{R}^n$". This volume, 109-130.

DIEWERT, W.E. (1974). "Applications of Duality Theory." In *Frontiers of Quantitative Economics.* (Intriligator M.D. and Kenduck D.A., Editors), Vol.II. North-Holland Publishing Company, Amsterdam.

DIEWERT, W.E. (1978). "Duality Approaches to Microeconomic Theory." Discussion paper 78-09, Dept. of Economics, University of British Columbia, Vancouver.
DIEWERT, W.E. (1981). "Generalized Concavity and Economics." This volume, 511-541.
EL QORTOBI, A. (1980). "Contributions à la Théorie de la Dualité pour les Fonctionnelles Quasiconvexes." Thése de 3ème Cycle, Université de Toulouse.
FLACHS, J. (1979). "Global Saddle-Point Theorems Involving Conjugate Operators for General and Quasiconvex Programs." Technical Report, National Research Institute for Mathematical Sciences, Pretoria.
GLOVER, F. (1975). "Surrogate Duality Constraint in Mathematical Programming." *Operations Research 23*, 434-451.
GOL'STEIN, E.G. (1972). "Theory of Convex Programming." Translations of mathematical monographs, American Math. Society.
GREENBERG, H.J. (1970). "Surrogate Mathematical Programs." *Operations Research 18*, 924-939.
GREENBERG, H.J. (1973). "The Generalized Penalty Function Surrogate Model." *Operations Research 21*, 162-178.
GREENBERG, H.J. and PIERSKALLA, W.P. (1973). "Quasiconjugate Functions and Surrogate Duality." *Cahiers du Centre d'Etude de Recherche Operationnelle 15*, 437-448.
JOLY, J.L. and LAURENT, P.J. (1971). "Stability and Duality in Convex Minimization Problems." *Revue Française d'Informatique et de Recherche Operationnelle R2*, 3-42.
KEMENY, J.G., MORGENSTERN, O. and THOMPSON, G.L. (1956). "A Generalization of the von Neumann Model of an Expanding Economy." *Econometrica 24*, 115-135.
LAU, L.J. (1970). "Duality and the Structure of Utility Functions." *Journal of Economic Theory 1*, 374-396.
LUENBERGER, D.G. (1968). "Quasiconvex Programming." *SIAM Journal of Applied Mathematics 16*, 1090-1095.
VON NEUMANN, J. (1945). "A Model of General Economic Equilibrium." *Review of Economic Studies 13*, 1-9.
PASSY, U. and KESLASSY, A. (1980). "Pseudo Duality and Duality for Explicitly Quasiconvex Functions." Mimeograph Series No.249, Technion, Haifa.
ROCKAFELLAR, R.T. (1970). *Convex Analysis*. Princeton University Press, Princeton, N.J.
ROCKAFELLAR, R.T. (1974). "Conjugate Duality and Optimization." *C.B.M.S. Series No.16, SIAM.*
SCHAIBLE, S. (1976a). "Minimization of Ratios." *Journal of Optimization Theory and Applications 19*, 347-352.
SCHAIBLE, S. (1976b). "Fractional Programming I, Duality." *Management Science 22*, 858-867.
SCHAIBLE, S. (1976c). "Duality in Fractional Programming: A Unified Approach." *Operations Research 24*, 452-461.

OPTIMALITY CONDITIONS INVOLVING GENERALIZED CONVEX MAPPINGS

W. Oettli

For optimization problems involving quasiconvex functions and quasiconvex mappings the classical Kuhn-Tucker conditions are inappropriate. In the following a type of optimality conditions, which has been introduced originally by Luenberger, is presented in the general framework of topological vector spaces. These conditions are then exploited for studying duality and biduality.

This note is concerned with the programming problem

$$\inf \{f(x) \mid x \in X,\ 0_Y \in \Gamma(x)\}, \tag{1}$$

where: X, Y are real topological vector spaces,

$f : X \to [-\infty, +\infty]$ is an extended real-valued function,

$\Gamma : X \to 2^Y$ is a multivalued mapping.

The symbol 0_Y, of course, means the origin of Y, henceforth denoted simply by 0. Ordinary programming problems of the form

$$\inf \{f(x) \mid x \in A,\ G(x) \in P\},$$

where $P \subset Y$, and $G : A \to Y$ is a single-valued mapping, can be subsumed under (1) by setting

$$\Gamma(x) := \begin{cases} \{G(x)\} - P & \text{if } x \in A \\ \emptyset & \text{if } x \notin A. \end{cases} \tag{2}$$

In the following we continue our exposition of optimality conditions for programming problems involving multivalued mappings (see Oettli (1980)), extending it from the convex case to the quasiconvex case.

Recall that a multivalued mapping $\Gamma : X \to 2^Y$ is called *convex* iff the set

$$\operatorname{graph} \Gamma := \{(x,y) \in X \times Y \mid y \in \Gamma(x)\}$$

is convex. And a function $f : X \to [-\infty, +\infty]$ is called *convex* iff the set

$$\operatorname{epi} f := \{(x,r) \in X \times \mathbb{R} \mid r \geq f(x)\}$$

is convex.

ISBN 0-12-621120-5

If the given vector space Y bears no topology originally, then for application of the following results we may endow Y with the "convex core topology", which occurs also later on in this note. In this topology, by definition, $y \in \text{int } A$, the topological interior of A, iff there exists a convex subset $C \subset A$ such that $C-y$ is absorbing in Y, i.e., for every $\eta \in Y$ there exists $\lambda_0 > 0$ such that $y+\lambda\eta \in C$ for all $\lambda \in [0,\lambda_0]$. It is the strongest locally convex, separated vector space topology on Y. Under the convex core topology every linear functional is continuous, and every convex function, in the interior of the domain where it is $\neq +\infty$, is upper semicontinuous (usc). Furthermore under this topology the topological interior of any convex set coincides with its algebraic interior (the core; see Holmes (1975)). In $\mathbb{R}^n$ the convex core topology coincides with the natural topology.

We recall now some background material about optimality conditions (see Oettli (1980)). Throughout this paper we use the perturbation function $\sigma : Y \to [-\infty,+\infty]$ of (1), namely

$$\sigma(y) := \inf \{f(x) \mid x \in X,\ y \in \Gamma(x)\} \tag{3}$$

(we use the convention that the infimum over an empty set is $+\infty$). In particular $\sigma(0)$ equals the infimum value of problem (1). We assume once and for all that this infimum is finite, $-\infty < \sigma(0) < +\infty$. We use the abbreviation

$$\text{dom } \sigma := \{y \in Y \mid \sigma(y) < +\infty\}.$$

For problem (1) the so-called Kuhn-Tucker condition may be stated succinctly as

$$\hat{f} \leq \sigma(y) + l(y) \quad \forall y \in Y, \tag{4}$$

where $l : Y \to \mathbb{R}$ is a continuous linear functional, and $\hat{f} \in \mathbb{R}$. The number $\hat{f}$ usually equals the infimum value of (1) or a minorant thereof, $\hat{f} \leq \sigma(0)$. Condition (4) can be written in the more explicit, but obviously equivalent form

$$\hat{f} \leq f(x) + l(y) \ \forall x \in A, \ y \in \Gamma(x), \tag{5}$$

where

$$A := \{x \in X \mid f(x) < \infty \ \& \ \Gamma(x) \neq \emptyset\}.$$

It is clear that

$$\operatorname{dom} \sigma = \Gamma(A) := \cup\{\Gamma(x) \mid x \in A\}.$$

Note that the inequality in (4) is trivially satisfied for $y \notin \Gamma(A)$, since then the right-hand side becomes $+ \infty$.

If Γ is given by (2), then condition (5) is easily seen to be equivalent to

$$\begin{cases} \hat{f} + \alpha \leq f(x) + l(G(x)) \ \forall x \in A, \\ l(y) \leq \alpha \ \forall y \in P, \end{cases} \tag{6}$$

where $\alpha \in \mathbb{R}$. It is obvious that (6) implies (5), and in order to derive (6) from (5) one sets $\alpha := \sup\{l(y) \mid y \in P\}$, this sup being finite because of (5). If P is a cone (i.e., $\lambda P \subset P \ \forall \lambda > 0$), then α may be set equal to zero. Incidentally, the often encountered requirement that a single valued mapping $G(\cdot)$ should be P-convex means exactly that $\Gamma(\cdot)$, as given by (2), should be a convex multivalued mapping.

Condition (4) obviously implies $\hat{f} \leq \sigma(0)$, and hence with $\hat{f} := f(\hat{x})$ is sufficient for the optimality of a feasible point $\hat{x}$. The validity of (4) with $\hat{f} := \sigma(0)$, i.e., as a necessary condition of optimality, is strongly related to the convexity and continuity properties of the perturbation function σ. If the function f is convex and the mapping Γ is convex, then σ, as given by (3), is again convex. And if σ is convex, then the upper semicontinuity of σ at $y = 0$ implies the existence of $l \in Y^*$ such that (4) holds with $\hat{f} := \sigma(0)$.

Finally, if $\hat{f} := \sigma(0)$, then the validity of (4) is equivalent to

$$\hat{f} = \max\{\sigma^*(y^*) \mid y^* \in Y^*\} = \sigma^*(l) \ \text{(duality)}, \tag{7}$$

where Y^* is the space of all continuous linear functionals over Y, and $\sigma^* : Y^* \to [-\infty, +\infty]$ is given by

$$\begin{aligned}\sigma^*(y^*) &:= \inf\{\sigma(y) + y^*(y) \mid y \in Y\} \\ &= \inf\{f(x) + y^*(y) \mid x \in A,\ y \in \Gamma(x)\}.\end{aligned} \tag{8}$$

This is easily verified; we refer to Oettli (1980).

Now a version of an optimality condition, which is only dependent on quasiconvexity of σ, has been introduced by Luenberger (1968) (see also Horst (1974), Singer (to appear)); it is given by (10) below. In fact this condition is of interest even in the convex case, since it shows almost all advantages of the Kuhn-Tucker condition, yet holds under weaker regularity assumptions. Hence it may replace very well the Kuhn-Tucker condition in many instances.

Let us insert here the relevant definitions: A function $f: X \to [-\infty, +\infty]$ is called *quasiconvex*, iff the level sets

$$\{x \in X \mid f(x) < \alpha\}$$

are convex for all $\alpha \in \mathbb{R}$. The function f is called *quasiconcave*, iff $-f$ is quasiconvex. A multivalued mapping $\Gamma : X \to 2^Y$ is called *quasiconvex* iff

$$\lambda\Gamma(x_1) + (1-\lambda)\Gamma(x_2) \subset \cup\{\Gamma(\xi) \mid \xi \in [x_1, x_2]\}$$

for all $x_1, x_2 \in X$ and all $\lambda \in [0,1]$.

Obviously a convex mapping is also quasiconvex, and a convex function is also a quasiconvex function. Moreover, $\Gamma : X \to 2^Y$ is quasiconvex if and only if $\Gamma(C) := \cup\{\Gamma(x) \mid x \in C\}$ is convex for every convex subset $C \subset X$.

PROPOSITION 1. If $f : X \to [-\infty, +\infty]$ *is quasiconvex, and* $\Gamma : X \to 2^Y$ *is quasiconvex, then* $\sigma(\cdot)$*, as given by (3), is quasiconvex.*

PROOF: Let $\sigma(y_i) < \alpha (i = 1,2)$. By definition of σ there exist $x_i \in X (i = 1,2)$ with $y_i \in \Gamma(x_i)$, $f(x_i) < \alpha$. For any $\eta \in [y_1, y_2]$ one has $\eta \in \Gamma(\xi)$ for some $\xi \in [x_1, x_2]$ by quasiconvexity of Γ, and by the quasiconvexity of f follows then $\sigma(\eta) \leq f(\xi) < \alpha$. ||

Now we have the following simple result, which applies in particular if σ is quasiconvex.

PROPOSITION 2. Let $\hat{f} \in \mathbb{R}$, *and let* $D := \{y \in Y | \sigma(y) < \hat{f}\}$ *be convex and open in* Y. *Then*

$$\hat{f} \leq \sigma(0) \tag{9}$$

if and only if there exists a continuous linear functional $l : Y \to \mathbb{R}$ *such that*

$$\hat{f} \leq \sigma(y) \text{ for all } y \in Y \text{ satisfying } l(y) \leq 0. \tag{10}$$

PROOF: That (10) implies (9) is trivial. Let (9) be satisfied. Then $0 \notin D$. If D is empty, then $\hat{f} \leq \sigma(y)$ $\forall y \in Y$, and (10) holds with $l = 0$. If D is not empty, then by the separation theorem (see Rudin (1973)) there exists a continuous linear functional $l \in Y^*$ such that $0 < l(y)$ $\forall\, y \in \text{int}\, D = D$. Then $l(y) \leq 0$ implies $y \notin D$ from which follows $\hat{f} \leq \sigma(y)$. Consequently (10) holds.||

In explicit form (9) says

$$\hat{f} \leq \inf \{f(x) | x \in X,\ 0 \in \Gamma(x)\}, \tag{11}$$

and (10) says

$$\hat{f} \leq f(x) \text{ for all } x \in X \text{ for which there exists } y \in \Gamma(x) \text{ satisfying } l(y) \leq 0. \tag{12}$$

Obviously condition (10) is implied by the Kuhn-Tucker condition (4). If $\hat{x}$ is feasible for (1), and (10) is satisfied with $\hat{f} := f(\hat{x})$, then via (11) $\hat{x}$ is optimal for (1). Thus condition (10) becomes a sufficient optimality condition.

It is possible to derive (10) from (9) under somewhat different assumptions which in essence go back to Luenberger (1968). This is the content of the next proposition.

PROPOSITION 3. Let the following assumptions hold:

i) $D := \{y \in Y | \sigma(y) < \sigma(0)\}$ *is convex, and if* $D \neq \emptyset$, *then* $\text{int } D \neq \emptyset$;

ii) $\text{dom } \sigma$ *is convex, and* 0 *is in the interior of* $\text{dom } \sigma$ *with regard to the convex core topology;*

iii) D *is open in* $\text{dom } \sigma$ *with regard to the convex core topology.*

Then there exists $l \in Y^*$ *such that*

$$\sigma(0) \leq \sigma(y) \text{ for all } y \in Y \text{ satisfying } l(y) \leq 0. \tag{13}$$

PROOF: If $D = \emptyset$, then the statement is true for $l = 0$. Assume $D \neq \emptyset$. By i) the separation theorem is applicable, since $0 \notin D$, and furnishes $l \in Y^*$, $l \neq 0$, such that $l(0) \leq l(y) \ \forall y \in D$, i.e., $l(y) < 0$ implies $y \notin D$. We have to settle the case $l(y) = 0$. If $y \in \text{dom } \sigma$ is such that $l(y) = 0$, then by ii) there exists $\eta_o \in \text{dom } \sigma$ with $l(\eta_o) < 0$, and then for all $\eta \in [\eta_o, y)$ we have $l(\eta) < 0$ and consequently $\eta \notin D$, which by iii) implies $y \notin D$, too. Hence $l(y) \leq 0$ implies $y \notin D$, which is (13).||

Assumption iii) is satisfied if $\sigma|_{\text{dom } \sigma}$ is usc with regard to the convex core topology. Note that for a quasiconvex function upper semicontinuity with regard to the convex core topology is equivalent to upper semicontinuity along lines. In $\mathbb{R}^n$ it is equivalent to upper semicontinuit with regard to the natural topology. Also note that an interior point of a convex set under the given topology is interior under the convex core topology, too. Assumption iii) can be replaced by the requirement iii'): $\sigma|_{\text{dom } \sigma}$ is usc along lines. If σ is convex, this is automatically satisfied, and conditions i) and ii) are then implied by the requirement "$0 \in \text{int dom } \sigma$, and σ is usc at $y = 0$"; the latter is the standard regularity assumption used in the convex case to establish the Kuhn-Tucker condition (4).

If $Y = \mathbb{R}^n$, then the requirement $\text{int } D \neq \emptyset$ in i) becomes superfluous, since it is no longer needed among the premises of the separation theorem.

Turning now to the specific form of σ as given by (3), we see easily that i) - iii') are fulfilled for instance if: Γ is quasiconvex, f is quasiconvex, $f|_{\text{dom } f}$ is usc along lines, int $\Gamma(x) \neq \emptyset$ $\forall x \in A$, $\exists x^o \in X$ with $0 \in \text{int } \Gamma(x^o)$.

DUALITY. Let Y* again denote the topological dual space of Y. Define the dual function $\sigma^* : Y^* \to [-\infty, +\infty]$ by means of

$$\begin{aligned} \sigma^*(y^*) &:= \inf \{\sigma(y) \mid y \in Y,\ y^*(y) \le 0\} \\ &= \inf \{f(x) \mid x \in X,\ y \in \Gamma(x),\ y^*(y) \le 0\} \quad (y^* \in Y^*). \end{aligned} \tag{14}$$

The function σ^* is positively homogeneous of degree zero, i.e., $\sigma^*(\lambda y^*) = \sigma^*(y^*)$ $\forall \lambda > 0$.

For σ^* as given by (14) the following propositions are valid without reference to the particular form of σ:

PROPOSITION 4. The function σ^* is quasiconcave.

PROOF: Let $\sigma^*(y_i^*) > \alpha$ for i = 1,2, and let $\eta^* \in [y_1^*, y_2^*]$. We have to show that $\sigma^*(\eta^*) > \alpha$. We have $\sigma^*(y_i^*) > \alpha + \varepsilon$ for some $\varepsilon > 0$ (i = 1,2). From (14) we see then that $\sigma(y) \le \alpha + \varepsilon$ implies $y_i^*(y) > 0$ (i = 1,2) and therefore also $\eta^*(y) > 0$. Hence $\eta^*(y) \le 0$ implies $\sigma(y) > \alpha + \varepsilon$. By (14) then $\sigma^*(\eta^*) \ge \alpha + \varepsilon > \alpha$.||

Note that we have used only the fact that $y^*(y)$ is quasiconcave in $y^* \in Y^*$ for fixed $y \in Y$.

PROPOSITION 5 (duality). Condition (13) holds if and only if

$$\sigma(0) = \sigma^*(1) = \max \{\sigma^*(y^*) \mid y^* \in Y^*\}. \tag{15}$$

PROOF: From (13) follows by the definition of σ^* that

$$\sigma(0) \le \sigma^*(1).$$

On the other hand for any $y^* \in Y^*$ we have from the definition of σ^* that

$$\sigma^*(y^*) \leq \sigma(0).$$

These two facts together imply (15). Conversely, if (15) is satisfied, then

$$\sigma(0) = \sigma^*(1) = \inf \{\sigma(y) | y \in Y, 1(y) \leq 0\},$$

and this implies (13).||

PROPOSITION 6. If σ is upper semicontinuous on Y with regard to the convex core topology, then σ^* is upper semicontinuous on $Y^* \setminus \{0\}$ with regard to the weak topology.*

PROOF: We have to show that for any $\alpha \in \mathbb{R}$ the set

$$\Lambda^*_\alpha := \{y^* \in Y^* \setminus \{0\} | \sigma^*(y^*) < \alpha\}$$

is weak* open. We know that

$$\Lambda_\alpha := \{y \in Y | \sigma(y) < \alpha\}$$

is open with regard to the convex core topology. Fix $\eta^* \in \Lambda^*_\alpha$. Then there exists $\eta \in Y$ with $\sigma(\eta) < \alpha$, $\eta^*(\eta) \leq 0$.

Case I: $\eta^*(\eta) < 0$. Set

$$U := \{y^* \in Y^* | y^* (\eta) < 0\}.$$

Then U is weak* open, and $\eta^* \in U \subset \Lambda^*_\alpha$.

Case II: $\eta^*(\eta) = 0$. Then there exists $t \in Y$ with $\eta^*(t) < 0$ and $\sigma(\eta+t) < \alpha$, because Λ_α is open in the convex core topology. Set

$$U := \{y^* \in Y^* | y^*(\eta+t) < 0\}.$$

Again U is weak* open, and $\eta^* \in U \subset \Lambda^*_\alpha$.||

If Y^* is endowed with some vector space topology, and if Y^{**} is the space of all continuous linear functionals on Y^*, then we may define the bidual function $\sigma^{**} : Y^{**} \to [-\infty, +\infty]$ by means of

$$\sigma^{**}(y^{**}) := \sup \{\sigma^*(y^*) | y^* \in Y^*, y^{**}(y^*) \leq 0\}. \tag{16}$$

It is quasiconvex by Proposition 4, and positively homogeneous of degree zero. Since every element $y \in Y$ induces on Y^* a linear functional $y(\cdot)$ by means of

$$y(y^*) := y^*(y) \quad \forall y^* \in Y^*,$$

it is then meaningful to compare σ and σ^{**} on $Y \cap Y^{**}$. It is easily seen that

$$\sigma^{**}(y) \leq \sigma(y) \quad \forall y \in Y \cap Y^{**},$$

since by (16) $\sigma^{**}(y) = \sup \{\sigma^*(y^*) | y^*(y) \leq 0\}$, and since $y^*(y) \leq 0$ implies, by (14), that $\sigma^*(y^*) \leq \sigma(y)$. Furthermore we have

$$\sigma^{**}(0) = \sigma(0) \iff \sigma(0) = \sup \{\sigma^*(y^*) | y^* \in Y^*\},$$

since by (16) $\sigma^{**}(0) = \sup \{\sigma^*(y^*) | y^* \in Y^*\}$. Since in particular (13) implies by Proposition 5 that

$$\sigma(0) = \max \{\sigma^*(y^*) | y^* \in Y^*\},$$

we have arrived at the following result:

PROPOSITION 7 (biduality). $\sigma^{**}(y) \leq \sigma(y)$ $\forall y \in Y \cap Y^{**}$, *and if (13) is satisfied, then* $\sigma^{**}(0) = \sigma(0)$.

In this situation $\sigma^{**}(y)$ is a quasiconvex (and positively homogeneous) support functional to $\sigma(y)$ at the origin, and one could term σ^{**} a *quasiconvex subgradient* of $\sigma|_{Y \cap Y^{**}}$.

PROPOSITION 8. If Y *is locally convex, and if* $\sigma : Y \to [-\infty, +\infty]$ *is quasiconvex and lower semicontinuous at* $y = 0$, *then* $\sigma^{**}(0) = \sigma(0)$.

PROOF: Choose $\varepsilon > 0$ arbitrary. Then there exists in Y a convex neighborhood U of zero such that U is disjoint from the convex set $D := \{y \in Y | \sigma(y) < \sigma(0) - \varepsilon\}$. Hence U and D can be separated, and we obtain a continuous linear functional $y^*_\varepsilon \in Y^*$ such that $y^*_\varepsilon(y) > 0$ for all $y \in D$, i.e.,

$$\sigma(0) - \varepsilon \leq \sigma(y) \text{ for all } y \in Y \text{ satisfying } y^*_\varepsilon(y) \leq 0.$$

Consequently from (14) it follows

$$\sigma(0) - \varepsilon \leq \sigma^*(y^*_\varepsilon).$$

Since ε was arbitrary we obtain then

$$\sigma(0) \leq \sup \{\sigma^*(y^*) \mid y^* \in Y^*\} = \sigma^{**}(0).$$

The proposition follows, because $\sigma^{**}(0) \leq \sigma(0)$ is already known to hold.||

Example. Since we have argued up to now mainly in terms of the function σ, let us briefly go through an example. We consider the problem

$$\text{(P)} \quad \inf \{F(x,y) \mid (x,y) \in X \times Y,\ y = Tx\},$$

where $T : X \to Y$ is a continuous linear operator. Let $(\xi,\eta) \in X \times Y$ with $\eta = T\xi$ be an optimal solution of (P). We assume ad hoc that for (P) condition (13) is satisfied with a suitable "multiplier" functional $\eta^* \in Y^*$ (formerly denoted by 1). With regard to (P), condition (13) can equivalently be written in explicit form as

$$F(\xi,\eta) \leq F(x,y) \text{ for all } (x,y) \in X \times Y \text{ satisfying } \eta^*(y-Tx) \leq 0. \quad (17)$$

By Proposition 5, η^* solves the dual problem $\sup \{\sigma^*(y^*) \mid y^* \in Y^*\}$, where

$$\sigma^*(y^*) := \inf \{F(x,y) \mid y^*(y-Tx) \leq 0\},$$

and $\sigma^*(\eta^*) = F(\xi,\eta)$. To make the dual more suggestive we define for all $(x^*,y^*) \in X^* \times Y^*$ the function

$$F^*(x^*,y^*) := \inf \{F(x,y) \mid x^*(x) + y^*(y) \leq 0\}.$$

Then $F^*(x^*,y^*) = \sigma^*(y^*)$ if $x^* = -T^*y^*$ (the adjoint map $T^* : Y^* \to X^*$ satisfies $(T^*y^*)(x) := y^*(Tx)\ \forall x \in X,\ y^* \in Y^*$). Hence the dual can be written as

$$\text{(P*)} \quad \sup \{F^*(x^*,y^*) \mid (x^*,y^*) \in X^* \times Y^*,\ x^* = -T^*y^*\}.$$

(ξ^*,η^*) with $\xi^* := -T^*\eta^*$ is then a solution to (P*), and $F^*(\xi^*,\eta^*) = F(\xi,\eta)$. Problem (P*) has the same structure as (P), and quasiconcave objective function.

Defining now for all $(x^{**},y^{**}) \in X^{**} \times Y^{**}$

$$F^{**}(x^{**},y^{**}) := \sup \{F^*(x^*,y^*) \mid x^{**}(x^*) + y^{**}(y^*) \leq 0\},$$

it follows as in Proposition 7 that $F^{**} \leq F$ on the intersection of the respective domains. We must assume now that X^*,Y^* have been topologized in such a way that $\xi \in X^{**}, \eta \in Y^{**}$ (but otherwise the bidual spaces may be larger or smaller than the original spaces). Then $\eta = T^{**}\xi$, and it follows without further assumption that (ξ,η) solves the bidual problem

$$(P^{**}) \quad \inf \{F^{**}(x^{**},y^{**}) \mid (x^{**},y^{**}) \in X^{**} \times Y^{**}, y^{**} = T^{**}x^{**}\},$$

and that $F^{**}(\xi,\eta) = F(\xi,\eta)$. Indeed: From

$$(x^* + T^*y^*)(\xi) \leq 0$$

follows by $\eta = T\xi$ that

$$x^*(\xi) + y^*(\eta) \leq 0,$$

and this implies by the definition of F^* that

$$F^*(x^*,y^*) \leq F(\xi,\eta) = F^*(\xi^*,\eta^*).$$

We have altogether

$$F^*(\xi^*,\eta^*) \geq F^*(x^*,y^*) \text{ for all } (x^*,y^*) \text{ satisfying } \xi(x^* + T^*y^*) \leq 0. \tag{18}$$

Now condition (18) (with $\xi \in X^{**}$ considered as the multiplier functional) plays for (P^*) exactly the same role as (17) played for (P), and a repetition of the argument which led from (17) to (P^*), leads from (18) to (P^{**}): We obtain that (ξ,η) solves (P^{**}), and that $F^{**}(\xi,\eta) = F^*(\xi^*,\eta^*)$.||

REFERENCES

HOLMES, R.B. (1975). *Geometrical Functional Analysis and its Applications*, Springer-Verlag, New York, p.7.

HORST, R. (1974). "Ein Beitrag zur quasikonvexen Optimierung in topologischen linearen Räumen," *Z. Operations Research*, 18, 19-26.

LUENBERGER, D.G. (1968). "Quasi-Convex Programming," *SIAM J. Appl. Math.*, 16, 1090-1095.

OETTLI, W. (1980). "Optimality Conditions for programming problems involving multivalued mappings," in B. Korte (editor), *Modern Applied Mathematics: Optimization and Operations Research*, North-Holland, Amsterdam.

RUDIN, W. (1973). *Functional Analysis*, McGraw-Hill, New York, p.58.

SINGER, I. Duality theorems for linear systems and convex systems, *J. Math. Anal. Appl.* (to appear).

PSEUDO DUALITY AND NON-CONVEX PROGRAMMING

Ury Passy

In this paper non-convex programs are analyzed via Legendre transforms. The first part includes definitions and a classification of the programs that can be handled by the transformation. Pseudo duality is defined and the pseudo duality theorem for non-convex programs is given. In the latter parts pseudo duality and duality for non-convex programs is derived together with several concrete examples.

1. INTRODUCTION

One of the most important concepts in the theory and practical applications of mathematical programming is that of duality. A powerful and elegant analysis of duality was done by Rockafellar (1970) by applying Fenchel's (1949) conjugate correspondence for convex functions (concave functions). This correspondence is a generalization of the classical Legendre transform. The conjugacy correspondence is intimately connected with convex (concave) functions. However, Legendre correspondence may exist even for non-convex (concave) differentiable functions and is the main tool used in the present paper.

Ekeland (1977) generalized the definition of such transform by replacing the notion of a function with that of a multi-valued function. In the present paper the approach is along the classical definition and the difficulty is bypassed by generating equivalent problems; its definition is given in Section 2.

Since convexity (concavity) will not be required, the notion of a minimum (maximum) point of a real valued function is replaced with that of a stationary point. Stationarity of a given real function f is associated with the properties of its gradient ∇f. It is therefore necessary to introduce the following definition.

ISBN 0-12-621120-5

2. PRELIMINARIES

Most of the material in this section can be found in Passy & Yutav (1978; 1980; 1981) and Passy (1979; 1981).

DEFINITION 2.1: Let h *be a real function which is defined on a set* $(S \cup s) \subset E^n$ *and is continuously differentiable on* S*, where* S *is an open set and* $s \subset Fr(S)$ *(the frontier of* S*). The gradient in the limit at a point* $x \in S \cup s$ *is a set denoted by* $\delta h(x)$ *and defined by:* $u \in \delta h(x)$ *if* $\exists$ *a converging sequence* $x^k \in S$ *such that* $x^k \to x \in S \cup s$ *and* $u = \lim_{k\to\infty} \nabla h(x^k)$. $\delta h(x)$ *may be empty.*

If h is as above and is also convex (concave) and $S \cup s$ is a convex set, then $\delta h(x) \subset \partial h(x)$, see Rockafellar (1970; Corrollary 24.51). Here $\partial h(x)$ denotes the subgradient of h evaluated at x. In what follows the symbol ∇h may be used instead of $y \in \delta h$.

Given a program

$$P : \text{Min } \{Q(x) \mid M_\alpha(x) \leq 0 \quad \alpha = 1,\ldots,\beta \quad x \in L \cap O\}$$

where

(i) $O \subseteq E^n$ is an m dimensional subspace of E^n

(ii) $Q(\cdot)$, $M_\alpha(\cdot)$ $\alpha = 1,\ldots,\beta$ are continuous functions defined on L and differentiable on $L^o \underline{\Delta}$ Int $L \neq \phi$.

Clearly, any set L can be written as

(iii) $L = L^0 \cup \ell \qquad L^0 = \text{Int } L \neq \phi \qquad \ell \subseteq Fr(L^0)$.

DEFINITION 2.2: A point x^* *is a stationary point of type minimal (Passy (1979)) of* P *if:*

(i) $M_\alpha(x^*) \leq 0 \qquad \alpha = 1,\ldots,\beta \qquad x^* \in L \cap O$ *(Feasibility)*

(ii) $\exists \quad \pi_0 \in \delta(Q(x^*)) \qquad \pi_\alpha \in \delta(M_\alpha(x^*)) \qquad \alpha = 1,\ldots,\beta$

such that $\forall\ \theta \in O$

$$\pi_0^T \theta < 0 \to \text{Not}[\pi_\alpha^T \theta \leq 0 \ \forall\ \alpha \in \{j \mid M_j(x^*) = 0\}] \ .$$

Condition (ii) can be replaced, using the Farkas lemma, with (iia) $\exists\ \pi_0 \in \delta(Q(x^*))$, $\pi_\alpha \in \delta(M_\alpha(x^*))$ $\alpha = 1,\ldots,\beta$

and $\lambda^*_\alpha \geq 0$ $\alpha = 1,\ldots,\beta$

such that (Kuhn Tucker conditions):

$$\pi_0 + \sum_{\alpha=1}^{\beta} \lambda^*_\alpha \pi_\alpha \in 0^\dagger \quad \textit{(the orthogonal complement of } 0\textit{)}$$

$$\lambda^* M_\alpha(x^*) = 0$$

A point that satisfies the above conditions is called a *minimal* point.

Let x* be a *minimal* point. If

(a) $\delta[Q(x^*)] = \nabla Q(x^*)$ $\delta[M_\alpha(x^*)] = \nabla M_\alpha(x^*)$ $\alpha = 1,\ldots,\beta$

then x* is a (regular) minimal point; otherwise

(b) x* is a minimal point in the limit.

If (i) the functions $Q(x)$ and $M_\alpha(x)$ $\alpha = 1,\ldots,\beta$ are convex

(ii) L is a convex set

Then a minimal point is a solution of *P*, i.e., it is a minimizer of P

$$\text{Min } \{Q(x) \mid G_\alpha(x) \leqslant 0 \quad \alpha = 1,\ldots,\beta \ , \quad x \in L \cap 0\} \ .$$

DEFINITION 2.3: Two programs P and D are a pair of pseudo duals, denoted P ↔ D, if

(i) There is a 1:1 correspondence between the collection A *of sets of minimal points of P and the collection* B *of sets of minimal points of D.*

(ii) Let $X_i \subseteq A$ *and* $Y_i \subseteq B$ *be two corresponding sets of primal and dual minimal points, then* $\forall\ x \in X_i\ \forall\ y \in Y_i$

$$Q(x) + \psi(y) = 0$$

where Q *and* ψ *are the objective functions of P and D respectively.*

If *P* and *D* are both convex programs then the only minimal points are the true minima. In this case

$$(P \leftrightarrow D) \Rightarrow \min_{x \in X} Q(x) + \min_{y \in Y} \psi(y) = 0 \Rightarrow [Q(x) + \psi(y) \geq 0 \quad \forall y \in Y \ x \in X]$$

where X is the feasible set of *primal* points and Y is the feasible set of *dual* points.

Accordingly, two programs P and D are a pair of *dual programs* if

(a) $P \leftrightarrow D$ (P and D are pseudo duals)

(b) $Q(x) + \psi(y) \geq 0$ holds for $\forall\, x \in X \quad \forall\, y \in Y$.

The problem of finding the minimal stationary points of a given program will be denoted

$$\underline{\text{Stat}}\ \{Q(x) \mid M_\alpha(x) \leq 0 \quad \alpha = 1,\ldots,\beta\ , \quad x \in L \cap 0\}\ .$$

It is understood that in some cases, for example the convex case, $\underline{\text{Stat}}$ may be replaced with Min.

A point that satisfies the necessary conditions for a maximum is called *maximal*. The problem of finding maximal stationary points of a given program will be denoted by

$$\overline{\text{Stat}}\ \{Q(x) \mid M_\alpha(x) \leq 0 \quad \alpha = 1,\ldots,\beta\ , \quad x \in L \cap 0\}\ .$$

Stationary points can be of different types:

(1) *Minimal* points - stationary points that satisfy the necessary conditions of a minimum.

(2) *Maximal* points - those points that satisfy the necessary conditions of a maximum.

(3) *Saddle* type - those points that satisfy the necessary conditions of a saddle points.

This classification is not exclusive. A stationary point can, for example, be a minimal and a maximal. Unless it is otherwise specified only minimal points are analyzed here.

2.1 Legendre Transform

Let f and w be differentiable functions defined on the open sets S and T respectively. The pair of two tuples (f,S) and (w,T) is called Legendre conjugates if there is a 1:1 correspondence between S and T such that any two corresponding points $x \in S$ and $y \in T$ satisfy $y = \nabla f(x)$, $x = \nabla w(y)$ and $f(x) + w(y) = x^T y$. The function w (f) is called the Legendre transform of f (w).

A differentiable function f has a Legendre transform over an open set $S \subset E^n$ if the map $\nabla f\colon S \to T$ is a 1:1 map; the transform w is given by

$$w(y) = x^T y - f(x) \quad x = \nabla f^{-1}(y) .$$

This is an extremely strong requirement. Ekeland (1977) generalized the definition of such a transform by replacing the notion of a function with that of a multi-valued one. In the present paper the approach is along the classical definition. Properties of Legendre transform can be found in Rockafellar (1970). If the function f is convex or concave then the Legendre transform is closely related to Fenchel's conjugate transform.

2.2 Partial Legendre Transform

Let $f(x,\mu)$ be a twice differentiable function defined on an open set $S \subseteq E^{\rho 1} \times E^{\rho 2}$. Let

(i) $S_\mu = \{x \mid (x,\mu) \in S\}$

(ii) $S_x = \{\mu \mid S_\mu \neq \phi\}$

(iii) $f_\mu(x) = f(x, \mu)$.

(iv) $T_\mu = \{y \mid y = \nabla f_\mu(x) \quad x \in S_\mu\}$

(v) $T = \{(y,\mu) \mid \mu \in S_x \quad y \in T_\mu\}$.

Assume that

(a) $(f_{\mu 1}, S_{\mu 1})$ and $(\omega_{\mu 1}, T_{\mu 1})$ is a pair of Legendre conjugates.

(b) Such a pair as in (a) exists for every $\mu \in S_x$ except possibly for a set $S_2 \subseteq S_x$ with the following property:

Let (i) $T(S_2) \underline{\Delta} \{(y, \mu) \mid \mu \in S_2 , y \in T_\mu\} \subseteq T$

(ii) $(\bar{y},\bar{\mu}) \in T(S_2)$,

then, for $\forall\, y \neq \bar{y}$ $(y,\bar{\mu}) \notin T$. Then, the function

$$w(y,\mu) = y^T x - f(x,\mu) \quad y = \nabla_x f(x,\mu) \tag{2.1}$$

is the x *partial Lengendre transform* of $f(x,\mu)$. The pair $(f(x,\mu),S)$ and $(w(y,\mu),T)$ are a pair of Legendre conjugates. The function $w(y,\mu_1)$ agrees with $\omega_{\mu_1}(y)$ on the set $T - T(S_2)$.

RESULT 2.1: If $(g(x),S)$ *and* $(w(y)\mathcal{T})$ *are a pair of Legendre conjugates. Then, the x partial Legendre conjugate of* $(\lambda g(x),\ x \in S \quad \lambda \geq 0)$ *is* $(\Omega(\lambda,y),T)$, *where*

$$\hat{T} = \{(y,\lambda) \mid \lambda > 0,\ y/\lambda \in \mathcal{T}\}, \quad \hat{T} = T \cup \{0,0\}, \quad \text{and}$$

$$\Omega(\lambda,y) = \begin{cases} \lambda w(y/\lambda) & \lambda > 0 \\ 0 \quad y = 0, & \lambda = 0 . \end{cases}$$

The proof follows directly from the definition of a partial transform and from equation (2.1).

RESULT 2.2: The function $\Omega(\lambda,y)$ *has a gradient in the limit at* $(0,0)$. *Any vector of the form* $(x,-g(x))$ *belongs to* $\delta\Omega(0,0)$.

PROOF: See Passy and Yutav (1980).

Given a pair of programs

$$\Gamma : \underline{\text{Stat}}\ \{f(x_0) \mid g_j(x_j) \leq 0,\ j = 1,\ldots,q;\ x \in Q \cap S \qquad x_j \in E^{n_j}, \qquad j = 0,\ldots,q\} \tag{2.2}$$

$$T : \underline{\text{Stat}}\ \{\phi(z,\lambda) \mid z \in Q^{\dagger},\ \lambda_j \geq 0,\ j = 1,\ldots,q,\ (z,\lambda) \in T\} , \tag{2.3}$$

where

(i) $n_j \quad j = 0,\ldots,q$ are given integers and $n = \sum_{j=1}^{q} n_j$,

(ii) S is an open subset of E^n,

(iii) Q is an m dimensional subspace of E^n,

(iv) $Q^{\dagger}$ is the orthogonal complement of Q,

(v) $f,\ g_j \quad j = 1,\ldots,q$ are differentiable functions defined on S,

(vi) The pairs (f,S), $(g_j,S) \quad j = 1,\ldots,q$ possess Legendre conjugates (ω_0,T_0), $(\omega_j,\mathcal{T}_j) \quad j = 1,\ldots,q$,

(vii) $T_0 = \{z_0 \mid \exists\ x \in S \ni z_0 = \nabla f(x_0)\}$

$\mathcal{T}_j = \{z_j \mid \exists\ x \in S \ni z_j = \nabla g_j(x_j)\}, \quad j = 1,\ldots,q,$

(viii) $\tilde{T}_j = \{(z_j,\lambda_j) \mid \lambda_j > 0 \quad z_j/\lambda_j \in \mathcal{T}_j\} \quad j = 1,\ldots,q,$

(ix) $T_j = \tilde{T}_j \cup \{0,0\}$,

(x) $T = \underset{j=0}{\overset{q}{\times}} T_j$,

(xi) $\phi(z,\lambda) = \omega_0(z_0) + \sum_{j=1}^{q} \Omega_j(\lambda_j, z_j)$, and

$$\Omega_j(\lambda_j, z_j) = \begin{cases} \lambda_j \ \omega_j(z_j/\lambda_j) & (z_j/\lambda_j) \in \mathcal{T}_j \\ 0 & z_j = 0, \ \lambda_j = 0 \ . \end{cases}$$ □

Remark 2.1. Program T, equation (2.3) was generated by the x *partial Legendre transform* of the associated Lagrangian, that is,

$$L(x,\lambda) = f(x_0) + \sum_{j=1}^{q} \lambda_j g_j(x_j). \qquad \text{(See Result 2.1)}$$

THEOREM 2.1 (Passy and Yutav (1980)): The two programs Γ *and* T *(equations (2.2) and (2.3)) are a pair of pseudo duals, that is,*

(a) *If* x^* *is a minimal point of* Γ *then* $\exists \ \lambda^* \geq 0 \ni$

$(z_0^* = \nabla f\ (x_0^*);\ z_j^* = \lambda_j^* \nabla g_j(x_j^*) \quad j = 1,\ldots,q;\ \lambda^*)$ *is a minimal point of* T *(possibly a minimal in the limit).*

It is implicitly assumed that Γ *satisfies the constraint qualification.*

(b) *If* (z^*,λ^*) *is a minimal point of* T *and if* $\bar{J} = \{j \mid \lambda_j^* = 0\}$, $J = \{j \mid \lambda_j^* > 0\}$ *then* $\exists \ \theta_j \in E^{\rho_j} \quad j \in \bar{J}$ *that satisfies*

(i) $y = [\nabla w_0(z_0^*);\ \lambda_j^* \nabla_{z_j} w_j(z_j^*/\lambda_j^*) \ \ j \in J;\ \nabla w_j(\theta_j) \ \ j \in \bar{J}] \in Q^{\dagger}$

(ii) $\eta = [w_j(z_j^*/\lambda_j^*) - \nabla w_j(z_j^*/\lambda_j^*)(z_j^*/\lambda_j^*) \ \ j \in J;$

$w_j(\theta_j) - \nabla w_j(\theta_j)\theta_j \quad j \in \bar{J}] \geq 0$

(iii) $(\lambda^*)^T \eta = 0$.

(In this case $(y,\eta) \in \delta\phi(z^*,\lambda^*)$, *see Result 2.2.)*

The point:

$x_0^* = \nabla w_0(z_0^*) \qquad x_j^* = \lambda_j \nabla w_j(z_j^*/\lambda_j^*) \quad j \in J \qquad x_j^* = \nabla w_j(\theta_j) \quad j \in \bar{J}$

is a minimal point of Γ *and*

(c) $f(x^*) + \phi(z^*,\lambda^*) = 0.$

Programs Γ and T are called *Legendre pseudo duals* (see Remark 2.1).

Remark 2.2. The following notation is adopted throughout this paper:

$\nabla w(\cdot)$ - The gradient with respect to the argument

$\nabla_z w(\cdot)$ - The gradient with respect to the variable z.

Legendre pseudo duals exist if each of the primal functions f and g_j $j = 1,\ldots,q$ have such a transform. Recall that a differentiable function has a transform over a given open set if the gradient is a 1:1 map (Rockafellar (1970)). This strong requirement can be bypassed by generating equivalent programs.

2.3 Equivalent Programs and Augmentation

Given a program Stat $\{x^2 \mid x \in E^1\}$, by a trivial change of variables it is possible to generate a new equivalent program Stat $\{xy \mid x = y\}$.

Two programs are said to be equivalent if one is obtained from the other by transforming the variables. In the present paper only linear transformations are considered.

DEFINITION 2.4: Consider two programs:

$\Gamma 1$: Stat $\{f_1(x) \mid x \in Q_1 \cap S_1\}$ where Q_1 is an m_1 dimensional subspace of E^{n_1} and S_1 is an open subset of E^{n_1}

$\Gamma 2$: Stat $\{f_2(x) \mid x \in Q_2 \cap S_2\}$ where Q_2 is an m_2 dimensional subspace of E^{n_2} and S_2 is an open subset of E^{n_2}.

Suppose

(i) R *is a linear map from* $Q_1 \to Q_2$,

(ii) $S_1 \cap Q_1 = R^{-1}(S_2 \cap Q_2)$, *and*

(iii) $f_1(z) = f_2(R(z))$.

Then if $n_2 > n_1$ $\Gamma 2$ *is an augmentation of* $\Gamma 1$ ($\Gamma 2 = \text{Aug}\ \Gamma 1$) *and if* $n_1 = m_1$ $\Gamma 1$ *is a reduction of* $\Gamma 2$ ($\Gamma 1 = \text{Red}\ \Gamma 2$).

DEFINITION 2.5: Two programs $\Gamma 1$ *and* $\Gamma 2$ *are equivalent* ($\Gamma 1 \approx \Gamma 2$) *if a reduction exists, such that* $\text{Red}\ \Gamma 1 = \text{Red}\ \Gamma 2$.

It can be easily verified that this relation is symmetric, reflexive and transitive.

LEMMA 2.1: Given two programs $\Gamma 1$ *and* $\Gamma 2$. $\Gamma 1 \approx \Gamma 2$ *iff a linear map* $R : Q_1 \to Q_2$ *exists, such that*

(i) $(S_1 \cap Q_1) = R^{-1}(S_2 \cap Q_2)$ *(ii)* $f_1(x_1) = f_2(Rx_1) \ \forall\ x_1 \in Q_1$.

LEMMA 2.2: If $\Gamma 1 \approx \Gamma 2$ *and* x_1 *is a solution to* $\Gamma 1$ *then* Rx_1 *is a solution to* $\Gamma 2$.

The proof of the two lemmas can be found in Passy and Yutav (1980).

Augmentation may be viewed as a generalized perturbation (Rockafellar (1970)). Given a primal program Γ if the defining functions f and g_j $j = 1,\ldots,q$ each have a Legendre transform then T can be constructed via Theorem 2.1. Otherwise it is possible, in many situations, to find an augmented program Γ^* such that:

(i) $\Gamma^* \approx \Gamma$

(ii) Γ^* has a Legendre pseudo dual T^*.

Γ^* and T^* are a pair of Legendre pseudo duals; Γ and T^* are pseudo duals by virute of the equivalence $\Gamma \approx \Gamma^*$. In that case:

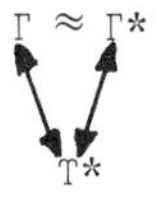

Given a pair (h,S). If h is differentiable on S then (h,S) has a Legendre conjugate if

(a) h is strictly convex; or

(b) h is strictly concave; or

(c) E^n can be decomposed into a direct sum $E^n = Q_1 \times Q_2$ such that $h(x) = h_1(x_1) + h_2(x_2)$ $x_1 \in Q_1$ $x_2 \in Q_2$ and h_1 $(-h_2)$ are strictly convex on Q_1 (Q_2).

Given a program with non-convex (concave) functions, it is impossible to generate an augmented program with convex or concave functions. However, in many cases it is possible to generate an augmented program with functions f and g_j $j = 1,\ldots,q$ satisfying each condition (c). Conditions under which such augmentation is possible appear in Passy and Yutav (1980).

It was shown in Theorem 2.1 that a pseudo dual to a given primal program Γ can be generated by applying an x partial Legendre transform to the Lagrangian function (see Remark 2.1), provided, of course, that it has such a transform.

Thus a pseudo dual can be generated in three steps:

Step 1: Generation of a Lagrangian program $\Gamma_L(\lambda)$ defined by:

$$\Gamma_L(\lambda) : \underline{S}\text{tat } \{L(x,\lambda) \underline{\Delta} f(x) + \sum_{j=1}^{q} \lambda_j g_j(x) \mid x \in Q \cap S\}, \tag{2.4}$$

where λ is a vector of q non-negative parameters.

Recall that if for a given λ^* the point (x^*,λ^*) satisfies the Kuhn Tucker conditions then x^* is a minimal point of both $\Gamma_L(\lambda^*)$ and Γ.

Step 2: Generating an augmentation $\Gamma_L^*(\lambda)$ of $\Gamma_L(\lambda)$ such that $\Gamma_L^*(\lambda)$ has a Legendre pseudo dual.

A constructive method for generating the required augmentation was given in Passy and Yutav (1980). However, in concrete situations the required augmentation can usually be recognized immediately.

Step 3: Generating the Legendre pseudo dual, T^*, of $\Gamma_L^*(\lambda^*)$ by applying an x partial Legendre transform to the objective function of $\Gamma_L^*(\lambda)$.

The existing relations between these four programs are

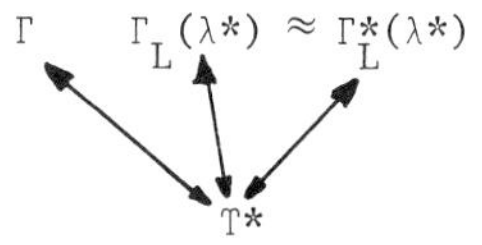

$\Gamma^*_L(\lambda) = \text{Aug } \Gamma_L(\lambda)$

According to Definition 2.5 Γ and $\Gamma_L(\lambda^*)$ are not equivalent.

3. PSEUDO DUALITY IN FRACTIONAL PROGRAMMING

For the purposes of the present analysis the following standard representation is adopted:

$$P: \quad \text{Stat}\{y_1/y_2 \mid (x,y) \in G\} \qquad G \subseteq S .$$

The set G is defined by the following conditions:

$$g_j(x) \le 0, \quad j = 1,\ldots,m-2 \tag{3.1}$$

$$g_{m-1}(x)-y_1 \le 0, \tag{3.2}$$

$$y_2-g_m(x) = 0, \tag{3.3}$$

$$y_2 > 0, \text{ and} \tag{3.4}$$

$$(x,y) \in E^{n+2} . \tag{3.5}$$

Another set denoted by G_1 is defined by the same conditions, but the equality sign, equation (3.3), is replaced with an inequality. The set S is given by $S = S_1 \times \{(y_1,y_2) \mid y_2 > 0\} \subseteq E^{n+2}$.

The following symbols will also be used:

(i) GL or [GL_1] - if $g_j(x) = \sum_{i=1}^{m} a_{ji}x_i - b_j \quad j=1,\ldots,m$ [if $y_2-g_m(x) \le 0$]

(ii) GQ or [GQ_1] - if $g_{m-1}(x) = \frac{1}{2}x^TQx + c^Tx + e$ and all other functions are linear as in (i) [if $y_2-g_m(x) \le 0$]

(iii) G(Q/Q) or [$G(Q/Q)_1$] - if $g_j(x) \quad j=1,\ldots,m-2$ are linear functions while $g_{m-1}(x)$ and $g_m(x)$ are quadratic functions [if $y_2-g_m(x) \le 0$].

Every program P is associated with a Lagrangian program:

$$LP: \quad \text{Stat}\{y_1/y_2-\mu_1y_1+\mu_2y_2 + \sum_{j=1}^{m-2} \lambda_jg_j(x)+\mu_1g_{m-1}(x)$$
$$-\mu_2g_m(x) \mid (\lambda,\mu) \ge 0,\ (\lambda,\mu) \in E^m,\ x \in S,\ (y,x) \in E^{2+n}\} .$$

It is well known that if (y^*,x^*) is a minimal point of P and the constraint qualification is satisfied, then $\exists\ (\mu^*,\lambda^*)$ such that $[(y^*,x^*),\ (\mu^*,\lambda^*)]$ is a minimax saddle type stationary point of LP.

LEMMA 3.1: x^* *solves* P_1 *iff* $(x^*, y_1^* = g_{m-1}(x^*), y_2^* = g_m(x^*))$ *solves* P.

Pseudo duals are generated via Legendre transform. It will therefore be assumed that the g_j's have such transform over the domain of definition S. In most cases if a transform does not exist it is possible to find an augmented program (Passy and Yutav (1980)) that has such transform (see for example section 4).

The corresponding Legendre transforms will respectively be denoted by ω_j $j=1,\ldots,m$. Define the following m+1 functions:

(a) $\dfrac{z_{o2}-\mu_2}{z_{o1}+\mu_1}$ $z_o \in E^2$, $\mu \in E^2$. This function is defined on the set $T_o = \{(z_o,\mu)\,|\,z_{o1}+\mu_1 > 0\}$. It is the y partial Legendre transform of the function $y_1/y_2 - \mu_1 y_1 + \mu_2 y_2$.

(b) m homogeneous functions of degree 1.

$$\Omega_j:\ E^{n+1} \to E^1 \quad j=1,\ldots,m$$

$$\Omega_j(v_o,v) = \begin{cases} v_o\omega_j(v/v_o) & v_o > 0 \\ 0 & (v_o,v) = 0 \end{cases} \qquad \begin{matrix} v_o \in E^1 \\ v \in E^n \end{matrix}$$

These functions are

(i) defined on $T_j = \hat{T}_j \cup \{(0,0)\}$ $j=1,\ldots,m$ where $\hat{T}_j = \{(v_o,v)\,|\,\exists\, x\in S \ni v/v_o = \nabla g_j(x),\ v_o > 0\}$ $j=1,\ldots,m$

(ii) are differentiable on $\hat{T}_j$, and

(iii) have a gradient in the limit on T_j .

If $(v_o,v) = (0,0)$ then $\delta\Omega_j(0,0) = \{(q_1,q_2)\,|\,q_1 = \omega_j(0) - \nabla\omega_j(\theta)\theta;$
$q_2 = \nabla\omega_j(\theta),\quad \forall\ \theta \in S\}$.

Define the following sets and function:

(i) $T = T_o \times T_1 \times \ldots \times T_m$

(ii) $\hat{I}_1 = \{(z,\lambda,\mu) | z = (z_o,\ldots,z_m) \in E^{nm+2},\ \lambda \in E^{m-2},\ \mu \in E^2,\ z_o = 0\ ,$

$\sum_{j=1}^{m} z_j = 0,\ ((z_o,\mu),(z_1,\lambda_1)\ldots(z_{m-2},\lambda_{m-2}),(z_{m-1},\mu_1),(z_m,\mu_2)) \in T\}.$

In what follows the abridged notation $(z,\lambda,\mu) \in T$ is used instead of the long but more accurate one. Observe that if $z_o = 0\ \ (0,\mu) \in T_o \rightarrow \mu_1 > 0$.

(iii) $\hat{I}_2 \subset \hat{I}_1$. The only additional requirement is that $\mu_2 \geq 0$.

(iv) $\Omega(\lambda,\mu,z) = \sum_{j=1}^{m-2} \Omega_j(\lambda_j z_j)+\Omega_{m-1}(\mu_1,z_{m-1})-\Omega_m(\mu_2,-z_m)$.

This function is defined on T and is differentiable on $\hat{T}$.

(v) $I_1 = \hat{I}_1 \cap \{(z,\lambda,\mu) | \Omega(\lambda,\mu,z) = 0\}$.

(vi) $I_2 = \hat{I}_2 \cap \{(z,\lambda,\mu) | \Omega(\lambda,\mu,z) \leq 0\}$ $\qquad I_1 \subseteq I_2$.

Consider the following program:

$$D_1:\ \ \text{Stat}\{\frac{z_{02}-\mu_2}{z_{01}+\mu_1} + \Omega(\lambda,\mu,z) | (z,\lambda,\mu) \in \hat{I}_1\}.$$

Since $z_o = 0$ it is possible to write D_1 differently:

$$D_1:\ \ \text{Stat}\{-\mu_2/\mu_1+\Omega(\lambda,\mu,z) | (z,\lambda,\mu) \in \hat{I}_1\}.$$

<u>Notation</u>:

$$\nabla_z \omega_j(\alpha,z) = [\frac{\partial \omega_j(\alpha,z)}{\partial z_i}\quad i=1,\ldots,n]$$

$$\nabla \omega_j(\alpha,z) = [\frac{\partial(\alpha\ \omega_j(\alpha,z))}{\partial z_i}\quad i=1,\ldots,n]$$

∇ is a row vector.

LEMMA 3.2: The two programs P *and* D_1 *are a pair of pseudo duals* ($P \leftrightarrow D_1$), *that is,*

(a) If (y^*,x^*) *is a minimal point of* P *then* $\exists\ \lambda^* \geq 0,\ \mu_1^* > 0$ *and* μ_2^* *such that:*

$$[\mu^*,\lambda^*,z_j^* = \lambda_j^*\nabla g_j(x^*)\ \ j=1,\ldots,m-2;\ z_{m-1}^*=\mu_1^*\nabla g_{m-1}(x^*),\ \ z_m^* = \mu_2^*\nabla g_m(x^*)]$$

is a minimal point of D_1 *(possibly in the limit).*

(b) If (z^*,λ^*,μ^*) *is a minimal point of* D_1 *(possibly in the limit) and if* $\bar{J} = \{j \mid \lambda_j^*=0\}$ *then every vector* $\theta_j \in S$ $j \in \bar{J}$ *that satisfies*

(i) $u = [\lambda_j^* \nabla_z \omega_j(z_j^*/\lambda_j^*)\ j \in J;\ \nabla_z \omega_j(\theta_j)\ j \in \bar{J},\ \mu_1^* \nabla_z \omega_{m-1}(z_{m-1}^*/\mu_1^*),$

$-\mu_2^* \nabla_z \omega_m(z_m^*/\mu_2^*)] \in L^{\dagger}$. $L^{\dagger}$ *is the space orthogonal to*

$$L = \{z \mid \sum_{j=1}^{m} z_j = 0\}$$

(ii) $\eta = [\omega_j(z_j^*/\lambda_j^*) - \nabla\omega_j(z_j^*/\lambda_j^*)\ j \in J,\ \omega_j(\theta_j) - \nabla\omega_j(\theta_j)\theta_j\ j \in \bar{J}] \geq 0$

(iii) $\xi = [\frac{1}{\mu_1^{*2}} + \omega_{m-1}(z_{m-1}^*/\mu_1^*) - \nabla\omega_{m-1}(z_{m-1}^*/\mu_1^*)(z_{m-1}^*/\mu_1^*);$

$- \frac{1}{\mu_1^*} - \omega_m(-z_m^*/\mu_2^*) - \nabla\omega_m(-z_m^*/\mu_2^*)(-z_m^*/\mu_2^*)] = 0$

(iv) $y^* = \nabla_{z_o} \left(\frac{z_{02}-\mu_2^*}{z_{01}+\mu_1^*}\right)\Bigg|_{z_o=0}$

(v) $(\lambda^*)^T \eta = 0 \qquad (\mu^*)^T \xi = 0$

(In this case $(y^*,u,\xi,\eta) \in \delta[\frac{z_{02}-\mu_2^*}{z_{01}+\mu_1^*} + \Omega(\lambda^*,\mu^*,z^*)]_{z_o=0}$*)*

induces a minimal point: (y^*,x^*), *where* x^* *can be calculated from any of the following equations:*

$x^* = \lambda_j^* \nabla_{z_j} \omega_j(z_j^*/\lambda_j^*) \quad \forall\ j \in J \qquad x^* = \mu_1^* \nabla_{z_{m-1}} \omega_{m-1}(z_{m-1}^*/\mu_1^*)$

$x^* = \nabla\omega_j(\theta_j) \quad \forall\ j \in \bar{J} \qquad x^* = \mu_2^* \nabla_{z_m} \omega_m(z_m^*/\mu_2^*)$

(c) $y_1^*/y_2^* + \Omega(\lambda^*,\mu^*,z^*) - \mu_2^*/\mu_1^* = 0$.

For the proof see Passy and Yutav (1980).

LEMMA 3.2a: The two programs D_1 *and* LP, *the Lagrangian program, are pseudo-duals* ($D_1 \leftrightarrow$ LP). *Here, the pseudo-duals relate the minimal points* (λ^*,μ^*, z^*) *of* D_1 *with the minimaxi saddle type points of* LP, $((y^*,x^*), (\lambda^*,\mu^*))$.

LEMMA 3.3: If (z^*,λ^*,μ^*) *solves* D_1 *(possibly a minimal point in the limit) then* $\Omega(\lambda^*,\mu^*,z^*) = 0$.

PROOF: The function Ω is a homogeneous function of degree 1. Therefore, if (z^*,λ^*,μ^*) is a minimal point of D_1, then for $r > 0$ the point $r(z^*,\lambda^*,\mu^*) \in \hat{I}_I$. Assume that $\Omega(\lambda^*,\mu^*,z^*) \neq 0$, let $\Omega^*(r) \equiv \Omega(r(\lambda^*,\mu^*,z^*))$ then $d\Omega^*(r)/dr = \Omega(\lambda^*,\mu^*,z^*) \neq 0$. Without loss of generality let $d\Omega^*(r)/dr < 0$, then $\exists \chi \in \delta[-\mu_2^*/\mu_1^* + \Omega(\lambda^*,\mu^*,z^*)]$ such that

$\chi^T(z^*,\lambda^*,\mu^*) = \left.\frac{d\Omega^*(r)}{dr}\right|_{r=1} < 0$ contradicting the assumption that (z^*,λ^*,μ^*) is a minimal point. □

It follows from lemma 3 that if D_1 has a solution then

$$D: \quad \text{Stat}\{-\mu_2/\mu_1 \mid (z,\lambda,\mu) \in I_1\}$$

is equivalent to D_1 and in that case $P \leftrightarrow D$.

Remark 3.1. It was shown that $P \leftrightarrow D_1$, that is, P has a solution iff D_1 has a solution. If P has a solution then and only then $P \leftrightarrow D$. If P does not possess a solution $P \not\leftrightarrow D$ (see example 4.1).

It will be assumed that P has a solution. In that case, both programs, P and D, have a fractional objective function. Consider the following program:

$$PE: \quad \text{Stat}\{y_1/y_2 \mid (x,y) \in G_1\}$$

In what follows it will be assumed that

$$g_{m-1}(x) \geq 0, \qquad g_m(x) > 0 \qquad \forall x \in S. \tag{3.6}$$

In this case the pseudo dual of PE is given by:

$$DE: \quad \text{Stat}\{-\mu_2/\mu_1 \mid (z,\lambda,\mu) \in I_1, \ \mu_2 \geq 0\} .$$

COROLLARY 3.1: If equation (3.6) holds, then (x^*,y^*) *solves* P *if* (x^*,y^*) *solves* PE.

PROOF: The stationarity conditions for P are defined by:

$$\forall \pi_j \in \delta[g_j(x^*)] \quad j=1,\ldots,m \ \nexists \ (\theta,\eta) \in E^{n+2} \text{ such that}$$

$$\pi_j^T\theta \leq 0 \quad j \in J=\{k \mid g_k(x^*) = 0 \quad k=1,\ldots,m-2\} , \tag{3.7}$$

$$\pi_{m-1}^T\theta - \eta_1 \leq 0 \ , \tag{3.8}$$

$$\eta_2 - \pi_m^T\theta = 0 \ , \quad \text{and} \tag{3.9}$$

$$1/y_2^*[\eta_1 - y_1^*/y_2^*\eta_2] < 0 \rightarrow [\eta_1 - y_1^*/y_2^*\eta_2] < 0 \ . \tag{3.10}$$

The stationary condition for PE are defined by Equations (3.7)-(3.10) but the equality sign in equation (3.9) is replaced with an inequality. In this case $y_1^*/y_2^* \geq 0$ and the proof of Corollary 3.1 is therefore the same as that of Lemma 3.1. □

In the following paragraph it will be assumed that g_j, j=1,...,m-2 are convex functions, g_{m-1} is convex and non-negative and g_m is a positive concave function and equation (3.6) holds. In this case the constrained set G_1 is convex and any minimal point of PE is the true global minimum.

Remark 3.2. If g_j j=1,...,m-1 are convex functions, g_m is linear and positive. In that case the set G is convex, equations (3.1)-(3.5) and the non-negativity of g_{m-1}, equation (3.6), are not required.

COROLLARY 3.2: If equation (3.6) holds and G_1 *is as above, then* Ω_j j=1,...,m-1 *and* $(-\Omega_m)$ *are respectively convex functions on* T_j j=1,...,m *and* Ω *is a convex function on* T.

The proof can be found in Passy and Yutav (1980).

LEMMA 3.4: If the conditions defined by equation (3.6) hold then the two programs:

DE: $\text{Stat}\{-\mu_2/\mu_1 | (z,\lambda,\mu) \in I_1, \ \mu_2 \geq 0\}$ *and*

DC: $\text{Stat}\{-\mu_2/\mu_1 | (z,\lambda,\mu) \in I_2\}$ *are equivalent.*

If G_1 is convex, that is, g_j j=1,...,m-1 are convex and g_m is concave, and equation (3.6) holds, then program DC has a convex constrained set and a pseudo convex objective function (see Corollary (3.2)). The minimal points of PE and of DC are the true global minimum. Therefore, PE and (-DC) are a pair of dual programs where *Stat* is replaced with *Min* and *Max*.

$$\text{PE:} \quad \text{Min}\{y_1/y_2 \,|\, (x,y) \in G_1\} \tag{3.11}$$

$$\text{(-DC):} \quad \text{Max}\{\mu_2/\mu_1 \,|\, (z,\lambda,\mu) \in I_2\} \tag{3.12}$$

If program (PE) is feasible and has a finite solution then dual program (-DC) is feasible and has a finite solution. In that case (PE) $\geq$ (-DC) and Min PE = Max(-DC).

However, if primal program (PE) is not feasible or if it does not possess a solution then the dual is not given by (-DC) (see remark 3.1). In that case the dual is given by

$$(-\widetilde{DC}) \qquad \text{Max}\{\mu_2/\mu_1 - \Omega(\lambda,\mu,z) \,|\, (z,\lambda,\mu) \in \hat{I}_1,\ \mu_2 \geq 0\} \tag{3.13}$$

and it does not possess a solution (see example 4.1). If program (PE) does not possess a solution then the following situations are possible.

$(-\widetilde{DC}) \leftrightarrow$ (PE)		(-DC)
DO NOT		Not feasible
HAVE A	$\Longrightarrow$	Not bounded, but feasible
SOLUTION		Has a feasible solution

For the various possibilities see Examples (3.1), (4.1) and (4.2). A more complete relation between (-DC) and (PE) is given in Schaible (1976).

Example 3.1.

$$\text{PE:} \quad \text{Min}\{y_1/y_2 \,|\, \tfrac{1}{4}x_1^2 - y_2 \leq 0, \quad y_2 - 1 + \tfrac{1}{4}x_1^2 \leq 0 \quad x_1 \leq a \quad x_1 > b\}$$

The dual program in this example was calculated by the technique described in section 6

$$\text{(-DC):} \quad \text{Max}\Big\{\mu_2/\mu_1 \;\Big|\; \frac{z_1^2}{\mu_1} + \frac{(z_2-\lambda_1+\lambda_2)^2}{\mu_2} + \lambda_1 a - \lambda_2 b + \mu_2/4 \leq 0,$$
$$z_1 + z_2 = 0, \quad (\lambda_1 \lambda_2 \mu_2) \geq 0 \quad \mu_1 > 0\Big\} .$$

The point $z_1 = z_2 = 0$, $\lambda_1 = 0$, $\lambda_2 = \frac{b}{2}\mu_2$, $\mu_1 = 1$ is dual feasible if $2 - \sqrt{3} \leq b \leq 2 + \sqrt{3}$. Therefore, if

(i) $2 + \sqrt{3} \geq b \geq 2 - \sqrt{3}$ and

(ii) $b > a$

then primal program PE is not feasible and dual program (-DC) is unbounded. Program $(-\tilde{DC})$ is also unbounded.

In the following sections an explicit expression for DC is derived for various special primal programs. It should be noted that:

(1) the objective function is constant on rays emanating from the origin,

(2) Ω is a homogeneous function.

Therefore, it is possible to assume that $\mu_1 = 1$. Alternatively, it can be shown that if PE has a solution then the constraint $\mu_1 > 0$ can be replaced with $\mu_1 \geq 0$ (see Examples 3.1, 4.1 and 4.2).

4. PSEUDO DUALITY AND LINEAR FRACTIONAL PROGRAMMING

Consider the following program

$$P(L): \quad \text{Stat}\{y_1/y_2 \mid (x,y) \in GL\} .$$

To avoid confusion and complicated notation, it will be assumed that GL is defined by m+2 linear functions instead of m such functions. In the linear case, since the functions do not possess a Legendre transform, it is not possible to directly apply Lemma 3.2. Here, a pseudo dual will be generated in two steps:

4.1 Generating a Lagrangian Program

The constraints, in this case, are given by:

$$g_j(x) = \sum_{i=1}^{n} a_{ji}x_i - b_j \quad j=1,\dots,m ;$$

$$g_{m+1}(x) = \sum_{i=1}^{n} c_i x_i + d ; \quad \text{and} \quad g_{m+2}(x) = \sum_{i=1}^{n} f_i x_i + e .$$

The associated Lagrangian program, PL(L), is defined by

$$\text{PL(L)}: \ \text{Stat}\{y_1/y_2-\mu_1 y_1+\mu_2 y_2+\sum_{j=1}^{n} x_j(\sum_{i=1}^{m} a_{ij}\lambda_i+c_j\mu_1-f_j\mu_2)$$

$$-\lambda^T b+\mu_1 d-\mu_2 e \,|\, y_2 > 0,$$

$$(x,y)\in E^{n+2}, \quad (\lambda,\mu_1) \geq 0, \quad (\lambda,\mu)\in E^{m+2}\}$$

It is well known that if (y^*,x^*) solves P(L) then $\exists$ (μ^*,λ^*) such that $(y^*,x^*,\mu^*,\lambda^*)$ solves PL(L). This stationary point is *minimal* with respect to (y^*,x^*) and *maximal* with respect to (μ^*,λ^*) (Passy and Yutav (1980)). Program PL(L) does not possess a Legendre transform with respect to (x,y).

4.2 Generating an Augmented Program

Let $\phi_j(\cdot)$: $E^1 \to E^1$ $j=1,\ldots,n$ be n differentiable functions that have a Legendre transform ω_j, $j=1,\ldots,n$ over E^1. Such functions exist, i.e., $\phi_j = \frac{1}{2}(\theta_j)^2$. Define the following augmentation:

$$\text{PL*(L)} = \text{Aug(PL(L))}: \ \text{Stat}\{\psi(y,x,v,\mu,\lambda)\,|\,x = v, \quad y_2 > 0,$$

$$(x,v,y) \in E^{2n+2}, \quad (\lambda,\mu_1) \geq 0, \quad (\lambda,\mu)\in E^{m+2}\} .$$

The function ψ is given by:

$$\psi(y,x,v,\mu,\lambda) = y_1/y_2+\mu_2 y_2-\mu_1 y_1+\sum_{j=1}^{n} [\phi_j(x_j)+x_j(\sum_{i=1}^{m} a_{ij}+c_j\mu_1-f_j\mu_2)$$

$$-\phi_j(v_j)]-b^T\lambda+\mu_1 d-\mu_2 e .$$

Observe that if $x = v$ then PL*(L) is indeed equivalent to PL(L). The function ψ has an (x,v,y) *partial Legendre transform* and the following program, DL*(L), is therefore a pseudo dual to both PL*(L) and P(L).

$$\text{DL*(L)}: \ \text{Stat}\left\{\frac{z_{02}-\mu_2}{z_{01}+\mu_1} + \sum_{j=1}^{n} [\omega_j(z_j-\sum_{i=1}^{m} a_{ij}\lambda_i-c_j\mu_1+f_j\mu_2)-\omega_j(-w_j)]+\right.$$

$$+\lambda^T b-\mu_1 d+\mu_2 e\,|\,z_o = 0, \ z+W = 0, \ (\lambda,\mu_1) \geq 0,$$

$$\left.(z_o,z,W,\lambda,\mu) \in E^{2n+m-4}\right\}$$

The two pairs of programs DL*(L) ↔ PL(L) and DL*(L) ↔ P(L) are pseudo duals. The pseudo duality relates:

(i) The minimal points of DL*(L) with the minimaxi saddle type points of PL*(L), and

(ii) The minimal points of DL*(L) with those of P(L).

By setting $z_o = 0$ and eliminating W one obtains:

$$\text{DL*(L)}: \text{Stat}\{-\mu_2/\mu_1 + \sum_{j=1}^{m} \omega_j(z_j - \sum_{i=1}^{m} a_{ij}\lambda_i - c_j\mu_1 + f_j\mu_2) - \omega_j(z_j) + b^T\lambda - \mu_1 d + \mu_2 e \,|\, z \in E^n,\ (\lambda,\mu_1) \geq 0,\ (\lambda,\mu) \in E^{m+2}\}.$$

At a stationary point the gradient with respect to z vanishes

$$\frac{\partial\omega_j}{\partial z_j}(z_j - \sum_{i=1}^{m} a_{ij}\lambda_1 - c_j\mu_1 + f_j\mu_2) - \frac{\partial\omega_j}{\partial z_j}(z_j) = 0, \quad j=1,\ldots,n. \quad (4.1)$$

Since ω_j has a Legendre transform, equation 4.1 holds only if

$$\sum_{i=1}^{m} a_{ij}\lambda_i + c_j\mu_1 - f_j\mu_2 = 0, \quad j=1,\ldots,n .$$

Thus DL*(L) reduces to:

$$\text{DL*(L)}: \text{Stat}\{-\mu_2/\mu_1 + \sum_{i=1}^{m} \lambda_i b_i - \mu_1 d + \mu_2 e \,|\, \lambda^T A + \mu_1 c^T - \mu_2 f^T = 0,\ (\lambda,\mu_1) \geq 0\} .$$

Finally, by applying the results of Lemmas 3.3 and 3.4, one obtains:

$$\text{D(L)}: \text{Stat}\{-\mu_2/\mu_1 \,|\, A^T\lambda + c\mu_1 - f\mu_2 = 0,\ \lambda^T b - \mu_1 d + \mu_2 e \leq 0,\ (\lambda,\mu) \geq 0\} .$$

Remark 4.1.

(i) If the conditions of remark 3.1 hold then μ_2 is not restricted to be non-negative.

(ii) The dual objective function is a sum of two functions that are:

(a) homogeneous of degree zero $(-\mu_2/\mu_1)$; and

(b) homogeneous of degree one; usually, it is the function $\Omega(\lambda,\mu,z)$ except in some special cases where z is eliminated. If this function is also linear then the corresponding inequality constraint is replaced with an equality constraint, e.g., $\Omega(\lambda,\mu,z) \leq 0 \rightarrow \Omega(\lambda,\mu,z) = 0$.

It is possible therefore to write D(L) as

$$D(L)_1:\ \text{Stat}\{-\mu_2/\mu_1 | A^T\lambda+c\mu_1-f\mu_2 = 0,\ \lambda^Tb-\mu_1 d+\mu_2 e = 0,\ \lambda \geq 0,\ \mu_1 > 0\} .$$

The constrained set of D(L) is convex and the objective function is pseudo convex, thus:

$$\max\{\mu_2/\mu_1 | A^T\lambda+c\mu_1-\mu_2 f = 0,\ \lambda^Tb-\mu_1 d+\mu_2 e \leq 0,\ \ \mu_1>0,\ \ (\lambda,\mu) \geq 0\}$$

$$= \min\{y_1/y_2 | Ax \leq b,\ c^Tx+d-y_1 \leq 0,\ \ y_2-f^Tx-e \leq 0,\ \ y_2 > 0,\ (x,y) \in E^{n+2}\} .$$

Consider the following two examples (Schaible (1976)):

Example 4.1.

$$P_1:\ \text{Min}\{y_1/y_2 | x_1+x_2 \leq 1,\ \ -x_1-x_2 \leq -2,\ \ -x_1+\tfrac{1}{2}x_2-y_1 \leq 0,$$
$$y_2-x_1-2x_2-1 = 0,\ \ y_2 > 0\} .$$

The set of primal feasible points is empty.

The dual problem (see remark 4.1) is given by:

$$D_1:\ \text{Max}\{\mu_2/\mu_1 | \lambda_1-\lambda_2-\mu_1-\mu_2 = 0,\ \ \lambda_1-\lambda_2+\tfrac{1}{2}\mu_1-2\mu_2 = 0,$$
$$\lambda_1-2\lambda_2+\mu_2 = 0,\ \ \mu_1 > 0,\ \ \lambda \geq 0\} .$$

The point $\mu_2=3/2\mu_1$, $\lambda_2 = 8/2\mu_1$, $\lambda_1 = 13/2\mu_1$, $\mu_1 > 0$ is dual feasible and D_1 is constant on the feasible set $\mu_2/\mu_1 = 3/2$. However in that case the dual is (see remark 3.1)

$$\text{Max}\{\mu_2/\mu_1-[\lambda_1-2\lambda_2+\mu_2] | \lambda_1-\lambda_2-\mu_1-\mu_2 = 0,\ \ \lambda_1-\lambda_2+\tfrac{1}{2}\mu_1-2\mu_2 = 0,$$
$$\mu_1 > 0 , \lambda \geq 0\}$$

and it is unbounded.

Example 4.2.

$$P_2: \quad \text{Min}\{y_1/y_2 \mid -x \le -1, \quad 1-y_1 \le 0, \quad y_2-x=0, \ y_2>0, \ (y,x)\in E^3\} .$$

P_2 does not have a solution.

The dual (see remark 4.1) is given by:

$$D_2: \quad \text{Max}\{\mu_2/\mu_1 \mid -\lambda_1-\mu_2 = 0, \quad -\lambda_1-\mu_1 = 0, \quad \lambda \ge 0, \quad \mu_1 > 0\} .$$

The dual does not have a feasible solution. However, in that case the dual is (see remark 3.1)

$$\text{Max}\{\mu_2/\mu_1-[-\lambda_1-\mu_1] \mid -\lambda_1-\mu_2 = 0, \ \lambda_1 \ge 0, \quad \mu_1 > 0\}$$ and is unbounded.

4.3 Quadratic Programs

In this section the results for pseudo duality of quadratic fractional programs are summarized. The detailed derivation is similar to that in the linear case and is given in Passy (1980).

(i) Quadratic programs (numerator)

$$P(Q): \quad \text{Min}\{\frac{x^TQx+c^Tx+e}{d^Tx+h} \mid x\in GQ\} .$$

It is assumed that GQ is defined by m+2 linear constraints and not only by m_1 (see the first part of Section 4). It and that $\det Q \neq 0$ (otherwise see Passy (1981)).

The dual is

$$D(Q): \text{Max}\{\mu_2/\mu_1 \mid \frac{1}{2\mu_1} \ell^TQ^{-1}\ell+b^T\lambda-\mu_1e+\mu_2b\le 0, (\lambda,\mu)\ge 0, (\lambda,\mu)\in E^{m+2}\}$$

where $\ell_j = \sum_{i=1}^{m} a_{ij}\lambda_j+\mu_1c_j-\mu_2d_j$, $j=1,\dots,n$.

(ii) Quadratic programs (both numerator and denominator)

$$P(Q_1/Q_2): \quad \text{Min}\ \frac{x^TQ_1 \ \ +c^Tx}{n^TQ_2n-d^Tx+h} .$$

The dual is

$$D(Q_1/Q_2): \ \text{Max}\{\mu_2/\mu_1 \,|\, \frac{1}{2\mu_1}\ell^T Q(\mu_1,\mu_2)+b^T\lambda-\mu_1 e+\mu_2 b \leq 0, (\lambda,\mu) \geq 0\}$$

where $Q(\mu_1,\mu_2) = Q_1^{-1}[I-(\frac{Q_1^{-1}}{\mu_1} - \frac{Q_2^{-1}}{\mu_2})^{-1} \frac{Q_1^{-1}}{\mu_1}]$ and

$$\ell_1 = \sum_{i=1}^{m} a_{ij}\lambda_i+\mu_1 c_j-\mu_2 d_j, \quad j=1,\ldots,n$$

It is assumed that $\det Q_1 \neq 0$, $\det Q_2 \neq 0$ (otherwise see Passy (1981)). Using the ideas presented here one can obtain pairs of dual programs for:

(1) products and ratios of functions (Passy (1981)); and

(2) functions which are strictly quasiconvex and fractional (Passy (1981)).

REFERENCES

EKELAND, I. (1977). "Legendre Duality in Nonconvex Optimization and Calculus of Variations." *SIAM J. Control and Optimization 15*, 905-934.

FENCHEL, W. (1940). "On Conjugate Convex Functions." *Canadian J. Mathematics 1*, 73-77.

PASSY, U. (1979). "Duality and Pseudo Duality in Mathematical Programs with Quotients and Products of Finitely Many Functionals." Operations Research, Statistics and Economics Mimeograph Series No.218. Faculty of Industrial Engineering and Management, Technion, Haifa, Israel.

PASSY, U. (1981). "Pseudo Duality in Prgorams with Quotients and Products." *Journal of Optimization Theory and Applications* (forthcoming).

PASSY, U. and YUTAV, S. (1978). "Pseudo Duality and Saddle Points." Operations Research, Statistics and Economics Mimeograph Series No.214. Faculty of Industrial Engineering and Management, Technion, Haifa, Israel.

PASSY, U. and YUTAV, S. (1980). "Pseudo Duality-Inequality Constraints and Extended Geometric Programming." Operations Research, Statistics and Economics Mimeograph Series No.210. Faculty of Industrial Engineering and Management, Technion, Haifa, Israel.

PASSY, U. and YUTAV, S. (1981). "Pseudo Duality in Mathematical Programming Unconstrained Problems and Problems with Equality Constraints." *Mathematical Programming 18*, 248-273.

ROCKAFELLAR, R.T. (1970). *Convex Analysis*, Princeton University Press, Princeton, N.J.

SCHAIBLE, S. (1976). "Duality in Fractional Programming: A Unified Approach." *Operations Research 24*, 452-461.

GENERALIZED CONCAVITY AND DUALITY

B. Mond and T. Weir

A number of different duals distinct from the Wolfe dual are proposed for the nonlinear programming problem. These duals allow for the weakening of the usual convexity conditions required for duality to hold. A pair of symmetric and self-dual programs under weaker convexity conditions are also given.

1. INTRODUCTION

Consider the nonlinear programming problems

(P) Minimize $f(x)$ subject to $g(x) \geqslant 0$.

(PE) Minimize $f(x)$ subject to $g(x) \geqslant 0$, $h(x) = 0$ where f, g and h are functions with continuous first partial derivates from $\mathbb{R}^n$ into $\mathbb{R}$, $\mathbb{R}^m$ and $\mathbb{R}^k$ respectively.

The Wolfe duals of (P) and (PE) are, respectively,

(D) Maximize $f(u) - y^t g(u)$

subject to $\nabla y^t g(u) = \nabla f(u)$, $y \geqslant 0$,

(DE) Maximize $f(u) - y^t g(u) - z^t h(u)$

subject to $\nabla y^t g(u) + \nabla z^t h(u) = \nabla f(u)$, $y \geqslant 0$.

Here we propose a number of different duals to (P) and (PE) as well as to nonlinear programming problems requiring non-negative variables.

For example, another dual of (P) is

(D1) Maximize $f(u)$

subject to $\nabla y^t g(u) = \nabla f(u)$, $y \geqslant 0$, $y^t g(u) \leqslant 0$.

The advantage of (D1) over (D) is that the objective function of the dual is the same as that of the primal, and, more importantly, the convexity requirements for duality can be relaxed. Wolfe (1961) establishes his duality results on the assumption that f is convex and g concave. A number of authors (see e.g., Bector et al. (1977) and Mahajan and Vartak (1977)) have recently shown that weak and strong duality hold for the Wolfe dual (D) if only the Lagrangean $f-y^t g$ is pseudo-convex. We show

ISBN 0-12-621120-5

that the duality of (D1) to (P) holds under still weaker convexity conditions so that there are problems for which duality theorems do not hold for (D) but can be applied to the proposed dual (D1). A number of such examples are given below.

2. DUALITY

We prove the following results for the pair (P), (D1).

THEOREM 1 (Weak Duality): If, for all feasible (x,u,y), f *is pseudo-convex and* $y^t g$ *is quasi-concave, then* $\inf(P) \geqslant \sup(D1)$.

PROOF: Let x be feasible for (P) and (u,y) feasible for (D1). Since $y^t g(x) \geqslant 0$ and $y^t g(u) \leqslant 0$, we have

$$y^t g(x) - y^t g(u) \geqslant 0 \Rightarrow (x-u)^t \nabla y^t g(u) \geqslant 0$$

(by the quasi-concavity of $y^t g$). Thus

$$0 \leqslant (x-u)^t \nabla f(u) \Rightarrow f(x) \geqslant f(u)$$

(by the constraints of (D1) and the pseudo-convexity of f). □

THEOREM 2 (Strong Duality): If x_0 *is a local or global optimum of* (P) *at which a constraint qualification is satisfied, then there exists* $y \in \mathbb{R}^m$ *such that* (x_0, y) *is feasible for* (D1) *and the corresponding values of* (P) *and* (D1) *are equal. If, also, for all feasible* (x,u,y), f *is pseudo-convex and* $y^t g$ *is quasi-concave, then* x_0 *and* (x_0, y) *are global optima for* (P) *and* (D1) *respectively.*

PROOF: Assuming that a constraint qualification (Mangasarian (1969)) is satisfied at x_0, then by the Kuhn-Tucker necessary conditions (Kuhn and Tucker (1951)) there exists $y \geqslant 0$ such that

$$\nabla y^t g(x_0) = \nabla f(x_0), \quad y^t g(x_0) = 0.$$

Thus, (x_0, y) is feasible for (D1). Equality follows since the objective functions of the primal and dual are the same. Optimality follows, given the pseudo-convexity of f and quasi-concavity of $y^t g$, from weak duality. □

Mangasarian (1965, 1969) points out that weak and strong duality (unlike converse duality) do not hold between (P) and (D) for pseudo-convex f and quasi-concave g. His counterexample for which duality theorems do not apply to the Wolfe dual is the following:

Example 1: Minimize $-e^{-x^2}$ subject to $x-1 \geqslant 0$.

The optimum is attained at x=1, whereas the Wolfe dual (see Mangasarian (1965) or Mangasarian (1969; 158)) has no maximum, only a supremum equal to zero. Duality results, however, do hold between (P) and (D1) which is the problem:

$$\begin{aligned} &\text{Maximize} \quad -e^{-u^2} \\ &\text{subject to} \quad 2ue^{-u^2} - y = 0 \\ &\qquad\qquad\quad y(u-1) \leqslant 0,\ y \geqslant 0. \end{aligned}$$

The maximum is achieved at $u = x = 1$, $y = 2e^{-1}$. Weak duality holds, and the optimal values of the primal and the dual are equal.

An even simpler example of a problem with a pseudo-convex objective function and a concave constraint for which weak and strong duality do not hold for the Wolfe dual but for which the duality given here is applicable is the following:

Example 2: Minimize $x^3 + x$ subject to $x \geqslant 1$.

The optimum is attained at x = 1. While the Wolfe dual is unbounded, corresponding to (D1) we have the problem:

$$\begin{aligned} &\text{Maximize} \quad u^3 + u \\ &\text{subject to} \quad 3u^2 + 1 = y,\ y(u-1) \leqslant 0,\ y \geqslant 0 \end{aligned}$$

for which the maximum is attained at $u = 1$, $y = 4$.

Two other examples of problems with pseudo-convex objective functions and concave constraints for which weak and strong duality do not hold for the Wolfe dual are the following, given by Schaible (1976a, 1976b):

Example 3: $\underset{x>0}{\text{Min}} \; -\frac{1}{x^2}$, subject to $x \geqslant 1$ (1)

Example 4: $\underset{x>0}{\text{Min}} \; -\frac{1}{x}$, subject to $1 \leqslant x \leqslant 2$, (2)

In both cases, the optimal value -1 is achieved at x = 1. The duals corresponding to (D1) are, respectively,

$$\underset{u>0}{\text{Max}} \; -\frac{1}{u^2}, \text{ subject to } y = \frac{2}{u^3}, \; y(u-1) \leqslant 0, \; y \geqslant 0 \tag{3}$$

$$\underset{u>0}{\text{Max}} \; -\frac{1}{u}, \quad \text{subject to } y_1 - y_2 = \frac{1}{u^2},$$

$$y_1(u-1) + y_2(-u+2) \leqslant 0, \; y \geqslant 0. \tag{4}$$

In (3) the optimal value is achieved at u = 1, y = 2. In (4), the optimal value is achieved at $u = 1$, $y_1 = 1$, $y_2 = 0$. Thus in both examples, weak duality holds and the dual problems (3) and (4) achieve the same optimal values as the primal, whereas the corresponding Wolfe duals are unbounded. Another dual for (1) and (2), involving fractional programming, other than (D1), is given by Schaible (1976a, 1976b).

As is well known, duality theorems requiring concavity of the constraints are not applicable to (PE), since h(x) = 0 is equivalent to $h(x) \geqslant 0$ and $-h(x) \geqslant 0$ and if h is nonlinear, h and -h cannot both be concave. On the other hand, duality results requiring only pseudo-convexity of the Lagrangean are applicable to (PE). Thus, under the assumption that $f - y^t g - z^t h$ is pseudo-convex, (DE) is a dual to (PE) (Bector et al. (1977) and Mahajan and Vartak (1977)).

Analogous to (D1) and (P), the following problem is a dual to (PE), under the assumptions that f is pseudo-convex and $y^t g + z^t h$ is quasi-concave for all feasible (x,u,y,z).

(DE1) Maximize $f(u)$

subject to $\nabla y^t g(u) + \nabla z^t h(u) = \nabla f(u)$, $y \geqslant 0$

$y^t g(u) + z^t h(u) \leqslant 0$.

Other duals to (P) and (PE) are possible depending on the convexity conditions of f, g and h. We now state a general, if complicated, dual problem to (PE) (and thus to (P)).

Let $M = \{1,2,\ldots,m\}$, $K = \{1,2,\ldots,k\}$, $I_\alpha \subseteq M$, $\alpha=0,1,2,\ldots,r$, with $I_\alpha \cap I_\beta = \emptyset, \alpha \neq \beta$ and $\bigcup_{\alpha=0,1,2,\ldots r} I_\alpha = M$, and $J_\alpha \subseteq K$, $\alpha = 0,1,2,\ldots,r$, with $J_\alpha \cap J_\beta = \emptyset$, $\alpha \neq \beta$ and $\bigcup_{\alpha=0,1,2,\ldots r} J_\alpha = K$.

Note that any particular I_α or J_α may be empty. Thus if M has r_1 disjoint subsets and K has r_2 disjoint subsets, $r = \text{Max}[r_1,r_2]$. So that if $r_1 > r_2$, then J_α, $\alpha > r_1$, is empty.

(DEG) Maximize $f(u) - \sum_{i\in I_0} y_i g_i(u) - \sum_{j\in J_0} z_j h_j(u)$

subject to $\nabla y^t g(u) + \nabla z^t h(u) = \nabla f(u)$, $y \geqslant 0$

$$\sum_{i\in I_\alpha} y_i g_i(u) + \sum_{j\in J_\alpha} z_j h_j(u) \leqslant 0, \quad \alpha=1,2,\ldots,r.$$

THEOREM 3 (Weak Duality): Let x *be feasible for* (PE) *and* (u,y,z) *feasible for* (DEG). *If* $f - \sum_{i\in I_0} y_i g_i - \sum_{j\in J_0} z_j h_j$ *is pseudo-convex for all feasible* (x,u,y,z) *and if* $\sum_{i\in I_\alpha} y_i g_i + \sum_{j\in J_\alpha} z_j h_j$, $\alpha=1,2,\ldots,r$ *is quasi-concave for all feasible* (x,u,y,z), *then inf*(PE) $\geqslant$ *sup*(DEG).

PROOF: $\sum_{i\in I_\alpha} y_i g_i(x) + \sum_{j\in J_\alpha} z_j h_j(x) - \sum_{i\in I_\alpha} y_i g_i(u) - \sum_{j\in J_\alpha} z_j h_j(u) \geqslant 0$,

$\alpha=1,2,\ldots,r.$

Since $\sum_{i\in I_\alpha} y_i g_i + \sum_{j\in J_\alpha} z_j h_j$ is quasi-concave, $\alpha=1,2,\ldots,r$, then

$$(x-u)^t\{\nabla \sum_{i\in I_\alpha} y_i g_i(u) + \nabla \sum_{j\in J_\alpha} z_j h_j(u)\} \geqslant 0, \quad \alpha=1,2,\ldots,r.$$

Thus $(x-u)^t \left\{ \sum_{\substack{i\in \cup I_\alpha \\ \alpha=1,\ldots,r}} \nabla y_i g_i(u) + \sum_{\substack{j\in \cup J_\alpha \\ \alpha=1,\ldots,r}} \nabla z_j h_j(u) \right\} \geqslant 0.$

From the equality constraint of (DEG), we now have

$$(x-u)^t\{\nabla f(u) - \sum_{i\in I_0} \nabla y_i g_i(u) - \sum_{j\in J_0} \nabla z_j h_j(u)\} \geqslant 0$$

and since $f - \sum_{i\in I_0} y_i g_i - \sum_{j\in J_0} z_j h_j$ is pseudo-convex, then

$$f(x) - \sum_{i\in I_0} y_i g_i(x) - \sum_{j\in J_0} z_j h_j(x) \geqslant f(u) - \sum_{i\in I_0} y_i g_i(u) - \sum_{j\in J_0} z_j h_j(u).$$

Thus by the constraints of (PE),

$$f(x) \geqslant f(u) - \sum_{i\in I_0} y_i g_i(u) - \sum_{j\in J_0} z_j h_j(u).$$ □

THEOREM 4 (Strong Duality): If x_0 *is a local or global optimum of* (PE) *at which a constraint qualification is satisfied, then there exists* (y,z) *such that* (x_0,y,z) *is feasible for* (DEG) *and the corresponding values of* (PE) *and* (DEG) *are equal. If also* $f - \sum_{i\in I_0} y_i g_i - \sum_{j\in J_0} z_j h_j$ *is pseudo-convex for all feasible* (x,u,y,z) *and* $\sum_{i\in I_\alpha} y_i g_i + \sum_{j\in J_\alpha} z_j h_j$, $\alpha=1,2,\ldots,r$, *is quasi-concave for all feasible* (x,u,y,z), *then* x_0 *and* (x_0,y,z) *are global optimal solutions of* (PE) *and* (DEG) *respectively.*

PROOF: Assuming that a constraint qualification is satisfied at x_0, then by the Kuhn-Tucker necessary conditions there exists z and $y \geqslant 0$, such that

$$\nabla f(x_0) = \nabla y^t g(x_0) + \nabla z^t h(x_0),\ y^t g(x_0) = 0.$$

Thus (x_0,y,z) is feasible for (DEG). Equality holds since $y_i g_i(x_0) = 0$, $i=1,2,\ldots,m$ and $h(x_0) = 0$. Optimality then follows, if $f - \sum_{i\in I_0} y_i g_i - \sum_{j\in J_0} z_j h_j$ is pseudo-convex and $\sum_{i\in I_\alpha} y_i g_i + \sum_{j\in J_\alpha} z_j h_j$, $\alpha=1,2,\ldots,r$, is quasi-concave, from weak duality. □

We now consider some special cases of the dual (DEG) and Theorems 3 and 4.

If $K = \emptyset$, $I_0 = M$, then (PE) becomes (P) and (DEG) becomes (D). As already mentioned and also from Theorems 3 and 4, (D) is a dual to (P) if $f - y^t g$ is pseudo-convex for all feasible (x,u,y); in particular $f - y^t g$ is pseudo-convex if f is convex, g concave and $y \geqslant 0$.

If $K = \emptyset$, $I_0 = \emptyset$, $I_\alpha = M$ (for some $\alpha \in \{1,2,\ldots,r\}$), then (PE) becomes (P) and (DEG) becomes (D1). From Theorems 1 and 2 or Theorems 3 and 4, (D1) is a dual to (P) if f is pseudo-convex and $y^t g$ is quasi-concave for all feasible (x,u,y).

If $K = \emptyset$, $I_0 = \emptyset$, $I_1 = \{1\}$, $I_2 = \{2\},\ldots,$ $I_m = \{m\}$, $(r=m)$ then (PE) becomes (P) and (DEG) becomes

(D2) Maximize $f(u)$

subject to $\nabla y^t g(u) = \nabla f(u)$, $y \geqslant 0$

$y_i g_i(u) \leqslant 0$, $i=1,2,\ldots,m$.

From Theorems 3 and 4, (D2) is a dual to (P) if f is pseudo-convex and each $y_i g_i$, $i=1,2,\ldots,m$, is quasi-concave for all feasible (x,u,y). Note that if g_i is quasi-concave and $y_i \geqslant 0$, then $y_i g_i$ is quasi-concave; thus (D2) is a dual to (P) if f is pseudo-convex and g (i.e., each component of g) is quasi-concave.

Duality for (D1) required the quasi-concavity of $y^t g$, whereas only the quasi-concavity of g was required for duality in (D2). It may happen that some but not all of the components of g can be combined into one quasi-concave function while the others are individually but not collectively quasi-concave. In this case it is possible to find a dual somewhere between (D1) and (D2) as follows. If $K = \emptyset$, $I_0 = \emptyset$, $\Omega = \bigcup_{\alpha\in\Lambda} I_\alpha \subseteq M$ where $\Lambda \subseteq \{1,2\ldots,r\}$ then (DEG) becomes

(D3) Maximize $f(u)$

subject to $\nabla y^t g(u) = \nabla f(u)$, $y \geqslant 0$

$\sum_{i\in\Omega} y_i g_i(u) \leqslant 0$, $y_j g_j(u) \leqslant 0$, $j \in M \setminus \Omega$.

From Theorems 3 and 4, (D3) is a dual to (P) if, for all feasible (x,u,y), f is pseudo-convex, $\sum_{i\in\Omega} y_i g_i$ is quasi-concave and g_j, $j\in M\setminus\Omega$ are all quasi-concave.

As already noted duality holds between (P) and the Wolfe dual (D) if the Lagrangean $f - y^t g$ is pseudo-convex for all feasible (x,u,y). However if only part of the Lagrangean is pseudo-convex, it is possible to find a dual somewhere between the Wolfe dual (D) and some of the other duals suggested here, as follows. If $K=\emptyset$, $\Omega = \bigcup_{\alpha\in\Lambda} I_\alpha \subseteq M$ where $\Lambda \subseteq \{1,2,\ldots,r\}$ then (DEG) becomes

(D4) Maximize $f(u) - \sum_{i\in I_0} y_i g_i(u)$

subject to $\nabla y^t g(u) = \nabla f(u)$, $y\geqslant 0$

$\sum_{i\in\Omega} y_i g_i(u) \leqslant 0$, $y_j g_j(u) \leqslant 0$, $j\in M\setminus\Omega$.

From Theorems 3 and 4, (D4) is a dual to (P) if $f - \sum_{i\in I_0} y_i g_i$ is pseudo-convex, $\sum_{i\in\Omega} y_i g_i$ is quasi-concave and g_j, $j\in M\setminus\Omega$ are all quasi-concave, for all feasible (x,u,y).

If $I_0 = M$, $J_0 = K$, then (DEG) becomes (DE). From Theorems 3 and 4, as already mentioned, (DE) is a dual to (PE) if $f - y^t g - z^t h$ is pseudo-convex for all feasible (x,u,y,z).

If $I_0 = \emptyset$, $J_0 = \emptyset$, $I_1 = M$, $J_1 = K$, then (DEG) becomes (DE1). From Theorems 3 and 4 (DE1) is a dual to (PE) if f is pseudo-convex and $y^t g + z^t h$ is quasi-concave for all feasible (x,u,y,z).

If $I_0 = \emptyset$, $J_0 = K$, $I_\alpha = M$ (some $\alpha \in \{1,2,\ldots,r\}$), then (DEG) becomes

(DE2) Maximize $f(u) - z^t h(u)$

subject to $\nabla f(u) = \nabla y^t g(u) + \nabla z^t h(u)$, $y \geqslant 0$

$y^t g(u) \leqslant 0$.

From Theorems 3 and 4, (DE2) is a dual to (PE) if $f - z^t h$ is pseudo-convex and $y^t g$ is quasi-concave for all feasible (x,u,y,z).

If $J_0 = \emptyset$, $I_0 = M$, $J_1 = K$, then (DEG) becomes

(DE3) Maximize $f(u) - y^t g(u)$

subject to $\nabla f(u) = \nabla y^t g(u) + \nabla z^t h(u),\ y \geqslant 0$

$z^t h(u) \leqslant 0.$

From Theorems 3 and 4, (DE3) is a dual to (PE) if $f - y^t g$ is pseudo-convex and $z^t h$ is quasi-concave for all feasible (x,u,y,z).

3. NON-NEGATIVE VARIABLES

Consider the problem

(P') Minimize $f(x)$

subject to $g(x) \geqslant 0,\ x \geqslant 0.$

The Wolfe dual of (P') is the problem

(D') Maximize $f(u) - y^t g(u) - u^t[\nabla f(u) - \nabla y^t g(u)]$

subject to $\nabla f(u) \geqslant \nabla y^t g(u),\ y \geqslant 0.$

Duality holds if f is convex and g concave (Mangasarian (1962))

One can also show that if $f - y^t g - v^t x$, for all $v \geqslant 0$, is pseudo-convex, then duality also holds. By applying some of our earlier results to (P'), or directly, one can establish different dual problems to (P') that will hold under different convexity conditions. We state three such duals and corresponding theorems although many other combinations are also possible.

(D1') Maximize $f(u)$

subject to $\nabla y^t g(u) - \nabla f(u) \leqslant 0,\ y \geqslant 0$

$y^t g(u) + u^t[\nabla f(u) - \nabla y^t g(u)] \leqslant 0,$

(D2') Maximize $f(u) - u^t[\nabla f(u) - \nabla y^t g(u)]$

subject to $\nabla f(u) - \nabla y^t g(u) \geqslant 0,\ y \geqslant 0$

$y^t g(u) \leqslant 0,$

(D3') Maximize $f(u) - y^t g(u)$

subject to $\nabla f(u) - \nabla y^t g(u) \geqslant 0,\ y \geqslant 0$

$u^t[\nabla f(u) - \nabla y^t g(u)] \leqslant 0.$

COROLLARY 5: (a) If f *is pseudo-convex and* $y^t g + v^t[.]$ *is quasi-concave for all feasible* (x,u,y) *and* $v \geqslant 0$, *then inf* (P') $\geqslant$ *sup* (D1').

(b) If $f - v^t[.]$ *is pseudo-convex and* $y^t g$ *is quasi-concave for all feasible* (x,u,y) *and* $v \geqslant 0$, *then inf*(P') $\geqslant$ *sup* (D2').

(c) If $f - y^t g$ *is pseudo-convex for all feasible* (x,u,y), *then inf*(P') $\geqslant$ *sup*(D3').

Observe that $x^t v$, for all v, will be concave, and hence quasi-concave, in x, and so this condition does not have to be stated in (c).

The corollary, as well as the corresponding strong duality result, can be obtained directly, or by applying Theorems 3 and 4, respectively, to (P') and then eliminating from the dual problems the multiplier corresponding to the constraints $x \geqslant 0$.

4. CONVERSE DUALITY

THEOREM 6: Let $(\bar{x},\bar{y},\bar{z})$ *be a local or global optimum of* (DEG). *Let* $f - \sum_{i\in I_0} y_i g_i - \sum_{j\in J_0} z_j h_j$ *be pseudo-convex for all feasible* (x,u,y,z), *and let* $\sum_{I\in I_\alpha} y_i g_i + \sum_{j\in J_\alpha} z_j h_j$ *be quasi-concave for all feasible* (x,u,y,z), $\alpha=1,2,\ldots,r$.

If the $n \times n$ *Hessian matrix*

$$\nabla^2 f(\bar{x}) - \nabla^2 \bar{y}^t g(\bar{x}) - \nabla^2 \bar{z}^t h(\bar{x}) \tag{5}$$

is positive or negative definite and if the set

$$\left\{ \sum_{i\in I_\alpha} \nabla \bar{y}_i g_i(\bar{x}) + \sum_{j\in J_\alpha} \nabla \bar{z}_j h_j(\bar{x}) \ :\ \alpha=1,2,\ldots,r \right\} \tag{6}$$

is linearly independent, then $\bar{x}$ *is an optimal solution to* (PE).

PROOF: By the generalized Fritz John Theorem (Mangasarian and Fromovitz (1967)), there exist $\tau \in \mathbb{R}$, $v \in \mathbb{R}^n$, $w_\alpha \in \mathbb{R}$, $\alpha = 1,2,\ldots,r$, $s \in \mathbb{R}^m$ such that

$$\tau(\nabla f(\bar{x}) - \sum_{i\in I_0} \nabla \bar{y}_i g_i(\bar{x}) - \sum_{j\in J_0} \nabla \bar{z}_j h_j(\bar{x})) + (\nabla^2 \bar{y}^t g(\bar{x}) + \nabla^2 \bar{z}^t h(\bar{x}) - \nabla^2 f(\bar{x}))v$$

$$- \sum_{\alpha=1}^{r} w_\alpha (\sum_{i\in I_\alpha} \nabla \bar{y}_i g_i(\bar{x}) + \sum_{j\in J_\alpha} \nabla \bar{z}_j h_j(\bar{x})) = 0 \quad (7)$$

$$\tau g_i(\bar{x}) - v^t \nabla g_i(\bar{x}) - s_i = 0,\ i \in I_0 \quad (8)$$

$$v^t \nabla g_i(\bar{x}) - w_\alpha g_i(\bar{x}) + s_i = 0,\ i \in I_\alpha,\ \alpha=1,2,\ldots,r \quad (9)$$

$$\tau h_j(\bar{x}) - v^t \nabla h_j(\bar{x}) = 0 \quad , \quad j \in J_0 \quad (10)$$

$$v^t \nabla h_j(\bar{x}) - w_\alpha h_j(\bar{x}) = 0 \quad , \quad j \in J_\alpha,\ \alpha=1,2,\ldots,r \quad (11)$$

$$w_\alpha (\sum_{i\in I_\alpha} \bar{y}_i g_i(\bar{x}) + \sum_{j\in J_\alpha} \bar{z}_j h_j(\bar{x})) = 0 \quad , \quad \alpha=1,2,\ldots,r \quad (12)$$

$$s^t \bar{y} = 0 \quad (13)$$

$$(\tau,\ v,\ s,\ w_1,\ w_2,\ldots,w_r) \neq 0 \quad (14)$$

$$(\tau,\ s,\ w_1,\ w_2,\ldots,w_r) \geqslant 0\,. \quad (15)$$

Multiplying (9) by $\bar{y}_i \geqslant 0$, and using (13) gives

$$v^t \nabla \bar{y}_i g_i(\bar{x}) - w_\alpha \bar{y}_i g_i(\bar{x}) = 0,\ \alpha=1,2,\ldots,r,\quad i \in I_\alpha.$$

Thus

$$\sum_{i\in I_\alpha} (v^t \nabla \bar{y}_i g_i(\bar{x}) - w_\alpha \bar{y}_i g_i(\bar{x})) = 0,\quad \alpha=1,2,\ldots,r. \quad (16)$$

Multiplying (11) by $\bar{z}_j$ gives

$$v^t \nabla \bar{z}_j h_j(\bar{x}) - w_\alpha \bar{z}_j h_j(\bar{x}) = 0,\quad \alpha=1,2,\ldots,r,\quad j \in J_\alpha.$$

Thus

$$\sum_{j\in J_\alpha} (v^t \nabla \bar{z}_j h_j(\bar{x}) - w_\alpha \bar{z}_j h_j(\bar{x})) = 0,\quad \alpha=1,2,\ldots,r. \quad (17)$$

From (12), (16) and (17) it follows that

$$\sum_{i\in I_\alpha} (v^t \nabla \bar{y}_i g_i(\bar{x})) + \sum_{j\in J_\alpha} (v^t \nabla \bar{z}_j h_j(\bar{x})) = 0,\quad \alpha=1,2,\ldots,r. \quad (18)$$

Hence multiplying (7) by v^t, then using (18) and the equality constraint of (DEG) we get

$$v^t(\nabla^2 f(\bar{x}) - \nabla^2 \bar{y}^t g(\bar{x}) - \nabla^2 \bar{z}^t h(\bar{x}))v = 0.$$

Since (5) is assumed positive or negative definite, then v = 0. Using v = 0, and the equality constraint of (DEG) in (7) we get

$$\sum_{\alpha=1}^{r} (\tau - w_\alpha)\Big(\sum_{i\in I_\alpha} \nabla \bar{y}_i g_i(\bar{x}) + \sum_{j\in J_\alpha} \nabla \bar{z}_j h_j(\bar{x})\Big) = 0. \tag{19}$$

Since the set (6) is assumed linearly independent, it follows that

$$\tau = w_\alpha,\ \alpha=1,2,\dots,r.$$

If $\tau=0$, then $w_\alpha = 0$, $\alpha=1,2,\dots,r$; and so from (8) and (9) and v=0, it follows that s=0. But $(\tau, v, s, w_1, w_2,\dots,w_r) = 0$ contradicts (14), hence $\tau > 0$. Thus (8), (9) give $g(\bar{x}) \geqslant 0$ and (10), (11) give $h(\bar{x}) = 0$. Thus $\bar{x}$ is feasible for (PE) and the objective functions of (PE) and (DEG) are equal, since $h(\bar{x}) = 0$ and by (8) and (9), $\bar{y}_i g_i(\bar{x}) = 0$, $i \in I_0$. If $f - \sum_{i\in I_0} y_i g_i - \sum_{j\in J_0} z_j h_j$ is pseudo-convex and if $\sum_{i\in I_\alpha} y_i g_i + \sum_{j\in J_\alpha} z_j h_j$, $\alpha=1,2,\dots,r$, are quasi-concave, then by Theorem 3, $\bar{x}$ is an optimal solution for (PE). □

5. SYMMETRIC DUALITY

Following Dorn (1960), a program and its dual are said to be *symmetric* if the dual of the dual is the original problem, i.e. if, when the dual program is recast in the form of the primal, its dual is the primal. Dantzig et al. (1965) formulate a pair of symmetric dual programming problems involving a scalar function f(x,y), $x \in \mathbb{R}^n$, $y \in \mathbb{R}^m$ that is required to be convex in x for fixed y and concave in y for fixed x. Here we give a different pair of symmetric dual nonlinear programming problems in which the convexity and concavity assumptions have been reduced to pseudo-convexity and pseudo-concavity.

Primal (PS)

$$\text{Minimize} \quad f(x,y)$$

$$\text{subject to} \quad f_y(x,y) \leqslant 0 \tag{20}$$

$$y^t f_y(x,y) \geqslant 0 \tag{21}$$

$$x \geqslant 0 \tag{22}$$

Dual (DS)

$$\text{Maximize} \quad f(u,v)$$

$$\text{subject to} \quad f_x(u,v) \geqslant 0 \tag{23}$$

$$u^t f_x(u,v) \leqslant 0 \tag{24}$$

$$v \geqslant 0 \tag{25}$$

Here f is a twice differentiable real-valued function of x and y where $x \in \mathbb{R}^n$ and $y \in \mathbb{R}^m$. f_x and f_y denote gradient (column) vectors with respect to x and y respectively. Subsequently f_{yy} and f_{yx} will denote, respectively the m x m and n x m matrices of 2nd partial derivatives.

THEOREM 7 (Weak Duality): Let $f(\cdot,y)$ *be pseudo-convex (for fixed* y*) and let* $f(x,\cdot)$ *be pseudo-concave (for fixed* x*). Let* (x,y) *be feasible for (PS) and let* (u,v) *be feasible for (DS). Then* inf(PS) $\geqslant$ sup(DS).

PROOF: From (22), (23) and (24), we have

$$(x-u)^t f_x(u,v) \geqslant 0.$$

Since $f(\cdot,v)$ is pseudo-convex, it follows that

$$f(x,v) \geqslant f(u,v).$$

From (20), (21) and (25) we have

$$(v-y)^t f_y(x,y) \leqslant 0.$$

Since $f(x,\cdot)$ is pseudo-concave, it follows that $f(x,v) \leqslant f(x,y)$. Thus, $f(x,y) \geqslant f(u,v)$. □

THEOREM 8 (Strong Duality): Let $f(\cdot,y)$ *be pseudo-convex (for fixed* y*) and let* $f(x,\cdot)$ *be pseudo-concave (for fixed* x*). Let* (x_0,y_0) *be a local or global optimal solution of the primal problem,* $f_{yy}(x_0,y_0)$ *be positive or negative definite and* $f_y(x_0,y_0) \neq 0$. *Then* (x_0,y_0) *is an optimal solution to the dual problem.*

PROOF: Since (x_0,y_0) is an optimal solution of the primal problem there exists $\tau \in \mathbb{R}$, $r \in \mathbb{R}^m$, $w \in \mathbb{R}$, $s \in \mathbb{R}^n$ that satisfy the following Fritz John conditions (John (1948)): (for simplicity of notation, we write f_x, f_y, f_{yx}, f_{yy} instead of $f_x(x_0,y_0)$ $f_y(x_0,y_0)$ $f_{yx}(x_0,y_0)$ and $f_{yy}(x_0,y_0)$)

$$\tau f_x - f_{yx}(wy_0-r) - s = 0 \tag{26}$$

$$(\tau-w)f_y - f_{yy}(wy_0-r) = 0 \tag{27}$$

$$r^t f_y = 0 \tag{28}$$

$$wy_0{}^t f_y = 0 \tag{29}$$

$$s^t x_0 = 0 \tag{30}$$

$$(\tau, r, w, s) \neq 0 \tag{31}$$

$$(\tau, r, w, s) \geqslant 0. \tag{32}$$

Multiplying (27) by $(wy_0-r)^t$ and applying (28) and (29) gives

$$(wy_0-r)^t f_{yy}(wy_0-r) = 0.$$

Since it is assumed that f_{yy} is positive or negative definite, then

$$wy_0 = r. \tag{33}$$

Thus from (27), $(w-\tau)f_y = 0$, and since, by assumption $f_y \neq 0$, then

$$w = \tau. \tag{34}$$

If $\tau = 0$, $w = 0$ by (34), $r = 0$ by (33) and $s = 0$ by (26); but this contradicts (31) so that $\tau > 0$, $w > 0$. Thus by (33) $y_0 \geqslant 0$ and by (26) and (33) $f_x \geqslant 0$, and (26), (30) give $x_0^t f_x = 0$. Thus (x_0, y_0) is feasible for (DS) and the objective functions are equal. Optimality follows by weak duality. □

6. SELF DUALITY

A mathematical programming problem is said to be *self dual* if, when the dual is recast in the form of the primal, the new program thus constructed is the same as the primal problem.

A function $f(x,y)$, $x \in R^n$, $y \in R^n$, is said to be *skew-symmetric* if

$$f(x,y) = -f(y,x)$$

for all x and y in the domain of f. Mond and Cottle (1966) showed that for the symmetric dual program of Dantzig et al. (1965), if $f(x,y)$ is skew-symmetric, the program is self dual. A similar result holds here for (PS). Indeed, the proof for the following theorem is simpler than the corresponding self duality result in Mond and Cottle (1966).

THEOREM 9: Let f *be skew-symmetric. Then* (PS) *is self dual. If also* (PS) *and* (DS) *are dual programs, and* (x_0, y_0) *is a joint optimal solution, then so is* (y_0, x_0) *and*

$$f(x_0, y_0) = f(y_0, x_0) = 0.$$

PROOF: (DS) can be written as

$$\begin{aligned} \text{Minimize} \quad & -f(u,v) \\ \text{subject to} \quad & -f_x(u,v) \leqslant 0 \\ & -u^t f_x(u,v) \geqslant 0 \\ & v \geqslant 0. \end{aligned}$$

Since f is skew-symmetric, $f_x(u,v) = -f_y(v,u)$ and (DS) becomes

$$\begin{aligned} \text{Minimize} \quad & f(v,u) \\ \text{subject to} \quad & f_y(v,u) \leqslant 0 \\ & u^t f_y(v,u) \geqslant 0 \\ & v \geqslant 0 \end{aligned}$$

which is just (PS).

Thus (x_0,y_0) optimal for (DS) implies (y_0,x_0) optimal for (PS) and, by symmetric duality, also for (DS). Therefore

$$f(x_0,y_0) = f(y_0,x_0) = -f(x_0,y_0) = 0. \qquad \square$$

7. REMARKS

One application of the results given here that will soon be more fully explored elsewhere is fractional programming. Consider the problem

$$\text{Minimize } \hat{f}(x)/\hat{g}(x) \quad \text{subject to } \hat{h}(x) \geqslant 0$$

where $\hat{f}$, $\hat{g}$ and $\hat{h}$ are differentiable functions from R^n into R, R and R^m respectively. If $\hat{f} \geqslant 0$ and convex, $\hat{g} > 0$ and concave, then (Mangasarian (1969; 148)) $\hat{f}/\hat{g}$ is pseudo-convex. If, also, $\hat{h}$ is concave, many duality results can be found in the literature, see e.g., Bector et al. (1977), Mahajan and Vartak (1977), Schaible (1976a, 1976b)); for a unified approach that relates different duality results in fractional programming to each other, see Schaible (1976a). The theorems given here, however, are applicable if $\hat{h}$ or $y^t\hat{h}$ or some of the components of $\hat{h}$ are quasi-concave, rather than concave.

Extensions of some of the results given here to generalized convex and invex functions of Craven (1981), Craven and Mond (1980), Hanson and Mond (1980), Hanson (1980), and Mond and Hanson (1980) are also possible and will be discussed in a subsequent paper.

REFERENCES

BECTOR, C.R., BECTOR, M.K. and KLASSEN, J.E. (1977). "Duality for a Nonlinear Programming Problem." *Utilitas Mathematica 11*, 87-99.

CRAVEN, B.D. (1981). "Duality for Generalized Convex Fractional Programs." This volume, 473-489.

CRAVEN, B.D. and MOND, B. (1980). "On Fractional Programming Duality with Generalized Convexity." Submitted for publication.

DANTZIG, G.G., EISENBERG, E. and COTTLE, R.W. (1965). "Symmetric Dual Nonlinear Programs." *Pacific J. Math. 15*, 809-812.

DORN, W.S. (1960). "A Symmetric Dual Theorem for Quadratic Programs." *J. Oper. Res. Soc. Japan 2*, 93-97.

HANSON, M.A. (1980). "On Sufficiency of the Kuhn-Tucker Conditions." *Pure Mathematics Research Paper No. 80-1*, March 1980, La Trobe University, Bundoora, Melbourne, Australia. To appear in J. Math. Anal. Appl.

HANSON, M.A. and MOND, B. (1980). "Further Generalizations of Convexity in Mathematical Programming." *Pure Mathematics Research Paper No. 80-6*, June 1980, La Trobe University, Bundoora, Melbourne, Australia.

JOHN, F. (1948). "Extremum Problems with Inequalities as Side Conditions" In *Studies and Essays, Courant Anniversary Volume* (K.O. Friedrichs, O.L. Neugebauer and J.J. Stoker, eds.). Interscience, New York, 187-204.

KUHN, H.W. and TUCKER, A.W. (1951). "Nonlinear Programming." *Proceedings of the 2nd Berkeley Symposium on Mathematical Statistics and Probability*. University of California Press, 481-492.

MAHAJAN, D.G. and VARTAK, M.N. (1977). "Generalization of Some Duality Theorems in Nonlinear Programming." *Mathematical Programming 12*, 293-317.

MANGASARIAN, O.L. (1965). "Pseudo-convex Functions." *SIAM J. Control 3*, 281-290.

MANGASARIAN, O.L. (1969). *Nonlinear Programming*. McGraw-Hill, New York.

MANGASARIAN, O.L. (1962). "Duality in Nonlinear Programming." *Quart. Appl. Math 20*, 300-302.

MANGASARIAN, O.L. and FROMOVITZ, S. (1967). "The Fritz John Necessary Optimality Conditions in the Presence of Equality and Inequality Constraints." *J. Math. Anal. Appl. 17*, 37-47.

MOND, B. and COTTLE, R.W. (1966). "Self-Duality in Mathematical Programming." *SIAM J. Appl. Math 14*, 420-423.

MOND, B. and HANSON, M.A. (1980). "On Duality with Generalized Convexity." *Pure Mathematics Research Paper No. 80-3*, April 1980, La Trobe University, Bundoora, Melbourne, Australia.

SCHAIBLE, S. (1976a). "Duality in Fractional Programming: A Unified Approach." *Operations Research 24*, 452-461.

SCHAIBLE, S. (1976b). "Fractional Programming. I, Duality." *Management Science 22*, 858-867.

WOLFE, P. (1961). "A Duality Theorem for Nonlinear Programming." *Quart. Appl. Math 19*, 239-244.

ESTIMATES OF THE DUALITY GAP FOR DISCRETE AND QUASICONVEX OPTIMIZATION PROBLEMS

F. Di Guglielmo

Estimates of the duality gap for discrete optimization problems in terms of the lack of convexity of the objective function and the constraints are derived. In addition we obtain estimates of the duality gap for nonlinear programs in continuous variables in terms of the lack of quasiconvexity of the objective function and constraints.

1. ESTIMATES OF THE DUALITY GAP FOR DISCRETE OPTIMIZATION PROBLEMS

Integer programming problems generally have a nonzero duality gap, that is the difference between the optimal values of the primal and of the dual problem is different from zero because of the discrete character of the feasible region.

The aim of the present paper is to give estimates of the duality gap by using an appropriate measure of the nonconvexity of the data involved.

A natural measure of the nonconvexity of a function f defined on a convex set X is given by Aubin-Ekeland (1974) in the form:

$$\rho_X(f) = \sup_{\substack{x_i \in X,\ \alpha_i \geqslant 0 \\ \sum_{i\varepsilon I} \alpha_i = 1}} \left[f\left(\sum_{i\varepsilon I} \alpha_i x_i\right) - \sum_{i\varepsilon I} \alpha_i f(x_i) \right] .$$

Since for a finite set X the barycentric combinations $\sum_{i\varepsilon I} \alpha_i x_i$ do not necessarily belong to X, one has to introduce the lack of γ-convexity of f, where γ is a mapping which associates with every discrete probability measure $m = \sum_{i\varepsilon I} \alpha_i \delta(x_i)$ on X a point $\gamma m \in X$

$$\rho_{\gamma,X}(f) = \sup_m \left[f(\gamma m) - \sum_{i\varepsilon I} \alpha_i f(x_i) \right] .$$

Let us consider a discrete optimization problem of the following form

$$\text{(P)} \qquad \begin{array}{l} \text{Minimize } f(x) \\ \text{subject to } x \in X\ ,\ A(x) \leqslant b \end{array}$$

ISBN 0-12-621120-5

where $X = \{x_1, x_2, \ldots, x_R\} \subset \mathbb{R}^n$, A is a given mapping from $\mathbb{R}^n$ to $\mathbb{R}^m$ and f is a real valued function defined on X. The dual problem of (P) is defined by

$$\sup_{p \in \mathbb{R}^m_+} \inf_{x \in X} \{f(x) + \langle p, A(x) - b \rangle\} .$$

We shall prove that there exists a mapping γ such that

$$o \leqslant v - v^* \leqslant \rho_{\gamma, X}(f) .$$

Here v resp. v* denotes the optimal value of the primal and the dual problem respectively. The preceding estimate can be applied to the study of Lagrangean relaxations of discrete optimization problems of the form

$$(Q) \quad \begin{cases} \text{Minimize } f(x) \\ \text{subject to} \\ A(x) \leqslant b \\ C(x) \leqslant d \\ x \in X . \end{cases}$$

We shall begin by studying discrete optimization problems of the form (P). We first derive a particular form of the dual of (P) well suited for the study of the duality gap.

Let us denote by

$$L(x,p) = f(x) + \langle p, A(x) - b \rangle \quad , \quad p \in \mathbb{R}^m$$

the Lagrangean function of (P). We can formulate the dual problem as

$$\sup_{p \in \mathbb{R}^m_+} \inf_{x \in X} L(x,p) = \sup_{p \in \mathbb{R}^m_+} \{-\langle p,b \rangle + \inf_{x \in X} [f(x) + \langle p, A(x) \rangle]\} . \qquad (1)$$

Since $X = \{x_1, x_2, \ldots, x_R\}$ the dual problem can be written explicitly as

$$(P^*) \quad \begin{cases} \text{Maximize } z \\ \text{subject to} \\ z \leqslant \langle p, A(x^r) - b \rangle + f(x^r) \qquad r = 1,2,\ldots,R \\ p \geqslant o . \end{cases}$$

With (P*) being a linear programming problem, its optimum must coincide with that of the second dual

$$\operatorname*{Sup}_{\mu,q \geqslant o} \ \operatorname*{Inf}_{z,p} \ L(z,p,\mu,q)$$

where $\mu = (\mu_1,\ldots,\mu_R)$ and $q \in \mathbb{R}^m_+$ are Lagrange multipliers and

$$L(z,p,\mu,q) = -z + \sum_{r=1}^{R} \mu_r [z - \langle p, A(x^r) - b\rangle - f(x^r)] - \langle q,p\rangle \qquad (2)$$

is the Lagrangean function of (P*).

This second dual can be rewritten in the following form (Fisher-Shapiro (1974))

$$(P^{**}) \begin{cases} \text{Minimize } \sum_{r=1}^{R} \mu_r f(x^r) \\ \text{subject to} \\ \sum_{r=1}^{R} \mu_r A(x^r) \leqslant b \\ \sum_{r=1}^{R} \mu_r = 1 , \quad \mu_r \geqslant o . \end{cases}$$

We shall derive an estimate of the duality gap of (P) in terms of the lack of γ-convexity $\rho_{\gamma,X}(f)$ of the objective function

$$\rho_{\gamma,X}(f) = \operatorname*{Sup}_{m \in M(X)} [f(\gamma m) - f^{\Delta}(m)]$$

where $M(X)$ denotes the set of discrete probability measures on X i.e. measures of the form $m = \sum_{i=1}^{R} \alpha_i \delta(x^i)$ with $\alpha_i \geqslant o$, $\sum_{i=1}^{R} \alpha_i = 1$, $\delta(x^i)$ being the Dirac measure at the point x^i and γ denoting a given mapping from $M(X)$ into X. If A is an arbitrary mapping from X into a vector space V, we shall denote by A^{Δ} its extension to $M(X)$ defined by

$$A^{\Delta} m = \sum_{i=1}^{R} \alpha_i A(x^i) \text{ for } m = \sum_{i=1}^{R} \alpha_i \delta(x^i) \in M(X) .$$

We can now prove the following result.

THEOREM 1: Let X *be a finite subset of* $\mathbb{R}^n$, f *a realvalued function defined on* X *and* A *a mapping from* $\mathbb{R}^n$ *into* $\mathbb{R}^m$. *Let us suppose that one of the following assumptions is satisfied:*

$$(H)\begin{cases} A \text{ is a linear operator and there exists a mapping } \gamma \text{ from} \\ M(X) \text{ into } X \text{ such that for } \mu \in M(X) \\ A(\beta\mu) \leqslant b \quad \text{implies} \quad A(\gamma\mu) \leqslant b \\ \text{where } \beta \text{ is the barycentric operator,} \end{cases}$$

$$(H')\begin{cases} A \text{ is a nonlinear operator and there exists a mapping } \gamma \text{ from} \\ M(X) \text{ into } X \text{ such that for } \mu \in M(X) \\ A^{\Delta}(\mu) \leqslant b \quad \text{implies} \quad A(\gamma\mu) \leqslant b \text{ .} \end{cases}$$

Then, if we denote by v *and* v* *the optimum of* (P) *and* (P*) *respectively and by* γ *a mapping satisfying either* (H) *or* (H'), *we have*

$$v \leqslant \rho_{\gamma,X}(f) + v^* \text{ .} \tag{3}$$

If γ *is an arbitrary mapping from* $M(X)$ *into* X *we have the following estimate*

$$v(\rho_{\gamma,X}(A),\ \rho_{\gamma,X}(-A)) = \operatorname*{Inf}_{x\in K(\rho_{\gamma,X}(A),\rho_{\gamma,X}(-A))} f(x) \leqslant \rho_{\gamma,X}(f) + v^* \tag{4}$$

where

$$v(c,d) = \operatorname*{Inf}_{x\in K(c,d)} f(x) \quad \text{and}$$

$$K(c,d) = \{x \in X \,|\, [A(x) - c,\ A(x) + d] \cap (b - \mathbb{R}^m_+) \neq \emptyset\} \text{ .}$$

PROOF: Let us consider the second dual (P**) and introduce the discrete probability measure $\mu \in M(X)$ defined by

$$\mu = \sum_{r=1}^{R} \mu_r\, \delta(x^r) \text{ with } \mu_r \geqslant 0, \quad \sum_{r=1}^{R} \mu_r = 1, \quad x^r \in X \text{ .}$$

If we denote by $M_A(X)$ the subset of $M(X)$ of the probability measures satisfying $\sum_{r=1}^{R} \mu_r A(x^r) \leqslant b$, we may write (P**) as

$$v^* = \operatorname*{Inf}_{\mu\in M_A(X)} \sum_{r=1}^{R} \mu_r\, f(x^r) \text{ .} \tag{5}$$

It is then clear that for any $\varepsilon > 0$ there exists $\mu \in M(X)$ such that

$$\begin{cases} \sum_{r=1}^{R} \mu_r\, A(x^r) \leqslant v^* + \varepsilon \\ \sum_{r=1}^{R} \mu_r\, A(x^r) \leqslant b \text{ .} \end{cases} \tag{6}$$

Let us denote by γ a mapping from $M(X)$ into X and let us consider the vector

$$x = \gamma\mu = \gamma\left(\sum_{r=1}^{R} \mu_r\, \delta(x^r)\right) .$$

Then we may write

$$f(\gamma\mu) \leqslant f(\gamma\mu) - \sum_{r=1}^{R} \mu_r\, f(x^r) + v^* + \varepsilon = f(\gamma\mu) - f^{\Delta}(\mu) + v^* + \varepsilon .$$

Hence it results that

$$f(\gamma\mu) \leqslant \sup_{m\in M(X)} [f(\gamma m) - f^{\Delta}(m)] + v^* + \varepsilon = \rho_{\gamma,X}(f) + v^* + \varepsilon . \qquad (7)$$

By definition of μ we also have

$$y = A^{\Delta}(\mu) = \sum_{r=1}^{R} \mu_r\, A(x^r) \leqslant b .$$

Now it follows from the definition of the lack of convexity that

$$\begin{aligned} A(\gamma\mu) - \rho_{\gamma,X}(A) &= A(\gamma\mu) - \sup_{m\in M(X)} [A(\gamma m) - A^{\Delta}(m)] \\ &\leqslant A(\gamma\mu) - A(\gamma\mu) + A^{\Delta}(\mu) \\ &\leqslant A(\gamma\mu) + \sup_{m\in M(X)} [-A(\gamma m) - (-A)^{\Delta}(m)] \\ &= A(\gamma\mu) + \rho_{\gamma,X}(-A) \end{aligned} \qquad (8)$$

where the supremum is taken for each component of A. This means that the parallelotope $[A(\gamma\mu) - \rho_{\gamma,X}(A),\ A(\gamma\mu) + \rho_{\gamma,X}(-A)]$ intersects the orthant $b - \mathbb{R}^m_+$. We shall set

$$K(c,d) = \{x \in X \mid [A(x) - c,\ A(x) + d] \cap (b - \mathbb{R}^m_+) \neq \emptyset\}$$

and

$$v(c,d) = \inf_{x\in K(c,d)} f(x) .$$

Then it follows from (7) and (8) that

$$v(\rho_{\gamma,X}(A),\ \rho_{\gamma,X}(-A)) \leqslant f(\gamma\mu) \leqslant \rho_{\gamma,X}(f) + v^* + \varepsilon .$$

Since $\varepsilon > o$ is arbitrary this yields

$$v(\rho_{\gamma,X}(A),\ \rho_{\gamma,X}(-A)) \leqslant \rho_{\gamma,X}(f) + v^* . \qquad (9)$$

If we choose for γ a mapping from $M(X)$ into X satisfying either (H) or (H') we deduce from (7)

$$\inf_{\substack{x\in X \\ A(x) \leqslant b}} f(x) \leqslant f(\gamma\mu) \leqslant \rho_{\gamma,X}(f) + v^* + \varepsilon \ . \tag{10}$$

Since $\varepsilon > o$ is arbitrary this implies

$$v \leqslant v^* + \rho_{\gamma,X}(f) \ .$$ □

We shall now show that the assumption (H) resp.(H') can be satisfied if we take $\gamma = t \circ \beta$ where $\beta : M(X) \to Co(X)$ is the barycentric operator defined by

$$\beta m = \sum_{i=1}^{R} \alpha_i x^i \quad \text{for} \quad m = \sum_{i=1}^{R} \alpha_i \delta(x^i) \in M(X) \tag{11}$$

and t is a mapping from the convex hull Co(X) into X such that it maps the points of $A^{-1}(b + \rho_{\beta,X}(A) - \mathbb{R}^m_+)$ into $A^{-1}(b - \mathbb{R}^m_+)$ where

$$\rho_{\beta,X}(A) = \sup_{m\in M(X)} [A(\beta m) - A^{\Delta}(m)] \ . \tag{12}$$

We shall assume that A is defined on $Co(X) \subset \mathbb{R}^n$ and satifies

$$\{x \in X \mid A(x) \leqslant b\} \neq \emptyset \ . \tag{13}$$

We have then the following

PROPOSITION 2: Let X *be an arbitrary subset of* $\mathbb{R}^n$ *and* t *a mapping from* Co(X) *into* X *which satisfies the following condition:*

$$A \ t \ A^{-1} \ \textit{maps} \ \ b + \rho_{\beta,X}(A) - \mathbb{R}^m_+ \ \textit{into} \ \ b - \mathbb{R}^m_+ . \tag{14}$$

Then, if we set $\gamma = t \circ \beta$, *we have for any* $\mu \in M(X)$

$$A^{\Delta}(\mu) \leqslant b \ \ \textit{implies} \ \ A(\gamma\mu) \leqslant b \ . \tag{15}$$

If the mapping A *is linear, then condition (14) becomes*

$$A \ t \ A^{-1} (b - \mathbb{R}^m_+) \subset b - \mathbb{R}^m_+ \ . \tag{14'}$$

PROOF: The definition of $\rho_{\beta,X}(A)$ implies that

$$A(\beta\mu) \leqslant \rho_{\beta,X}(A) + A^{\Delta}(\mu) \qquad \forall \mu \in M(X) \ . \tag{16}$$

Since $A^{\Delta}(\mu) \leqslant b$ we have

$$A(\beta\mu) \leqslant b + \rho_{\beta,X}(A)$$

or equivalently

$$\beta\mu \in A^{-1}[b + \rho_{\beta,X}(A) - \mathbb{R}^m_+] \ . \qquad (17)$$

This relation together with assumption (14) implies that

$$A(\gamma\mu) = A([t \circ \beta]\mu) \in A\ t\ A^{-1}[b + \rho_{\beta,X}(A) - \mathbb{R}^m_+] \subset b - \mathbb{R}^m_+$$

or equivalently

$$A(\gamma\mu) \leqslant b \ .$$

If A is a linear mapping we have

$$A^{\Delta}(\mu) = \sum_{i=1}^{R} \alpha_i\ A(x^i) = A\ (\sum_{i=1}^{R} \alpha_i\ x^i) = A(\beta\mu) \ .$$

Hence it results that $\rho_{\beta,X}(A) = o$, and the assumption (14) takes the simpler form

$$A\ t\ A^{-1}(b - \mathbb{R}^m_+) \subset b - \mathbb{R}^m_+ \ . \qquad \square$$

Remark. 1. Such a mapping t exists if $\{x\in X \mid A(x) \leqslant b\} \neq \emptyset$, i.e. $X \cap A^{-1}(b - \mathbb{R}^m_+) \neq \emptyset$. It suffices to take t with range in the set $X \cap A^{-1}(b - \mathbb{R}^m_+)$. □

Concerning the lack of γ-convexity of the function f, it can be shown that when $\gamma = t \circ \beta$ where β is the barycentric operator, one can replace the lack of γ-convexity of the function f defined on an arbitrary set X by the usual lack of β- convexity on the convex set Co(X).

PROPOSITION 3: Let X *be an arbitrary subset of* $\mathbb{R}^n$ *and let* t *be a mapping from* Co(X) *into* X *such that*

$$t(x) = x \qquad \forall x \in X \ . \qquad (18)$$

Then, if we set $\gamma = t \circ \beta$ where $\beta : M(X) \to Co(X)$ *is the barycentric operator, we have*

$$i)\quad \rho_{\gamma,X}(f) = \rho_{t \circ \beta,X}(f) = \rho_{\beta,X}(f \circ t) \qquad (19)$$

$$ii)\quad \rho_{\gamma,X}(f) \leqslant \rho_{\beta,Co(X)}(f \circ t) \ . \qquad (20)$$

PROOF: The definition of the lack of γ-convexity and the fact that $t(x) = x$ for any $x \in X$ imply that

$$\rho_{\gamma,X}(f) = \sup_{m\in M(X)} [f(\gamma m) - \sum_{i\in I} \alpha_i f(x^i)]$$

$$= \sup_{m\in M(X)} [f(t(\beta m)) - \sum_{i\in I} \alpha_i f(t(x^i))]$$

$$= \sup_{m\in M(X)} \{[f \circ t](\beta m) - \sum_{i\in I} \alpha_i [f \circ t](x^i)\} = \rho_{\beta,X}(f \circ t) .$$

Since $M(X) \subset M(Co(X))$ we have

$$\rho_{\beta,X}(f \circ t) \leqslant \sup_{m\in M(Co(X))} \{[f \circ t](\beta m) - \sum_{i\in I} \alpha_i [f \circ t](x^i)\}$$

$$= \rho_{\beta,Co(X)} (f \circ t) . \qquad \square$$

Since for a discrete set $X = \{x^1, x^2,\ldots,x^R\}$ the mapping $t : Co(X) \to X$ takes only a finite number of values, it is clear that the mapping $f \circ t$ is piecewise constant, and we have

$$[f \circ t]\,(x) = f(tx) = f(x^i) \quad \forall x \in E_i = t^{-1}(\{x^i\}) \text{ with } \bigcup_{i=1}^{R} E_i = Co(X)$$

so that the lack of γ-convexity of f on the discrete set X can be replaced by the lack of convexity of the piecewise constant function $f \circ t$ defined on the convex set Co(X).

We will now give an application of the preceding estimates to the study of Lagrangean relaxations of discrete optimization problems.

Let X be a finite subset of $\mathbb{R}^n$ and let us consider the following discrete optimization problem with two sets of inequality constraints

$$(Q) \quad \begin{cases} \text{Minimize } f(x) \\ \text{subject to} \\ A(x) \leqslant b \\ C(x) \leqslant d \\ x \in X \end{cases}$$

where f is a realvalued function defined on X, A and C are mappings from $\mathbb{R}^n$ into $\mathbb{R}^m$ and $\mathbb{R}^q$ respectively, b and d are vectors of $\mathbb{R}^m$ and $\mathbb{R}^q$ respectively.

Let λ be a vector of $\mathbb{R}^q$. Let us associate with problem (Q) the following Lagrangean relaxation:

$$(Q_\lambda)\quad \begin{cases} \text{Minimize } f(x) + \langle\lambda, C(x) - d\rangle \\ \text{subject to} \\ A(x) \leqslant b \\ x \in X\ . \end{cases}$$

We shall estimate the difference between the optima of problems (Q) and (Q_λ). Denoting by $v(Q_\lambda)$ the optimum of problem (Q_λ) we may also introduce the following optimization problem

$$(D)\qquad \operatorname*{Max}_{\lambda \geqslant 0} v(Q_\lambda) = \operatorname*{Max}_{\lambda \geqslant 0} \operatorname*{Min}_{x \in Y} [f(x) + \langle\lambda, C(x) - d\rangle]$$

where $\qquad Y = \{x \in X \mid A(x) \leqslant b\}$.

Problem (D) may be interpreted as the dual problem of an optimization problem

$$\begin{cases} \text{Minimize } f(x) \\ \text{subject to} \\ C(x) \leqslant d \\ x \in Y \end{cases}$$

which by definition of the set Y coincides with (Q).

By applying Theorem 1 to problems (Q) and (D) we obtain first the following inequality

$$v(Q) \leqslant \rho_{\gamma,Y}(f) + \operatorname*{Max}_{\lambda \geqslant 0} v(Q_\lambda) \tag{21}$$

where γ is a mapping from $M(Y)$ into Y satisfying the condition

$$C^\Delta(\mu) \leqslant d \ \text{ implies } \ C(\gamma\mu) \leqslant d \qquad \text{for any } \mu \in M(Y)\ .$$

For example one can take $\gamma = t \circ \beta$ where $t : Co(Y) \to Y$ has its range contained in $Y \cap C^{-1}(d - \mathbb{R}^q_+)$.

We shall now prove two preliminary lemmas.

LEMMA 4: Let $x^* \in X$ *be an optimal solution of* (Q). *Then the following inequality is satisfied for any* $\lambda \in \mathbb{R}^q$

$$- \langle\lambda, C(x^*) - d\rangle \leqslant v(Q) - v(Q_\lambda) \ . \tag{22}$$

PROOF: From the definition of $v(Q_\lambda)$ we know that

$$v(Q_\lambda) \leqslant f(y) + \langle\lambda, C(y) - d\rangle \tag{23}$$

for any $y \in X$ such that $A(y) \leqslant b$. Taking for y the optimal solution x^* of problem (Q) and replacing $f(x^*)$ by $v(Q)$ yields the desired inequality. □

LEMMA 5: Let $x_{\bar\lambda}$ *be an optimal solution of problem* (Q_λ) *for* $\lambda = \bar\lambda$. *Then for any* $\lambda \in \mathbb{R}^q$ *we have*

$$v(Q_\lambda) + \langle\bar\lambda - \lambda, C(x_{\bar\lambda}) - d\rangle \leqslant v(Q_{\bar\lambda}) \ . \tag{24}$$

PROOF: Inequality (23) with $y = x_{\bar\lambda}$ implies

$$v(Q_\lambda) \leqslant f(x_{\bar\lambda}) + \langle\lambda, C(x_{\bar\lambda}) - d\rangle \ . \tag{25}$$

By definition of $x_{\bar\lambda}$ we have

$$v(Q_{\bar\lambda}) = f(x_{\bar\lambda}) + \langle\bar\lambda, C(x_{\bar\lambda}) - d\rangle \ . \tag{26}$$

Subtracting (26) from (25) yields the desired result. □

We can now prove the following

PROPOSITION 6: Let x_λ, x^* *and* λ^* *be optimal solutions of the problems* (Q_λ), (Q) *and* (D) *respectively. Then for any* $\lambda \in \mathbb{R}^q$ *we have*

$$- \langle\lambda, C(x^*) - d\rangle \leqslant v(Q) - v(Q_\lambda) \leqslant \rho_{\gamma,Y}(f) - \langle\lambda - \lambda^*, C(x_\lambda) - d\rangle \ . \tag{27}$$

PROOF: The left hand inequality is implied by Lemma 4. We deduce from Lemma 5 with $\lambda = \lambda_*$ that

$$v(Q_{\lambda^*}) + \langle\lambda - \lambda^* \ , \ C(x_\lambda) - d\rangle \leqslant v(Q_\lambda) \ . \tag{28}$$

Now the duality gap estimate (21) shows that

$$v(Q) \leqslant \rho_{\gamma,Y}(f) + v(Q_{\lambda^*}) \tag{29}$$

and by using inequalities (28) and (29) we get

$$v(Q) - v(Q_\lambda) \leqslant \rho_{\gamma,Y}\,(f) - \langle \lambda-\lambda^*,\ C(x_\lambda) - d\rangle$$

which is the right hand side of (27). □

By using the estimate (27) it is possible to bound the difference between the optima of problems (Q) and (Q_o) where (Q_o) is defined by dropping the constraint $C(x) \leqslant d$ from (Q)

$$(Q_o)\quad \begin{cases} \text{Minimize } f(x) \\ \text{subject to} \\ A(x) \leqslant b \\ x \in X\ . \end{cases}$$

Taking $\lambda = 0$ in (27) yields

$$0 \leqslant v(Q) - v(Q_o) \leqslant \rho_{\gamma,Y}(f) + \langle \lambda^*,\ C(x_o) - d\rangle \qquad (30)$$

where λ^* is an optimal solution of the dual problem (D).

2. ESTIMATES OF THE DUALITY GAP FOR QUASICONVEX OPTIMIZATION PROBLEMS

We consider now the optimization problem

$$(P)\qquad \begin{array}{l} \text{Minimize } f(x) \\ \text{subject to } x \in X,\ A(x) \leqslant 0 \end{array}$$

where f is an arbitrary function defined on $\mathbb{R}^n$, X is a subset of $\mathbb{R}^n$ and A a mapping from $\mathbb{R}^n$ to $\mathbb{R}^m$. We associate with problem (P) its quasi-convex or surrogate dual

$$(S)\quad \left\{ \sup_{p \geqslant o}\ \inf_{\substack{x\in X \\ \langle p,\ A(x)\rangle \leqslant o}} f(x) \right.$$

in which the constraints $A(x) \leqslant o$ have been replaced by a single scalar constraint obtained as a convex combination of the original constraints.

A theory of quasiconvex duality is developed in Luenberger (1968) where the existence of an optimal multiplier $\bar{p}$ for the quas convex dual in the case of a quasiconvex objective function and of convex constraints is established.

We consider here the case where the objective function is not quasiconvex. We derive estimates of the duality gap between the optima of (P) and (S) by using a measure of the lack of quasiconvexity of the objective function.

We will first recall some well-known properties of quasiconvex functions and quasiconjugates (Greenberg-Pierskalla (1973)). A function $f : R^n \to \bar{R}$ is called quasiconvex if and only if its level sets

$$L_c(f) = \{x \in \mathbb{R}^n \mid f(x) \leqslant c\}$$

are convex for all real numbers c. An equivalent definition is that for any $x_1, x_2 \in \mathbb{R}^n$ the following inequality is satisfied

$$f(\alpha x_1 + [1-\alpha]x_2) \leqslant \text{Sup}\ \{f(x_1),\ f(x_2)\} \quad \text{for all } \alpha \in [o,1]\ .$$

If $f : \mathbb{R}^n \to \bar{\mathbb{R}}$ is an arbitrary function, its quasiconjugate is defined by

$$f_z^*(y) = z - \inf_{\langle x,y\rangle \geqslant z} f(x)\ ,$$

and it is easy to verify that the mapping $y \to f_z^*(y)$ is quasiconvex.

Then the second quasiconjugate of f in the sense of Greenberg-Pierskalla (1973) is defined by

$$f^{**}(x) = \sup_z\ (f_z^*)_z^*(x)\ .$$

It satisfies

$$f^{++} \leqslant f^{**} \leqslant f \tag{31}$$

where f^{++} denotes the second convex conjugate of f defined by

$$f^{++}(x) = \sup_y\ [\langle x,y\rangle - f^+(y)]$$

with $f^+(y) = \sup_w\ [\langle w,y\rangle - f(w)]$.

Further one can also verify that f^{**} can alternatively be defined by the following formula

$$f^{**}(x) = \sup_y\ \inf_w\ \{f(w) \mid \langle w-x,y\rangle \geqslant o\}\ .$$

We can now introduce the following definition of the lack of quasiconvexity of an arbitrary function f on the convex subset X

$$\sigma_X(f) = \sup_{x \in X}\ [f(x) - f^{**}(x)]\ . \tag{32}$$

From (32) it is clear that if $\sigma_X(f) = o$ then f is quasiconvex on X. On the other hand if f is quasiconvex and lower semi-continuous on the set X then $f(x) = f^{**}(x)$ for any $x \in X$ and thus $\sigma_X(f) = o$. If we take the following definition for the lack of convexity of f in X

$$\bar{\rho}_X(f) = \operatorname*{Sup}_{x \in X} [f(x) - f^{++}(x)] ,$$

then it results immediately from (31) that $\sigma_X(f) \leqslant \bar{\rho}_X(f)$ for any f.

We now recall some properties of the quasiconvex dual of an optimization problem. If we set

$$s(p) = \operatorname*{Inf}_{\substack{x \in X \\ \langle p, A(x) \rangle \leqslant o}} f(x)$$

then the quasiconvex dual (S) can be written in the form

$$v^* = \operatorname*{Sup}_{p \geqslant o} s(p) .$$

One can easily check that the function s(p) is quasiconcave. Further we have the following result of Luenberger (1968) concerning the existence of quasiconvex multipliers for (P):

THEOREM 7: Assume $\operatorname*{Inf}_{\substack{x \in X \\ A(x) \leqslant o}} h(x)$ *is finite. Furthermore let the function* h *satisfy the following regularity assumption:*

for any $x_1, x_2 \in X$ $\alpha : [o,1] \to h(\alpha x_1 + (1-\alpha) x_2)$ *is an upper-semicontinuous function in* α [0 1].

Let $A : \mathbb{R}^n \to \mathbb{R}^m$ *be a convex mapping. If there exists* $x_1 \in X$ *such that* $A(x_1) < o$, *then there exists* $\bar{p} \in \mathbb{R}^m_+$, $\bar{p} \neq o$ *such that*

$$\operatorname*{Inf}_{\substack{x \in X \\ A(x) \leqslant o}} h(x) = \operatorname*{Inf}_{\substack{x \in X \\ \langle \bar{p}, A(x) \rangle \leqslant o}} h(x) .$$

We will first derive an estimate for the duality gap in the case of convex constraints, that is we shall assume that X is a convex subset of $\mathbb{R}^n$ and A a convex mapping from $\mathbb{R}^n$ into $\mathbb{R}^m$.

We shall introduce the following auxiliary problem

$$(P_A) \qquad w = \operatorname*{Inf}_{\substack{x \in X \\ A(x) \leqslant o}} f^{**}(x)$$

where f^{**} denotes the second quasiconjugate of f.

We can now prove the following result.

THEOREM 8: Let us assume that f *is such that* f^{**} *satisfies the upper-semicontinuity assumption of Theorem 7,* A *is a convex mapping from* $\mathbb{R}^n$ *into* $\mathbb{R}^m$, $X \subset \mathbb{R}^n$ *is convex and there exists* $x_1 \in X$ *such that* $A(x_1) < o$. *Then we have the following estimate for the duality gap of* (P)

$$v - v^* \leqslant \sigma_X(f) \ . \tag{33}$$

PROOF: We first notice that, by Theorem 7 applied to (P_A), there exists $\bar{p} \in \mathbb{R}^m_+$ such that

$$w = \operatorname*{Inf}_{\substack{x \in X \\ A(x) \leqslant o}} f^{**}(x) = \operatorname*{Inf}_{\substack{x \in X \\ <\bar{p}, A(x)> \leqslant o}} f^{**}(x) \ .$$

Since $f^{**}(x) \leqslant f(x)$ for all $x \in X$, this implies that

$$w \leqslant \operatorname*{Inf}_{\substack{x \in X \\ <\bar{p}, A(x)> \leqslant o}} f(x) \leqslant \operatorname*{Sup}_{p \geqslant o} \ \operatorname*{Inf}_{\substack{x \in X \\ <p, A(x)> \leqslant o}} f(x) = v^* \ . \tag{34}$$

Let now $\varepsilon > o$ be given. By definition of w there exists $\xi \in X$ such that

$$\begin{aligned} & f^{**}(\xi) \leqslant w + \varepsilon \\ & A(\xi) \leqslant o \end{aligned} \tag{35}$$

and we have

$$f(\xi) \leqslant f(\xi) - f^{**}(\xi) + w + \varepsilon \quad \operatorname*{Sup}_{x \in X} [f(x) - f^{**}(x)] + w + \varepsilon$$

$$\leqslant \sigma_X(f) + w + \varepsilon \ .$$

Since ξ satisfies the constraints of problem (P) this implies that

$$\operatorname*{Inf}_{\substack{x \in X \\ A(x) \leqslant o}} f(x) \leqslant f(\xi) \leqslant \sigma_X(f) + w + \varepsilon$$

Then (33) follows since $w \leqslant v^*$ and $\varepsilon > o$ is arbitrary. □

We will now consider the case of nonconvex constraints. We will first assume that the mapping A is nonconvex and the set X is convex. We denote by

$$\bar{\rho}_X(A) = \sup_{x\in X} [A(x) - A^{++}(x)] \tag{36}$$

the lack of convexity of A where A^{++} denotes the second convex conjugate of A. We shall introduce the auxiliary problem

$$(P'_A) \quad w = \inf_{\substack{x\in X \\ A^{++}(x) \leqslant \bar{\rho}_X(-A)}} f^{**}(x) .$$

We then have the following.

THEOREM 9: Assume that f *is such that* f^{**} *satisfies the upper semi-continuity assumption of Theorem 7, that* X *is a convex subset of* $\mathbb{R}^n$ *and* A *is an arbitrary mapping from* $\mathbb{R}^n$ *into* $\mathbb{R}^m$ *such that* $A(x_1) < o$ *for some* $x_1 \in X$. *Then we have the following estimate*

$$v(\bar{\rho}_X(A) + \bar{\rho}_X(-A)) \leqslant \bar{\rho}_X(f) + v^* \tag{37}$$

where $v(c) = \inf_{\substack{x\in X \\ A(x) \leqslant c}} f(x)$

PROOF: Let us notice first that the assumption $A(x_1) < o$ implies $A^{++}(x_1) < o$. By applying Theorem 7 to Problem (P'_A) we deduce that there exists $\bar{p} \in \mathbb{R}^m_+$ such that

$$\inf_{\substack{x\in X \\ A^{++}(x) \leqslant \bar{\rho}_X(-A)}} f^{**}(x) = \inf_{\substack{x\in X \\ \langle \bar{p}, A^{++}(x)\rangle \leqslant \langle \bar{p}, \bar{\rho}_X(-A)\rangle}} f^{**}(x) .$$

From the definition of the lack of convexity of -A we have

$$A^{++}(x) \leqslant A(x) + \bar{\rho}_X(-A)$$

This implies

$$\langle \bar{p}, A^{++}(x)\rangle \leqslant \langle \bar{p}, A(x)\rangle + \langle \bar{p}, \bar{\rho}_X(-A)\rangle .$$

Since

$$\{x\in X \mid \langle \bar{p}, A(x)\rangle \leqslant o\} \subset \{x \in X \mid \langle \bar{p}, A^{++}(x)\rangle \leqslant \langle \bar{p}, \bar{\rho}_X(-A)\rangle\} \tag{38}$$

we have

$$w = \operatorname*{Inf}_{\substack{x\in X \\ \langle\bar{p},A^{++}(x)\rangle \leqslant \langle\bar{p},\ \bar{\rho}_X(-A)\rangle}} f^{**}(x) \leqslant \operatorname*{Inf}_{\substack{x\in X \\ \langle\bar{p},A(x)\rangle \leqslant o}} f^{**}(x) \leqslant \operatorname*{Inf}_{\substack{x\in X \\ \langle\bar{p},A(x)\rangle \leqslant o}} f(x) \leqslant v^* .$$

We can now estimate the difference between w and v.

By definition of (P'_A) for arbitrary $\varepsilon > o$ there exists $\xi \in X$ such that

$$\begin{cases} f^{**}(\xi) \leqslant w + \varepsilon \\ A^{++}(\xi) \leqslant \bar{\rho}_X(-A) \ . \end{cases} \tag{39}$$

We deduce from the first inequality (39) that

$$f(\xi) \leqslant f(\xi) - f^{**}(\xi) + w + \varepsilon \leqslant \operatorname*{Sup}_{x\in X}[f(x) - f^{**}(x)] + w + \varepsilon \leqslant \sigma_X(f) + w + \varepsilon \ .$$

On the other hand it results from the definition of $\bar{\rho}_X(A)$ that

$$A(x) \leqslant \bar{\rho}_X(A) + A^{++}(x) \quad \text{for any } x \in X \ .$$

Since $A^{++}(\xi) \leqslant \bar{\rho}_X(-A)$ we have

$$A(\xi) \leqslant \bar{\rho}_X(A) + \bar{\rho}_X(-A) \ .$$

This finally yields

$$\operatorname*{Inf}_{x\in X} f(x) \leqslant f(\xi) \leqslant \sigma_X(f) + w + \varepsilon$$

$$A(x) \leqslant \bar{\rho}_X(A) + \bar{\rho}_X(-A)$$

which implies inequality (37) since $w \leqslant v^*$ and $\varepsilon > o$ is arbitrary. □

We can now consider the case where the mapping A and the set X are not convex. We shall associate with (P) the following auxiliary problem

$$(P''_A) \qquad w = \operatorname*{Inf}_{\substack{x\in Co(X) \\ A^{++}(x) \leqslant \bar{\rho}_{Co(X)}(-A)}} f^{**}(x)$$

where $Co(X)$ denotes the convex hull of the set X and A^{++} is the second convex conjugate of A.

We shall prove the following result:

THEOREM 10: Assume that f *is such that* f^{**} *satisfies the uppersemicontinuity assumption of Theorem 7. Let* $X \subset \mathbb{R}^n$ *be an arbitrary subset and*

$A : \mathbb{R}^n \to \mathbb{R}^m$ *an arbitrary mapping such that there exists* $x_1 \in X$ *with* $A(x_1) < o$. *Denote by* t *a mapping from* Co(X) *into* X *such that*

$$A\, t\, A^{-1}\, [\bar{\rho}_{Co(X)}(A) + \bar{\rho}_{Co(X)}(-A) - \mathbb{R}^m_+] \subset \mathbb{R}^m_- \ . \tag{40}$$

Then we have the following estimate

$$v - v^* \leqslant \sigma_{t,Co(X)}(f) \tag{41}$$

where

$$\sigma_{t,Co(X)}(f) = \sup_{x \in Co(X)} [f(tx) - f^{**}(x)] \ .$$

PROOF: By Theorem 7 there exists $\bar{p} \geqslant o$ such that

$$w = \inf_{\substack{x \in Co(X) \\ A^{++}(x) \leqslant \rho_{Co(X)}(-A)}} f^{**}(x) = \inf_{\substack{x \in Co(X) \\ <\bar{p}, A^{++}(x)> \leqslant <\bar{p}, \rho_{Co(X)}(-A)>}} f^{**}(x) \ .$$

Since in view of (38)

$$\{x \in Co(X) \mid <\bar{p}, A(x)> \leqslant o\} \subset \{x \in Co(X) \mid <\bar{p}, A^{++}(x)> \leqslant <\bar{p}, \bar{\rho}_{Co(X)}(-A)>\}$$

we have

$$w = \inf_{\substack{x \in Co(X) \\ <\bar{p}, A^{++}(x)> \leqslant <\bar{p}, \bar{\rho}_{Co(X)}(-A)>}} f^{**}(x) \leqslant \inf_{\substack{x \in X \\ <\bar{p}, A(x)> \leqslant o}} f(x) \leqslant \sup_{p \geqslant o} \inf_{\substack{x \in X \\ <p, A(x)> \leqslant o}} f(x) = v^*.$$

In view of the definition of (P''_A) there exists a vector $\xi \in Co(X)$ for $\varepsilon > o$ such that

$$\begin{cases} f^{**}(\xi) \leqslant w + \varepsilon \\ A^{++}(\xi) \leqslant \bar{\rho}_{Co(X)}(-A) \ . \end{cases} \tag{42}$$

If t is any mapping from Co(X) into X we deduce from the first inequality that

$$f(t\xi) \leqslant f(t\xi) - f^{**}(\xi) + w + \varepsilon \leqslant \sup_{x \in Co(X)} [f(tx) - f^{**}(x)] + w + \varepsilon \ ,$$

or equivalently

$$f(t\xi) \leqslant \sigma_{t,Co(X)}(f) + w + \varepsilon \ .$$

From the definition of $\bar{\rho}_{Co(X)}(A)$ and the second inequality in (42) we deduce that

$$A(\xi) \leqslant A^{++}(\xi) + \bar{\rho}_{Co(X)}(A) \leqslant \bar{\rho}_{Co(X)}(-A) + \bar{\rho}_{Co(X)}(A) \ ,$$

or equivalently

$$\xi \in A^{-1} [\bar{\rho}_{Co(X)}(-A) + \bar{\rho}_{Co(X)}(A) - \mathbb{R}^m_+] . \tag{43}$$

If the mapping $t : Co(X) \to X$ satisfies (40), then (43) implies $A(t\xi) \leqslant o$.

Finally we have

$$f(t\xi) \leqslant \sigma_{t,Co(X)}(f) + w + \varepsilon$$

$$A(t\xi) \leqslant o \qquad \text{with } t\xi \in X$$

These inequalities imply

$$v = \inf_{\substack{x \in X \\ A(x) \leqslant o}} f(x) \leqslant f(t\xi) \leqslant \sigma_{t,Co(X)}(f) + w + \varepsilon \leqslant \sigma_{t,Co(X)}(f) + v^* + \varepsilon .$$

Since $\varepsilon > o$ is arbitrary this yields

$$v \leqslant \sigma_{t,Co(X)}(f) + v^* .$$ □

The existence of a mapping t satisfying (40) results from the fact that there exists at least one $x \in X$ such that $A(x) \leqslant o$ or equivalently that $A^{-1} \mathbb{R}^m_- \cap X \neq \emptyset$.

By the nonnegativity of the lack of convexity of A we have

$$A(x) \leqslant \bar{\rho}_{Co(X)}(A) + \bar{\rho}_{Co(X)}(-A) \qquad \text{for some } x \in X ,$$

or equivalently

$$X \cap A^{-1} [\bar{\rho}_{Co(X)}(A) + \bar{\rho}_{Co(X)}(-A) - \mathbb{R}^m_+] \neq \emptyset .$$

Then we can choose a mapping t from Co(X) into X such that

$$t\{A^{-1} [\bar{\rho}_{Co(X)}(A) + \bar{\rho}_{Co(X)}(-A) - \mathbb{R}^m_+] \cap X\} \subset A^{-1} \mathbb{R}^m_- \cap X$$

and this implies (40).

REFERENCES

AUBIN, J.P. and EKELAND, I. (1974). "Estimates of the duality gap of non-convex optimization problems." *University of Wisconsin, M.R.C. Report # 1491.*

CROUZEIX, J.P. (1976). "Dualité quasiconvexe." *Séminaire d'Analyse Numérique, Université de Clermont.*

FISHER, M. and SHAPIRO, J. (1974). "Constructive Duality in Integer Programming." *SIAM J. for Applied Mathematics, 27,* 31-52.

GREENBERG, H.J. and PIERSKALLA, W.P. (1973). "Quasiconjugate functions and surrogate duality." *Cahiers Centre d'Etudes de Recherche Opérationnelle, 15,* 437-448.

LUENBERGER, D. (1968). "Quasiconvex Programming." *SIAM J. for Applied Mathematics, 16,* 1090-1095.

PART IV

NEW CLASSES OF GENERALIZED CONCAVE FUNCTIONS

F-CONVEX FUNCTIONS: PROPERTIES AND APPLICATIONS

A. Ben-Tal[†] and A. Ben-Israel

For a given family of functions F, a function is F-convex if its epigraph is supported at each point by a member of F. We introduce notions such as F-subgradients and F-conjugate functions, and generalized results associated with their classical counterparts. In particular we derive monotonicity results for the F-subgradients, and unimodality results for differences of F-convex and F-concave functions. A Fenchel-type duality theory for primal problems involving such differences is given. Other augmented Lagrangean type duals, involving closely related notions of generalized convexity, are briefly sketched. We also outline a numerical method for a root-finding problem, which is based on "F-approximation" rather than "linearization" as in Newton's method.

1. INTRODUCTION

Convex functions are generated by the family of affine functions in the sense that a (closed) convex function f is the pointwise supremum of affine functions, or equivalently that the epigraph of f is supported at each point by the graph of an affine function.

A natural generalization of convex functions is obtained by replacing the family of affine functions with an appropriate family F. We introduced *F-convex functions* in Ben-Tal (1973) and Ben-Tal and Ben-Israel (1976), the latter dealt mainly with the issue of *characterizations*. These included generalizations of the classical characterizations such as the gradient inequality, the monotonicity of the gradient mapping and the positive-definiteness of the Hessian. For such characterizations, and other results in F-convexity, to be concrete, we consider families F of continuous functions $F: X \subset R^n \to R$, depending continuously on n+1 parameters

$$\{x^*, \eta\} \in X^* \times Y \subset R^n \times R.$$

The general member of F is a function

$$F(\cdot) = F(x^*, \eta; \cdot) \quad , \quad (x^* \in X^*, \eta \in Y).$$

An example is the family F_k discussed in section 3 whose general member evaluated at $x = (x_1, \ldots, x_n)$ is (k being a fixed constant)

$$F(x) = k \sum_{i=1}^{n} \cosh(x_i^* + x_i) - \eta.$$

[†]Research partly done while visiting University of Delaware and University of British Columbia.

ISBN 0-12-621120-5

Any function $f:R^n \to R$ with second derivatives bounded above [below] is shown in section 3 to be F_k-concave [F_k-convex] for suitably chosen k. The family F_k consists of separable functions, a fact which greatly simplifies the treatment of F_k-convexity.

Unlike convex functions, F-convex functions need not possess unimodal properties, and may in fact have several local extrema. We show in section 4 however that differences of F-convex and F-concave functions are unimodal (Theorem 4).

The building blocks of F-convex analysis are laid in sections 5-6. These are the *conjugate family* F^* of a given family F[1], the *F-subgradient* and the *F-conjugate function*. Under suitable assumptions on F, the basic results of convex analysis (see Rockafellar (1970)) are naturally extended to the F-convex case. In particular, the F-subdifferential mapping has monotonicity properties (Theorem 5, corollary 7); the F-conjugate f^F of any $f:R^n \to R$ is F^*-convex (Theorem 8), and dually, when f is F-convex, $f=(f^F)^{F^*}$. F-conjugate functions can be computed using a "Legendre transform" (50). Examples 3-5 illustrate that the computation of F-conjugates is not much more demanding than the computation of conjugate functions in the classical case.

These results are applied in section 7 to give a Fenchel-type duality theorem (Theorem 10). As in the Fenchel duality scheme, the primal problem is to minimize the difference of an F-convex function and an F-concave function, and the dual problem is to maximize the difference of associated F^*-concave and F^*-convex functions. Under suitable assumptions on F (further elaborated in Lemmas 12-14), strong duality (i.e. no duality gaps) is established, and explicit relations between the optimal solutions of the dual problems are developed.

Another approach to duality, in the framework of F-convexity, due

[1]Absent in classical convexity by the fact that the family of affine functions is self conjugate.

to Balder (1978), and Dolecki and Kurcyusz (1978), is briefly reviewed in section 8.

An illustrative application of $\mathcal{F}$-convexity is given in section 9. Here an iterative method for solving a nonlinear equation, is based on "$\mathcal{F}$-approximation" rather than "linearization" as in Newton's method.

2. PRELIMINARIES AND NOTATION

We collect here the main definitions from [2], and additional ones needed in this paper.

Let $\mathcal{F}$ be a family of functions $F:X\to R$, where $X\subset R^n$, with range

$$Y \triangleq \cup\{\text{range } F: F\in\mathcal{F}\}. \tag{1}$$

Let f be a function $:R^n\to R$ with domain:[2]

$$\text{dom } f\subset X, \tag{2}$$

and let S be an open subset of dom f. Then f is called *$\mathcal{F}$-convex in* S if for every $x\in S$, there exists an $F\in\mathcal{F}$ such that

$$f(x) = F(x) \text{ and } f(z)\geq F(z) \text{ for all } x \neq z\in S, \tag{3}$$

in which case F is a *support of* $f:S$ *at* x. The function f is called *strictly $\mathcal{F}$-convex in* S if strict inequality holds in (3) for all $x\neq z\in S$.

If there is no need to specify S, for example if S=dom f, the above names are abbreviated by omitting S, e.g., *$\mathcal{F}$-convex, support of* f *at* x, etc.

If the reverse inequality holds in (3), f is called *$\mathcal{F}$-concave*.

Note that in general, the $\mathcal{F}$-concavity of f is not equivalent to the $\mathcal{F}$-convexity of -f. Also, the sum of $\mathcal{F}$-convex functions is not necessarily $\mathcal{F}$-convex.

In this paper we consider families $\mathcal{F}$ of continuous functions $F:X\subset R^n\to R$ depending continuously on n+1 parameters

$$\{x^*,\eta\}\in X^*\times Y\subset R^n\times R.$$

The general member of $\mathcal{F}$ is denoted by

[2] The domain of f, dom f, here denotes dom $f\triangleq\{x:f(x)<\infty\}$.

$$F(\cdot) = F(x^*,\eta;\cdot) \quad (x^*\in X^*,\ \eta\in Y).$$

We assume that the correspondence

$$\{x^*,\eta\}\leftrightarrow F(x^*,\eta;\cdot\}$$

between $X^*\times Y$ and F is one to one. A family F is called *complete* if for every convergent sequence $\{x^*_\nu,\eta_\nu\}\subset X^*\times Y$, $(x^*_\nu,\eta_\nu)\to(x^*_\infty,\eta_\infty)$:

either $\lim_{\nu\to\infty} F(x^*_\nu,\eta_\nu;x) = \infty$ for every x

or $F(x^*_\infty,\eta_\infty;\cdot)\in F$.

An example of a non-complete family is the family of non-horizontal affine functions

$$F_0 = \{<x^*,x>-\eta : 0\neq x^*\in R^n,\ \eta\in R\}.$$

The undesirability of non-complete families is shown by the example

$$f(x) = x^2$$

which is the pointwise supremum of its minorants in F_0, but not F_0-convex since at x=0 f is not supported by any $F\in F_0$.

If a family F is complete, then a function f is F-convex if and only if f is the pointwise supremum of its minorants in F. Note that a family F is complete if the parameter sets X^*,Y are closed.

A family F of functions: $R\to R$ is a *Beckenbach family* if for every two points (x_1,y_1), (x_2,y_2); $x_1<x_2$; there is a unique function $F_{1,2}\in F$ such that $F_{1,2}(x_i) = y_i$, i = 1,2 i.e., there is a unique solution (x^*,η) to

$$F(x^*,\eta;x_i) = y_i,\ i = 1,2.$$

For Beckenback families, F-convexity is equivalent, [Ben-Tal and Ben-Israel (1970), Proposition 2.4], to the notion of sub-F functions studied in Beckenbach (1937), Peixoto (1949); see also Beckenbach and Bellman (1965), Roberts and Varberg (1973).

A family F of functions: $R^n\to R$ is called a *Beckenbach family* if every $F\in F$ is of the form

$$F(x) = \sum_{i=1}^{n} F_i(x_i)$$

and each F_i is a member of a Beckenbach family: $R\to R$.

A family F of differentiable functions is said to be in *class A*, if the system

$$\begin{aligned} \alpha &= F(x^*,\eta;x) \\ a &= \nabla F(x^*,\eta;x) \end{aligned} \qquad (4)$$

of n+1 equation in the n+1 unknowns (x^*,η) has a unique solution for every $x\in X$ and $\binom{\alpha}{a}$ in the appropriate range; ∇F is the gradient of F with respect to x. Further, if F is continuously differentiable, and the solutions of (4) are continuous in (α,a,x) then F is in *class* $\overline{A}$.

For $F\in A$, and a function f, the solution of the system

$$f(z) = f(x^*,\eta;x) \qquad (5)$$

$$\nabla f(x) = F(x^*,\eta;x) \qquad (6)$$

is denoted by $x_f^*(x)$, $\eta_f(x)$. The vector $x_f^*(x)$ is called the *F-gradient* of f. For $F\in A$ and a differentiable function f which is F-convex in an open subset S of dom f, the parameters $\{x_f^*(x),\eta_f(x)\}$ determine the unique support $F(x_f^*(x),\eta_f(x);\cdot)$ at $x\in S$. Thus F-convexity is characterized by the F-gradient inequality

$$f(z) \geqslant F(x_f^*(x),\eta_f(x);z) \qquad \forall\ z\in S. \qquad (7)$$

A continuity argument shows that for $f\in C^1[S]$ and $F\in\overline{A}$, the inequality (7) holds for all $z\in c\ell S$. It should be noted that strict F-convexity is characterized by a strict inequality in (7) for all $z\neq x$.

A family F is said to be in *class C* if for every $x^*\in X^*$, $x\in X$, the function $F(x^*,\cdot;x):Y\to R$ is strictly decreasing. The *inverse* of $F(x^*,\cdot;x)$ is then denoted by $F^I(x,\cdot,;x^*)$. A family F is *separable* if every $F\in F$ is of the form

$$F(x) = \sum_{i=1}^{n} \phi_i(x_i^*,x_i) - \eta \qquad (8)$$

where $\frac{\partial}{\partial x_i}\phi_i(\cdot,x_i)$ are strictly increasing functions (of x_i^*), i=1,...,n. Note that a separable F is in $A\cap C$. The equations (5), (6), defining the F-gradient x_f^* reduce here to

$$\frac{\partial}{\partial x_i} f(x) = \frac{d}{dx_i}\phi_i(x_i^*,x_i) \qquad (i=1,\ldots,n) \qquad (9)$$

whose solution x_i^* is the ith component of $x_f^*(x)$.

The class of *affine functions:*

$$\mathcal{F} = \{F(x)=\langle x^*,x\rangle-\eta : x^*\in R^n, \eta\in R\}$$

is separable, and all the above notions reduce to their ordinary versions. In particular, the $\mathcal{F}$-gradient: $x_f^*(x) = \nabla f(x)$.

Example 1. For i=1,...,n let the differentiable functions

$u^i : R\to R$ be strictly increasing

$v^i : R\to R$ be strictly decreasing.

Then the family

$$\{F(x) = \sum_{i=1}^{n} (x_i^* u^i(x_i) + \frac{1}{x_i^*} v^i(x_i)) - \eta : x_i^*\in R^+, \eta\in R, \quad i=1,\dots,n\} \tag{10}$$

is separable. Here (9) reduces to

$$\frac{\partial f}{\partial x_i} = x_i^* \frac{du^i}{dx_i} + \frac{1}{x_i^*}\frac{dv^i}{dx_i} \quad (i=1,\dots,n) \tag{11}$$

so that the ith component of x_f^* is given as the positive solution of the quadratic equation (11), i.e.,

$$(x_f^*(x))_i = \frac{1}{2\frac{du^i}{dx_i}} \left(\frac{\partial f}{\partial x_i} + \sqrt{\left(\frac{\partial f}{\partial x_i}\right)^2 - 4\frac{du^i}{dx_i}\frac{dv^i}{dx_i}} \right) \tag{12}$$

3. AN EXAMPLE OF $\mathcal{F}$-CONVEXITY

For any real k consider the family of functions: $R^n\to R$

$$\mathcal{F}_k = \{F(x) = k\sum_{i=1}^{n} \cosh(\theta_i^*+x_i) - \eta : \theta_i^*\in R,\ i=1,\dots,n, \eta\in R\} \tag{13}$$

$\mathcal{F}_k$ is a separable family of the type given in Example 1 with

$$u^i(x_i) = \frac{k}{2}e^{x_i},\ v^i(x_i) = \frac{k}{2}e^{-x_i}$$

$$\theta_i^* = \log x_i^* \quad .$$

Let f be an arbitrary function: $R^n\to R$ with continuous second derivatives, whose Hessian $\nabla^2 f(x)$ is uniformly bounded, i.e. there is an M > 0 such that for all $x \in R^n$

$$y^T \nabla^2 f(x) y \leqslant M y^T y, \quad \forall\, y \in R^n \ . \tag{14}$$

Then f is strictly F_k-concave for any $k \geqslant M$. Similarly, if the Hessian is bounded below in the sense that there is an $m < 0$ such that for all $x \in R^n$

$$m y^T y \leqslant y^T D^2 f(x) y, \quad \forall\, y \in R^n \tag{15}$$

then f is F_k-convex for any $k \leqslant m$. (The cases where $M < 0$ or $m > 0$ are not interesting because then f is concave or convex respectively).

In particular if Q is an indefinite symmetric matrix with

maximal eigenvalue $\lambda_{max} > 0$

minimal eigenvalue $\lambda_{min} < 0$

then the quadratic form $x^T Q x$ (in fact any quadratic function $x^T Q x + p^T x + \alpha$) is $F_{\lambda_{max}}$-concave and $F_{\lambda_{min}}$-convex.

We prove that a function f satisfying (14) is strictly F_k-concave, for $k \geqslant M$. Using (8), the F_k-gradient of f at x, is the solution $\theta^*(x) = (\theta_1^*, \ldots, \theta_n^*)$ of the system

$$\frac{\partial f}{\partial x_i}(x) = k \sinh(\theta_i^* + x_i), \ i=1,\ldots,n. \tag{16}$$

In particular,

$$\theta^*(x) + x = 0 \Leftrightarrow \nabla f(x) = 0. \tag{17}$$

We show that f is supported (from above) at any point x by some $F \in F_k$. We distinguish two cases for the point x.

Case 1: $\nabla f(x) = 0$. By Taylor's theorem,

$$\begin{aligned} f(z) &= f(x) + (z-x)^T \nabla f(x) + \tfrac{1}{2}(z-x)^T \nabla^2 f(\bar{x})(z-x), \text{ for some } \bar{x} \\ &\leqslant f(x) + \tfrac{1}{2} k (z-x)^T (z-x), \text{ by (3.2) and } k \geqslant M \\ &< f(x) + k \sum_{i=1}^{n} (\cosh(z_i - x_i) - 1), \quad \forall\, z \neq x. \end{aligned}$$

The latter inequality holds since $\cosh(t) > 1 + \frac{1}{2}t^2$, $\forall\, t \neq 0$. The above inequality can be written as

$$f(z) < k \sum_{i=1}^{n} \cosh(\hat{\theta}_i + z_i) - \hat{\eta}, \quad \forall\ z \neq x$$

where $\hat{\theta}_i = -x_i$, $\hat{\eta} = nk - f(x)$. This shows that

$F(z) = k \sum_{i=1}^{n} \cosh(\hat{\theta}_i + z_i) - \hat{\eta}$, which belongs to $\mathcal{F}_k$, supports f at x.

Case 2: $\nabla f(x) \neq 0$. By Ben-Tal and Ben-Israel (1976) the strict $\mathcal{F}$-concavity of f follows from the negative-definiteness of the matrix

$$H(x) = \nabla^2 f(x) - \nabla_x^2 \phi(\theta^*(x), x) \tag{18}$$

where

$$\phi(\theta, x) = k \sum_{i=1}^{n} \cosh(\theta_i + x_i). \tag{19}$$

Now by (17), $\theta^*(x) + x \neq 0$,

and therefore

$$\cosh(\theta_i^* + x_i) \geqslant 1 \quad i=1,\dots,n \tag{20}$$

with at least one strict inequality.

From (18)

$$\nabla_x^2 \phi(\theta^*(x), x) = k \operatorname{diag}(\cosh(\theta_1^* + X_1), \dots, \cosh(\theta_n^* + X_n))$$

so by (20) for every $y \in R^n$,

$$y^T \nabla_x^2 \phi(\theta^*(x), x) y > k y^T y \geqslant y^T \nabla^2 f(x) y, \text{ by (14)},$$

showing that H(x) of (18) is negative definite. ■

The proof that f satisfying (15) is $\mathcal{F}_k$-convex for $k \leqslant m$ follows from

f is $\mathcal{F}$-convex $\Leftrightarrow$ $-$ f is $(-\mathcal{F})$-concave.

4. UNIMODALITY

A differentiable function $h: R^n \to R$ is said to be *unimodal* in a set $S \subset \operatorname{dom} h$ if every critical point in int S is a global minimizer of h on S, i.e.

$$x \in \text{int } S,\ \nabla h(x) = 0 \Rightarrow h(z) \geqslant h(x), \quad \forall\ z \in S. \tag{21}$$

The function h is *pseudoconvex at $x \in S$* if

$$\langle \nabla h(x), z - x \rangle \geqslant 0 \Rightarrow h(x) \geqslant h(x), \quad \forall\ z \in s. \tag{22}$$

If this holds at every $x \in S$, h is *pseudoconvex in S*. h is said to be *pseudoconcave* if (-h) is pseudoconvex, and *pseudomonotonic* if both pseudo-convex and pseudoconcave. For detailed dicussion see e.g. Mangasarian (1969), Martos (1975) and Greenberg and Pierskalla (1971).

Clearly, any pseudoconvex function is unimodal, and the two notions coincide for functions of a single variable. In the sequel, unimodality appears to be more useful than pseudoconvexity: Consider a separable function

$$h(x) = \sum_{i=1}^{n} h_i(x_i)$$

with pseudoconvex (hence unimodal) components h_i. Then h is unimodal, but not necessarily pseudoconvex.

Unlike convex functions, *F*-convex functions need not be unimodal. However an *F*-convex function inherits unimodality from the family *F*.

THEOREM 1. Let $F \in A$ *be a family of unimodal [pseudoconvex] functions* {F}. *Then every F-convex function is unimodal [pseudoconvex].*

PROOF. Let f be *F*-convex, which by Ben-Tal and Ben-Israel (1976) is characterized by the *F*-gradient inequality

$$f(z) \geqslant F(x_f^*(x), \eta_f(x); z) \quad \forall\ x, z \tag{7}$$

where $x_f^*(x)$ and $\eta_f(x)$ are the solutions of (5), (6).

Unimodality: The unimodality of *F* implies

$$\nabla_x F(x^*, \eta, x) = 0 \Rightarrow F(x^*, \eta, z) \geqslant F(x^*, \eta, x) \quad \forall\ z, x\ .$$

In particular, for $x^* = x_f^*(x)$, $\eta = \eta_f(x)$ we get from (21), (22) and the *F*-gradient inequality

$$\nabla f(z) = 0 \Rightarrow f(z) \geqslant f(x_f^*(x), \eta_f(x); z) \geqslant F(x_f^*(x), \eta_f(x); x) = f(x),$$

showing that f is unimodal.

Pseudoconvexity: Similarly proved ■

An important class of separable unimodal families is the following special case of Example 1

$$F(x) = \sum_{i=1}^{n} \left(x_i^* \psi^i(x_i) + \frac{1}{x_i^* \psi^i(x_i)}\right) - \eta \tag{23}$$

where $x_i^* > 0$, $i=1,\ldots,n$ and each ψ^i is a positive, strictly increasing differentiable function. To show the unimodality of (23), it suffices to show the pseudoconvexity of each component. A slightly more general result is the following:

LEMMA 2. Let $\psi\colon R\to R$ *be a positive, strictly increasing, differentiable function, and let* $\alpha,\beta > 0$. *Then the function*

$$\gamma(x) = \alpha\psi(x) + \beta\frac{1}{\psi(x)} \tag{24}$$

is pseudoconvex.

PROOF. Let $\gamma'(x)(z-x) \geq 0$. Then by (24), $\psi'(x) > 0$ and $\psi(x) > 0$, implies that

$$\begin{aligned} 0 &\leq (z-x)(\alpha\psi(x)^2-\beta) = \\ &= (z-x)[\alpha\psi(x)^2-\alpha\psi(x)\psi(z)+\alpha\psi(x)\psi(z)-\beta] \\ &= \alpha\psi(x)(z-x)[\psi(x)-\psi(z)] + (z-x)\ [\alpha\psi(x)\psi(z)-\beta]\ . \end{aligned}$$

Now $\alpha\psi(x)(z-x)(\psi(x)-\psi(z)) \leq 0$ by $\alpha > 0$, $\psi > 0$ and the monotonicity of ψ. Therefore $(z-x)[\alpha\psi(x)\psi(z)-\beta] \geq 0$, but $\text{sign}(z-x) = \text{sign}(\psi(z)-\psi(x))$ so that

$$0 \leq (\psi(z)-\psi(x))[\alpha\psi(x)\psi(z)-\beta] = \frac{1}{\psi(z)\psi(x)}(\gamma(z)-\gamma(x)) \text{ by } (24)$$

proving the implication

$$\gamma'(x)(z-x) \geq 0 \ \Rightarrow \gamma(z) \geq \gamma(x).$$ ■

Having established the unimodality of the separable family (24), we prove now the analogous result for non-separable families.

COROLLARY 3. Let $\phi\colon R^n\to R$ *be a positive, pseudomonotonic function. Then for any* $\alpha,\beta \geq 0$ *the function* $F(x) = \alpha\phi(x) + \frac{\beta}{\phi(x)}$ *is pseudoconvex.*

PROOF. Follows from Lemma 2 by the fact that a function is pseudomonotonic if its restrictions to lines are either constants or strictly monotone (see Martos (1975), result 3.30, p. 55 and result 7.30, p. 114). ■

Even families F which are not unimodal, give rise to unimodal functions, namely differences of F-convex and F-concave functions. Consider the *class* ϕ of families F of the form

$$F = \{\phi(x^*,x)-\eta : x^* \in X^*, \eta \in R\}. \tag{25}$$

THEOREM 4. Let $F \in \phi \cap A$, let f be a differentiable F-convex function and g be a differentiable F-concave function in an open set S. Then f-g is unimodal in S.

PROOF. Let $x \in S$ be a critical point of f-g, i.e.

$$\nabla f(x) = \nabla g(x) \tag{26}$$

The F-convexity of f implies, by the F-gradient inequality,

$$f(z) \geqslant f(x) + \phi(x_f^*(x),z) - \phi(x_f^*(x),x), \quad \forall\ z \in S \tag{27}$$

where $x_f^*(x)$, the F-gradient of f at x, is the unique solution of

$$\nabla f(x) = \nabla_x \phi(x^*,x). \tag{28}$$

The F-concavity of g implies

$$g(z) \leqslant g(x) + \phi(x_g^*(x),z) - \phi(x_g^*(x),x), \quad \forall\ z \in S \tag{29}$$

where $x_g^*(x)$ is the unique solution of

$$\nabla g(x) = \nabla_x \Phi(x^*,x)\ . \tag{30}$$

Therefore (26), (28) and (30) give

$$x_f^*(x) = x_g^*(x) \tag{31}$$

substituting (31) in (27) and (29), one obtains by subtraction

$$f(z) - g(z) \geqslant f(x) - g(x), \quad \forall\ z \in S$$

which shows the critical point x to be a global minimizer in S. ■

Note that in ordinary convexity the unimodality of f-g follows from the convexity of f-g, which follows from two facts:

(i) g concave $\Leftrightarrow$ (-g) convex

(ii) The sum of convex functions is convex.

Neither one of these properties holds in $\mathcal{F}$-convexity, and yet f-g is unimodal (though in general not $\mathcal{F}$-convex.)

5. $\mathcal{F}$-SUBGRADIENTS

From here on we consider only families $\mathcal{F} = \{F(x^*,\eta;\cdot):x^*\in X^*,\eta\in Y\}$ in class $\mathcal{C}$. The *conjugate family* of a given $\mathcal{F}$ is the family $\mathcal{F}^*$ of functions $F^*:X^*\subset R^n\to R$,

$$\mathcal{F}^* = \{F^*(\cdot) = F^*(x,\zeta;\cdot):x\in X,\zeta\in Z\} \tag{32}$$

defined for each $(x,\zeta)\in X\times Z$ by

$$F^*(x,\zeta;x^*) = F^I(x,\zeta;x^*), \quad \forall\ x^* \tag{33}$$

where the inverse function F^I is defined in section 2. It is obvious that $\mathcal{F}^*$ is also in class $\mathcal{C}$ and

$$F^{**} = F, \quad \forall\ F\in\mathcal{F}\ . \tag{34}$$

The family $\mathcal{F}$ of the affine functions is self-conjugate, i.e., $\mathcal{F}^* = \mathcal{F}$. So is every family $\mathcal{F}$ of the form (25) with $\phi(x^*,x) = \phi(x,x^*)$.

Example 2. The conjugate family of

$$\mathcal{F} = \{F(x^*,\eta;x) = x^*e^{x} - \eta e^{-x}:x^*,\ \eta\in R\} \tag{35}$$

is

$$\begin{aligned}\mathcal{F}^* &= \{F^*(x,\zeta;x^*) = e^{2x}x^* - \zeta e^{x}:\ x,\zeta\in R\} \\ &= \{F^*(\alpha,\beta;x^*) = \alpha x^* - \beta:\alpha > 0,\beta\in R\}\end{aligned} \tag{36}$$

i.e. the lines with positive slope.

Let f be a function: $R^n\to R$ and let $x\in\text{dom } f$. A vector $x^*\in X^*$ is called an *$\mathcal{F}$-subgradient of* f *at* x if

$$f(z) \geqslant F(x^*,F^*(x,f(x);x^*);z), \quad \forall\ z\in\text{dom } f. \tag{37}$$

The *set of all $\mathcal{F}$-subgradients of* f *at* x is denoted by $\partial_{\mathcal{F}}f(x)$, and the multivalued mapping

$$x \to \partial_{\mathcal{F}}f(x)$$

is called the *$\mathcal{F}$-subdifferential mapping*.

If f is F-convex it follows from (37) and (33) that for every $x \in \text{dom } f$ and $x^* \in \partial_F f(x)$, the function

$$F(x^*, F^*(x, f(x); x^*), \cdot)$$

is a support of f at x. Therefore $\partial_F f(x) \neq \phi$ for all x. Further, if $F \in A$, then f has a unique support at every x where f is differentiable (by Ben-Tal and Ben-Israel (1976), Lemma 4.1 and Theorem 4.2) and at such x

$$\partial_F f(x) = \{x_f^*(x)\}. \tag{38}$$

For convex function, the monotonicity properties of subgradients are well known and widely used, e.g. Rockafellar (1970), Ortega and Rheinboldt (1970). The monotonicity of the F-gradient x_f^*, for separable F, was studied in Ben-Tal and Ben-Israel (1976). We now extend these results to F-subdifferentials, and improve some of them for the differentiable case as well.

The monotonicity notions useful here are the following ones. Let $T: S \subset R^n \to R^n$ be a multivalued mapping. Then T is a *P_o-function* [*P-function*] in S if for every $x^1, x^2 \in S; x^1 \neq x^2;$ $y^1 \in T(x^1)$, $y^2 \in T(x^2)$ there is an integer $k = k(x^1, x^2) \in \{1, 2, \ldots, n\}$ such that $x_k^1 \neq x_k^2$ and $(x_k^1 - x_k^2) \cdot (y_k^1 - y_k^2) \geqslant 0$ [>0], e.g. Moré and Rheinboldt (1973).

THEOREM 5. Let F be a Beckenbach separable family[1] *of the form (8), and let* f *be F-convex* [*strictly F-convex*]. *Then the F-subdifferential mapping is a* P_0-*function* [P-*function*].

PROOF. Since each F_k is a Beckenbach family, the system

$$y^1 = \phi_k(x^*, x_k^1) - \eta \qquad\qquad y^2 = \phi_k(x^*, x_k^2) - \eta$$

has a unique solution (x^*, η), for every (x_k^1, y^1), $(x_k^2, y^2); x_k^1 \neq x_k^2$ i.e. the equation

[1] i.e. $F \in \mathcal{F} \Leftrightarrow F(x) = \Sigma F_k(x_k)$ where $F_k \in \mathcal{F}_k = \{\phi_k(x_k^*, \cdot) - \eta\}$ and $\mathcal{F}_k$ is a Beckenbach family of functions: $R \to R$.

$$y = \phi_k(x^*, x_k^1) - \phi_k(x^*, x_k^2)$$

has a unique solution x^* for every $x_k^1 > x_k^2$ and y. This implies the strict monotonicity of the function

$$\text{i.e.} \quad h(\cdot) = \phi_k(\cdot, x_k^1) - \phi_k(\cdot, x_k^2) \text{ for } x_k^1 > x_k^2, \tag{39}$$

$$(x_k^1 - x_k^2)(x_k^{*1} - x_k^{*2})(h(x_k^{*1}) - h(x_k^{*2})) > 0$$

or, using (39),

$$(x_k^1 - x_k^2)(x_k^{*1} - x_k^{*2})([\phi_k(x_k^{*1}, x_k^1) - \phi_k(x_k^{*2}, x_k^1)] - $$
$$-[\phi_k(x_k^{*1}, x_k^2) - \phi_k(x_k^{*2}, x_k^2)]) > 0, \quad \forall k. \tag{40}$$

Now, let $x^1, x^2 \in R^n$ and let $x^{*j} \in \partial_F f(x^j)$ $j=1,2$.

i.e., for every z

$$f(z) \geq f(x^j) + \sum_i \phi_i(x_i^{*j}, z_i) - \sum_i \phi_i(x_i^{*j}, x_i^j), \quad j=1,2. \tag{41}$$

substituting $(z=x^1, j=2)$ and $(z=x^2, j=1)$ in (41) gives

$$f(x^1) \geq f(x^2) + \sum_i \phi_i(x_i^{*2}, x_i^1) - \sum_i \phi_i(x_i^{*2}, x_i^2)$$

$$f(x^2) \geq f(x^1) + \sum_i \phi_i(x_i^{*1}, x_i^2) - \sum_i \phi_i(x_i^{*1}, x_i^1).$$

Adding and canceling equal terms yields

$$\sum_i \{[\phi_i(x_i^{*1}, x_i^1) - \phi_i(x_i^{*2}, x_i^1)] - [\phi_i(x_i^{*1}, x_i^2) - \phi_i(x_i^{*2}, x_i^2)]\} \geq 0$$

hence, for at least one index, say k

$$[\phi_k(x_k^{*1}, x_k^1) - \phi_k(x_k^{*2}, x_k^1)] - [\phi_k(x_k^{*1}, x_k^2) - \phi_k(x_k^{*2}, x_k^2)] \geq 0. \tag{42}$$

Comparing (42) and (40) we deduce

$$(x_k^1 - x_k^2)(x_k^{*1} - x_k^{*2}) \geq 0$$

which is the inequality required to prove that $\partial_F f(\cdot)$ is a P_0-function. The proof that $\partial_F f(\cdot)$ is a P-function, under the condition of strict F-convexity, is similar. ∎

As a byproduct of the above proof we have the following.

COROLLARY 6. A separable family of differentiable functions is a Beckenbach family if and only if it is in class A.

PROOF. Let F be a Beckenbach family of differentiable functions of the form (8). By (40), each component ϕ_k satisfies

$$(x^{*1}-x^{*2})\left(\frac{\partial}{\partial x}\phi_k(x^{*1},x)-\frac{\partial}{\partial x}\phi_k(x^{*2},x)\right) > 0 \tag{43}$$

for every x; $x^{*1} > x^{*2}$.

This is equivalent to the strict monotonicity of $\frac{\partial}{\partial x}\phi_k(\cdot,x)$, $k=1,\ldots,n$ which is equivalent to property A. Conversely, property A implies (43) which can be rewritten as

$$\frac{\partial}{\partial x}[(x^{*1}-x^{*2})(\phi_k(x^{*1},x)-\phi_k(x^{*2},x))] > 0$$

showing that the function

$$(x^{*1}-x^{*2})(\phi_k(x^{*1},\cdot) - \phi_k(x^{*2},x))$$

is monotone increasing in x. This implies (40), which in turn was shown (in the proof of Theorem 5) to be the Beckenbach family property. ■

COROLLARY 7. Let $F\in A$ be a separable family. Then the F-gradient of a differentiable F-convex [strictly F-convex] function is a (single valued) P_0-*function* [P-*function*].

PROOF. Follows from Corollary 6, Theorem 5 and (38). ■

This result has been proved in Ben-Tal and Ben-Israel (1976) Theorem 6.4 under stronger hypotheses.

6. F-CONJUGATE FUNCTIONS

From here on we consider only families F for which both F and its conjugate family F^* are complete.

Let $F\in C$ and let $f:R^n\to R$. The *F-convex conjugate of* f, denoted by f^F, is a function: $R^n \to R$ defined by

$$f^F(x^*) \triangleq \sup_{x \in \text{dom } f} \{F^*(x, f(x); x^*)\} \tag{44}$$

for $x^* \in X^*$.

The *F-concave conjugate of* f, f_F, is similarly defined by

$$f_F(x^*) \triangleq \inf_{x \in \text{dom } f} \{F^*(x, f(x); x^*)\}. \tag{45}$$

Since f^F is used more often than f_F, the former is simply called *F-conjugate*. For the family F of affine functions: $R^n \to R$, (44) reduces to the classical definition of the (convex) conjugate function

$$f^*(x^*) \triangleq \sup_{x \in \text{dom } f} \{\langle x^*, x\rangle - f(x)\},$$

see Rockafellar (1970), section 12. In this case f^* is a closed convex function. If f itself is closed and convex then $f^{**} = f$. Analogous results for F-conjugate functions are given below.

THEOREM 8. Let $F \in C$ and let $f: R^n \to R$. Then

(a) f^F *is F^*-convex*

(b) f *is F - convex if and only if*

$$(f^F)^{F^*} = f . \tag{46}$$

PROOF.

(a) For every $x^* \in \text{dom } f^F$

$$\begin{aligned} f^F(x^*) &= \sup_{x \in \text{dom } f} F^*(x, f(x); x^*) \\ &= \sup_{\substack{x, \zeta \\ \zeta \geq f(x)}} F^*(x, \zeta, x^*) \end{aligned} \tag{47}$$

since $F \in C \Rightarrow F^* \in C$, which implies that, at supremum, $\zeta = f(x)$. Now $\zeta \geqslant f(x)$ is equivalent to

$$F^*(x, f(x); x^*) \geqslant F^*(x, \zeta, x^*), \quad \forall\, x, x^*$$

which in turn is equivalen to

$$f^F(x^*) \geq F^*(x,\zeta;x^*).$$

This and (47) show that f^F is pointwise supremum of its minorants in F^* which, since F^* is complete, shows that f^F is F^*-convex.

(b) If (46) holds then f is F^{**}-convex, by part (a). Then f is F-convex since $F = F^{**}$. Conversely, let f be F-convex, i.e. the pointwise supremum of the functions $F \in F$ which are majorized by f

$$\begin{aligned} f(x) &= \sup_{x^*,\eta}\{F(x^*,\eta;x):f(z) \geq F(x^*,\eta;z),\ \forall\ z\} \\ &= \sup_{x^*,\eta}\{F(x^*,\eta;x):\eta \geq F^*(z,f(z);x^*),\ \forall\ z\} \\ &= \sup_{x^*,\eta}\{F(x^*,\eta;x):\eta \geq f^F(x^*)\} \\ &= \sup_{x^*} F\,(x^*,f^F(x^*);x) \quad \text{since F is decreasing in } \eta \\ &= \sup_{x^*} F^{**}(\,x^*,f^F(x^*)\cdot,x) \\ &= (f^F)^{F^*}(x). \end{aligned}$$

■

The following theorem reduces, for the family F of affine functions, to a classical result on convex functions (Rockafellar (1970), Theorem 23.5).

THEOREM 9. Let $F \in C$, $f:R^n \to R$. *Then for every* $x \in \text{dom } f$ *and* $x^* \in \text{dom } f^F$ *the following two conditions are equivalent*

(a) $x^* \in \partial_F f(x)$

(b) $f^F(x^*) = F^*(x,f(x);x^*)$

If f *is F-convex, then a third equivalent condition is*

(c) $x \in \partial_{F^*} f^F(x^*)$.

PROOF.

(a) means that

$$f(z) \geq F(x^*,F^*(x,f(x);x^*);z), \quad \forall\ z \in \text{dom } f$$

which, by $F \in C$ and the definition (33), is equivalent to

$$F^*(x,f(x);x^*) \geq F^*(z,f(z);x^*), \quad \forall\ z \in \text{dom } f$$

and, by (44), to (b).

Now, (b) is equivalent to

$$f(x) = F(x^*,f^{F}(x^*);x)$$

by (33) and (34). But if f is F-convex, Theorem 8 (b) gives

$$f(x) = \sup_{z^* \in \text{dom } f^F}\{F(z^*,f^{F}(z^*);x)\} = f^{FF^*}(x)$$

hence (b) is equivalent to

$$F(x^*,f^{F}(x^*);x) \geq F(z^*,f^{F}(z^*);x) \quad \forall\ z^* \in \text{dom } f^{F}$$

which is the same as

$$f^{F}(z^*) \geq F^*(x,F(x^*,f^{F}(x^*);x);z^*), \ \forall\ z^* \in \text{dom } f^{F}$$

proving (c). ■

Remarks.

1. The equivalence of (a), (c)

$$x^* \in \partial_{F} f(x) \Leftrightarrow x \in \partial_{F^*} f^{F}(x^*)$$

which holds for F-convex functions, means that the two (multivalued) mappings

$$x \to \partial_{F} f(x), \ x^* \to \partial_{F^*} f^{F}(x),$$

are inverses of each other.

2. From definition (44) follows the inequality

$$f^{F}(x^*) \geq F^*(x,f(x);x^*) \tag{48}$$

which in the classical case reduces to the *Fenchel-Young inequality*. The equivalence of Theorem 9 (a) ⇔ (b) shows that equality holds in (48) if and only if

$$x^* \in \partial_{F} f(x) \ . \tag{49}$$

3. If f is a differentiable strictly F-convex function where $F \in A \cap C$, then by (38) and Ben-Tal and Ben-Israel (1976), Corollary 4.4 (b) the mapping

$$x \to x^*_f(x)$$

is one to one. Therefore Theorem 9 implies that

$$f^{F}(x^*) = F^*(x_f^{*\,-1}(x^*), f[x_f^{*\,-1}(x^*)]; x^*) \quad \forall\, x^* \in \text{range } x_f^* . \tag{50}$$

This formula is useful in computing conjugates. The right-hand side of (50) reduces in the classical case to the *Legendre Transform*, (Rockafellar (1970), section 26).

Example 3. Consider the family F and its conjugate F^* from Example 2. The F-conjugate of $f:R\to R$ is here

$$f^{F}(x^*) = \sup_{x\in R}\{x^* e^{2x} - f(x)e^{x}\} \tag{51}$$

and the second conjugate is

$$f^{FF^*}(x) = \sup_{x^*\in R}\{x^* e^{x} - f^{F}(x^*)e^{-x}\} . \tag{52}$$

The function $f(x) = e^{x}(x-\frac{1}{2})$ which is neither convex nor concave, is strictly F-convex (as can be shown by using the differential inequality characterization in Ben-Tal and Ben-Israel (1976), Corollary 6.6). By solving (5) and (6) we verify that $x_f^*(x) = x$ which is substituted in (50) to give

$$f^{F}(x^*) = \frac{1}{2}e^{2x^*}, \quad x^*\in R . \tag{53}$$

Observe that f^{F} is an increasing convex function, as can be expected from the fact that its support family, F^* of (36), is the family of lines with positive slope. Substituting (53) in (52) one recovers f, in agreement with Theorem 8 (b). The "Fenchel-Younge inequality" (48) here becomes

$$x^* e^{2x} - e^{x}(x-\frac{1}{2})e^{x} \leqslant \frac{1}{2}e^{2x^*}$$

which can be reduced to the familiar

$$e^{t} \geqslant 1 + t.$$

Example 4. Consider a family of functions:R→R of the type considered in section 3

$$F = \{F(x)=\cosh(x^*+x)-\eta : x^*\in R, \eta\in R\}. \tag{54}$$

Then the strictly *convex* function $f(x) = e^x$ is strictly *F-concave*. Indeed its F-gradient is

$$x_f^*(x) = \ln(1+\sqrt{1+e^{-2x}}), \text{ by } (6) \tag{55}$$

which is strictly decreasing, proving the F-concavity of f by Ben-Tal and Ben-Israel (1976), Corollaries 4.4(b) and 6.5.

Note that the family F is self-conjugate (see remark following (34)). Therefore the conjugate of f is, by (45),

$$f_F(x^*) = \inf_x\{\cosh(x^*+x)-e^x\}$$

$$= \begin{cases} \sqrt{1-2e^{-x^*}} & x^* \geqslant \ln 2 \\ -\infty & \text{otherwise.} \end{cases}$$

This result, derived by elementary calculus, could also be obtained by computing $(x_f^*)^{-1}$ from (55) and substituting in (50).

Example 5. H-ϕ convex functions were defined in Ben-Tal (1977) in terms of the "generalized means" introduced by Hardy, Littlewood and Polya (1934). Let $\phi:R\to R$ be continuous strictly increasing function, and let $H:R^n\to R^n$ be an invertible operator. A function $f:R^n\to R$ is H-ϕ *convex* if

$$f(H^{-1}[\lambda H(x)+(1-\lambda)H(y)]) \leqslant \phi^{-1}[\lambda\phi(f(x)) + (1-\lambda)\phi(f(y))]$$
$$\forall \; x,y \text{ and } 0 \leqslant \lambda \leqslant 1 .$$

Using the results of Ben-Tal (1977) it can be shown that the H-ϕ convex functions are F-convex with

$$F = \left\{ \phi^{-1}[\langle H(x^*),H(x)\rangle-\phi(\eta)] : \begin{matrix} X=X^*=\text{dom } H \\ Y=\text{dom } \phi=R \end{matrix} \right\}.$$

Note that this is a self-conjugate family, so

$$f^F(x^*) = \sup_{x \in \text{dom } H}\{\phi^{-1}[\langle H(x^*),H(x)\rangle-\phi(f(x))]\}$$

$$= \phi^{-1}[\sup_{u \in \text{range } H}\{\langle H(x^*),u\rangle-\phi f[H^{-1}(u)]\}]$$

$$= \phi^{-1}[(\phi f H^{-1})^*(H(x^*))]$$

where $(\phi f H^{-1})^*$ is the classical conjugate of $\phi f H^{-1}$.

Here (46) can be illustrated directly:

$$f^{FF*}(x^*) = \sup_{x^*} F(x^*,f^F(x^*);x) =$$

$$= \sup_{x^* \in \text{dom } H}\{\phi^{-1}[\langle H(x^*),H(x)\rangle-\phi[\phi^{-1}[(\phi f H^{-1})^*(H(x^*))]]]\}$$

$$= \phi^{-1}[\sup_{u^* \in \text{range } H}\{\langle u^*,H(x)\rangle-(\phi f H^{-1})^*(u^*)\}]$$

$$= \phi^{-1}[(\phi f H^{-1})^{**}(H(x))] \ .$$

Now $\phi f H^{-1}$ is convex by Ben-Tal (1977), Corollary 3. If it is also closed then

$$f^{FF*}(x) = \phi^{-1}[\phi f H^{-1}(H(x))] = f(x).$$

7. A DUALITY THEORY FOR *F*-CONVEX FUNCTIONS

Let $F \in \Phi$, i.e.

$$F = \{\phi(x^*,\cdot)-\eta : x^* \in X^*, \eta \in R\}$$

and let $F \in \bar{A}$, see definition following (4).

For any two continuously differentiable functions $f,g:R^n \to R$ and a convex subset S of dom f $\cap$ dom g in which f is *F*-convex and g is *F*-concave, we consider the *primal program*

(P) $\inf\{f(x)-g(x):x \in S\}$.

The problem (P) with the underlying assumptions is called a *unimodal program* (a name justified by Theorem 4). A program dual to (P) is given in terms of the *F-conjugates of* f *and* g *relative to* S

$$f_S^F(x^*) = \sup_{x \in S}\{\phi(x,x^*)-f(x)\} \tag{56}$$

$$g_{\mathcal{F}}^{S}(x^*) = \inf_{x\in S}\{\phi(x,x^*)-g(x)\} \tag{57}$$

over the intersection

$$S^* = \text{dom } f_S^{\mathcal{F}} \cap \text{dom } g_{\mathcal{F}}^{S} \tag{58}$$

$$= \{x^*: f_S^{\mathcal{F}}(x^*)<\infty \text{ and } g_{\mathcal{F}}^{S}(x^*)>-\infty\}.$$

The problem

$$\text{(D)} \qquad \sup\{g_{\mathcal{F}}^{S}(x^*)-f_S^{\mathcal{F}}(x^*):x^*\in S^*\}$$

is called the *dual of* (P).

Note that even if $S = \text{dom } f \cap \text{dom } g$ and $\mathcal{F}$ is the family of affine functions, (D) is not the Fenchel dual of (P), (See Rockafellar (1970), Stoer and Witzgall (1970)) unless dom f = dom g. This follows since for all $x^*\in S^*$

$$f_S^{\mathcal{F}}(x^*) \leqslant f^{\mathcal{F}}(x^*) \quad \text{(defined by (44))} \tag{59}$$

$$g_{\mathcal{F}}^{S}(x^*) \geqslant g_{\mathcal{F}}(x^*) \quad \text{(defined by (45))}. \tag{60}$$

An immediate relation between (P) and (D) is the *weak duality* inequality

$$\inf\{f-g:x\in S\} \geqslant \sup\{g_{\mathcal{F}}^{S}-f_S^{\mathcal{F}}:x^*\in S^*\} \tag{61}$$

which follows from the definitions (56) - (58).

We proceed directly to the *strong duality* theorem.

THEOREM 10. Let

(A) (P) *be a unimodal program and* $\bar{x}\in S$ *a local minimizer of* (f-g).

(B) the function

$$h(\cdot) = \phi(x_f^*(\bar{x}),\cdot) - \phi(x_g^*(\bar{x}),\cdot) \tag{62}$$

be pseudoconvex at $\bar{x}$ *(see definition (4.1)). Then*

(a) $\bar{x}$ *is a global minimizer of* f-g *on* S.

(b) $\max\{f-g:x\in S\} = \min\{g_{\mathcal{F}}^{S}-f_S^{\mathcal{F}}:x^*\in S^*\}$

(c) $x_1^* = x_f^*(\bar{x})$ *and* $x_2^* = x_g^*(\bar{x})$ *are optimal solutions of* (D).

PROOF. For $\bar{x}$ to be a local minimizer of f-g on S it is necessary that

$$\langle\nabla f(\bar{x})-\nabla g(\bar{x}),x-\bar{x}\rangle \geqslant 0 \quad \forall\ x\in S. \tag{63}$$

But from (6)

$$\nabla f(\bar{x}) = \nabla\phi(x_f^*(\bar{x}),\bar{x})$$

$$\nabla g(\bar{x}) = \nabla\phi(x_g^*(\bar{x}),\bar{x})$$

and therefore (63) is, by (62),

$$\langle\nabla h(\bar{x}),x-\bar{x}\rangle \geqslant 0 \quad \forall\ x\in S$$

which by the pseudoconvexity of h at $\bar{x}$ implies

$$h(x) \geqslant h(\bar{x}) \quad \forall\ x\in S.$$

The last inequality can be rearranged to read

$$\phi(x_g^*(\bar{x}),\bar{x}) - \phi(x_g^*(\bar{x}),x) \geqslant \phi(x_f^*(\bar{x}),\bar{x}) - \phi(x_f^*(\bar{x}),x), \quad \forall\ x\in S. \tag{64}$$

Combining (64) and the *F*-gradient inequality (see (27))

$$\phi(x_f^*(\bar{x}),\bar{x}) - \phi(x_f^*(\bar{x}),x) \geqslant f(\bar{x}) - f(x), \quad \forall\ x\in S$$

one gets

$$\phi(x_g^*(\bar{x}),\bar{x}) - \phi(x_g^*(\bar{x}),x) \geqslant f(\bar{x}) - f(x), \quad \forall\ x\in S$$

or

$$\phi(x_g^*(\bar{x}),\bar{x}) - f(\bar{x}) \geqslant \phi(x_g^*(\bar{x}),x) - f(x), \quad \forall\ x\in S$$

so that by (56)

$$f_S^F(x_g^*(\bar{x})) = \phi(x_g^*(\bar{x}),\bar{x}) - f(\bar{x}). \tag{65}$$

Now from Theorem 9 (a), (b)

$$g_F^S(x_g^*(\bar{x})) = \phi(x_g^*(\bar{x}),\bar{x}) - g(\bar{x})$$

which subtracted from (65) gives

$$f(\bar{x}) - g(\bar{x}) = g_F^S(x_g^*(\bar{x})) - f_S^F(x_g^*(\bar{x})). \tag{66}$$

From (66) and the weak duality inequality (61) follow conclusions (a), (b) and half of (c). The other half, namely, that $x_1^* = x_f^*(\bar{x})$ is also an optimal solution of (D), follows from

$$f(\bar{x}) - g(\bar{x}) = g_F^S(x_f^*(\bar{x})) - f_S^F(x_f^*(\bar{x}))$$

which is proved analogously to (66). ∎

Remarks.

1. If F is the affine family, then $h(\cdot)$ in (62) is the affine function $h(\cdot)=\langle\nabla f(\bar{x})-\nabla g(\bar{x}),\cdot\rangle$ and Assumption (B) holds.

2. If the optimal solution of (P) is at a critical point of f-g (i.e. the gradient of (f-g) vanishes at $\bar{x}$), in particular if $\bar{x}\in \text{int}\, S$, then, by the proof of Theorem 4, $x_f^*(\bar{x}) = x_g^*(\bar{x})$. Consequently the function h of (62) is identically zero and assumption (B) is trivially satisfied.

3. Even if the optimal solution $\bar{x}$ is not a critical point then assumption (B) is not too restrictive. Indeed the function h of (62) is at least *unimodal*, by Theorem 4, since h is the difference of the F-convex function $\phi(x_f^*(\bar{x}),\cdot)$ and the F-concave function $\phi(x_g^*(\bar{x}),\cdot)$. This takes care of the one-dimensional case since then unimodality $\Leftrightarrow$ pseudoconvexity. However the one-dimensional case is of independent interest and merits a direct proof, given in Theorem 11. We return to the n-dimensional case in Lemmas 12-14.

THEOREM 11. Let f,g *be differentiable functions:*$R\to R$ *which are F-convex and F-concave respectively where* $F = \{\phi(x^*,x)-\eta : x^*\in R, \eta\in R\}\in A$. *Let* $S = [a,b]\subset \text{dom } f \cap \text{dom } g$. *Then*

(a) $$\min_{a\leq x\leq b}\{f(x)-g(x)\} = \max_{x^*}\{g_F^S(x^*)-f_S^F(x^*)\}$$

(b) If $\bar{x}$ is a minimizer of f-g *in* [a,b], *then every* x^* *in the interval*

$$S^* = [x_g^*(\bar{x}),x_f^*(\bar{x})\} \tag{67}$$

is an optimal solution of the dual problem.

PROOF. The case where (f-g) has a critical point in [a,b] is covered by remark 1 above, so we assume, without loss of generality, that

$$f'(x) - g'(x) > 0, \quad a \leq x \leq b \tag{68}$$

in which event the solution of (P) is

$$\min_{a\leqslant x\leqslant b}(f-g) = f(a) - g(a). \tag{69}$$

Since $F\in A$ it follows that $\frac{d}{dx}\phi(x^*,x)$ is a monotone function of the argument x^*. We assume without loss of generality that

$$\frac{d}{dx}\phi(x^*,x) \text{ is an increasing function of } x^* \text{ for all } a\leqslant x\leqslant b. \tag{70}$$

From the definition of the F-gradient (6)

$$f'(x) = \frac{d}{dx}\phi(x_f^*(x),x) \tag{71}$$

$$g'(x) = \frac{d}{dx}\phi(x_g^*(x),x),$$

so by (68) and (70)

$$x_f^*(x) > x_g^*(x), \quad a\leqslant x\leqslant b\ . \tag{72}$$

By Corollary 7 (specialized to one dimension) $x_f^*(\cdot)$ is an increasing function, and $x_g^*(\cdot)$ is decreasing in [a,b]. These facts and (72) give

$$x_g^*(b) \leqslant x_g^*(a) < x_f^*(a) \leqslant x_f^*(b)\ . \tag{73}$$

The monotonicity of $x_f^*(\cdot)$ also means that x_f^* maps [a,b] into $[x_f^*(a),x_f^*(b)]$. Therefore for $x^* < x_f^*(a)$,

$$\frac{d}{dx}\phi(x^*,x) - f'(x) \neq 0 \quad \text{for all } a\leqslant x\leqslant b. \tag{74}$$

For such x^*, from (70)

$$\frac{d}{dx}\phi(x^*,a) - f'(a) < \frac{d}{dx}\phi(x_f^*(a),a) - f'(a) = 0, \quad \text{by (71)}$$

which shows that $\phi(x^*,x) - f(x)$ is decreasing in [a,b]. Therefore

$$f_S^F(x^*) = \sup_{a\leqslant x\leqslant b}\{\phi(x^*,x)-f(x)\} = \phi(x^*,a) - f(a) \tag{75}$$

for $x^*\leqslant x_f^*(a)$.

Similarly, it can be shown that

$$g_F^S(x^*) = \inf_{a\leqslant x\leqslant b}\{\phi(x^*,x)-g(x)\} = \phi(x^*,a) - g(a) \tag{76}$$

for $x^*\geqslant x_g^*(a)$.

Combining (75) and (76) it follows that

$$g_F^S(x^*) - f_S^F(x^*) = f(a) - g(a), \quad x_g^*(a) \leqslant x^* \leqslant x_f^*(a)$$

which by (69) and the weak duality inequality (61), proves (a) and (b). ∎

The following two lemmas give conditions for Assumption (B) (Theorem 10) to hold in separable families.

LEMMA 12. Let F be a separable family as defined in (8), and let S *be an* n-*dimensional interval* $I_1 \times I_2 \times \ldots \times I_n$. *Then Assumption (B) in Theorem 10 is satisfied.*

PROOF. The optimality of $\bar{x}$ implies that

$$\langle \nabla f(\bar{x}) - \nabla g(\bar{x}), x - \bar{x} \rangle \geqslant 0, \quad \forall\, x \in S \tag{63}$$

in particular for x differing from $\bar{x}$ in only one component, say the kth,

$$\left(\frac{\partial f}{\partial x_k}(\bar{x}) - \frac{\partial g}{\partial x_k}(\bar{x}) \right) (x_k - \bar{x}_k) \geqslant 0 \ .$$

As in the proof that (68) ⇒ (72), it follows here that:

$$\operatorname{sign} \left(\frac{\partial f}{\partial x_k}(\bar{x}) - \frac{\partial g}{\partial x_k}(\bar{x}) \right) = \operatorname{sign}(x_f^*(\bar{x})_k - x_g^*(\bar{x})_k) \tag{77}$$

and consequently

$$(x_f^*(\bar{x})_k - x_g^*(\bar{x})_k)(x_k - \bar{x}_k) \geqslant 0.$$

Combining this with (40) we obtain

$$\phi_k(x_f^*(\bar{x})_k, x_k) - \phi_k(x_g^*(\bar{x})_k, x_k) \geqslant \phi_k(x_f^*(\bar{x})_k, \bar{x}_k) - \phi_k(x_g^*(\bar{x})_k, \bar{x}_k)$$

which, summed over all k, gives

$$h(x) \geqslant h(\bar{x}), \quad \forall \quad x \in S. \tag{78}$$

In the beginning of the proof of Theorem 10 we showed that (63) is equivalent to

$$\langle \nabla h(\bar{x}), x - \bar{x} \rangle \geqslant 0, \quad \forall\, x \in S \tag{79}$$

thus (79) ⇒ (78), proving the pseudoconvexity of h. ∎

LEMMA 13. Let F be a separable family (8) such that, for each i,

$$\frac{d^2}{dx_i^2} \phi_i(\cdot, x_i) \text{ is increasing} \tag{80}$$

and let S *be a bounded polyhedron*

$$S = \{x : Ax \leqslant b\} \tag{81}$$

where A *is a full row-rank nonnegative matrix. Then Assumption (B) in Theorem 10 is satisfied.*

PROOF. Assumption (80) means that the family $\{\sum_{i=1}^{n} \frac{d}{dx_i}\phi_i(x_i^*, x_i) - \eta\}$ is in class *A*. Applying (40) to this family we get for k=1,...,n and every x^1, x^2

$$(x_k^1 - x_k^2)(x_k^{*1} - x_k^{*2})\left\{\left[\frac{d}{dx_k}\phi_k(x_k^{*1}, x_k^1) - \frac{d}{dx_k}\phi_k(x_k^{*2}, x_k^1)\right] - \left[\frac{d}{dx_k}\phi_k(x_k^{*1}, x_k^2) - \frac{d}{dx_k}\phi_k(x_k^{*2}, x_k^2)\right]\right\} > 0 \tag{82}$$

where

$$x_k^{*1} \triangleq x_f^*(\bar{x})_k$$
$$x_k^{*2} \triangleq x_g^*(\bar{x})_k \quad . \tag{83}$$

The definition (62) reduces to

$$h(x) = \sum_{k=1}^{n} h_k(x) = \sum_{k=1}^{n} \left(\phi_k(x_k^{*1}, x_k) - \phi_k(x_k^{*2}, x_k)\right). \tag{84}$$

Suppose (79) $\not\Rightarrow$ (78), i.e. (79) holds and $\bar{x}$ is not a global minimizer in S. Then there exists $\hat{x} \neq \bar{x}$, $\hat{x} \in S$ such that $\hat{x}$ is a minimizer of h in S. It follows from the necessary optimality condition at $\hat{x}$ that

$$\langle \nabla h(\hat{x}), \bar{x} - \hat{x} \rangle \geqslant 0 \ . \tag{85}$$

Combining (85) and (79) at $x = \hat{x}$ we get

$$\langle \nabla h(\bar{x}) - \nabla h(\hat{x}), \hat{x} - \bar{x} \rangle \geqslant 0 \quad . \tag{86}$$

Using (82) with $x^1 = \hat{x}, x^2 = \bar{x}$ we get from (86) and (83)

$$x_f^*(\bar{x})_k > x_g^*(\bar{x})_k$$

for some k=1,...,n; hence by (77)

$$\frac{\partial f}{\partial x_k}(\bar{x}) > \frac{\partial g}{\partial x_k}(\bar{x}) \ . \tag{87}$$

Since $\bar{x}$ is a minimizer for (P), it satisfies the Fritz John condition

$$\lambda_0(\nabla f(\bar{x})-\nabla g(\bar{x})) + A^T\lambda = 0, \text{ for some } 0 \neq (\lambda_0,\lambda) \geqslant 0 .$$

Since A has full row-rank and $\lambda_0 \neq 0$, the above equation (in view of the nonnegativity of A) implies $\nabla f(\bar{x}) - \nabla g(\bar{x}) \leqslant 0$, contradicting (87). ■

The last result concerns the (nonseparable) family

$$F(a,\beta) = \{F(x)=\frac{<x^*,x>}{<a,x>+\beta}-\eta : x\in X=\{x\in R^n : <a,x>+\beta>0\} x^*\in R^n, \eta\in R\} \qquad (88)$$

parametrized by $a\in R^n$, $0\neq \beta\in R$. Note that for a=0, β=1 this family reduced to affine functions, so convexity is a special case of $F(a,\beta)$-convexity.

LEMMA 14. For the family $F(a,\beta)$ *in (88), Assumption (B) of Theorem 10 is satisfied.*

PROOF. It is well known that fractional-linear functions, as those in (85) are pseudoconvex and pseudoconcave, and hence also the function h is such, and in particular pseudoconvex. It is only left to show that $F(a,\beta)$, $a\in R^n$, $\beta\neq 0$ is of class *A*. This however was shown in Ben-Tal and Ben-Israel (1976). ■

8. OTHER APPROACHES TO DUALITY

In this section we briefly review some of the important results of Dolecki and Kurcyusz (1978).

Let U be a vector space and Φ a family of functions $y:U\to R$. A function $h:U\to R$ is Φ-*convex* if

$$h(u) = \sup_{y\in\Phi'} y(u) \quad , \text{ for some } \Phi'\subset\Phi .$$

A set $A\subset U$ is Φ-*convex* if it can be expressed as

$$A = \bigcap_{y\in\Phi'} \{u:y(u)\leqslant\alpha_y\}, \quad \alpha_y\in R, \quad \forall\ y\in\Phi'.$$

In the sequel the family Φ is assumed to satisfy

$$y\in\Phi \Rightarrow \alpha y + \eta\in\Phi \quad \forall\ \alpha > 0, \quad \eta\in R .$$

Thus Φ-convexity is closely related to F_α-convexity where F_α is a complete family of the form

$$F_\alpha = \{F(x)=\alpha y(x)-\eta:\ \alpha>0,\ \eta\in R,\ y\in\Phi\}.$$

Let X be another vector space and f a function from X to the extended real line $\bar{R}$. For each $u\in U$ let Su be a subset of X. In term of a fixed element $u_0\in U$, the *primal problem* is

(P) $\inf\{f(x):\ x\in Su_0\}$.

The class of *perturbed problems* is

(fu) $\inf\{f(x):\ x\in Su\} \equiv p(u),\quad u\in U.$

Let $S_x^{-1} = \{u:\ x\in Su\}$ and $F(x,u) = f(x) + \delta(u|S_x^{-1})$ (here $\delta(\cdot|T)$ is the indicator function of T). The *lagrangean* of (P) is

$$L(x,y) = y(u_0) - \sup_{u\in U}\{y(u)-F(x,u)\}.$$

This can be expressed as $y(u_0) - F^\Phi(x,y)$, where $F^\Phi(x,u)$ is the Φ-conjugate function of $F(x,\cdot)$ evaluated at y. The *dual problem* of (P) is

(D) $\sup_{y\in\Phi} g(y) = \sup_y \inf_x L(x,y)$.

It is always true that

$$\inf_x \sup_y L(x,y) \leqslant \inf(P)$$

but for equality to hold, certain Φ-convexity properties must hold. In particular this is the case if

$$\bigcap_{u\in U} Su = \emptyset \text{ and } S_x^{-1} \text{ is a } \Phi\text{-convex set},\quad \forall \in X$$

Lack of duality gap i.e. $\sup(D) = \inf(P)$, is equivalent to the Φ-convexity of the perturbation function $p(\cdot)$ at u_0. A stronger result

$$\inf(P) = \mathrm{Max}(D)$$

holds if and only if $p(\cdot)$ is Φ-subdifferentiable at u_0.

These results are given in terms of the function $p(u)$. The verfication, in terms of the data, that a given primal problem satisfies the above condition, is by no means an easy task. Specific results are obtained in Dolecki and Kurcyusz (1978) for specific type of families $\mathfrak{L}$, the so-called "metric-like", and under certain growth and stability assumptions. These authors also review and compare similar types of conditions given in Balder (1976).

9. AN ILLUSTRATIVE APPLICATION OF $\mathcal{F}$-CONVEXITY IN NUMERICAL ANALYSIS

Let $f:R\to R$ be a twice differentiable strictly convex and increasing function, and assume that f has a root at $\bar{x}$, i.e. $f(\bar{x}) = 0$. The classical Newton Method attempts to find $\bar{x}$ as a limit point of the sequence

$$\bar{x}_{n+1} = \bar{x}_n - \frac{f(\bar{x}_n)}{f'(\bar{x}_n)} \qquad n = 0,1,2,\ldots . \tag{89}$$

The geometric interpretation of (89) is the following. At the current point x_n, an affine support L_n is constructed, its equation being

$$L_n(x) = f(\bar{x}_n) + (x-\bar{x}_n)f'(\bar{x}_n) \tag{90}$$

and (89) is nothing else but

$$L_n(\bar{x}_{n+1}) = 0. \tag{91}$$

Let $\mathcal{F}\in\mathcal{A}$ be a family with respect to which f is $\mathcal{F}$-convex, and assume further that the members of $\mathcal{F}$ are convex functions. Let $F_n\in\mathcal{F}$ denote the support of f at x_n, i.e.

$$F_n(x) = F(x_n^*,\eta_n;x) \tag{92}$$

where x_n^* and η_n are solutions of the equations

$$F(x_n^*,\eta_n,x_n) = f(x_n)$$

$$\frac{d}{dx}F(x_n^*,\eta_n,x_n) = f'(x_n). \tag{93}$$

At the nth iteration, the root of the "$\mathcal{F}$-approximation" F_n is used as the $(n+1)^{st}$ approximation of $\bar{x}$, i.e. x_{n+1} is determined by

$$F_n(x_{n+1}) = 0 \ . \tag{94}$$

Since f is convex, and so is F_n, it follows that

$$f_n(x) \leqslant F_n(x) \leqslant L_n(x).$$

Moreover (see Figure 1), by the monotonicity of f, the new point x_{n+1}, generated by (94), will lie closer to the root $\bar{x}$ than the Newton point $\bar{x}_{n+1}$ generated by (91).

As a specific example consider the family

$$\mathcal{F}_k = \{x^*\operatorname{arcsinh}(e^{kx}) - \eta : x^*, \eta \in R\}$$

where k is a positive parameter. Here (93) can be solved to yield

$$x_n^* = \frac{1}{k} f'(x_n)\sqrt{1+e^{-2kx_n}} \tag{95}$$

$$\eta_n = x_n^*\operatorname{arcsinh}(e^{kx_n}) - f(x_n)$$

and so

$$F_n(x) = x_n^*\operatorname{arcsinh}(e^{kx}) - \eta_n.$$

and the iteration (94) becomes, after some algebraic manipulations

$$x_{n+1} = \frac{1}{k}\log\left[\frac{1}{2}(\alpha_n - \frac{1}{\alpha_n})\right] \tag{96}$$

where

$$\alpha_n = \exp\left(-\frac{f(x_n)}{x_n^*}\right) \cdot \left[e^{kx_n} + \sqrt{1+e^{2kx_n}} \right] .$$

Note that since $f' > 0$ and $k > 0$ it follows from (95) that $x_n^* > 0$ and consequently the support F_n is indeed convex. For f to be $\mathcal{F}_k$-convex it is necessary and sufficient that the second order differential inequality

$$f''(x) \geqslant \frac{d}{dx^2} F(x_f^*(x), \eta_f(x), x)$$

holds for all x. For the family F_k this becomes

$$f''(x) \geqslant kf'(x)\frac{1}{1+e^{2kx}} \quad . \tag{97}$$

Since $f'' > 0$, (97) will be satisfied for $k > 0$ sufficiently small. In particular it suffices to take

$$k \leqslant \min_x \frac{f''(x)}{f'(x)} \quad . \tag{98}$$

Numerical results obtained for several test functions (satisfying the above assumptions) revealed that, in most cases, the F_k-approximation method (96), with k chosen appropriately, performs only slightly better then Newton's method. But in some cases it might perform considerably better. As a somewhat extreme example , consider the function

$$f(x) = e^{1-x} - 1$$

whose root is $\bar{x} = 1$. Starting from the initial point $x_0 = 4$ one obtained:

	Newton's method	*F_k-approximation method (k=1)*
x_1	-15.08553692	0.97771439
x_2	-14.08553703	0.99996944
x_3	-13.08553731	0.99999999
.		
.		
.		
x_{10}	-6.08602398	
x_{15}	-1.15646158	
x_{20}	0.99996926	

Starting from $x_0 = 10$, Newton's method failed due to overflow, while the F_k-approximation method still behaved excellently as above. Note that the choice $k = 1$ here corresponds to (98).

Remarks. (a) The assumptions $f' > 0$, $f'' > 0$ were made only to simplify the discussion. If $f' < 0$, $f'' < 0$ then one can still use the family F_k above to get better approximations than in Newton methods. Similarly if sign f' ≠ sign f" one could use F_k with $k < 0$. (b) It is not necessary to use the *same* family F at each point in the sequence $\{x_n\}$. In fact F can be made to depend on x_n. As an example, at x_n one might consider the family F_{k_n} where, based on (98),

$$k_n = \frac{f''(x_n)}{f'(x_n)} .$$

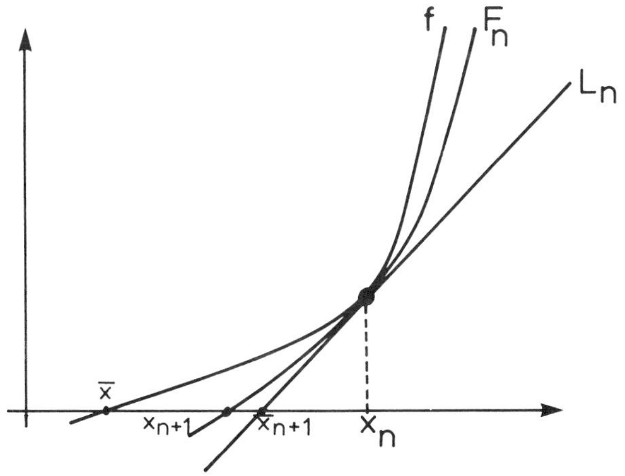

REFERENCES

BALDER, (1976). "An extension of duality-stability relations to nonconvex optimization problems." *SIAM Journal Control and Optimization, 15,* 329-343.

BEN-TAL, A. (1973). "Contributions to generalized convexity and geometric programming." Doctoral Dissertation in Applied Mathematics, Northwestern University, Evanston, Illinois, U.S.A.

BEN-TAL, A. (1977). "On generalized means and generalized convex functions." *Journal Optimization Theory and Applications, 21,* 1-13.

BEN-TAL, A. and BEN-ISRAEL, A. (1976). "A generalization of convex functions via support properties." *The Journal of The Australian Mathematical Society, 21,* (Series A), 341-361.

BECKENBACH, E.F. (1937). "Generalized convex functions." *Bulletin American Mathematical Society, 43,* 363-371.

BECKENBACH, E.F. and BELLMAN, R. (1965). *Inequalities.* Springer-Verlag, New York, 2nd revised printing.

DOLECKI, S. and KURCYUSZ, S. (1978). "On Φ-convexity in extremal problems." *SIAM Journal Control and Optimization, 16,* 277-300.

GREENBERG, H.J. and PIERSKALLA, W.P. (1971). "A review of quasi-convex functions." *Operations Research, 19,* 1553-1570.

HARDY, G.H., LITTLEWOOD, J.E. and POLYA, G. (1934). *Inequalities,* Cambridge University Press, Cambridge, England.

MANGASARIAN, O.L. (1969). *Nonlinear Programming.* McGraw-Hill, New York.

MARTOS, B. (1975). *Nonlinear Programming: Theory and Methods.* North-Holland, Amsterdam.

MORÉ, J. and RHEINHOLDT, E.C. (1973) "On P- and S-functions and related classes of n-dimensional nonlinear mappings." *Linear Algebra and Its Applications, 6,* 45-68.

ORTEGA, J.M. and RHEINHOLDT, W.C. (1970). *Iterative Solution of Nonlinear Equations In Several Variables,* Academic Press, New York.

PEIXOTO, M.M. (1949). "Generalized convex functions and second order differential inequalities." *Bulletin American Mathematical Society, 55,* 563-572.

ROBERTS, A.W. and VARBERG, D.E. (1973). *Convex Functions,* Academic Press, New York.

ROCKAFELLAR, R.T. (1970). *Convex Analysis.* Princeton University Press, Princeton, N.J.

STOER, J. and WITZGALL, C. (1970). *Convexity and Optimization in Finite Dimensions.* Vol. I, Springer-Verlag, Berlin and New York.

CONVEX RELATIONS IN ANALYSIS AND OPTIMIZATION

J.M. Borwein

In this paper our aim is to show some of the ways in which the use of convex relations simplifies, unifies and strengthens the study of convex constrained optimization problems with vector objectives. First we sketch the topological and analytic properties of convex relations which are of particular use in optimization and many of which are of considerable independent analytic interest. Subsequently we apply these considerations to the study of constrained optimization problems.

1. ANALYSIS

1.1 Motivation

Let X be a (*real, separated, topological*) *vector space* and consider the *ordinary convex program*

$$(P) \quad p = \inf_{x \in C} \left\{ f(x) \;\middle|\; \begin{matrix} g_i(x) \le 0 \ (i=1,\dots,m) \\ h_j(x) = 0 \ (j=1,\dots,k) \end{matrix} \right\} \tag{1.1}$$

where $f, g_i, h_j : X \to \bar{R}$ and the f, g_i are *convex* while the h_j are *affine*, on the convex set C. Set

$$H_g(x) = \begin{cases} g(x) + R_+^m , & x \in C \\ \emptyset , & x \notin C \end{cases} , \tag{1.2}$$

$$H_h(x) = \{h(x)\} , \tag{1.3}$$

where $g = (g_1, \dots, g_m)$, $h = (h_1, \dots, h_k)$ and R_+^m is the non-negative orthant in R^m. Now set

$$H(x) = (H_g(x), H_h(x)) . \tag{1.4}$$

Then H is a (*multifunction, set-valued map*) *relation* between X and $Y = R^{m+k}$ and it is clear that we may rewrite (P) in the multivalued equation form

$$(P) \quad p = \inf \{f(x) \mid 0 \in H(x)\} . \tag{1.5}$$

This will be the form in which we will study (P) in the sequel. The reasons for this are various: it unifies the study of equality and

ISBN 0-12-621120-5

inequality constraints via (1.4); it often simplifies proofs and strengthens results; also it is the author's contention that many results about constraint structure are intrinsically results about H and *not* about g and h.

Let us observe that H will turn out to be a convex relation as defined below exactly when g is convex and h is affine on a convex set C.

1.2 Some Basic Relationships

Before turning to convex relations let us summarize the central properties of relations with which we will be concerned. The reader is referred to Berge (1959), Dolecki (1981), Jameson (1972), Huard (1979), Robinson (1972; 1976a), and Rockafellar (1970; 1976) for more details.

Throughout the paper all spaces are supposed to be real separated topological vector spaces. If one wishes to study purely algebraic notions one may use the *finest locally convex* or *core* topology (see Oettli (1979) and Robertson and Robertson (1964)). Let $H:X \to 2^Y$ be a relation between X and Y. We will reserve lower case letters for relations and upper case letters for functions. Thus we will not usually distinguish between h and $\{h\}$ and also we will consider $H:X \to Y$ although this is an abuse of notation. The *domain* of H, D(H), is defined by

$$D(H) = \{x \in X \mid H(x) \neq \emptyset\} . \tag{1.6}$$

Given any set C in X we define

$$H(C) = \cup\{H(x) \mid x \in C\} . \tag{1.7}$$

Then $H(X) = R(H)$ is the *range* of H. The *inverse* relation $H^{-1}:Y \to X$ is defined by

$$H^{-1}(y) = \{x \in X \mid y \in H(x)\} . \tag{1.8}$$

The *graph* of H is given by

$$\mathrm{GrH} = \{(x,y) \mid y \in H(x)\} . \qquad (1.9)$$

Thus H and H^{-1} essentially share their graph since (x,y) lies in GrH exactly if (y,x) lies in GrH^{-1}. Then one may identify relations between X and Y and sets in XxY via (1.9).

If $H:X \to Y$ and $K:Y \to Z$ the composite relation is defined by

$$(HK)(x) = H(K(x)) , \qquad (1.10)$$

using (1.7). It is easily verified that

$$(HK)^{-1} = K^{-1}H^{-1} . \qquad (1.11)$$

Now $H:X \to Y$ is said to be *lower semi-continuous (LSC) at* (x_0,y_0) if given any neighbourhood V of zero in Y one may find a neighbourhood U of zero in X with

$$H(x) \cap (V + y_0) \neq \emptyset \qquad x \in U + x_0 . \qquad (1.12)$$

It is implicit in (1.12) that $y_0 \in \mathrm{cl}\ H(x_0)$. If H is LSC at (x_0,y_0) for all y_0 in $H(x_0)$ we say that H is *LSC at* x_0.

Similarly H is *open at* (x_0,y_0) if given any neighbourhood U of zero in X one can find a neighbourhood V of zero in Y with

$$V + y_0 \subset H(x_0 + U). \qquad (1.13)$$

If this holds for all x_0 in $H^{-1}(y_0)$ we say H is *open at* y_0.

Since (1.12) is equivalent to $U + x_0 \subset H^{-1}(V + y_0)$ we see that the following proposition holds.

PROPOSITION 1:

(a) H *is open at* (x_0,y_0) *if and only if* H^{-1} *is* LSC *at* (y_0,x_0).

(b) H *is open at* y_0 *if and only if* H^{-1} *is* LSC *at* y_0.

(c) *If* H *is open at* (x_0,y_0) *and* K *is open at* (y_0,z_0) *then* HK *is open at* (x_0,z_0).

(d) *If* H *is* LSC *at* (x_0,y_0) *and* K *is* LSC *at* (y_0,z_0) *then* HK *is* LSC *at* (x_0,z_0).

PROOF: (a) and (b) are clear from the previous discussion. Now (c) follows easily from (1.10) and (1.13) whence (d) follows from (1.11) (a) and (c). □

1.3 Convex Relations

A relation $H:X \to Y$ is *convex* if its graph is convex. Equivalently, one has

$$t\, H(x_1) + (1-t)\, H(x_2) \subset H(tx_1 + (1-t)x_2) \tag{1.14}$$

whenever $0 \leq t \leq 1$ and x_1, x_2 lie in X. It is immediate from the previous section that the inverse and composition of convex relations are convex and that the domain and range of a convex relation are convex. Moreover there is a one-to-one identification of convex relations between X and Y and convex sets in XxY.

Example 1:

(a) Let $g:C \subset X \to Y$ be a single valued function and $P \subset Y$ a convex cone (or just a convex set). Then

$$H_g(x) = \begin{Bmatrix} g(x) + P & x \in C \\ \emptyset & x \notin C \end{Bmatrix} \tag{1.15}$$

is a convex relation exactly when g is P-convex on C, (Borwein (1977a)). In particular this is true for H as in (1.2) or (1.4). The empty set here plays the role of $(+\infty)$ in convex analysis, the whole space the role of $(-\infty)$. Note that GrH_g is exactly the P-*epigraph* of g on C.

(b) Let $A:X \to Y$ be a linear operator and let C and D be convex sets

$$H_A(x) = \begin{Bmatrix} A(x) + D & x \in C \\ \emptyset & x \notin C \end{Bmatrix} . \tag{1.16}$$

Then H_A is a convex relation. This is a particularly useful special case of (1.15).

(c) With H_g as in (1.15)

$$L_g(y) = H_g^{-1}(y) = \{x \in X \mid y \in g(x) + P, \quad x \in C\} \tag{1.17}$$
$$= g^{-1}(y-P) \cap C$$

is the "level set" relation and is convex exactly when H_g is. Thus in this framework L_g and H_g have entirely symmetric roles. For example, if g is continuous at x_0 in int C, H_g is LSC at x_0 and hence L_g is open at x_0.

(d) Let $H_1:X \to Y_1$, $H_2:X \to Y_2$ be convex. Then so is H given by

$$H(x) = (H_1,H_2)(x) = (H_1(x),H_2(x)) \ . \tag{1.18}$$

The inverse of H is given by

$$H^{-1}(y_1,y_2) = H_1^{-1}(y_1) \cap H_2^{-1}(y_2) \ . \tag{1.19}$$

(e) Let H_1, $H_2:X \to Y$ be convex. So also is the *sum* $H_1 + H_2$ given by

$$(H_1 + H_2)(x) = H_1(x) + H_2(x) \tag{1.20}$$

and the *convolution* $H_1 \square H_2$ given by

$$H_1 \square H_2(x) = \{y_1 + y_2 \mid y_1 \in H(x_1),\ y_2 \in H(x_2),\ x_1 + x_2 = x\}. \tag{1.21}$$

Note that convolution corresponds to adding graphs and sum to adding images. As for convex function one may define partial convolution. In the real valued case we have

$$f \square g = \inf H_f \square H_g . \tag{1.22}$$

However, (1.21) makes sense in many situations in which (1.22) is not well defined.

(f) A convex relation is a *convex process* if its graph is a cone. For any convex relation H we may define the *convex process* P *generated by* H by

$$\mathrm{Gr}\ P = \mathrm{cone}\ (\mathrm{Gr}\ H) \tag{1.23}$$

or equivalently

$$P(x) = \bigcup_{\lambda>0} \{\lambda\ H(\tfrac{x}{\lambda})\}. \tag{1.24}$$

The reader is referred to Berge (1959), Rockafellar (1967; 1970), Robinson (1972) and Makarov and Rubinov (1977) for information on convex processes and to Borwein (1977a; 1979), Robinson (1972; 1976a) and Jameson (1970; 1972) for more information on convex relations.

Any other pieces of convex terminology are consistent with that of Ekeland and Temam (1976) or Robertson and Robertson (1964).

1.4 Convexity and Continuity

In a fashion analogous to that of convex functions the continuity properties of convex relations are considerably simpler than those of arbitrary relations.

PROPOSITION 2:

(a) *Suppose* $H:X \to Y$ *is convex and open* (LSC) *at* (x_0,y_0). *Then* H *is open* (LSC) *at* (x_1,y_1) *whenever* $y_1 \in H(x_1)$ *and* $y_1 \in \text{core } R(H)$ $(x_1 \in \text{core } D(H))$.

(b) *In particular if* H *is open* (LSC) *at* (x_0,y_0) *then* H *is open* (LSC) *at* $y_0(x_0)$.

(c) *If* $H_1(x) \subset H_2(x) \qquad \forall\, x \in X \qquad (1.25)$

where H_1 *is open* (LSC) *at* (x_0,y_0) *and* H_2 *is convex then* H_2 *is open* (LSC) *at* $y_0(x_0)$.

PROOF: We prove only the unbracketed assertions. The parenthetic assertions follow on reversing the roles of X and Y and using H^{-1}, as in Proposition 1.

(a) Since y_1 is in core R(H) one may find $y_2 \in H(x_2)$ and $0 < \varepsilon < 1$ with $y_1 = \varepsilon y_0 + (1-\varepsilon)y_2$. Set $\bar{x}_1 = \varepsilon x_0 + (1-\varepsilon)x_2$. Then suppose that a neighbourhood of zero in X,W, is given and that there exists a neighbourhood V with

$$y_0 + V \subset H(x_0 + U) \qquad (1.26)$$

for a balanced U with $U + U \subset W$. Then using convexity and (1.26)

$$y_1 + \varepsilon V = \varepsilon(y_0 + V) + (1-\varepsilon)y_2 \subset \varepsilon H(x_0 + U) + (1-\varepsilon)H(x_2) \subset H(\bar{x}_1 + \varepsilon U). \tag{1.27}$$

Now pick $0 < \delta < 1$ with $\delta(\bar{x}_1 - x_1) \in \varepsilon U$. Then

$$y_1 + \delta\varepsilon V = (1-\delta)y_1 + \delta(y_1 + \varepsilon V) \subset (1-\delta)\, H(x_1) + \delta H(\bar{x}_1 + \varepsilon U) \subset H(x_1 + \delta(\bar{x}_1 - x_1) + \delta\varepsilon U) \tag{1.28}$$

and thus

$$y_1 + \delta\varepsilon V \subset H(x_1 + \varepsilon W) \tag{1.29}$$

and H is open at (x_1, y_1). Note that one actually now gets

$$y_1 + \lambda\delta V \subset H(x_1 + \lambda W) \tag{1.30}$$

for $0 < \lambda \leq \varepsilon$. In Dolecki's terms (1981) H is *open at linear rate*.

If X is locally convex the previous argument may be a little simplified.

(b) is now immediate since y_0 lies in core R(H).

(c) Since H_1 is open at (x_0, y_0) it follows that H_2 is open at (x_0, y_0) and (b) now applies. □

In a normed or semi-normed setting the constants in (1.30) can be quantified as in Robinson (1976). Proceeding as above one can show that a convex relation is locally uniformly open or LSC at any point at which it is open or LSC: i.e. for (x_1, y_1) near (x_0, y_0) in Gr H, (Dolecki (1981)). In particular one derives the following.

PROPOSITION 3: Let $H: Y \to Y$ *be a convex relation between normed spaces. Let* H *be* LSC *at* x_0. *For all* $\eta > 0$ *sufficiently large there exist* ε, $k > 0$ *such that*

$$D(H(x) \cap \eta B, H(x') \cap \eta B) \leq k\, ||x-x'|| \tag{1.31}$$

whenever $||x-x_0||, ||x'-x_0|| \leq \varepsilon$.

Here B is the unit ball in Y and D is the Hausdorff metric. A homogeneous form of the above is given by Robinson (1972) and in Tuy and Du'ong (1978). One can derive the corresponding Lipschitzness result for convex functions from (1.31). It is more convenient to establish the next lemma.

Recall that a cone S in Y is said to be *normal* if there is a base at zero of neighbourhoods V with

$$(V - S) \cap (S - V) \subset V. \tag{1.32}$$

Most commonly occurring cones are normal but by no means all. The reader is referred to Peressini (1967), Schaefer (1971), Jameson (1970), or Borwein (1980a) for more details.

LEMMA 4: Let $f:X \to Y$ *be* S*-convex. Let* $H_f(x) = f(x) + S$.

(a) Then H_f *is* LSC *at* x_0, *whenever* f *is continuous at* x_0.

(b) Conversely, if S *is normal and* H_f *is* LSC *at* x_0, f *is continuous at* x_0.

PROOF:

(a) is immediate.

(b) Let us suppose $x_0 = 0$, $f(x_0) = 0$. Since H_f is LSC at x_0 one can find for each neighbourhood V of zero a neighbourhood U of zero with

$$(f(x) + S) \cap V \neq \emptyset \qquad x \in U. \tag{1.33}$$

Since f is S-convex

$$f(x) + f(-x) \subset 2f(0) + S = S. \tag{1.34}$$

Hence for x in U

$$f(-x) \subset S - f(x) \subset (S + S - V) = (S - V)$$

using (1.33) and (1.34). Thus if x lies in -u

$$f(x) \subset (S - V) \cap (V - S) \subset V \tag{1.35}$$

on using (1.33) again and the normality of S. Thus f is actually continuous at x_0. □

It follows as in Borwein (1980a) and in Robinson (1976) that, in the normed case, f is actually locally Lipschitz throughout int(domf). When S is not normal it is usual that H_f is LSC while f is not continuous. In these cases one is much better advised to study H_f than f itself. A case of interest which is not normal occurs when Y = D[0,1] is the space of continuously differentiable functions on the unit interval with

$$||f|| = |f(0)| + \sup_{0\leq t\leq 1} |f'(t)| \tag{1.36}$$

and S is the cone of non-negative functions. Then S is not normal since it takes no account of derivative behavior. This is a prototype for Sobelov space behavior. Note also that in the core topology any pointed cone is weakly normal.

For the final result in this section we need one more topological definition. A relation $H:X \to Y$ is *strongly open* at (x_0,y_0) if

$$y_0 \in \text{int } H(x_0) \ . \tag{1.37}$$

PROPOSITION 5: Let $H_1:X \to Y_1$, $H_2:X \to Y_2$ *be convex. Let* $y_1 \in H_1(x_1)$, $y_2 \in H_2(x_1)$. *Suppose that*

(a) H_1 *is* LSC *at* x_1 *and strongly open at* (x_1,y_1),

(b) H_2 *is open at* y_2.

Then

$$H(x) = (H_1(x),H_2(x))$$

is open at x_1.

A proof of this result is given in Borwein (1980b) and a special case may be found in Robinson (1976a). A dual result may be given for H as in (1.19) involving semi-continuity.

Example 2: If H_g is given by (1.15) H_g is strongly open at $(x_1,0)$ exactly when x_1 lies in C and

$$g(x_1) \in - \text{int } P \quad (\textit{Slater's condition}). \tag{1.38}$$

Notice that while we know from Proposition 2 that H_g is then open at zero it is not generally strongly open at every point in $H_g^{-1}(0)$.

Proposition 5 allows us to replace various constraints by one product constraint while maintaining openness. For example, if in (1.2), (1.3), (1.4) g is continuous and satisfies Slater's condition at some point in the interior of C while h is open (in its range) then H given by (1.4) is open at zero. If one relativizes these conditions as one may in R^n one discovers that the standard constraint qualification given in Rockafellar (1970) is equivalent to the relative openness of H at 0.

In general, even for open continuous linear operators T_1 and T_2 between Banach spaces the product (T_1, T_2) need not have closed range and so cannot be open in its range.

Example 3: Let X be a reflexive Banach space. Let $K \subset X$ be a closed convex cone and K^+ its positive *dual* cone. Let $T:X \to X^*$ be a continuous linear operator and suppose T is *coercive* on K

$$(T(x),x) \geq c\,||x||^2 \qquad (\forall x \in K) \tag{1.39}$$

for some $c > 0$. Let

$$H(x) = \begin{Bmatrix} T(x)-K^+ & x \in K \\ \emptyset & x \notin K \end{Bmatrix}. \tag{1.40}$$

Then the theorem of Lions and Stampacchia given in Ekeland and Temam (1976) may be used to show that H is surjective. It will follow from Theorem 8 that H is open although in general T will not be surjective and K will have no interior. This has applications in variational inequality theory. See Borwein (1980a), Cryer and Dempster (1980), and Ekeland and Temam (1976).

1.5 CS-closed Relations and the Open Mapping Theorem

We now introduce CS-closed sets and prove a general form of the

Open-Mapping-Closed Graph theorem for convex relations. We then give two easy applications.

DEFINITION 1: A set C *in a linear space* X *is* <u>*CS-closed*</u> *(convergent series-closed) if whenever* $\{\lambda_n\}$ *is a non-negative sequence with sum 1 and* $\{c_n\}$ *lies in* C *one has*

$$\sum_{n=1}^{\infty} \lambda_n c_n \in C \tag{1.41}$$

whenever $\sum^{\infty} \lambda_n c_n$ *exists in* X. If $\sum^{\infty} \lambda_n c_n$ *always exists,* C *is said to be* <u>*CS-compact*</u>.

A relation is CS-closed if its graph is. This definition is due to Jameson (1970; 1972). A similar notion of *ideal convexity* can be found in Holmes (1975).

It is obvious that CS-closed sets are convex. It is less obvious but true that if (1.41) holds for one infinitely positive sequence of λ_n it holds for all. We now gather up some of the signal properties of CS-closed sets.

PROPOSITION 6:

(a) Finite dimensional, closed and open sets are CS-*closed.*

(b) The intersection of CS-*closed sets is* CS-*closed; if in addition one of the sets is* CS-*compact so is the intersection.*

(c) A convex G_δ *in a Fréchet space is* CS-*closed.*

(d) A bounded complete CS-*closed subset of a normed space is* CS-*compact.*

(e) If H *is a* CS-*closed relation and* C *is* CS-*compact then* H(C) *is* CS-*closed.*

All of the above are easy and can be found in Jameson (1972) except for (c) which is much harder and is proven in Fremlin and Talagrand (1979).

Example 4: Let $H(x) = Tx + C_2$ where T is continuous and C_2 is CS-closed. Then part (e) shows $T(C_1) + C_2$ is CS-closed whenever C_1 is CS-compact.

In particular $C_1 + C_2$ is CS-closed.

The central topological property of CS-closed sets is that they are *semi-closed* (have the same interior as their closure) at least in a metrizable setting.

THEOREM 7 (Jameson 1972): Let Y *be a metrizable topological vector space. Let* $C \subset Y$ *be* CS-*closed. Then*

$$\text{int } C = \text{int cl}C. \tag{1.42}$$

This fails more generally. Let Y be B[0,1] the bounded functions in the product topology. Let C be the subspace of functions of countable support. Then C is sequentially closed and so CS-closed. However, clC = Y and (1.42) fails. The easiest way of showing a set is not CS-closed in a metrizable setting is showing (1.42) fails. Consider the convex hull of a countable dense subset of the unit ball in a separable Banach space.

We can now state simultaneously the Closed Graph and Open Mapping theorems.

THEOREM 8:

(a) *Let* $H:X \to Y$ *be a* CS-*closed relation between complete metrizable spaces. Then* H *is* LSC *throughout core* D(H) *and open throughout core* R(H).

(b) *Let* H *be a closed convex relation. Suppose* X *is barreled and* Y *is complete, metrizable and locally convex. Then* H *is* LSC *throughout core* D(H).

(c) *Let* H *be a closed convex relation. Suppose* X *is complete, metrizable and locally convex and* Y *is barreled. Then* H *is open throughout core* R(H).

PROOF: As before it is only necessary to establish the open version. Thus even in the linear case the standard proofs are simplified. Before proceeding let us indicate various special cases. Parts (b) and (c) may be found in Ursescu (1975), albeit with slightly different terminology. The Banach space case of (b) is in Robinson (1976), while Jameson (1972) gives the Banach space case of (a) and other related results. When H is single valued these results reduce to the classical ones (Robertson & Robertson (1964), Rudin (1973)). When H is a convex process the condition that $0 \in \text{core } R(H)$ is equivalent to $R(H) = Y$. In this case one can weaken the requirement in (a) to having $R(H) \cap - R(H)$ second category.

1.6 A Proof of the Metrizable Case

We may suppose that $0 \in \text{core } R(H)$ and $0 \in H(0)$, by translation. We wish to show H is open at (0,0). Let us pick closed balanced neighbourhood bases $\{U_n\}$ at zero in X and $\{V_n\}$ at zero in Y such that

$$U_n + U_n \subset U_{n-1} \; ; \qquad V_n \subset V_{n-1} \; . \tag{1.43}$$

Since $0 \in \text{core } R(H)$ and each U_n is absorbing it follows easily from the convexity of H that

$$D_n = \overline{-H(U_n) \cap H(U_n)} \qquad (n=1,2,\ldots) \tag{1.44}$$

is absorbing. The Baire category theorem shows that some multiple of D_n has interior and by homothety D_n itself does. Let x_n be in int D_n. Then we may suppose that

$$\pm\, x_{n+1} + 2V_{n+1} \subset \overline{H(U_{n+1})}, \qquad (n=1,2,\ldots) \tag{1.45}$$

on selecting a subsequence from V_n if need be. This does not disturb (1.43). By convexity and (1.43)

$$2V_n \subset \overline{H(\tfrac{1}{2}U_{n+1} + \tfrac{1}{2}U_{n+1})} \subset \overline{H(U_n)} \subset H(U_n) + V_{n+1} \tag{1.46}$$

for each $n \geq 1$. Repeated use of (1.46) produces

$$V_1 \subset \tfrac{1}{2}H(U_1) + \tfrac{1}{4}H(U_2) + \ldots \frac{1}{2^{n-1}} H(U_{n-1}) + \frac{1}{2^{n-1}} V_n \; . \tag{1.47}$$

Let v_1 lie in V_1. Then there exist $y_i \in H(u_i)$, $u_i \in U_i$ such that

$$v_1 - \sum_{i=1}^{n-1} \frac{y_i}{2^i} \in V_n \ . \tag{1.48}$$

Since Y is metrizable one has

$$\sum_{i=1}^{\infty} \frac{y_i}{2^i} = v_1 \ . \tag{1.49}$$

Also

$$\sum_{i=m}^{k} \frac{u_i}{2^i} \in \sum_{i=m}^{k} \frac{1}{2^i} U_i \subset \frac{1}{2^m} U_{m-1} \tag{1.50}$$

on using (1.43) repeatedly. This shows that

$$\sum_{i=1}^{\infty} \frac{u_i}{2^i} = u_0 \tag{1.51}$$

exists since $\sum_{i=1}^{k} \frac{u_i}{2^i}$ is a Cauchy sequence and X is complete.

Moreover, $u_0 \in \frac{1}{2}U_0$ on using (1.50) with m = 1. Thus as $y_i \in H(u_i)$ and H is CS-closed (1.49) and (1.51) show that

$$v_1 \in H(u_0) \subset H(U_0) \ , \tag{1.52}$$

and as v_1 is arbitrary V_1 lies in $H(U_0)$. The identical considerations show that

$$V_{n+1} \subset H(U_n) \ , \tag{1.53}$$

for each $n \geq 0$ and so H is open at (0,0).

1.7 The Barreled Case

If actually H is closed and Y is barreled one proceeds in much the same fashion. Now we assume each U_n is convex and D_n turns out to be a barrel and so has zero in its interior. Then (1.46) will still hold except that V_n will not actually form a base at zero. Then one may fix n and pick an arbitrary neighbourhood V to replace V_n in (1.47) and (1.48). Now (1.48), (1.51) and the convexity of H in conjunction with $0 \in H(0)$ imply that

$$(u_0,v_1) = \sum_{i=1}^{n-1} \frac{1}{2^i}(u_i,y_i) + \left(\sum_{i=n}^{\infty} \frac{u_i}{2^i} , v_1 - \sum_{i=1}^{n-1} \frac{y_i}{2^i} \right) \quad (1.54)$$

and so

$$(u_0,v_1) \in \mathrm{Gr}(H) + (U_{n-1},V) . \quad (1.55)$$

Then (1.55) shows that (u_0,v_1) lies in $\overline{\mathrm{Gr}(H)}$ since U_{n-1} is in a neighbourhood base for zero in X and V is arbitrary. Since Gr(H) is closed the openness of H at (0,0) follows as before. □

Let us observe that while a CS-closed linear operator in a metrizable space is closed a CS-closed convex relation is generally not as is shown by Proposition 6 (a),(b),(c). Thus the generality of Theorem 8 is not spurious. Indeed it conveniently allows us to deal with constraint structures involving intersections of open, closed or finite dimensional sets and many others which may occur in infinite dimensional programming. Notice also that in the metrizable case neither X nor Y need be locally convex. This is useful in applications to partially ordered topological vector spaces. Let us derive some easy consequences.

COROLLARY 9: Let $f:X \to Y\cup\{\infty\}$ *be* S-*convex with* S *a normal cone. Suppose that either*

(a) f *has a* CS-*closed epigraph while* X *and* Y *are complete metrizable linear spaces,*

or

(b) f *has a closed epigraph while* X *is barreled and* Y *is a complete metrizable locally convex space.*

Then f *is continuous throughout the core of its domain.*

PROOF: Since the epigraph of f is just the graph of the associated relation with $H_f(x) = f(x) + S$, it follows from Theorem 8 that H_f is LSC throughout core $D(H_f)$ = core(domf). Lemma 4 finishes the result. □

Again in the normed case f is actually locally Lipschitz at core points of its domain. When Y is the real line and $S = R_+$ (b) is a well-known result. In the Banach space setting (b) is due to Robinson (1976a). Part (a) is new.

As a sample of how the open mapping theorem is used in partially ordered vector space theory we give the following:

COROLLARY 10: Let Y *be a complete metrizable linear space and suppose that* C_1, C_2 *are* CS-*closed sets in* Y *with*

$$0 \in \text{core } (C_1 + C_2) \ . \tag{1.56}$$

Then

$$0 \in \text{int } (C_1 \cap U) + (C_2 \cap U) \tag{1.57}$$

for any neighbourhood U *of zero.*

PROOF: Let $X = Y \times Y$ in the product topology, let $C = C_1 \times C_2$ and set

$$H(x) = \begin{cases} x_1 + x_2 & x_1 \in C_1, x_2 \in C_2 \\ \emptyset & \text{else} \end{cases} \ . \tag{1.58}$$

Then H is CS-closed and (1.56) shows $0 \in \text{core } R(H)$. Since (1.57) merely says H is open at 0 we are done. □

When C_1 and C_2 are cones a similar result is given in Jameson (1977) and can be used to derive and extend normal -B-cone duality Jameson (1970) and Schaefer (1971).

2. SOME FURTHER ANALYTIC RESULTS

2.1 The Principle of Uniform Boundedness

We suppose for simplicity in this development that X and Y are Banach spaces.

THEOREM 11: Let $H_i : X \to Y$ $(i \in I)$ *be a family of convex relations each* LSC *at* x_0. *Suppose that*

$$\sup_I \inf ||H_i(x)|| = m(x) < \infty \tag{2.1}$$

for each x *in some set* A *with* x_0 *in core* A. *Then if for* $c > 0$

$$y_{i_0} \in H_i(x_0)\ ,\ ||y_{i_0}|| \le c \tag{2.2}$$

the relations H_i *are locally equi-Lipschitz at* (x_0, y_{i_0}).

PROOF: Let us set

$$K_i(x) = (H_i(x+x_0) - y_{i_0}) \cap B. \tag{2.3}$$

Then $0 \in K_i(0)$ and each K_i is still convex, is LSC at 0 and satisfies (2.1) for some set A with 0 in core A. Let

$$U = \bigcap_I \overline{K_i^{-1}(\bar{B})} \tag{2.4}$$

where B is the unit open ball in Y. Since (2.1) holds and K_i is convex it is easy to show that U is absorbing and being convex, closed in a Banach space is a neighbourhood of zero.

Since each $K_i^{-1}(B)$ actually has interior, being LSC at 0, we have

$$\operatorname{int} U \subset \operatorname{int} \overline{K_i^{-1}(\bar{B})} \subset K_i^{-1}(\bar{B}). \tag{2.5}$$

Now we may assume U is the closed unit ball in X and pick x_1 in int $\frac{1}{2}$ U and any y_{i_1} in $K_i(x_1)$. Thus y_{i_1} is in B and

$$x_1 + \frac{U}{2} \subset \operatorname{int} U \subset K_i^{-1}(\bar{B}) \subset K_i^{-1}(y_{i_1}+2\bar{B})\ . \tag{2.6}$$

Since K_i^{-1} is convex and $x_1 \in K_i^{-1}(y_{i_1})$ we must have

$$x_1 + \lambda U \subset K_i^{-1}(y_{i_1}+4\lambda\bar{B}) \tag{2.7}$$

whenever $0 < \lambda < \frac{1}{2}$. This shows that for i in I and any x in $x_1 + \lambda U$

$$H_i(x+x_0) \cap (B + y_{i_0}) \cap (y_{i_1}+y_{1_0} + 4\lambda\bar{B}) \ne \emptyset. \tag{2.8}$$

Since $y_{i_1} + y_{i_0}$ is an arbitrary point of $H_i(x_1+x_0) \cap (B + y_{i_0})$ one has

$$H_i(x_1+x_0) \cap (B + y_{i_0}) \subset H_i(x+x_0) \cap (B + y_{i_0}) + 4||x-x_1||\bar{B}\ . \tag{2.9}$$

Since x and x_1 play symmetric roles we see that actually

$$D(H_i(x_1) \cap (\bar{B} + y_{i_0})\ ,\ H_i(x) \cap (\bar{B} + y_{i_0})) \le 4||x-x_1|| \tag{2.10}$$

whenever $||x-x_0||, ||x_1-x_0|| < \frac{1}{2}$. The choice of the unit ball in (2.10) is merely one of convenience. □

A slightly more extended argument allows one to replace (2.10) by (1.31) for each i. There is a more cumbersome dual form of this result involving open mappings.

It is convenient to have conditions which ensure that each H_i is LSC at x_0 and that (2.1) holds. One such condition is as follows. We say H_i is *image semicontinuous at* (x_0,i_0) if for each sufficiently small neighbourhood U of zero in X and each $\varepsilon > 0$ one has

$$H_{i_0}(x) \subset H_i(x) + \varepsilon B \qquad x \in U + x_0 , \tag{2.11}$$

for i sufficiently near i_0. Of course this supposes that I is topologized. This notion is due to Dolecki and Rolewicz (1979).

PROPOSITION 12: Suppose that H_i *is image semi-continuous at* (x_0,i_0) *and that each* H_i *has a* CS-*closed graph. Then for* i *near* i_0 *(2.1) holds. If* x_0 *is a core point of* $D(H_{i_0})$ *then actually all the* H_i *are* LSC *at* x_0 *for* i *near* i_0.

PROOF: Fix any U and ε for which (2.11) holds. Then for x in $U + x_0$ it is clear that (2.1) holds if i is near i_0. Moreover if x_0 is a core point of $D(H_{i_0})$, Theorem 8(a) shows that H_{i_0} is LSC at x_0. Let $x_0 + U$ be an open set inside $D(H_{i_0})$. Then (2.11) shows that for U sufficiently small and i near i_0 actually $x_0 + U$ lies in $D(H_i)$. Since each H_i has a CS-closed graph it is actually LSC at x_0. □

Example 5: Let $I = \{1,2,...\}$ and let $g_i:X \to Y$ be S_i-convex on some open set C with

$$\lim_{i\to\infty} g_i(x) = g_0(x), \qquad (\forall x \in C) . \tag{2.12}$$

Suppose that for $i = 0,1,2,...,g_i + S_i$ is LSC. It is immediate that (2.1) holds on C and that the result of Theorem 11 obtains. Thus $g_i(\cdot) + S_i$

satisfy (2.10) with $y_{i_0} = g_i(x_0)$. If in addition the S_i are *equi-normal* we may obtain from (2.10) that the g_i are equi-Lipschitz locally at x_0. This is Kosmol's central theorem (1977). He then applies this to the study of perturbed optimization problems of the form

$$h(i) = \inf \{f_i(x) \mid g_i(x) \in -S_i, x \in C\} . \tag{2.13}$$

We may use Theorem 11 and Proposition 12 to produce similar results without assuming normality of the S_i or even such a specific constraint form.

Finally, observe that if each $g_i + S_i$ is CS-closed ($i=1,2,\ldots,$) and g_0 is continuous and if the convergence in (2.12) is uniform near x_0, then Proposition 12 applies. Note that Theorem 11 does include the classical Banach-Steinhaus result (Rudin (1973)).

2.2 The Dieudonné Closure Condition

Let us begin by recalling that the *recession cone* of a convex set C in X is defined by

$$\text{rec } C = \{x \in X \mid C + \lambda x \subset C, \quad \lambda \geq 0\} . \tag{2.14}$$

We refer to Holmes (1975) for details of the recession cone's properties in infinite dimensions. The central one is that a locally compact closed convex set is compact exactly when rec C = 0.

THEOREM 13: Let H *be a closed convex relation between topological vector spaces* X *and* Y. *Suppose that* C *is a closed convex set in* X *and* $0 \in R(H)$. *If*

(i) $D(H) \cap C$ *is relatively locally compact,* (2.15)

(ii) $\text{rec } C \cap \text{rec } H^{-1}(0)$ *is linear* (2.16)

then H(C) *is closed.*

PROOF: Let P be the projection of $X \times Y$ on Y given by $P(x,y) = y$. Now

$$H(C) = P(C \times Y \cap \text{Gr}H) \tag{2.17}$$

and since P is open and linear it follows much as in Holmes (1975) that H(C) is closed exactly when

$$P^{-1}(0) - (C\times Y \cap \mathrm{Gr}H)$$

is closed. Let $A = X\times 0 = P^{-1}(0)$ and $B = C\times Y \cap \mathrm{Gr}H$. Then A and B are closed convex sets and calculation shows

$$\begin{aligned}\mathrm{rec}\,A \cap \mathrm{rec}\,B &= (X\times 0) \cap (\mathrm{rec}\,C\times Y) \cap (\mathrm{rec}\,\mathrm{Gr}H) \\ &= (\mathrm{rec}\,C \cap \mathrm{rec}\,H^{-1}(0),\ 0)\ .\end{aligned} \tag{2.18}$$

Thus all common recession directions for A and B are *lineality directions*. Dieudonné's theorem in Holmes (1975) would apply if either A or B were locally compact. This fails in the present case. However, the following extension of that theorem is valid and completes the proof of the present result.

LEMMA 14: Let A *and* B *be closed convex subsets of a product space* X×Y *such that*

(i) rec A ∩ rec B *is linear,* (2.19)

(ii) $P_X A$ *and* $P_Y B$ *are locally compact (relatively).* (2.20)

Then A - B *is closed.*

PROOF: The proof is a slight complication of that of Dieudonné's theorem which is the case that X = 0 above. □

Note that Lemma 14 covers a variety of cases in which Dieudonné's theorem is inapplicable.

Gwinner (1977) gives a similar result for upper semi-continuous relations and indicates applications to duality theory, splines and best approximation. It should be clear that Theorem 12 has applications whenever one needs closure of an image set as is frequently the case in convex analysis. One can also derive in a straight forward fashion conditions for the composition of closed relations to be closed since

$$R(H_1, H_2) = \mathrm{Gr}(H_1 H_2^{-1}) \ . \qquad (2.21)$$

Application of those considerations yields the following condition for $H_1 H_2^{-1}$ to be closed.

(i) $D(H_1) \cap D(H_2)$ is relatively locally compact, (2.22)

(ii) $\mathrm{rec}\ H_1^{-1}(0) \cap \mathrm{rec}\ H_2^{-1}(0)$ is linear. (2.23)

In finite dimensions (2.22) is, of course, always satisfied.

Often one has in mind the case where $H_1(x) = f(x) + R_+$ and $H_2(x)$ is an arbitrary convex relation. Then if the *value function* h is defined by

$$h(u) = \inf\{f(x) \mid u \in H_2(x)\}, \qquad (2.24)$$

$$\mathrm{Epi}\,h = \overline{\mathrm{Gr}(H_1 H_2^{-1})}\ ; \qquad (\text{i.e. } h(u) + R_+ = \mathrm{cl}(H_1 H_2^{-1})(u)) \quad (2.25)$$

and (2.22) and (2.23) give conditions for h to be (single-valued) lower semi-continuous and to have the infimum in (2.24) attained. This has of course, the usual implications for duality theory. See Gwinner (1977) and Levine and Pomerol (1979) for further details and in particular discussion of the relationships between the closure of the epigraph of the dual value function and the existence of Lagrange multipliers for (P) in (1.5).

One may apply (2.22) and (2.23) to semi-infinite programming pairs such as in Charnes et al (1963) and derive conditions ruling out a *duality gap* (Duffin and Karlovitz (1965)) either from the primal or the dual.

Note that in (2.16) or (2.23) 0 may be replaced by any point y_0 in R(H) and the hypothesis that $0 \in R(H)$ abandoned. Indeed it is shown in Borwein (1977a) that when H is closed $\mathrm{rec}\ H^{-1}(y_0)$ is independent of y_0. Thus one could define a recession relation in analogy to the recession function of convex analysis.

2.3 The Inverse Function Theorem for Approximately Convex Relations

Let us, in keeping with the terminology in Michel (1974) and Pomerol (1979), call a relation Θ *approximately convex* at x_0 if θ satisfies

$$\Theta(x) = f(x) + H(x) , \tag{2.26}$$

where f is a single-valued continuously Fréchet differentiable mapping and H is a LSC (on D(H)) convex relation, in a neighbourhood of x_0. In this case there is an associated "derivative"

$$\nabla\Theta(x_0)(h) = \nabla f(x_0)(h) + f(x_0) + H(x_0+h) \tag{2.27}$$

where $\nabla f(x_0)$ is the derivative of f at x_0. These approximately convex relations admit much simpler convex approximates than is usual for more general relations. Indeed the heart of the next theorem can be established, using considerably more terminology, in the general framework of Dolecki (1981) and Dolecki and Rolewicz (1979). The more direct relationships indicated below seem to be limited to approximately convex relations.

Example 6:

(i) If $H(x) = h(x) + P$ where h is P convex one is in the setting of Michel (1974) and Pomerol (1979).

(ii) If $H(x) = \begin{cases} K & x \in C \\ \emptyset & x \notin C \end{cases}$ and K is a convex cone and C a convex set one is in Robinson's (1976a)and Du'ong and Tuy's (1978) setting, while if K is only a convex set one is in a setting that various authors have considered.

(iii) If f is zero one is again considering arbitrary convex relations.

THEOREM 15: Suppose Θ *is a closed approximately convex relation at* x_0 *and that* $\nabla\Theta(x_0)$ *is open at* $y_0 \in \Theta(x_0)$. *Then there exist* $K > 0$, $\varepsilon > 0$ *such that*

$$d(y \mid \Theta(x)) \geq K\, d(x \mid \Theta^{-1}(y)) \tag{2.28}$$

whenever $||x-x_0|| \leq \varepsilon$ *and* $||y-y_0|| \leq \varepsilon$.

PROOF: The proof is a little delicate. We sketch its outline. Fix $y \in Y$. We may assume that $y_0 = 0, x_0 = 0$ on translating Θ. Define

$$P(h) = \nabla\Theta(x_0)^{-1}(\nabla f(x_0)h + f(x_0) + y - f(x_0 + h)) \tag{2.29}$$

and, for $\delta > 0$, set

$$Q(h) = P(h) \cap B_\delta(0) \ . \tag{2.30}$$

Since $\nabla\Theta(x_0)$ is open at 0, $\nabla\Theta(x_0)^{-1} \cap B_\delta$ is k-locally Lipschitz at 0 using (1.31) or (2.10) and, for sufficiently small $\varepsilon > 0$, Q is a multivalued contraction mapping on $B_\varepsilon(0)$ since f is strictly differentiable. Also, one can show that for x in $D(\Theta)$

$$d(y \mid \Theta(x)) \to 0 \quad \text{as } x \to x_0 \ , \ y \to y_0 \tag{2.31}$$

since Θ is LSC on $D(\Theta)$ at x_0, and as $\nabla f(x_0)x + H(x) \subset \nabla\Theta(x_0)(x)$,

$$d(x \mid Q(x)) \leq k\, d(y \mid \Theta(x)) \ . \tag{2.32}$$

It follows from (2.30) and (2.32) that for δ sufficiently small and $||x-x_0||, ||y-y_0|| < \varepsilon$ one has a fixed point x(y) with

$$x(y) \in Q(x(y)) \quad , \quad d(x(y), x) \leq K\, d(y \mid \Theta(x)) \quad , \tag{2.33}$$

for some $K > 0$ (independent of x and y) as promised by the multivalued contraction mapping theorem (Robinson (1976b or Tuy and Du'ong (1978)). Note that P, and so Q, does have closed convex images since the continuity of f and closure of Θ forces closure of H and hence of $\nabla\Theta(x_0)$. Now a fixed point x(y) of Q satisfies

$$y \in f(x_0 + x(y)) + H(x_0 + x(y)) = \Theta(x_0 + x(y)) \tag{2.34}$$

and in conjunction with (2.32) this shows that (since $x_0 = 0$)

$$d(x \mid \Theta^{-1}(y)) \leq K\, d(y \mid \Theta(x)) \tag{2.35}$$

as required. □

Let us recall that the *tangent cone* to a set E at x_0 is defined by

$$T(E,x_0) = \{x \mid t_n(x_n - x_0) \to x,\ x_n \to x_0\ \ x_n \in E,\ t_n \geq 0\}. \quad (2.36)$$

Let us call Θ *regular* at (x_0, y_0) if (2.28) holds. Then this corresponds with the usage in Ioffe (1972) and we can derive the following extended Ljusternik theorem (Luenberger (1969)).

THEOREM 16: Suppose Θ *is approximately convex and regular at* $(x_0,0)$. *Then if* $0 \in \Theta(x_0)$

$$\nabla\Theta(x_0)^{-1}(0) \subset T(\Theta^{-1}(0),\ x_0)\ . \quad (2.37)$$

PROOF: Suppose $0 \in \nabla\Theta(x_0)(h)$. Since $0 \in \nabla\Theta(x_0)(0)$ and $\nabla\Theta(x_0)$ is convex one has

$$0 \in \nabla\Theta(x_0)(th) \quad (2.38)$$

for $0 \leq t \leq 1$. Also

$$\begin{aligned} d(0 \mid \Theta(x_0 + th)) &\leq \|f(x_0 + th) - f(x_0) - \nabla f(x_0)(th)\| \\ &\quad + d(o \mid \nabla\Theta(x_0)(th)) \\ &= o(t) \end{aligned}$$

since f is (strictly) differentiable at x_0. By the regularity of Θ at $(x_0,0)$ one has

$$d(x_0 + th \mid \Theta^{-1}(0)) = o(t)\ , \quad (2.39)$$

and it is easily verified that

$$h \in T(\Theta^{-1}(0),\ x_0)\ .$$

□

This type of approximation result can be extended as in Oettli (1980), Dolecki (1981) and Tuy and Du'ong (1978) to cases where $\nabla f(x_0)(\cdot)$ is a convex approximate to f in a sufficiently strong sense.

3. MATHEMATICAL PROGRAMMING

3.1 A General Theorem of the Alternative

Let $K:X \to Y$, $H:X \to Z$ be convex relations. Let Y^* and Z^* be the topological duals of Y and Z.

THEOREM 17: Suppose that H *is open at* 0, *that* $x_0 \in H^{-1}(0)$ *and that* K *is* LSC *at* x_0. *Suppose also that for some* x_1, $\text{int } K(x_1) \neq \emptyset$. *Then of the following exactly one has a solution:*

$$\text{(i)} \quad 0 \in H(x) \quad , \quad 0 \in \text{int } K(x) \; ; \qquad (3.1)$$

$$\text{(ii)} \quad y' \, K(x) + z' \, H(x) \geq 0 \; , \qquad (3.2)$$

for all x *in* X *and some* y' *in* Y*, z' *in* Z* *with* y' *non-zero.*

PROOF: Let $G(x) = \text{int } K(x)$. It is straightforward that G is a convex relation. Also $\text{int } D(K) \subset D(G)$ since for $0 < t \leq 1$ and any x in X

$$t \text{ int } K(x_1) + (1 - t) K(x) \subset \text{int } K(tx_1 + (1 - t)x). \qquad (3.3)$$

Now given x_2 in int D(K) we may find x in D(K) and $0 < t < 1$ with $x_2 = tx_1 + (1 - t)x$. Then (3.3) shows that x_2 is in D(G). Let y_0 lie in $G(x_0)$. Since G is LSC at x_0 and strongly open at (x_0, y_0) Proposition 5 shows that

$$F = (G, H) \qquad (3.4)$$

is open at the image point $(y_0, 0)$. Thus R(F) is a convex set with non-empty interior and the failure of (3.1) says that (0,0) does not lie in R(F). The Mazur separation theorem guarantees the existence of y' in Y* and z' in Z*, not both zero, such that

$$y'(y) + z'(z) \geq 0 \quad \forall\, y \in G(x) \; , \; z \in H(x) \; . \qquad (3.5)$$

Since x_0 lies in int D(G) and H is open at 0 a standard argument shows y' is non-zero. It remains to show that G can be replaced by K in (3.5). Let y lie in K(x) and z in H(x) for some fixed x. Set

$$z_\varepsilon = (1-\varepsilon)z \; , \; y_\varepsilon = \varepsilon y_0 + (1-\varepsilon)y \; , \; x_\varepsilon = \varepsilon x_0 + (1-\varepsilon)x, \qquad (3.6)$$

for $0 < \varepsilon < 1$. Then z_ε lies in $H(x_\varepsilon)$ since H is convex while y_ε lies in $G(x_\varepsilon)$ on applying (3.3) with x_0 replacing x_1. Thus (3.5) shows that

$$y'(y_\varepsilon) + z'(z_\varepsilon) \geq 0 \; . \qquad (3.7)$$

If we let ε tend to zero we establish (3.3). The converse is easy. □

We say that a relation M is S-*convex* if M + S is a convex relation, Borwein (1977a). An immediate application of Theorem 17 is with K = M + S when int S is non-empty. Then (3.1) and (3.2) become, when S is a cone,

$$\text{(i)}' \quad 0 \in M(x) + \text{int } S \quad , \quad 0 \in H(x) \quad , \tag{3.8}$$

$$\text{(ii)}' \quad s^{+}M(x) + z'H(x) \geq 0 \tag{3.9}$$

for some non-zero s^+ in S^+ and z′ in z*. Here $S^+ = \{s^+ \in Y^* | s^+(s) \geq 0, s \in S\}$ is the *dual cone* to S.

This alternative was established in Borwein (1977a) and similar results may be found in Oettli (1980); and used to prove weak Lagrange multiplier theorems extending those in Oettli (1980) and in Craven and Mond (1973).

One may also treat Pareto alternatives in this framework, extending results in Borwein (1977b) and elsewhere.

3.2 The Lagrange Multiplier Theorem

We next establish a Lagrange multiplier theorem for vector convex programs and derive various consequences.

PROPOSITION 18 (Borwein (1980b): Let K:X → Y *be a convex relation. Suppose that* Y *is partially ordered by an order complete normal cone* S. *Suppose that* K *is* LSC *at* 0 *and*

$$K(0) \geq 0 \ . \tag{3.10}$$

Then there exists a continuous linear operator T:X → Y *such that*

$$T(x) \leq K(x) \qquad (\forall x \in X) \ . \tag{3.11}$$

This result is established in Borwein (1980b) from the Hahn-Banach theorem (Day (1973)) which holds exactly when S is order-complete (Elster and Nehse (1978)).

THEOREM 19: Let H:X → Z *be a convex relation. Let* f:X → Y ∪{∞} *be* S-*conv with respect to an order complete normal cone* S. *Consider*

$$\text{(P)} \quad \mu = \inf_S \{f(x) \mid 0 \in H(x)\} \quad (\mu \in Y). \tag{3.12}$$

Suppose H *is open at zero and* f *is continuous at* x_1 *for some* x_1 *in* $H^{-1}(0)$. *Then there exists a continuous linear operator* $T:Z \to Y$ *such that*

$$\mu \leq f(x) + T\,H(x) \qquad (\forall x \in X). \tag{3.13}$$

Moreover the set of all such T *is equicontinuous. Also* h *given by*

$$h(u) = \inf_S \{f(x) \mid u \in H(x)\} \tag{3.14}$$

is well defined convex and continuous in some neighbourhood of 0.

PROOF: This is a specialization of the Lagrange multiplier result given by the author in Borwein (1980b). Define

$$K(u) = H_f\, H^{-1}(u) - \mu \tag{3.15}$$

where $H_f(x) = f(x) + S$. Then $K:Z \to Y$ is a convex relation and $K(0) \geq 0$ by (3.12). Since H is open at $(x_1, 0)$ H^{-1} is LSC at $(0, x_i)$and also H_f is LSC at $(x_1, f(x_1))$. Then Proposition 1 shows that K is LSC at $(0, f(x_1) - \mu)$ and hence at 0. Proposition 18 now guarantees the existence of a continuous linear T with

$$K(u) \geq -T(u). \tag{3.16}$$

Equivalently,

$$H_f(x) + Tu \geq \mu \quad (x,u) \in G(H) \tag{3.17}$$

or

$$f(x) + T\,H(x) \geq \mu \quad x \in X. \tag{3.18}$$

Since K is LSC at 0 (3.16) shows that

$$h(u) = \inf K(u) \tag{3.19}$$

is well defined and finite (and so convex) on some neighbourhood U of zero. Moreover Propositions 1 and 2 show that H_h is LSC at zero and thus Lemma 4 shows h is continuous at zero since S is normal. Finally, the solutions T of (3.13) are exactly

$$-\partial h(0) = -\{T \in B[Z,Y] \mid T(u) \leq h(u) - h(0)\}. \tag{3.20}$$

This last set is the *subgradient* of h at zero. Again the proof of Lemma 4 shows that $\partial h(0)$ is equi-continuous. □

Define the *vector Lagrangian* L by

$$L(T) = \inf f(x) + T\,H(x). \qquad (3.21)$$

Then, as in the scalar case, L is S-concave and one always has *weak duality*

$$\sup_T L(T) \leq h(0). \qquad (3.22)$$

The considerations of the last proof show that the openness of H at zero in conjunction with some continuity assumption on f is a *constraint qualification* which ensures that *strong duality* holds:

$$\sup_T L(T) = h(0) \qquad (3.23)$$

with attainment on the left hand side of (3.23). Indeed T is a solution in (3.23) exactly when

$$T \in -\,\partial h(0). \qquad (3.24)$$

It seems much harder than it is in the scalar case to give semi-continuity conditions for (3.23) to hold rather than (3.24).

Equation (3.24) (Borwein, Penot and Thera (1980)) shows that multipliers may still be interpreted as marginal prices for the program (P). In many situations the multiplier is actually unique and is the Gateaux or Fréchet derivative of -h at zero (Borwein (1980a).

Let us also observe that if the infimum in (3.12) is attained by a feasible point x_0 one actually has the *complementary slackness* condition

$$\min T\,H(x_0) = 0, \qquad (3.25)$$

which reduces to the standard slackness condition when H is given by (1.4) or (1.15).

Example 6: The Hahn-Banach theorem is a special case of Theorem 19 in the algebraic core topology. Let X be a vector space, M a subspace of X, p a sublinear operator and T a linear operator dominated by p on M. Then

$$0 = \inf_S \{p(x) - T(x) \mid 0 \in x - M\} \qquad (3.26)$$

and application of Theorem 19 provides an appropriate multiplier. Note that the constraint is open because in the algebraic core topology a convex relation is, not surprisingly, open at any core point of its range.

Remark: Suppose in Theorem 19 that Z is replaced by Z_0 equal to span R(H). Then one may establish the existence of a Lagrange multiplier assuming only that H is open at zero in Z_0 or equivalently that H is *relatively open* at zero. The multiplier T will only be defined on Z_0. If S has non-empty interior (as is true in the scalar case) T will have a continuous extension to all of Z (Peressini (1967)). In any case T will have a linear extension to all of Z. Thus in the algebraic case it suffices that zero be a relative core point of R(H) as in Kutateladze (1979) or Oettli (1980).

Finally, some remark is in order about which cones are order complete. Considerable detail may be found in Jameson (1970), Peressini (1967) or Schaefer (1971). In short, however, the only finite dimensional cones which are order complete are those with linearly independent generators (Peressini (1967). In an infinite dimensional setting all dual Banach lattices are order complete as are many other spaces.

3.3 Vector Perturbational Duality

Now we examine the relationship between our Lagrange duality and the vector version of the Fenchel-Rockafellar bifunctional duality given in Ekeland and Temam (1976) or in Rockafellar (1970, 1974b). As we will see below nothing is lost in our framework and certain results may be strengthened. Moreover, it is the author's contention that the constraint structure of (3.12) is more intuitive and easier to work with than that of (3.27).

Let $f: X \times Z \to Y \cup \{\infty\}$ be an S-convex (bi)function.

Define

$$h(u) = \inf_S f(x,u), \quad \mu = \inf f(x, 0) \tag{3.27}$$

where $h(u) = -\infty$ if the infimum fails to exist. We may rewrite (3.12) in this form by setting

$$f(x,u) = \begin{Bmatrix} f(x) & u \in H(x) \\ +\infty & u \notin H(x) \end{Bmatrix}. \tag{3.28}$$

Conversely if we set

$$u \in K(x, y) \qquad f(x, u) \leq y \tag{3.29}$$

we may write

$$h(u) = \inf_S \{y \mid u \in K(x, y)\}. \tag{3.30}$$

Then K is a convex relation since its graph is essentially the epigraph of f. The usual regularity condition on f is that $f(x_0, \cdot)$ should be continuous at zero for some x_0 (Ekeland and Temam (1976)). This is easily seen to be equivalent (when S is normal) to the openness of $K(x_0, \cdot)$ at zero (Propositions 1, 2 and Lemma 4). In turn this implies that K is open at zero, but in general the converse is false. Thus the openness of K is a weaker requirement to impose in (3.27) or (3.29). A related result is given in Robinson (1976a) and Rockafellar (1974b). Now (3.13) becomes

$$\begin{aligned} h(0) &= \inf_S \{y + Tu \mid f(x, u) \leq y\} \\ &= -\sup_S \{-Tu - f(x, u)\} = -f^*(0, -T), \end{aligned} \tag{3.31}$$

where f* is by the definition the *vector conjugate* of f defined in Zowe (1975a). Now (3.31) and (3.23) give the conjugate duality result obtained by Zowe and others under stronger regularity conditions. We will have more to say about regularity conditions below.

Let us rederive the ordinary Lagrange multiplier theorem in our framework.

COROLLARY 20: Consider the ordinary convex program given by (1.1). Suppose that g *is continuous at some point* x_0 *in* int C *with* $h(x_0) = 0$ *and* $g(x_0) < 0$. *Then there exist non-negative numbers* $\lambda_1, \ldots, \lambda_m$ *and real numbers* $\mu_1, \ldots, \mu_k$ *such that*

$$f(x) + \sum_{i=1}^{m} \lambda_i g_i(x) + \sum_{j=1}^{k} \mu_j h_j(x) \geq p \qquad \forall x \in C. \tag{3.32}$$

PROOF: We have supposed implicitly that h is everywhere defined and that C lies in the domain of each function. Then the associated relation H of (1.4) is open considered as a mapping between X and $R^m xR(h)$ by Proposition 5. Indeed H_h is open since R(h) is a finite dimensional subspace and h is surjective while H_g is LSC at x_0 and strongly open. Then (3.13) becomes

$$f(x) + \sum_{i=1}^{m} \lambda_i(g_i(x) + R_+) + \mu\, h(x) \geq p \qquad \forall x \in C; \tag{3.33}$$

for some $\lambda_1, \ldots, \lambda_m$ and some $\mu \in R(h)^*$. Since we may suppose that μ is extended to R^k, $\mu = (\mu_1, \ldots, \mu_k)$ for some μ_j in R and now (3.33) is easily seen to show that $\lambda_i \geq 0$ and (3.32) holds. □

The analogous considerations apply to more general mixtures of inequality and equality constraints. If $X = R^n$ the continuity assumptions on g and C may be relativized.

Various other duality calculations are given in Borwein (1980b). One may also consider *augmented Lagrangians* (Rockafellar (1974a)) and *limiting Lagrangians* (Duffin (1973); Duffin and Jeroslow (1978)) profitably in this framework.

3.4 The Krein-Rutman Theorem

We prove a form of the Krein-Rutman theorem given in Peressini (1967) which will be applied in section 4 to produce Kuhn-Tucker conditions.

Let us define the S-*dual* cone to a set C in X by

$$C^S = \{T \in B[X,Y] \mid T(C) \subset S\}. \tag{3.34}$$

When $S = R_+$, $Y = R$, C^S is just C^+.

THEOREM 21 (Borwein (1980b)): Let A *be a linear mapping between* X *and* Z *and let* K_1 *and* K_2 *be convex cones in* X *and* Z *respectively. Set*

$$H_A(x) = \begin{Bmatrix} A(x) - K_2 & x \in K_1 \\ \emptyset & x \notin K_1 \end{Bmatrix} . \tag{3.35}$$

Suppose that S *is an order complete normal cone in* Y. *Then*

$$(K_1 \cap A^{-1}(K_2))^S = K_1^S + K_2^S A , \tag{3.36}$$

whenever H_A *is open at zero.*

PROOF: The right hand side of (3.36) always lies in the left hand side. The converse is established by considering the convex program

$$0 = \inf \{T_0(x) \mid 0 \in H_A(x)\} \tag{3.37}$$

where T_0 is any member of $(K_1 \cap A^{-1}(K_2))^S = (H_A^{-1}(0))^S$. The Lagrange multiplier T then satisfies

$$T_0 + TA \in K_1^S \quad , \quad - T \in K_2^S \tag{3.38}$$

and (3.36) follows. □

In particular H_A is open if either (a) $R(H_A) = A(K_1) - K_2$ is Z in a situation in which Theorem 8 holds or (b) in more general spaces if $A(K_1) \cap \text{int } K_2 \neq \emptyset$. The scalar version of (a) is obtained in Banach space by Kurcuysz (1976), while (b) is the classic Krein-Rutman result given by Peressini (1967).

A similar result can be proven in similar ways assuming that K_1 and K_2 are only convex sets. This extends a result in Ben Tal and Zowe (1980) and elsewhere.

3.5 A Gauvin-Tolle Type Result

We show that for convex programming openness constraints is the weakest constraint qualification with various pleasant properties. In the next section this will be seen to imply the same type of behavior for differentiable programming as in Gauvin and Tolle (1977).

THEOREM 22: Consider the scalar convex program

$$(P) \quad h(u) = \inf \{f(x) \mid u \in H(x)\} \tag{3.39}$$

where $f:X \to R$ *is a continuous function and* $H:X \to Y$ *is a convex relation. Consider the following statements.*

(a) H *is open at zero and* h(o) *is finite.*

(b) The value function h *is continuous at zero.*

(c) The Lagrange multiplier set at zero is equicontinuous and non-empty.

Then the following relationships hold:

(i) (a) => (b) => (c).

(ii) If X *is locally convex, complete, metrizable and either* H *has a* CS-*closed graph and* Y *is complete, metrizable or* H *has a closed graph and* Y *is barreled then (a) and (b) are equivalent.*

(iii) If in addition R(H) *is supportable at boundary points, as occurs if* R(H) *is finite dimensional or if* H *is open at some point* y_0*, then (a), (b) and (c) are equivalent.*

PROOF:

(i) Follows from Theorem 19 and (3.20).

(ii) In either case the openness of H is equivalent to $0 \in \text{int } R(H)$ which since dom f = X and h(o) is finite is equivalent to $0 \in \text{int}(\text{dom } h)$.

(iii) By (ii), if H is not open at zero, zero lies in the boundary of R(H). Let $\Phi \in Y^*$ be a non-zero vector with

$$\Phi(R(H)) \geq 0. \tag{3.40}$$

If λ is a Lagrange multiplier for (P) at zero one has

$$f(x) + \lambda\, H(x) \geq h(0). \tag{3.41}$$

It follows from (3.40) and (3.41) that

$$f(x) + (\lambda + n\Phi)\, H(x) \geq h(0)\ , \tag{3.42}$$

and as $\lambda + n\Phi$ lies in the Lagrange multiplier set for each $n \geq 0$ this set is unbounded. Thus (c) implies (a). □

If R(H) is finite dimensional one may add to the above equivalences the condition that h is finite in some neighbourhood of zero.

Suppose that H has a closed graph, that D(H) is locally compact, and that the ε-solution mapping S_ε given by

$$S_\varepsilon(u) = \{x \mid u \in H(x), \quad f(x) \leq h(u) + \varepsilon\} \tag{3.43}$$

has $S_\varepsilon(o)$ compact. Then the continuity of h forces *upper semi-continuity* of S_ε at zero (in ε and u). Thus the upper semi-continuity of S_ε may be added as an equivalence in (iii).

If actually H is strongly open at $(0,x_0)$ $S_\varepsilon(u)$ can be shown to be lower semi-continuous (in ε and u) at zero.

4. VECTOR KUHN-TUCKER THEORY

4.1 Karush-Kuhn-Tucker Conditions in Guignard's Form

In this section we consider the non-linear program

$$\text{(P)} \quad \min_S\{f(x) \mid g(x) \in B \ , \ x \in C\} \ , \tag{4.1}$$

where $f:X \to Y$, $g:X \to Z$ are Fréchet or Hadamard differentiable at X_0 (as in Craven (1978)) and B and C are arbitrary sets. Let $A = g^{-1}(B) \cap C$ and suppose x_0 lies in A and $f(x_0)$ minimizes (vectorially) f over A.

For an arbitrary set E recall that the *tangent cone* to E at x_0 is defined by (2.35) and the *pseudo-tangent* cone to E at x_0 by

$$P(E, x_0) = \overline{\text{co}}\ T(E, x_0) \ . \tag{4.2}$$

Let the *linearizing cone* K be defined by

$$K = \nabla g(x_0)^{-1}(P(B, g(x_0))) \ , \tag{4.3}$$

and let G be any closed convex cone in X such that

$$G \cap K \subset P(A, x_0) \ . \tag{4.4}$$

Define H_G by

$$H_G(x) = \begin{Bmatrix} \nabla g(x_0)(x) - P(B,g(x_0)), & x \in G \\ \emptyset & x \in G \end{Bmatrix} \ . \tag{4.5}$$

THEOREM 23:

(a) If H_G *is open at zero and* S *is normal and order complete then*

$$(G \cap K)^S = G^S + P(B, g(x_0))^S \nabla g(x_0) \quad (4.6)$$

(b) If (4.6) *holds and* S *is closed then one obtains the necessary condition*

$$\nabla f(x_0) - T \nabla g(x_0) \in G^S \quad (4.7)$$

for some T *in* $P(B, g(x_0))^S$.

PROOF:

(a) This is the Krein-Rutman theorem for H_G. Thus in a complete metrizable setting it holds if $\nabla g(x_0)(G) - P(B, g(x_0)) = Z$.

(b) Now (4.4) shows that

$$P(A, x_0)^S \subset G^S + P(B, g(x_0))^S \nabla g(x_0) \quad (4.8)$$

Thus, once we show that $\nabla f(x_0)$ lies in $P(A, x_0)^S$ we will obtain (4.7). Let h lie in $T(A, x_0)$. Then

$$h_n \to h,\ h_n = t_n(a_n - x_0), t_n > 0,\ a_n \to x,\ a_n \in A.$$

It follows that $x_0 + t_n^{-1} h_n$ is feasible and so

$$t_n[f(x_0 + t_n^{-1} h_n) - f(x_0)] \subset S. \quad (4.9)$$

The differentiability assumptions imply that

$$\nabla f(x_0)(h) = \lim t_n[f(x_0 + t_n^{-1} h_n) - f(x_0)] \quad (4.10)$$

and since S is closed

$$\nabla f(x_0)(T(A, x_0)) \subset S. \quad (4.11)$$

Since $\nabla f(x_0)$ is linear and continuous one may replace $T(A, x_0)$ by $P(A, x_0)$ in (4.11). □

For more details about the tangent cones and the scalar conditions in this form we refer to Guignard (1969) or Borwein (1978). Even in the scalar case Theorem 23(b) slightly extends the necessary results in these papers while (a) is entirely new.

Ideally one wishes to take $G = P(C, x_0)$ or even larger. The Kuhn-Tucker constraint qualification of Kuhn and Tucker (1951) ensures that this is possible.

Remark: Suppose we take f and g linear in (4.1) and take B a closed convex cone with C equal to X and $x_0 = 0$. Then one verifies that one may take G = X in (4.4) and that (4.6) becomes

$$(g^{-1}(B))^S = B^S g . \tag{4.12}$$

Then (4.1) becomes

$$g(x) \in B \quad \text{implies} \quad f(x) \in S \tag{4.13}$$

and (4.7) is

$$f(x) = Tg(x) \quad , \quad T(B) \subset S . \tag{4.14}$$

Thus one sees that Theorem 23 includes the Farkas lemma in Craven (1978; 1980) and in Elster and Nehse (1978). It is shown in Nehse (1978) that, even assuming Slater's condition holds for g and B, the Farkas lemma only holds when S is order complete (Day (1973)). The same, therefore, is true of the Kuhn-Tucker conditions unless one places more restrictions such as surjectivity on the constraints (Craven (1980) and Ritter (1970)). This is also true of the Krein-Rutman Theorem.

In the case that $S = R_+$, Y = R (4.6) is equivalent to the closure of $G^+ + P(B, g(x_0))^+ \nabla g(x_0)$. This can be ensured either by making G and $P(B, g(x_0))$ polyhedral or by applying the Dieudonné closure condition of Theorem 13.

Sufficiency conditions may be given for (4.7) to guarantee the minimality of a feasible x_0. These are essentially unchanged from those given in Borwein (1978) and in Guignard (1969) for the scalar case.

4.2 Karush-Kuhn-Tucker Conditions in Robinson's Form

We now consider a program in Banach space of the form

$$\text{(P)} \quad \min_S \{f(x) + m(x) \mid 0 \in g(x) + H(x)\} \tag{4.15}$$

where $\Theta(x) = g(x) + H(x)$ is approximately convex at x_0 as in Theorem 15 and $f(x) + m(x) + S$ is also approximately convex at x_0. We assume that $0 \in \theta(x_0)$, that the minimum occurs at x_0 and that H has a closed graph.

THEOREM 24: Suppose that S *is order-complete closed and normal and that*

$$\nabla\Theta(x_0)(h) = \nabla g(x_0)(h) + g(x_0) + H(x_0 + h) \tag{4.16}$$

is open at zero.

Then the following necessary condition for a minimum holds at x_0:

$$\begin{aligned} 0 \leq_S \nabla f(x_0)(h) + m(x_0 + h) - m(x_0) + \\ T(\nabla g(x_0)(h) + g(x_0) + H(x_0 + h)) \end{aligned} \tag{4.17}$$

for some continuous linear T.

PROOF: The extended Ljusternik theorem of Theorem 16 shows that if $0 \in \nabla\Theta(x_0)(h)$ then $h \in T(\Theta^{-1}(0), x)$. Much as in the last theorem we have $h_n \to h$, $h_n = t_n^{-1}(x_n - x_0)$; $0 \in \Theta(x_n)$ and $x_n \to x_0$, $t_n > 0$. Then

$$\frac{f(x_n) - f(x_0)}{t_n} + \frac{m(x_n) - m(x_0)}{t_n} \geq_S 0 \ . \tag{4.18}$$

Now as before

$$\frac{f(x_n) - f(x_0)}{t_n} \to \nabla f(x_0)(h) \ , \tag{4.19}$$

while, for large n,

$$\frac{m(x_0 + t_n h_n) - m(x_0)}{t_n} \leq_S m(x_0 + h_n) - m(x_0) \tag{4.20}$$

since m is S-convex. Since S is closed and m is continuous we may derive from (4.18), (4.19) and (4.20) that

$$\nabla f(x_0)(h) + m(x_0 + h) - m(x_0) \geq_S 0 \ . \tag{4.21}$$

Thus one has

$$0 = \min_S \{\nabla f(x_0)(h) + m(x_0 + h) - m(x_0) \mid 0 \in \nabla\Theta(x_0)(h)\}. \tag{4.22}$$

Now (4.22) is a convex program which has an open constraint and a continuous objective. Theorem 19 now produces the desired result. □

In addition, one has as a consequence of Theorem 22 that openness of $\nabla\Theta(x_0)$ at 0 is essentially the weakest constraint qualification yielding a bounded set of multipliers in (4.17) and a continuous value function for (P) of (15.8). This extends some of the results in Gauvin & Tolle (1977), Kurcyusz & Zowe (1979), Lempio & Zowe (1980) and elsewhere.

For $S = R^+$, $Y = R$ the result with H induced by a convex function is due to Pomerol (1979). For m = 0 and H a convex cone it is due to Robinson (1976). One may think of (4.16) as a generalized *Mangasarian-Fromovitz* constraint qualification (Mangasarian-Fromovitz (1967)).

One may also use Theorem 16 to derive second order conditions for (P) of (4.15) and analogous to those in Ben Tal and Zowe (1980) and extending those in Maurer and Zowe (1979), Lempio and Zowe (1980). The *second order variation* for θ looks something like

$$\nabla g(x_0)(d) + r\,\nabla g(x_0)(d_1) + \frac{1}{2}\nabla^2 g(x_0)(d_1, d_1) + M(x_0 + rd_1 + d)\ . \tag{4.23}$$

Here d and r are variable while x_0 and d_1 are fixed.

4.3 Constraint Regularization

We conclude with an example showing a new interpretation of the LSC hull of the value function h of (3.14). Suppose that in

$$(P) \quad h(u) = \inf_S \{f(x) \mid u \in H(x)\}, \tag{4.24}$$

H is not open at zero. Let U be an arbitrary neighbourhood of zero in Z and define

$$H_U(x) = H(x) - U\ . \tag{4.25}$$

Then H_U is strongly open at zero. Consider

$$(P_U) \quad P_U = \inf_S \{f(x) \mid 0 \in H_U(x)\}. \tag{4.26}$$

This has infimal value

$$P_U = \inf_{u \in U} h(u)\ . \tag{4.27}$$

Assume that P_U is finite for some U. Then Theorem 19 applies to (P_U) and we derive that

$$\inf_{u\in U} h(u) = \inf[f(x) + T_U(H_U(x))] \tag{4.28}$$

for some continuous linear T_U. Thus

$$\begin{aligned}\sup_U \inf_{u\in U} h(u) &= \sup_U \inf[f(x) + T_U(H_U(x))] \\ &\le \sup_T \inf_X [f(x) + T\,H(x)] \\ &= \sup_T L(T)\ .\end{aligned} \tag{4.29}$$

The left hand side of (4.29) is the *order lower semi-continuous hull* of h at zero, $\bar{h}(0)$, while the right hand side is easily seen to be equal to the *second conjugate* of h at zero, $h^{**}(0)$, as in Zowe (1975a). Thus we have shown that

$$\bar{h}(0) \le h^{**}(0) \le h(0) \tag{4.30}$$

whenever $\bar{h}(0)$ is finite. In restricted cases such as the scalar case equality actually holds in (4.30). More generally, however, linear continuous operators T may exist with

$$\bar{T}(0) \equiv -\infty,\ T(0) = T^{**}(0)\ , \tag{4.31}$$

as the example of the Hilbert matrix in Peressini (1967) shows. The positive result is as follows.

THEOREM 25: Suppose that $h: X \to Y$ *is* S-*convex and finite at zero. Suppose that* S *is an order-complete, normal, Daniell lattice ordering with interior. Then*

$$\bar{h}(0) = h^{**}(0)\ . \tag{4.32}$$

PROOF: By (4.30) $h^{**}(0)$ is finite. Also as above

$$\sup_T L(T) = h^{**}(0)\ . \tag{4.33}$$

For each finite subset F of B[X,Y] let

$$Y_F = \max_F L(T)\ . \tag{4.34}$$

Then $\{Y_F\}$ is an increasing net with supremum $h^{**}(0)$. A Daniell order is one in which the supremum is actually the topological limit (Borwein (1980a); Penot (1975-6)). Hence one can find F such that

$$Y_F \geq h^{**}(0) - s_0 \tag{4.35}$$

where s_0 is an interior point of S. Since each T in F is continuous one can find a neighbourhood U with

$$T(U) \leq s_0 \tag{4.36}$$

for each T in F. Then, for these T and that U,

$$\inf_{x,u} f(x) + T(H(x) - U) \geq L(T) - s_0 \tag{4.37}$$

and so combining (4.35) and (4.37)

$$\sup_T \inf_{x,u} [f(x) + T(H(x) - U)] \geq h^{**}(0) - 2s_0 \ . \tag{4.38}$$

Finally (4.29) shows that

$$\bar{h}(0) \geq h^{**}(0) - 2s_0 \ . \tag{4.39}$$

Since S is linearly closed and s_0 is arbitrary in int S we have (4.32). □

Related results may be found in Borwein, Penot and Thera (1980).

Similar considerations allow us to extend the Transposition theorem of Theorem 17 to include what Holmes calls "Tuy's inconsistency condition" (Holmes (1975)).

REFERENCES

BERGE, C. (1959). *Espaces Topologiques.* Dunod, Paris.

BLAIR, C.E., BORWEIN, J., and JEROSLOW, R.G. (September 1978). "Convex Programs and Their Closures." Management Science Series, GSIA, Carnegie-Mellon University, and Georgia Institute of Technology.

BEN-TAL, A. and ZOWE, J. (1980). "A Unified Theory of First and Second Order Optimality Conditions for Extremum Problems in Topological Vector Spaces." (To appear.)

BORWEIN, J.M. (1977a). "Multivalued Convexity: A Unified Approach to Equality and Inequality Constraints." *Math. Programming 13*, 163-180.

BORWEIN, J.M. (1977b). "Proper Efficient Points for Maximizations with Respect to Cones." *SIAM J. Control Optimization 15*, 57-63.

BORWEIN, J.M. (1978). "Weak Tangent Cones and Optimization in Banach Spaces." *SIAM J. Control 16 (3)*, 512-522.

BORWEIN, J.M. (1979). "Convex Relations and Optimization." Unpublished manuscript.

BORWEIN, J.M. (1980a). "Continuity and Differentiability of Convex Operators." Dalhousie Research Report. *Proc. L.M.S.* (To appear.)
BORWEIN, J.M. (1980b). "A Lagrange Multiplier Theorem and a Sandwich Theorem for Convex Relations." *Math. Scand.* (To appear.)
BORWEIN, J.M., PENOT, J.-P., and THERA, M. (1980). "Conjugate Vector-Valued Convex Mappings." (To appear.)
BRONSTED, A. (1972). "On the Subdifferential of the Supremum of Two Convex Functions." *Math. Scand. 31*, 225-230.
CHARNES, A., COOPER, W.W., and KORTANEK, K.O. (1963). "Duality in Semi-Infinite Programs and Some Works of Haar and Caratheodory." *Management Science 9*, 209-229.
CRAVEN, B.D. (1978). *Mathematical Programming and Control Theory.* Chapman and Hall, London.
CRAVEN, B.D. (1980). "Strong Vector Minimization and Duality." *ZAMM 60 (1)*, 1-5.
CRAVEN, B.D. and MOND, B. (1973). "Transposition Theorems for Cone-Convex Functions." *SIAM J. Applied Math. 24*, 603-612.
CRYER, C.W. and DEMPSTER, M.A.H. (1980). "Equivalence of Linear Complementarity Problems and Linear Programs in Vector Lattice Hilbert Spaces." *SIAM J. Control Optimization.* (To appear.)
DAY, M.M. (1973). *Normed Linear Spaces.* 3rd Edition, Springer-Verlag, New York.
DEWILDE, M. (1978). *Closed Graph Theorems and Webbed Spaces.* Pitman, London.
DOLECKI, S. (1981). "A General Theory of Optimality Conditions." *J. Math. Anal. Appl.* (To appear.)
DOLECKI, S. and ROLEWICZ, S. (1979). "Exact Penalties for Local Minima." *SIAM J. Control Optimization 17*, 596-606.
DUFFIN, R.J. (1973). "Convex Analysis Treated by Linear Programming." *Mathematical Programming 4*, 125-143.
DUFFIN, R.J. and KARLOVITZ, L.A. (1965). "An Infinite Linear Program with a Duality Gap." *Management Science 12*, 122-134.
DUFFIN, R.J. and JEROSLOW, R.G. (November 1978). "Lagrangean Functions and Affine Minorants." Preliminary Report.
EKELAND, I. and TEMAM, R. (1976). *Convex Analysis and Variational Problems.* North-Holland, Amsterdam.
ELSTER, K. and NEHSE, R. (1978). "Necessary and Sufficient Conditions for the Order-Completeness of Partially Ordered Vector Spaces." *Math. Nachr. 81*, 301-311.
FREMLIN, D.H. and TALAGRAND, M. (1979). "On CS_Closed Sets." *Mathematika 26*, 30-32.
GAUVIN, J. and TOLLE, J.W. (1977). "Differential Stability in Nonlinear Programming." *SIAM J. Control Optimization 15 (2)*, 296-311.
GUIGNARD, M. (1969). "Generalized Kuhn-Tucker Conditions for Mathematical Programming in a Banach Space." *SIAM J. Control*, 232-241.
GWINNER, J. (1977). "Closed Images of Convex Multivalued Mappings in Linear Topological Spaces with Applications." *J. Math. Anal. Appl. 60*, 75-86.
HIRIART-URRUTY, J.B. (1979). "Lipschitz r-Continuity of the Approximate Subdifferential of a Convex Function." Preprint (Clermont).
HOLMES, R.B. (1975). *Geometric Functional Analysis.* Springer-Verlag, New York.
HUARD, P. (1979). "Point to Set Maps and Mathematical Programming." *Math. Programming Study 10.*
IOFFE, A.D. (1979). "Regular Points of Lipschitz Functions." *Trans. A.M.S. 251*, 61-69.
IOFFE, A.D. and LEVIN, V.L. (1972). "Subdifferentials of Convex Functions." *Trans. Moscow Math. Soc. 26*, 1-72.

IOFFE, A.D. and TIKHOMIROV, V.M. (1979). *Theory of Extremum Problems.* North-Holland.
JAMESON, G.J. (1970). *Ordered Linear Spaces.* Lecture Notes in Math.141. Springer-Verlag, New York.
JAMESON, G.J. (1972). "Convex Series." *Proc. Camb. Phil. Soc. 72*, 37-47.
JEROSLOW, R.G. (April 1978 revised June and July 1978). "A Limiting Lagrangean for Infinitely-Constrained Convex Optimization in Rn." Management Science Research Report No.417r, GSIA, Carnegie-Mellon University.
KELLEY, J.L. and NAMIOKA, I. (1963). *Topological Spaces.* Springer-Verlag, New York.
KOSMOL, P. (1977). "On Stability of Convex Operators." In *Lecture notes in Economics and Optimization 157*, Springer-Verlag.
KORTANEK, K.O. (1977). "Constructing a Perfect Duality in Infinite Programming." *Applied Mathematics and Optimization 3*, 357-372.
KUHN, H.W. and TUCKER, A.W. (1951). "Nonlinear Programming." *Proceedings Second Berkeley Symposium on Mathematics, Statistics and Probability, Volume 5.* University of California Press, Berkeley, 481-492.
KURCYUSZ, S. (1976). "On Existence and Nonexistence of Lagrange Multipliers in Banach Space." *J.O.T.A. 20*, 81-110.
KURCYUSZ, S. and ZOWE, J. (1979). "Regularity and Stability for the Mathematical Programming Problem in Banach Spaces." *Appl. Math. Optim. 5*, 49-62.
KUTATELADZE, S. (1979). "Convex Operators." *Russian Math. Surveys 34 (1)*, 181-214.
LEMPIO, F. and ZOWE, J. (1980). "Higher Order Optimality Conditions." (To appear.)
LEVINE, P. and POMEROL, J. CH. (1979). "Sufficient Conditions for Kuhn-Tucker Vectors in Convex Programming." *SIAM J. Control Optimization 17.*
LUENBERGER, D.G. (1968). "Quasi-Convex Programming." *SIAM J. Appl. Math. 16*, 1040-1095.
LUENBERGER, D.G. (1969). *Optimization by Vector Space Methods.* John Wiley, New York.
MAKAROV, V.L. and RUBINOV, A.M. (1977). *Mathematical Theory of Economic Dynamics and Equilibrium.* Springer-Verlag, New York.
MANGASARIAN, O.L. (1969). *Nonlinear Programming.* McGraw-Hill, New York.
MANGASARIAN, O.L. and FROMOVITZ, S. (1967). "The Fritz John Necessary Optimality Condition in the Presence of Equality and Inequality Constraints." *J. Math. Anal. Appl. 17*, 37-47.
MAURER, H. and ZOWE, J. (1979). "Second Order Necessary and Sufficient Conditions for Infinite Dimensional Programming Problems." *Math. Programming 16*, 641-652.
MICHEL, P. (1974). "Probleme d'Optimisation Défini par des Fonctions qui Sont Sommes de Fonctions Convexes et de Fonction Dérivable." *J. Math. Pures et App. 53*, 321-330.
NEHSE, R. (1978). "The Hahn-Banach Property and Equivalent Conditions." *Comment. Math. Univ. Carolinae 19*, 165-177.
OETTLI, W. (1980). "Optimality Conditions for Programming Problems Involving Multivalued Mappings." In *Modern Applied Mathematics: Optimization and Operations Research.* (B. Korte, ed.) North-Holland, Amsterdam.
OETTLI, W. (1981). "Optimality Conditions Involving Generalized Convex Mapping". This volume, 227-238.
PENOT, J.-J. (1975-6). Extremisation et Optimisation. Lecture notes (Univ. Pau).
PERESSINI, A.L. (1967). *Ordered Topological Vector Spaces.* Harper and Row, New York.

PETERSON, E.L. (1976). "Geometric Programming." *SIAM Review 18*, 1-51.

POMEROL, J.-CH. (1979). "Application de la Programmation Convexe à la Programmation Differèntiable." *C.R. Acad. Sc. Paris 228*, 1041-1044.

RITTER, K. (1970). "Optimization in Linear Spaces, III." 184, 133-154.

ROBERTSON, A.P. and ROBERTSON, W.J. (1964). *Topological Vector Spaces*. Cambridge University Press, Cambridge.

ROBINSON, S. (1972). "Normed Convex Processes." *Trans. A.M.S. 174*, 127-140.

ROBINSON, S. (1976a). "Regularity and Stability for Convex Multivalued Functions." *Math. of Oper. Res. 1*, 130-143.

ROBINSON, S. (1976b). "Stability Theory for Systems of Inequalities, Part II: Differentiable Nonlinear Systems". *SIAM J. NUMER. ANAL. 13*, 497-513.

ROCKAFELLAR, R.T. (1967). Monotone Processes of Convex and Concave Type. American Math Society, Memoir No.77.

ROCKAFELLAR, R.T. (1970). *Convex Analysis*. Princeton University Press, Princeton.

ROCKAFELLAR, R.T. (1974a). "Augmented Lagrange Multiplier Functions and Duality in Nonconvex Programming." *SIAM J. Control 12*, 268-275.

ROCKAFELLAR, R.T. (1974b). *Conjugate Duality and Optimization*. SIAM Publications, Philadelphia.

RUDIN, W. (1973). *Functional Analysis*. McGraw-Hill, New York.

SCHAEFER, H.H. (1971). *Topological Vector Spaces*. Springer-Verlag, Berlin.

TUY, H. and DU'ONG, P.C. (1978). "Stability, Surjectivity and Local Invertibility of Nondifferentiable Mappings." *Acta Math. Vietnamica 3*, 89-103.

URSESCU,C. (1975). "Multifunctions with Convex Closed Graph." *Czech. Math. J. 7 (25)*, 438-441.

ZOWE, J. (1975a). "A Duality Theorem for a Convex Programming Problem in an Order Complete Vector Lattice." *J. Math. Anal. and Appl. 50*, 273-287.

ZOWE, J. (1975b). "Linear Maps Majorized by a Sublinear Map." *Arch. Math. 34*, 637-645.

ZOWE, J. (1978). "Sandwich Theorems for Convex Operators with Values in an Ordered Vector Space." *J. Math. Anal. and Appl. 66*, 292-296.

CONE-CONVEX PROGRAMMING: STABILITY AND AFFINE CONSTRAINT FUNCTIONS

J.M. Borwein* and H. Wolkowicz†

We present strengthened optimality conditions for the abstract convex program. In particular, we consider the special case when the range space of the constraint is finite dimensional and when the constraint function is affine. Applications include a sensitivity theorem and a generalization of Farkas' Lemma.

1. INTRODUCTION

In this paper we consider both: the general abstract convex program

$$(P_g) \quad \mu = \inf\{p(x): g(x) \in -S,\ x \in \Omega\}, \tag{1}$$

where $p: X \to R$ is a convex functional, $g: X \to Y$ is an S-convex function, S is a convex cone in Y and Ω is a convex set in X; and the special abstract convex program

$$(P_A) \quad \mu = \inf\{p(x): Ax \in -S,\ x \in \Omega\}, \tag{2}$$

where now $Ax = Lx - b$ is an affine operator and Ω is a polyhedral set, i.e., L is a linear operator, b is a vector and Ω is a finite intersection of closed half spaces.

Such programs arise in several situations. For example, the semi-infinite program

$$\mu = \inf\{p(x): h(x,t) \leq 0, \text{ for all } t \text{ in } T \text{ and } x \text{ in } \Omega\}$$

can be written in the form (P_g) if we set $g(x) = h(x,\cdot)$, Y some subspace of R^T and S the cone of nonnegative functions in Y. If T is compact, $h(\cdot,t)$ convex for each t and $h(x,\cdot)$ continuous for each x, then we can choose $Y = C[T]$, the continuous functions on T. The Lagrange multipliers are then taken from the dual space of Borel measures on T, see e.g. Borwein and Wolkowicz (1979b).

---

*Research partially supported by NSERC A4493.

†Research partially supported by NSERC A3388.

ISBN 0-12-621120-5

Optimal control problems with linear dynamics, an initial condition, a final target set and a control set can be rewritten in the form (P_A). The constraint $Ax \in -S$ represents the state hitting the target for a given control x, see e.g. Luenberger (1969).

Many problems in stability of differential equations can be phrased in terms of the feasibility of an operator equation in the space of linear operators endowed with the symmetric ordering, i.e. the ordering induced by the cone of positive semidefinite operators, see e.g. Berman (1973).

Characterizations of optimality for (P_g) with constraint qualification have been given by many authors, see e.g. Luenberger (1969). The usual constraint qualification used is Slater's condition, i.e. there exists a feasible point $\hat{x}$ whose image under g is in the topological interior of -S. This condition has been weakened by Craven and Zlobec (1980) to require only nonempty relative algebraic interior of the feasible set and nonempty interior of the cone S, while a characterization of optimality for (P_g) without any constraint qualification has been given by Borwein and Wolkowicz (1979b).

Massam (1979) considered (P_g) in the special case that Y is finite dimensional. She used the "minimal exposed face" of -S which contained the image of the feasible set to get optimality conditions which held when the minimal exposed face actually coincided with the minimal face. This was pointed out and strengthened in Borwein and Wolkowicz (1981a). In Borwein and Wolkowicz (1981b) we presented a Lagrange multiplier theorem for (P_g) which holds without any constraint qualification. This result differs from the standard Lagrange multiplier theorem in two ways. First the Lagrange multiplier is chosen from the dual cone of the minimal face (or "minimal cone") of (P_g), denoted S^f, rather than the (smaller) dual cone of S; second, the Lagrangian is restricted to $x \in \Omega \cap g^{-1}(S^f - S^f)$.

In Borwein and Wolkowicz (1979a) we presented an algorithm which "regularizes" (P_g) by finding S^f and $g^{-1}(S^f-S^f)$.

In this paper we restrict ourselves to the special case when Y is finite dimensional. We first recall and strengthen the Lagrange multiplier characterization of optimality for (P_g) given in Borwein and Wolkowicz (1981b). This is then applied to the program (P_A) to yield a Lagrange multiplier result which differs from the standard case only in that the multiplier is chosen from a larger dual cone. This differs from previous results where the multiplier is chosen from a larger dual cone *and* the variable x is restricted to a smaller set than Ω.

These results are then applied to extend the sensitivity theorem for (P_g) given in Luenberger (1969) and derive a generalization of the Farkas' lemma given in Ben-Israel (1969). These results hold without any constraint qualification.

2. PRELIMINARIES

Let us first consider the *general abstract convex program*

$$(P_g) \quad \mu = \inf\{p(x): g(x) \in -S,\ x \in \Omega\}$$

where $p: X \to R$ is a continuous convex functional (on Ω); $g: X \to R^m$ is a continuous S-*convex function* (on Ω), i.e.

$$tg(x) + (1-t)g(y) - g(tx+(1-t)y) \in S$$

for all $0 \leq t \leq 1$ and x, y in Ω; S is a convex cone, i.e. $S + S \subset S$ and $tS \subset S$ for all $t \geq 0$; Ω is a convex set in X (not necessarily polyhedral); and X is a locally convex (Hausdorff) space. The *feasible set* of (P_g) is

$$F = \Omega \cap g^{-1}(-S). \tag{3}$$

We assume throughout that

$$F \neq \emptyset. \tag{4}$$

The *polar cone* of a set K in R^m is

$$K^+ = \{\phi \in R^m: \phi x \geq 0 \text{ for all } x \text{ in } K\}, \tag{5}$$

where ϕx denotes the inner product in R^m. The *annihilator* of a set K is

$$K^{\perp} = K^{+} \cap -K^{+}.$$

Recall that for two closed convex cones K and L,

$$(K \cap L)^{+} = \overline{K^{+} + L^{+}}\,, \tag{6}$$

where $\bar{\cdot}$ denotes closure, e.g. Luenberger (1969).

We will need to use the smallest face of S which contains $-g(F)$.

Definition 1: (i) K is a *face* of a convex cone S if K is a convex cone and

$$x,\ y \in S,\ x + y \in K \quad \text{implies} \quad x,\ y \in K. \tag{7}$$

(ii) The *minimal cone* for (P_g), denoted S^f, is defined to be the smallest face of S which contains $-g(F)$. (Note that S^f is the intersection of all faces of S which contain $-g(F)$).

The following characterization of optimality for (P_g) was given in Borwein and Wolkowicz (1981b) Theorem 4.1.

THEOREM 1: For the program (P_g),

$$\mu = \inf\{p(x) + \lambda g(x) : x \in F^f\} \tag{8}$$

for some λ *in* $(S^f)^+$ *and* $F^f = \Omega \cap g^{-1}(S^f - S^f)$. *In addition, if* $\mu = p(a)$ *with* a *in* F, *then*

$$\lambda g(a) = 0 \qquad \text{(complementary slackness)} \tag{9}$$

and (8) and (9) characterize optimality of a *in* F. □

To prove the above result one shows that

$$F^f = \Omega \cap g^{-1}(S^f - S^f) = \Omega \cap g^{-1}(S^f - S), \tag{10}$$

$$g(F) \cap -\mathrm{ri}S^f \neq \emptyset \tag{11}$$

where ri denotes *relative interior*, and that

$$g \text{ is } S^f - \text{convex on } F^f.$$

Then, the standard Lagrange multiplier theorem is applied to the equivalent program

$$\mu = \inf\{p(x) : g(x) \in -S^f,\ x \in F^f\}.$$

The details are given in Borwein and Wolkowicz (1981b). Note that by the standard Lagrange multiplier theorem we mean Theorem 1 with $F^f = \Omega$ and $(S^f)^+$ replaced by S^+, see e.g. Luenberger (1969). This theorem requires that Slater's condition hold, i.e. that

$$\Omega \cap g^{-1}(-\text{int } S) \neq \emptyset , \tag{12}$$

where int denotes interior.

We will now present a version of Theorem 1 which is stronger in the sense that the Lagrange multiplier relation holds on a larger set. A similar result was given in Borwein and Wolkowicz (1979a), Theorem 6.2 and Remark 6.3. This result will be applied in Section 3. First we need the following lemma. Note that a *polyhedral function* is the maximum of a finite number of affine functions.

LEMMA 1: Consider the ordinary convex program

$$(P_o) \qquad \mu = \inf\{p(x): g^k(x) \leq 0, \ k = 1,\ldots,m, \ x \in V\}$$

where $g^k: X \to R$, $k = 1,\ldots,m$, *are continuous convex functionals* (on V) *and* V *is a polyhedral set in* X. *Suppose that there exists an* $\hat{x} \in V$ *such that* $g^k(\hat{x}) \leq 0$, $k = 1,\ldots,m$, *with strict inequality if* g^k *is not polyhedral on* V. *Then*

$$\mu = \inf\{p(x) + \lambda g(x): x \in V\} \tag{13}$$

for some $\lambda = (\lambda_k) \geq 0$. *Moreover, if* $\mu = p(a)$ *for some* $a \in F$, *then*

$$\lambda g(a) = 0 \tag{14}$$

and (13) and (14) characterize optimality of a *in* F.

PROOF. Let $I = \{k: g^k \text{ is polyhedral}\}$, $J = \{1,\ldots,m\}\setminus I$ and

$$W = \{x \in X: g^k(x) \leq 0, \text{ for all } k \text{ in } I\}.$$

Then (P_o) is equivalent to the program

$$\mu = \inf\{p(x): g^k(x) \leq 0, \text{ for all } k \text{ in } J, \ x \in V \cap W\}. \tag{15}$$

The point $\hat{x}$ now satisfies Slater's condition for this program. Therefore there exist nonnegative scalars α_k, k in J, such that

$$\mu = \inf\{p(x) + \sum_{k\in J} \alpha_k g^k(x) : x \in V \cap W\}.$$

Now since V is polyhedral,

$$V = \{x \in X: \psi^k(x) = \phi^k x - b_k \leq 0, \ k = 1,\ldots,t$$

for some ϕ^k in X*, the dual of X, and b_k in R. Moreover, for each k in I, $g^k(x) = \max_{i\in I^k} \eta_i^k(x)$ for some affine functions η_i^k and *finite* index sets I^k.

Therefore

$$\mu = \inf\{p(x) + \sum_{k\in J} \alpha_k g^k(x): \ \eta_i^k(x) \leq 0, \ i \text{ in } I^k, \ k \text{ in } I,$$

$$\psi^k(x) \leq 0, \ k = 1,\ldots,t\}.$$

This program is linearly constrained which implies that there exist further nonnegative scalars α_i^k, $i \in I^k$, k in I, and a nonnegative vector λ in R^t such that

$$\mu = \inf\{p(x) + \Sigma\alpha_i^k \eta_i^k(x) + \lambda\psi(x)\} \leq \inf\{p(x) + \alpha g(x) + \lambda\psi(x)\}$$

where $\alpha_k = \sum_{i\in I^k} \alpha_i^k$, $\alpha = (\alpha_k)$ and $\psi = (\psi^k)$. Since $\lambda\psi(x) \leq 0$ for all x in V,

$$\mu \leq \inf\{p(x) + \alpha g(x): \ x \in V\}. \tag{16}$$

The reverse inequality follows from the definition of (P_o), since $\alpha g(x) \leq 0$ for all feasible x. This proves (13). The rest of the proof, i.e. complementary slackness and sufficiency, is standard. □

If $\hat{x} \in$ ri V, then the above lemma holds with V convex but not necessarily polyhedral, see Rockafellar (1970a, Theorem 28.1).

We now present the strengthened version of Theorem 1.

THEOREM 2: Suppose that

$$g \ \text{ is } \ S^f\text{-convex on } F^H, \tag{17}$$

where

$$F \subset F^H \subset \Omega \ . \tag{18}$$

Moreover, suppose that F^H *is polyhedral. Then Theorem 1 holds with* F^f *replaced by* F^H.

PROOF. We can rewrite (P_g) as the equivalent program

$$(P_H) \qquad \mu = \inf\{p(x): g(x) \in -S^f,\ x \in F^H\}.$$

By (11) and (18), a generalized Slater's condition holds for this program and thus the standard Lagrange multiplier theorem holds. Let us show this by applying Theorem 1 and Lemma 1. Now, by Theorem 1 applied to (P_H), we get

$$\mu = \inf\{p(x) + \bar{\lambda} g(x): x \in F^H \cap g^{-1}(S^f - S^f)\} \tag{19}$$

for some $\bar{\lambda}$ in $(S^f)^+$. Since $Y = R^m$ is finite dimensional, we can find ϕ_i, $i = 1,\ldots,t$, in $(S^f)^\perp$ such that

$$(S^f - S^f) = \bigcap_{i=1}^{t} \{\phi_i\}^+.$$

Thus $x \in g^{-1}(S^f - S^f)$ if and only if $\phi_i g(x) \leq 0$, $i = 1,\ldots,t$, and (19) is equivalent to

$$\mu = \inf\{p(x) + \bar{\lambda} g(x): \phi_i g(x) \leq 0,\ i = 1,\ldots,t,\ x \in F^H\}. \tag{20}$$

Since g is S^f-convex on F^H and $\{\phi_i\} \subset (S^f)^\perp \subset (S^f)^+$, we conclude that both $\phi_i g$ and $-\phi_i g$ are convex (on F^H), which in turn implies that

$$\phi_i g \text{ is affine (on } F^H),\ i = 1,\ldots,t$$

and, without loss of generality, we can assume them affine on X. Now since F^H is polyhedral, we can apply Lemma 1 to (20). Thus, there exist nonnegative scalars α_i such that

$$\mu = \inf\{p(x) + \bar{\lambda} g(x) + \sum_{i=1}^{t} \alpha_i \phi_i g(x): x \in F^H\}. \tag{21}$$

We can now let

$$\lambda = \bar{\lambda} + \sum_{i=1}^{t} \alpha_i \phi_i .$$

Then clearly λ is in $(S^f)^+$ and this shows that (8) holds with F^f replaced by F^H. That (9) holds follows, for if $a \in F$, then

$$\lambda g(a) = \bar{\lambda} g(a) \tag{22}$$

since $\sum_{i=1}^{t} \alpha_i \phi_i$ is in $(S^f)^{\perp}$. The sufficiency of (8) and (9) is clear. □

The above two theorems differ from the standard case in two ways. First, x is restricted to a set smaller than Ω while the multiplier λ is chosen from the larger cone $(S^f)^+$ rather than S^+. Theorem 2 shows how to weaken the restriction on x. This strengthens the optimality conditions. We now show that we can further strengthen the conditions by replacing $(S^f)^+$ with a smaller cone.

COROLLARY 1: Suppose that in Theorem 2 we have

$$S^f \subset L \cap -H; \; F^H = \Omega \cap g^{-1}(H). \tag{23}$$

Then Theorem 2 holds with L^+ *replacing* $(S^f)^+$ *if*

$$L^+ - H^+ = (S^f)^+ \tag{24}$$

or equivalently, when equality holds in (23) and H *and* L *are closed convex cones, if*

$$L^+ - H^+ \text{ is closed.} \tag{25}$$

Moreover, if equality holds in (23) and H *is a subspace, then the above condition (24) or (26) is also necessary (independent of the particular functions* p *and* g*). In particular, we can choose* $H = S^f - S^f$ *and* $L = S$.

PROOF. Note that when equality holds in (23) then (24) and (25) are equivalent since, by (6)

$$(S^f)^+ = (L \cap -H)^+ = \overline{L^+ - H^+},$$

when H and L are closed convex cones. Now, if (24) holds and λ satisfies (8) and (9) (with F^H instead of F^f), then one can solve $\lambda = \phi - h$ with ϕ in L^+ and h in H^+. Hence for any x in F^H we get

$$\begin{aligned} \lambda g(x) &= \phi g(x) - hg(x) \\ &\leq \phi g(x), \end{aligned} \tag{26}$$

since $g(x) \in H$ when $x \in F^H$. Thus

$$\begin{aligned} \mu &= \inf\{p(x) + \lambda g(x) : x \in F^H\} \\ &\leq \inf\{p(x) + \phi g(x) : x \in F^H\}. \end{aligned} \tag{27}$$

The reverse inequality holds also. For, $\phi g(x) \leq 0$, for all $x \in F \subset F^H$, since $g(F) \subset -S^f \subset -L$. The complementary slackness condition is proved in the usual way, i.e. if $\mu = p(a)$ with $a \in F$, then

$$\begin{aligned} p(a) &= \mu \\ &= \inf\{p(x) + \phi g(x): x \in F^H\} \\ &\leq p(a) + \phi g(a), \text{ since } a \in F \subset F^H \\ &\leq p(a), \qquad \text{since } g(F) \subset -S^f \subset -L. \end{aligned}$$

Conversely, suppose ϕ lies in $(S^f)^+$, equality holds in (23) and H is a subspace. Assuming that Theorem 2 holds with L^+ replacing $(S^f)^+$, we need to show that (24) holds. Let P be the orthogonal projection on H. Consider the program

$$\mu = \inf\{\phi P(y): -Py \in -\text{ cone } L,\ y \in R^m\}$$

where cone L denotes the convex cone generated by L. Then pp^{-1} (cone L) $\subset pp^{-1}(L \cap -H) = S^f$ so $\mu = 0$. Also $-P^{-1}(H) = R^m$ so that Theorem 2 yields

$$0 = \mu = \inf\{\phi Py + \lambda(-Py): y \in R^m\}.$$

Since we now assume that $\lambda \in L^+$, we get $\phi P - \lambda P = 0$ and

$$\begin{aligned} \phi &= \phi - (\phi P - \lambda P) \\ &= (\phi - \lambda)(I-P) + \lambda \in -H^+ + L^+. \end{aligned}$$

□

Remark 1. As an example of the above theorem, let us consider the case $S = R^m_+$, i.e. (P_g) is now equivalent to the ordinary finite dimensional convex program with convex (on Ω) constraints $g^k(x) \leq 0$, $k \in P = \{1,\ldots,m\}$, $x \in \Omega$. Let $P^= = \{k \in P: g^k(x) = 0, \text{ for all } x \in F\}$ be the minimal indexing set of binding constraints, e.g. Abrams and Kerzner (1978). Then it is easy to see that

$$S^f = \{y = (y_k) \in R^m_+: y_k = 0, \text{ for all } k \in P^=\}$$

is the minimal cone of (P_g) and Theorem 1 yields

$$\mu = \inf\{p(x) + \lambda g(x): x \in \Omega \text{ and } g^k(x) = 0, \text{ for all } k \in P^=\},$$

for some $\lambda = (\lambda_k)$ in $(S^f)^+$, i.e. $\lambda_k \geq 0$ for $k \in P \setminus P^=$, λ_k arbitrary for $k \in P^=$. This result extends the characterization of optimality for the ordinary convex program given in e.g. Abrams and Kerzner (1978), Ben-Israel et al. (1976) and Ben-Tal and Ben-Israel (1976). The extension is in the sense that the infimum may be unattained and the constraint $x \in \Omega$ is included. (Note that one might also be able to use the approach of Abrams (1975) for problems with unattained infima.) Since S is polyhedral, Corollary 1 implies we can assume that $\lambda \in S^+$, i.e. $\lambda_k \geq 0$ for all k in P.

Now suppose that Ω is polyhedral and the constraints g^k, $k \in P^=$, are analytic convex (or only piecewise faithfully convex, see Ben-Israel et al. (1976) and Rockafellar (1970b)) on Ω. In Theorem 2 set $H = \{\alpha\}^\perp$, where $\alpha = (\alpha_k)$ with $\alpha_k > 0$ if $k \in P^=$ and g^k is not polyhedral (on Ω), $\alpha_k \geq 0$ if $k \in P^=$ and g^k is polyhedral (on Ω) and $\alpha_k = 0$ otherwise. Then

$$F^H = \Omega \cap g^{-1}(H)$$
$$= \Omega \cap \{x \in X: \sum_{k \in P^=} \alpha_k g^k(x) = 0\}.$$

Let x be any feasible point for (P) and

$$D_h = \{d \in X: \text{ there exists } \bar{\alpha} > 0 \text{ with } h(x+\alpha d) = h(x), \text{ for all } 0 < \alpha \leq \bar{\alpha}\}$$

be the cone of directions of constancy at x of h, e.g. Ben-Israel et al. (1976), where $h = \Sigma \alpha_k g^k$. Then D_h is a subspace (or polyhedral cone) independent of the point x (see e.g. Ben-Tal and Ben-Israel (1979) for X finite dimensional and Wolkowicz (1980a) for the general case). Note that Theorem 2 holds with F^H replaced by any polyhedral set G such that

$$F \subset G \subset F^H$$

since the program (P_H) remains equivalent to the original program (P_g). We now set

$$G = \Omega \cap (\hat{x}+D_h) .$$

To apply Theorem 2 with F^H replaced by G, we need only show that g is

S^f-convex on G. Now if $x,y \in G$ and $0 \leq \lambda \leq 1$, then $\lambda x + (1-\lambda)y \in G$, since D_h is a subspace and so convex, and

$$g^k(\lambda x+(1-\lambda)y) - \lambda g^k(x) - (1-\lambda)g^k(y) \leq 0 \qquad (28)$$

for all $k \in P$, since the constraints g^k are convex on Ω. To show that g is S^f-convex on G we need to show that equality holds in (28) for all $k \in P^=$. In fact, it is sufficient to show that g^k is affine on G for all $k \in P^=$. Now suppose that $k_o \in P^=$ and g^{k_o} is not affine (on Ω), then $\alpha_{k_o} > 0$, $h(x) = h(\hat{x}) = 0$ for all x in G and so

$$-\alpha_{k_o} g^{k_o}(x) = \sum_{k \in P\setminus\{k_o\}} \alpha_k g^k(x)$$

for all $x \in G$. Thus $-g^{k_o}$ as well as g^{k_o} are convex functions on G which implies that g^{k_o} is affine on G. Theorem 2 now yields

$$\mu = \inf\{p(x) + \lambda g(x): \ x \in G\}$$

for some λ in $(S^f)^+$. Moreover, by Corollary 1 we can assume that $\lambda \in S^+$, i.e. $\lambda \geq 0$. This result yields the optimality conditions given in Wolkowicz (1980b). Note that, if $\{x \in X: \ h(x) = 0\}$ is convex, then $F^H = G$.

The above set G can be found computationally in the analytic convex (or faithfully convex) case by calculating the cones of directions of constancy of the appropriate functions, see Wolkowicz (1978).

In the case that Slater's condition (12) holds, the above theorems reduce to the standard Lagrange multiplier theorem (see e.g. Luenberger (1969)) i.e. in this case $g^{-1}(S^f-S^f)$ becomes all of X, while $(S^f)^+$ becomes S^+.

Note that A is S-convex (on Ω) since it is affine (on Ω). Thus the above preliminaries all hold for the program (P_A), see (2).

Remark 2. We can extend the functions $p: \ X \to R \cup \{\infty\}$ and $g: \ X \to R^m \cup \{\infty\}$ and remove the continuity assumptions on p and g.

Theorem 1 (see Wolkowicz and Borwein (1981b)) would then require that dom p $\supset$ F, where dom p denotes the essential domain of p. Lemma 1 needs x $\in$ dom p and p continuous at some feasible point and g continuous on V. Theorem 2 needs dom p $\supset$ F and $p(x) + \bar{\lambda}g(x)$ continuous at some point in $F^H \cap g^{-1}(S^f - S^f)$.

3. A LAGRANGE MULTIPLIER THEOREM FOR (P_A)

If the cone S is polyhedral, then the program (P_A) is linearly constrained and the standard Lagrange multiplier theorem always holds. (Recall that Ω is a polyhedral set in (P_A).) However, we now see that even when S is an arbitrary convex cone, then program (P_A) allows a Lagrange multiplier theorem which differs from the standard case only in that the multiplier λ is chosen from a larger cone containing S^+, namely from $(S^f)^+$. Moreover, when A is a linear operator and Ω is a subspace, we get a geometric weakest constraint qualification.

THEOREM 3: For the program (P_A),

$$\mu = \inf\{p(x) + \lambda Ax\colon x \in \Omega\}, \tag{29}$$

for some λ *in* $(S^f)^+$. *In addition, if* $\mu = p(a)$ *with* a *in* F, *then*

$$\lambda g(a) = 0 \tag{30}$$

and (29) and (30) characterize optimality of a *in* F.

PROOF: By Theorem 1, we get that

$$\mu = \inf\{p(x) + \bar{\lambda}Ax\colon x \in \Omega \cap A^{-1}(S^f - S^f)\} \tag{31}$$

for some $\bar{\lambda}$ in $(S^f)^+$ and if $\mu = p(a)$ with a in F, then

$$\bar{\lambda}g(a) = 0 \tag{32}$$

and (31) and (32) characterize optimality of a in F. Now since $Y = R^m$ is finite dimensional, we see that

$$A^{-1}(S^f - S^f) = \{x \in X\colon \phi_i Ax = 0,\ i = 1,\ldots,k\}, \tag{33}$$

for some $\phi_i \in (S^f - S^f)^\perp = (S^f)^\perp$ with

$$S^f - S^f = \bigcap_{i=1}^{k} \{\phi_i\}^\perp. \tag{34}$$

Lemma 1 now implies that

$$\mu = \inf\{p(x) + \bar{\lambda}Ax + \sum_{i=1}^{k} \alpha_i \phi_i Ax : x \in \Omega\}, \tag{35}$$

for some α_i in R_+. We now let

$$\lambda = \bar{\lambda} + \sum_{i=1}^{k} \alpha_i \phi_i . \tag{36}$$

Since the functionals ϕ_i are in (S^f), we get that

$$\lambda \in (S^f)^+ \tag{37}$$

and moreover, since $g(a) \in g(F) \subset -S^f$,

$$\lambda g(a) = \bar{\lambda} g(a). \tag{38}$$

The conclusion now follows by substituting λ into (35) and noting that the program (35) is equivalent to (31) and that, by (38), $\lambda g(a) = 0$ if and only if $\bar{\lambda} g(a) = 0$. □

Remark 3. In (31) we can assume $\bar{\lambda}$ is in S^+ rather than $(S^f)^+$ if

$$S^+ + (S^f)^\perp = (S^f)^+ \tag{39}$$

(see Corollary 1). Thus if

$$A(\Omega) \subset (S^f - S^f) \tag{40}$$

then $A^{-1}(S^f - S^f)$ is redundant in (31) and (29) holds with λ in S^+. Thus (39) and (40) is a constraint qualification for (P_A).

We can further strengthen the above theorem by using Theorem 2. In fact, in this case we see that regularity of the problem, in the sense that the standard Lagrange multiplier theorem holds, independent of p, depends solely on the condition (39), i.e. solely on the geometry of the cone S.

THEOREM 4: First, for the program (P_A), *there exists a polyhedral cone* H *in* R^m *such that*

$$S^f = S \cap -H \text{ and } A(\Omega) \subset H. \tag{41}$$

Now if K *and* H *in* R^m *satisfy*

$$\text{(i)}\quad S^f \subset K \cap -H \text{ and } \quad \text{(ii)}\quad A(\Omega) \subset H \tag{42}$$

with H *polyhedral, then Theorem 3 holds with* $(S^f)^+$ *replaced by* K^+, *if*

$$K^+ - H^+ = (S^f)^+, \tag{43}$$

or equivalently, when equality holds in (42)(i) *and* H *and* K *are closed convex cones, if*

$$K^+ - H^+ \text{ is closed.} \tag{44}$$

Moreover, if H *is a subspace and equality holds in (42)*(i), *then the above conditions are also necessary for* K^+ *to replace* $(S^f)^+$.

PROOF: If $S^f = S$, we can choose $H = R^m$. If S^f is a proper face of S, then $A(\Omega) \cap -\text{ri } S = \emptyset$ and the Hahn-Banach Theorem implies that there exists $\phi_1 \in S^+$ such that

$$\phi_1(A(\Omega)) \geq 0, \quad \phi_1(\text{ri } S) > 0. \tag{45}$$

This implies that $-S^f \subset -S \cap \{\phi_i\}^+$. If equality holds, then we can set $H = \{\phi_1\}^+$. If not, then we repeat the same process but with the cone $-S \cap \{\phi_1\}^+$ replacing the cone $-S$. Since R^m is finite dimensional this process must stop in a finite number of steps. We then set

$$H = \bigcap_{i=1}^{t} \{\phi_1\}^+ .$$

This H then satisfies (41).

Now suppose that (42) holds. Since A is affine, we see that A is S^f-convex on the polyhedral set

$$F^H = \Omega \cap A^{-1}(H) \tag{46}$$

and Theorem 2 implies that we can replace F^f by F^H in (8), i.e.

$$\mu = \inf\{p(x) + \lambda Ax: \quad x \in \Omega \cap A^{-1}(H)\}, \tag{47}$$

for some λ in $(S^f)^+$. By Corollary 1, we see exactly when $(S^f)^+$ can be replaced by K^+, but with $x \in \Omega \cap A^{-1}(H)$ rather than $x \in \Omega$ as desired. But since $A(\Omega) \subset H$, we see that $x \in \Omega$ if and only if $x \in \Omega \cap A^{-1}(H)$. □

Slater's condition (12) is a *constraint qualification* for (P_g), i.e. it is a sufficient condition which guarantees that the standard Lagrange multiplier theorem holds for (P_g), independent of the objective function p. A *weakest constraint qualification* is a necessary and sufficient constraint qualification. Weakest constraint qualifications for the ordinary convex program, i.e. for the program (P_g) with S polyhedral and $\Omega = X$, have been given in Wolkowicz (1980a). The nonconvex case is treated in Gould and Tolle (1972). See also Bazaraa et al. (1976). We now see that the above theorem yields a very elegant weakest constraint qualification for (P).

COROLLARY 2: Consider the program (P_A) *when* A *is linear and* Ω *is a subspace. Then there exists a subspace* H *satisfying (41) and moreover, the condition*

$$S^+ - H^+ = (S^f)^+ \tag{48}$$

or equivalently, if S *is closed,*

$$S^+ - H^+ \text{ is closed} \tag{49}$$

is a weakest constraint qualification.

PROOF: Since $A(\Omega)$ is now a subspace, let us choose H to be any subspace satisfying (41), e.g. $H = A(\Omega) + S^f - S^f$. The result now follows from Theorem 4. □

4. STABILITY AND A GENERALIZATION OF FARKAS' LEMMA

We now apply the results in the previous section. First, we have the following sensitivity theorem for a perturbed program $(P_{g,\varepsilon})$, when the perturbation ε is restricted to $S^f - S$.

THEOREM 5: Suppose that μ *and* $\mu(\varepsilon)$ *are the optimal values of the program* (P_g) *and the perturbed program*

$$(P_{g,\varepsilon}) \qquad \begin{array}{l} \textit{minimize} \quad p(x) \\ \textit{subject to} \quad g(x) - \varepsilon \in -S \textit{ and } x \in \Omega \end{array}$$

respectively, where $\varepsilon \in S^f - S$ *(resp.* $\in S^f_\varepsilon - S$*). Suppose that* λ *and* λ_ε *are the (restricted) Lagrange multipliers for* (P_g) *and* $(P_{g,\varepsilon})$ *respectively, found using Theorem 1. Then*

$$\mu - \mu(\varepsilon) \leq \lambda\varepsilon \text{ (resp } \geq -\lambda_\varepsilon\varepsilon). \tag{50}$$

PROOF: Suppose that S^f_ε is the minimal cone for the perturbed program $(P_{g,\varepsilon})$ with $\varepsilon \in S^f - S$. Let us first show that $S^f_\varepsilon \subset S^f$. Suppose not, i.e. suppose that there exists $x \in \Omega$ such that $g(x) - \varepsilon = -s \in -S$ but $s \notin S^f$. Then $g(x) = -s + \varepsilon \in S^f - S$ but $-s + \varepsilon \notin S^f - S^f$, since $(S-S^f) \cap (S^f-S) = S^f - S^f$. Now by (11), there exists $\hat{x} \in \Omega$ such that $g(\hat{x}) = -\hat{s} \in -\text{ri } S^f$. Thus for $0 < \lambda < 1$ and λ sufficiently small, we see that $(1-\lambda)\hat{x} + \lambda x \in \Omega$, $-\hat{s} + \lambda(\hat{s}+\varepsilon) \in -\text{ri } S^f - S$ and

$$\begin{aligned} g((1-\lambda)\hat{x}+\lambda x) &= (1-\lambda)g(\hat{x}) + \lambda g(x) - s_1, \quad \text{for some } s_1 \in S \\ &= -\hat{s} + \lambda(\hat{s}+\varepsilon) - \lambda s - s_1 \in -S\backslash -S^f, \\ &\qquad \text{since } s \in S^f. \end{aligned} \tag{51}$$

This contradicts the definition of the minimal cone S^f. Thus

$$S^f_\varepsilon \subset S^f. \tag{52}$$

Now by Theorem 1

$$\begin{aligned} \mu - p(x) &\leq \lambda g(x), \quad \text{for all } x \text{ in } F^f = \Omega \cap g^{-1}(S^f-S) \\ &\leq \lambda\varepsilon, \quad \text{for all } x \text{ in } F_\varepsilon, \text{ the feasible set of} \\ &\qquad (P_{g,\varepsilon}), \text{ by (52) and since } \lambda \in (S^f)^+. \end{aligned}$$

The first inequality of (50) now follows by taking the infimum over x in F_ε. The second inequality follows symmetrically. □

This result reduces to the standard stability result if Slater's Condition holds, i.e. in this case the perturbation ε is no longer restricted since $S^f - S = S - S = R^m$, e.g. Luenberger (1969; 222) or Geoffrion (1971). Thus $\mu(\varepsilon)$ is a continuous function of ε. (In Geoffrion (1971) it was shown that μ is a continuous function of ε when restricted to the subspace $S^f - S^f$. This follows from the above since $S^f = S^f_\varepsilon$ if $\varepsilon \in S^f - S^f$.)

We now derive a generalization of Farkas' lemma. Recall that the usual Farkas' lemma (e.g. Mangasarian (1969)) holds with the assumption that S is polyhedral. In the following $A^t\colon\ R^m \to R^n$ is the transpose of the m×n matrix A and S^f is the minimal cone for the constraint $Ax \in S$.

THEOREM 6: Suppose that A *is an* m×n *matrix,* ϕ *is a vector in* R^n, H *is a subspace (whose existence is promised by Corollary 2) and* K *is a closed convex cone which satisfies*

$$S^f = K \cap -H,\ A(R^n) \subset H \quad \text{and } (44). \tag{53}$$

Then the following are equivalent:

i) The system

$$A^t\lambda = \phi\ ,\quad \lambda \in (S^f)^+ \tag{54}$$

is consistent.

ii) $Ax \in S \implies \phi x \geq 0.$

iii) The system

$$A^t\lambda = \phi\ ,\quad \lambda \in K^+ \tag{55}$$

is consistent

iv) $Ax \in K \implies \phi x \geq 0.$

PROOF: Consider the abstract convex program

$$\text{(P)}\qquad \mu = \inf\{\phi x\colon\ -Ax \in -S,\ \ x \in \Omega = X\}.$$

Then (ii) is equivalent to the fact that $x^* = 0$ solves (P). By Theorem 3, this is equivalent to

$$0 = \mu = \inf\{\phi x - \lambda Ax\colon\ \ x \in R^n\}$$

for some λ in $(S^f)^+$. This in turn is equivalent to

$$0 = \phi - A^t\lambda \ , \quad \text{for some } \lambda \in (S^f)^+,$$

since the gradient

$$\nabla(\phi x - \lambda Ax) = \phi - A^t\lambda.$$

Thus (i) is equivalent to (ii). The remaining equivalences follow from Theorem 4 and Corollary 2. □

Note that if (39) holds, then (iii) and (iv) yield the extension of Farkas' Lemma, with $K^+ = S^+$ (see e.g. Ben-Israel (1969))

$$A^t\lambda = \phi \ , \quad \lambda \in S^+ \text{ is consistent}$$

$$\text{if and only if}$$

$$Ax \in S \Longrightarrow \phi x \geq 0.$$

This result holds if and only if $A(S^+)$ is closed, or equivalently, $N(A) + S$ is closed which is equivalent to (39).

REFERENCES

ABRAMS, R.A. (1975). "Projections of Convex Programs With Unattained Infima." *SIAM J. Control 13*, 706-718.

ABRAMS, R.A. and KERZNER, L. (1978). "A Simplified Test for Optimality." *J. of Optimization Theory and Applications 25*, 161-170.

BAZARAA, M.S., SHETTY, C.M., GOODE, J.J. and NASHED, M.Z. (1976). "Nonlinear Programming Without Differentiability in Branch Spaces: Necessary and Sufficient Constraint Qualification." *Applicable Analysis 5*, 165-173.

BEN-ISRAEL, A. (1969). "Linear Equations and Inequalities on Finite Dimensional, Real or Complex, Vector Spaces: A Unified Theory." *J. of Mathematical Analysis and Its Applications 27*, 367-389.

BEN-ISRAEL, A., BEN-TAL, A. and ZLOBEC, S. (1976). "Optimality Conditions in Convex Programming." The IX International Symposium on Mathematical Programming, Budapest.

BEN-TAL, A. and BEN-ISRAEL, A. (1979). "Characterizations of Optimality in Convex Programming: The Nondifferential Case." *Applicable Analysis 9*, 137-156.

BERMAN, A. (1973). "Cones, Matrices and Mathematical Programming." Lecture Notes in Economics and Mathematical Systems 79, Springer-Verlag, New York.

BORWEIN, J. and WOLKOWICZ, H. (1981a). "Facial Reduction for a Cone-Convex Programming Problem." *J. of the Australian Mathematical Society. 30*, 369-380.

BORWEIN, J. and WOLKOWICZ, H. (1981b). "Characterizations of Optimality for the Abstract Convex Program With Finite Dimensional Range." *J. of the Australian Mathematical Society 30*, 390-411.

BORWEIN, J. and WOLKOWICZ, H. (1979a). "Regularizing the Abstract Convex Program." *J. of Mathematical Analysis and Its Applications*, (forthcoming).

BORWEIN, J. and WOLKOWICZ, H. (1979b). "Characterizations of Optimality Without Constraint Qualification for the Abstract Convex Program." *Mathematical Programming Study*, (forthcoming).

CRAVEN, B.D. and ZLOBEC, S. (1980). "Complete Characterization of Optimality for Convex Programming in Banach Spaces." *Applicable Analysis 11*, 61-78.

GEOFFRION, A.M. (1971). "Duality in Nonlinear Programming: A Simplified Applications-Oriented Development." *SIAM Review 13*, 1-37.

GOULD, F.J. and TOLLE, J.W. (1972). "Geometry of Optimality Conditions and Constraint Qualifications." *Mathematical Programming 2*, 1-18.

LUENBERGER, D.G. (1969). *Optimization by Vector Spaces Methods*. John Wiley & Sons Inc., New York.

MANGASARIAN, O. (1969). *Nonlinear Programming*. McGraw-Hill, New York.

MASSAM, H. (1979). "Optimality Conditions for a Cone-Convex Programming Problem." *J. of the Australian Mathematical Society (Series A), 27*, 141-162.

MOREAU, J.J. (1966). "Convexity and Duality." In *Functional Analysis and Optimization* (E.R. Caranells, ed.), Academic Press, New York, 145-169.

PERESSINI, A.L. (1967). *Ordered Topological Vector Spaces*. Harper and Row, New York.

ROCKAFELLAR, R.T. (1970a). *Convex Analysis*. Princeton University Press, Princeton, N.J.

ROCKAFELLAR, R.T. (1970b). "Some Convex Programs Whose Duals are Linearly Constrained." In *Nonlinear Programming* (J.B. Rosen, O.L. Mangasarian and K. Ritter, eds.), Academic Press, New York, 293-322.

WOLKOWICZ, H. (1978). "Calculating the Cone of Directions of Constancy." *J. of Optimization Theory and Applications 25*, 451-457.

WOLKOWICZ, H. (1980a). "Geometry of Optimality Conditions and Constraint Qualifications: The Convex Case." *Mathematical Programming 19*, 32-60.

WOLKOWICZ, H. (1980b). "A Strengthened Test for Optimality." *J. of Optimization Theory and Applications*, (forthcoming).

NEWTON'S METHOD FOR W-CONVEX OPERATORS

Shelby L. Brumelle and Martin L. Puterman

Under general conditions, W-convex operators have supports which are analogous to subgradients of convex functions. A Newton's method using supports in place of derivatives is defined. The sequence of points generated by this method are shown to converge monotonically. Rates of convergence and error bounds are provided.

1. PRELIMINARY DEFINITIONS

Let X be a real linear space. If $W \subseteq X$ is a convex set such that $tW \subseteq W$ for $t \geq 0$, then W is a *wedge*, and (X,W) is a *partially ordered linear space* (PL space). The partial order on a PL space (X,W) is defined by $x \geq y$ if and only if $x - y \in W$. The set $W \cup -W$ consisting of the positive and negative elements is denoted by $W^{\pm}$. Two elements, x and y, are *comparable* if $x \geq y$ or $y \geq x$. If each pair of elements from X are comparable (i.e., $X = W^{\pm}$), then the order is *total*. Any space X can be totally ordered by taking the positive wedge to be X. If A and B are subsets of a PL space (X,W), then $A \leq B$ means $a \leq b$ for each $a \in A$ and $b \in B$. An *order-interval* is a set of points of the form $\{x \in X: x_1 \leq x \leq x_2\}$ and is written $[x_1,x_2]$. A subset $D \subseteq X$ is *order-convex* if $[x_1,x_2] \subseteq D$ whenever $x_1, x_2 \in D$. If $W \cap -W = \{0\}$, then W is a *cone*. It follows that W is a cone if and only if $x = y$ whenever $x \geq y$ and $y \geq x$. If each subset of X which is bounded above has a least upper bound in (X,W), then X is *order complete* and W is a *fully minihedral* wedge. A wedge W is *generating* if $X = W - W$. If (X,W) is a PL space and $\mathcal{T}$ is a topology on X such that $(X,\mathcal{T})$ is a topological linear space, then $(X,W,\mathcal{T})$ is a *partially ordered topological linear space* (PTL space). These definitions follow Day (1962) and Wong and Ng (1973).

The real line is denoted by R, and the non-negative numbers by R^+. Inequalities between numbers will, as usual, refer to the PL space (R,R^+).

A set D in a linear vector space X is *absorbing* at $x \in D$, if for each $y \in X$, there exists some $\varepsilon > 0$ such that $x + \alpha(y-x) \in D$ for $0 \leq \alpha \leq \varepsilon$.

ISBN 0-12-621120-5

Define D^o as the set of points for which D is absorbing. If X has a topology such that scalar multiplication is continuous, then the interior of D is included in D^o (Taylor (1964;124)).

2. W-CONVEXITY AND SUPPORTS

An operator F mapping a convex subset D of a linear space X into a PL space (Z,V) is *convex* if

$$F(\alpha x + (1-\alpha)y) \leqslant \alpha F(x) + (1-\alpha)F(y) \tag{1}$$

whenever $x, y \in D$ and $0 \leqslant \alpha \leqslant 1$. It follows by induction that F is convex if and only if $F(x) \leqslant \sum_{i=1}^{n} \alpha_i F(y_i)$ for any finite subsets $\{y_i\colon i=1,2,\ldots,n\} \subseteq D$ and $\{\alpha_i \in [0,1]\colon i=1,2,\ldots,n\}$ such that $x = \sum_{i=1}^{n} \alpha_i y_i$ and $1 = \sum_{i=1}^{n} \alpha_i$. The inequality in this characterization of convexity is named after Jensen (1906).

Brumelle (1978), introduced a more general notion of convexity based on Jensen's inequality. Suppose that (X,W) and (Z,V) are PL spaces, and that $F\colon D \subseteq X \to Z$. The operator F is *W-convex at* $x \in D$ if, whenever x is a convex combination of elements in D comparable to x, say

$$x = \sum_{i=1}^{n} \alpha_i y_i, \quad \sum_{i=1}^{n} \alpha_i = 1,$$

and $y_i \in D$, $y_i - x \in W^{\pm}$, $\alpha_i \geqslant 0$ for $i = 1,2,\ldots,n$, then $F(x) \leqslant \sum_{i=1}^{n} \alpha_i F(y_i)$. If F is W-convex at each $x \in D$, then F is *W-convex*. If F is convex then, by definition, Jensen's inequality must hold for all convex combinations of points which equal x; whereas if F is W-convex, then Jensen's inequality only needs to hold for those convex combinations consisting of points comparable to x.

PROPOSITION 1.

Suppose that F *maps a convex subset* D *of a linear space* X *into a* PL *space* (Z,V). *If* F *is convex, then* F *is W-convex for any wedge* $W \subseteq X$.

Conversely, if F *is W-convex and* $W^{\pm} = X$, *then* F *is convex. In particular,* F *is convex if and only if* F *is X-convex* .

An example, due to Ortega and Rheinboldt (1967), of a function which is W-convex but not convex is xAx: $R^n \to R$, where R^n is the set of n-dimensional real vectors, A is an n x n matrix which has non-negative components but is not positive semi-definite, and W is the set of vectors with non-negative components. Thus W-convexity generalizes the notion of convexity. Even X-convexity is more general than convexity, since the domain of an X-convex operator is not necessarily convex.

The definition of convexity (1) is in terms of the behavior of the operator on line segments. Ortega and Rheinboldt (1967), (1970) (also see Vandergraft (1967)) generalized convexity by requiring the defining inequality (1) to hold only on line segments oriented in a "positive direction". More precisely, let X and Z be PL spaces and $F: D \subseteq X \to Z$ for some order-convex set D. The function F is *order-convex* if $F(\alpha x+(1-\alpha)y) \leqslant \alpha F(x) + (1-\alpha)F(y)$ whenever x, $y \in D$ are comparable and $0 \leqslant \alpha \leqslant 1$. Order-convex functions are further developed by Ortega and Rheinboldt (1970) and Vandergraft (1967).

Real-valued convex functions have the important property that subgradients exist in the interior of their domain (Rockafellar (1970)). This property is used in a central way in many mathematical programming and numerical analysis algorithms.

Brumelle (1978) showed that W-convex operators (and hence convex operators) have an analogous property. Suppose that (X,W) and (Z,V) are PL spaces, and that $F: D \subseteq X \to Z$. If $x \in D^o$, then an affine operator $A[x]: X \to Z$ with the property that $A[x](x) = F(x)$ and $A[x](y) \leqslant F(y)$ for each $y \in D$ comparable to x is called a *W-support of* F *at* x.

If F has a W-support A[x] at each $x \in D^o$, then we say that F *has a support on* D and the map $A: x \to A[x]$ is a *support of* F. Theorem 2

(Brumelle (1978)) states that if F has a W-support at $x \in D$, then F is W-convex at x. Conversely, if Z is order complete and F is W-convex at $x \in D^o$, then F has a W-support at x. If W=X and $(Z,V)=(R,R^+)$, then the converse portion of this result reduces to the statement that convex functions have subgradients on the interior of their domain.

It is not known whether or not order-convex operators have an analogous property. Whenever order-convex operators are used in the literature (e.g. Ortega and Rheinboldt (1967, 1970); Vandergraft (1967)) they are accompanied by other assumptions such as Gateux differentiability which actually make them W-convex.

In this paper, we develop a Newton's method for solving F(v) = 0, where F has a support. Given an operator F, a sequence $\{y_k\}$ is a *Newton sequence* if, for each k, y_{k+1} solves $D_k y = 0$, where D_k is an affine operator obtained by "linearizing" F about y_k. A *Newton's method* is an iterative procedure which at the k-th iteration constructs D_k and solves the affine equation. The objective is to find a Newton's method which generates a sequence which rapidly converges to a root of F(y). The usual Newton's method generates y_{k+1} by solving $F'(y_k)(y-y_k) = -F(y_k)$. So we are using the adjective "Newton" in a generalized sense, for which the alternative "quasi-Newton" is frequently used.

If X and Z are Euclidean spaces with the usual component-wise partial order, and if $F: D \subseteq X \to Z$ has a support on D, then Ortega and Rheinboldt (1967) provide sufficient conditions to obtain a convergent and monotone Newton sequence. Vandergraft (1967, Theorems 5.1, 5.4) has similar result but with X and Z being more general spaces and with the assumption that F is Gâteaux differentiable and order-convex. To obtain quadratic convergence of the Newton sequence, Vandergraft assumes that F is twice uniforml differentiable.

In this paper we provide sufficient conditions, somewhat weaker than Vandergraft's, to obtain a convergent and monotone Newton sequence for an operator F. Throughout the paper, F is assumed to have a support (and is thus W-convex), but is not assumed to be differentiable. The hypotheses of our convergence theorem are modelled after those of Ortega and Rheinboldt (1967), which also do not require differentiability. However, our method of generating a Newton sequence differs from theirs. Let $L(X,Z)$ be the space of linear operators from X to Z, with the usual topology. To obtain the rate of convergence, it is sufficient that the support A, viewed as an operator from X to $L(X,Z)$, be right Lipschitz continuous of order p at the limit point y^* of the Newton sequence. Convergence is then of order $1 + p$. The Lipschitz condition on A is, of course, analogous to the common assumption that F' is Lipschitz continuous as for example in Ortega (1968), Rheinboldt (1968), Gragg and Tapia (1974), and Rall (1974). However, it should be noted that our continuity condition is imposed only at the root, y^*.

This paper was motivated by problems arising in control theory where F is convex, and thus has a support, but is usually not differentiable at every $x \in D$. Further, the global Lipschitz continuity condition of the supports is usually not satisfied but the weaker conditions presented here hold. The theory developed here is specialized to dynamic programs in Puterman and Brumelle (1979). Puterman and Shin (1978) report on their numerical experience in implementing a modification of the algorithm. Brumelle (1978) relaxes the convexity assumption.

3. MONOTONE CONVERGENCE

Before proceeding to the convergence theorems, it is necessary to give several definitions concerning relationships between the partial order and the topology in a PTL space. Vandergraft (1967) has a nice discussion of some of these relationships relevant to the implementation of Newton's method. Our definitions generally follow those of Wong and Ng (1973).

A PTL space is *regular* if $y_0 \geqslant y_1 \geqslant \ldots \geqslant x_0$ implies that $\{y_n\}$ converges. For example, with the cone of nonnegative functions, L^p, $0 < p < \infty$ are regular, but $C[0,1]$, $C^n[0,1]$, and $L^\infty[0,1]$ are not (cf. Wong and Ng, (1973;409)).

A PTL space $(X,W,\mathcal{T})$ is a *locally order-convex* space if $\mathcal{T}$ admits a neighborhood base at 0 consisting of order-convex sets. From Wong and Ng (1973, Theorem 5.1), $(X,W,\mathcal{T})$ is locally order-convex if and only if the net $\{x_n : n \in D\}$ converges to 0 whenever there exists a net $\{y_n : n \in D\}$ converging to 0 and $0 \leqslant x_n \leqslant y_n$ for each $n \in D$. A PTL space $(X,W,\mathcal{T})$ is a *locally o-convex* space if $\mathcal{T}$ admits a neighborhood base at 0 consisting of convex and order-convex sets. The L^p spaces $1 \leqslant p \leqslant \infty$, R^n, and $C[0,1]$ are locally o-convex. L^p spaces, $0 < p < 1$,are locally order-convex, but are not locally convex and hence not locally o-convex .

If X and Z are PL spaces, an operator $B: X \to Z$ is *positive* (written $B \geqslant 0$) if $B(x) \geqslant 0$ whenever $x \geqslant 0$. If A also maps X into Z then we write $A \geqslant B$ if $A - B \geqslant 0$. If $A: Z \to X$ is linear, then a linear operator $B: X \to Z$ is a *sub-inverse* of A provided $AB \leqslant I$ and $BA \leqslant I$ where I is, in each, an identity operator. An operator F mapping a PTL space X into a PTL space Z is *right continuous* at x if the sequence $\{F(x_n)\}$ converges to $F(x)$ for any decreasing sequence $\{x_n\}$ converging to x. It is *right closed* at x if $\lim F(x_n) = F(x)$ whenever $\{x_n\}$ is a decreasing sequence converging to x and $\{F(x_n)\}$ converges.

The first theorem is a generalization of the Ortega and Rheinboldt (1967, Theorem 4.1) and Vandergraft (1967, Theorem 5.1) convergence theorems for Newton iteration. The Newton sequence is defined in a similar manner to Vandergraft; however we have followed Ortega and

Rheinboldt in allowing the B_k's to be distinct and not necessarily invertible.

THEOREM 1. Let X *and* Z *be PTL spaces with closed wedges and* X *regular. Let* $F: D \subseteq X \to Z$ *be right continuous and suppose that there exists points* $x_0, y_0 \in D$ *such that* $x_0 \leq y_0$, $[x_0,y_0] \subseteq D$, *and* $F(x_0) \leq 0 \leq F(y_0)$. *Assume for each* $x \in [x_0,y_0]$ *that there exists a linear operator* $A[x] : X \to Z$ *such that*

$$F(y) \geq F(x) + A[x](y-x) \quad \textit{for } y \in [x_0,y_0] \tag{2}$$

and comparable to x, *and that* $A[x]$ *has a positive subinverse* $B[x]$ *such that* $B[x]A[x]$ *is right closed. Then there exists a monotone sequence* $y_0 \geq y_1 \geq y_2 \geq \ldots$ *satisfying*

$$B_k[A_k(y_{k+1}-y_k) + F(y_k)] = 0, \quad k = 0,1,2,\ldots, \tag{3}$$

where $A[y_k] = A_k$ *and* $B[y_k] = B_k$. *This sequence* $\{y_k\}$ *has a limit* y^* *such that* $y_k \geq y^* \geq x_0$ *for each* k. *Moreover, any solution of* $F(x) = 0$ *in* $[x_0,y_0]$ *is contained in* $[x_0,y^*]$.

If in addition there exists a positive one-to-one linear operator $B: Z \to X$ *such that* $B_k \geq B$, $k = 0,1,2,\ldots$ *and if* BF *is right closed at* y^* *and if* X *has a locally order-convex Hausdorff topology, then* $F(y^*) = 0$.

PROOF.

Define $V: X \to X$ by $V(x) = x - B_0(F(y_0) + A_0x)$ and $\{v_n: n = 0,1,2,\ldots\}$ by $v_0 = 0$ and $v_{n+1} = V(v_n)$. It follows as in Theorem 1 of Kantorovich (1939) (cf. Vandergraft (1967)) that $\{y_0+v_n\}$ is a decreasing sequence in $[x_0,y_0]$ which converges to a limit $y_0^* \in [x_0,y_0]$ satisfying

$$B_0(F(y_0) + A_0(y_0^*-y_0)) = 0.$$

From the definition of v_n it follows that

$$v_n = - \sum_{j=0}^{n-1} (I-B_0A_0)^j B_0 F(y_0) \qquad n = 1,2,3\ldots .$$

Hence

$$A_0 v_n = - A_0 B_0 \sum_{j=0}^{n-1} (I-A_0B_0)^j F(y_0). \tag{4}$$

Using (4) together with the identity $H \sum_{j=0}^{n-1} (I-H)^j = I - (I-H)^n$ gives

$$F(y_0) + A_0 v_n = (I-A_0B_0)^n F(y_0).$$

From (1) and $I \geqslant A_0B_0$ we have

$$F(y_0+v_n) \geqslant F(y_0) + A_0v_n = (I-A_0B_0)^n F(y_0) \geqslant 0.$$

Hence the right continuity of F together with the closure of the positive wedge of Z, implies that $F(y_0^*) \geqslant 0$.

Similarly, for $x \in [x_0,y_0]$,

$$x - \sum_{j=0}^{n-1} (I-B_0A_0)^j B_0 F(x) \leqslant (y_0+v_n) - (y_0-x) + \sum_{j=0}^{n-1} (I-B_0A_0)^j B_0A_0 (y_0-x)$$

$$= y_0 + v_n - (I-B_0A_0)^n (y_0-x) \leqslant y_0 + v_n . \tag{5}$$

If $\hat{x} \in [x_0,y_0]$ solves $F(\hat{x}) = 0$, then $\hat{x} \leqslant y_0 + v_n$ for each n and $\hat{x} \leqslant y_0^* \leqslant y_0$. Letting $x = x_0$ in (5) verifies that $x_0 \leqslant y_0^*$.

Setting $y_1 = y_0^*$ and proceeding in the above manner, we inductively generate a sequence $\{y_k\}$ such that $y_{k-1} \geqslant y_k \geqslant x_0$, $F(y_k) \geqslant 0$, $y_k \geqslant \hat{x}$ for $k = 1,2,\ldots$ and any $\hat{x} \in [x_0,y_0]$ satisfying $F(\hat{x}) = 0$. Then since X is regular, the positive wedges are closed, and F is right continuous, it follows that $y_k \to y^*$, $y_k \geqslant y^* \geqslant x_0$ for each k, $\hat{x} \leqslant y^*$, and $F(y^*) \geqslant 0$.

Assume now that there exists a positive, one-to-one linear operator B such that $B_k \geqslant B$, $k = 0,1,2,\ldots$ and BF is right continuous. Since $F(y_k) \geqslant 0$ for all n we have

$$0 \leqslant BF(y_k) \leqslant B_k F(y_k) = B_k A[y_k](y_k - y_{k+1}) \leqslant (y_k - y_{k+1}).$$

Since convergent sequences are Cauchy and X is locally order-convex, $BF(y_k) \to 0$, and since BF is right closed $BF(y_k) \to BF(y^*)$. However, X is Hausdorff separable and B is one-to-one, so that $F(y^*) = 0$. □

The preceding theorem admits several useful modifications.

First, for numerical computation, it is important to note that the iterative procedure used in the proof to solve (3) need not be carried to completion. In fact, it is clear from the proof that (3) can be replaced by

$$y_{n+1} = y_n - \sum_{j=0}^{k_n} (I-B_nA_n)^j B_n F(y_n), \tag{6}$$

where $\{k_n\}$ is any sequence of nonnegative integers, without affecting the validity of the theorem. With this substitution, it is no longer necessary to assume that F and B_kA_k are right continuous and right closed; however, the right closure of BF at y* must be retained in the last paragraph. If each $k_n = 0$ then the sequence defined by (6) becomes

$$y_{n+1} = y_n - B_n F(y_n),$$

which is the same as the Newton sequence used by Ortega and Rheinboldt (1970). It might be noted that the increment $y_1 - y_0$ generated as in the proof of the Theorem is larger (i.e. $\geqslant$) than the increment $y_1 - y_0$ generated by (6), but it is also harder to construct.

Second, the linear equation at (3) was solved in the proof by finding a fixed point for a linear operator in X. In some cases, it might be more convenient to solve the linear equation (3) in Z. The expression (6) can also be written

$$y_{n+1} = y_n - B_n \sum_{0}^{k_n} (I-A_nB_n)^j F(y_n) . \tag{7}$$

As $k \to \infty$, the quantity $\sum_0^k (I-A_0B_0)^j F(y_0)$ converges to a fixed point w* of the function $W: Z \to Z$ defined by $W(w) = w - F(y_0) - A_0B_0w$, provided that A_0B_0 (instead of B_0A_0 as in the theorem) is right closed, that Z is regular, and that the other assumptions (except for the right continuity of F) in the theorem are satisfied. In this case, $y_1 = y_0 + B_0w^*$ has the following properties: 1) $x_0 \leqslant y_1 \leqslant y_0$; 2) $F(y_1) \geqslant 0$; and 3) $y_1 \geqslant \hat{x}$ whenever $F(\hat{x}) = 0$ and $\hat{x} \in [x_0,y_0]$. Properties 1) and 3) can be established by arguments similar to those in the theorem. Property 2) follows since $F(y_1) \geqslant F(y_0) + A_0B_0w^* = 0$, and does not require F to be right continuous as in the theorem. A sequence $\{y_n\}$ satisfying

$$A_n(y_{n+1}-y_n) = -F(y_n) \tag{8}$$

can now be generated recursively, assuming that A_nB_n is right closed for each n. This sequence has all of the properties attributed to $\{y_n\}$ in the theorem. Vandergraft (1967) used (8) with F' in place of A to define a Newton sequence.

Third, in the modification described in the previous paragraph, the assumption that X is regular can be replaced by the assumptions that X is locally order-convex and that there exists a linear operator $T: X \to Z$ such that $\{T(y_n)\}$ is topologically bounded and T^{-1} is completely continuous. This device is due to Vandergraft (1967, Section 6) and insures the existence of a limit of $\{y_n\}$.

4. RATE OF CONVERGENCE AND ERROR BOUNDS

A semi-norm $\|\cdot\|$ on the PL space (X,W) is *k-normal* if $x \leqslant y \leqslant z$ implies $\|y\| \leqslant k \max\{\|x\|,\|z\|\}$. If $\|\cdot\|$ is a k-normal semi-norm on a PL space (X,W), then by Wong and Ng (1973, Proposition 5.6) the PTL space $(X,W,\|\cdot\|)$ is locally o-convex. Conversely, if the topology of a PTL space $(X,W,\mathcal{T})$ is semi-normable and locally o-convex, then there exits a semi-norm on X which is 1-normal and induces the topology $\mathcal{T}$. It follows that any continuous semi-norm on a locally o-convex PTL space is k-normal, for some k.

Throughout the rest of the paper $(X,W,\mathcal{T})$ and $(Z,V,\mathcal{S})$ will be locally o-convex PTL spaces and $\|\cdot\|_1$ and $\|\cdot\|_2$ will be continuous semi-norms on $(X,\mathcal{T})$ and $(Z,\mathcal{S})$, respectively. Thus there exists numbers k_1 and k_2 such that $\|\cdot\|_i$ is k_i-normal, $i = 1,2$. If A is a linear operator from X to Z, then

$$\|A\| = \sup_{\{x:\|x\|_1 \leqslant 1\}} \|A(x)\|_2$$

and if A is a linear operator from Z to X, then

$$\|A\| = \sup_{\{z:\|z\|_2 \leqslant 1\}} \|A(z)\|_1 .$$

In order to explore the rate of convergence of $\{y_k\}$ to y^*, and error bounds, we use the following mean value theorem for operators having supports. Ortega and Rheinboldt (1970, 3.2.12) use an analogous result to develop error bounds with F a continuously differentiable function.

THEOREM 2. Suppose that $F\colon D \subseteq X \to Z$ *with* D *an order-convex subset of* X. *Assume that* F *has a support* A *on* D. *Then*

(i) $0 \leqslant F(y) - F(x) - A(x)(y-x) \leqslant [A(y)-A(x)](y-x)$ *for comparable* $x, y \in D^o$.

(ii) *If* $\|A(y)-A(x)\|_2 \leqslant K\|y-x\|_1^p$ *for some comparable* $x, y \in D^o$, *then* $\|F(y)-F(x)-A(x)(y-x)\|_2 \leqslant k_2K\|y-x\|_1^{p+1}$.

PROOF.

(ii) follows immediately from (i). To verify (i), subtract $A(x)(y-x) + F(x)$ from each of the three terms in the inequality

$$A(x)(y-x) + F(x) \leqslant F(y) \leqslant A(y)(y-x) + F(x),$$

which holds since A is a support of F. □

We note as a corollary that if, in addition to the hypotheses of Theorem 2(ii), $X = R$, $D = [0,1]$ and $\|\cdot\|_2$ is a norm then

$$F(1) - F(0) = \int_0^1 A(t)dt, \tag{9}$$

where the integral is defined as a $\|\cdot\|_2$ - limit of Riemann sums as in Kantorovich and Akilov (1964, p.665). An analogous result, with Lipschitz continuity of A replaced by uniform differentiability of F, is due to Vandergraft (1967, Theorem 4.2). The equation at (9) follows since for any partition $\{t_i: i=0,1,2,\ldots,n\}$ of $[0,1]$ with $0 = t_0 < t_1 < \ldots < t_n = 1$,

$$0 \leqslant F(1) - F(0) - \sum_0^{n-1} A[t_k](t_{k+1}-t_k) \leqslant \sum_0^{n-1} (A[t_{k+1}]-A[t_k])(t_{k+1}-t_k);$$

and since $\|\cdot\|_2$ is k_2-normal,

$$\|F(1) - F(0) - \sum_0^{n-1} A[t_k](t_{k+1}-t_k)\|_2 \leqslant k_2 \max_k \|A[t_{k+1}] - A[t_k]\|$$
$$\leqslant k_2 K \max_k (t_{k+1}-t_k).$$

Consequently, the Riemann sums converge to $F(1) - F(0)$ as the mesh of $\{t_n\}$ goes to 0.

In the following theorem we study the rate of convergence of the algorithm described in Theorem 1. We obtain the rate of convergence in the absence of the usual differentiability conditions placed on F(x).

THEOREM 3. Suppose that $F: D \subseteq X \to Z$ *with* D *an order-convex subset of* X, *that* A *is a support of* F, *that* $\{y_k\} \subseteq D^o$ *is a decreasing sequence satisfying (3), that there exists a* $y^* \in D^o$ *such that* $y_k \geqslant y^*$ $k = 0,1,2,\ldots$ *and* $F(y^*) = 0$, *and that for some* k

$$\|A_k - A[y^*]\| \leqslant K\|y_k - y^*\|_1^p, \quad k = 0,1,2,\ldots .$$

In addition, assume that there exists linear operators $C_k: X \to X$ *such that*

$$C_k B_k A_k \geqslant I \quad \textit{for} \quad k = 0,1,2,\ldots \tag{10}$$

and a constant M *independent of* k *such that*

$$\|C_kB_k\| \leqslant M \quad \text{for } k = 0,1,2,\ldots \,. \tag{11}$$

Then

$$\|y_{k+1} - y^*\|_1 \leqslant L\|y_k - y^*\|_1^{1+p} \quad \text{for } k = 0,1,2,\ldots$$

where $L = k_1k_2KM$.

PROOF.

From (10),

$$0 \leqslant y_{k+1} - y^* \leqslant C_kB_kA_k(y_{k+1}-y^*).$$

However, since $F(y^*) = 0$ and (3) holds, the right hand side equals $C_kB_k[F(y^*) - F(y_k) - A_k(y^*-y_k)]$, and so

$$\|y_{k+1} - y^*\|_1 \leqslant k_1\|C_kB_k[F(y^*) - F(y_k) - A_k(y^*-y_k)]\|_1$$

$$\leqslant k_1\|C_kB_k\| \; \|F(y^*) - F(y_k) - A_k(y^*-y_k)\|_2.$$

An application of (11) and Theorem 2(ii) gives the desired result. □

The B_k's referred to in the previous theorem need not be the ones used to generate the Newton sequence $\{y_k\}$. For example, suppose that $\{y_k\}$ is constructed using the algorithm in Theorem 1 with each B_k invertible. Then $\{y_k\}$ actually satisfies (8), and so

$$B[F(y_k) + A_k(y_{k+1}-y_k)] = 0$$

for any B. Theorem 3 can then be applied if there exists $\{C_k\}$ and some B such that $\|C_kB\| \leqslant M$ and $C_kBA_k \geqslant I$ for each k.

In order to develop error bounds, observe that if $C_kB_kA_k \geqslant I$, and $\{y_k\}$ satisfies (3), then

$$\begin{aligned} 0 &\leqslant y_k - y_{k+1} \leqslant C_kB_kA_k(y_k-y_{k+1}) \\ &= C_k(B_k-B_{k-1})F(y_k) + C_kB_{k-1}[F(y_k) - F(y_{k-1}) - A_{k-1}(y_k-y_{k-1})]. \end{aligned} \tag{12}$$

If conditions are imposed sufficient to insure that $\|\cdot\|_1$ evaluated at the right hand side of (12) is bounded by $L\|y_k - y_{k-1}\|_1^{1+p}$, then an error bound can be developed as in the following theorem.

THEOREM 4. Suppose that $F: D \subseteq X \to Z$ *with* D *an order-convex subset of* X, *that* A *is a support of* F, *that* $\{y_k\}$ *is a decreasing sequence satisfying (3) with each* $B_k = B$ *and with* $\|A_{k+1} - A_k\| \leqslant K\|y_{k+1} - y_k\|_1^p$ *for each* k *and some* $p > 0$, *and that there exists a* y^* *such that* $\|y_k - y^*\|_1 \to 0$.

In addition, assume that there exists linear operators C_k *as in Theorem 3 and that* $h = L^{\frac{1}{p}}\delta < 1$ *where* $\delta = \|y_1 - y_0\|_1$ *and* $L = k_1 k_2 KM$. *Then*

$$\|y^* - y_k\|_1 \leqslant L^{-\frac{1}{p}} \sum_{n=k}^{\infty} h^{(1+p)^n} \leqslant \frac{L^{-\frac{1}{p}}}{\ln(1+p)} \int_{a_k}^{\infty} \frac{1}{v} e^{-v} dv \tag{13}$$

for $k = 1,2,\ldots$ *where* $a_k = -(1+p)^{k-1}(\ln h)$.

PROOF. Since $B_{k+1} = B_k = B$ the first term of (12) is zero, and so from Theorem 2(ii), it follows that

$$\|y_k - y_{k+1}\|_1 \leqslant L\|y_{k-1} - y_k\|_1^{1+p} . \tag{14}$$

Define the real valued sequence $\{t_k\ ;\ k = 0,1,2,\ldots\}$ by $t_0 = 0$, $t_1 = \delta$ and $t_{k+1} - t_k = L(t_k - t_{k-1})^{1+p}$ for $k = 2,3,\ldots$. Hence by (14) we have that

$$\|y_{k+1} - y_k\|_1 \leqslant t_{k+1} - t_k \qquad k = 0,1,2,\ldots$$

or that $\{t_k\}$ is a majorizing sequence for $\{y_k\}$ (cf. Ortega and Rheinboldt 1970, p. 414). This implies that for $m > k$,

$$\|y_{m+1} - y_k\|_1 \leqslant t_{m+1} - t_k = \sum_{n=k}^{m} (t_{n+1} - t_n) . \tag{15}$$

Following Ortega and Rheinboldt (1970, 12.4.3) we can pass to the limit in (15). Thus

$$\|y^* - y_k\|_1 \leqslant \sum_{n=k}^{\infty} (t_{n+1}-t_n) \ . \tag{16}$$

The equality

$$(t_{n+1}-t_n) = L^{-1/p} h(1+p)^n$$

together with (16) gives the first inequaltiy in (13).

To bound the sum, note that

$$\sum_{n=k}^{\infty} h^{(1+p)^n} \leqslant \int_{k-1}^{\infty} h^{(1+p)^u} du = \frac{1}{\ln(1+p)} \int_{a_k}^{\infty} \frac{e^{-v}}{v} dv$$

where the last equality follows by the change in variable $v = -(\ln h)(1+p)^u$.

□

Note that if y_1 is constructed as in the proof of Theorem 1 then

$$\delta = \left\| \sum_{j=0}^{\infty} (I-B_0A_0(y_0))^j B_0F(y_0) \right\|_1 \ .$$

Note also that the integral in (13) is called the exponential integral, Ei, and is tabulated in Spiegel (1968).

REFERENCES

BRUMELLE, S.L. (1978). "Convex Operators and Supports." *Math. of Operations Research*, Vol. 3, No. 2, pp. 171-175.

BRUMELLE, S.L. (1977). "Generalized Policy Improvement for Monotone Dynamic Programs." Working Paper 503, Faculty of Commerce, University of British Columbia.

DAY, M.M. (1962). *Normed Linear Spaces*. Springer-Verlag, Berlin.

DENARDO, E.V. and MITTEN, L.G. (1967). "Elements of Sequential Decision Processes." *Journal of Industrial Engineering XVIII*, 1, pp. 106-112.

DENNIS, J.E., Jr. (1970). "Toward a Unified Convergence Theory for Newton-Like Methods." *Nonlinear Functional Analysis and Applications*, ed. L.B. Rall, Academic Press, New York.

GRAGG, W.B. and TAPIA, R.A. (1974). "Optimal Error Bounds for the Newton-Kantorovich Theorem." *SIAM J. Numer. Anal.*, Vol. 11, pp. 10-13.

KANTOROVICH, L.V. (1939). "The Method of Successive Approximations for Functional Equations." *Acta Math.*, Vol. 71, pp. 63-77.

KANTOROVICH, L.V. and AKILOV, G.P. (1964). *Functional Analysis in Normed Spaces*. Macmillan, New York.

ORTEGA, J.M. (1968). "The Newton-Kantorovich Theorem." *American Math. Monthly*, Vol. 75, pp. 658-660.

ORTEGA, J.M. and RHEINBOLDT, W.C. (1967). "Monotone Iterations for Nonlinear Equations with applications to Gauss-Seidel Methods." *SIAM J. Numer. Anal.*, Vol. 4, pp. 171-190.

ORTEGA, J.M. and RHEINBOLDT, W.C. (1970). *Iterative Solutions of Nonlinear Equations in Several Variables*. Academic Press, New York.

PUTERMAN, M.L. and BRUMELLE, S.L. (1979). "On the Convergence of Policy Iteration in Stationary Dynamic Programming." *Math. of Operations Research*, Vol. 4, No. 1, pp. 60-69.

PUTERMAN, M.. and SHIN, M. (1978). "Modified Policy Iteration Algorithms for Discounted Markov Decision Problems." *Management Science*, 24, pp. 1127-1137.

RALL, L.B. (1974). "A Note on the Convergence of Newton's Method." *SIAM J. Numer. Anal.*, Vol. 11, pp. 34-36.

RHEINBOLDT, W.C. (1968). "A Unified Convergence Theory for a Class of Iterative Processes." *SIAM J. Numer. Anal.*, Vol. 5, pp. 42-63.

ROCKAFELLAR, R.T. (1970). *Convex Analysis*. Princeton University Press, Princeton, New Jersey.

SPIEGEL, M.R. (1968). *Mathematical Handbook*. McGraw-Hill, New York.

TAYLOR, A.E. (1964). *Introduction to Functional Analysis*. John Wiley & Sons, New York.

VANDERGRAFT, J.S. (1967). "Newton's Method for Convex Operators in Partially Ordered Spaces." *SIAM J. Numer. Anal.*, Vol. 4, pp. 406-432.

WONG, Y.C. and NG, K.F. (1973). *Partially Ordered Topological Vector Spaces*, Clarendon Press, Oxford.

PART V

FRACTIONAL PROGRAMMING

A SURVEY OF FRACTIONAL PROGRAMMING*

Siegfried Schaible†

Fractional programs have been treated in a considerable number of papers. In this survey we first give a detailed outline on major applications of fractional programming. Following a review of some basic theoretical results including duality relations we then describe different algorithmic approaches: a primal, a dual and a parametric solution method.

1. INTRODUCTION

Since the early paper of Isbell and Marlow (1956) dealing with a constrained ratio optimization problem (fractional program) at least 400 articles have appeared in this field. In an extensive treatment of the subject in Schaible (1978) it is tried to relate the majority of these papers to each other. There applications, theory and algorithms of fractional programming are discussed in detail.

The purpose of this paper is to outline some of the major developments in fractional programming. In the first part we survey major applications of fractional programs. A more detailed presentation is given in Schaible (1978). We then briefly review duality results in fractional programming. Papers on duality that appeared until 1975 are already related to each other in Schaible (1976c). For more recent results see Schaible (1978). Finally we discuss solution methods in fractional programming. A primal, a dual and a parametric approach are considered.

In our presentation we mainly concentrate on applications and solution methods of fractional programming since theoretical results have already been surveyed to some extent earlier in Schaible (1976a, 1976c). Furthermore, we shall stress developments in nonlinear frac-

---

† I would like to thank Professor W.T. Ziemba for helpful comments.
* This work was supported by Grant A 4534 from Natural Sciences and Engineering Research Council of Canada and by Grant 79-41 from the University of Alberta.

ISBN 0-12-621120-5

tional programming. Many of the results in linear fractional programming were outlined before; see for example Abrham and Luthra (1977), Schaible (1976a), Wagner and Yuan (1968). Therefore we put more emphasis on the nonlinear case.

The bibliography in the end is rather selective. It contains only material referred to in this survey. A more detailed bibliography is given in Schaible (1978).

We utilize the following notation. An optimization problem is called a *fractional program* if it is of the form

(P) $$\operatorname{Max} \{q(x) = \frac{n(x)}{d(x)} \mid x \in S\}$$

where $S = \{x \in R^n \mid h_i(x) \leq b_i \ , \ i=1,\ldots,m\}$ and $d(x) > 0$ on S.

More specifically, we differentiate between the following types of fractional programs:

(P) is called a *linear fractional program* if all functions $n(x)$, $d(x)$ and $h_i(x)$ are affine, i.e., the sum of a linear function and a constant.

(P) is said to be a *quadratic fractional program* if $n(x)$ and $d(x)$ are quadratic functions and $h_i(x)$ are affine.

(P) is called a *concave-convex fractional program* if $n(x)$ is concave and $d(x)$ as well as $h_i(x)$ are convex. Here we assume nonnegativity of $n(x)$ for nonaffine $d(x)$. This condition turns out to be necessary from a theoretical as well as algorithmic point of view and is satisfied in many applications.

Linear, quadratic and concave-convex fractional programs (P) are generalizations of linear, quadratic and concave programs. They are obtained from the latter ones by dividing by an affine, quadratic or convex function, respectively.

In addition, concave-concave fractional programs, convex-concave fractional programs and convex-convex fractional programs (P) can be introduced. However, very few results have been obtained for these models.

More recently *generalized fractional programs* have been discussed too; see Crouzeix-Ferland-Schaible (1981), Passy-Keslassy (1979). Here the objective function is not just one ratio. Instead, it is the minimum of several ratios.

In the next section we shall survey decision problems that give rise to the different types of fractional programs that were mentioned before.

2. APPLICATIONS

Applications of fractional programming are widely spread over the literature. Grunspan (1971) already reviewed many of the earlier applications known at that time. In Schaible (1978) it is tried to give an extensive presentation of decision problems that give rise to various fractional programs.

2.1 Optimizing Efficiency

Ratios to be optimized often describe some kind of an efficiency measure for a system. Examples of this type will be reviewed in this section.

Gilmore and Gomory (1963) discuss a stock cutting problem in the paper industry. They show that under the given circumstances it makes sense only to minimize the ratio of wasted and used amount of raw material instead of just minimizing the amount of wasted material. This stock cutting problem is formulated as a linear fractional program. Optimizing productivity of material is an objective often met in industrial plants; see Gutenberg (1975; 29).

According to Heinen (1971; 62) maximizing the return on investment is sometimes persued in resource allocation problems. This means that the ratio of profit and capital is to be maximized. Recently Mjelde (1978) discussed resource allocation problems where the ratio of return and cost is to be maximized. In this paper the term cost may also stand for the amount of pollution or the probability of a disaster in nuclear energy production for example. Depending on the nature of functions describing return, profit, cost or capital in the aforementioned examples we obtain linear, quadratic or concave-convex fractional programs. If, for instance, the price per unit depends linearly on the output, and cost and capital are linear functions then maximizing the return on investment gives rise to a concave-convex quadratic fractional program; see Pack (1962, 1965).

In a portfolio selection problem with risk free borrowing and lending and normally distributed random returns Ziemba, Parkan and Brooks-Hill (1974) obtain as a major subproblem the constrained maximization of the ratio of expected return and risk. The numerator is an affine function and the denominator a (nonlinear) convex function. Hence the portfolio selection problem gives rise to a concave-convex fractional program. It is not a quadratic fractional program. Ziemba (1974) generalizes the results from normal distributions to stable distributions for the return. Similar fractional programs which are not linear or quadratic are met in portfolio theory as well (Dexter, Yu and Ziemba (1980), Ohlson and Ziemba (1976)). They arise from lognormally distributed assets. Generalized concavity properties of these models are investigated in Schaible and Ziemba (1981).

Ratios involving the time occur frequently. Dantzig, Blattner and Rao (1966) discuss a routing problem for ships or planes where a cycle in the network is to be determined that minimizes the cost-to-time ratio. The same ratio appears in papers by Fox (1969) and Lawler (1972) dealing with related problems. In a cargo-loading problem considered by Kydland

(1969) profit per unit time is to be maximized. Both loading cost and loading time depend on the cargo chosen. In case of linear functions for revenue as well as linear functions for loading cost and time one obtains a linear fractional program; see Kydland (1969). If loading cost and loading time are convex and quadratic then a concave-convex quadratic fractional program occurs. On the other hand, if these functions are concave then a convex-concave fractional program arises.

Stochastic processes also give rise to the minimization of cost per unit time as demonstrated in a paper by Derman (1962); see also Hordijk (1974). An example of such a stochastic process is discussed by Klein (1963) who formulates a maintenance problem as a Markov decision process that then leads to a linear fractional program. Here the ratio of the expected cost for inspection, maintenance and replacement and the expected time between two inspections is to be minimized. An inventory problem where again the expected cost per unit time is to be minimized is discussed in Barlow and Proschan (1965; 115). The same objective is used in Brender (1963). Fox (1966) considers a Markov renewal programming problem where the objective is the minimization of the expected cost-to-time ratio. Depending on the nature of the cost function one has to deal with linear, quadratic or concave-convex fractional programs. In addition, convex-concave fractional programs are to be considered if the cost function is not convex but concave.

Another ratio optimization problem also involving time appears in a paper by Arisawa and Elmaghraby (1972). The authors consider a GERT-network. By additional investments the duration of the project can be reduced. The available funds are to be allocated to each of the activities in such a way that the reduction of duration per unit cost is maximized. The authors formulate this problem as a linear fractional program.

Charnes, Cooper and Rhodes (1978) recently used a linear fractional program to evaluate the activities of not-for profit entities participating in public programs. Given a collection of decision-making units the efficiency of any unit is obtained from the maximization of a ratio of weighted outputs and weighted inputs subject to the condition that the similar ratios for every decision-making unit be less than or equal to unity. The variable weights are then the efficiency of each member relative to that of the others. A linear fractional program has to be solved to determine these efficiencies.

We briefly mention that there are games that give rise to linear fractional programs; see for instance Isbell and Marlow (1956).

So far we have described fractional programs that arise when the efficiency of a system is to be optimized. Theoretical work by Eichhorn (1972, 1978) argues that the terms technical or economical "effectiveness" generally will be ratios of two functions. Only under more restrictive assumptions these terms are expressed as differences of two functions. Hence it is not surprising that the optimization of ratios comes up when the effectiveness (or efficiency) of a system is to be maximized.

In the management science literature there recently has been an increasing interest in optimizing relative terms such as relative profit; see Kern (1971). No longer these terms are merely used to control past economic behavior. Instead the optimization of rates is receiving more and more attention in decision-making processes for future projects; see Heinen (1970, 1971), Kern (1971), Meyer (1976).

There are a number of problems that give rise to a generalized fractional program, i.e., the maximization of the smallest of several ratios. An early application is J. von Neumann's model of an expanding economy; see von Neumann (1945). More recently generalized fractional programs have been discussed in goal programming and multicriteria optimization where a finite number of ratios is considered and Chebychev's norm is used. Examples can be found in Charnes-Cooper (1975) and Ashton-Atkins (1979).

2.2 Non-Economic Applications

Meister and Oettli (1967) consider the capacity of a discrete, constant communication channel. This term is defined as the maximum value of the transmission rate which is a ratio of a concave and a linear function. Hence a concave-convex fractional program has to be solved in order to determine the capacity of a communication channel. It is not a quadratic fractional program.

Fractional programs also occur in numerical analysis. The eigenvalues of a matrix can be calculated as constrained maxima of the Rayleigh quotient, a ratio of two quadratic forms (see Noble (1969)). Hence the eigenvalue problem gives rise to quadratic fractional programs.

Generalized fractional programs are met in discrete rational approximation where the Chebychev norm is used (Barrodale, 1973).

2.3 Indirect Applications

There are a number of operations research problems that indirectly give rise to a fractional program. Here it comes up as a surrogate problem or a subproblem.

Nonlinear objective functions in optimization models are often approximated by a linear or a quadratic function. However, there are problems where an approximation by a ratio of two functions seems to be more appropriate. This is the case for example if the objective function has poles, i.e., tends to infinity in finite points. Hence fractional programs may come up as approximations of a nonlinear program.

Concave-convex fractional programs which are quadratic arise in location theory in an indirect way. As Elzinga, Hearn and Randolph (1976) show the dual of certain minimax multifacility location problems is this type of model.

A good source of fractional programs is also large scale mathematical programming. Often some of the constraints in such a large model have a special structure and many coefficients are zero. Examples of some basic structures are the multidivisional or the multitime period problem; see Hillier and Lieberman (1979), Müller-Merbach (1973). Applying a decomposition principle to a large scale linear program this can be reduced to a finite number of smaller problems with linear fractional objective functions. Decomposition methods of that type are suggested by Lemke and Powers (1961), Abadie and Williams (1963) and by Bell (1965). The ratio in these fractional programs originates in the minimum-ratio-rule of the simplex method (second simplex criterion). The feasible region in the sequence of linear fractional programs is the same. It is determined by the specially structured constraints of the large scale linear program. Hence a sequence of linear fractional programs with specially structured constraints is obtained. An extension of the decomposition methods in linear programming to large scale nonlinear programs involving also fractional subproblems has been suggested by Whinston (1966).

Fractional programs occur in stochastic linear and nonlinear programming as first shown by Charnes and Cooper 1963) and Bereanu (1965).

Consider the optimization problem

$$\text{Max } \{c^T x \mid x \in S\} \tag{1}$$

where c is given by a joint normal distribution, T denotes the transpose and $S \subset R^n$ is a convex feasible region. There are different deterministic equivalents of the stochastic program (1). One is the maximum probability model associated with (1) (Charnes and Cooper (1963); Bereanu (1965)).

$$\text{Max } \{P(c^T x \geq k) \mid x \in S\} \tag{2}$$

where k is a given constant. An example of (2) would be the maximization of the probability that a certain profit k is achieved.

Problem (2) reduces to the nonlinear fractional program

$$\operatorname{Max}\left\{ q(x) = \frac{\bar{c}^T x - k}{\sqrt{x^T V x}} \;\middle|\; x \in S \right\} \tag{3}$$

as shown in Charnes and Cooper (1963), Bereanu (1965). Here $\bar{c}$ is the expected value of c and the matrix V is the variance-covariance matrix of c. The objective function in (3) is the ratio of an affine and a convex function. Hence for the stochastic convex program (1) the maximum probability model reduces to a concave-convex fractional program. We mention that the same nonquadratic fractional program is obtained in the portfolio selection problem by Ziemba, Parkan and Brooks-Hill (1974) discussed before. Ziemba (1974) shows that with c stable one gets a variety of fractional programs like (3) where $\sqrt{x^t V x}$ is replaced by a positive convex function.

In Schaible (1978) it is shown that the maximum probability model of nonlinear fractional programs leads to different types of nonlinear fractional programs (see also Stancu-Minasian (1976)). Since stochastic mathematical programs may arise in many different decision problems linear and nonlinear fractional programs come up indirectly in a variety of contexts.

3. CONCAVE-CONVEX FRACTIONAL PROGRAMS

There are a considerable number of decision problems that directly or indirectly give rise to various classes of fractional programs. In many cases the numerator and denominator of the objective function $q(x) = n(x)/d(x)$ is either affine or quadratic. On the other hand, other functions have to be considered too as we see from (3). Frequently concavity of n(x) and convexity of d(x) is found. For these concave-convex fractional programs a local maximum is a global one since the objective function q(x) is semistrictly quasiconcave in this case

(Mangasarian (1969a), Martos (1975), Schaible (1974a)). For the definition of semistrictly quasiconcave functions see Avriel, Diewert, Schaible and Ziemba (1981). Also for generalized concave-convex fractional programs the objective function, a minimum of finitely many concave-convex ratios, is semistrictly quasiconcave and therefore a local maximum is a global one.

If in a concave-convex fractional program the numerator n(x) is even strictly concave or the denominator is strictly convex then n(x)/d(x) is even strictly quasiconcave. This can easily be seen from the definition of strict quasiconcavity (see Avriel, Diewert, Schaible and Ziemba (1981)) using the fact that n(x)/d(x) is semistrictly quasiconcave in any concave-convex fractional program. For strictly quasiconcave functions we know that a maximum is unique if there exists one at all. Hence for concave-convex fractional programs where the numerator is strictly concave or the denominator is strictly convex there exists at most one maximum (see also Ziemba (1974)).

For differentiable functions n(x), d(x) and $h_i(x)$ a Kuhn-Tucker point of a concave-convex fractional program is a maximum since the objective function q(x) = n(x)/d(x) is pseudoconcave; see Mangasarian (1969a).

As we see concave-convex fractional programs, although not concave in general, still have some important properties in common with concave programs. Many algorithms of concave programming solve also nonlinear programs with a pseudoconcave objective function; see Avriel (1976), Martos (1975). Therefore in principle many concave-convex fractional programs can be solved by various standard methods of concave programming, for instance by the Frank-Wolfe (1956) algorithm as pointed out by Mangasarian (1969b).

At this point one may ask the question why there has been so much emphasis on deriving additional theoretical and algorithmic results in fractional programming. Discussions of this point as found in Mangasarian (1969b) are scarce in the literature. The author sees two major reasons for deriving theoretical and algorithmic results specially designed for fractional programs. Firstly, a very useful tool of concave programming, namely duality (Geoffrion (1971), Rockafellar (1970), Stoer and Witzgall (1970)) cannot be applied in its classical form to concave-convex fractional programs, not even to linear fractional programs as shown in Schaible (1973, 1976c). Hence dual solution methods of concave programming are not applicable in concave-convex fractional programming. Secondly, primal methods of concave (pseudoconcave) programming do not always exploit the structure of the objective function $q(x) = n(x)/d(x)$. Note that often $n(x)$ and $d(x)$ are affine or quadratic or they are of a related form as in (3). There is felt a need for methods that suitably use the structure of $n(x)$ and $d(x)$.

As far as duality is concerned we shall only briefly outline some basic results and then move on to algorithms in fractional programming. In Schaible (1976c) already a unified approach to duality in nondifferentiable and differentiable fractional programming was given. Results that appear until 1975 are related to each other in that approach. In Schaible (1978) additional results that have been obtained recently by different authors are considered.

There were suggested various approaches to define duality in concave-convex fractional programming in a meaningful way (see Bector (1973), Bitran and Magnanti (1976), Gol'stein (1967), Jagannathan (1973), Schaible (1973; 1976a; 1976c)).

The duality approach of the author is based on the following theorem. More special cases of this theorem appear in Bradley and Frey (1974), Charnes and Cooper (1962, 1963), Manas (1968), Mond and Craven

(1973). An application of it is given in Kallberg and Ziemba (1981). The theorem is due to Schaible (1973) (see also Schaible (1976a)).

THEOREM 1: A concave-convex fractional program (P) *can be reduced to the concave program*

(P') $\quad \text{Max } \{t\cdot n(y/t) \mid t > 0,\ t\cdot h_i(y/t) - t\cdot b_i \leqslant 0 \quad i=1,\dots, m,\ t\cdot d(y/t) \leqslant 1\}$

applying the variable transformation $y = (1/d(x))x$, $t = 1/d(x)$. *If* (y',t') *solves* (P') *then* $x' = y'/t'$ *solves* (P).

This theorem enables us to define duality in a simple and straightforward way: we consider a (classical) dual of the concave program (P') as the dual fractional program; see Schaible (1973, 1976a, 1976c). Using this approach we gain access to every duality theory in concave programming (Geoffrion (1971), Mangasarian (1969a), Rockafellar (1970), Stoer and Witzgall (1970)). Weak, strong, converse and other duality theorems including those on the existence of optimal primal or dual solutions have been obtained for concave-convex fractional programs.

The other approaches to duality in linear and nonlinear fractional programming lead to essentially one of the classical duals of the equivalent concave program (P') (Schaible (1978)). This is true for nondifferentiable fractional programs in case of Gol'stein's (1967) and Bitran-Magnanti's (1976) dual and it has also been shown for differentiable fractional programs in case of duals by Bector (1973) and Jagannathan (1973) (for details of these equivalences see Schaible (1976a; 1976c; 1978)). Meanwhile new concepts of duality in (nonlinear) fractional programming have been proposed by Borwein (1976), Craven (1981), Flachs and Pollatschek (1978), Mond and Weir (1981), and Passy (1981). Some of these papers deal with nondifferentiable fractional programs.

Without going into any detail we mention that additional duality relations can be obtained for quadratic fractional programs; see Schaible

(1976c, 1978). In this way classical duality results in quadratic programming are extended.

In case of linear fractional programming the equivalent problem (P') is a linear program as it was already shown by Charnes and Cooper (1962). Hence the dual of a linear fractional program is a linear program. It can be seen that almost all duality approaches in linear fractional programming yield essentially this dual. Other duals that were suggested are nonlinear programs that seem to be less useful. For a survey on duality in linear fractional programming see Abrham and Luthra (1977), Schaible (1976a). More recent results are considered in Schaible (1978).

Very recently duality for generalized concave-convex fractional programs has been investigated more thoroughly (Crouzeix (1981), Crouzeix, Ferland and Schaible (1981), Passy and Keslassy (1979)). In contrast to fractional programs involving only one ratio these problems cannot be reduced to a concave program in general. The duals are introduced in a different way. They are again generalized fractional programs. They appear in the Min-Max format whereas the primal is a Max-Min problem.

Sensitivity analysis and parametric programming in fractional programming has been discussed by Bitran and Magnanti (1976), Schaible (1973, 1975, 1978), Craven (1981). Earlier results on sensitivity for rather special cases are reviewed in Bitran and Magnanti (1976). Still more is to be done.

4. ALGORITHMS

We now turn to solution methods in fractional programming. The theorem above helps us to devise new algorithms for concave-convex fractional programs.

Strategy 1: Solve the equivalent concave program (P')*.*

In case of a linear fractional program the equivalent problem (P') is a linear program. Hence any solution method of linear programming can be used to solve (P). Wagner and Yuan (1968) in their survey on methods in linear fractional programming showed that other algorithms like the method by Martos (1964) and Swarup (1965) are algorithmicly equivalent to solving (P') by the simplex method. For further details on methods in linear fractional programming see Wagner and Yuan (1968). Recently Bitran (1978) compared various methods on a sample of randomly generated linear fractional programs. As the study shows Martos' method seems to be particularly good.

The theorem above proves that concave-convex fractional programs can be solved by any method of concave programming (Avriel (1976)). As we shall see below, for some fractional programs there will be other methods that are more appropriate however. These algorithms will not make use of the equivalent program (P'). On the other hand, there are nonlinear fractional programs where strategy 1 is recommendable. For instance the maximum probability model (3) reduces to a convex quadratic program if (P') is used (see Bergthaller (1970), Schaible (1973)). Similarly, the portfolio selection model discussed by Ziemba, Parkan and Brooks-Hill (1974) also is transformed into a convex quadratic program. For these models strategy 1 seems to be particularly useful. Here the structure of the objective function is exploited. Applying strategy 1 these problems can be solved by any method of quadratic programming.

The theorem above also suggests a second method for concave-convex fractional programs:

Strategy 2: Solve the dual of the equivalent concave program (P').

The advantage of this method is that we get also some insight into the sensitivity of the optimal value $\bar{q}(b) = \text{Max}\{q(x) \mid h_i(x) \leq b_i, i=1,\ldots,m\}$ with respect to right-hand side changes in the constraints. From an optimal solution of the dual of (P') the marginal values $\partial \bar{q}(b)/\partial b_i$ can be calculated as shown in Schaible(1975). Examples of this analysis are discussed in Schaible (1978). For sensitivity analysis in linear fractional programming see Bitran and Magnanti (1976), Kornbluth and Salkin (1972), Kydland (1972), and Schaible(1978).

Strategy 2 seems to be particularly appropriate in case of a quadratic fractional program with an affine denominator (numerator). Here the (Wolfe-) dual of (P') (Geoffrion (1971)) turns out to be a linear program with one additional convex quadratic constraint (Schaible(1974b)). If the matrix defining the quadratic function in the numerator (denominator) is nonsingular the dual fractional program essentially reduces to an optimization problem with a linear objective and one convex quadratic constraint. In any case the dual of these quadratic fractional programs can be solved by van de Panne's method (van de Panne (1966)) designed for linear programs with one additional quadratic constraint.

There is also a third algorithmic approach in fractional programming that actually is the oldest one. Jagannathan (1966) and Dinkelbach (1967) consider a parametric concave program related to a concave-convex fractional program (P) (for a similar parametric problem see Geoffrion (1967b)):

$$(P_q) \qquad \text{Max } \{n(x) - qd(x) \mid x \in S\}, \quad q \in R \text{ parameter.}$$

If the unique zero $\hat{q}$ of the strictly decreasing function $F(q) = \text{Max } \{n(x) - q\, d(x) \mid x \in s\}$ is determined then an optimal solution $\hat{x}$ of $(P_{\hat{q}})$ solves also (P) and we have $\hat{q} = q(\hat{x})$. For convenience compactness of S and continuity of n(x) and d(x) on S may be assumed although these conditions can be relaxed considerably (see Schaible (1976b)). In view of the properties of (P_q) a third way of solving (P) is the following:

Strategy 3: Calculate $\hat{q}$ and $\hat{x}$ for the parametric concave program (P_q).

One way of computing $\hat{q}$ is to solve the program (P_q) parametricly. Another method is suggested by Dinkelbach (1967). In his approach a sequence of parameter values $\{q_i\}$ is determined that converges to $\hat{q}$. He shows that the formula $q_{i+1} = q(x_i)$ generates such a sequence with x_i denoting an optimal solution of (P_{q_i}).

Again we ask the question: For which fractional programs is strategy 3 appropriate? One may even go a step further here. After strategy 3 had been suggested the theorem above was proved demonstrating that a concave-convex fractional program can be solved by a *single* concave program. Now calculating $\hat{q}$ and $\hat{x}$ involves a continuum of concave programs (P_q) or at least a finite number of problems (P_{q_i}) if Dinkelbach's method is used. There must be additional reasons to solve (P) by a parametric approach rather than by a single program (P'). Certainly, if n(x) and d(x) are just any functions then it would be not recommendable to solve (P) parametricly via (P_q). On the other hand, there may well be special cases where the structure of n(x) and d(x) is not well exploited if one solves (P') whereas it is if (P_q) is solved. This is the case if n(x) and d(x) are quadratic functions. Here (P_q) is a parametric quadratic program while the objective function in (P') is nonquadratic and a constraint is added that possibly is nonlinear. In case of quadratic fractional programming it may still be advantageous to solve (P) parametricly via (P_q). Using Dinkelbach's approach, it makes sense here to solve finitely many (concave) problems (P_q) instead of (P') if the number of subproblems (P_{q_i}) is small. Equivalently, this means that the sequence $\{q_i\}$ must converge fast. Recently it has been shown that the sequence $\{q_i\}$ converges superlinearly, i.e., $\lim_{i\to\infty} (\hat{q} - q_{i+1})/\hat{q} - q_i) = 0$ (see Schaible (1976b)). In addition one can prove that often $\{q_i\}$ converges locally at least quadratically. For quadratic fractional programs this is the case if the matrix in the numerator or denominator is nonsingular. The properties of Dinkelbach's method are extensively analysed in Schaible (1976b).

Since in quadratic fractional programming (P_q) exploits the structure and in addition $\{q_i\}$ converges fast, strategy 3 may still be appropriate. Some limited computations show (Mazzoleni (1973), Meyer (1975)) that only a few subproblems (P_{q_i}) have to be solved in quadratic fractional programming as predicted in Schaible (1976b). There also, a combination of a section method and Dinkelbach's method is suggested. Other modifications of Dinkelbach's algorithm have been presented in Bitran and Magnanti (1976), Ibaraki et al. (1976), Schaible (1976b). For a survey see Ibaraki (1981).

Ibaraki et al. (1976) compared their modified version of Dinkelbach's method with a parametric method for solving (P_q) on a sample of randomly generated quadratic fractional programs. The latter method turned out to be slightly faster. The parametric solution should gain more attention in the future than it has had in the past; see for instance Cottle (1972). Ritter (1967) suggested a parametric procedure that solves a class of concave-convex fractional programs. More recently Cambini, Martein and Pellegrini (1978), Cambini (1981) suggested another method by which linear and special classes of nonlinear fractional programs are solved parametricly.

Little work has been done to solve fractional programs where the ratio of two concave, two convex or the ratio of a convex and a concave function is to be maximized. As seen in Section 2, these problems are met for instance if cost functions are concave. Some first results concerning these fractional programs are presented in Schaible (1976d).

Also not much attention has been given to nonlinear programs where the sum of ratios is to be maximized. Almogy and Levin (1971) and

Ritter (1967) presented some algorithmic results. The special case of a sum of a linear and linear fractional function has been treated in Hirche (1975), Ritter (1967), Schaible (1977), Teterev (1969). In contrast to what was claimed by Teterev (1969) for these problems a local maximum is not a global one and an optimal solution is not attained at an extreme point of a polyhedron in general; see Hirche (1975). This is due to the fact that the sum of a linear and linear fractional function is neither quasiconcave nor quasiconvex in general as shown in Schaible (1977). Only under very restrictive assumptions on the linear and linear fractional function the sum is (semistrictly) quasiconcave or quasiconvex.

In the applications of fractional programming sometimes specially structured constraints are met (see for example Abadie and Williams (1963), Arisawa and Elmaghraby (1972), Bell (1965), Lemke and Powers (1961)). Some work has been done in this field already. Recently Charnes and Cooper (1973), Charnes, Granot and Granot (1976) suggested solution methods for linear fractional programs with two-sided linear inequality constraints.

In some applications the variables of a fractional program are restricted to be integers. For instance in investment planning such fractional programs are met. The variables are either any integers or they are to be 0 or 1. Solution methods for integer or 0-1 fractional programs have been suggested by a number of different authors; see Anzai (1974), Beale (1979), Bitran and Magnanti (1976), Florian and Robillard (1970), Granot and Granot (1976, 1977), Grunspan and Thomas (1973), Hammer and Rudeanu (1968), Ishii, Ibaraki and Mine (1976, 1977), Robillard (1971), Williams (1974). Recently Rhode (1978) gave a survey of these methods.

REFERENCES

ABADIE, J.M. and WILLIAMS, A.C. (1963). "Dual and Parametric Methods in Decomposition," in Graves, R. and Wolfe, P. (Eds.), *Recent Advances in Mathematical Programming*, McGraw-Hill, New York, 149-158.

ABRHAM, J. and LUTHRA, S. (1977). "Comparison of Duality Models in Fractional Linear Programming." *Zeitschrift für Operations Research 21*, 125-130.

ALMOGY, Y. and LEVIN, O. (1971). "A Class of Fractional Programming Problems." *Operations Research 19*, 57-67.

ANZAI, Y. (1974). "On Integer Fractional Programming." *Journal of the Operations Research Society of Japan 17*, 49-66.

ARISAWA, S. and ELMAGHRABY, S.E. (1972). "Optimal Time-cost Trade Offs in GERT-networks." *Management Science 18*, 589-599.

ASHTON, D.J. and ATKINS, D.R. (1979). "Multi-criteria Programming for Financial Planning." *Journal of the Operational Research Society 30*, 259-270.

AVRIEL, M. (1976). *Nonlinear Programming: Analysis and Methods*. Prentice-Hall, Englewood Cliffs, New Jersey.

AVRIEL, M., DIEWERT, W.E., SCHAIBLE, S. and W.T. ZIEMBA (1981). "Introduction to Concave and Generalized Concave Functions." This Volume, 21-50.

BARLOW, R.E. and PROSCHAN, F. (1965). *Mathematical Theory of Reliability*. Wiley, New York.

BARRODALE, I. (1973). "Best Rational Approximation and Strict Quasiconvexity." *SIAM Journal of Numerical Analysis 10*, 8-12.

BEALE, E.M.L. (1979). "Fractional Programming with Zero-One Variables," In *Proceedings of International Symposium in Extremal Methods and Systems Analysis*. (A. Fiacco, ed.). Springer, New York.

BECTOR, C.R. (1973). "Duality in Nonlinear Fractional Programming." *Zeitschrift für Operations Research 17*, 183-193.

BELL, E.J. (1965). "Primal-dual Decomposition Programming." Ph.D. Thesis, Operations Research Center, University of California at Berkeley.

BEREANU, B. (1965). "Decision Regions and Minimum Risk Solutions in Linear Programming," In *Colloquium on Applications of Mathematics to Economics*. (A. Prekopa, ed.). Publication House of the Hungarian Academy of Sciences, Budapest. 37-42.

BERGTHALLER, C.A. (1970). "Quadratic Equivalent of the Minimum Risk Problem." *Rev. Roum. Math. Pures et Appl. XV*, 17-23

BITRAN, G.R. (1978). "Experiments with Linear Fractional Problems". Technical Report 149, Operations Research Center, Massachusetts Institute of Technology.

BITRAN, G.R. and MAGNANTI, T.L. (1976). "Duality and Sensitivity Analysis for Fractional Programs." *Operations Research 24*, 675-699.

BORWEIN, J.M. (1976). "Fractional Programming Without Differentiability." *Mathematical Programming 11*, 283-290.

BRADLEY, S.P. and FREY, S.C. (1974). "Fractional Programming With Homogeneous Functions." *Operations Research 22*, 350-357.

BRENDER, D.M. (1963). "A Surveilance Model for Recurrent Events." IBM Watson Research Center Report.

CAMBINI, A, MARTEIN, L., and PELLEGRINI, L. (1978). "A Decomposition Algorithm for a Particular Class of Non-linear Programs *Proceedings of the AFCET-SMF Meeting*, Ecole Polytechnique, Palaiseau (Paris).

CAMBINI, A. (1981). "An Algorithm for a Special Class of Fractional Programs." This Volume, 491-508.

CHARNES, A. and COOPER, W.W. (1962). "Programming with Linear Fractional Functionals." *Naval Research Logistics Quarterly 9,* 181-186.

CHARNES, A. and COOPER, W.W. (1963). "Deterministic Equivalents for Optimizing and Satisficing Under Chance Constraints." *Operations Research 11,* 18-39.

CHARNES, A. and COOPER, W.W. (1973). "An Explicit General Solution in Linear Fractional Programming." *Naval Research Logistics Quarterly 20,* 449-467.

CHARNES, A. and COOPER, W.W. (1975). "Goal Programming and Multi-objective Optimization, Part I." Research Report CCS 250, Center for Cybernetic Studies, University of Texas at Austin, November.

CHARNES, A., GRANOT, D. and GRANOT, F. (1976). "An Algorithm for Solving General Fractional Interval Programming Problems." *Naval Research Logistics Quarterly 23,* 53-65.

CHARNES, A., COOPER, W.W., and RHODES, E. (1978). "Measuring the Efficiency of Decision Making Units." *European Journal of Operational Research 2,* 429-444.

COTTLE, R.W. (1972). "Monotone Solutions of the Parametric Linear Complementarity Problem." *Mathematical Programming 3,* 210-224.

CRAVEN, B.D. (1981). "Duality for Generalized Convex Fractional Programs. This volume, 473-489.

CROUZEIX, J.P. (1981). " A Duality Framework in Quasiconvex Programming." This Volume, 207-225.

CROUZEIX, J.P., FERLAND, J.A., and SCHAIBLE, S. (1981). "Duality in Generalized Linear Fractional Programming." Technical Report No. 399, Département d'informatique et de recherche opérationelle. Université de Montréal, Montréal.

DANTZIG, G.B., BLATTNER, W., AND RAO, M.R. (1966). "Finding a Cycle in a Graph with Minimum Cost to Time Ratio with Applications to a Ship Routing Problem," in: *Theory of Graphs,* Dunod, Paris and Gordon and Breach, New York, 77-83.

DERMAN, C. (1962). "On Sequential Decisions and Marcov Chains." *Management Science 9,* 16-24.

DEXTER, N.S., YU, J.N.W., and ZIEMBA, W.T. (1980). "Portfolio Selection in a Lognormal Market when the Investor Has a Power Utility Function: Computational Results," in Dempster, M.A.H. (Ed.), *Stochastic Programming,* Academic Press, New York, 507-523.

DINKELBACH, W. (1962). "Die Maximierung eines Quotienten zweier linearer Funktionen unter linearen Nebenbedingungen." *Zeitschrift für Wahrscheinlichkeitstheorie und Verwandte Gebiete 1,* 141-145.

DINKELBACH, W. (1967). "On Nonlinear Fractional Programming." *Management Science 13,* 492-498.

EICHHORN, W. (1972). "Effektivität von Produktionsverfahren." *Operations Research Verfahren 12,* 98-115.

EICHHORN, W. (1978). *Functional Equations in Economics.* Addison Wesley, New York.

ELZINGA, J., HEARN, D. and RANDOLPH, W.D. (1976). "Minimax Multifacility Location with Euclidian Distances." *Transportation Science 10,* 321-336.

FLACHS, J. and POLLATSCHEK, M. (1978). "Equivalence Between a Generalized Fenchel Duality Theorem and a Saddle-Point Theorem for Fractional Programs." Technical Report 114, Faculty of Industrial and Management Engineering, Technion, Haifa.

FLORIAN, M. and ROBILLARD, P. (1970). "Hyperbolic Programming with Bivalent Variables." Département d'Informatique, Université de Montréal, No. 41, August.

FOX, B. (1966). "Markov Renewal Programming By Linear Fractional Programming." *SIAM Journal of Applied Mathematics 14,* 1418-1432.

FOX, B. (1969). "Finding Minimum Cost-time Ratio Circuits." *Operations Research 17*, 546-550.
FRANK, M. and WOLFE, P. (1956). "An Algorithm for Quadratic Programming." *Naval Research Logistics Quarterly 3*, 95-110.
GEOFFRION, A.M. (1967a). "Strictly Concave Parametric Programming, Part I: Basic Theory." *Management Science 13*, 244-253; "Part II: Additional Theory and Computational Considerations." *Management Science 13*, 359-370.
GEOFFRION, A.M. (1967b). "Solving Bi-criterion Mathematical Programs." *Operations Research 15*, 39-54.
GEOFFRION, A.M. (1971). "Duality in Nonlinear Programming: A Simplified Applications-oriented Development." *SIAM Review 13*, 1-37.
GILMORE, P.C. and GOMORY, R.E. (1963). "A Linear Programming Approach to the Cutting Stock Problem - Part II." *Operations Research 11*, 863-888.
GOL'STEIN, E.G. (1967). "Dual Problems of Convex and Fractionally-Convex Programming in Functional Spaces." *Soviet Math. Dokl. 8*, 212-216.
GRANOT, D. and GRANOT, F. (1976). "On Solving Fractional (0,1) - Programs by Implicit Enumeration." *INFOR 14*, 241-249.
GRANOT, D. and GRANOT, F. (1977). "On Integer and Mixed Integer Programming Problems." *Annals of Discrete Mathematics 1*, 221-231.
GRUNSPAN, M. (1971). "Fractional Programming: A Survey." Technical Report 50, Department of Industrial and Systems Engineering, University of Florida.
GRUNSPAN. M. and THOMAS, M.E. (1973). "Hyperbolic Integer Programming." *Naval Research Logistics Quarterly 20*, 341-356.
GUTENBERG, E. (1975). *Einführung in die Betriebswirtschaftslehre.* Gabler, Wiesbaden.
HAMMER, P.L. and RUDEANU, S. (1968). *Boolean Methods in Operations Research and Related Areas.* Springer, Berlin - Heidelberg - New York.
HEINEN, E. (1970). "Betriebliche Kennzahlen, Eine Organisationstheoretische und Kybernetische Analyse," in: H. Linhardt, P. Penzkofer und P. Scherpf (eds.), *Dienstleistungen in Theorie und Praxis.* Stuttgart, 227-236.
HEINEN, E. (1971). *Grundlagen Betriebswirtschaftlicher Entscheidungen, das Zielsystem der Unternehmung.* Gabler, Wiesbaden.
HILLIER, F.S. and LIEBERMAN, G.J. (1979). *Introduction to Operations Research*, 3rd Ed. Holden Day, San Francisco.
HIRCHE, J. (1975). "Zur Extremwertannahme und Dualität bei Optimierungsproblemen Mit Linearem und Gebrochen-linearem Zielfunktionsanteil." *Zeitschrift für Angewandte Mathematik und Mechanik 55*, 184-185.
HORDIJK, A. (1974). "Dynamic Programming and Marcov Potential Theory." *Mathematical Centre Tracts 51*, Amsterdam.
IBARAKI, T. (1981). "Solving Mathematical Programming Problems With Fractional Objective Functions." This Volume, 441-472.
IBARAKI, T., ISHII, H., IWASE, J., HASEGAWA, T., and MINE, H. (1976). "Algorithms for Quadratic Fractional Programming Problems." *Journal of the Operations Research Society of Japan 19*, 174-191.
ISHII, H., IBARAKI, T. and MINE. H. (1976). "A Primal Cuttting Plane Algorithm for Integer Fractional Programming Problems." *Journal of the Operations Research Society of Japan 19*, 228-244.
ISHII, H., IBARAKI, T. and MINE, H. (1977). "Fractional Knapsack Problems." *Mathematical Programming 13*, 255-271.
ISBELL, J.R. and MARLOW, W.H. (1956). "Attrition Games." *Naval Research Logistics Quarterly 3*, 71-93.
JAGANNATHAN, R. (1966). "On Some Properties of Programming Problems in Parametric Form Pertaining to Fractional Programming." *Management Science 12*, 609-615.

JAGANNATHAN, R. (1973). "Duality for Nonlinear Fractional Programs." *Zeitschrift für Operations Research 17*, 1-3.
KALLBERG, J.G. and ZIEMBA, W.T. (1981). "Generalized Concave Functions in Stochastic Programming and Portfolio Theory." This Volume, 719-767.
KERN, W. (1971). "Kennzahlensysteme als Niederschlag Interdependenter Unternehmensplanung." *Zeitschrift für Betriebswirtschaftliche Forschung 23*, 701-718.
KLEIN, M. (1963). "Inspection-maintenance-replacement Schedule Under Marcovian Deterioration." *Management Science 9*, 25-32.
KORNBLUTH, J.S. and SALKIN. G.R. (1972). "A Note on the Economic Interpretation of the Dual Variables in Linear Fractional Programming." *Zeitschrift für Angewandte Mathematik und Mechanik 52*, 175-178.
KYDLAND, F. (1969). "Simulation of Linear Operators." Institute of Shipping Research, Norwegian School of Economics and Business Administration, Bergen; translated from *Sosialoekonomen 23*.
KYDLAND, F. (1972). "Duality in Fractional Programming." *Naval Research Logistics Quarterly 19*, 691-697.
LAWLER, E.M. (1972). "Optimal Cycles in Graphs and the Minimal Cost-to-time Ratio Problem." Technical Report ERL-M343, Department of Electrical Engineering, University of California, Berkeley.
LEMKE, C.E. and POWERS, T.J. (1961). "A Dual Decomposition Principle." *RPI Math. Rep. 48*, Troy, New York.
MANAS, M. (1968). "On Transformations of Quasiconvex Programming Problems." *Econ. Mat. Obzor 4*, 93-99.
MANGASARIAN, O.L. (1969a). *Nonlinear Programming*. McGraw-Hill, New York.
MANGASARIAN, O.L. (1969b). "Nonlinear Fractional Programming." *Journal of the Operations Research Society of Japan 12*, 1-10.
MANGASARIAN, O.L. (1970). "Convexity, Pseudo-convexity and Quasi-convexity of Composite Functions." *Cahiers du Centre d'Etudes de Recherche Operationelle 12*, 114-122.
MARTOS, B. (1964). "Hyperbolic Programming." *Naval Research Logistics Quarterly 11*, 135-155; originally published in *Math. Institute of Hungarian Academy of Sciences 5*, (1960) 383-406 (in Hungarian).
MARTOS, B. (1975). *Nonlinear Programming: Theory and Methods*. North-Holland, Amsterdam.
MAZZOLENI, P. (1973). "Some Experience on a Moving Truncation Method Applied to a Nonlinear Programming Problem with a Fractional Objective Function." Technical Report, Department of Mathematics, University of Venice.
MEISTER, B. and OETTLI, W. (1967). "On the Capacity of a Discrete, Constant Channel." *Information and Control 11*, 341-351.
MEYER, C. (1976). *Kennzahlen und Kennzahlensysteme*. Poeschel, Stuttgart.
MEYER, K.H. (1975). "Untersuchung eines Verfahrens zur Optimierung von Quotienten Quadratischer Funktionen." *Diplomarbeit*. Universität Köln.
MJELDE, K.M. (1978). "Allocation of Resources According to a Fractional Objective." *European Journal of Operational Research 2*, 116-124.
MOND, B. and CRAVEN, B.D. (1973). "A Note on Mathematical Programming With Fractional Objective Functions." *Naval Research Logistics Quarterly 20*, 577-581.
MOND, B. and WEIR, T. (1981). "Generalized Concavity and Duality." This Volume, 263-279.
MÜLLER-MERBACH, H. (1973). *Operations Research*, 3. Anfl. Vahlen, München.
NOBLE, B. (1969). *Applied Linear Algebra*. Prentice-Hall, Englewood-Cliffs, N.J.

OHLSON, J.A. and ZIEMBA, W.T. (1976). "Portfolio Selection in a Lognormal Market When the Investor has a Power Utility Function." *Journal of Financial and Quantitative Analysis 11*, 57-71.
PACK, L. (1962). "Maximierung der Rentabilität als Preispolitisches Ziel," in: H. Koch (ed.), *Zur Theorie der Unternehmung*, Festschrift für E. Gutenberg, Gabler, Wiesbaden, 73-135.
PACK, L. (1965). "Rationalprinzip, Gewinnprinzip und Rentabilität." *Zeitschrift für Betriebswirtschaft 35*, 525-551.
PASSY, U. (1981). "Pseudo Duality in Mathematical Programs with Quotients and Ratios." *Journal of Optimization Theory and Applications*, to appear.
PASSY, U. and KESLASSY, A. (1979). "Pseudo Duality and Duality for Explicitly Quasiconvex Functions." Mimeograph Series No. 249, Faculty of Industrial Engineering and Management, Technion, Haifa.
RHODE, P. (1978). "Verfahren zur Ganzzahligen Linearen Quotieten-Programmierung." *Diplomarbeit*, Köln.
RITTER, K. (1962). "Verfahren zur Lösung Parameterabhängiger Nichtlinearer Maximum-probleme." *Unternehmensforschung 6*, 149-166.
RITTER, K. (1967). "A Parametric Method for Solving Certain Nonconcave Maximization Problems." *Journal of Computer and System Sciences 1*, 44-54.
ROBILLARD, P. (1971). "(0,1) Hyperbolic Programming Problems." *Naval Research Logistics Quarterly 18*, 47-57.
ROCKAFELLAR, R.T. (1970). *Convex Analysis*. Princeton University Press, Princeton, N.J.
SCHAIBLE, S. (1971). "Beiträge zur Quasikonvexen Programmierung." Doctoral Dissertation, Universität Köln.
SCHAIBLE, S. (1972). "Quasi-convex Optimization in General Real Linear Spaces." *Zeitschrift für Operations Research 16*, 205-213.
SCHAIBLE, S. (1973). "Fractional Programming: Transformations, Duality and Algorithmic Aspects." Technical Report 73-9, Department of Operations Research, Stanford University, Stanford, CA.
SCHAIBLE, S. (1974a). "Maximization of Quasiconcave Quotients and Products of Finitely Many Functionals." *Cahiers du Centre d'Etudes de Recherche Operationelle 16*, 45-53.
SCHAIBLE, S. (1974b). "Parameter-free Convex Equivalent and Dual Programs of Fractional Programs." *Zeitschrift für Operations Research 18*, 187-196.
SCHAIBLE, S. (1975). "Marginalwerte in der Quotientenprogrammierung." *Zeitschrift für Betriebswirtschaft 45*, 649-658.
SCHAIBLE, S. (1976a). "Fractional Programming. I, Duality." *Management Science 22*, 858-867.
SCHAIBLE, S. (1976b). "Fractional Programming. II, On Dinkelbach's Algorithm." *Management Science 22*, 868-873.
SCHAIBLE, S. (1976c). "Duality in Fractional Programming: A Unified Approach." *Operations Research 24*, 452-461.
SCHAIBLE, S. (1976d). "Minimization of Ratios." *Journal of Optimization Theory and Applications 19*, 347-352.
SCHAIBLE, S. (1977). "On the Sum of a Linear and Linear-fractional Function." *Naval Research Logistics Quarterly 24*, 691-693.
SCHAIBLE, S. (1978). *Analyse und Anwendungen von Quotientenprogrammen*. Hain-Verlag, Meisenheim.
SCHAIBLE, S. and ZIEMBA, W.T. (1981). "On the Generalized Concavity of a Fractional Function Arising in Portfolio Theory." To appear.
STANCU-MINASIAN, I.M. (1976). "Asupra Problemei Lui Kataoka." *Studii si Cercetari Matematice 28*, 95-111.

STOER, J. and WITZGALL, Ch. (1970). *Convexity and Optimization in Finite Dimensions I."* Springer, Berlin-Heidelberg-New York.

SWARUP, K. (1965). "Linear Fractional Functionals Programming." *Operations Research 13,* 1029-1036.

TETEREV, A.G. (1969). "On a Generalization of Linear and Piece-wise Linear Programs" (Russian). *Ekon. i. Mat. Met. 5,* 440-447; English translation in *Matekon* (1970), 246-259.

VAN DE PANNE, C. (1966). "Programming with a Quadratic Constraint." *Management Science 12,* 798-815.

VON NEUMANN, J. (1945). "A Model of General Economic Equilibrium." *Review of Economic Studies 13,* 1-9.

WAGNER, H.M. and YUAN, J.S.C. (1968). "Algorithmic Equivalence in Linear Fractional Programming." *Management Science 14,* 301-306.

WHINSTON, A. (1966). "A Decomposition Algorithm for Quadratic Programing." *Cahiers du Centre d'Etudes de Recherche Operationnelle 8,* 112-131.

WILLIAMS, H.P. (1974). "Experiments in the Formulation of Integer Programming Problems." *Mathematical Programming Studies 2,* 180-197.

ZIEMBA, W.T. (1974). "Choosing Investment Portfolios When the Returns Have a Stable Distribution," in Hammer, P.L. and Zoutendijk (eds.), *Mathematical Programming in Theory and Practice.* North-Holland Publishing Company, 443-482.

ZIEMBA, W.T., PARKAN, C. and BROOKS-HILL, R. (1974). "Calculation of Investment Portfolios with Risk Free Borrowing and Lending." *Management Science 21,* 209-222.

SOLVING MATHEMATICAL PROGRAMMING PROBLEMS WITH FRACTIONAL OBJECTIVE FUNCTIONS

Toshihide Ibaraki[†]

This paper surveys algorithms for solving nonlinear programming problems where the objective function is the ratio of two given functions. There are basicly three types of approaches: (i) variable transformations, (ii) direct nonlinear programming approaches, and (iii) use of a related parameterized nonlinear program. Various strategies in these approaches are discussed putting emphasis on the third approach. The case of integer variables is also considered.

1. INTRODUCTION

This paper briefly surveys the algorithms for solving mathematical programming problems with fractional objective functions:

$$\begin{aligned} P:\ &\text{maximize} \quad f(x)/g(x) \\ &\quad \text{subject to } x \in X, \end{aligned} \tag{1}$$

where $f:R^n \to R$, $g:R^n \to R$ and $X \subset R^n$ (R denotes the set of real numbers). For simplicity $g(x) > 0$ is assumed for all $x \in X$. It is also assumed that X is nonempty and a finite optimal objective value exists. Depending upon whether X is a continuous set or a discrete set, P becomes a nonlinear programming problem or a nonlinear integer programming problem. (Existence of optimal solutions under less restrictive assumptions is discussed in Charnes-Cooper (1962a) and Martos (1964).)

A fractional programming problem arises whenever the optimization of ratios such as performance/cost, income/investment and cost/time is required. Typical applications include attrition games (Isbell-Marlow (1956)), transportation scheduling (Charnes-Cooper (1962b), Pollack-Novaes-Frankel (1965), Dantzig-Blattner-Rao (1967), Lawler (1976)), Markov decision processes resulting from maintenance and repair problems

[†]The author wishes to thank Professors H. Mine and T. Hasegawa of Kyoto University for their support, and Dr. H. Ishii of Osaka University for his careful reading of the earlier version of the manuscript and helpful comments. He is also indebted to Professor S. Schaible of the University of Alberta for many invaluable comments.

ISBN 0-12-621120-5

(Derman (1962), Klein (1962)), the cutting stock problem (Gilmore-Gomory (1963)), portfolio selection theory (Tobin (1958), Ziemba-Brooks-Hill-Parkan (1974)), Chebyshev maximization problems (Blau (1973)), decomposition algorithms for linear programming (Abadie-Williams (1963), Bell (1965), Lasdon (1970)), and the maximum probability models of stochastic programming (Charnes-Cooper (1963), Bereanu (1965), Vajda (1972)).

Pioneered by Isbell and Marlow (1956), and Charnes and Cooper (1962a), a number of solution algorithms have been proposed so far. They may be roughly classified into the following three categories.

(i) *Variable transformation:* By introducing new variables, transform problem P into an equivalent problem P' which is more tractable than P. Then solve P'.

(ii) *Direct nonlinear programming approach:* Regard problem P as a nonlinear programming problem and apply a suitable nonlinear programming algorithm.

(iii) *Use of the parameterized problem:* Define the auxiliary problem $Q(\lambda)$ by

$$Q(\lambda):\ \text{maximize}\quad f(x) - \lambda g(x) \qquad \text{subject to } x \in X. \tag{2}$$

Based on the relationship between P and $Q(\lambda)$, described in Section 4, P can be solved via a series of $Q(\lambda)$.

Each of these approaches will be explained in the subsequent sections. In particular, various strategies in approach (iii) are discussed in some detail.

An extensive bibliography of fractional programming, including over 300 references, was compiled by Stancu-Minasian (1977). More recently, Schaible (1978) published a monograph on fractional programming in which he relates the major contributions in this area to each other in a unified framework. For a comprehensive bibliography of applications, theory and algorithms in fractional programming, see Schaible (1978, 1981).

2. APPROACHES BY VARIABLE TRANSFORMATION

By using the new variables

$$t = 1/g(x)$$
$$y = xt, \qquad (3)$$

problem P can be transformed to the following problem P', as first introduced by Charnes-Cooper (1962a) for the linear fractional programming problem (defined in (5) below), and discussed by Bradley-Frey (1974), Schaible (1974, 1976c) and others for nonlinear fractional programming problems:

$$\begin{aligned} P'\text{: maximize} \quad & f(y/t)t \\ \text{subject to} \quad & g(y/t)t = 1 \qquad (4) \\ & y/t \in X, \ t > 0. \end{aligned}$$

This transformation is very useful in the particular case where f(x) and g(x) are affine functions, and X is given by linear inequalities, i.e.

$$\begin{aligned} f(x) &= c_0 + \textstyle\sum_{j=1}^n c_j x_j \\ g(x) &= d_0 + \textstyle\sum_{j=1}^n d_j x_j \qquad (5) \\ X &= \{x \in R^n \mid Ax \leqslant b, \ x \geqslant 0\}. \end{aligned}$$

(A is an $m \times n$ matrix, $b \in R^m$ and $g(x) > 0$ for all $x \in X$.) In this case, P is called the *linear fractional programming problem*. P' is then an ordinary linear programming problem:

$$\begin{aligned} P'\text{: maximize} \quad & c_0 t + \textstyle\sum_{j=1}^n c_j y_j \\ \text{subject to} \quad & d_0 t + \textstyle\sum_{j=1}^n d_j y_j = 1 \qquad (6) \\ & Ay - bt \leqslant 0 \\ & y, \ t \geqslant 0. \end{aligned}$$

Since P' can be efficiently solved by the simplex method, this is probably the most efficient method for the linear fractional programming problem.

If P is not linear, however, P' is no longer a linear programming problem. This is true if either f(x) or g(x) is nonlinear, even if X is defined by linear inequalities. Then we must resort to nonlinear programming algorithms.

Another example of the variable transformation is discussed in Williams (1974) for the *linear fractional 0-1 programming problem:*

$$\begin{aligned} \text{P: maximize} \quad & (c_0 + \textstyle\sum_{j=1}^{n} c_j x_j)/(d_0 + \textstyle\sum_{j=1}^{n} d_j x_j) \\ \text{subject to } & x \in X \\ & x_j = 0 \text{ or } 1,\ j=1,2,\ldots,n, \end{aligned} \tag{7}$$

where $d_0 + \sum_{j=1}^{n} x_j d_j > 0$ for all $x \in X$. Introduce new variables

$$y = (c_0 + \textstyle\sum_{j=1}^{n} c_j x_j)/(d_0 + \textstyle\sum_{j=1}^{n} d_j x_j),$$

namely

$$d_0 y + \textstyle\sum_{j=1}^{n} d_j x_j y - (c_0 + \textstyle\sum_{j=1}^{n} c_j x_j) = 0, \tag{8}$$

and

$$z_j = x_j y. \tag{9}$$

The last nonlinear equation can be replaced by the following linear inequalities:

$$\begin{aligned} & z_j - Ux_j \leqslant 0 \\ & z_j - y \leqslant 0 \\ & -z_j + y + Ux_j \leqslant U \\ & z_j \geqslant 0, \end{aligned} \tag{10}$$

where U is an upper bound of y. Then problem (7) is transformed to the following 0-1 mixed integer programming problem:

$$\begin{aligned} \text{P': maximize} \quad & y \\ \text{subject to } & x \in X \\ & x_j = 0 \text{ or } 1,\ j=1,2,\ldots,n \\ & d_0 y + \textstyle\sum_{j=1}^{n} d_j z_j - (c_0 + \textstyle\sum_{j=1}^{n} c_j x_j) = 0 \\ & \text{(10) for } j=1,2,\ldots,n. \end{aligned} \tag{11}$$

This problem contains n 0-1 variables x_j, n continuous variables z_j and one continuous variable y. Any algorithm for the 0-1 mixed integer programming problem can then be used to solve P'.

3. TREATMENT AS NONLINEAR PROGRAMMING

If f(x) and g(x) of (1) satisfy certain convexity or concavity assumptions, problem P becomes a rather tractable nonlinear programming problem.

THEOREM 1 (Mangasarian (1969a)): Let f(x) *and* g(x) *of (1) be differentiable on* X, f(x) *be concave on* X, g(x) *be convex on* X *and* $f(x) \geqslant 0$, $g(x) > 0$ *on* X *where* X *is a convex set. Then* f(x)/g(x) *is pseudoconcave and hence semistrictly quasiconcave on* X*.

A notable characteristic of a semistrictly quasiconcave function is that *every local maximum is a global maximum* (e.g., Mangasarian (1969b)). Therefore, many nonlinear programming algorithms such as gradient methods can be directly applied to concave-convex fractional programming problems to obtain global optimal solutions. The adjacent vertex method proposed for the linear fractional programming problem by generalizing the simplex method (Martos (1965)) may be included in this class of gradient methods. However, it will be explained in Section 8 as an algorithm in another category. Closely related to the gradient method is the *linearization technique* due to Kortanek-Evans (1967). Let h(x) be pseudoconcave and X be convex, and consider the following two problems.

$$
\begin{aligned}
&P: \text{maximize} \quad h(x) \\
&\qquad \text{subject to } x \in X. \\
&R(x^*): \text{maximize} \quad \nabla_x h(x^*)x \\
&\qquad\qquad \text{subject to } X \in X.
\end{aligned}
\qquad (12)
$$

Usually R(x*) is easier to solve than P because the objective function is

* For definitions of pseudoconcavity and semistrict quasiconcavity see Avriel-Diewert-Schaible-Ziemba (1981).

linear. P and $R(x^*)$ are related as follows.

THEOREM 2: In the above P *and* $R(x^*)$, x^* *is an optimal solution of* P *if and only if it is an optimal solution of* $R(x^*)$.

Thus P can be solved by finding an x^* which solves $R(x^*)$. For example, Bector-Jolly (1977) proposes a search strategy for such x^*, considering the case of integer variables. The approach using $R(x^*)$ is also related to the parametric approach discussed in the next section because $R(x^*)$ is equivalent to $Q(\lambda)$ of (2) with $\lambda = f(x^*)/g(x^*)$ if P is the linear fractional programming problem. But $Q(\lambda)$ and $R(x^*)$ are generally different in other cases.

Characterization of $f(x)/g(x)$ under other assumptions may be found in Mangasarian (1969a), Bector (1973) and Schaible (1976b).

4. USE OF THE PARAMETERIZED PROBLEM $Q(\lambda)$

The auxiliary problem $Q(\lambda)$ of (2) is closely related to P of (1), as the following theorem shows.

THEOREM 3 (Isbell-Marlow (1956), Jagannathan (1966), Dinkelbach (1967)): Let P *have an optimal solution* $\bar{x}$ *with* $\bar{\lambda} = f(\bar{x})/g(\bar{x})$. *Let* $z(\lambda)$ *denote the optimal value of* $Q(\lambda)$. *Then*

(A) $z(\lambda) > 0$ *if and only if* $\lambda < \bar{\lambda}$,

(B) $z(\lambda) = 0$ *if and only if* $\lambda = \bar{\lambda}$,

(C) $z(\lambda) < 0$ *if and only if* $\lambda > \bar{\lambda}$.

PROOF: We prove (A) only; (B) and (C) are similarly treated. $z(\lambda) > 0$ implies that a solution x' of $Q(\lambda)$ satisfies $f(x') - \lambda g(x') > 0$. Since $g(x') > 0$ by assumption, this implies $f(x')/g(x') > \lambda$. $\bar{\lambda} \geqslant f(x')/g(x')$ is also obvious since x' is a feasible solution of P. Conversely, $\lambda < \bar{\lambda}$ implies that there is a solution x'' of P satisfying $f(x'')/g(x'') > \lambda$ (e.g., consider the case of $x'' = \bar{x}$). By $g(x'') > 0$, this implies

$$z(\lambda) \geqslant f(x'') - \lambda g(x'') > 0. \qquad \square$$

COROLLARY 4: If $z(\lambda) = 0$ *in Theorem 3, then an optimal solution of* $Q(\lambda)$ *is also optimal in* P.

PROOF: $z(\lambda) = 0$ implies $\lambda=\bar{\lambda}$ by Theorem 3. Also an optimal solution x' of $Q(\lambda)$ satisfies $f(x') - \bar{\lambda}g(x') = 0$, implying $f(x')/g(x') = \bar{\lambda}$ by $g(x') > 0$. Thus x' is an optimal solution of P. □

By the above results, P can be solved by finding $\lambda=\bar{\lambda}$ satisfying $z(\lambda) = 0$; an optimal solution of $Q(\bar{\lambda})$ is optimal in P. The curve $z(\lambda)$ has rather nice properties for this purpose.

THEOREM 5 (Jagannathan (1966), Dinkelbach (1967)): The above $z(\lambda)$ *is continuous, strictly decreasing and convex over* R. *(Therefore* $z(\lambda) = 0$ *has the unique solution if there is any.)*

PROOF: Only the convexity of $z(\lambda)$ is proved for simplicity. Let x^t be an optimal solution of $Q(t\lambda' + (1-t)\lambda'')$, where $\lambda' < \lambda''$ and $0 \leqslant t \leqslant 1$. Then

$$\begin{aligned} z(t\lambda' + (1-t)\lambda'') &= f(x^t) - (t\lambda' + (1-t)\lambda'')g(x^t) \\ &= t(f(x^t) - \lambda' g(x^t)) + (1-t)(f(x^t) - \lambda'' g(x^t)) \\ &\leqslant tz(\lambda') + (1-t)z(\lambda''). \end{aligned}$$

This proves the convexity of z. □

A typical $z(\lambda)$ is illustrated in Fig. 1. Various procedures for solving P may result according as how the equation $z(\lambda) = 0$ is solved. Some details of such procedures will be given in the following sub-sections.

4.1 Direct Trace of $z(\lambda)$

In the rest of this paper, we assume $f(x) \geqslant 0$ for some $x \in X$. Combining with the previous assumption $g(x) > 0$ for all $x \in X$, we have

$$z(0) \geqslant 0 \text{ and } \bar{\lambda} \geqslant 0, \tag{13}$$

where $\bar{\lambda}$ is the optimal value of P. The simplest way to find a λ with

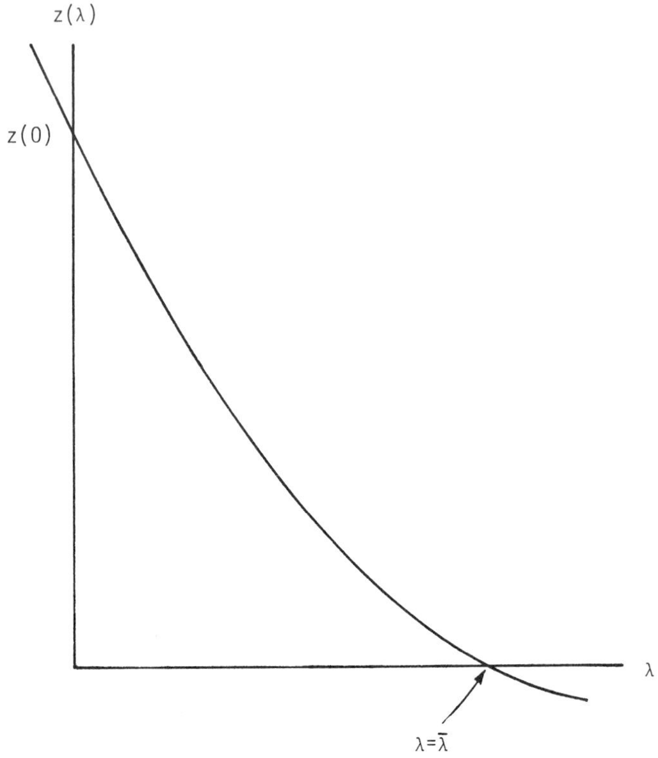

FIGURE 1. Illustration of curve $z(\lambda)$.

$z(\lambda) = 0$ would be to directly trace the curve $z(\lambda)$ starting from $\lambda = 0$.

Procedure DIRECT TRACE

Step 1: $\lambda \leftarrow 0$ (or any λ satisfying $\lambda \leqslant \bar{\lambda}$ may be used).

Step 2: Solve $Q(\lambda)$. If $z(\lambda) = 0$, halt. Otherwise go to Step 3.

Step 3: If $z(\lambda) > 0$, let $\lambda \leftarrow \lambda + \Delta$ and return to Step 2. Otherwise (i.e., $z(\lambda) < 0$), let $\lambda \leftarrow \lambda - \Delta$ and return to Step 2.

The increment (or decrement) Δ (> 0) in the above procedure is to be determined according to the current values of λ and $z(\lambda)$ so that the generated sequence of λ converges to $\bar{\lambda}$. This approach would be attractive if $Q(\lambda + \Delta)$ (or $Q(\lambda - \Delta)$) can be easily solved by using the knowledge about $Q(\lambda)$ (i.e., parametric programming technique).

As such an example, consider the *quadratic fractional programming problem:*

$$\begin{aligned} \text{P: maximize} \quad & (\tfrac{1}{2}x^T Cx + r^T x + s)/(\tfrac{1}{2}x^T Dx + p^T x + q) \\ \text{subject to} \quad & Ax \leqslant b, \end{aligned} \tag{14}$$

where C is an n×n symmetric negative definite matrix, D is an n×n symmetric positive semidefinite matrix, A is an m×n matrix, $r, p \in R^n$, $b \in R^m$, $s, q \in R$, $x = (x_1, x_2, \ldots x_n)^T$ is an n-vector of variables and T denotes the transpose. The numerator of the objective function is strictly concave, while the denominator is convex. It is also assumed that

$$\begin{aligned} & \tfrac{1}{2}x^T Cx + r^T x + s \geqslant 0 \text{ for some } x \in X \\ & \tfrac{1}{2}x^T Dx + p^T x + q > 0 \text{ for all } x \in X, \end{aligned} \tag{15}$$

where $X = \{x \mid Ax \leqslant b\}$. $Q(\lambda)$ for this P is given by

$$\begin{aligned} Q(\lambda)\text{: maximize} \quad & \tfrac{1}{2}x^T K(\lambda)x + c(\lambda)^T x + d(\lambda) \\ \text{subject to} \quad & Ax \leqslant b, \end{aligned} \tag{16}$$

where

$$\begin{aligned} K(\lambda) &= C - \lambda D \\ c(\lambda) &= r - \lambda p \\ d(\lambda) &= s - \lambda q. \end{aligned} \tag{17}$$

Note that $K(\lambda)$ is negative definite for $\lambda \geqslant 0$, and finite algorithms are known to solve such quadratic programming problems (e.g., Wolfe (1959), Lemke (1965), van de Panne (1975)).

The Kuhn-Tucker theorem (e.g., Mangasarian (1969b)) tells that an optimal solution of $Q(\lambda)$ is obtained by solving the following problem:

$$Q'(\lambda): \quad n+m\left\{\overbrace{\begin{bmatrix} K(\lambda) & A^T & 0 \\ & & \\ A & 0 & I_m \end{bmatrix}}^{n+m}\right. \begin{pmatrix} x \\ u \\ y \end{pmatrix} = \begin{pmatrix} -c(\lambda) \\ \\ b \end{pmatrix} \tag{18}$$

$$y^T u = 0$$
$$y \geqslant 0,\ u \leqslant 0, \tag{19}$$

where I_m is the unit matrix of order m, and y, u are m-vectors of variables. Let B be a basis of (18) (i.e., an $(n+m) \times (n+m)$ nonsingular submatrix), and let x_B, u_B and y_B denote the basic variables corresponding to the columns of B. Then

$$\begin{pmatrix} x_B \\ u_B \\ y_B \end{pmatrix} = B^{-1} \begin{pmatrix} -c(\lambda) \\ \\ b \end{pmatrix} \tag{20}$$

nonbasic variables = 0

is the basic solution corresponding to B. It is *feasible* if (19) is also satisfied, and the x-part of such a solution provides an optimal solution of $Q(\lambda)$. For each λ, there is the interval $[\lambda^{\ell}, \lambda^{u}]$ containing λ, for which the same basis provides an optimal solution of $Q(\lambda)$. Thus the curve $z(\lambda)$ is partitioned corresponding to such intervals. This is illustrated in Fig. 2.

The theory of parametric programming says that two bases corresponding to adjacent intervals can be obtained by one pivot operation (i.e., by changing one column of B). Thus the trace of $z(\lambda)$ from $\lambda = 0$ is carried out by setting Δ equal to the size of the current interval. Initially $Q(0)$ is solved by an existing algorithm for the quadratic programming problem. But each $Q(\lambda + \Delta)$ can be solved from $Q(\lambda)$ by applying only one pivot operation. The details of such computation are discussed in Wolfe (1959) and Ritter (1967) (the first reference treats the case of $D = 0$). Furthermore, if $z(\lambda) > 0$ and $z(\lambda + \Delta) < 0$, the equation $z(\lambda) = 0$ is equal to

$$\tfrac{1}{2}x^T K(\lambda) x + c(\lambda)^T x + d(\lambda) = 0$$
$$x \text{ is given by (20).} \tag{21}$$

This is a quadratic equation if $D = 0$. The solution of this equation which lies in interval $[\lambda, \lambda + \Delta]$ provides $\bar{\lambda}$. Thus the case of

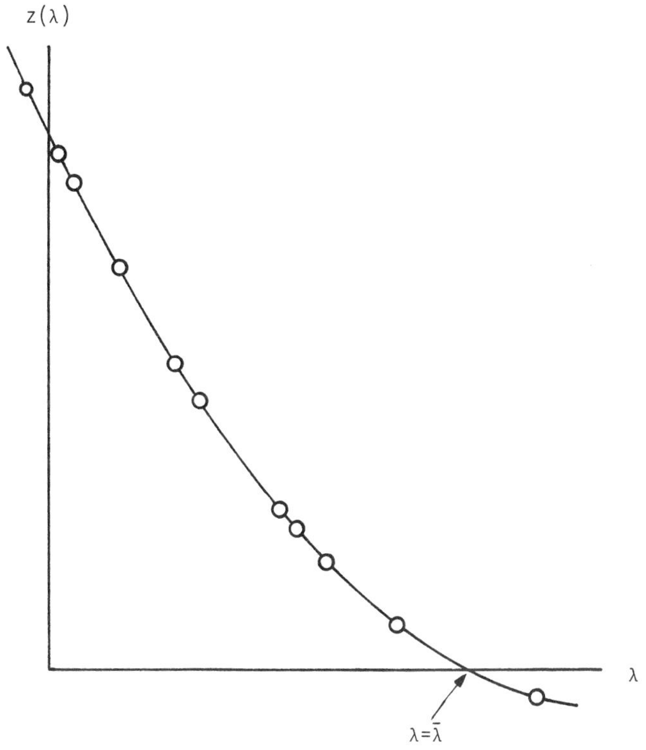

FIGURE 2. Partition of the curve $z(\lambda)$ corresponding to optimal bases (○ denotes a point of basis change)

$z(\lambda) < 0$ in Step 3 of DIRECT TRACE is not necessary.

The proof for the finite convergence of the above procedure and some computational results are given in Ibaraki et al. (1976). Part of the computational results will be cited in the next subsection.

4.2 Application of the Newton-like Method

The nonlinear equation $z(\lambda) = 0$ may be solved by means of a Newton-like method well known in numerical analysis. Starting with $\lambda_0 = 0$, a sequence of $\lambda_0, \lambda_1, \lambda_2, \ldots$ is generated; λ_{i+1} is determined as the point where the straight line tangent to $z(\lambda)$ at $\lambda = \lambda_i$ crosses the λ-axis. This is illustrated in Fig. 3. Note here that the tangent line can be defined even if $z(\lambda)$ is not differentiable but has a subgradient at $\lambda = \lambda_i$.

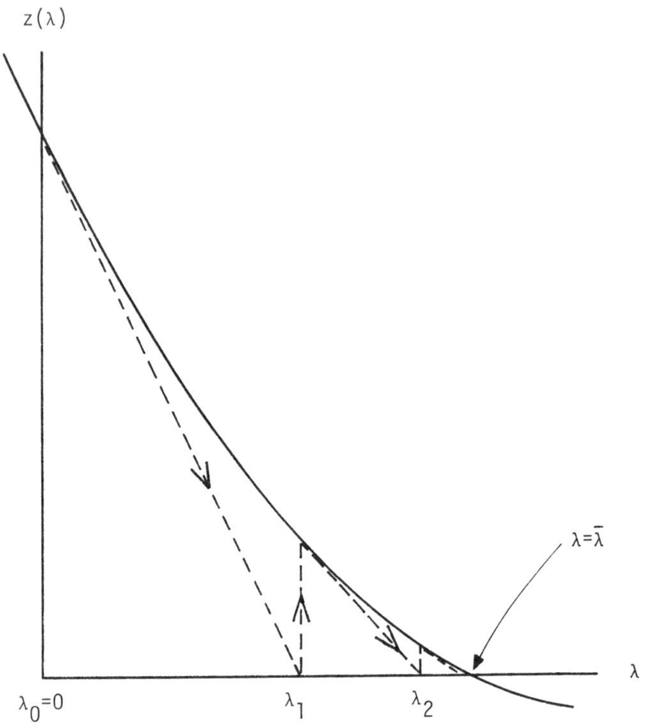

FIGURE 3. The Newton method for solving $z(\lambda)=0$.

Although the Newton method is usually considered only for the differentiable case, we use the term "Newton" because of its obvious similarity. This procedure is also called Dinkelbach's method after his paper in 1967.

To carry out the Newton method, it is important to see that the following line is tangent to $z(\lambda)$ at $\lambda = \lambda_i$.

$$y = f(x^{(i)}) - \lambda g(x^{(i)}),$$

where $x^{(i)}$ is an optimal solution of $Q(\lambda_i)$. At $\lambda=\lambda_i$, y is equal to $z(\lambda)$ by definition. For other λ, $z(\lambda) \geqslant f(x^{(i)}) - \lambda g(x^{(i)})$ is also obvious since $x^{(i)}$ is a feasible solution of $Q(\lambda)$.

The above line crosses the λ-axis at

$$\lambda_{i+1} = f(x^{(i)})/g(x^{(i)}) \tag{22}$$

and hence this value may be used in the iterations of the Newton method.

As a general characteristic of the Newton method, the sequence $\{\lambda_i\}$ converges to $\bar{\lambda}$ rather quickly. In fact it is shown in Schaible (1976d) that the convergence is superlinear. In the actual computation, the iteration is terminated if λ_i comes sufficiently close to $\bar{\lambda}$.

Procedure NEWTON

Step 1: $\lambda \leftarrow 0$ (or an appropriate initial value).

Step 2: Solve $Q(\lambda)$. If $z(\lambda) \leqslant \delta$ for a given nonnegative constant δ, halt. Otherwise go to Step 3.

Step 3: Let $\lambda \leftarrow f(x')/g(x')$, where x' is an optimal solution of $Q(\lambda)$ obtained in Step 2, and return to Step 2.

This type of procedure was first proposed by Isbell and Marlow (1956) for the linear fractional programming problem, and then by Dinkelbach (1967) for the general nonlinear fractional programming problem.

If $\delta > 0$, the procedure terminates after a finite number of iterations. In this case, the obtained approximate optimal solution x' satisfies

$$z(\lambda') = |\, z(\lambda') - z(\bar{\lambda}) | \leqslant \delta, \tag{23}$$

where $\lambda' = f(x')/g(x')$. Another rule proposed by Bitran-Magnanti (1976) is to use

$$\lambda \leftarrow (f(x')/g(x')) + \varepsilon$$

in Step 3, where $\varepsilon \geqslant 0$ is a given constant (δ in Step 2 is set to 0). In this case, the obtained approximate optimal solution x' satisfies

$$\bar{\lambda} \leqslant (f(x')/g(x')) + \varepsilon. \tag{24}$$

Schaible (1976d) contains yet another rule which makes use of an optimal solution x'' of $Q(\lambda'')$ with $\lambda'' > \bar{\lambda}$.

As the initial $\lambda = \lambda_0$ of Step 1, any $\lambda_0 \leqslant \bar{\lambda}$ may be used in place of $\lambda_0 = 0$. For example, $\lambda_0 = f(x')/g(x')$ for a feasible solution x' of P (i.e., $x' \in X$) always satisfies $\lambda_0 \leqslant \bar{\lambda}$ since $f(x')/g(x') \leq f(\bar{x})/g(\bar{x})$, where $\bar{x}$ is an optimal solution of P. In this case, it is proved in Schaible (1976d) that the i-th solution x^i obtained in Step 2 of NEWTON initiated with such λ_0 satisfies

$$\bar{\lambda} - (f(x^i)/g(x^i)) \leqslant (\lambda_1 - \lambda_0)K^i/(1-K), \tag{25}$$

where $K = (g(x')/g(\bar{x})) - 1$. Therefore if we start NEWTON with λ_0 which is close enough to $\bar{\lambda}$ (i.e., K is close to 0), the convergence speed of NEWTON becomes much faster.

Specializing P in the above discussion, a number of papers have been published; Florian-Robillard (1970), Grunspan-Thomas (1973) and Anzai (1974) treat the fractional integer programming problem, Bitran-Novaes (1973) the linear fractional programming problem, and Ibaraki et al. (1976) the quadratic fractional programming problem. Here we shall discuss the last problem in some detail.

The initial $Q(\lambda)$ in Step 1 of NEWTON may be solved by any existing algorithm for the quadratic programming problem. The subsequent $Q(\lambda)$, however, can be solved by using the sensitivity analysis technique (Wolfe (1959), Ritter (1967))from the data obtained for the previous $Q(\lambda)$. Although this is much faster than solving $Q(\lambda)$ from scratch at each iteration, it still requires a considerable amount of computation if compared with the one pivot operation required by one iteration of DIRECT TRACE. The number of iterations of Step 2 and Step 3 required by NEWTON is however much smaller than that required by DIRECT TRACE.

The termination rule may also be modified in the case of the quadratic fractional programming problem. As discussed in Section 4.1, the interval $[\lambda^{\ell}, \lambda^{u}]$ containing a given λ can be computed, for which the basis B solving (18)(19) does not change. If $z(\lambda^u) \leqslant 0$ is found in Step 2 of NEWTON, therefore, $\lambda = \bar{\lambda}$ satisfying $z(\lambda) = 0$ can be directly computed from (21). With this mechanism, NEWTON becomes a finite algorithm even if $\delta = 0$ (and $\varepsilon = 0$) is used, i.e., an exact optimal solution is sought. The formal proof for the finiteness and further elaboration of the algorithm are found in Ibaraki et al. (1976).

Table 1 lists the computational results of some randomly generated quadratic fractional programming problems (14), in which $D = 0$ is assumed. Both DIRECT TRACE and NEWTON are tested. The computation times required by two algorithms are about the same; DIRECT TRACE seems to be slightly faster than NEWTON. It should be emphasized that the major part of the entire computation is spent on solving the initial quadratic programming problem. The rest of computation consumes only about 10% of the total time. This shows that the parametric programming and sensitivity analysis techniques are quite powerful in reducing the computation time at each iteration.

4.3 Use of Heuristic Algorithms in NEWTON

In procedure NEWTON, the auxiliary problem $Q(\lambda)$ is solved for many values of λ. If $Q(\lambda)$ is difficult to solve, this is quite a computational burden. Since the main purpose of solving $Q(\lambda)$ is to know whether $z(\lambda) > 0$ or $z(\lambda) = 0$, the computation of the exact value $z(\lambda)$ is not always necessary. Therefore, a heuristic algorithm for $Q(\lambda)$ (i.e., an algorithm for obtaining a good feasible solution quickly) may be used in place of an exact algorithm in the following manner.

(1) Solve $Q(\lambda)$ by a heuristic algorithm. If a feasible solution x' with $f(x') - \lambda g(x') > 0$ is obtained, $z(\lambda) > 0$ is concluded.

(2) Otherwise, solve $Q(\lambda)$ by an exact algorithm.

Usually a heuristic algorithm consumes much less computation time than an exact algorithm, and hence this approach may greatly save the total computation time.

To use the above idea in NEWTON, we need also the next property.

THEOREM 6: Let x', x''$\in$X *and let* λ = f(x'')/g(x''). *Then* f(x')- λg(x')> 0 *holds if and only if* f(x')/g(x') >f(x'')/g(x'').

PROOF: Obvious by g(x') > 0 and g(x'') > 0. □

In other words, whenever a feasible solution x' of Q(λ) with f(x') - λg(x') > 0 is obtained, the new value λ' = f(x')/g(x') satisfies $\lambda < \lambda' \leqslant \bar{\lambda}$.

TABLE 1. Computational Results for Quadratic Fractional Programming Problems

Problem Size	Algorithm	Initial Q(λ)		Other Q(λ)		Iteration[d]	Total[c,e] Time (Sec.)
		Pivot[a]	Time[c] (Sec.)	Pivot[b]	Time[c] (Sec.)		
n=40	DIRECT-TRACE	344	31.8	25	2.7	25	34.5
m=40	NEWTON	344	31.8	22	2.5	2	34.3
n=50	DIRECT-TRACE	586	85.3	25	4.2	26	89.5
m=50	NEWTON	586	85.3	53	9.6	2	94.9
n=50	DIRECT-TRACE	606	83.2	27	4.5	28	87.7
m=50	NEWTON	606	83.2	68	12.5	2	95.7
n=50	DIRECT-TRACE	575	71.1	24	3.9	25	75.0
m=50	NEWTON	575	71.1	39	6.7	2	77.8
n=50	DIRECT-TRACE	694	110.6	29	4.8	30	115.4
m=50	NEWTON	694	110.6	57	9.5	3	120.1

Notes

(a) The number of pivots required to solve the initial Q(λ) by Wolfe's method.

(b) The number of pivots required to solve all the other Q(λ).

(c) Machine is FACOM 230/75 (roughly corresponds to IBM 370/165).

(d) The number of times Step 2 is executed.

(e) This includes the time for Step 3.

Procedure NEWTON WITH HEURISTICS

Step 1: $\lambda \leftarrow 0$ (or an appropriate initial value)[†].

Step 2: Solve $Q(\lambda)$ by a heuristic algorithm. If a feasible solution x' with $f(x') - \lambda g(x') > \delta$ is obtained, where δ is a given nonnegative constant, go to Step 3; otherwise solve $Q(\lambda)$ by an exact algorithm. If $z(\lambda) \leqslant \delta$, halt. Otherwise let x' be an optimal solution of $Q(\lambda)$ and go to Step 3.

Step 3: Let $\lambda \leftarrow f(x')/g(x')$ and return to Step 2.

This type of approach was first used in Ishii-Ibaraki-Mine (1977) for the *fractional knapsack problem:*

$$\begin{aligned} \text{P: maximize} \quad & (c_0 + \textstyle\sum_{j=1}^{n} c_j x_j)/(d_0 + \sum_{j=1}^{n} d_j x_j) \\ \text{subject to} \quad & \textstyle\sum_{j=1}^{n} a_j x_j \leqslant b \\ & x_j\text{: nonnegative integer, } j=1,2,\ldots,n. \end{aligned} \tag{26}$$

For this problem, $Q(\lambda)$ is the ordinary knapsack problem.

$$\begin{aligned} Q(\lambda)\text{: maximize} \quad & c_0(\lambda) + \textstyle\sum_{j=1}^{n} c_j(\lambda) x_j \\ \text{subject to} \quad & \textstyle\sum_{j=1}^{n} a_j x_j \leqslant b \\ & x_j\text{: nonnegative integer, } j=1,2,\ldots,n, \end{aligned} \tag{27}$$

where

$$c_j(\lambda) = c_j - \lambda d_j, \; j=1,2,\ldots,n. \tag{28}$$

A good feasible solution is provided by the greedy algorithm (Hu-Lenard (1973), Magazine-Nemhauser-Trotter (1975)):

$$\begin{aligned} x_1' &= \lfloor b/a_1 \rfloor \\ x_k' &= \lfloor (b - \textstyle\sum_{j=1}^{k-1} a_j x_j')/a_k \rfloor \quad k=2,3,\ldots,n, \end{aligned} \tag{29}$$

where $\lfloor\cdot\rfloor$ denotes the integer part and the indices are arranged in such a way that

[†] If a heuristic algorithm for solving P is available, it may be used to obtain the initial $\lambda = f(x')/g(x')$, where x' is the obtained feasible solution.

$$\frac{c_1(\lambda)}{a_1} \geqslant \frac{c_2(\lambda)}{a_2} \geqslant \cdots \geqslant \frac{c_n(\lambda)}{a_n} \qquad (30)$$

Although the knapsack problem $Q(\lambda)$ can be rather efficiently solved by dynamic programming or by branch-and-bound (even though it is known to be NP-complete), the greedy solution (29) is much faster to compute. The finite convergence and further properties are discussed in Ishii-Ibaraki-Mine (1977).

4.4 Iterative Improvement of Feasible Solutions

The primal algorithms such as the primal simplex method generate a sequence of feasible solutions converging to an optimal solution. Based on the idea of Subsection 4.3, the generation of feasible solutions of $Q(\lambda)$ for a given λ may be cut off whenever a feasible solution x' with $f(x') - \lambda g(x') > 0$ is obtained. The computation is resumed after updating the value of λ to $f(x')/g(x')$. The difference between this method and NEWTON WITH HEURISTICS is that a primal algorithm obtains an optimal solution of $Q(\lambda)$ upon termination, while a heuristic algorithm does not (hence another exact algorithm should be called upon).

Procedure PRIMAL

Step 1: $\lambda \leftarrow 0$ (or an appropriate initial value).

Step 2: Apply a primal method to $Q(\lambda)$. As soon as a feasible solution x' with $f(x') - \lambda g(x') > \delta$ is obtained in this computation, where δ is a given nonnegative constant, go to Step 3. Otherwise (i.e., the computation is carried out until termination and $z(\lambda) \leqslant \delta$ is proved), halt.

Step 3: Let $\lambda \leftarrow f(x')/g(x')$ and return to Step 2.

By Theorem 6, the value of λ strictly increases at each iteration.

This approach was first applied to linear fractional programming problems (6)by Martos (1964; 1965) (Swarup (1965) also contains almost the same algorithm). In this case, $Q(\lambda)$ is the linear programming problem and the primal simplex method may be used in Step 2.

Assume that $\delta = 0$ is used in PRIMAL. Let u_i $(i=1,2,\ldots,m)$ be the current basic variables and t_j $(j=1,2,\ldots,n)$ be the current nonbasic variables. Expanding basic variables and two functions $z_1 = f(x)$ and $z_2 = g(x)$ by nonbasic variables, the following simplex tableau results.

$$\begin{aligned} z_1 &= \beta_{00} + \textstyle\sum_{j=1}^{n} \beta_{0j} \, (-t_j) \\ z_2 &= \gamma_{00} + \textstyle\sum_{j=1}^{n} \gamma_{0j} \, (-t_j) \\ u_i &= \alpha_{i0} + \textstyle\sum_{j=1}^{n} \alpha_{ij} \, (-t_j), \ i=1,2,\ldots,m \\ u_i&, \ t_j \geqslant 0. \end{aligned} \tag{31}$$

The objective function of $Q(\lambda)$ is given by

$$(\beta_{00} - \lambda\gamma_{00}) + \textstyle\sum_{j=1}^{n}(\beta_{0j} - \lambda\gamma_{0j}) \, (-t_j). \tag{32}$$

Assume that λ is currently set to

$$\lambda = \beta_{00}/\alpha_{00} \tag{33}$$

so that $(\beta_{00} - \lambda\alpha_{00})$ of (32) is equal to 0. The current tableau (31) and (32) represents an optimal solution of $Q(\lambda)$ if and only if

$$(\beta_{0j} - \lambda\gamma_{0j}) \geqslant 0, \ j=1,2,\ldots,n. \tag{34}$$

If this is the case, the algorithm halts by indicating $z(\lambda) = 0$. On the other hand, if some j satisfies

$$(\beta_{0j} - \lambda\gamma_{0j}) < 0, \tag{35}$$

a pivot operation is applied according to the primal simplex method, and the new tableau is computed. The above computation is then repeated with the new tableau and the new λ given by (33).

The proof for the finite convergence can be done in a manner similar to the ordinary simplex method (under an appropriate assumption of non-degeneracy).

As discussed in Martos (1965) and mentioned in Section 3, this approach may also be viewed as a nonlinear programming algorithm directly applied to the linear fractional programming problem (5). This approach is also related to problem (6) obtained by variable transformation. In fact the sequence of basic solutions generated by the above procedure is exactly the same as the one generated by the primal simplex algorithm applied to (6), as proved in Wagner-Yuan (1968).

Bitran (1979) reports some computational results of NEWTON, PRIMAL and their variants applied to linear fractional programming problems. PRIMAL seems to give better performance than others.

PRIMAL can also be used for the *linear fractional integer programming problem*.

$$\begin{aligned} \text{P: maximize} \quad & \left(c_0 + \textstyle\sum_{j=1}^{n} c_j x_j\right) / \left(d_0 + \textstyle\sum_{j=1}^{n} d_j x_j\right) \\ \text{subject to} \quad & \textstyle\sum_{j=1}^{n} a_{ij} x_j \leqslant b_i, \ i=1,2,\ldots,m \\ & x_j : \text{nonnegative integer}, \ j=1,2,\ldots,n. \end{aligned} \tag{36}$$

As pointed out by Ishii-Ibaraki-Mine (1976) and Granot-Granot (1977), the primal cutting plane method for the linear integer programming problem such as proposed by Young (1968) and Glover (1968) may be used. The computational details are similar to the above linear programming case, except that a cut is generated at each iteration and a pivot operation is executed on this cut-row.

4.5 Megiddo's Approach to Solve $z(\lambda) = 0$

Assume that we have an algorithm A to solve $Q(\lambda)$. As stated in Corollary 4, an optimal solution of $Q(\bar{\lambda})$ is also an optimal solution of P. Thus we want to apply Algorithm A to $Q(\bar{\lambda})$. However, the exact value of $\bar{\lambda}$ is not known unless an optimal solution of P is obtained. Megiddo (1979) found an ingeneous way to resolve this dilemma.

Let Algorithm A applied to $Q(\bar{\lambda})$ contain conditional jump operations to select a proper computation path according to conditions involving the value of $\bar{\lambda}$, i.e.,

$$\bar{\lambda} > \lambda^*,\ \bar{\lambda} = \lambda^* \text{ or } \bar{\lambda} < \lambda^*, \tag{37}$$

where λ^* are constants specific to conditional jump operations. Also assume that only conditional jump operations are relevant to $\bar{\lambda}$. To know which of (37) holds at each conditional jump operation, $Q(\lambda^*)$ is solved by Algorithm A. Based on the value of $z(\lambda^*)$, a proper computation path is determined by Theorem 3 even if the exact value of $\bar{\lambda}$ is not known.

If Algorithm A executes at most $O(a(n))$[†] conditional jump operations and $O(b(n))$ other operations (i.e., A requires $O(a(n) + b(n))$ time), where n denotes the size of the given problem, the total number of operations required by the above algorithm is

$$O(a(n)[a(n) + b(n)] + b(n)), \tag{38}$$

since Algorithm A is called at most a(n) times to determine condition (37).

This algorithm is interesting if $Q(\lambda)$ has an efficient Algorithm A. Various examples of this type are included in Meggido (1979), e.g., the minimum ratio spanning tree problem and the minimum ratio cycle problem.

To illustrate how such algorithms are implemented, we treat here the fractional resource allocation problem and try to improve the algorithm efficiency by modifying the standard procedure.

The *fractional resource allocation problem* is a simple integer programming problem defined by

$$\begin{aligned} \text{P: minimize} \quad & z(x) = \textstyle\sum_{j=1}^{n} f_j(x_j) / \sum_{j=1}^{n} g_j(x_j) \\ \text{subject to} \quad & \textstyle\sum_{j=1}^{n} x_j = N \\ & x_j: \text{nonnegative integer, } j=1,2,\ldots,n, \end{aligned} \tag{39}$$

where $\sum_{j=1}^{n} g_j(x_j) > 0$ is assumed for any feasible x. The parameterized problem $Q(\lambda)$ is then given by

---

† $T = O(f(n))$ indicates that T is bounded above by cf(n), where c is a constant.

$$Q(\lambda):\quad \text{minimize} \quad \sum_{j=1}^{n} (f_j(x_j) - \lambda g_j(x_j))$$
$$\text{subject to} \quad \sum_{j=1}^{n} x_j = N \tag{40}$$
$$x_j: \text{nonnegative integer}, \ j=1,2,\ldots,n.$$

The standard technique of dynamic programming (e.g., Bellman (1957)), Dreyfus-Law (1977)) can be applied to solve $Q(\bar{\lambda})$. Let

$$h^{(k)}(y) = \min\{\sum_{j=1}^{k}(f_j(x_j) - \bar{\lambda}g_j(x_j)) \mid \sum_{j=1}^{k}x_j = y,\ x_j: \text{nonnegative integer}\}, \tag{41}$$
$$k = 1,2,\ldots,n,\ y=0,1,\ldots,N.$$

Based on the principle of optimality of dynamic programming, it is not difficult to show that $h^{(k)}(y)$ are computed by the boundary condition

$$h^{(k)}(0) = \sum_{j=1}^{k} f_j(0) - \bar{\lambda}\sum_{j=1}^{k} g_j(0),\ k=1,2,\ldots,n$$
$$h^{(1)}(y) = f_1(y) - \bar{\lambda}g_1(y),\ y=0,1,\ldots,N \tag{42}$$

and the recurrence formula

$$h^{(k)}(y) = \min\{h^{(k-1)}(y-\ell) + (f_k(\ell) - \bar{\lambda}g_k(\ell)) \mid \ell = 0,1,\ldots,y\}, \tag{43}$$
$$k=2,3,\ldots,n,\ y=1,2,\ldots,N.$$

Obviously $h^{(n)}(N)$ gives the optimal vaue of $Q(\bar{\lambda})$, which is 0.

If $\bar{\lambda}$ is a given constant, this formula is directly computed for $k=1,2,\ldots,n$ (in this order), each with all values of y. The required number of elementary operations (such as additions, subtractions and comparisons of two numbers) is $O(nN^2)$, provided that each evaluation of $f_j(y)$ or $g_j(y)$ is done in a constant number of operations.

Since the exact value of $\bar{\lambda}$ is not known, however, each $h^{(k)}(y)$ is represented in the form of

$$h^{(k)}(y) = f^{(k)}(y) - \bar{\lambda}g^{(k)}(y),$$

and $f^{(k)}(y)$ and $g^{(k)}(y)$ are stored instead of $h^{(k)}(y)$.

To compute the minimum in (43), let

$$h^{(k)}(y, \ell) = (f^{(k-1)}(y-\ell) - \bar{\lambda} g^{(k-1)}(y-\ell)) + f_k(\ell) - \bar{\lambda} g_k(\ell).$$

Then the condition

$$h^{(k)}(y, \ell) \begin{pmatrix} \leqslant \\ \geqslant \end{pmatrix} h^{(k)}(y, m) \tag{44}$$

is written as follows:

$$f^{(k-1)}(y-\ell) + f_k(\ell) - f^{(k-1)}(y-m) - f_k(m)$$
$$\begin{pmatrix} \leqslant \\ \geqslant \end{pmatrix} \bar{\lambda} \, (g^{(k-1)}(y-\ell) + g_k(\ell) - g^{(k-1)}(y-m) - g_k(m)).$$

In other words, (44) is determined by

$$\lambda^{(k)}(y, \ell, m) \begin{pmatrix} \leqslant \\ \geqslant \end{pmatrix} \bar{\lambda}, \tag{45}$$

where

$$\lambda^{(k)}(y, \ell, m) = \frac{f^{(k-1)}(y-\ell)+f_k(\ell)-f^{(k-1)}(y-m)-f_k(m)}{g^{(k-1)}(y-\ell)+g_k(\ell)-g^{(k-1)}(y-m)-g_k(m)} . \tag{46}$$

(The sign of the denominator of (46) should be taken into consideration to determine the direction of (44).) Thus the minimum in (43), i.e., $h^{(k)}(y)$, can be computed by checking condition (45) at most N times.

To facilitate this computation, we obtain

$$\lambda_{max}^{(k)} = \max\{\lambda^{(k)}(y, \ell, m) \mid \lambda^{(k)}(y, \ell, m) \leqslant \bar{\lambda}, \quad y=0,1,\ldots,N, \ \ell=0,1,\ldots,y, \ m=0,1,\ldots,y\}, \tag{47}$$

before computing $h^{(k)}(y)$ (a similar technique is already suggested in Megiddo (1979) for other problems). Then we have obviously

$$\begin{aligned} &\lambda^{(k)}(y, \ell, m) > \bar{\lambda} \text{ if } \lambda^{(k)}(y, \ell, m) > \lambda_{max}^{(k)} \\ &\lambda^{(k)}(y, \ell, m) \leqslant \bar{\lambda} \text{ if } \lambda^{(k)}(y, \ell, m) \leqslant \lambda_{max}^{(k)} , \end{aligned} \tag{48}$$

and this is used to determine the direction of (45).

The computation of $\lambda_{max}^{(k)}$ can be done by applying binary search over the set

$$S^{(k)} = \{\lambda^{(k)}(y, \ell, m) \mid y=0,1,\ldots,N,\ \ell=0,1,\ldots,y,\ m=0,1,\ldots,y\}.$$

To carry out this binary search, it is necessary (i) to compute the medians of the selected subsets of $S^{(k)}$ and (ii) to check whether $\lambda^{(k)}(y, \ell, m) \leqslant \bar{\lambda}$ holds or not for the obtained medians $\lambda^{(k)}(y, \ell, m)$. It is shown by Blum et al. (1972) that the median of $S^{(k)}$ can be computed in $O(N^3)$ time since $|S^{(k)}| = O(N^3)$. Since the subsets considered by binary search are halved each time, the time spent to find medians in binary search is

$$O(N^3 + \tfrac{1}{2}N^3 + \tfrac{1}{4}N^3 + \ldots) = O(N^3).$$

(This trick is first proposed by Balas-Zemel (1976) and Lawler (1977).) The checking of the condition $\lambda^{(k)}(y, \ell, m) \leqslant \bar{\lambda}$ in (ii) is done by solving $Q(\lambda^{(k)}(y, \ell, m))$ by dynamic programming (e.g., by (42) and (43) with $\bar{\lambda}$ replaced by $\lambda^{(k)}(y, \ell, m)$). Since this dynamic programming computation requires $O(nN^2)$ time for each problem and at most $O(\log N^3) = O(\log N)$ problems are solved during the execution of binary search, the time required to carry out the computation of (ii) is $O(nN^2 \log N)$ for each k.

The entire algorithm is now summarized as follows.

Procedure FRACTIONAL RESOURCE ALLOCATION

Note: P of (39) is solved.

Step 1: Compute $f^{(1)}(y)$ and $g^{(1)}(y)$ of $h^{(1)}(y)$ for $y=0,1,\ldots,N$ by (42). Let $k \leftarrow 2$.

Step 2: Compute $\lambda^{(k)}_{max}$ as described above.

Step 3: Compute $f^{(k)}(y)$ and $g^{(k)}(y)$ of $h^{(k)}(y)$ for $y=0,1,\ldots,N$ as described above.

Step 4: If $k=n$, halt. The solution realizing $h^{(n)}(N)$ is an optimal solution. Otherwise, let $k \leftarrow k + 1$ and return to Step 2. □

The solution $\bar{x}$ realizing $h^{(n)}(N)$ of Step 4 can be easily retrieved if some additional data are stored during the computation of $h^{(k)}(y)$. The details are not given here since it is a standard technique of dynamic programming.

The time required by FRACTIONAL RESOURCE ALLOCATION is analyzed below. Step 2 requires $O(N^3 + nN^2 \log N)$ time for each k, as noted previously; thus $O(nN^3 + n^2N^2 \log N)$ time for all k. Step 3 requires $O(nN^2)$ time in total, since condition (45) (by using (48)) is checked $O(N)$ times to compute each $h^{(k)}(y)$ and there are $O(nN) h^{(k)}(y)$'s. Other computation time is obviously dominated by these two. Consequently FRACTIONAL RESOURCE ALLOCATION requires

$$O(nN^3 + n^2N^2 \log N) \tag{49}$$

computation time. This time bound is better than $O(n^2N^4)$ directly obtained from (38).

4.6 Other Approaches to Solve $z(\lambda) = 0$

There are other approaches to solve $z(\lambda) = 0$, some of which are briefly described below.

Assume that two problems $Q(\lambda')$ and $Q(\lambda'')$ with $\lambda' < \bar{\lambda} < \lambda''$ are solved (i.e., $z(\lambda') > 0$ and $z(\lambda'') < 0$). As illustrated in Fig. 4, the convexity of $z(\lambda)$ (Theorem 5) implies that the point $\lambda = v(\lambda', \lambda'')$ at which the straight line connecting two points $(\lambda', z(\lambda'))$ and $(\lambda'', z(\lambda''))$ crosses $z(\lambda) = 0$ satisfies

$$\bar{\lambda} \leqslant v(\lambda', \lambda'') < \lambda''. \tag{50}$$

The same procedure may then be repeated by using λ' and $v(\lambda', \lambda'')$ as initial values.

Procedure OUTER

Step 1: Find λ' and λ'' such that $z(\lambda') > 0$ and $z(\lambda'') < 0$. $\lambda \leftarrow v(\lambda', \lambda'')$ (v is defined above).

Step 2: Solve $Q(\lambda)$. If $|z(\lambda)| \leqslant \delta$ for a given nonnegative constant δ, halt. Otherwise go to Step 3.

Step 3: Let $\lambda \leftarrow v(\lambda', \lambda)$ and return to Step 2.

With this procedure, a sequence starting from λ'' $(\geqslant \bar{\lambda})$ and converg-

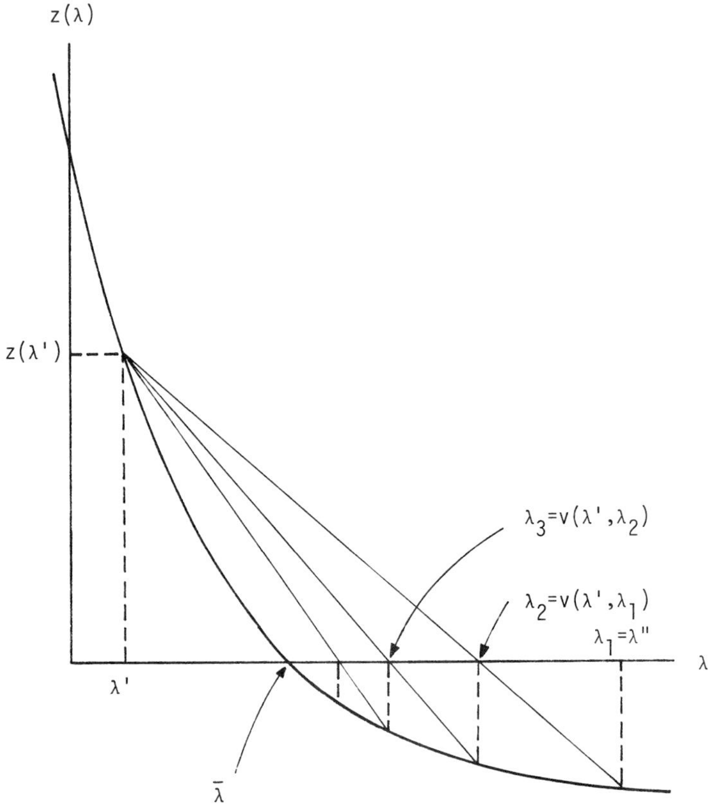

FIGURE 4. Convergence sequence $\{\lambda_i\}$ generated by OUTER.

ing to $\bar{\lambda}$ from above is generated, as indicated in Fig. 4. Note that each λ generated in Step 3 is infeasible in the sense that no feasible solution x realizes $\lambda=f(x)/g(x)$. However the obtained solution is feasible because $Q(\lambda)$ and P have the same feasible region. This method was proposed by Schaible (1976d).

When an interval $[\lambda', \lambda'']$ containing $\bar{\lambda}$ is known, another common approach is to apply binary search in order to find $\bar{\lambda}$. In general, let $[\lambda^{\ell}, \lambda^{u}]$ denote an interval containing $\bar{\lambda}$. The middle value $\lambda = (\lambda^{\ell} + \lambda^{u})/2$ is then tested. If $z(\lambda) \geqslant 0$, $\lambda \leqslant \bar{\lambda}$ holds by Theorem 3 and the interval $[\lambda, \lambda^u]$ is considered in the next step; otherwise the other half $[\lambda^\ell, \lambda]$ is considered.

Procedure BINARY

Step 1: Find λ' and λ'' such that $z(\lambda') > 0$ and $z(\lambda'') < 0$. Let $\lambda^\ell \leftarrow \lambda'$ and $\lambda^u \leftarrow \lambda''$.

Step 2: Let $\lambda \leftarrow (\lambda^\ell+\lambda^u)/2$ and solve $Q(\lambda)$. If $|z(\lambda)| \leqslant \delta$ for a given nonnegative constant δ, halt. Otherwise go to Step 3.

Step 3: If $z(\lambda) > 0$, let $\lambda^\ell \leftarrow \lambda$ and return to Step 2. Otherwise (i.e., $z(\lambda) < 0$), let $\lambda^u \leftarrow \lambda$ and return to Step 2.

Since the interval $[\lambda^\ell, \lambda^u]$ is halved at each iteration, the convergence rate of BINARY is 1/2, i.e., linear convergence. In this respect, NEWTON seems to be superior to BINARY because NEWTON converges superlinearly. BINARY has been sometimes used to derive polynomial upper bounds on the number of steps required to solve the fractional versions of network type problems (Lawler (1976) and Chandrasekaran (1977)). (Megiddo's approach however seems to provide better bounds for these problems.)

A two-stage procedure was suggested by Schaible (1976d). With BINARY a good approximation λ' of $\bar{\lambda}$ is calculated and then NEWTON is started with λ'. The switchover from one to the other method is controlled by a parameter which can be chosen in such a way that the advantages of both methods are fully exploited.

If two problems $Q(\lambda')$ and $Q(\lambda'')$ corresponding to $[\lambda', \lambda'']$ containing $\bar{\lambda}$ are solved, we may obtain a smaller interval containing $\bar{\lambda}$:

$$f(x')/g(x') \leqslant \bar{\lambda} \leqslant v(\lambda', \lambda''), \tag{51}$$

where x' is an optimal solution of $Q(\lambda')$. (51) is obvious from the previous discussion. This observation gives rise to the following variation of BINARY.

Procedure MODIFIED BINARY

Step 1: Find λ' and λ'' such that $z(\lambda') > 0$ and $z(\lambda'') < 0$. Let $\lambda^{\ell} \leftarrow f(x')/g(x')$ where x' is an optimal solution of $Q(\lambda')$, and $\lambda^{u} \leftarrow v(\lambda', \lambda'')$.

Step 2: Let $\lambda \leftarrow (\lambda^{\ell} + \lambda^{u})/2$ and solve $Q(\lambda)$. If $|z(\lambda)| \leqslant \delta$ for a given nonnegative constant δ, halt. Otherwise go to Step 3.

Step 3: If $z(\lambda) > 0$, let $\lambda' \leftarrow \lambda$ and $\lambda^{\ell} \leftarrow f(x^*)/g(x^*)$, where x^* is an optimal solution of $Q(\lambda)$, and return to Step 2. Otherwise, let $\lambda^{u} \leftarrow v(\lambda', \lambda)$ and return to Step 2.

5. DISCUSSION

Various algorithms for solving fractional programming problems have been surveyed, putting emphasis on those algorithms based on the auxiliary problem $Q(\lambda)$. But this does not intend to be exhaustive. For example, there are other interesting approaches, which may be categorized as direct nonlinear programming treatment of Section 3, e.g., Hammer-Rudeanu (1968) and Granot-Granot (1976) for the fractional integer programming problem.

Sensitivity analysis and parametric programming techniques of fractional programming are also not mentioned.

Another aspect of fractional programming not discussed in this paper is the *duality theory*. It is pointed out by Craven-Mond (1973) and Schaible (1974, 1976a, 1976c, 1978) that the Lagrangean dual to problem (4) obtained by variable transformation shares many nice properties with classical Lagrangean duality for concave programming problems. Other types of duality are discussed in Jagannathan (1973), Bector (1973) and Bitran-Magnanti (1976). Further exploration to use the duality theory for developing efficient algorithms may turn out to be fruitful. For some first results see Schaible (1974, 1976d, 1981a).

As a concluding remark of this paper, we may state that, if a mathematical programming problem can be solved with certain computational complexity, its fractional version can be usually solved with only a slight additional effort. This is clearly true in the case of the linear fractional programming problem, and confirmed by the actual computational experiment for the quadratic fractional programming problem.

REFERENCES

ABADIE, J.M. and WILLIAMS, A.C. (1963). "Dual and Parametric Methods in Decomposition", in *Recent Advances in Mathematical Programming*, Graves, R.L. and Wolfe, P. (eds.), 149-158, McGraw-Hill, New York.

ANZAI, Y. (1974). "On Integer Fractional Programming." *J. Operations Research Society of Japan 17*, 49-66.

AVRIEL, M., DIEWERT, W.E., SCHAIBLE, S. and ZIEMBA, W.T. (1981). "Introduction to Concave and Generalized Concave Functions." This volume, 21-50.

BALAS, E. and ZEMEL, E. (1976). "Solving Large Zero-One Knapsack Problems." *Management Science Research Report No. 408*, Carnegie-Mellon University.

BECTOR, C.R. (1973). "Duality in Nonlinear Fractional Programming." *Zeitschrift für Operations Research 17*, 183-193.

BECTOR, C.R. and JOLLY, P.L. (1977). "Pseudo-Monotonic Integer Programming," in *Proceedings of the 7th Manitoba Conference on Numerical Mathematics and Computing*, 211-218.

BELL, E.J. (1965). "Primal-Dual Decomposition Programming." Ph.D. Thesis, Operations Research Center, University of California at Berkeley, Report ORC 65-23.

BELLMAN, R.E. (1957). *Dynamic Programming*. Princeton University Press, Princeton, N.J.

BEREANU, B. (1965). "Decision Regions and Minimum Risk Solutions in Linear Programming," in *Colloquim on Applications of Mathematics to Economics*, Prekopa, A. (ed.), pp. 37-42, Hungarian Academy of Science, Budapest.

BITRAN, G.R. and NOVAES, A.G. (1973), "Linear Programming with a Fractional Objective Function." *Operations Research 21*, 22-29.

BITRAN, G.R. and MAGNANTI, T.L. (1976). "Duality and Sensitivity Analysis for Fractional Programs." *Operations Research 24*, 675-699.

BITRAN, G.R. (1979). "Experiments with Linear Fractional Problems." *Naval Res. Log. Quarterly 26*, 689-693.

BLAU, R.A. (1973). "Decomposition Techniques for the Chebyshev Problem." *Operations Research 21*, 1163-1167.

BLUM, M., Floyd, R.W., PRATT, R., RIVEST, R.L. and TARJAN, R.E. (1972). "Time Bounds for Selection." *Journal of Computer and System Sciences 7*, 448-461.

BRADLEY, S.P. and FREY, S.C., Jr. (1974). "Fractional Programming with Homogeneous Functions." *Operations Research 22*, 350-357.

CHANDRASEKARAN, R. (1977). "Minimum Ratio Spanning Trees." *Networks 7*, 335-342.

CHARNES, A. and COOPER, W.W. (1962a). "Programming with Linear Fractional Functionals." *Naval Res. Log. Quarterly 9*, 181-186.

CHARNES, A. and COOPER, W.W. (1962b). "Systems Evaluation and Repricing Theorems." *Management Science 9*, 33-49.
CHARNES, A. and COOPER, W.W. (1963). "Deterministic Equivalents for Optimizing and Satisfying under Chance Constraints." *Operations Research 11*, 18-39.
CRAVEN, B.D. and MOND, B. (1973). "The Dual of a Fractional Linear Program." *Journal of Mathematical Analysis and Applications 42*, 507-512.
DANTZIG, G.B., BLATTNER, W. and RAO, M.R. (1967). "Finding a Cycle in a Graph with Minimum Cost to Time Ratio with Application to a Ship Routing Problem." In *Theory of Graphs* (P. Rosenstiehl, ed.). Dunod, Paris, and Gordon and Breach, New York, 78-84.
DERMAN, C. (1962). "On Sequential Decisions and Markov Chains." *Management Science 9*, 16-24.
DINKELBACH, W. (1967). "On Nonlinear Fractional Programming." *Management Science 13*, 492-498.
DREYFUS, S.E. and LAW, A.W. (1977). *The Art and Theory of Dynamic Programming.* Academic Press, New York.
FLORIAN, M. and ROBILLARD, P. (1970). "Hyperbolic Programming with Bivalent Variables." Département d'Informatique, Université de Montréal, Publication No. 41.
FRANK, M. and WOLFE, P. (1956). "An Algorithm for Quadratic Programming." *Naval. Res. Log. Quarterly 3*, 95-110.
GILMORE, P.C. and GOMORY, R.E. (1963). "A Linear Programming Approach to the Cutting Stock Problem - Part II." *Operations Research 11*, 863-888.
GLOVER, F. (1968). "A New Foundation for a Simplified Primal Integer Programming Algorithm." *Operations Research 16*, 727-740.
GRANOT, D. and GRANOT, F. (1976). "On Solving Fractional (0,1) Programs by Implicit Enumeration." *INFOR 14*, 241-249.
GRANOT, D. and GRANOT, F. (1977). "On Integer and Mixed Integer Fractional Programming Problems." *Annals of Discrete Mathematics 1*, 221-231.
GRUNSPAN, M. and THOMAS, M.E. (1973). "Hyperbolic Integer Programming." *Naval Res. Log. Quarterly 20*, 341-356.
HAMMER, P.L. and RUDEANU, S. (1968). *Boolean Methods in Operations Research and Related Areas.* Springer-Verlag, Berlin.
HU, T.C. and LENARD, M.L. (1973). "A Study of a Heuristic Algorithm." MRC Technical Report No. 1370, University of Wisconsin-Madison.
IBARAKI, T., ISHII, H., IWASE, J., HASEGAWA, T. and MINE, H. (1976). "Algorithms for Quadratic Fractional Programming Problems." *Journal of the Operations Research Society of Japan 19*, 174-191.
ISBELL, J.R. and MARLOW, W.H. (1956). "Attrition Games." *Naval Res. Log. Quarterly 3*, 71-93.
ISHII, H., IBARAKI, T. and MINE, H. (1976). "A Primal Cutting Plane Algorithm for Integer Fractional Programming Problem." *Journal of the Operations Research Society of Japan 19*, 228-244.
ISHII, H., IBARAKI, T. and MINE, H. (1977). "Fractional Knapsack Problems." *Mathematical Programming 13*, 255-271.
JAGANNATHAN, R. (1966). "On Some Properties of Programming Problems in Parametric Form Pertaining to Fractional Programming." *Management Science 12*, 609-615.
JAGANNATHAN, R. (1973). "Duality for Nonlinear Fractional Programs." *Zeitschrift für Operations Research 17*, 1-3.
KLEIN, M. (1962). "Inspection-Maintenance-Replacement-Schedule under Markovian Deterioration." *Management Science 9*, 25-32.

KORTANEK, K.O. and EVANS, J.P. (1967). "Pseudo-Concave Programming and Lagrange Regularity." *Operations Research 15*, 882-891.

LASDON, L.S. (1970). *Optimization Theory for Large Systems.* MacMillan,

LAWLER, E.L. (1976). *Combinatorial Optimization: Networks and Matroids.* Holt, Rhinehart, and Winston, New York, 94-97.

LAWLER, E.L. (1977). "Fast Approximation Algorithms for Knapsack Problems." Memorandum UCB/ERL M77/45, University of California, Berkeley.

LEMKE, C.E. (1965). "Bimatrix Equilibrium Points and Mathematical Programming." *Management Science 11*, 681-689.

MAGAZINE, M.J., NEMHAUSER, G.L. and TROTTER, L.E. (1975). "When the Greedy Solution Solves a Class of Knapsack Problems." *Operations Research 23*, 207-217.

MANGASARIAN, O.L. (1969a). "Nonlinear Fractional Programming." *Journal of the Operations Research Society of Japan 12*, 1-10.

MANGASARIAN, O.L. (1969b). *Nonlinear Programming.* McGraw-Hill, New York.

MARTOS, B. (1964). "Hyperbolic Programming (translated by A. and V. Whinston)." *Naval Res. Log. Quarterly 11*, 135-155.

MARTOS, B. (1965). "The Direct Power of Adjacent Vertex Programming Methods." *Management Science 17*, 241-252.

MEGIDDO, N. (1979). "Combinatorial Optimization with Rational Objective Functions." *Mathematics of Operations Research 4*, 414-424.

POLLACK, E.G., NOVAES, G.N. and FRANKEL, E.G. (1965). "Optimization and Integration of Shipping Ventures." *Ship Building and Marine Eng. Monthly 1*, 267-281.

RITTER, K. (1967). "A Method for Solving Nonlinear Maximum-Problems Depending on Parameters (translated by M. Meyer)." *Naval Res. Log. Quarterly 14*, 147-162.

SCHAIBLE, S. (1974). "Parameter-Free Convex Equivalent and Dual Programs of Fractional Programming Problems." *Zeitschrift für Operations Research 18*, 187-196.

SCHAIBLE, S. (1976a). "Duality in Fractional Programming: A Unified Approach." *Operations Research 24*, 452-461.

SCHAIBLE, S. (1976b). "Minimization of Ratios." *Journal of Optimization Theory and Applications 19*, 347-352.

SCHAIBLE, S. (1976c). "Fractional Programming I, Duality." *Management Science 22*, 858-867.

SCHAIBLE, S. (1976d). "Fractional Programming II, on Dinkelbach's Algorithm." *Management Science 22*, 868-873.

SCHAIBLE, S. (1978). *Analyse und Anwendungen von Quotientenprogrammen.* Hain-Verlag, Meisenheim.

SCHAIBLE, S. (1981). "A Survey of Fractional Programming." This volume, 417-440.

STANCU-MINASIAN, I.M. (1977). *Bibliography of Fractional Programming 1960-1976.* Academy of Economic Studies.

SWARUP, K. (1965). "Linear Fractional Functionals Programming." *Operations Research 13*, 1029-1036.

TOBIN, J. (1958). "Liquidity Preference as Behavior Toward Risk." *Rev. Economic Studies 26*, 65-86.

VAJDA, S. (1972). *Probabilistic Programming.* Academic Press, New York.

VAN DE PANNE, C. (1975). *Methods for Linear and Quadratic Programming,* North-Holland, Amsterdam.

WAGNER, H.M. and YUAN, J.S.C. (1968). "Algorithmic Equivalence in Linear Fractional Programming." *Management Science 14*, 301-306.

WILLIAMS, H.P. (1974). "Experiments in the Formulation of Integer Programming Problems." *Mathematical Programming Study 2*, 180-197.

WOLFE, P. (1959). "The Simplex Method for Quadratic Programming." *Econometrica 27*, 382-398.

YOUNG, R. (1968). "A Simplified (All-Integer) Integer Programming Algorithm." *Operations Research 16*, 750-782.

ZIEMBA, W.T., BROOKS-HILL, F.J. and PARKAN, C. (1974). "Calculating of Investment Portfolios with Risk Free Borrowing and Lending." *Management Science 21*, 209-222.

DUALITY FOR GENERALIZED CONVEX FRACTIONAL PROGRAMS

B.D. Craven

Various dual problems are obtained for nonlinear fractional programming problems, under generalized convexity hypotheses, in place of the convexity usually assumed to obtain weak duality. The dual problems provide a measure of sensitivity of the given problems to perturbations. The results are applied to fractional programs arising in portfolio selection, agricultural planning, and information transfer.

1. INTRODUCTION

A large part of mathematical programming theory (see e.g. Rockafellar, (1970), Craven (1978, section 4.2)) assumes that the *objective function* being minimized is a *convex* function, and that the constraints are convex. These assumptions allow a *duality* theorem to hold. Consider a given *(primal)* problem (P): Minimize $\phi(x)$ subject to $x \in Q$. The related problem (D): Maximize $\psi(z)$ subject to $z \in N$, is called a *dual* of (P) if there hold

(i) *(weak duality)* $\phi(x) \geq \psi(z)$ whenever $x \in Q$ and $z \in N$; and

(ii) *(zero duality gap)* If (P) reaches a minimum at some point $\bar{x} \in Q$, then (D) reaches a maximum at some point $\bar{z} \in N$, and $\phi(\bar{x}) = \psi(\bar{z})$.

(Similarly (P) is a dual of (D) if (i) holds, and if (D) reaches a maximum at some $\hat{z}$, then (P) reaches a minimum at some $\hat{x}$, with $\phi(\hat{x}) = \psi(\hat{z})$.) Note that (D) is a dual of (P) if (i) holds, and $\phi(x') = \psi(z')$ for some $x' \in Q$ (in particular, if x' minimizes (P)) and some $z' \in N$; for then z' maximizes (D), from (i). The relevance of the dual problem is that (a) (D) gives a bound, which may be computationally useful, for the minimum of (P), thus $\phi(\bar{x}) \geq \psi(z)$ for each $z \in N$, and this bound can be a good approximation; (b) the dual variable $\bar{z}$ has often a significant economic interpretation, often as a price to be charged; and (c) the optimal dual variable $\bar{z}$ provides a measure of the *sensitivity* of the optimal primal

ISBN 0-12-621120-5

objective function, $\phi(\bar{x})$, to small changes in the specification of the primal problem (P).

If (P) is a linear program, there is a unique linear program which serves as a dual (D) to (P). For nonlinear programs, there may be several nonlinear duals - if a dual exists at all. Weak duality depends on some convex-like property assumed for (P). The objective function f(x)/g(x) for a *fractional program* is *not* usually convex; it is pseudo-convex (Mangasarian (1969)) if f and g are affine functions, or if f is convex and g is concave positive. *It will be shown that even weaker generalized-convexity properties will suffice for weak duality in fractional programming.* Dual problems and sensitivity are then discussed, in sections 5 and 6, for several fractional programs arising in applications.

For references on applications of fractional programming, see Schaible (1976a; 1978; 1981), Craven (1978, chapter 6; 1979), Bereanu (1963), Meister and Oettli (1967), Gilmore and Gomory (1961; 1963), Kallberg and Ziemba (1981), Pollack, Novaes and Frankel (1965), Ziemba, Parkan and Brooks-Hill (1974).

2. DEFINITIONS AND NOTATION

Denote by $\mathbb{R}^n$ the Euclidean space of n dimensions. Vectors will usually be considered (in matrix notation) as columns. The transpose of a column v is denoted v^T. All functions will be considered to map between spaces of finite dimensions (although most of the results will work also with infinite dimensions). Inequalities are understood pointwise (if h maps $\mathbb{R}^n$ into $\mathbb{R}^m$, then $h(x) \leq 0$ means $h_j(x) \leq 0$ for $j=1,2,\ldots,m$). The function which takes x to f(x)/g(x) will sometimes be denoted f(.)/g(.). If m is a function of two variables, the partial derivative of m(u,v) with respect to u will be denoted by $m_u(u,v)$. The norm (or length) of $x \in \mathbb{R}^n$ will be denoted $|x|$.

A differentiable function f is *convex* if and only if, for all x and u in the domain of f, $f(x) - f(u) \geq f'(u)(x-u)$, where $f'(u)$ denotes the Fréchet derivative. Generalizing this, a real function will be called *invex* if, for some function η,

$$f(x) - f(u) \geq f'(u)\eta(x,u); \tag{1}$$

quasiinvex if $$f(x) \leq f(u) \Longrightarrow f'(u)\eta(x,u) \leq 0; \tag{2}$$

pseudoinvex if $$f'(u)\eta(x,u) \geq 0 \Longrightarrow f(x) \geq f(u). \tag{3}$$

These definitions (though not the names) are due to Hanson (1980), who showed that various duality results remained true when *convex* hypotheses were weakened to *invex* hypotheses (with the same η for all the functions), and pointed out that similar statements would apply when *quasiconvex* and *pseudoconvex* were weakened to (2) and (3). Theorem 4 proves this for a class of fractional programs. For definitions of quasiconvexty and pseudoconvexity, see Avriel, Diewert, Schaible and Ziemba (1981).

To relate *invex* to *convex*, let $q : \mathbb{R}^n \to \mathbb{R}$ be differentiable; let $\phi : \mathbb{R}^r \to \mathbb{R}^n$ be differentiable surjective (with $r \geq n$), and $u = \phi(y)$, $x = \phi(z)$; denote by g' the composite function $q \circ \phi$. From the chain rule for differentiation,

$$g'(y)\eta(z,y) = q'(u)\phi'(y)\eta(z,y) = q'(u)(x-u) \tag{4}$$

provided $\phi'(y)\eta = x-u$ can be solved for η, and this is so if $\phi'(y)$ has maximal rank n. Assuming this, if q is convex (pseudoconvex, quasiconvex), then g is invex (pseudoinvex, quasiinvex). This holds in particular when ϕ is bijective; thus *invex* (unlike *convex*) is a property invariant to bijective coordinate transformations. (The name *invex* is here proposed from *invariant convex*.) One example of a function which is invex but not convex is

$$g(y_1,y_2) = 3(y_1 - ay_1^3)^2 - 2(y_1 - ay_1^3)(y_2 + by_2^3) + 2(y_2 + by_2^3)^2,$$

where a and b are positive constants.

The inequality $h(x) \leq 0$ (when h maps $\mathbb{R}^n$ into $\mathbb{R}^m$) is *locally solvable* at $x = a$ if (i) $h(a) \leq 0$, and (ii) whenever the direction d satisfies $h(a) + h'(a)d \leq 0$, for $|d|$ small enough, $h(x) \leq 0$ has a solution $x = a + \alpha d + r(\alpha)$ with $r(\alpha)/\alpha \to 0$ as $\alpha \downarrow 0$. *Locally solvable* is equivalent to the "Kuhn-Tucker constraint qualification", but extends also to more general systems (see Craven (1978, pages 34 and 61)). Suppose now that f(x) is minimized, subject to $h(x) \leq 0$, at $x = a$, where $h(x) \leq 0$ is locally solvable there. The *Kuhn-Tucker theorem* then shows that a *Lagrange multiplier* λ exists, satisfying the Kuhn-Tucker conditions:

$$f'(a) + \lambda^T h'(a) = 0, \quad \lambda^T h(a) = 0, \quad \lambda \geq 0. \tag{5}$$

3. DUALITY THEOREMS

Consider the pair of nonlinear fractional programming problems:

(P) Minimize $f(x)/g(x)$ subject to $h(x) \leq 0$; (6)

(D) Maximize $f(u)/g(u)$ subject to $v \geq 0$, $v^T h(u) \geq 0$,

$$0 = g(u)f'(u) - f(u)g'(u) + g(u)v^T h'(u) - v^T h(u)g'(u). \tag{7}$$

Here f, g and h are differentiable functions (with derivatives $f'(x)$, etc.) on an open convex domain X_o in $\mathbb{R}^n$.

Under suitable hypotheses, problem (D) will be shown to be a dual of problem (P). Observe that this dual has the special feature that the dual objective function has the same functional form as the primal objective function. Such a form of dual, first given (for a linear fractional program) by Sharma and Swarup (1972) is *not* the only form of dual possible; see e.g. Schaible (1976a,b) for other forms of dual, under different assumptions.

THEOREM 1 (Duality Theorem): For problem (6), assume that

(i) $f : X_o \to \mathbb{R}$, $g : X_o \to \mathbb{R}$, $h : X_o \to \mathbb{R}^m$ *are differentiable functions;*

(ii) *for all* $x \in X_o$, $g(x) > 0$;

(iii) *for each* $u \in X_o$, *the function* $f(.) - [f(u)/g(u)]g(.)$ *is pseudoconvex;*

(iv) *when* $v \geq 0$, *the function* $v^T h(.)/g(.)$ *is quasiconvex;*

(v) *problem (6) reaches a minimum at* $x = \bar{x}$, *and the constraint* $h(x) \leq 0$ *is locally solvable there.*

Then the problem (7) is a dual to problem (6).

PROOF: Weak duality requires that $f(x)/g(x) \geq f(u)/g(u)$ whenever x and (u,v) are feasible for (6) and (7) respectively. To prove this, let x be feasible for (6), and let (u,v) be feasible for (7). Then $h(x) \leq 0$, $v \geq 0$, $g(x) > 0$, $g(u) > 0$, $v^T h(u) \geq 0$. Define the function $\theta(.) = v^T h(.)/g(.)$. Then $v^T h(x) \leq 0$, so $\theta(x) \leq 0$; also $\theta(u) \geq 0$ from $v^T h(u) \geq 0$ and $g(u) > 0$. Since $\theta(x) \leq \theta(u)$ and θ is quasiconvex, by hypothesis (iv), it follows that $\theta'(u)(x - u) \leq 0$. Hence

$$0 \leq -\theta'(u)\,[g(u)]^2\,(x - u)$$
$$= [-g(u)v^T h'(u) + v^T h(u)g'(u)]\,(x - u). \tag{8}$$

Consider u temporarily as fixed, and define the function ϕ by $\phi(.) = g(u)f(.) - f(u)g(.)$. Then, from (3.3) and the third constraint of (3.2),

$$0 \leq \phi'(u)(x - u). \tag{9}$$

From hypothesis (iii) and $g(u) > 0$, ϕ is pseudoconvex. Therefore (9) implies that $\phi(x) \geq \phi(u)$. Consequently $g(u)f(x) \geq f(u)g(x)$; since $g(x) > 0$ and $g(u) > 0$, there results

$$f(x)/g(x) \geq f(u)/g(u), \tag{10}$$

which proves weak duality.

From hypothesis (v), the Kuhn-Tucker theorem, applied to the minimum of (6) at $x = \bar{x}$, shows that a Lagrange multiplier λ exists, satisfying

$$[g(\bar{x})]^{-2}[g(\bar{x})f'(\bar{x}) - f(\bar{x})g'(\bar{x})] + \lambda^T h'(\bar{x}) = 0;$$

$$\lambda^T h(\bar{x}) = 0; \ \lambda \geq 0. \tag{11}$$

Then $(u,v) = (\bar{x}, g(\bar{x})\lambda)$ is feasible for (7), and the objective functions of (6) and (7) are equal there. Then weak duality implies that (7) reaches its maximum at $(\bar{x}, g(\bar{x})\lambda)$. It follows that (7) is a dual to(6) □

Hypotheses (iii) and (iv) of Theorem 1 hold, in particular, when *either* g is concave, f is convex, and $x \in X_o$ implies $g(x) > 0$ and $f(x) \geq 0$; *or* when g is convex, f is convex, and $x \in X_o$ implies $g(x) > 0$ and $f(x) < 0$. Duality for (7) was discussed under these stronger assumptions in Craven and Mond (1979b;1980); there also the case of non-differentiable convex or concave functions was discussed. Schaible (1976a; 1976b) gave a unified approach that yields duality results of various authors.

Assume now that hypothesis (ii) of Theorem 1 is strengthened to:

$$\text{For all } x \in X_o,\ g(x) > 0 \text{ and } f(x) > 0. \tag{12}$$

Then problem (6) is equivalent to problem

$$\text{Maximize } g(x)/f(x) \quad \text{subject to } h(x) \leq 0, \tag{13}$$

in the sense that (6) and (13) both reach an optimum at the same $x = \bar{x}$. (The optimum values of the objective functions are reciprocal.) This fact enables results for the maximization problem

$$\text{Maximize } f(x)/g(x) \quad \text{subject to } h(x) \leq 0 \tag{14}$$

to be deduced from those for the minimization problem (6) using Theorem 1. The dual obtained is

$$\text{Minimize } f(u)/g(u) \quad \text{subject to } v \geq 0,\ v^T h(u) \geq 0,$$

$$0 = f(u)g'(u) - g(u)f'(u) + f(u)v^T h'(u) - v^T h(u)f'(u). \tag{15}$$

THEOREM 2: For problem (14), let f,g,h *be differentiable functions, satisfying (12). Let (14) reach a maximum at* $x = \bar{x}$, *and let the constraint* $h(x) \leq 0$ *be locally solvable there. Let* $g(.) - [g(u)/f(u)]f(.)$ *be pseudo-*

convex for each $u \in X_o$, *and let* $v^T h(.)/f(.)$ *be quasiconvex when* $v \geq 0$. *Then (15) is a dual to (14).*

PROOF: From Theorem 1, interchanging the roles of f and g. □

A different approach to duality for (14) is as follows. For constant k, problem (14) is equivalent to the problem

$$\text{Minimize } [k - [f(x)/g(x)]] = [kg(x) - f(x)]/g(x)$$
$$\text{subject to } h(x) \leq 0, \tag{16}$$

in the sense that (14) and (16) both reach their optima at the same value of x (and the optimal objective functions are related in an obvious way). It follows that, given appropriate hypotheses of generalized convexity, etc., Theorem 1 (with f replaced by kg - f) may be applied to find a dual problem for (16), and hence for (1). The constant k cancels out of the dual problem and the pseudoconvex hypothesis. This gives the result:

THEOREM 3: For problem (14), let f,g,h *be differentiable functions, satisfying* $g(x) > 0$ *for all* $x \in X_o$, *and* $f(x)/g(x) \leq k < \infty$ *whenever* $h(x) \leq 0$. *Let (14) reach a maximum at* $x = \bar{x}$, *and let* $h(x) \leq 0$ *be locally solvable there. Let* $-f(.) + [f(u)/g(u)]g(.)$ *be pseudoconvex for each* $u \in X_o$, *and let* $v^T h(.)/g(.)$ *be quasiconvex when* $v \geq 0$. *Then a dual to (14) is the problem*

$$\text{Minimize } f(u)/g(u) \quad \text{subject to } v \geq 0,\ v^T h(u) \geq 0,$$
$$0 = -g(u)f'(u) + f(u)g'(u) + g(u)v^T h'(u) - v^T h(u)g'(u). \tag{17}$$

The pseudoconvex hypothesis holds, in particular, when f is convex, g is concave positive, and $f(u)/g(u) \leq k$ for all $u \in X_o$.

These pseudoconvex and quasiconvex properties can be further weakened to pseudoinvex and quasiinvex, as follows.

THEOREM 4: Let problem (6) satisfy the hypotheses of Theorem 1, except that pseudoconvex and quasiconvex are replaced respectively by pseudoinvex and quasiinvex with respect to the same function η. *Then (7) is a dual to (6).*

PROOF: As in the proof of Theorem 1, it follows that $\theta(x) \leq \theta(u)$. From the quasiinvex hypothesis, it follows that $\theta'(u)\eta(x,u) \leq 0$. Hence $0 \leq \phi'(u)\eta(x,u)$. Since ϕ is assumed pseudoinvex, with respect to this *same* η, there follows $\phi(x) \geq \phi(u)$. The rest of the proof is unchanged. □

The results of section 3 are based on those of Craven and Mond (1979b; 1980), and Hanson (1980). The sign of the last term of (17) is given incorrectly in Craven and Mond (1979b), and corrected here. The concepts of invex, pseudoinvex and quasiinvex functions, though not the names here given to them, were introduced by Hanson (1980), who pointed out that these concepts could replace the usual convex and generalized convex concepts in various duality theorems. For the precise relations between these concepts and convex, pseudo- and quasi-convex, see section 2 above. Note also the special case of invex, when $\eta(x,u) = \beta(x,u)(x - u)$ with β a real positive function (*strong pseudoconvex*, Chandra (1972)).

4. EXAMPLES AND COMPARISON WITH OTHER DUALS

Consider, in particular, the linear program

$$\text{Minimize } c^T x/1 \quad \text{subject to } Ax \geq b, \tag{18}$$

where c and b are constant (column) vectors, A is a constant matrix, and the objective function has been written as a quotient, in order to apply the theory of fractional programming. Then Theorem 1 gives a dual of (18) as the nonlinear program

$$\text{Maximize } c^T u/1 \quad \text{subject to } v \geq 0,\ v^T(Au - b) \leq 0,$$
$$0 = c^T - v^T A. \tag{19}$$

Now (19) is equivalent to

$$\text{Maximize } c^T u \text{ subject to } c^T u \le b^T v,\ v \ge 0,\ c^T = v^T A,$$

(since each constraint set implies the other), and so to the usual linear programming dual

$$\text{Maximize } b^T v \text{ subject to } v \ge 0,\ c^T = v^T A. \tag{20}$$

Of course, a dual problem, from its definition (in section 1) remains a dual problem when a further constraint (such as $v^T(Au - b) \le 0$) is added, provided that this constraint does not exclude the dual optimum. (A more constrained dual may be more useful for bounding the minimum of the primal.)

Similarly, Theorem 3 gives a dual to the linear program

$$\text{Maximize } b^T x \text{ subject to } A^T x \le c, \tag{21}$$

the (nonlinear) dual problem

$$\text{Minimize } b^T u \text{ subject to } Av = b,\ v \ge 0,\ v^T(A^T u - c) \ge 0. \tag{22}$$

Consider now the dual problem to (19) given by Theorem 3. This is

$$\begin{aligned} &\text{Minimize } c^T x \text{ subject to } p \ge 0,\ q \ge 0,\\ &\quad -p^T v + qy^T(Ax - b) + r^T(A^T y - c) \ge 0,\ 0 = -c^T + qy^T A,\\ &\quad 0 = -p^T + q(Ax - b)^T + r^T A^T. \end{aligned} \tag{23}$$

This is obtained by rewriting the constraints of (19) as inequalities:

$$-v \le 0,\ v^T(Au - b) \le 0,\ A^T v - c \le 0,\ -A^T v + c \ge 0.$$

The analog, for this problem, of " $v^T h(u)$ " in (17) is then

$$-p^T y + qy^T(Ax - b) + r_1(A^T y - c) + r_2(-A^T y + c),$$

where $p, q, r_1, r_2 \ge 0$; set $r = r_1 - r_2$. From the constraints of (23) $c = qA^T y$, whence $q \ne 0$; by linearity, since also $q \le 0$, it suffices to take $q = 1$. Then it follows that $-r^T c \ge 0$;
$Ax - b = -A(q^{-1} r) + q^{-1} p \ge -A(q^{-1} r)$; writing $z = -Ar$ (with $q = 1$), the constraints of (23) reduce to

$$Ax - b \ge z,\ y^T z \ge 0,\ c = A^T y. \tag{24}$$

Hence, with the duality theorems currently discussed, a "dual of a dual" does *not* reduce to the original program (though clearly (23) relates closely to the constraint $Ax \geq b$ of (18)).

If the original ("primal") problem has the form

$$\text{Minimize } \Phi(x) \quad \text{subject to } h(x) \leq 0, \tag{25}$$

then the "Wolfe dual" of (25) has the form

$$\begin{aligned} &\text{Maximize } \Phi(u) + v^T h(u) \quad \text{subject to } v \geq 0, \\ &\qquad \Phi'(u) + (v^T h)'(u) = 0. \end{aligned} \tag{26}$$

Now (26) will be in fact a dual for (25) if the Kuhn-Tucker theorem can be applied to a minimum of (25) (leading to zero duality gap), and if also (25) has enough convex-like properties (see Theorem 5 below) to make weak duality work. Assuming weak duality, and that (26) reaches a maximum at $(u,v) = (\bar{u},\bar{v})$, it is then also true that (26) is also a dual of (25), provided that a certain *solvability* property is assumed for the constraint of (26), namely (see Craven (1978); 64) that, whenever $v = \bar{v} + t$ and $|t|$ is sufficiently small, $\Phi'(u) + (v^T h)(u) = 0$ has a solution $u = \xi(t)$ with $\xi(t) \to \bar{u}$ as $t \to 0$. The property (as here) that the "dual of the dual" reduces to the primal is called *converse duality*.

In order to obtain a "Wolfe dual" to a fractional program, weak duality must be proved, without assuming convexity of $\Phi(x) = f(x)/g(x)$. For the differentiable problem

$$\text{Maximize } f(x)/g(x) \quad \text{subject to } h(x) \leq 0, \tag{27}$$

Schaible (1976a, 1978) obtained the following dual for (27).

$$\begin{aligned} &\text{Minimize } \pi \text{ subject to } v \geq 0, \ \pi \geq 0, \\ &-f'(u) + v^T h'(u) + \pi g'(u) = 0, \ -f(u) + v^T h(u) + \pi g(u) \geq 0, \end{aligned} \tag{28}$$

assuming that $g(u) > 0$ always. Denote

$$m(u,v) = [-f(u) + v^T h(u)]/g(u).$$

From the last constraint of (28), there follows $\pi \geq -m(u,v)$. If (28) is further restricted, to replace this inequality by an equality, then the preceding constraint reduces to $g(u)m_u(u,v) = 0$. So the modified problem (28) reduces to

$$\text{Minimize } -m(u,v) = -\frac{-f(u) + v^T h(u)}{g(u)} \quad \text{subject to } v \geq 0,$$

$$m_u(u,v) = 0 \tag{29}$$

together with further constraint $-m(u,v) \geq 0$. Observe that (29) is the Wolfe dual for the problem

$$\text{Maximize } f(x)/g(x) \quad \text{subject to } h(x)/g(x) \leq 0. \tag{30}$$

Since $g(x) > 0$, (30) and (27) describe the same problem.

For the problem (25), define the Lagrangian $L(u,v) = \Phi(u) + v^T h(u)$. The relevant weak duality theorem (Bector et. al (1977)) is as follows.

THEOREM 5: Let x *and* (u,v) *satisfy the constraints of (25) and (26) respectively. Assume that* L(.,v) *is pseudoconvex, whenever* $v \geq 0$. *Then weak duality holds, namely* $\Phi(x) \geq L(u,v)$.

PROOF: $L_u(u,v) = 0 \Longrightarrow L_u(u,v)(x-u) = 0$

$\Longrightarrow L(x,v) - L(u,v) \geq 0$ since L(.,v) is pseudoconvex

$\Longrightarrow \Phi(x) - L(u,v) \geq - v^T h(x) \geq 0.$ □

Consequently, for (30) and (29), it suffices for weak duality if h is convex, and if m(.,v) is pseudoconvex whenever $v \geq 0$. The latter happens, in particular, if -f, h and g are convex, and $g(.) > 0$. □

5. SOME APPLICATIONS

Consider the following fractional programs.

$$\text{Maximize } c^T x/(x^T A x)^{\frac{1}{2}} \quad \text{subject to } Kx \geq b,\ x \in X_o \subset \mathbb{R}^n. \tag{31}$$

$$\text{Maximize } (c^T x - t \log t)/(d^T x) \quad \text{subject to } Kx + kt \geq b,$$
$$(x,t) \in X_o \subset \mathbb{R}^n \times \mathbb{R}. \tag{32}$$

$$\text{Minimize } [-f(x) + s(x|C_1)]/[g(x) + s(x|C_2)]$$
$$\text{subject to } h(x) \leq 0,\ x \in X_o. \tag{33}$$

Here, c,b,d,k are constant vectors, A and K are matrices (with A positive definite), and $s(.|C_i)$ is the support function of the closed convex set $C_i \subset \mathbb{R}^n$, namely

$$s(x|C_i) = \sup \left\{ x^T v : v \in C_i \right\} . \tag{34}$$

Program (31) arises both in an agricultural application (Bereanu (1963)), where the probability that a certain set of linear inequalities (involving random elements) holds is to be maximized, and also as part of the analysis of an optimum portfolio selection model (Ziemba et al (1974), see also Kallberg and Ziemba (1981)). In both these applications, the denominator $(x^TAx)^{\frac{1}{2}}$ represents a standard deviation, with A a positive definite matrix. Program (32) arises in calculating the maximum rate of information transferred through an information channel (Meister and Oettli (1967)).

For (31) and (32), both Theorem 3 and Theorem 2 provide dual problems, under different hypotheses. These duals are as follows.

For (31): Minimize $c^Tu/(u^TAu)^{\frac{1}{2}}$ subject to $v \geq 0$, $v^T(Ku - b) \leq 0$,

$$u \in X_o, \ (K^Tv + c)(u^TAu) = (c^Tu + v^TKu - v^Tb)Au. \tag{35}$$

(Hypothesis: $x \in X_o \implies 0 \neq x$ and $c^Tx \geq 0$.)

Or Minimize $c^Tu/(u^TAu)^{\frac{1}{2}}$ subject to $v \geq 0$, $v^T(Ku - b) \leq 0$,

$$u \in X_o, \ 0 = (c^Tu)Au(u^TAu)^{\frac{1}{2}} - (u^TAu)^{\frac{1}{2}}c - (c^Tu)K^Tv + v^T(Ku - b)c. \tag{36}$$

(Same hypothesis as for (35).)

For (32): Minimize $(c^Tu - s \log s)/(d^Tu)$ subject to $v \geq 0$,

$(u,s) \in X_o$, $v^T(Ku + ks - b) \leq 0$,

$$0 = (d^Tu)(1 + \log s) - (d^Tu)(v^TK),$$

$$0 = (-cd^T + dc^T)u - (s \log s)d - (d^Tu)(K^Tv) - v^T(b - Ku - ks)d. \tag{37}$$

(Hypothesis: $(x,t) \in X_o \implies t > 0$, $d^Tx > 0$.)

Or $\quad$ Minimize $(c^Tu - s \log s)/(d^Tu)$ subject to $v \geq 0$,

$$(u,s) \in X_o,\ v^T(Ku + ks - b) \leq 0,$$

$$0 = (c^Tu - s \log s)(d - K^Tv) - (d^Tu)c + v^T(Ku + ks - b)c,$$

$$0 = (1 + \log s)(d^Tu - v^T(Ku + ks - b)) - (c^Tu - s \log s)(v^TK). \qquad (38)$$

(Hypothesis: $(x,t) \in X_o \Longrightarrow d^Tx > 0$ and $c^Tx - t \log t > 0$.)

These detailed results are quoted from Craven and Mond (1979b ; 1980).

In (33), the support functions are not always differentiable. For example, if C is the closed unit ball in $\mathbb{R}^n$, with centre 0, then $s(x|C) = (x^Tx)^{\frac{1}{2}}$, and the derivative $s'(x|C) = x^T(x^Tx)^{-\frac{1}{2}}$ exists whenever $x \neq 0$, but not at $x = 0$. For the present, assume that the support functions are differentiable at the points required. Then, from Theorem 1, problem (33) has a dual problem

$$\text{Maximize } [-f(u) + s(u|C_1)]/[g(u) + s(u|C_2)]$$

$$\text{subject to } v \geq 0,\ v^Th(u) \geq 0,$$

$$0 = [g(u) + s(u|C_2)][-f'(u) + s'(u|C_1) + v^Th'(u)] - [-f(u) + s(u|C_1) + v^Th(u)][g'(u) + s'(u|C_2)]. \qquad (39)$$

Assuming differentiability, it can be shown that $s'(u|C_i) \in C_i$. The hypotheses for (39) to be a dual to (33) are that f,g.h and the support functions are differentiable at the points considered, $g(x) + s(x|C_2) > 0$ for all $x \in X_o$, $-f(.) + s(.|C_1) - \{[-f(u)+s(u|C_1)]/[g(u)+s(u|C_2)]\}[g(.)+s(.|C_2)]$ is pseudoconvex for each $u \in X_o$, $v^Th(.)/[g(.) + s(.|C_2)]$ is quasiconvex for $v \geq 0$.

For an alternative dual problem, analogous to (28), see Mond and Craven (1979). The dual problem to (33) obtained there is equivalent to

$$\text{Maximize } -z \quad \text{subject to } z \geq 0,\ v \geq 0,\ p \in C_1,\ q \in C_2,$$

$$0 = (-f + zg)'(u) + (v^Th)'(u) + p + zq\ ;$$

$$- f(u) + zg(u) + v^Th(u) + (p + zq)^Tu \geq 0\ . \qquad (40)$$

The assumptions made include convexity of $-f, g, h$, and (here) also differentiability. Examination of the proof shows that the last inequality in (40) may be replaced by an equality, and a dual is still obtained. Thus modified, (40) becomes

$$\text{Maximize } -z = [-f(u) + v^T h(u) + p^T u]/[g(u) + q^T u]$$

$$\text{subject to } v \geq 0,\ p \in C_1,\ q \in C_2,$$

$$z \geq 0,\ [-f'(u) + (v^T h)'(u) + p^T] + z[g'(u) + q^T] = 0. \quad (41)$$

Appropriate hypotheses here for weak duality for (33) and (41) are that the denominator of $-z$ is > 0, h is convex, and that the function

$$[-f(.) + s(.|C_1)] - [z\ g(.) + s(.|C_2)]$$

is pseudoconvex. These dual problems are mainly useful for considering sensitivity to perturbations.

6. SENSITIVITY

For a differentiable primal problem

$$\text{Minimize } p(x) \quad \text{subject to } h(x) \leq 0,\ k(x) = 0, \quad (42)$$

consider a *perturbed problem* of the form

$$\text{Minimize } p(x,q) \quad \text{subject to } h(x,q) \leq 0,\ k(x,q) = 0. \quad (43)$$

Here $q \in \mathbb{R}^s$ is a small perturbation parameter, and (43) is minimized over $x \in X_o$, the (open) domain of the functions $p(.,q)$, $h(.,q)$, $k(.,q)$. For (6.2), define the *Lagrangian*

$$L(x;q;\lambda,\mu) = p(x,q) + \lambda^T h(x,q) + \mu^T k(x,q), \quad (44)$$

with Lagrange multipliers λ and μ. Let $L_x(\ )$ denote $\partial L(\ )/\partial x$; assume that $p(x) = p(x,0)$, $h(x) = h(x,0)$, and $k(x) = k(x,0)$; denote $L_q(\ ) = \partial L(\ )/\partial q$ (assumed to exist).

Assume now that (43) reaches a minimum, say at $x = \bar{x}(q)$, and that the objective function there equals $F(q) = p(\bar{x}(q),q)$. Assume that the constraint system of (42) is locally solvable at $x = \bar{x} \equiv \bar{x}(0)$. Then, from the Kuhn-Tucker theorem, Lagrange multipliers $\bar{\lambda}$ and $\bar{\mu}$ exist for which

$$L_x(\bar{x};0;\bar{\lambda},\bar{\mu}) = 0; \quad \bar{\lambda}^T h(\bar{x},0) = 0; \quad \bar{\lambda} \geq 0. \tag{45}$$

Denote by $K(\bar{x})$ the set of $L_q(\bar{x};0;\bar{\lambda},\bar{\mu})$ for which $\bar{\lambda}$, $\bar{\mu}$ satisfy (6.4). Also let

$$\rho(q) = \sup \left\{ z^T q : z \in K(\bar{x}) \right\} . \tag{46}$$

From a theorem of Maurer (1977a,b), it follows that, for any $q \neq 0$ and any $\beta > \rho(q)$,

$$F(\alpha q) \geq F(0) - \alpha\beta, \tag{47}$$

whenever α is sufficiently small positive. It follows that $\rho(q)$ gives a measure of the sensitivity of the primal problem to perturbations "in the direction q". Since Maurer's theorem makes no kind of convexity assumptions, these estimates can be applied to fractional programming problems whose objective functions are *not* convex. If, in particular, the constraints are $h(x) - q = 0$, and if also $p(x,q) = p(x)$, then $\rho(q) = -\bar{\lambda}^T q$, in case the multiplier $\bar{\lambda}$ is unique. Here $\bar{\lambda}$ is analogous to the *shadow cost* in linear programming.

Consider now the perturbed fractional program

$$\text{Minimize } p(x,q) = f(x,q)/g(x,q) \quad \text{subject to } h(x,q) \leq 0. \tag{48}$$

For this problem, the number $\rho(q)$, representing sensitivity to perturbation, is calculated as

$$\rho(q) = \bar{\lambda}^T h_q + g^{-2}(gf_q - fg_q), \tag{49}$$

with the functions and derivatives evaluated at $\bar{x} = \bar{x}(0)$. The Lagrange multiplier $\bar{\lambda}$ is some constant multiple of an appropriate optimal dual variable. For the dual problems of section 3, the dual problem is optimized at $(\bar{u},\bar{v}) = (\bar{x},g(\bar{x})\bar{\lambda})$; hence $\bar{\lambda} = [g(\bar{x})]^{-1}\bar{v}$. This result is also discussed by Schaible (1975) and (1978; 187-202). Here the optimum $(\bar{u},\bar{v})$ is assumed unique; if $\bar{v}$ is not unique, then (46) must be used. The formula $\bar{\lambda} = g(\bar{x})^{-1}\bar{v}$ applies to the duals (35) through (38), and also (29).

Such sensitivity measures may be applied to other forms of fractional programming dual. Another dual for (14) is (Craven and Mond 1976, 1979b) is

$$\text{Minimize } f(u)/g(u) \quad \text{subject to } v \geq 0,\ v^T[h'(u)u - h(u)] \leq 0,$$
$$v^T h'(u) = g(u)f'(u) - f(u)g'(u). \tag{50}$$

This dual is obtained using a Lagrangian $f(u)/g(u) - [g(\bar{x})]^{-2}v^T h(x)$, and it is assumed that f,g,h are differentiable, with f concave, g convex, h convex, f and g homogeneous of the same degree, and $x \in X_o$ implying $f(x) \geq 0$, $g(x) \geq 0$, with $g(x) = 0 \Longrightarrow f(x) \neq 0$. From this Lagrangian, formula (46) for $\rho(q)$ applies here, but with $\bar{\lambda} = [g(\bar{x})]^{-2}\bar{v}$. (For comparison, the dual objective function in (7) may be derived from a different Lagrangian:

$$f(u)/g(u) - [g(\bar{x})]^{-2}v^T g(u)h(u). \tag{51}$$

For references, see Craven (1979) and Craven and Mond (1980).

ACKNOWLEDGEMENT

I am indebted to Professor B. Mond for a number of amendments to these lecture notes.

REFERENCES

AVRIEL, M., DIEWERT, W.E., SCHAIBLE, S. and ZIEMBA, W.T. (1981). "Introduction to Concave and Generalized Concave Functions." This volume, 21-50.

BECTOR, C.R., BECTOR, M.H., and KLASSEN, J.E. (1977). "Duality for a Nonlinear Programming Problem." *Utilitas Math. 11*, 87-99.

BEREANU, B. (1963). "Distribution Problems and Minimum Risk Solutions in Stochastic Programming." In *Colloquium on Applications of Mathematics to Economics*, (A. Prékopa, ed.). Akadémiai Kiadó, Budapest, 37-42.

CHANDRA, S. (1972). "Strong Pseudo-Convex Programming." *Indian J. of Pure and Applied Math. 3*, 178-182.

CRAVEN, B.D. (1978). *Mathematical Programming and Control Theory*. Chapman and Hall, London.

CRAVEN, B.D. (1979). "Duality and Sensitivity for Some Fractional Programs." Paper presented at Applied Math. Conference. Leura, N.S.W., Australia (February 1979).

CRAVEN, B.D. and MOND, B. (1973, 1976). "The Dual of a Fractional Linear Program." *J. Math. Anal. Appl. 42,* 507-512, and *55,* 807.

CRAVEN, B.D. and MOND, B. (1976). "Duality for Homogeneous Fractional Programming." *Cahiers du Centre D'Etude de Recherche Opérationelle 18,* 413-417.

CRAVEN, B.D. and MOND, B. (1979a). "A Note on Duality in Homogeneous Fractional Programming." *Naval. Res. Log. Quart. 26,* 153-155.

CRAVEN, B.D. and MOND, B. (1979b). "On Duality for Fractional Programming." *Zeits. Angew. Math.Mech. 59,* 278-279.

CRAVEN, B.D. and MOND, B. (1980). "On Fractional Programming Duality with Generalized Convexity." (Submitted for publication.)

GILMORE, P.C. and GOMORY, R.E. (1961, 1963). "A Linear Programming Approach to the Cutting Stock Problem." *Operations Research 9,* 849-859 and *11,* 863-888.

HANSON, M.A. (1980). "On Sufficiency of the Kuhn-Tucker Conditions." *Pure Math. Research Paper 80-1.* La Trobe University, Australia.

KALLBERG, J.G. and ZIEMBA, W.T. (1981). "Generalized Concave Functions in Stochastic Programming and Portfolio Theory." This volume, 719-767.

MANGASARIAN, O.L. (1969). *Nonlinear Programming.* McGraw-Hill, New York.

MAURER, H. (1977a). "A Sensitivity Result for Infinite Nonlinear Programming Problems, Part I: Theory." Math. Institut Univ. Wörzburg, West Germany, Preprint 21.

MAURER, H. (1977b). "Zur Störungstheorie in der Infiniten Optimierung." *Zeits. Angew. Math. Mech. 57,* 340-341.

MEISTER, B. and OETTLI, W. (1967). "On the Capacity of a Discrete, Constant Channel." *Inf. and Control 11,* 341-351.

MOND, B. and CRAVEN, B.D. (1979). "A Duality Theorem for a Nondifferentiable Nonlinear Fractional Programming Problem." *Bull. Austral. Math. Soc. 23,* 397-406.

POLLACK, E.G., NOVAES, G.N. and FRANKEL G. (1965). "On the Optimization and Integration of Shipping Ventures." *International Shipbuilding Progress 12,* 267-281.

ROCKAFELLAR, R.T. (1970). *Convex Analysis.* Princeton Univ. Press.

SCHAIBLE, S. (1975). "Marginalwerte in der Quotientenprogrammierung." *Zeitschrift für Betriebswirtschaft 45,* 649-658.

SCHAIBLE, S. (1976a). "Fractional Programming I: Duality." *Management Science 22,* 858-867.

SCHAIBLE, S. (1976b). "Duality in Fractional Programming - A Unified Approach." *Operations Research 24,* 452-461.

SCHAIBLE, S. (1978). "Analyse und Anwendungen von Quotientprogrammen." Verlag Anton Hain, Meisenheim.

SCHAIBLE, S. (1981). "A Survey of Fractional Programming." This volume, 417-440.

SHARMA, I G. and SWARUP, K. (1972). "On Duality in Linear Fractional Functionals Programming." *Zeitschrift für Operations Research 16,* 91-100.

ZIEMBA, W.T., PARKAN, C. and BROOKS-HILL, R. (1974). "Calculation of Investment Portfolios with Risk Free Borrowing and Lending." *Management Science 21,* 209-222.

AN ALGORITHM FOR A SPECIAL CLASS OF GENERALIZED CONVEX PROGRAMS

Alberto Cambini

This paper aims to study a maximization problem with linear constraints, in which the objective function f(x) is the product of a concave function N(x) and an integer power of an affine function D(x). General properties and optimality conditions are given. Moreover, algorithms are proposed for solving such problems in the case where a) N(x) is affine and b) N(x) is a strictly concave quadratic function.

1. INTRODUCTION

Let us consider the following problem

$$(P) \qquad \max_{x \in S} \; [f(x) = N(x).[D(x)]^p]$$

where

$S = \{x \in \mathbb{R}^n : Ax \geq b,\ x \geq 0\}$; A is a $m \times n$ matrix and $b \in \mathbb{R}^m$

$N : S \to \mathbb{R}$ is a concave function

$D : S \to \mathbb{R}$ is an affine function which is assumed to be positive if $p < 0$

p is a non-zero integer.

Problem (P) becomes a linear fractional program if $p = -1$ and N is affine; (P) is a concave-convex fractional program if $p < 0$ and N is non-negative. Let S* denote the set of optimal solutions of (P). Furthermore, let $S_1 = \{x \in S : f(x) \geq 0\}$ and f_1 be the restriction of f to the set S_1.

The results of this section and those of section 2 are in a more general form than the ones given in Cambini (1977) and in Cambini, Martein, and Pellegrini (1978).

GENERALIZED CONCAVITY
IN OPTIMIZATION AND ECONOMICS

ISBN 0-12-621120-5

THEOREM 1: If D *is positive in* S *then*

(a) S_1 *is a convex set and* f_1 *is a semistrictly quasiconcave function*(*)

(b) a local maximum point for f_1 *is also a global maximum point for problem* (P)

(c) if $S_1 \neq \emptyset$, S^* *is a convex set.*

PROOF. Since N is a concave function, the set $N^+ = \{x \varepsilon S : N(x) \geq 0\}$ is convex and $S_1 = N^+$. (a), (b), (c) are direct consequences of known properties (Mangasarian (1969), Schaible (1972)). □

THEOREM 2: Let N *be an affine function. If* $S^* \neq \emptyset$ *then there exists* $x_o \varepsilon S^*$ *such that* x_o *belongs to an edge of* S*; when* $p = -1$ *there exists a vertex of* S *belonging to* S^*.

PROOF. Let $x^* \varepsilon S^*$. It is sufficient to note that problem (P) is equivalent to the linear problem

$$(P') \qquad \max_{x \varepsilon S'} [h(x) = N(x)[D(x^*)]^p]$$

where

$S' = \{x \varepsilon S : D(x) = D(x^*)\}$. An optimal solution of problem (P') is reached at a vertex x_o of S' and such a vertex belongs to an edge of S.

When $p = -1$, it can be seen that the restriction of f to an edge of S is an increasing, decreasing or a constant function by differentiating $f(x_o + \alpha d)$ with respect to α where x_o is on the edge of S, α is a scalar and d points in the direction of the edge of S. Thus, x_o is a vertex of S or α vertex on the line $x_o + \alpha d$ exists that is an optional solution. □

(*) For the definition of a semistrictly quasiconcave function see Avriel-Diewert-Schaible-Ziemba (1981).

2. A FUNDAMENTAL LEMMA

In this section we assume $D > 0$. Let I be the range of $D(x)$, $x \varepsilon S$. For any $\xi \varepsilon I$ consider the problem obtained from P by adding the constraint $D(x) = \xi$, that is the problem

$$P(\xi): \quad \max_{x \varepsilon S} [N(x) . \xi^{p}] \triangleq z(\xi) \qquad D(x) = \xi$$

Let S_ξ and $x(\xi)$ denote the feasible region and an optimal solution of problem $P(\xi)$, respectively. Let us suppose that problem $P(\xi_k)$, $\xi_k \varepsilon I$, has an optimal solution, say $x(\xi_k)$. Setting

$$D(x) = d^t x + d_o, \quad \text{where } d \varepsilon \mathbb{R}^n, \ d_o \text{ is a scalar}$$

$$F = \left[A^t \vdots I \vdots d \vdots -d \right]^t, \quad f = (b^t, 0^t, \xi_k - d_o, -\xi_k + d_o)^t$$

problem $P(\xi_k)$ can be written in this way:

$$\begin{array}{l} \max [N(x) . \xi_k^p] \\ Fx \geq f \end{array} \tag{1}$$

Denote by:

μ the vector of Lagrange multipliers associated with $x(\xi_k)$

$\mu_\ell, \mu_{\ell+1}$ the last two components of μ.

Moreover, set

$$z_k = z(\xi_k) = N(x(\xi_k)) \cdot \xi_k^p \qquad \mu_o = \mu_{\ell+1} - \mu_\ell .$$

We are interested in finding a new optimal solution $x(\xi_k + \theta)$ of problem $P(\xi_k + \theta)$, such that $z(\xi_k + \theta) \triangleq z(\theta) > z_k$. To this end we state, first of all, the following theorem.

THEOREM 3 (Giannessi (1979)): Consider a set $X \subseteq \mathbb{R}^n$ *and functions* $h: X \to \mathbb{R}$, $g: X \to \mathbb{R}^m$. *Assume* h *and* g *are concave and* X *is convex. The system*

$$\begin{cases} h(x) > 0 \\ g(x) \geq 0 \\ x \varepsilon X \end{cases}$$

is inconsistent iff there exists $\lambda \in \mathbb{R}_+^{(*)}$ *and* $u \in \mathbb{R}_+^{m(*)}$, *with* $(\lambda,u) \neq 0$ *such that*

$$\lambda\, h(x)+u^t g(x) \leq 0 \qquad \forall x \in X.$$

Now we are able to prove the following fundamental Lemma.

LEMMA 4: Let us consider the function

$$Q(\theta) = z_k(\xi_k+\theta)^{-p}-z_k\xi_k^{-p}-\mu_o\theta$$

and the set $Q^+ = \{\theta\in\mathbb{R} : \xi_k+\theta > 0,\ Q(\theta) \geq 0\}$. *Then*

(a) for any $\theta\in Q^+$ *such that* $S_{\xi_k+\theta}\neq\emptyset$ *we have* $z(\theta) \leq z_k$

(b) $Q(\theta)$ *is increasing or decreasing at* $\theta=0$ *according to* $\mu_o < -pz_k\xi_k^{-p-1}$ *or* $\mu_o > -pz_k\xi_k^{-p-1}$.

If $\mu_o = -pz_k\xi_k^{-p-1}$ *and* $z_k \geq 0$, *then* $\theta=0$ *is a local minimum point for* $Q(\theta)$.

PROOF:

(a) For any θ, such that $\xi_k+\theta > 0$, consider the system

$$\begin{cases} N(x)\cdot(\xi_k+\theta)^p > z_k \\ x \in S \\ D(x)=\xi_k+\theta \end{cases} \tag{2}$$

Set $f_o = (0,0,\ldots,0,1,-1)^t$; then system (2) can be written in the following way

$$\begin{cases} N(x) - z_k(\xi_k+\theta)^{-p} > 0 \\ Fx-f-\theta f_o \geq 0 \\ x \in S \end{cases} \tag{3}$$

According to Theorem 3, system (3) is inconsistent iff there exists a vector (λ,u) such that

$$\lambda(N(x)-z_k(\xi_k+\theta)^{-p})+u^t(Fx-f-\theta f_o) \leq 0 \qquad \forall x\in S \tag{4}$$

where $(\lambda,\mu) \neq 0$.

(*) $\mathbb{R}_+$ is the set of nonnegative real numbers and $\mathbb{R}_+^m$ is the nonnegative orthant of $\mathbb{R}^m$.

Substituting in (4) $u=\mu$, $\lambda=1$, we have

$$N(x)-z_k(\xi_k+\theta)^{-p}+\mu^t(Fx-f)-\theta(\mu_\ell-\mu_{\ell+1})\leq 0 \qquad \forall x\varepsilon S$$

$$N(x)-N(x(\xi_k))+N(x(\xi_k))-z_k(\xi_k+\theta)^{-p}+\mu^t(Fx-f)+\mu_o\theta\leq 0 \qquad \forall x\varepsilon S.$$

Set

$$\psi(x)=N(x)-N(x(\xi_k))+\mu^t(Fx-f).$$

Then the last inequality becomes

$$\psi(x) - Q(\theta) \leq 0 \qquad \forall x\varepsilon S. \tag{5}$$

Since $x(\xi_k)$ is the optimal solution of problem

$\max N(x)$

$Fx-f \geq 0$

we have $N(x)+\mu^t(Fx-f) \leq N(x(\xi_k)$ $\forall x\varepsilon S$, that is $\psi(x) \leq 0$ $\forall x\varepsilon S$. Consequently, condition (5) holds for any θ satisfying $Q(\theta) \geq 0$, and this implies that (4) is feasible and system (3) is inconsistent. This completes the proof of (α).

(b) It turns out that

$$Q'(0)=-pz_k\xi_k^{-p-1}-\mu_o \ , \ Q''(0)=p(p+1)z_k\xi_k^{-p-1} \ .$$

Hence, $Q'(0) > 0$ iff $\mu_o < -pz_k\xi_k^{-p-1}$ and $Q'(0) < 0$ iff $\mu_o > -pz_k\xi_k^{-p-1}$. Consider now the case $Q'(0)=0$, that is $\mu_o = -pz_k\xi_k^{-p-1}$. If $z_k=0$ or $p=-1$, $Q(\theta)$ becomes the null function; otherwise for $z_k > 0$ the inequality $Q''(0) > 0$ holds. In both cases $\theta = 0$ is a local minimum point for $Q(\theta)$. □

Remark 1. Obviously $Q(0)=0$. If $Q'(0) > 0$ ($Q'(0) < 0$), that is $\mu_o < -pz_k\xi_k^{-p-1}$ ($\mu_o > -pz_k\xi_k^{-p-1}$), then (b) of Lemma 4 implies that $Q(\theta)$ is increasing (decreasing) at $\theta=0$. Thus, in a neighbourhood of 0, on the right (left), we have $Q(\theta) > 0$. In order to find a better solution than $x(\xi_k)$ we must consider $\theta < 0$ ($\theta > 0$) according to (a) of Lemma 4.

3. OPTIMALITY CONDITIONS FOR A SUBCLASS OF PROBLEMS (P)

In this section we will consider problem (P) under the assumption that N is affine; we assume also non-degeneracy of all basic solutions which will be met. By adjoining slack variables, problem $P(\xi_k)$

$$P(\xi_k):\quad \begin{aligned} &\max\ N(x)\cdot\xi_k^p \\ &Ax \geq b \\ &d^t x=\xi_k-d_o \\ &x \geq 0 \end{aligned}$$

can be converted into the standard form

$$\begin{aligned} &\max\ (c^t x+c_o)\xi_k^p \\ &A^*\begin{pmatrix}x\\y\end{pmatrix} = b^* \\ &\begin{pmatrix}x\\y\end{pmatrix} \geq 0 \end{aligned} \tag{6}$$

where

$$A^* = \begin{bmatrix} A & \vdots & -I \\ d^t & \vdots & 0 \end{bmatrix} \quad , \quad b^*=(b^t,\xi_k-d_o)^t \quad .$$

The first n components of x are the initial variables of problem $P(\xi_k)$ and the others are slack variables. Without any loss of generality we can assume $\xi_k > 0$ (if $\xi_k=0$ problem $P(\xi_k)$ is trivial; if $\xi_k < 0$ we replace $N(x)$ by $(-1)^p N(x)$ and $D(x)=d^t x+d_o$ by $-D(x)$).

Assume that problem (6) has an optimal basic solution $x(\xi_k)$ in a vertex of S, with corresponding basis B. We partition c^t as $c^t=(c_B^t,c_N^t)$. In the vertex $x(\xi_k)$ n constraints of S are binding as well as the constraint $D(x)=\xi_k$ and thus $x(\xi_k)$ is a degenerate basic solution.

To point out the dependence of the vector of Lagrange multipliers and that of the function $Q(\theta)$ on the basis (actually, it is $\mu=c_B^t B^{-1}$ and μ_o is the last component of μ), we set

$$\mu_B = \mu_o \quad , \quad Q_B(\theta)=Q(\theta) \quad , \quad Q_B^+(\theta)=Q^+ \; .$$

The following theorem gives a fundamental optimality condition.

THEOREM 5 (Local Optimality):

(a) *If there are two different bases* B_1 *and* B_2 *corresponding to an optimal solution* $x(\xi_k)$, *such that*

$$\mu_{B_1} < -pz_k\xi_k^{-p-1} \text{ and } \mu_{B_2} > -pz_k\xi_k^{-p-1}$$

or

$$\mu_{B_1} > -pz_k\xi_k^{-p-1} \text{ and } \mu_{B_2} < -pz_k\xi_k^{-p-1}$$

then $x(\xi_k)$ *is a local maximum point for problem* (P).

(b) *If* $\mu_B = -pz_k\xi_k^{-p-1}$ *and* $z_k \geq 0$, *then* $x(\xi_k)$ *is a local maximum point for problem* (P).

PROOF:

(a) Since $\mu_{B_1} < -pz_k\xi_k^{-p-1}$ and $\mu_{B_2} > -pz_k\xi_k^{-p-1}$, from (b) of Lemma 4 we obtain $z(\theta) \leq z_k$ in a neighbourhood of 0, on the right and on the left, respectively. Consequently, $x(\xi_k)$ is a local maximum point for problem (P). The proof for the other case in (α) is analogous.

(b) According to (b) of Lemma 4, $\theta=0$ is a local minimum point for $Q_B(\theta)$ that is $Q_B(\theta) \geq 0$ in a neighbourhood U of 0. From (a) of Lemma 4 it follows $z(\theta) \leq z_k$, $\forall\theta\varepsilon U$. □

Remark 2. If D is positive and $z_k \geq 0$, the local optimality condition becomes a global optimality condition. (see Theorem 1).

When problem $P(\xi_k)$ is solved, we consider the parametric problem $P(\xi_k+\theta), \theta \in I_k \triangleq [-\xi_k, +\infty)$. Let e_{m+1} denote the unit (m+1)-vector $(0,\ldots,0,1)^t$, and set

$$H_B = \{\theta\varepsilon I_k : B^{-1}(b^*+\theta e_{m+1}) \geq 0\}$$

$$Q_B^- = \{\theta : Q_B(\theta) \leq 0\}$$

$$U_B = H_B \cap Q_B^-$$

$\overset{\circ}{U}_B$ the interior of U_B.

Since $z_k = (c_B^t B^{-1} b^*)\xi_k^p$ we have for any $\theta \in H_B$

$$z(\theta) = c_B^t B^{-1}(b^*+\theta e_{m+1})(\xi_k+\theta)^p = (z_k \xi_k^{-p} + \theta c_B^t B^{-1} e_{m+1})(\xi_k+\theta)^p$$

that is

$$z(\theta) = (z_k \xi_k^{-p} + \mu_B \theta)(\xi_k+\theta)^p \ . \qquad (7)$$

The following theorem holds.

THEOREM 6:

(a) *If* $\overset{\circ}{U}_B \neq \emptyset$ *then for any* $\theta \in \overset{\circ}{U}_B$ *we have* $z(\theta) > z_k$.

(b) *Let* $\theta_k = -\dfrac{pz_k + \xi_k^{p+1}\mu_B}{(p+1)\mu_B \xi_k^p}$, $(p+1)\mu_B \xi_k \neq 0$.

If $\theta_k \in \overset{\circ}{U}_B$, *then* $x(\xi_k+\theta_k)$ *is a local maximum point for problem* (P).

If $\theta_k \notin \overset{\circ}{U}_B$, *then* $z(\theta)$ *is an increasing function in* U_B.

PROOF:

(a) We have $(z_k \xi_k^{-p} + \mu_B \theta)(\xi_k+\theta)^p > z_k$ iff $z_k(\xi_k+\theta)^{-p} - z_k \xi_k^{-p} - \mu_o \theta < 0$, that is iff $Q_B(\theta) < 0$. Thus for any $\theta \in U$ we have $z(\theta) > z_k$.

(b) Differentiating (7), we have

$$z'(\theta) = \xi_k^{-p}(\xi_k+\theta)^{p-1}[(p+1)\xi_k^p \mu_B \theta + pz_k + \mu_B \xi_k^{p+1}].$$

Since the inequalities $\xi_k > 0$ and $\xi_k + \theta > 0$ hold, we have $z'(\theta) \gtrless 0$ according to $q(\theta) \gtrless 0$ where $q(\theta)$ is the first degree polynomial

$$q(\theta) = (p+1)\xi_k^p \mu_B \theta + pz_k + \mu_B \xi_k^{p+1}$$

whose unique root is θ_k when $(p+1)\mu_B \xi_k \neq 0$. There are two possibilities: either the sign of $z'(\theta)$ is constant or not. In the first case we have $z'(\theta) > 0 \ \forall \theta \in U$ because $z(\theta) > z_k = z(0) \ \forall \theta \in U$. For the same reason in the second case we have $z'(\theta) > 0 \ \forall \theta \in [0, \theta_k)$ and $z'(\theta) < 0 \ \forall \theta > \theta_k, \theta \in U$. Thus, θ_k is a local maximum point for $z(\theta)$ and, consequently, $x(\xi_k+\theta_k)$ is a local maximum point for problem (P). □

Remark 3. As a consequence of Theorem 6 $x(\xi_k)$ is not a local maximum point if $\overset{o}{U}_B \neq \emptyset$; in such a case we can find a better solution than $x(\xi_k)$ namely $x(\xi_k+\theta_k)$ if $\theta_k \varepsilon \overset{o}{U}_B$ or the vertex of U_B adjacent to $x(\xi_k)$ if $\theta_k \notin \overset{o}{U}_B$ and U_B is a compact interval.

When U_B is unbounded and $\theta_k \notin \overset{o}{U}_B$ we have:

if $p > 0$ $\quad \sup_{x\varepsilon S} f(x)=+\infty$ because in such a case $\lim_{\theta\to+\infty} z(\theta)=+\infty$

if $p=-1$ $\quad \sup_{x\varepsilon S} f(x)=\mu_B$ because $z(\theta)$ becomes

$$z(\theta) = \frac{z_k\xi_k+\mu_B\theta}{\xi_k+\theta} \quad \text{and} \quad \lim_{\theta\to+\infty} z\ (\theta)=\mu_B.$$

4. A SEQUENTIAL METHOD FOR A LINEAR FRACTIONAL PROGRAM

Let us consider problem (P) where N is affine and p=-1. The results obtained in Section 3 allow us to state a solution method for a linear fractional problem (Cambini (1977)).

Step 0. Let x_o be a feasible solution of problem (P) and set $D(x_o)=\xi_o$; replace index 0 by k and go to Step 1.

Step 1. Solve the linear problem $P(\xi_k)$. If $P(\xi_k)$ does not have optimal solutions, the algorithm terminates and problem (P) does not have an optimal solution. Otherwise, let B denote a basis corresponding to the optimal solution (*) $x(\xi_k)$.

If $\mu_B=z_k$ then $x(\xi_k)$ is an optimal solution for problem (P); if $\mu_B < z_k (\mu_B > z_k)$ go to Step 2.

Step 2. If $\overset{o}{U}_B = \emptyset$ go to Step 3; otherwise go to Step 4.

Step 3. Change the basis; let B_1 denote the new optimal basis and find μ_{B_1}. If $\mu_{B_1} > z_k (\mu_{B_1} < z_k)$ the algorithm terminates $x(\xi_k)$ is an optimal solution for problem (P); otherwise replace the basis B_1 by B and go to Step 2.

(*) -One of the n hyperplanes binding at $x(\xi_k)$ is $D(x)=\xi_k$.

Step 4. If $U_B=[0,+\infty)$ the algorithm terminates. Problem (P) does not have an optimal solution and $\sup_{x\in S} f(x)=\mu_B$. Otherwise U_B is an interval of vertices 0 and $\bar{\theta}$; set $\xi_{k+1}=\xi_k+\bar{\theta}$. Replace index k+1 by k and go to Step 1.

Example 1. Consider the problem

$$\max\ [f(x)=(2x_1+5x_2)(x_1+2x_2+1)^{-1}]$$
$$-x_1+x_2 \le 1$$
$$x_1+x_2 \le 5;\ x_1,x_2 \ge 0.$$

Step 0. $x_o=(0,0)$ is a feasible solution; set $\xi_o=D(x_o)=1$. Go to Step 1.

Step 1. Solve the linear problem

$$\begin{aligned} &\max\ 2x_1+5x_2\\ &-x_1+x_2+x_3 \quad =1\\ &\ x_1+x_2 \quad +x_4=5\\ &\ x_1+2x_2 \qquad =0 \quad ,\quad x_1,x_2 \ge 0\ . \end{aligned}$$

We have

$$x_B=(x_3,x_4,x_2)^t=(1,5,0)^t \quad , \quad c_B^t=(0,0,5),\ z_o=0$$

$$B^{-1}=\begin{bmatrix} 1 & 0 & -1/2\\ 0 & 1 & -1/2\\ 0 & 0 & 1/2 \end{bmatrix} \quad , \quad \mu_B=c_B^t\begin{pmatrix} -1/2\\ -1/2\\ 1/2 \end{pmatrix}=\frac{5}{2}\ .$$

It occurs $\mu_B=5/2 > z_o=0$. Go to Step 2.

Step 2. When $b=(1,5,0)^t$ is replaced by $b'=b+\theta(0,0,1)^t$ we have

$$x_B(\theta)\triangleq B^{-1}b+\theta B^{-1}\begin{pmatrix} 0\\ 0\\ 1 \end{pmatrix}=\begin{pmatrix} 1 & -1/2\ \theta\\ 5 & -1/2\ \theta\\ & 1/2\ \theta \end{pmatrix}\ .$$

It follows

$$H_B=\{\theta:x_B(\theta)\ge 0\}=[0,2],\ Q_B^-=\{\theta:Q_B(\theta)$$
$$=(z_{k_o}-\mu_B)\theta \le 0\}=[0+\infty);\ U_B=H_B\cap Q_B^-=[0,2].$$

Since $U_B \neq \emptyset$ go to Step 4.

Step 4. Since $\bar{\theta}=2$ put $\xi_1=1+2=3$ and go to Step 1.

Step 1. For $\bar{\theta}=2$ we have

$$x_B(\bar{\theta}=2)=(x_3,x_4,x_2)^t=(0,4,1), \quad z_1=5/3 ,$$

$$\mu_B = 5/2 .$$

Since $\mu_B=5/2 > z_1=5/3$ go to Step 2.

Step 2. $H_B=[-2,0]$, $Q_B^-=[0,+\infty)$, $U_B=\{0\}$.

Since $\overset{\circ}{U}_B=\emptyset$ go to Step 3.

Step 3. x_3 leaves the basis and x_1 enters the basis. We have

$$x_B^t=(x_1,x_4,x_2)=(0,4,1)^t, \quad z_1=5/3, \quad c_B^t=(2,0,5),$$

$$B^{-1} = \begin{bmatrix} -2/3 & 0 & 1/3 \\ 1/3 & 1 & -2/3 \\ 1/3 & 0 & 1/3 \end{bmatrix} , \quad \mu_B=c_B^t \begin{pmatrix} 1/3 \\ -2/3 \\ 1/3 \end{pmatrix} =7/3 .$$

It follows $\mu_B=7/3 > z_1=5/3$. Go to Step 2.

Step 2. Set $\xi=3+\theta$. We have

$$x_B(\theta)=(\tfrac{1}{3}\theta, 4-\tfrac{2}{3}\theta, 1+\tfrac{1}{3}\theta)^t \qquad H_B=[0,6],$$

$$U_B=[0,6] .$$

Since $\overset{\circ}{U}_B\neq\emptyset$ go to Step 4.

Step 4. Since $\bar{\theta}=6$ put $\xi_2=3+\bar{\theta}=9$ and go to Step 1.

Step 1. For $\bar{\theta}=6$ we have

$$x_B(\bar{\theta}=6)=(x_1,x_4,x_2)^t=(2,0,3)^t \quad z_2=19/9,$$

$$\mu_B=7/3 .$$

Since $\mu_B=7/3 > z_2=19/9$ go to Step 2.

Step 2. $H_B=[-1,0]$, $Q_B^-=[0,+\infty)$, $U_B=\{0\}$.

Since $\overset{\circ}{U}_B = \emptyset$, go to Step 3.

Step 3. x_4 leaves the basis and x_5 enters the basis. We have

$$x_B^t=(x_1,x_5,x_2)=(2,0,3)^t \ , \ c_B^t=(2,0,5) \ , \ z_2=19/9 \ ,$$

$$B^{-1}=\begin{bmatrix} -1/2 & 1/2 & 0 \\ -1/2 & -3/2 & 1 \\ 1/2 & 1/2 & 0 \end{bmatrix} \qquad \mu_B=c_B^t \begin{pmatrix} 0 \\ 1 \\ 0 \end{pmatrix} = 0 \ .$$

Since $\mu_B=0 < z_2=19/9$ the algorithm terminates: $\bar{x}=(2,3,0,0,0)$ is an optimal solution for the given problem.

5. A SEQUENTIAL METHOD FOR A SUBCLASS OF PROBLEM (P)

Now we shall describe an algorithm for problem (P) where N is affine, which generalizes the one considered in Section 4 (Martein and Pellegrini (1977); Cambini, Martein and Pellegrini (1978)). This algorithm allows us to find a local maximum point $x(\xi_k)$ for problem (P). If D is nonnegative and $z_k \geq 0$ then $x(\xi_k)$ is also an optimal solution for (P). Otherwise, we are interested to find a new local maximum point better than $x(\xi_k)$. According to Lemma 4 we must find a suitable root $\hat{\theta} \neq 0$ of $Q_B(\theta)$. With respect to the number of roots of $Q_B(\theta)$, the following theorem holds:

THEOREM 7:

(a) *Let* $p < 0$. $Q_B(\theta)$ *has at most one real root different from zero if* p *is even, and* $Q_B(\theta)$ *has at most two real roots different from zero if* p *is odd.*

(b) *Let* $p > 0$. $Q_B(\theta)$ *has at most two real roots different from zero if* p *is even, and* $Q_B(\theta)$ *has at most three real roots different from zero is* p *is odd.*

PROOF: It is sufficient to note that $Q'(\theta)=-pz_k(\xi_k+\theta)^{-p-1}-\mu_B$ has only one root if p is even and two roots if p is odd. Since $\theta=0$ is a root of $Q_B(\theta)$ and $Q_B(\theta)$has at most one discontinuity point the results (a), (b) follow. □

We define $\hat{\theta}$ in the following way:

$\hat{\theta}$ is the positive root of $Q_B(\theta)$ when $p < 0$,

$\hat{\theta}$ is the negative root of $Q_B(\theta)$ belonging to $[-\xi_k, 0]$, when $p > 0$ and $z_k < 0$,

$\hat{\theta}$ is the negative root of $Q_B(\theta)$ nearest to 0, belonging to $(-\infty, -\xi_k]$, when $p > 0$ and $z_k > 0$.

If $S_{\xi_k+\hat{\theta}} = \emptyset$, $x(\xi_k)$ is an optimal solution for problem P. Otherwise we solve the problem $P(\xi_k+\hat{\theta})$ and then we apply the next algorithm. In such a way we find an optimal solution of (P) by a finite sequence of local maximum points $x_1, \ldots, x_s$ such that $f(x_1)<f(x_2)<\ldots<f(x_s)$. This fact guarantees the convergence, in a finite number of steps, of the following algorithm:

Step 0. Let x_o be a feasible solution of problem (P) and set $D(x_o)=\xi_o$. We can suppose $\xi_o \geq 0$. (Otherwise we replace D by -D and N by -N, if p is even, D by -D if p is odd.) Replace index 0 by k and go to Step 1.

Step 1. Solve the linear problem $P(\xi_k)$. If $P(\xi_k)$ does not have optimal solutions the algorithm terminates: Problem (P) does not have optimal solutions. Otherwise let B denote a basis corresponding to the optimal solution $x(\xi_k)$. If $\mu_B < -pz_k\xi_k^{-p-1}$ $(\mu_B > -pz_k\xi_k^{-p-1})$ or if $\mu_B=-pz_k\xi_k^{-p-1}$ and $z_k < 0$, go to Step 2. If $\mu_B=-pz_k\xi_k^{-p-1}$ and $z_k \geq 0$, then $x(\xi_k)$ is a local maximum point (global if $p < 0$); go to Step 5 if $p > 0$.

Step 2. If $\overset{o}{U}_B=\emptyset$ go to Step 3. Otherwise go to Step 4.

Step 3. Change the basis; let B_1 denote the new optimal basis. If $\mu_{B_1} < -pz_k\xi_k^{-p-1}$ $(\mu_{B_1} < -pz_k\xi_k^{-p-1})$ then $x(\xi_k)$ is a local maximum point (global if $p < 0$ and $z_k \geq 0$), go to Step 5. Otherwise replace basis B_1 by B and go to Step 2. If B_1 does not exist $x(\xi_k)$ is an optimal solution.

Step 4. Set

$$\theta_k = -\frac{pz_k+\xi_k^{-p-1}\mu_B}{(p+1)\xi_k^p\mu_B}, \quad (p+1)\xi_k^p\,\mu_B \neq 0.$$

If $\theta_k \varepsilon \overset{o}{U}_B$ then $x(\xi_k+\theta_k)$ is a local maximum point (global if $p < 0$ and $z \geq 0$). Go to Step 5 replacing ξ_k by $\xi_k+\theta_k$. Otherwise the following cases occur:

a) $U_B=[0,+\infty)$. If $p > 0$ the algorithm terminates: problem P does not have optimal solutions. If $p < 0$ go to Step 5.

b) U_B is an interval of vertices $0,\theta^*$. Put $\xi_{k+1}=\xi_k+\theta^*$. If $\xi_{k+1} \geq 0$ replace index k+1 by k and go to Step 1. If $\xi_{k+1} < 0$ replace D by -D if p is even , or N by -N and D by -D if p is odd. Go to Step 1.

Step 5. Set $\xi_{k+1}=\xi_k+\hat{\theta}$. If $S_{\xi_{k+1}} \neq \emptyset$ replace index k+1 by k and go to Step 1. If $S_{\xi_{k+1}} = \emptyset$ the algorithm terminates: $x(\xi_k)$ is an optimal solution for problem P or $\sup_{x\varepsilon S} f(x) = 0$ if we are coming from (a) of Step 4.

Example 2. Consider the problem

$$\max\ (2x_1-x_2+1)(x_1+x_2-2)^2$$

$$x_1+2x_2 \leq 2,\ x_1,x_2 \geq 0\ .$$

Step 0. $x_o=(0,0)$ is a feasible solution and $D(x_o)=-2$. We replace D by -D and we set $\xi_o=2$. Go to Step 1.

Step 1. Solve the linear problem

$$4\max(2x_1-x_2+1)$$
$$x_1+2x_2+x_3 = 2$$
$$-x_1 - x_2 = 0$$
$$x_1,x_2,x_3 \geq 0.$$

We have

$$x_B=(x_1,x_3)=(0,2),\ z_o=4,\ \mu_B=-2,\ -pz_o\xi_o^{-p-1}=-1$$

Go to Step 2.

Step 2. When $b=(2,0)^t$ is replaced by $b'=b+\theta(0,1)^t$, we have

$$x_B(\theta)=(-\theta,2+\theta)^t \qquad Q_B(\theta)=\frac{\theta(2\theta^2+7\theta+4)}{(2+\theta)^2}\ .$$

It follows that

$$Q_B^- = \left[-\infty,\frac{-7-\sqrt{17}}{4}\right] \cup \left(0,\frac{-7+\sqrt{17}}{4}\right],\ H_B=[-2,0],$$

$$U_B = \left[0,\frac{-7+\sqrt{17}}{4}\right].$$

Since $\overset{\circ}{U}_B \neq \emptyset$ go to Step 4.

Step 4. $\theta_o=-\frac{1}{3}\in\overset{\circ}{U}_B$. Set $\xi_1=\xi_o+\theta_o=2-\frac{1}{3}=\frac{5}{3}$. Then $x_B(-\frac{1}{3})=(x_1,x_3)=(\frac{1}{3},\frac{16}{3})$ is a local maximum point and $z_1=125/27$. Go to Step 5.

Step 5. We have $Q_B(\theta)=\theta^2(2\theta+5)\cdot(\frac{5}{3}+\theta)^{-2}$ and $\hat{\theta}=-5/2$. We set $\xi_2=\xi_1+\hat{\theta}=\frac{5}{3}-\frac{5}{2}=-\frac{5}{6}$. Since $S_{\xi_2}=\emptyset$ $(x_1,x_3)=(\frac{1}{3},\frac{16}{3})$ is also a global maximum point for problem (P).

6. THE QUADRATIC FRACTIONAL CASE

Martein and Pellegrini (1978) have studied and proposed an algorithm for solving the problem (P) when p=-1 and N is a strictly concave quadratic function. Such a problem has a unique optimal solution. Let us mention briefly the main features of the algorithm. For this problem, $P(\xi_k)$ becomes

$$\xi_k^{-1} \max_{\substack{x\in S \\ D(x)=\xi_k}} N(x)$$

Using the same notation as in Section 2, the problem $P(\xi_k)$ can be written in the following way:

$$\xi_k^{-1} \max_{Fx\geq f} N(x) .$$

Now, $x(\xi_k)$ is not a vertex of S. Nevertheless $x(\xi_k)$ is binding for $D(x)=\xi_k$ and for s constraints of S where $0 \leq s \leq n-1$. If f is replaced by $f+\theta f_o$, the optimal solution $x(\xi_k+\theta) \triangleq x(\theta)$ of problem $P(\xi_k+\theta)$ is binding for the same constraints of $x(\xi_k)$ in an appropriate neighbourhood of $\theta=0$. Let $\mu=\mu(\theta)$ denote the vector of the Lagrange multipliers associated with $x(\theta)$ and let $\mu_o(\theta),\ldots,\mu_{s-1}(\theta)$ be the components of $\mu(\theta)$ which are not the null function. In particular $\mu_o(\theta)$ is the component corresponding to the constraint $D(x)=\xi_k$. Let us remark that $x(\theta)$, $\mu_i(\theta)$, $i=0,\ldots,s-1$ are affine functions. In our case, the function $Q(\theta)$ has the simple form

$$Q(\theta)=(z_k-\mu_o)\theta$$

If $z_k=\mu_o$ we have $Q(\theta)=0 \quad \forall\theta\in\mathbb{R}$, and thus $x(\xi_k)$ is the optimal solution of (P) according to Lemma 4. If $z_k \neq \mu_o$, the set $Q^-=\{\theta : Q(\theta) \leq 0\}$ is a half-line. Denote by:

$\bar{\theta}$ - the real number satisfying the following conditions: $\bar{\theta}\in Q^-$ and $\theta > \bar{\theta}$ implies $x(\theta)\notin S$;

$\hat{\theta}$ - the minimum of the roots of the functions $\mu_i(\theta)$, $i=0,\ldots,s-1$, belonging to Q^-.

It occurs that $\mu_i(\theta) \geq 0$ either for all $\theta\in[0,\hat{\theta}]$ if $\hat{\theta} > 0$ or for all $\theta\in[\hat{\theta},0]$ if $\hat{\theta} < 0$.

θ_o - the root of the equation $z(\theta)=\mu_o(\theta)$ belonging to Q^-.

Set $\theta^* = \min(\bar{\theta},\hat{\theta})$ and $H=[0,\theta^*]$ if $\theta^* > 0$, or $H=[\theta^*,0]$ if $\theta^* < 0$.

There are two cases:

(1) $\theta_o \varepsilon H$. Then $x(\theta_o)$ is the optimal solution of the problem (P) because $x(\theta_o)\varepsilon S$, $\mu(\theta_o) \geq 0$ and the function $Q(\theta)$ corresponding to the problem $P(\xi_k+\theta_o)$ becomes the null function

(2) $\theta_o \notin H$. Set $\xi_k'=\xi_k+\theta^*$. We find the vectors $\mu(\theta)$, $x(\theta)$ corresponding to the problem $P(\xi_k'+\theta)$. Replacing ξ_k' with ξ_k we can repeat the previous considerations.

In such a way we find a finite sequence of optimal solutions $x(\xi_{k_1}),\ldots,x(\xi_{k_s})$ such that $z_{k_1} < z_{k_2} <\ldots< z_{k_s}$ because $N(x)$ is a strictly concave quadratic function. This guarantees the convergence of the algorithm in a finite number of steps.

REFERENCES

AVRIEL, M., DIEWERT, W.E., SCHAIBLE, S., and ZIEMBA, W.T. (1981). "Introduction to Concave and Generalized Concave Functions." This volume, 21-50.

BENDERS, J.F. (1962). "Partitioning Procedures for Solving Mixed-Variables Programming Problems." *Numerische Mathematik 4*, 238-252.

BITRAN, G.R. and NOVAES, A.G. (1973). "Linear Programming with a Fractional Objective Function." *Operations Research 21*, 22-29.

CAMBINI, A., CARRARESI, P., and GIANNESSI, F. (1976). "Sequential Methods and Decomposition in Mathematical Programming." Paper A-38, Dept. of Operations Research, University of Pisa, Italy.

CAMBINI, A. (1977). "Un Algoritmo per il Massimo del Quoziente di Due Forme Affini con Vincoli Lineari." Paper n. A-42, Dept. of Operations Research, University of Pisa, Italy.

CAMBINI, A., MARTEIN, L., and PELLEGRINI, L. (1978). "Decomposition methods and Algorithms for a Class of Non-Linear Programming Problems." First Meeting AFCET-SMF Palaiseau, Ecole Polytechnique Palaiseau, (Paris), 2, 179-189.

CASTELLANI, G., and GIANNESSI, F. (1976). "Decomposition of Mathematical Programs by Means of Theorems of Alternative for Linear and Nonlinear Systems." *Proceeding IX Int. Symposium on Mathematical Programming, Budapest 1976.* (A. Prekopa, ed.). *Hungarian Academy of Sciences*, 423-439.

CHADDA, S.S. (1976). "A Decomposition Principle for Fractional Programming." *Opsearch 4*, 123-132.

CHARNES, A., and COOPER, W.W. (1962). "Programming with Linear Fractional Functionals." *Naval Research Logistics Quarterly 9*, 181-186.

GIANNESSI, F. (1979). "Theorems of Alternative, Quadratic Programs and Complementarity Problems." *Variational Inequalities and Complementarity Problems. Theory and Applications*. (R.W. Cottle, F. Giannessi and J.L. Lions, eds.), Wiley, New York, 151-186.

JAGANNATHAN, R. (1966). "On Some Properties of Programming in Parametric Form Pertaining to Fractional Programming." *Management Science 12*, 609-615.

JOKSCH, H.C. (1964). "Programming with Fractional Linear Objective Functions." *Naval Research Logistics Quarterly 11*, 197-204.

MANGASARIAN, O.L. (1969). *Nonlinear Programming*. McGraw-Hill, New York.

MARTEIN, L., and PELLEGRINI, L. (1977). "Un Algoritmo per la Determinazione del Massimo di una Particolare Funzione Razionale Fratta Soggetta a Vincoli Lineari." Paper n. A-45, Dept. of Operations Research, University of Pisa, Italy.

MARTEIN, L., and PELLEGRINI, L. (1977). "Su un'Estensione di una Particolare Classe di Problemi di Programmazione Frazionaria." Paper n. A-47, Dept. of Operations Research, University of Pisa, Italy.

MARTEIN, L., and PELLEGRINI, L. (1977). "Su una Classe di Problemi non Lineari e non Convessi." Paper n. A-48, Dept. of Operations Research, University of Pisa, Italy.

MARTOS, B. (1964). "Hyperbolic Programming." *Naval Research Logistics Quarterly 11*, 135-155.

MOND, B., and CRAVEN, B.D. (1973). "A Note on Mathematical Programming with Fractional Objective Functions." *Naval Research Logistics Quarterly 20*, 577-581.

MOND, B., and CRAVEN, B.D. (1975). "Non-Linear Fractional Programming." *Bull. Austr. Math. Soc. 3*, 391-397.

MOND, B., and CRAVEN, B.D. (1975). "On Fractional Programming and Equivalence." *Naval Research Logistics Quarterly 22*, 405-410.

SCHAIBLE, S. (1972). "Quasi-Concave, Strictly Quasi-Concave and Pseudo-Concave Functions." In *Methods of Operations Research, Vol.17*, 308-316.

SCHAIBLE, S. (1976). "Fractional Programming. I. Duality." *Management Science 22*, 858-867.

SCHAIBLE, S. (1977). "A Note on the Sum of a Linear and Linear-Fractional Function." *Naval Research Logistics Quarterly 24*, 691-693.

SCHAIBLE, S. (1981). "A Survey of Fractional Programming."This vol. 417-440

SWARUP, K. (1965). "Linear Fractional Functionals Programming." *Operations Research* , 1029-1036.

VAN DE PANNE, C. (1975). *Methods for Linear and Quadratic Programming*. North-Holland Publishing Company, Amsterdam.

ZIONTS, S. (1968). "Programming with Linear Fractional Functionals." *Naval Research Logistics Quarterly 15*, 449-452.

PART VI

APPLICATIONS OF GENERALIZED CONCAVITY IN MANAGEMENT SCIENCE AND ECONOMICS

GENERALIZED CONCAVITY AND ECONOMICS

W.E. Diewert†

The paper indicates where the various kinds of generalized concave functions are used in economics. The producer's cost minimization, profit maximization, and the consumer's utility maximization problems are discussed along with some applications to various areas of economics. The role of generalized concave functions is discussed in what is known in the economics literature as comparative statics analysis and in the applied mathematics literature as perturbation or stability theory.

1. OVERVIEW

The purpose of this paper is to indicate where generalized concave functions are used in economics. In section 2, we discuss the producer's cost minimization problem where the producer is subject to a production function or technological constraint and we indicate how the various types of concavity play a role in this theory. In Section 3, we undertake a similar analysis for the consumer's utility maximization problem. In both sections, we indicate applications of the basic theory to various areas of economics. In Section 4, we study briefly the producer's profit maximization and comparative statics analysis, one of the fundamental tools of economic theory pioneered by Hicks (1946) and Samuelson (1947). The latter topic is known as perturbation theory in the applied mathematics literature. Section 5 offers a few concluding remarks.

The reader may find it useful to consult the earlier paper in this volume by Avriel, Diewert, Schaible and Ziemba (1981) for definitions and alternative terminology for the various types of generalized concave functions discussed in this paper.

†The author thanks W. Eichhorn, Y. Kannai and W.T. Ziemba for helpful comments, the Social Sciences and Humanities Research Council of Canada for financial support, and the National Bureau of Economic Research at Stanford, California for providing office space.

ISBN 0-12-621120-5

2. THE COST OR EXPENDITURE FUNCTION

2.1 The Basic Problem

One of the fundamental paradigms in economics has a producer competitively minimizing costs subject to his technological constraints. *Competitive* means that the producer takes input prices as fixed and unchanging during the given period of time irrespective of the producer's demand for those inputs. This is not always a reasonable assumption, particularly if the producer is large relative to the size of the market, but it is often a reasonable first approximation.

Assume that only one output is produced using N inputs and that the producer's technology can be summarized by a production function F: $y = F(x)$, where $y \geqslant 0$ is the maximal amount of output that can be produced during a period given the nonnegative vector of inputs $x^T \equiv (x_1, x_2, \ldots, x_N) \geqslant 0_N$; T denotes transposition and all vectors are columns. We further assume that the cost of purchasing one unit of input i is $p_i > 0$, $i=1,2,\ldots,N$ and that the positive vector of input prices that the producer faces is $p \equiv (p_1, p_2, \ldots, p_N)^T >> 0_N$. For $y \geqslant 0$, $p >> 0_N$, the *producer's cost function* C is defined as the solution to the constrained minimization problem:

$$C(y,p) \equiv \min_x \{p^T x : F(x) \geqslant y\} \tag{1}$$

where $p^T x \equiv \sum_{i=1}^{N} p_i x_i$.

Of course, the minimum in (1) may not exist. However, if we impose the following very weak regularity condition on the production function F, it can be shown (e.g., see McFadden (1978)) that C will be well defined as a minimum (at least for $p >> 0_N$):

Assumption 1 on F: F is a real valued nonnegative function defined for all nonnegative input vectors $x \geqslant 0_N$ and F is *continuous from above;* i.e., for every $y \in$ Range F, the upper level set $U(y) \equiv \{x: F(x) \geqslant y\}$

is closed.

Assumption 1 above is very weak from an empirical point of view, since it cannot be contradicted by a finite set of data on the inputs and outputs of a producer. Define $\bar{y} \equiv \sup_x \{F(x): x \geqslant 0_N\}$. If the supremum is attained, define $Y \equiv \{y: 0 \leqslant y \leqslant \bar{y}\}$; if the sup is not attained, define $Y \equiv \{y: 0 \leqslant y < \bar{y}\}$.

THEOREM 1. (Diewert, 1978b): If F *satisfies Assumption 1, then* C *is well defined by (1) for* $y \in Y$ *and* $p >> 0_N$. *Moreover,* C *has the following seven properties:*

(2) $y \in Y,\ p >> 0_N \rightarrow C(y,p) \geqslant 0$ *(nonnegativity)*

(3) $y \in Y,\ p >> 0_N,\ \lambda > 0 \rightarrow C(y, \lambda p) = \lambda C(y,p)$ *(linear homogeneity in prices for fixed output)*

(4) $y \in Y,\ 0_N << p^0 < p^1$ (i.e., $p^0 \leqslant p^1$ but $p^0 \neq p^1$) $\rightarrow C(y, p^0) \leqslant C(y, p^1)$ *(nondecreasing in prices for fixed output)*

(5) $y \in Y,\ p^1 >> 0_N,\ p^2 >> 0_N,\ 0 \leqslant \lambda \leqslant 1 \rightarrow C(y, \lambda p^1 + (1-\lambda)p^2) \geqslant \lambda C(y, p^1) + (1-\lambda)C(y, p^2)$ *(concavity in prices for fixed output)*

(6) $y \in Y \rightarrow C(y, p)$ is continuous in p for $p >> 0_N$ *(continuity in prices for fixed output)*

(7) $p >> 0_N,\ y^0 \in Y,\ y^1 \in Y,\ y^0 < y^1 \rightarrow C(y^0, p) \leqslant C(y^1, p)$ *(nondecreasing in output for fixed prices)*, and

(8) $p >> 0_N,\ \alpha \in R^1 \rightarrow \{y: C(y, p) \leqslant \alpha\}$ is a closed set *(continuity from below in output for fixed prices).*

From an economic point of view, properties (2), (3), (4), and (7) are intuitively obvious. The most puzzling property to an economist is the concavity property (5). In order to explain why this property holds, we sketch a proof of it due to McKenzie (1956-7; 185).

Let $y \in Y$, $p^1 >> 0_N$, $p^2 >> 0_N$ and $0 \leqslant \lambda \leqslant 1$. Then

$$\begin{aligned} C(y, \lambda p^1 + (1-\lambda)p^2) &\equiv \min_x\{(\lambda p^1 + (1-\lambda)p^2)^T x: F(x) \geqslant y\} \\ &= (\lambda p^1 + (1-\lambda)p^2)^T x^* \text{ for some } x^* \text{ such that } \\ &\quad F(x^*) \geqslant y \\ &= \lambda(p^{1T}x^*) + (1-\lambda)(p^{2T}x^*) \\ &\geqslant \lambda C(y; p^1) + (1-\lambda)C(y; p^2) \end{aligned}$$

since x^* is feasible for the p^1 and p^2 cost minimization problems, but it is not necessarily optimal; i.e., $p^{iT}x^* \geqslant C(y, p^i)$ for $i = 1, 2$.

To a mathematician, the concavity property may not be so puzzling since $C(y, p)$ is the *support function* for the upper level set $U(y) \equiv \{x: F(x) \geqslant y\}$, the set of input vectors that can produce at least the output level y.

Properties (6) and (8) are somewhat technical in nature. Property (6) follows from the fact that a concave function is continuous over the interior of its domain of definition (see Fenchel (1953; 75) or Rockafellar 1970; 82)), while (8) follows from Berge's (1963; 111-112) Upper Semicontinuous Maximum Theorem.

Economists have been studying the cost function for a long time. Under somewhat stronger assumptions on the production function F, Shephard (1953; 14-15) obtained properties (2), (3), (4) and (5), Uzawa (1964; 217) deduced (6) and (7), and Shephard (1970; 83) obtained (8).

Theorem 1 has some important empirical implications. For example, economists often have data on cost, output and input prices for a firm during period t, C^t, y^t and p^t, respectively. A functional form for the firm's cost function, C^* say, is assumed and the unknown parameters which characterize C^* are estimated by minimizing the sum of squared errors $(e^t)^2$ where the error in period t is:

$$e^t \equiv C^t - C^*(y^t, p^t), \quad t = 1,2,\ldots,T. \tag{9}$$

Frequently, the functional form for C* is assumed to be linear:

$$C^*(y, p) \equiv a_0 + \sum_{i=1}^{N} a_i p_i + a_{N+1}\, y \qquad (10)$$

where the a_i are unknown parameters to be estimated. However, if the firm's production function satisfies the very weak regularity condition in Assumption 1 above and if the firm is competitively minimizing costs during the T periods, then property (3) of Theorem 1 implies that the firm's true cost function must be linearly homogeneous in prices p. Thus if C* is the firm's true cost function, then we must have $a_0 = 0 = a_{N+1}$. But then C* does not depend on the output level y, which is unrealistic. If C*(y, p) is a quadratic function in y, p, we encounter similar difficulties. For examples of functional forms for cost functions that are consistent with Theorem 1 but at the same time can approximate an arbitrary (differentiable) cost function to the second order, see Diewert (1974, 1978b) and Lau (1974, 1978).

We conclude this section by noting that the production function F completely determines the cost function C defined by (1). In the following section, we indicate how the process can be reversed.

2.2 Duality Between Cost and Production Functions

Given a production function F satisfying Assumption 1, for $y \in Y$ define the upper level set U(y) by

$$U(y) \equiv \{x: F(x) \geqslant y\}. \qquad (11)$$

The family of level sets (or production possibilities sets) L(y) completely determines the production function F. Also, the producer's cost function C can be defined by (1) or equivalently by

$C(y, p) \equiv \min_x \{p^T x: x \in U(y)\}$ for $y \in Y$, $p >> 0_N$.

For $y \in Y$, $p >> 0_N$, define the isocost plane for output level y and input price vector p as $\{x: p^T x = C(y, p)\}$. From the definitions of

$C(y, p)$ and $L(y)$, it follows that the set $L(y)$ lies *above* this isocost plane and is tangent to it; i.e., $L(y) \subset \{x: p^T x \geqslant C(y, p)\}$, where $\{x: p^T x \geqslant C(y, p)\}$ is a supporting halfspace for the true production possibilities set $L(y)$. Thus we may use the cost function in order to form an *outer approximation* $M(y)$ to the true set $U(y)$:

$$M(y) \equiv \bigcap_{p >> 0_N} \{x: p^T x \geqslant C(y, p)\}. \tag{12}$$

Since $L(y) \subset \{x: p^T x \geqslant C(y, p)\}$ for every $p >> 0_N$, $U(y) \subset M(y)$, where by (12) $M(y)$ is the intersection of the supporting halfspaces to the set $L(y)$. $M(y)$ is often called (e.g., McFadden (1966)) the *free disposal convex hull* of $U(y)$.

Since $M(y)$ is the intersection of a family of convex sets, $M(y)$ is also a convex set. $M(y)$ also has the following (free disposal) property: $x^0 \in M(y)$, $x^0 \leqslant x^1 \rightarrow x^1 \in M(y)$. Thus if we want $U(y)$ to coincide with $M(y)$ for each $y \in Y$, then $U(y)$ must be a convex set with the free disposal property for every $y \in Y$. However, if the family of sets $U(y)$ is convex for every $y \in Y$, then the production function F is *quasiconcave* (see Avriel, Diewert, Schaible and Zang). Similarly, if $U(y)$ satisfies the free disposal property for every $y \in Y$, then F must be a *nondecreasing* function.

Assumption 2 on F: F is a quasiconcave function.

Assumption 3 on F: F is nondecreasing; i.e., if $x^0 < x^1$, then $F(x^0) \leqslant F(x^1)$.

THEOREM 2: If the production function F *satisfies Assumption (1) then the cost function* C *defined by (1) satisfies properties (2) - (8). If in addition,* F *satisfies Assumptions (2) and (3) and we use the cost function* C *in order to define the function* F^* *for* $x \geqslant 0_N$ *as*

$$F^*(x) \equiv \max_y \{y: x \in M(y)\} \tag{13}$$

where M(y) *is defined in terms of* C *by (12), then* F = F*; *i.e., the production function* F *is completely characterized by its cost function* C.

This duality theorem between cost and production functions (essentially due to Shephard (1953, 1970)) has been established under a variety of regularity conditions. In the economics literature, see Samuelson (1953-4; 15), Uzawa (1964), McFadden (1966, 1978), Diewert (1971, 1974, 1978), and for local duality theorems, see Epstein (1981), Diewert (1978b) and Blackorby and Diewert (1979). In the mathematics literature, see Fenchel (1953; 122-124) and Crouzeix (1977; 222).

We have seen why *concavity* plays a central role in economic theory -- from Theorem 1, the cost function C(y, p) is concave in p, no matter what the functional form for the producer's production function F is (provided that F is continuous from above, an empirically harmless assumption). *Quasiconcavity* also plays a central role in economic theory: from an empirical point of view, it is harmless to assume that the producer's production function is quasiconcave (and nondecreasing), provided that the producer is competitively minimizing costs. Why is this? Assume that the true production function F satisfies only Assumption 1 and construct F's cost function C using definition (1). Then use C to construct the function F* by definition (13). It is straightforward to show that the cost function that corresponds to F* is also C. Thus the original production function F and the derived (from the cost function) production function F* both generate the same cost function, and so we might as well assume that F* is the true production function. However, F* is quasiconcave (and nondecreasing) even though the original F need not be.

To further explain the last point, we need an additional definition. For $y \in Y$, $p >> 0_N$, let $\gamma(y, p)$ denote the solution set of x's to the cost minimization problem (1). Thus if $x \in \gamma(y, p)$, then

$p^T x = C(y, p)$. Similarly, let $\gamma^*(y, p)$ denote the solution set to the cost minimization problem: $\min_x\{p^T x: F^*(x) \geq y\}$. It can be seen that if $y \in Y$, $p >> 0_N$ and $x \in \gamma(y, p)$, then $x \in \gamma^*(y, p)$ also. However, if $x \in \gamma^*(y, p)$, then it is not necessarily true that $x \in \gamma(y, p)$, unless the original production function F was quasiconcave. Thus the net effect of assuming that F^* is the true production function is to enlarge (incorrectly) the solution set to the cost minimization problem if the true F is not quasiconcave. However, this possibly enlarged solution set $\gamma^*(y, p)$ will always contain the true solution set $\gamma(y, p)$. Thus we can always assume that any observable data set on costs, output and inputs has been generated by the quasiconcave F^* rather than the true F.

2.3 Some Applications of Cost Functions

THEOREM 3. (Hicks (1946; 331), Samuelson (1947; 68), Shephard (1953: 11), Fenchel (1953; 104), McKenzie (1956-7), and Karlin (1959; 272): Suppose that the production function F *satisfies Assumption 1 and the cost function* C *is defined by (1). Let* $y^* \in Y$, $p^* >> 0_N$ *and suppose that* x^* *is a solution to (1), i.e.,* $x^* \in \gamma(y^*, p^*)$. *Finally, suppose that the partial derivaties of* C *with respect to input prices at* y^*, p^* *exist. Then*

$$x^* = \nabla_p C(y^*, p^*) \qquad (14)$$

and thus the solution to (1) is unique when $y = y^*$, $p = p^*$.

PROOF (Karlin (1959)): For every $p >> 0_N$, since x^* is feasible for the cost minimization problem $\min_x\{p^T x: F(x) \geq y^*\}$,

$$p^T x^* \geq C(y^*, p). \qquad (15)$$

For $p >> 0_N$, define $g(p) \equiv C(y^*, p) - p^T x^*$. From (15), $g(p) \leq 0$ for all $p >> 0_N$ but since $p^{*T} x^* = C(y^*, p^*)$, $g(p^*) = 0$. Thus $g(p)$ attains a (global) maximum at $p = p^*$ and the following first order necessary conditions must be satisfied:

$$\nabla g(p^*) = \nabla_p C(y^*, p^*) - x^* = 0_N. \tag{16}$$

Q.E.D.

Thus if $C(y^*, p^*)$ is differentiable with respect to prices, then the vector of first order partial derivatives with respect to input prices evaluated at (y^*, p^*), $\nabla_p C(y^*, p^*)$, is the unique solution to the cost minimization problem. This property is known in the economics literature (see McFadden (1978)) as *the derivative property for the cost function* or as *Shephard's Lemma*. Actually, Shephard proved a version of the following more difficult result:

THEOREM 4 (Shephard (1953; 11)): Suppose a cost function C *satisfies properties (2) - (8). If in addition the partial derivatives of* C *with respect to input prices exist at* $y^* \in Y$, $p^* >> 0_N$, *then* $x^* \equiv \nabla_p C(y^*, p^*)$ *is the unique solution to the cost minimization problem* $\min_x \{p^{*T}x: F^*(x) \geq y^*\}$ *where* F^* *is defined in terms of the given cost function* C *by (12) and (13).*

In Theorem 3, the production function F is given, while in Theorem 4, the cost function C is given, and the underlying production function F^* is defined in terms of the cost function.

One common application of Theorem 4 (see Diewert (1971, 1974) or Lau (1974, 1978)) is as an easy method of generating a theoretically valid system of cost minimizing input demand functions $x(y, p)$: simply postulate a functional form for the cost function C that is consistent with properties (2) - (8) and in addition is differentiable with respect to input prices and then estimate the unknown parameters occurring in the cost function by minimizing a function of the errors e^t in the equation system

$$x^t = \nabla_p C(y^t, p^t) + e^t; \qquad t=1,2,\ldots,T \tag{17}$$

where x^t, y^t and p^t are observed data on inputs, output and input prices, respectively, during period t. The alternative to the above method is to assume a functional form for the production function F and then solve the cost minimization problem (1) directly for the system of cost minimizing demand function's $x(y, p)$, assuming that the solution set is unique. The problem with this more direct method is that the unknown parameters which characterize the production function F usually occur in the demand system $x(y, p)$ in a highly nonlinear manner. On the other hand, using Theorem 4 in order to generate the system of demand functions $x(y, p)$, we can choose our functional form for C in such a way that the system of equation (17) is linear in the unknown parameters, and thus linear regression techniques can be applied to the estimation problem.

A second equally important application of Theorem 4 can be given. Suppose that C satisfies properties (2) - (8) and is twice continuously differentiable with respect to input prices at the point $y^* \in Y$, $p^* >> 0_N$. Then by Theorem 4, the input demand functions are given by $x(y^*, p) = \nabla_p C(y^*, p)$ for p close to p*. Differentiate each demand function $x_i(y^*, p^*) = \partial C(y^*, p^*)/\partial p_i$ with respect to input prices and denote the resulting matrix of partial derivatives as $\nabla_p x(y^*, p^*)$. From Theorem 4,

$$\nabla_p x(y^*, p^*) = \nabla^2_{pp} C(y^*, p^*) \tag{18}$$

where $\nabla^2_{pp} C(y^*, p^*) \equiv [\partial^2 C(y^*, p^*)/\partial p_i \partial p_j]$ denotes the matrix of second order partial derivatives of C with respect to the components of p. By property (5), $C(y^*, p)$ is concave in p. By Proposition 3 in Avriel, Diewert, Schaible and Ziemba (1981), $\nabla^2_{pp} C(y^*, p^*)$ is a negative semi-definite matrix, so that in particular, we must have

$$\partial x_i(y^*, p^*)/\partial p_i = \partial^2 C(y^*, p^*)/\partial p_i^2 \leqslant 0, \quad i=1,2,\ldots,N. \tag{19}$$

The inequalities in (19) admit a very simple economic interpretation: if the price of input i increases, then the cost minimizing demand for input i needed to produce a fixed output level y* will not increase.

If N = 2, so that there are only two inputs, then we may deduce that $\partial x_i(y^*, p^*)/\partial p_2 = \partial^2 C(y^*, p^*)/\partial p_1 \partial p_2 = \partial^2 C(y^*, p^*)/\partial p_2 \partial p_1$ (using the twice continuously differentiability assumption) $= \partial x_2(y^*, p^*)/\partial p_1 = -(p_2^*/p_1^*)\partial x_2(y^*, p^*)/\partial p_2 \geqslant 0$ since $p_1^* \partial x_2(y^*, p^*)/\partial p_1 + p_2^* \partial x_2(y^*, p^*)/\partial p_2 = 0$. The last equality follows from Euler's Theorem on homogeneous functions, since $C(y^*, p)$ is linearly homogeneous in p, $\partial C(y^*, p)/\partial p_2$ is homogeneous of degree zero in $(p_1, p_2) \equiv p^T$.

The above results were obtained (under somewhat stronger hypotheses) by Hicks (1946; 311 and 331) and Samuelson (1947; 69). The present derivation is taken from McFadden (1966, 1978) and Diewert (1978b). These results illustrate a second major application of the duality between cost and production functions: duality theory usually enables us to derive theoretical theorems about the solutions to various economic optimization problems in a comparatively effortless manner.

One question that we have not yet resolved is: under what conditions on the production function F will the cost function C be differentiable with respect to input prices? It is fairly easy to show that a sufficient condition for C to be once differentiable with respect to input prices is that F be *strictly quasiconcave*. However, the following example shows that strict quasiconcavity of F is not necessary. Let $a_1 > 0$, $a_2 > 0$, N = 2 and define $F(x_1, x_2) \equiv \min\{x_1/a_1, x_2/a_2\}$. This production function is known in the economics literature as a Leontief or fixed coefficient production function. It is not strictly quasiconcave nor is it directionally differentiable since the two sided directional derivatives of F, $D_v F(x) \equiv \lim_{t\to 0}[F(x + tv) - F(x)]/t$, do not exist for all directions v. However, the cost function which corresponds to F is $C(y, p_1, p_2) \equiv (a_1 p_1 + a_2 p_2)y$ which has partial derivatives of all orders. Thus simple necessary and sufficient conditions on F for the differentiability of C with respect to input prices do not seem to be known at present.

Another interesting question is: under what conditions on F will C be twice differentiable with respect to prices with a negative definite Hessian matrix $\nabla^2_{pp}C(y^*, p^*)$ (which would imply that the inequalities in (19) are strict)? Unfortunately, $\nabla^2_{pp}C(y^*, p^*)$ cannot be negative definite for any F, since property (3) on C (linear homogeneity in prices) implies, using Euler's Theorem on homogeneous functions, that

$$\nabla^2_{pp}C(y^*, p^*)p^* = 0_N \tag{20}$$

so that p* is an eigenvector of $\nabla^2_{pp}C(y^*, p^*)$ that corresponds to a zero eigenvalue. However, we can ask under what conditions will $\nabla^2_{pp}C(y^*, p^*)$ be negative definite in the subspace orthogonal to $p^* >> 0_N$, and what does the last condition on C imply about F?

Suppose:

C satisfies (2) - (8), $y^* \in$ Interior Y, $p^* >> 0_N$, C is twice continuously differentiable in a neighborhood around y*, p* with $x^* \equiv \nabla_p C(y^*, p^*) >> 0_N$ and $\nabla^2_{pp}C(y^*, p^*)$ satisifes the following property: (21)

$$z^T p^* = 0,\ z \neq 0_N \rightarrow z^T\nabla^2_{pp}C(y^*, p^*)z < 0. \tag{22}$$

First we show that (21) and (22) imply that the inequalities in (19) hold strictly provided that $N \geqslant 2$. By (5), $\nabla^2_{pp}C(y^*, p^*)$ is negative semidefinite while (3) implies (20); i.e., that p* is an eigenvalue of $\nabla^2_{pp}C(y^*, p^*)$ that corresponds to a zero eigenvalue. (22) implies that the other N-1 eigenvalues of $\nabla^2_{pp}C(y^*, p^*)$ are negative. Thus $\nabla^2_{pp}C(y^*, p^*)$ satisifies:

$$k \in R^1,\ z \neq kp^* \rightarrow z^T\nabla^2_{pp}C(y^*, p^*)z < 0. \tag{23}$$

From (19), $\partial^2 C(y^*, p^*)/\partial p_i^2 \leqslant 0$. Suppose $\partial^2 C(y^*, p^*)/\partial p_i^2 = 0$. Then $e_i^T \nabla_{pp}^2 C(y^*, p^*) e_i = \partial^2 C(y^*, p^*)/\partial p_i^2 = 0$ where e_i is the ith unit vector. Since $p^* >> 0_N$, $e_i \neq kp^*$ for any scalar k, and we have contradicted (23). Thus our supposition is false and

$$\partial x_i(y^*, p^*)/\partial p_i = \partial^2 C(y^*, p^*)/\partial p_i^2 < 0 \quad \text{for } i=1,2,\ldots,N. \tag{24}$$

Thus assumptions (21) and (22) on C are sufficient to imply that the ith input demand function $x_i(y, p)$ has a negative slope with respect to the ith price p_i for all y, p in a neighborhood of y^*, p^*.

The meaning of (22) will become clearer if we note that (21) and (22) are equivalent to (21) and (25):

$$z^T \nabla_p C(y^*, p^*) = 0,\ z \neq 0_N \rightarrow z^T \nabla_{pp}^2 C(y^*, p^*) z < 0. \tag{25}$$

Suppose (21) and (22) hold. Then (23) also holds. Let $z^T \nabla_p C(y^*, p^*) = 0$, $z \neq 0_N$ where $\nabla_p C(y^*, p^*) \equiv x^* >> 0_N$ by (21). If $z = kp^*$ for some $k \neq 0$, then $z^T \nabla_p C(y^*, p^*) < 0$ if $k < 0$ or $z^T \nabla_p C(y^*, p^*) > 0$ if $k > 0$ since $p^* >> 0_N$, which contradicts $z^T \nabla_p C(y^*, p^*) = 0$. Thus $z \neq kp^*$ for any scalar k and (23) implies $z^T \nabla_{pp}^2 C(y^*, p^*) z < 0$. Thus (25) holds. The proof that (21) and (25) imply (22) is similar.

From Proposition 17 in Avriel, Diewert, Schaible and Ziemba (1981) (21) and (25) imply that C(y, p) is *strongly pseudoconcave* (quasiconcave) with respect to p for y, p close to y^*, p^*. We now ask: what does local strong pseudoconcavity of the cost function C imply about the corresponding dual production function F^* defined by (12) and (13)? Using the material in Blackorby and Diewert (1979; 589-591), it can be shown that if C satisfies (21) and (25), then F^* satisfies Assumptions 1 - 3 and the following additional conditions (where $x^* \equiv \nabla_p C(y^*, p^*) >> 0_N$):

F^* is twice continuously differentiable in a neighborhood around x^* with $\nabla F^*(x^*) >> 0_N$; and (26)

$$z^T \nabla F^*(x^*) = 0, \; z \neq 0_N \rightarrow z^T \nabla^2 F^*(x^*) z < 0. \qquad (27)$$

Using Proposition 17 in Avriel, Diewert, Schaible and Ziemba (1981), it can be seen that the differentiability properties of F* and (27) imply that F* is *strongly pseudoconcave* (quasiconcave) in a neighborhood of x*.

It can also be shown that if a production function F satisfies Assumptions 1 - 3, (26) and (27), then the corresponding cost function C defined by (1) satisfies (2) - (8), (21) and (25). Thus local strong pseudoconcavity (quasiconcavity) of the production function is equivalent to local strong pseudoconcavity (quasiconcavity) of the cost function in the twice continuously differentiable case.

We have shown how concavity, quasiconcavity, strict quasiconcavity and strong quasiconcavity arise in the context of the producer's cost minimization problem. We conclude this section by indicating a few additional economic applications of cost functions.

First, the producer's cost minimization problem can be given an interpretation in the context of consumer theory: interpret F as the consumer's *utility function* (if $F(x^0) < F(x^1)$, then the consumer prefers the commodity vector x^1 to x^0), interpret $p >> 0_N$ as a vector of commodity prices and $y \in Y$ as a utility level rather than as an output level. Then the cost minimization problem (1), $\min_x\{p^Tx: F(x) \geq y\}$ can be interpreted as the problem of minimizing the cost or expenditure needed to achieve a certain utility level $y \in Y$ given that the consumer faces the vector of commodity prices $p >> 0_N$. In the economics literature, the resulting cost function C(y, p) is often called an *expenditure function* (e.g., Blackorby and Diewert (1979)).

Two applications of expenditure functions can readily be given. The expenditure function plays a central role in the economic theory of the cost of living index. Let the consumer face the commodity price vectors $p^0 >> 0_N$ and $p^1 >> 0_N$ during periods 0 and 1, let F be the consumer's utility function and C be the corresponding expenditure function defined by (1). Then the Könus (1939) *cost of living index* P corresponding to $x > 0_N$ (the reference vector of quantities), is defined as

$$P(p^0, p^1, x) \equiv C[F(x), p^1]/C[F(x), p^0]. \tag{28}$$

P is interpreted as follows: pick a reference indifference surface indexed by the quantity vector $x > 0_N$. Then $P(p^0, p^1, x)$ is the minimum cost of achieving the standard of living indexed by x when the consumer faces commodity prices p^1 relative to the minimum cost of achieving the same standard of living when the consumer faces period 0 prices p^0. Thus P can be interpreted as a level of prices in period 1 relative to a level of prices in period 0. The mathematical properties of P are completely determined by the mathematical properties of F and C.

The expenditure function also plays a central role in the *cost benefit analysis* which makes use of the *consumer surplus* concept. Let F, C, p^0 and p^1 be defined as in the previous paragraph and let x^0 and x^1 be the consumer's observed consumption vectors in periods 0 and 1, respectively. Then Hicks (1941-42; 128) has used the expenditure function in order to define two closely related measures of consumer surplus or welfare change: Hick's *compensating variation in income* is defined as $C[F(x^1), p^1] - C[F(x^0), p^1]$ while Hick's *equivalent variation in income* is defined as $C[F(x^1), p^0] - C[F(x^0), p^0]$. If the sign of either of these consumer surplus measures is positive (negative), then the consumer's utility or welfare has increased (decreased) going from period 0 to period 1. The mathematical properties of these consumer surplus measures are again completely determined by the mathematical properties of F and C.

3. CONSUMER THEORY

3.1 The Basic Problem

Another fundamental paradigm in economics is where a consumer's preferences over alternative (nonnegative) consumption vectors $x \geqslant 0_N$ are represented by a continuous *utility function* F and the consumer chooses his optimal consumption vector by maximizing utility F(x) with respect to the consumption vector x, subject to a budget constraint of the form $p^T x \leqslant m$ where $p >> 0_N$ is a vector of positive commodity prices that he faces and $m > 0$ is the amount of money that he can spend on the commodities during the period under consideration. Formally, the consumer's utility maximization problem is defined by (29) below for $p >> 0_N$:

$$G(p) \equiv \max_x \{F(x): p^T x \leqslant 1, \ x \geqslant 0_N\} \tag{29}$$

where we have replaced the consumer's original budget constraint, $\bar{p}^T x \leqslant m$ say, by the equivalent budget constraint $p^T x \leqslant 1$ where $p = (p_1, p_2, \ldots, p_N) \equiv (\bar{p}_1/m, \bar{p}_2/m, \ldots, \bar{p}_N/m)$ denotes a vector of *normalized prices* (the original commodity price vector divided by the consumer's positive income). In what follows, the price vector p is always interpreted as a vector of normalized commodity prices.

Note that we have not only defined the consumer's constrained utility maximization problem in the right hand side of (29), but we have also defined the function G as the maximized utility level as a function of the normalized prices that the consumer faces. The function G is known in the economics literature as the *indirect utility function*.

It is technically more convenient to assume that the utility function F is continuous rather than just being continuous from above as in Section 2.

Assumption 4 on F: F is a real valued continuous function defined for all nonnegative consumption vectors $x \geqslant 0_N$.

THEOREM 5 (Diewert (1974; 121-124)): Let F *satisfy Assumption 4. Then* G *is well defined by (29) for* $p >> 0_N$ *and satisfies:*

(30) G *is a continuous (finite) function,*

(31) G *is a nonincreasing (if* $0 << p^0 < p^1$, *then* $G(p^0) \geqslant G(p^1)$),

(32) G *is a quasiconvex function* ($\alpha \in R^1 \to \{p: G(p) \leqslant \alpha, p >> 0_N\}$ *is a convex set), and*

(33) if for $x >> 0_N$, *we define* $F^*(x) \equiv \min_p\{G(p): p^T x \leqslant 1, p \geqslant 0_N\}$, *then* F* *is continuous over* $x >> 0_N$ *and has a continuous extension to the nonnegative orthant* $x \geqslant 0_N$.

A few technical comments on the above theorem are in order. Property (33) requires G to be defined over the nonnegative orthant. We extend G from the positive orthant to the nonnegative orthant using the *Fenchel* (1953; 78) *closure operation;* i.e., define the epigraph of G over the positive orthant as $E \equiv \{(u, p): p >> 0_N, u \geqslant G(p)\}$, let $\bar{E}$ denote the closure of E and define $\bar{G}$ over the nonnegative orthant by $\bar{G}(p) \equiv \inf_u\{u: (u, p) \in \bar{E}\}$. For $p >> 0_N$, $\bar{G}(p) = G(p)$. Although $\bar{G}$ need not be finite on the boundary of the nonnegative orthant, $\bar{G}$ will be continuous from below and thus the minimum in (33) will exist using a result in Berge (1963; 76) (we have abused notation in (33) by letting $G = \bar{G}$).

From the perspective of the role of generalized concavity in economics, the interesting point to notice about the above theorem is that the indirect utility function G is *quasiconvex* (i.e., -G is *quasiconcave*) irrespective of the properties of the direct utility function F (provided only that F is continuous).

From the viewpoint of economics, properties (30) and (33) are technical in nature while property (31) is intuitively obvious: if the prices that a consumer faces increase, then his feasible consumption set $\{x: p^T x \leqslant 1, x \geqslant 0_N\}$ shrinks, and thus maximal utility must decrease or remain constant. However, property (32) is not intuitive and it may be useful to indicate a proof of it (see Diewert, (1974; 160)).

Let $0 \leqslant \lambda \leqslant 1$, $p^i >> 0_N$, and $G(p^i) \leqslant u$ for $i = 1, 2$. Define the sets $H^i \equiv \{x: p^{iT}x \leqslant 1, x \geqslant 0_N\}$, $i = 1, 2$ and $H^\lambda \equiv \{x: (\lambda p^1 + (1-\lambda)p^2)^T x \leqslant 1, x \geqslant 0_N\}$. Then

$$\begin{aligned} G(\lambda p^1 + (1-\lambda)p^2) &\equiv \max_x\{F(x): x \in H^\lambda\} \\ &\leqslant \max_x\{F(x): x \in H^1 \cup H^2\} \\ &\qquad \text{since } H^\lambda \subset H^1 \cup H^2 \\ &\leqslant u \end{aligned}$$

since $\max_x\{F(x): x \in H^i\} = G(p^i) \leqslant u$ for $i = 1, 2$.

Suppose that the utility function F satisfies Assumption 4. Then we may proceed as in the previous section and define the function F* using (1), (12) and (13). The resulting F* is the free disposal, quasiconave hull of the original function F. It can be verified that F and F* generate the same indirect utility function G by definition (29) and furthermore, that the F* defined in (33) coincides with the F* defined using (1), (12) and (13). Again, it is costless from an empirical point of view to assume that F satisfies Assumptions 2 and 3 in addition to Assumption 4; i.e., given Assumption 4 on F and the assumption of utility maximizing behavior on the part of the consumer, denote the solution set to the original utility maximization problem (29) as $\beta(p)$, denote the solution set to $\max_x\{F^*(x): p^T x \leqslant 1, x \geqslant 0_N\}$ as $\beta^*(p)$ where F* is the free disposal quasiconcave hull of F. Then for every $p >> 0_N$, $\beta(p) \subset \beta^*(p)$. Thus if p and x are observed price and quantity vectors that correspond to the original utility maximization problem (29) (i.e., $x \in \beta(p)$), then x is also a solution to $\max_x\{F^*(x): p^T x \leqslant 1, x \geqslant 0_N\}$; (i.e., $x \in \beta^*(p)$ also).

Hence in both the producer's competitive cost minimization problem and the consumer's maximization problem, it is completely harmless from an empirical point of view to assume that the producer's production function or the consumer's utility function is *quasiconcave.*

There is a complete *duality* between utility functions F satisfying Assumptions 2, 3 and 4 and indirect utility functions G satisfying (30) - (33). For references to such duality theorems in the economics literature, see Diewert (1974; 132) or Blackorby and Diewert (1979). For a similar duality theorem in the mathematics literature, see Crouzeix (1977, 222; 1981).

3.2 Semistrict Quasiconcavity and Consumer Theory

Suppose F satisfies Assumption 1 and we define the consumer's utility maximization problem by (29) for every (normalized) commodity price vector $p >> 0_N$. By a theorem due to Berge (1963; 76), the indirect utility function G(p) is well defined as a maximum in (29). Suppose $G(p) < \sup_x\{F(x): x \geqslant 0_N\}$ and denote the solution set of optimal consumption vectors for (29) as $\beta(p)$. If F is constant over a neighborhood (in economics terminology, some indifference surfaces are thick), then $\beta(p)$ can contain points x such that $p^T x < 1$; i.e., the consumer's entire budget is not spent. From the viewpoint of economic applications, it is useful to be able to rule out this kind of behavior. The following theorem gives sufficient conditions for this.

THEOREM 5: Suppose F *satisfies Assumption 1 and is semistrictly quasiconcave. Then* F *satisfies the following property:*

$$p >> 0_N,\ G(p) < \sup_x\{F(x): x \geqslant 0_N\},\ x \in \beta(p) \rightarrow p^T x = 1. \qquad (34)$$

PROOF: Suppose there exists p* such that

$$p^* >> 0_N,\ G(p^*) < \sup_x\{F(x): x \geqslant 0_N\},\ x^* \in \beta(p^*) \text{ and}$$
$$p^{*T} x^* < 1. \qquad (35)$$

Then there exists $x^2 \geqslant 0_N$ such that $G(p^*) = F(x^*) < F(x^2)$ with $p^{*T} x^2 > 1$. By the semistrict quasiconcavity of F (see Definition 7 in Avriel, Diewert,

Schaible and Ziemba):

$$F(x^*) < F(\lambda x^* + (1-\lambda)x^2) \quad \text{for } 0 \leqslant \lambda < 1. \tag{36}$$

Since $p^{*T}x^* < 1$ and $p^{*T}x^2 > 1$, there exists $0 < \lambda^* < 1$ such that $p^{*T}(\lambda^* x^* + (1-\lambda^*)x^2) = 1$. Thus $\lambda^* x^* + (1-\lambda^*)x^2$ is feasible for the $G(p^*)$ maximization problem, $\max_x\{F(x)\colon p^{*T}x \leqslant 1,\ x \geqslant 0_N\}$. But (36) shows that this feasible solution gives a higher level of utility than $F(x^*) = G(p^*)$, a contradiction to the definition of $G(p^*)$. Thus (35) is false and (34) follows.

Q.E.D.

Semistrict quasiconcavity of F is not a necessary condition for (34); the following function of two variables satisfies Assumptions 1 through 4 as well as property (34), but it is not semistrictly quasiconcave (consider the behavior of the function along the x_1 axis):

$$F(x_1,x_2) \equiv \begin{cases} 0 & \text{if } 0 \leqslant x_1 \leqslant 1,\ x_2 = 0 \\ t & \text{if } x_1 \geqslant 0,\ x_2 \geqslant 0,\ t > 0,\ tx_1 + (1+t)x_2 = t(1+t) \end{cases}$$

3.3 Strict Quasiconcavity and Consumer Theory

In most applications of consumer theory in economics, we require not only that property (34) hold, but also that the set of optimal consumption vectors $\beta(p)$ is a singleton. Thus we are interested in finding conditions on the direct utility function F or the indirect utility function G that will ensure that the consumer's demand correspondence, the set of maximizers $\beta(p)$ to (29), is an ordinary function, at least for a region of prices.

THEOREM 6: Suppose the direct utility function F *satisfies Assumptions 2 and 4 and the following (local) condition:*

$$p^* >> 0_N,\ G(p^*) < \sup_x\{F(x)\colon x \geqslant 0_N\},\ x^* \in \beta(p^*) \textit{ and } F \textit{ is} \tag{37}$$

strictly quasiconcave over $N_\delta(x^*) \equiv \{x\colon (x - x^*)^T(x - x^*) < \delta^2,$

$x \geqslant 0_N\}$, *the open ball of radius* δ *around* x^* *for some* $\delta > 0$.

Then (38) holds:

$$p^* >> 0_N,\ \beta(p) \textit{ is a continuous function in a neighborhood of normalized prices around } p^*,\ P \textit{ say, with } p^T\beta(p) = 1 \textit{ for } p \in P. \tag{38}$$

PROOF: Let p^* and x^* be defined in (37) and let x^2 be such that $G(p^*) = F(x^*) < F(x^2)$. Suppose $p^{*T}x^* < 1$. Then since F is quasiconcave, for $0 \leqslant \lambda \leqslant 1$, $F(\lambda x^* + (1-\lambda)x^2) \geqslant F(x^*)$. By the local strict quasiconcavity of F, there exists $0 < \lambda^* < 1$ such that

$$F(\lambda x^* + (1-\lambda)x^2) > F(x^*) \quad \text{for } \lambda^* \leqslant \lambda < 1. \tag{39}$$

For λ^1 sufficiently close to 1, $p^{*T}(\lambda^1 x^* + (1-\lambda^1)x^2) \leqslant 1$, which implies along with (39) that $G(p^*) \geqslant F(\lambda^1 x^* + (1-\lambda^1)x^2) > F(x^*)$, a contradiction. Thus our supposition is false, and $p^{*T}x^* = 1$.

Now suppose there exists $x^1 \neq x^*$ such that $x^1 \in \beta(p^*)$. Then the quasiconcavity and local strict quasiconcavity of F again imply a contradiction. Thus $\beta(p^*)$ is the singleton x^* with $p^{*T}x^* = 1$.

By the Debreu (1952; 889-890, 1959; 19), Berge (1963; 116) Maximum Theorem, $\beta(p)$ is an upper semicontinuous correspondence for $p >> 0_N$. Hence for p sufficiently close to p^*, $\beta(p) \cap N_\delta(x^*) \neq \phi$. Strict quasiconcavity of F in $N_\delta(x^*)$ will imply as before that $\beta(p)$ is single valued and $p^T\beta(p) = 1$ for p close to p^*. When β is single valued, upper semicontinuity reduces to continuity.

Q.E.D.

Versions of the above theorem are common in the economics literature (e.g., Debreu (1959)) except that strict quasiconcavity is generally assumed to hold globally rather than locally.

Another set of conditions that are sufficient to ensure that the consumer's demand correspondence $\beta(p)$ is single valued and continuous around $p^* >> 0_N$ can be phrased in terms of the indirect utility function

G. Sufficient conditions are that G satisfy (30) - (33) and the following condition:

$$G \text{ is once continuously differentiable in neighborhood around } p^* \text{ with } \nabla G(p^*) \neq 0_N. \quad (40)$$

If the above conditions on G are satisfied, then it can be shown (see Roy (1947; 222) or Diewert (1974; 125)) that $\beta(p) = (p^T \nabla G(p))^{-1} \nabla G(p)$ for p close to p*.

Neither set of sufficient conditions for the single valuedness and continuity of $\beta(p)$ implies the other. It appears to be difficult to find necessary and sufficient conditions for this problem, just as it is difficult to find necessary and sufficient conditions for the differentiability of the cost function with respect to input prices.

3.4 Strong Quasiconcavity and Consumer Theory

Suppose that we not only want the consumer's demand correspondence $\beta(p)$ to be single valued and continuous for a neighborhood of prices p, but also we want the demand functions to be once continuously differentiable.

As usual, it is difficult to find necessary and sufficient conditions on F for the above property to hold. However, the following theorem gives sufficient conditions.

THEOREM 7: Suppose the direct utility function F *satisfies:*

Assumptions 2,3, and 4, (41)

it is twice continuously differentiable in a neighborhood around $x^* >> 0_N$ *with* $\nabla F(x^*) >> 0$, *and* (42)

$$z^T z = 1,\ z^T \nabla F(x^*) = 0 \rightarrow z^T \nabla^2 F(x^*) z < 0. \quad (43)$$

Then the indirect utility function G *defined by (29) satisifes (30) - (33). Moreover, if we define* $p^* \equiv (x^{*T} \nabla F(x^*))^{-1} \nabla F(x^*)$, *then* G *satisfies*

$$G \text{ is twice continuously differentiable in a neighborhood around } p^* >> 0_N \text{ with } \nabla G(p^*) << 0_N, \text{ and} \tag{44}$$

$$z^T z = 1,\ z^T \nabla G(p^*) = 0 \rightarrow z^T \nabla G(p^*) z > 0. \tag{45}$$

Moreover, the solution to the utility maximization problem (29) reduces to $\beta(p) = (p^T \nabla G(p))^{-1} \nabla G(p)$ *for* p *close to* p^* *and is once continuously differentiable there.*

Conversely, given an indirect utility function G *satisfying (30) - (33), (44) and (45), then* F^* *defined in (33) satisfies (41) - (43) if we define* $x^* = (p^{*T} \nabla G(p^*))^{-1} \nabla G(p^*)$.

The above theorem can readily be proven using the techniques in Blackorby and Diewert (1979).

Our differentiability assumptions and (43) imply that F is *strongly quasiconcave* in a neighborhood around x^* while (45) implies that G is *strongly quasiconvex* around p^*.

Thus sufficient conditions for the local continuous differentiability of a consumer's system of demand functions can be obtained by assuming that the direct utility function F is strongly quasiconcave locally or equivalently, by assuming that the indirect function G is strongly quasiconvex locally.

3.5 Pseudoconcavity and Consumer Theory

To obtain a system of demand functions consistent with utility maximizing behavior, economists frequently postulate a functional form for direct utility function F and then they attempt to solve algebraically the utility maximization problem (29) for an explicit solution. As part of this exercise, economists often assume that necessary conditions for x^0 to solve (29) are also sufficient.

When F is not necessarily differentiable, if x^0 is a solution to

(29) when normalized prices $p >> 0_N$ prevail, it can be verified that the following necessary conditions must be satisfied:

$$D_v^{+u}F(x^0) \leqslant 0 \quad \text{for every direction } v \in S(x^0, p) \tag{46}$$

where $D_v^{+u}F(x^0) \equiv \limsup_{t \to 0^+} [F(x^0 + tv) - F(x^0)]/t$ is the upper Dini derivative in the direction v (see Diewert (1981)) and the set of feasible directions is defined as $S(x^0, p) \equiv \{v: v^Tv = 1$ and there exists at $t > 0$ such that $x^0 + tv \geqslant 0_N$ and $p^T(x^0 + tv) \leqslant 1\}$. The meaning of (46) is that for all feasible directions, F cannot increase locally.

If F is once continuously differentiable over the nonnegative orthant, (46) reduces to the following more familiar Kuhn-Tucker (1951) conditions: if x^0 solves (29), then there exists a multiplier λ^0 such that

$$\nabla F(x^0) - \lambda^0 p \leqslant 0_N,\ \lambda^0 \geqslant 0,\ x^0 \geqslant 0_N,\ x^{0T}[\nabla F(x^0) - \lambda^0 p] = 0, \tag{47}$$
$$p^Tx^0 - 1 \leqslant 0,\ \lambda^0[p^Tx^0 - 1] = 0.$$

Since the constraint set in (29) is polyhedral, constraint qualification conditions are satisfied.

From the definition of pseudoconcavity (see Diewert (1981), definition B), the *necessary* conditions (46) or (47) are also *sufficient* if F is *pseudoconcave;* i.e., if F is pseudoconcave and x^0 satisfies (46), then x^0 is a solution to the consumer's utility maximization problem (29). Similarly, if F is *strictly pseudoconcave* and x^0 satisfies (46), then x^0 is the *unique* solution to (29).

There are several other similar applications of the concept of strict pseudoconcavity to economics; e.g., see Donaldson and Eaton (1978; 1981).

Thus far, we have shown how seven different kinds of generalized concavity occur naturally in economics. In the following section, we indicate how an additional two types occur.

4. PROFIT MAXIMIZATION AND COMPARATIVE STATICS ANALYSIS

Consider the model of producer behavior presented in section 2.1 where the producer minimized costs subject to an output constraint. We now make a stronger assumption about producer behavior; namely, that he competitively maximizes profits. Denote the production function by F, and input vector by $x \geqslant 0_N$, an input price vector by $\bar{p} >> 0_N$ and a scalar price of output by $\bar{p}_0 > 0$. A competitive producer takes all prices as fixed and chooses x in order to maximize profits, $\bar{p}_0 F(x) - \bar{p}^T x$. Thus the producer's competitive profit maximization problem is

$$\sup_x \{F(x) - p^T x: x \geqslant 0_N\} \equiv \pi(p) \tag{48}$$

where the vector of normalized input prices is $p \equiv (\bar{p}_0)^{-1}\bar{p} \equiv (\bar{p}_1/\bar{p}_0, \bar{p}_2/\bar{p}_0, \ldots, \bar{p}_N/\bar{p}_0)$. It is obvious that maximizing profits is equivalent to solving (48), since the objective function in (48) is simply profits divided by the positive price of output $\bar{p}_0$. The maximized value in (48) is defined to be the producer's (normalized) *profit function* π (see Lau (1978) or Jorgenson and Lau (1974a, 1974b) for additional details on this model).

Mathematicians will recognize π as the (negative of the) *conjugate function* to F (see Fenchel (1953; 90) or Rockafellar (1970)).

Recall how we used the cost function C to define outer approximations to the true level of the production function F. Now we show how the profit function π can be used in order to provide another outer approximation to the true production function.

For $p \geqslant 0_N$, define the following halfspace in R^{N+1}: $H(p) \equiv \{(y, x): y - p^T x \leqslant \pi(p)\}$. Then an outer approximation to the true production possibilities set $S \equiv \{(y, x): y \leqslant F(x), x \geqslant 0_N\}$ can be defined as $\hat{S} \equiv \bigcap_{p \geqslant 0_N} H(p)$. Assuming that $\pi(p) < +\alpha$ for at least one $p >> 0_N$, an outer approximation to F can be defined for $x \geqslant 0_N$ as

$$\hat{F}(x) \equiv \max_y\{y: (y, x) \in \hat{S}\}. \tag{49}$$

Since $\hat{S}$ is a closed convex set, the approximating production function $\hat{F}$ is a *concave function*. Moreover, for any input price vector $p^* >> 0_N$ such that there exists an $x^* \geqslant 0_N$ which attains the sup in (48), we have $\max_x\{F^*(x) - p^{*T}x: x \geqslant 0_N\} = \max_x\{F(x) - p^{*T}x: x \geqslant 0_N\} = p^{*T}x^* = \pi(p^*)$. Hence from an empirical point of view, it is harmless to assume that the producer's production function is concave, provided that the producer is competitively maximizing profits. The present situation is analogous to the result we obtained in Section 2.2 where we indicated that the production function could be assumed to be quasiconcave provided that the producer was competitively minimizing costs. We now obtain the stronger concavity result because we assume the stronger hypothesis of competitive profit maximizing behavior.

For references to duality theorems between production functions, production possibilities sets and profit functions, see McFadden (1978), Lau (1978) and Diewert (1974).

However, our main concern in this section is not to show how concavity arises naturally in the context of production theory, but to show how strong concavity arises naturally in the content of comparative statics theory.

A typical application of the Hicks (1946) Samuelson (1947) comparative statics analysis can be illustrated by using the profit maximization model (48). The analysis given below follows Lau (1978; 147-149). Given a positive vector of (normalized) input prices $p^* >> 0_N$, it is assumed that $x^* >> 0_N$ is the unique solution to the profit maximization problem $\max_x\{F(x) - p^{*T}x: x \geqslant 0_N\} \equiv \pi(p^*)$. (A sufficient condition for this is that F be globally concave and locally *strictly concave* around a point $x^* >> 0_N$ such that $\nabla F(x^*) = p^*$.) It is also assumed that F is twice continuously differentiable in a neighborhood of x^*. Since

$x^* >> 0_N$ solves the profit maximization problem, the nonnegativity constraints $x \geqslant 0_N$ are not binding and hence the following *first order necessary conditions* must be satisifed:

$$\nabla F(x^*) - p^* = 0_N. \tag{50}$$

Since F is assumed to be twice continuously differentiable, the following *second order necessary conditions* must also be satisfied:

$$z \neq 0_N \rightarrow z^T \nabla^2 F(x^*) z \leqslant 0; \tag{51}$$

i.e., the matrix of second order partial derivatives of F evaluated at x^* must be negative semidefinite. Comparative statics analysis proceeds by assuming that F satisfies the following stronger conditions at x^* *(second order sufficient conditions).*

$$z \neq 0_N \rightarrow z^T \nabla^2 F(x^*) z < 0; \tag{52}$$

i.e., $\nabla^2 F(x^*)$ is a negative definite symmetric matrix, and hence $[\nabla^2 F(x^*)]^{-1}$ exists. Note that (51) implies that F is *strongly concave* in a neighborhood of x^* (see Proposition 5 in Avriel, Diewert, Schaible and Ziemba (1981)).

The reason for assuming (51) is that we may now apply the Implicit Function Theorem to (49) to deduce the existence of a unique once continuously differentiable solution to the profit maximization problem (48), x(p) say, for p close to p^*, and the matrix of partial derivatives of the functions $[x_1(p),\ldots,x_N(p)]^T \equiv x(p)$ evaluated at p^* is

$$\nabla x(p^*) \equiv [\partial x_i(p^*)/\partial p_j] = [\nabla^2 F(x^*)]^{-1}. \tag{53}$$

Since $\nabla^2 F(x^*)$ is a negative definite matrix, so is its inverse $[\nabla^2 F(x^*)]^{-1}$. Hence the input demand functions satisfy the following restrictions (the diagonal elements of a negative definite matrix are negative):

$$\partial x_i(p^*)/\partial p_i < 0, \quad i = 1,2,\ldots,N. \tag{54}$$

Thus if the price of the ith input increases, the demand for the ith input decreases. Hence the assumption of local strong concavity leads to economically meaningful theorems.

In the mathematical programming literature, the terms "stability theory" or "perturbation theory" are used in place of the economist's term "comparative statics analysis".

5. CONCLUSION

We have considered three models of economic behavior in this paper (the producer's cost minimization problem, the consumer's utility maximization problem, and the producer's profit maximization problem) and have shown how nine different kinds of generalized concavity and convexity arise in the context of these models. However, we have by no means exhausted the applications of generalized concavity in economics.

In many economic problems (e.g., see Kannai (1977)), it is extremely useful to be able to represent a consumer's preferences by means of a concave utility function rather than a quasiconcave function. Thus given a quasiconcave utility function F(x) defined for $x \geqslant 0_N$, we ask under what conditions does there exist an increasing function of one variable, f say, such that f[F] is a concave function? Under various regularity conditions, answers have been given by Fenchel (1953), Kannai (1977; 1981), Crouzeix (1977; 1980), Schaible and Zang (1979), and Zang (1981). In a related vein, Diewert (1973) using the work of Afriat (1967) shows how any finite set of price-quantity data generated by a consumer maximizing a continuous from above utility function can be generated by maximizing a concave utility function, while Kannai (1974) shows how arbitrary quasiconcave preferences can be approximated by concave preferences.

A problem which is related to the issue of concavifiability is:

given that the direct utility function is concave, what does this imply about the corresponding cost and indirect utility functions C and G? On the other hand, if G is convex, what does this imply about the corresponding F and C? Answers to these questions can be found in Fenchel (1953; 122-123), Crouzeix (1977; 110, 1980) and Diewert (1978a; 33).

Another interesting problem is: given that $F(x^1, x^2, \ldots, x^M) = \sum_{i=1}^{M} f^i(x^i)$ where $x^i \in S^i$, a convex subset of finite dimensional Euclidean space for each i and F is quasiconcave, what does this imply about the functions f^i? For answers to this question and some economic applications (particularly to consumer choice under uncertainty), see Yaari (1977; 1184), Blackorby, Davidson and Donaldson (1977) and Debreu and Koopmans (1978).

Finally, another area which is too large to be reviewed here is economic dynamics, where strong concavity plays a particulary important role (see, e.g., Rockafellar (1976; 75)).

REFERENCES

AFRIAT, S.N. (1967). "The Construction of Utility Functions from Expenditure Data." *International Economic Review 8*, 67-77.

AVRIEL, M., DIEWERT, W.E., SCHAIBLE, S., and ZIEMBA, W.T. (1981). "Introduction to Concave and Generalized Concave Functions." This volume, 21-50.

BERGE, C. (1963). *Topological Spaces*. Macmillan, New York.

BERSTEIN, B. and TOUPIN, R.A. (1962). "Some Properties of the Hessian Matrix of a Strictly Convex Function." *Journal für die Reine and Angewandte Mathematik 210*, 65-72.

BLACKORBY, C., DAVIDSON, R., and DONALDSON, D. (1977). "A Homiletic Exposition of the Expected Utility Hypothesis." *Economica 44*, 351-358.

BLACKORBY, C. and DIEWERT, W.E. (1979). "Expenditure Functions, Local Duality and Second Order Approximations." *Econometrica 47*, 579-601.

CROUZEIX, J.-P. (1977). "Contributions à l'étude des fonctions quasiconvexes." Thèse présentée à l'université de Clermont-Ferrand, France.

CROUZEIX, J.-P. (1980). "Conditions for Convexity of Quasiconvex Functions." *Mathematics of Operations Research*. Forthcoming.

CROUZEIX, J.-P. (1981). "A Duality Framework in Quasiconvex Programming." This volume, 207-225.

DEBREU, G. (1952). "A Social Equilibrium Existence Theorem." *Proceedings of the National Academy of Sciences 38*, 886-893.

DEBREU, G. (1959). *Theory of Value*. John Wiley and Sons, New York.

DEBREU, G. and KOOPMANS, T.C. (1978). "Additively Decomposed Quasiconvex Functions." Mimeo.

DIEWERT, W.E. (1971). "An Application of the Shephard Duality Theorem: A Generalized Leontief Production Function." *Journal of Political Economy 79*, 481-507.

DIEWERT, W.E. (1973). "Afriat and Revealed Preference Theory." *Review of Economic Studies 40*, 419-426.

DIEWERT, W.E. (1974). "Applications of Duality Theory." In *Frontiers of Quantitative Economics*, Vol.II. (M.D. Intriligator and D.A. Kendrick, eds.). North-Holland Publishing Company, Amsterdam.

DIEWERT, W.E. (1978a). "Hicks' Aggregation Theorem and the Existence of a Real Value Added Function." In *Production Economics: A Dual Approach to Theory and Applications*, Vol.2. (M. Fuss and D. McFadden, eds.). North-Holland Publishing Co., Amsterdam.

DIEWERT, W.E. (1978b). "Duality Approaches to Microeconomic Theory." Discussion Paper 78-09. Department of Economics, University of British Columbia, Vancouver, Canada.

DIEWERT, W.E. (1981). "Alternative Characterizations of Six Kinds of Quasiconcavity in the Nondifferentiable Case with Applications to Nonsmooth Programming." This volume, 51-93.

DONALDSON, D. and EATON, B.C. (1978). "Person Specific Costs of Production: Hours of Work, Rates of Pay, Labour Contracts." Mimeo, Department of Economics, University of British Columbia, Vancouver, Canada.

DONALDSON, D. and EATON, B.C.(1981). "Patience, More Than Its Own Reward: A Note on Price Discrimination." *Canadian Journal of Economics*, forthcoming.

EPSTEIN, L. (1979). "Generalized Duality and Integrability." Working Paper 7901, Institute for Policy Analysis, University of Toronto, Toronto, Canada.

FENCHEL, W. (1953). "Convex Cones, Sets and Functions." Lecture notes, Department of Mathematics, Princeton University.

HICKS, J.R. (1941-2). "Consumer's Surplus and Index-Numbers." *Review of Economic Studies 9*, 126-137.

HICKS, J.R. (1946). *Value and Capital*. Oxford: Clarendon Press.

JORGENSON, D.W. and LAU, L.J.(1974a). "The Duality of Technology and Economic Behavior." *The Review of Economic Studies 41*, 181-200.

JORGENSON, D.W. and LAU, L.J. (1947b). "Duality and Differentiability in Production." *Journal of Economic Theory 9*, 23-42.

KANNAI, Y. (1974). "Approximation of Convex Preferences." *Journal of Mathematical Economics 1*, 101-106.

KANNAI, Y. (1977). "Concavifiability and Constructions of Concave Utility Functions." *Journal of Mathematical Economics 4*, 1-56.

KANNAI, Y. (1981). "Concave Utility Functions - Existence, Constructions and Cardinality." This volume, 543-611.

KARLIN, S. (1959). *Mathematical Methods and Theory in Games, Programming and Economics*, Vol. I. Addison-Wesley, Palo Alto, California.

KÖNUS, A.A. (1939). "The Problem of the True Index of the Cost of Living." *Econometrica 7*, 10-29.

KUHN, H.W. and TUCKER, A.W.(1951). "Nonlinear Programming." In *Proceedings of the Second Berkeley Symposium on Mathematical Statistics and Probability*. (J. Neyman, ed.). University of California Press, Berkeley, California.

LAU, L.J. (1974). "Applications of Duality Theory: Comments." In *Frontiers of Quantitative Economics*, Vol. II. (M.D. Intriligator and D.A. Kendrick, eds.). 176-199. North-Holland Publishing Company, Amsterdam.

LAU, L.J. (1978). "Applications of Profit Functions." In *Production Economics: A Dual Approach to Theory and Applications*, Vol. 1. (M. Fuss and D. McFadden, eds.). 133-216. North-Holland Publishing Co

McFADDEN, D. (1966). "Cost, Revenue and Profit Functions: A Cursory Review." Working Paper No. 86, IBER, University of California at Berkeley.

McFADDEN, D. (1978). "Convex Analysis." In *Production Economics: A Dual Approach to Theory and Applications*, (Eds. M. FUSS and D. McFADDEN) Vol.1. 383-408. North-Holland Publishing Co., Amsterdam.

McKENZIE, L.W. (1956-7). "Demand Theory Without a Utility Index." *Review of Economic Studies 24*, 185-189.

ROCKEFELLAR, R.T. (1970). *Convex Analysis.* Princeton University Press, Princeton, N.J.

ROCKEFELLAR, R.T. (1976). "Saddle Points of Hamiltonian Systems in Convex Lagrange Problems Having a Nonzero Discount Rate." *Journal of Economic Theory 12*, 71-113.

ROY, R. (1974). "La distribution du revenu entre les divers biens." *Econometrica 15*, 205-225.

SAMUELSON, P.A. (1947). *Foundations of Economic Analysis.* Harvard University Press, Cambridge, Mass.

SAMUELSON, P.A. (1953-4). "Prices of Factors and Goods in General Equilibrium." *Review of Economic Studies 21*, 1-20.

SCHAIBLE, S. and ZANG, I. (1979). "On the Convexifiability of Pseudoconvex C^2 Functions." Mimeo.

SHEPHARD, R.W. (1953). *Cost and Production Functions.* Princeton University Press, Princeton, N.J.

SHEPHARD, R.W. (1970). *Theory of Cost and Production Functions.* Princeton University Press, Princeton, N.J.

UZAWA, H. (1964). "Duality Principles in the Theory of Cost and Production." *International Economic Review 5*, 216-220.

YAARI, M.E. (1977). "A Note on Separability and Quasiconcavity." *Econometrica 45*, 1183-1186.

ZANG, I. (1981). "Concavifiability of C^2 Functions: A Unified Exposition." This volume, 131-152.

CONCAVE UTILITY FUNCTIONS - EXISTENCE, CONSTRUCTIONS AND CARDINALITY

Yakar Kannai

We discuss the problem of concavifiability of convex preference orderings - i.e., the problem of existence of a *concave* function having the same level sets as a given continuous *quasi-concave* one. The conditions for existence are intimately related to constructions of special (least concave) utility representations. The importance of those least concave functions in several economic and bargaining situations is reviewed.

1. INTRODUCTION

It is well-known that a concave function has convex level surfaces. That the converse is not true was discovered by de Finetti (1949). Using the terminology of Economic Theory, one can ask, therefore, for the conditions that a convex preference ordering $\succsim$ defined on a convex subset K of R^m has to satisfy so that there exists a *concave* (not just a quasi-concave) utility function u representing $\succsim$ in K. If such a u exists, we say that the ordering $\succsim$ is concavifiable.

The author's work on concavifiability and constructions of concave utility functions was presented, and some earlier work (especially by Fenchel (1953, 1956)) was surveyed, in Kannai (1977). The invitation to lecture in the Nato Advanced Study Institute on Generalized Concavity has provided the author with the opportunity and motivation to rewrite that 1977 paper. While the broad outline and main ideas of that paper have been kept, numerous improvements and simplifications were incorporated in the present lecture notes. (I hasten to add that many of those are due to other people.)

The starting point here, as in Kannai (1977), is the observation that concavity of a function of a single variable can be regarded as (i) a condition relating the values of the function at three collinear points, (ii) a condition relating the values of the derivatives (if the

ISBN 0-12-621120-5

function itself is differentiable) in two points, and (iii) a pointwise condition on the second derivatives (if the function is twice differentiable). For several variables (i) is unchanged, and (iii) is a condition on the definiteness of the matrix of the second derivatives, whereas the analogue of (ii) is either a monotomicity condition on the (sub)-gradient mapping (this viewpoint was adopted in Kannai (1977)), or a monotonicity condition on the one-sided directional derivatives (this approach is suggested here and in Crouzeix (1977)). One attempts to see whether the given preference ordering $\succsim$ is compatible with the requirements that result from the representability of $\succsim$ by a concave utility, requirements involving one of these three types of conditions. It turns out that each of the three resulting types of conditions for concavifiability leads naturally to a construction of a concave utility function.

In section 2 we present 3 point conditions along lines similar to those appearing in Kannai (1977). The novelty here consists of a (hopefully) more transparent notation in the formulations of the main theorems. A new and more direct proof of Theorem 2, a proof which does not make any use of the minimum principle, is presented (in Appendix A). Also, a simple example (due to Z. Artstein) of a non-concavifiable $\succsim$ defined on a compact one-dimensional interval is exhibited.

Section 3 (which deals with 2 point conditions) has been entirely rewritten. In Kannai (1977) it was attempted to reconstruct the super gradient map from the knowledge of the normal cones to the indifference surfaces. First the "integrability" problem had to be solved, and then the "adapted" correspondences had to be modified further so as to become monotone. In the present paper super-gradients and normal cones do not appear at all. Instead, one observes that concavity is essentially a one-dimensional phenomenon, and that if $\succsim$ is at all concavifiable, then a suitably constructed auxiliary quasi-concave utility function *has* to possess finite and non-vanishing one-sided directional derivatives. This auxiliary function $v_{a,b}(p)$ is defined to be *linear* on the interval [a,b]. Concavifiability is then characterized in terms of the one-sided directional derivatives of $v_{a,b}$. Similar auxiliary functions have been introduced independently by Crouzeix, in his 1977 Thesis.

An unusual feature of concavifiability theory as presented in Kannai (1977) was the essential use of Perron's integral in expressing concavifiability in terms of one-point conditions involving twice-differentiable quasi-concave utility functions v. It turns out that if such a v exists at all, then the auxiliary $v_{a,b}$ constructed in Section 3 is also twice-differentiable, and the associated function, whose Perron integrability is equivalent to concavifiability, has a constant sign, hence Perron integrability is equivalent (in the construction performed on $v_{a,b}$) to *Lebesgue* integrability. (We note that Crouzeix (1977) replaced Perron by Lebesgue using a similar approach.) This matter is discussed in Section 5, as well as relationships between one, two, and three point-conditions, explained here in a less general (but easier to understand) manner than in Kannai (1977). (The changes in Section 4 - dealing with one-point conditions a-la Fenchel - are mostly stylistic.)

Following suggestions by a number of people (including a referee and an editor - to whom I hereby give my thanks) we have put some of the more complicated or technical proofs in Appendices. This applies - besides to the already mentioned Theorem 2 - to the proofs of the sufficiency parts of the two- and three-point conditions for concavifiability (Theorem 3 and 10), and to certain results concerning real variables, as well as to some computations, in Section 5.

In Section 6 a relatively simple one-dimensional example (developed in discussions with Z. Artstein) replaces the computationally more complicated two-dimensional example, exhibited in Kannai (1977), of concavifiable preference orderings $\succsim_n$ converging to a concavifiable $\succsim$ such that no sequence u_n of concave utility functions representing $\succsim_n$ can converge to a utility function representing $\succsim$.

If $\succsim$ is concavifiable, then certain concave utility functions representing $\succsim$ are "less concave" than all other utility functions representing $\succsim$. This concept of least concave utility functions was formulated in full clarity in Debreu (1976); de Finetti (1949) introduced those functions under the name "minimally concave functions" (this terminology was used also by Kannai (1977)). All the constructions in the present paper yield actually least concave utility functions. Those functions possess interesting economic properties (Kannai (1980)) and are behavioristically preferable to the others, in a number of bargaining situations. These and similar matters are discussed in Section 7.

It is hoped that the present lecture notes would make the material more accessible and would persuade a wider audience to study this fascinating subject. I am grateful to the editors and referees for suggestions whose implementation will improve - I hope - the readability of this survey.

In addition to the persons mentioned in the introduction to Kannai (1977) and in Kannai (1980), I am indebted to A.E. Roth, U.G. Rothblum and W. Thomson, for explaining to me their work on the bargaining problem - explanations which clarified some of the ideas presented in Section 7 below.

2. THREE POINT CONDITIONS

Let K be a convex subset of R^m. We assume that a continuous, complete and convex preference ordering $\succsim$ is given on K. By this we mean (as is customary in Economic theory - see e.g. Debreu (1959) and Hildenbrand (1974)) that $\succsim$ is a binary relation on K such that: (i) for all $p,q \in K$, either $p \succsim q$ or $p \precsim q$; (ii) $p \succsim p$ for all $p \in K$; (iii) for all $p,q,r \in K$, if $p \succsim q$ and $q \succsim r$, then $p \succsim r$; (iv) the set $\{(p,q) \in K \times K;\ p \succsim q\}$ is closed relative to $K \times K$; and (v) for all $q \in K$, $\{p : p \succsim q\}$ is convex. As is well-known, these conditions are equivalent to the statement that there exists a continuous, *quasi-concave* utility function g on K representing $\succsim$ (i.e., $p \succsim q$ if and only if $g(p) \geq g(q)$). Recall the usual notation: $p \sim q$ if $p \succsim q$ and $q \succsim p$, and $p \succ q$ if $p \succsim q$ but *not* $p \precsim q$. (Thus the level sets of g are the same as the "indifference surfaces" $\{q : q \sim p\}$ of the ordering $\succsim$.)

We are interested in knowing when $\succsim$ can be represented by a *concave* (and not just quasi-concave) utility function. Recall that a function f defined in K is said to be *concave* if for all $a,b \in K$, $0 \leq t \leq 1$,

$$f(ta + (1-t)b) \geq tf(a) + (1-t)f(b). \tag{1}$$

(On the other hand, if f is only quasi-concave, then the right-hand side of (1) is replaced by $\min\{f(a),f(b)\}$.) It is impossible to represent $\succsim$ by means of a concave utility function, unless one assumes, as we shall *always* do in the present paper (without explicitly mentioning it) that if $p \succ q$ and $0 < t < 1$, then $tp + (1-t)q \succ q$. It follows in particular that the set $\{q : q \sim p\}$ does not contain a non-empty open (relative to K) set, unless p is maximal. (Unfortunately, this condition is only necessary, but does not suffice for $\succsim$ to be representable by means of a concave function.)

The inequality (1), defining concavity, can be rewritten as

$$\frac{f(ta + (1-t)b) - f(a)}{f(b) - f(ta + (1-t)b)} \geq \frac{1-t}{t} \tag{2}$$

if $f(b) > f(ta + (1-t)b)$. We want to utilize (2) for investigating the problem of *concavifiability*, i.e., to see whether or not there exists a concave f defined on K such that f is a utility function representing $\succsim$. For this purpose, consider triplets p_1,p_2,p_3 with $p_i \in K$ for $i = 1,2,3$

and $p_1 \prec p_2 \prec p_3$. Let $q_1 \sim p_1$, $q_3 \sim p_3$ be arbitrary. There exists a real number t, $0 < t < 1$, such that $q_2 = tq_1 + (1-t)q_3$ satisfies $q_2 \sim p_2$. If f is an arbitrary concave utility function for $\succsim$, it follows from (2) that

$$\frac{f(p_2)-f(p_1)}{f(p_3)-f(p_2)} = \frac{f(q_2)-f(q_1)}{f(q_3)-f(q_2)} \geq \frac{1-t}{t} = \frac{|q_2-q_1|}{|q_3-q_2|} \tag{3}$$

where $|\ |$ denotes the Euclidean norm.

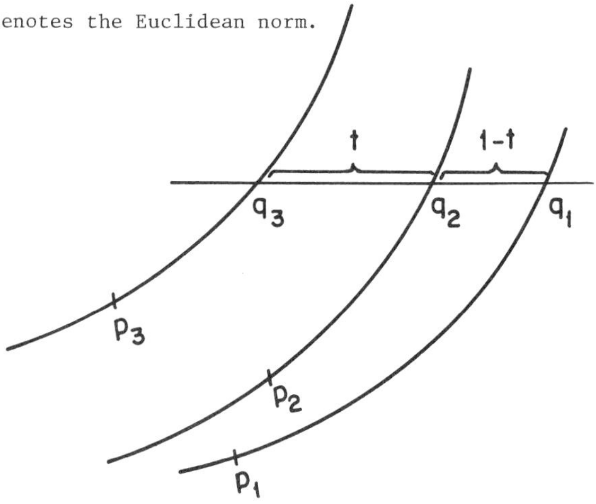

FIGURE 1: Three points conditions

Setting

$$\alpha(p_1,p_2,p_3) = \sup_{q_i \sim p_i} \frac{|q_2-q_1|}{|q_3-q_2|} \tag{4}$$

q_1,q_2,q_3 collinear with

q_2 between q_1 and q_3

we obtain

PROPOSITION 1: (i) If f is an arbitrary concave utility function for $\succsim$, *then*

$$\frac{f(p_2) - f(p_1)}{f(p_3) - f(p_2)} \geq \alpha(p_1,p_2,p_3) \tag{5}$$

for all triplets (p_1,p_2,p_3) *with* $p_1 \prec p_2 \prec p_3$. *(ii) A necessary condition for* $\succsim$ *to be concavifiable is that*

$$\alpha(p_1,p_2,p_3) < \infty \tag{6}$$

for all triplets (p_1,p_2,p_3) *with* $p_1 \prec p_2 \prec p_3$.

As a simple illustration of the meaning of (6), let us prove that there exists no concave function f in $\{(x_1,x_2) : x_1 > 0,\ x_2 > 0\}$ such that f is constant on the lines

$$\frac{x_2}{1+x_1} = \text{const.}$$

Indeed, let $0 < a_1 < a_2 < a_3$, $p_i = (1,2a_i)$, i=1,2,3, and let $q_3 = p_3$, q_i (i=1,2) - the intersection of the lines $a_i(1+x_1) = x_2$ with the line whose parametric representation is $(1+t,\ 2a_3+(a_1-\varepsilon)t)$ for $a_1 > \varepsilon > 0$ (see Fig.2).

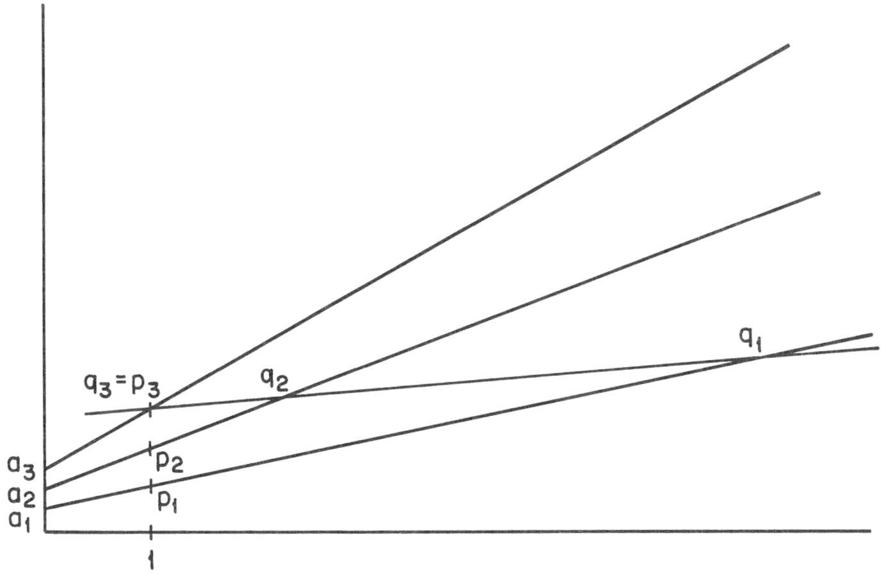

FIGURE 2. Non concavifiability for $x_2/(1+x_1)$

Then

$$\lim_{\varepsilon\to 0} \frac{|q_2-q_1|}{|q_3-q_2|} = \infty$$

and $\alpha(p_1,p_2,p_3) = \infty$. Thus, the preference ordering induced by $\frac{x_2}{1+x_1}$ is not concavifiable.

Let us relate $\alpha(p_1,p_2,p_3)$ to a ratio of differences of certain support functions, a ratio previously introduced in the investigations of concavifiability by Fenchel (1953) and approximability by concavifiable preferences (Kannai (1974)). Denote by H(u,i) the support function of the upper set of p_i, i.e.,

$$H(u,i) = \sup_{q \succsim p_i} \langle u,q\rangle \quad , \quad i=1,2,3$$

where $p_1 \prec p_2 \prec p_3$ and $\langle \ , \ \rangle$ denotes the inner (scalar) product in R^m. Assume that K is open and set

$$\beta(p_1,p_2,p_3) = \sup_{u\neq 0} \frac{H(u,1) - H(u,2)}{H(u,2) - H(u,3)} \quad . \tag{7}$$

If the set $\{q\colon \ q \succsim p_1\}$ is compact, then the supremum in the definition of the support functions is attained, and moreover

$$H(u,1) > H(u,2) > H(u,3)$$

for all $u \neq 0$. The ratio in the right hand side of (7) is positively homogeneous of order zero. Thus it suffices to consider this ratio only on the (compact) unit sphere, and there the supremum is actually a maximum. Hence the definition (7) makes sense, at least if the set $\{q\colon \ q \succsim p_1\}$ is compact, and we claim that in that case

$$\alpha(p_1,p_2,p_3) = \beta(p_1,p_2,p_3) \quad . \tag{8}$$

(A similar statement can be made in general; see Appendix A in Kannai (1977).)

To prove (8), consider first an arbitrary triplet (q_1,q_2,q_3) of collinear points with $q_i \sim p_i$, $i = 1,2,3$, and $q_2 \in (q_1,q_3)$ and choose

$u \neq 0$ such that

$$H(u,2) = \langle u,q_2\rangle.$$

(Such a u clearly exists since q_2 is a boundary point of $\{q: q \succcurlyeq p_2\}$.)

But

$$H(u,3) \geq \langle u,q_3\rangle$$

and

$$H(u,1) \geq \langle u,q_1\rangle.$$

Hence

$$\langle u,q_2-q_3\rangle \geq H(u,2)-H(u,3) > 0$$

and

$$H(u,1) - H(u,2) \geq \langle u,q_1-q_2\rangle > 0$$

(we use the fact that $q_1-q_2 = \lambda(q_2-q_3)$ with $\lambda > 0$). Thus

$$\frac{H(u,1)-H(u,2)}{H(u,2)-H(u,3)} \geq \frac{\langle u,q_1-q_2\rangle}{\langle u,q_2-q_3\rangle} = \frac{|q_2-q_1|}{|q_3-q_2|}$$

and so $\beta(p_1,p_2,p_3) \geq \alpha(p_1,p_2,p_3)$.

To prove the converse inequality, choose for a given $u \neq 0$ q_1,q_3 such that $q_i \sim p_i$ for $i=1,3$ and $H(u,i) = \langle u,q_i\rangle$ for these values of i. Let q_2 be the intersection point of the open interval (q_1,q_3) and the set $\{q:q \sim p_2\}$. Clearly

$$H(u,2) \geq \langle u,q_2\rangle.$$

It follows that

$$0 < H(u,1) - H(u,2) \leq \langle u,q_1-q_2\rangle$$

and

$$H(u,2) - H(u,3) \geq \langle u,q_2-q_3\rangle > 0.$$

Hence

$$\frac{H(u,1)-H(u,2)}{H(u,2)-H(u,3)} \leq \frac{\langle u,q_1-q_2\rangle}{\langle u,q_2-q_3\rangle} = \frac{|q_1-q_2|}{|q_2-q_3|}$$

which implies that $\beta(p_1,p_2,p_3) \leq \alpha(q_1,q_2,q_3)$.

After making these remarks concerning the meaning of the quantities $\alpha(p_1,p_2,p_3)$, we continue the consideration of convex preference orderings defined on (not necessarily open) convex sets K. First we determine the consequences following from iteration of the inequality (5). Thus, let $p_0 \prec p_1 \prec \cdots \prec p_k$ be a finite sequence of points of K, and let f be an arbitrary concave utility function for the ordering $\succsim$. Setting $\alpha_j = \alpha(p_{j-1},p_j,p_{j+1})$ for $1 \leq j \leq k-1$, we rewrite (5) for every triple p_{j-1},p_j,p_{j+1} as

$$f(p_1)-f(p_0) \geq \alpha_1[f(p_2)-f(p_1)]$$
$$f(p_2)-f(p_1) \geq \alpha_2[f(p_3)-f(p_2)] \qquad (5')$$
$$f(p_{k-1})-f(p_{k-2}) \geq \alpha_{k-1}[f(p_k)-f(p_{k-1})].$$

Substituting each inequality in the preceding one successively from the last inequality backwards, we get the following system of inequalities,

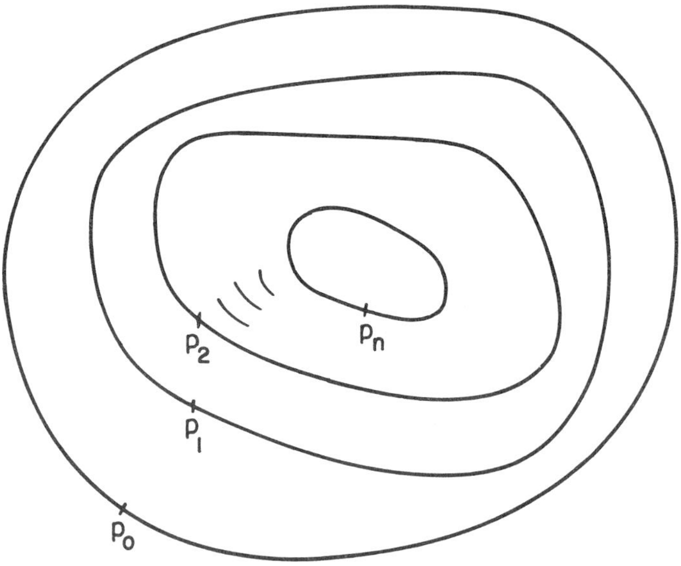

FIGURE 3. Finitely many indifference surfaces

$$f(p_1)-f(p_0) \geq \alpha_1\alpha_2 \cdots \alpha_{k-1} [f(p_k)-f(p_{k-1})]$$

$$f(p_2)-f(p_1) \geq \alpha_2 \cdots \alpha_{k-1} [f(p_k)-f(p_{k-1})] \tag{5''}$$

$$\vdots$$

$$f(p_{k-1})-f(p_{k-2}) \geq \alpha_{k-1} [f(p_k)-f(p_{k-1})]$$

which we supplement by the "inequality" $f(p_k)-f(p_{k-1}) \geq f(p_k)-f(p_{k-1})$. Adding those k inequalities, we see that

$$f(p_k)-f(p_0) \geq (1 + \alpha_{k-1} + \alpha_{k-2}\alpha_{k-1} + \cdots + \alpha_1\alpha_2 \cdots \alpha_{k-1}) \times [f(p_k)-f(p_{k-1})].$$

This inequality can be rewritten as

$$f(p_{k-1}) \geq \frac{f(p_0)}{1+\alpha_{k-1}+\alpha_{k-2}\alpha_{k-1} + \cdots + \alpha_1 \cdots \alpha_{k-1}} + \frac{(\alpha_{k-1}+\alpha_{k-2}\alpha_{k-1} + \cdots + \alpha_1 \cdots \alpha_{k-1})f(p_k)}{1+\alpha_{k-1}+\alpha_{k-2}\alpha_{k-1} + \cdots + \alpha_1 \cdots \alpha_{k-1}}. \tag{9}$$

Note that if k = 2 then (9) is just (5). If, however, k > 2, then we can draw further conclusions from further iterations of (9), as shown by the following result, the full proof of which is given in Appendix A.

THEOREM 2: Let $p_0 \prec p_1 \prec \cdots \prec p_n$ *be an otherwise arbitrary finite sequence of points of* K. *Let* f *be an arbitrary concave utility function for the preference ordering* $\succsim$. *Set*

$\alpha_j = \alpha(p_{j-1}, p_j, p_{j+1})$ *for* $1 \leq j \leq n-1$. *Then*

$$f(p_j) \geq \frac{1+\alpha_{n-1}+\alpha_{n-2}\alpha_{n-1} + \cdots + \alpha_{j+1} \cdots \alpha_{n-1}}{1+\alpha_{n-1}+\alpha_{n-2}\alpha_{n-1} + \cdots + \alpha_1 \cdots \alpha_{n-1}} f(p_0) + \frac{\alpha_j \cdots \alpha_{n-1} + \cdots + \alpha_1 \cdots \alpha_{n-1}}{1+\alpha_{n-1}+\alpha_{n-2}\alpha_{n-1} + \cdots + \alpha_1 \cdots \alpha_{n-1}} f(p_n) \tag{10}$$

for all $1 \leq j \leq n-1$.

Note that (10) reduces to (9) if j = k-1, n = k (as there is no term of the form $\alpha_n\alpha_{n-1}$ and the numerator of the fraction multiplying $f(p_0)$ in (10) is understood to be equal to 1).

Note also that the right hand side of (10) is a convex combination of $f(p_0)$ and $f(p_n)$, the weight of $f(p_0)$ decreasing and the weight of $f(p_n)$ increasing with j. Note also that the right hand sides of (10) are the (unique) solutions of the system (5'), when we (i) set k = n and treat $f(p_0)$ and $f(p_n)$ as given, and (ii) ask that equality holds everywhere in (5'), regarding $f(p_2),\dots,f(p_{n-1})$ as unknowns.

The right hand side of (10) can be used in order to approximate the convex preference ordering $\succsim$ by a concavifiable one. (This construction was used by Mas-Colell (1974) and Kannai (1974), and was discovered as early as 1969 by R. Mantel in an unpublished paper.) Setting $f(p_0) = 0$, $f(p_n) = 1$, we define a real function u on the set

$$A = \bigcup_{i=0}^{n} \{p \in K: p \sim p_i\}$$

by

$$u(p) = \frac{\alpha_j\cdots\alpha_{n-1} + \dots + \alpha_1 \cdots \alpha_{n-1}}{1+\alpha_{n-1} + \dots + \alpha_1 \cdots \alpha_{n-1}} \text{ if } p \sim p_j,\ 0 \leq j \leq n. \quad (11)$$

Then the function u is concave in its domain of definition, i.e., if $a,b \in A$, $0 \leq t \leq 1$, and $ta+(1-t)b \in A$, then

$$u(ta+(1-t)b) \geq tu(a)+(1-t)u(b). \quad (12)$$

In fact, if $p \succsim q$, $p,q \in A$, then $u(p) \geq u(q)$. If $a \sim b$ then $ta+(1-t)b \succsim a,b$ (by convexity of $\succsim$), and (12) follows in this case. Thus we may assume without loss of generality that $a \prec b$ and $a \sim p_i$, $b \sim p_k$ with $0 \leq i < k \leq n$. Then $ta+(1-t)b \sim p_j$ for some j, $i < j < k$. For every j, $i < j < k$, there exists a unique real number t_j with $0 \leq t_j \leq 1$ and $t_ja+(1-t_j)b \sim p_j$. We set also $t_i = 1$, $t_k = 0$ and note that t_j is a decreasing sequence for $i \leq j \leq k$. Then the points $t_{j-1}a+(1-t_{j-1})b$, $t_ja+(1-t_j)b$ and $t_{j+1}a+(1-t_{j+1})b$ are collinear. Hence

$$\alpha_j = \alpha(p_{j-1},p_j,p_{j+1}) \geq \frac{|t_ja+(1-t_j)b-t_{j-1}a-(1-t_{j-1})b|}{|t_{j+1}a+(1-t_{j+1})b-t_ja-(1-t_j)b|} = \frac{t_{j-1}-t_j}{t_j-t_{j+1}}.$$

But the definition (11) implies that

$$u(p_j) - u(p_{j-1}) = \alpha_j[u(p_{j+1}) - u(p_j)]$$

or

$$u(t_j a+(1-t_j)b)-u(t_{j-1}a+(1-t_{j-1})b) \geq$$

$$\frac{t_{j-1}-t_j}{t_j-t_{j+1}} [u(t_{j+1}a+(1-t_{j+1})b)-u(t_j a+(1-t_j)b)].$$

Extending the function $u(ta+(1-t)b)$ of the variable t from its domain $\{t_i,t_{i+1},\ldots,t_k\}$ to the full interval $[0,1]$ in a piecewise linear way (so as to be linear on the intervals $[t_j,t_{j-1}]$, $j = i+1,\ldots,k$) we see that the extended function is concave, as the slopes of the lines joining $(t_j,u(t_j a+(1-t_j)b)$ with $t_{j+1},u(t_{j+1}a+(1-t_{j+1})b)$ decrease (weakly) with t_j, and (12) follows.

The argument used in the proof of (12) can be used for extending u from A to $\text{conv}(A) = \{p \in K: p \succcurlyeq p_0\}$ (here and in the sequel $\text{conv}(A)$ denotes the convex hull of A). For every $p \in \text{conv}(A)$ there exist $a,b \in A$ and $t \in [0,1]$ such that $p = ta+(1-t)b$. We define $u(p)$ to be the supremum of $u(ta+(1-t)b)$, taken over all such a,b and t (see also Fig.4). Evidently, $u(p)$ is concave in its domain of definition. Alternatively, and perhaps more geometrically, consider the graph L of the (unextended) function u:

$$L = \bigcup_{j=0}^{n} \{p:p \sim p_j\} \times \{u(p_j)\},$$

and set $M = \text{conv}(L)$. Then M is the sub-graph of (the extended) u, as the preceding definition is equivalent to

$$u(p) = \max_{(p,t)\in M} t \quad . \tag{13}$$

(Note added in proof: see Appendix F.)

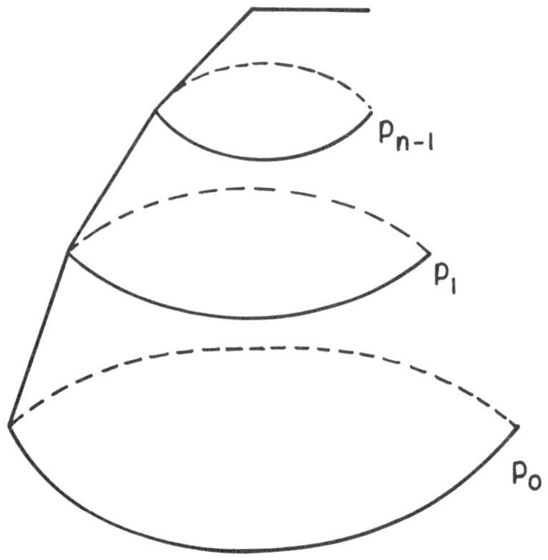

FIGURE 4. The extended approximating utility

After all these preparations we can state necessary and sufficient conditions for concavifiability. First consider the case where K is a convex body (i.e., a compact convex set with a non-empty interior).

THEOREM 3: Let K *be a convex body, and let* $\succsim$ *be a complete continuous and convex preference ordering on* K. *A necessary and sufficient condition for concavifiability of* $\succsim$ *is that for every* $p \in K$, p *not maximal with respect to* $\succsim$,

$$\sup \frac{\alpha_j \cdots \alpha_{n-1} + \cdots + \alpha_1 \cdots \alpha_{n-1}}{1+\alpha_{n-1} + \cdots + \alpha_1 \cdots \alpha_{n-1}} < 1 \tag{14}$$

where the supremum is taken over all finite sequences $p_0 \prec p_1 \prec \cdots \prec p_n$ *where* $p_i \in K$, $0 \leq i \leq n$, p_n *is maximal* w r.t. $\succsim$, $\alpha_j = \alpha(p_{j-1}, p_j, p_{j+1})$, *and* $p_j = p$. *Moreover, this supremum (taken as a function of* p*) is a concave utility.*

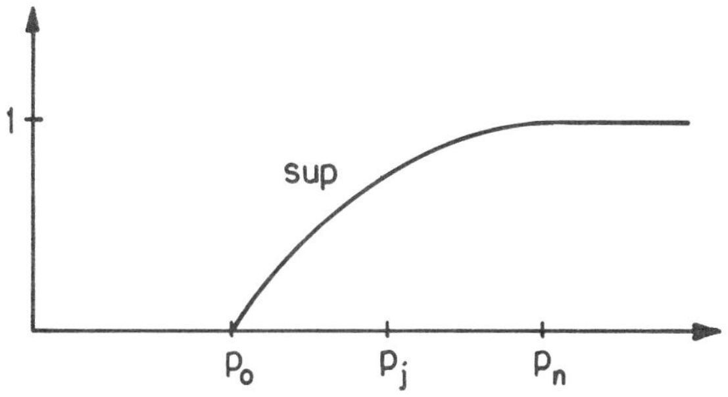

FIGURE 5. A concavifiable case

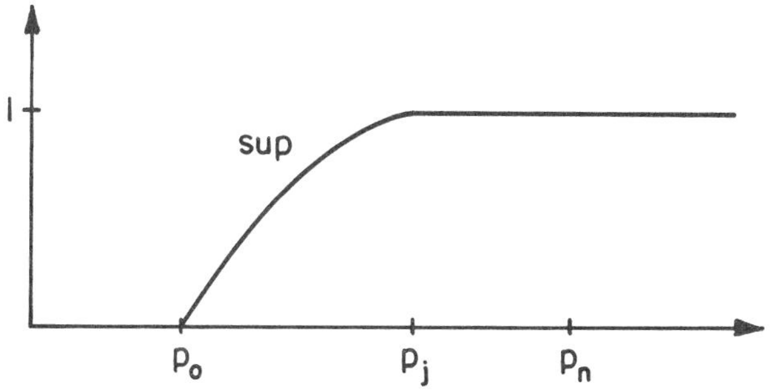

FIGURE 6. A nonconcavifiable case

PROOF: Let $\succsim$ be concavifiable and let f be any concave utility for $\succsim$. The compactness of K implies that we can choose f non-negative, and thus $f(p_n) > 0$. From Theorem 2 we learn that

$$\frac{\alpha_j \cdots \alpha_{n-1} + \cdots + \alpha_1 \cdots \alpha_{n-1}}{1+\alpha_{n-1} + \cdots + \alpha_1 \cdots \alpha_{n-1}}$$

$$\leq \frac{f(p_j)}{f(p_n)} = \frac{f(p)}{f(p_n)} < 1$$

which proves the necessity.

We will only outline roughly the proof of the sufficiency here, leaving the details to Appendix B. We begin by choosing a sequence of finite sequences $\{p_{i,n}\}$, $0 \leq i \leq n < \infty$, of points in K with $p_{0,n} \prec p_{1,n} \prec \cdots \prec p_{n,n}$, where $p_{n,n}$ is maximal and $p_{0,n}$ is minimal with respect to $\succsim$, and such that the sequence is dense with respect to $\succsim$. Then the concave functions u_n constructed according to (13) are defined over all of K, for n = 1,2,..., and $0 \leq u_n(x) \leq 1$ for all $x \in K$, n = 1,2,... . By a classical selection theorem for concave functions [e.g., Rockafellar (1970), Theorem 10.9)], we can select a converging subsequence converging to a continuous concave function u defined on int K. The condition (14) implies that u is indeed a utility function representing $\succsim$ on int K - that the situation exhibited in Fig.6, of the limit being equal to 1 on non-maximal points, does not occur. One then extends u to all of K.

REMARK 1. Note that the selection theorem, used in the proof of Theorem 3, is valid in non-compact sets as well as in compact sets. The compactness is used in Theorem 3 only to ensure the existence of maximal and minimal elements. Hence the theorem remains valid if K is an arbitrary closed convex subset of R^m and $\succsim$ is a convex preference ordering on $\succsim$ such that there exist in K maximal and minimal elements with respect to $\succsim$.

The utility function constructed in Theorem 3 (and Remark 1) is actually a minimally concave utility, in the sense of de Finetti's definition (1949) or, what amounts to the same thing, a least concave utility, in the sense of Debreu's definition (1976). Recall the definitions: a concave utility u is said to be *minimally concave* in the layer $\{x: u(a) \leq u(x) \leq u(b)\}$ if for all concave utilities of the same ordering satisfying $v(a) = u(a)$, $v(b) = u(b)$ we have $u(p) \leq v(p)$ for

$a \prec p \prec b$. It is said to be *least concave* in a convex set $X \subset K$ if every concave utility v defined on X can be represented in the form $v(p) = f(u(p))$, where f is a real valued concave function defined on the interval $u(X)$. (Debreu proves the existence of u by constructing u as a pointwise minimum. This should not mislead the reader: The terms "least" or "minimally" refer to the concavity of u which is least, and not to the minimality of the values of u.) (Note that while the existence of u is proved by Debreu by constructing u as a minimum, u is minimally concave - the concavity of u is "least", not just the values of u are minimal.) Clearly every least concave utility function is minimally concave in any sublayer of X. The converse follows at once from the following property of minimally concave utilities, a property which is also going to be useful in the sequel. This property was discovered by de Finetti; the first proof was given by Debreu (1976). (We suggest that the reader consult that paper for a more geometric argument than the one given here.)

PROPOSITION 4: A function u *which is minimally concave in a layer* $\{x: a \prec x \prec b\}$ *is minimally concave in any sublayer.*

PROOF: Let $a \precsim a' \prec b' \precsim b$, and let v be a concave utility with $v(a') = u(a')$, $v(b') = u(b')$. The conclusion $v(p) \geq u(p)$ for $a' \prec p \prec b'$ will follow, once we prove that the utility w defined by

$$w(p) = u(p) \qquad \text{if } p \precsim a' \text{ or } p \succsim b'$$

$$w(p) = \min\,(v(p),u(p)) \qquad \text{if } a' \prec p \prec b$$

is concave. We will prove that w satisfies the concavity inequality for the three collinear points c,p and d with $c \precsim a' \precsim p \precsim b' \precsim d$ (the other cases are similar or easier). Let a", b" be in the interval [c,d] with $a'' \sim a'$, $b'' \sim b'$. Hence

$$w(p) \geq \frac{|p-a''|}{|b''-a''|} w(b'') + \frac{|b''-p|}{|b''-a''|} w(a'') = \frac{|p-a''|}{|b''-a''|} u(b'') + \frac{|b''-p|}{|b''-a''|} u(a'') \tag{15}$$

The concavity of u implies that

$$u(b'') \geq \frac{|b''-c|}{|d-c|} u(d) + \frac{|d-b''|}{|d-c|} u(c) \quad \text{and}$$

$$u(a'') \geq \frac{|a''-c|}{|d-c|} u(d) + \frac{|d-a''|}{|d-c|} u(c).$$

Substituting in (15), we obtain the inequality

$$w(p) \geq \frac{|p-a''|\ |b''-c| + |b''-p|\ |a''-c|}{|b''-a''|\ |d-c|} u(d) + \frac{|p-a''|\ |d-b''| + |b''-p|\ |d-a''|}{|b''-a''|\ |d-c|} u(c) = \frac{|p-c|}{|d-c|} u(d) + \frac{|d-p|}{|d-c|} u(c).$$

Let now u be a minimally concave utility function representing $\succsim$ in the layer $\{x: a \precsim x \precsim b\}$. We wish to show that u is least concave there. Let v(p) be an arbitrary concave function representing $\succsim$ in the layer, so that we can write $v = f(u)$, where $f = (v \circ u^{-1})$ is a real function defined on the range of u. To prove the concavity of f, let $t_1 < t_2 < t_3$ be numbers in the domain of definition of f. There exist $p_1 \prec p_2 \prec p_3$ in the layer with $u(p_i) = t_i$, $i = 1,2,3$. Set

$$g(t) = \left[\frac{u(p_3)-u(p_1)}{v(p_3)-v(p_1)}\right] f(t) + \frac{u(p_1)v(p_3)-u(p_3)v(p_1)}{v(p_3)-v(p_1)} \quad \text{for } t \in [t_1,t_3],$$

and set $w(p) \equiv g(u(p))$ for $p_1 \precsim p \precsim p_3$. Then w(p) is a concave utility function representing $\succsim$ in the layer $\{p: p_1 \precsim p \precsim p_3\}$, and $w(p_1) = u(p_1)$, $w(p_3) = u(p_3)$. By Proposition 4, $w(p_2) \geq u(p_2)$, i.e., $g(t_2) \geq t_2$. This last inequality, along with the equations $g(t_1) = t_1$, $g(t_3) = t_3$ imply that

$$g(t_2) \geq \left(\frac{t_3-t_2}{t_3-t_1}\right) g(t_1) + \left(\frac{t_2-t_1}{t_3-t_1}\right) g(t_3).$$

Hence

$$f(t_2) \geq \left(\frac{t_3-t_2}{t_3-t_1}\right) f(t_1) + \left(\frac{t_2-t_1}{t_3-t_2}\right) f(t_3)$$

and f is concave.

It follows that if u is minimally concave in a layer, then $\alpha u+\beta$ is also least concave in the layer whenever α is a positive constant and β is an arbitrary real number, and that all minimally concave utility functions representing $\succsim$ in the layer are obtained in this manner. Hence a minimally concave utility is uniquely determined by its values on two not indifferent points.

Following Debreu (1976) we can use the properties of least concave utility functions in order to pass from the compact case (discussed in Theorem 2) or the layer case (discussed in Remark 1) to cases where K is not compact and $\succsim$ has no minimal points. We omit the details here, as they can be found in Kannai (1977, pp.13-15).

The following useful condition can be derived from (14).

PROPOSITION 5: If there exists a triple $p \prec q \prec r$ *and a sequence of finite sequences* $p_{0,n} = p \prec p_{1,n} \prec \cdots \prec p_{n-1,n} = q \prec p_{n,n} = r$ *such that*

$$\alpha_{1,n}\alpha_{2,n} \cdots \alpha_{n-1,n} \to \infty,$$

then (14) cannot be valid, and $\succsim$ *is not concavifiable.*

PROOF: The assumption implies that

$$\lim_{n\to\infty} \frac{\alpha_{j,n} \cdots \alpha_{n-1,n} + \ldots + \alpha_{1,n} \cdots \alpha_{n-1,n}}{1 + \alpha_{n-1,n} + \ldots + \alpha_{1,n} \cdots \alpha_{n-1,n}} = 1$$

for j = n-1, contradicting (14).

As an application of Proposition 5 let us prove that the preference relation induced by the utility $\frac{x_2}{1+x_1}$ is not concavifiable in the unit square. In fact, consider for every $n \geq 3$ the sequence $p_i = (0, \frac{i}{2n})$, and note that for each $3 \leq i \leq n$, $q_i = (1/i, \frac{i+1}{2n}) \sim p_i$, $\bar{q}_{i-1} = (\frac{2}{i-1}, \frac{i+1}{2n}) \sim p_{i-1}$, and the points $p_{i+1}, q_i, \bar{q}_{i-1}$ are collinear. Hence

$$\alpha(p_{i-1}, p_i, p_{i+1}) \geq \frac{|\bar{q}_{i-1} - q_i|}{|p_{i+1} - q_i|} = \frac{i+1}{i-1}$$

so that

$$\prod_{i=3}^{n-1} \alpha(p_{i-1}, p_i, p_{i+1}) = \frac{n(n-1)}{6} .$$

Since $\alpha(p_1,p_2,p_3) \geq 1$ and $\alpha(p_{n-1},p_n,r) \geq \frac{1}{n}$ where $r \in (0,1)$, we obtain a contradiction.

It was observed in Kannai (1977, p.17), that there exist non-concavifiable convex preference orderings even in the one dimensional case. In fact, it was shown that if $\succsim$ is the preference ordering induced on $(-1,\infty)$ by the utility function

$$v(x) = \begin{cases} x & \text{if } -1 < x < 1 \\ \frac{1}{x} & \text{if } 1 < x < \infty \end{cases}$$

Then $\succsim$ is not concavifiable, but $\succsim$ restricted to $(-1,n)$ is concavifiable for all n.

Let us present here a non-concavifiable convex preference ordering defined on a *compact* subset K of R^1. (I owe this example to Z. Artstein.) Let $K = [-1,3]$ and let $v(x)$ be defined as follows:

$$v(x) = \begin{cases} x & -1 \leq x < 0 \\ x^2 & 0 \leq x < 1 \\ 2-x & 1 \leq x \leq 3 \end{cases}$$

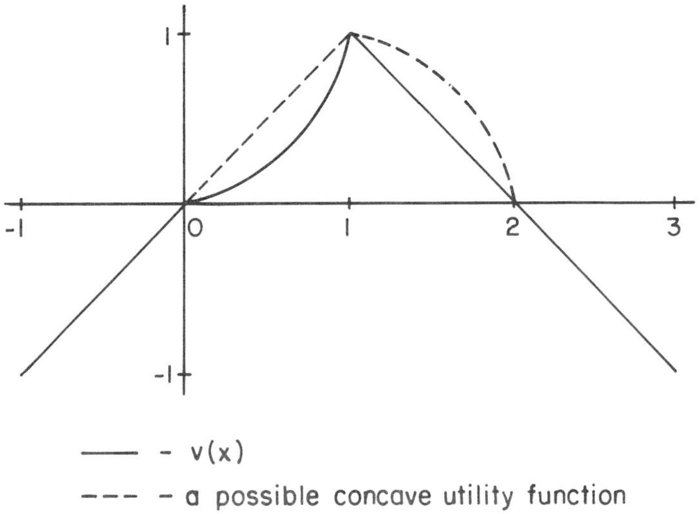

FIGURE 7. Non concavifiability in compact intervals

The preference ordering $\succsim$ is defined by $x \succsim y$ if and only if $v(x) \geq v(y)$. The function v is not concave on [0,1]. Note that $\succsim$ restricted to [0,2] is indeed concavifiable, but every concave utility has infinite slope at x = 2, hence cannot be extended beyong x = 2 (see Fig. 7). We shall prove directly from Theorem 2 that $\succsim$ is not concavifiable. In fact, set $p_k = 3 - \frac{2k}{n}$ for $0 \leq k \leq n$. Then $p_0 = 3$ - a minimal point, $p_n = 1$ - a maximal point, with respect to $\succsim$. Let n be even, $n = 2n'$. Then $p_{n'} = 2$. Note that for $n' \leq k \leq 2n'$, $p_k \sim x_k$ where $v(x_k) = x_k^2 = 2-p_k$. Hence

$$\alpha_k = \alpha(p_{k-1},p_k,p_{k+1}) = \frac{\sqrt{k-n'} - \sqrt{k-1-n'}}{\sqrt{k+1-n'} - \sqrt{k-n'}} \text{ for } n' < k \leq n-1$$

and $\alpha_k = 1$ for $1 \leq k \leq n'$. Then $\alpha_j \cdots \alpha_{n-1} = \dfrac{\sqrt{j-n'} - \sqrt{j-n'-1}}{\sqrt{n'} - \sqrt{n'-1}}$ if $j \geq n'+1$ and $\alpha_j \cdots \alpha_{n-1} = 1/(\sqrt{n'} - \sqrt{n'-1})$ if $j \leq n'$. Evaluating (14) at $j = n'$ we see that

$$\frac{\alpha_1 \cdots \alpha_{n-1} + \cdots + \alpha_{n'} \cdots \alpha_{n-1}}{1+\alpha_{n-1} + \cdots + \alpha_{n'} \cdots \alpha_{n-1} + \cdots + \alpha_1 \cdots \alpha_{n-1}} = \frac{n'}{n' + \sqrt{n'}} \to 1$$

so that the condition for concavifiability is violated.

3. TWO POINT CONDITIONS

We start by recalling several well-known concepts and results from the differential theory of concave functions [see Rockafellar (1970, part V); notice the change from convex to concave].

Let v be any real function defined on K. Let $p \in K$ and assume that $\{p+\lambda y : 0 \leq \lambda \leq \varepsilon\} \subset K$ for a sufficiently small ε. If the limit

$$\lim_{\lambda \to 0_+} (v(p+\lambda y) - v(p))/\lambda$$

exists ($+\infty$ and $-\infty$ are allowed as limits) we denote it by $v'(p,y)$, and call it the one-sided directional derivative of v at p with respect to y. (Clearly, $v'(p,\alpha y) = \alpha v'(p,y)$ if $\alpha > 0$.)

Note that if v is actually differentiable at p then $v'(p,y) = \langle \nabla v(p), y\rangle$ where $\nabla v(p)$ is the gradient of v at p and we denote the inner product in R^m by $<,>$. But $v'(p,y)$ exists in many cases in which $\nabla v(p)$ does not exist. Concave functions do possess directional derivatives, as summarized in the following theorem. The proofs can be found in Rockafellar (1970). We use the following notation: For $p \in K$, $y \in R^m$, set $I(p,y) = \{\lambda \in R^1 : p+\lambda y \in K\}$.

THEOREM 6: Let u *be a concave function defined in a convex subset* K *of* R^m. *If* $p \in \text{int } K$ *then* $u'(p,y)$ *exists and is finite for all* $y \in R^m$. *If* $p \in \partial K$ *and* $y \in K-\{p\}$, *then* $u'(p,y)$ *exists (but might be infinite). The (extended) real function* $u(\lambda;p,y)$, *defined in the real interval* $\lambda \in I(p,y)$ *by the equation*

$$u(\lambda;p,y) = u'(p+\lambda y,y) \tag{16}$$

is monotone non-increasing in $\lambda \in I(p,y)$.

Note that a converse for Theorem 6 can be stated in a number of ways (see Rockafellar (1970)). We will find the following formulation useful for proving the sufficiency part of Theorem 10.

PROPOSITION 7: Let $u(x)$ *be a continuous function of a single real variable defined in an interval* I. *If* $u'(x,1)$ *exists for all* $x \in I\backslash A$, *where* A *is an at most countable subset of* I, *and if* $u'(x,1)$ *is monotone non-increasing in* $I\backslash A$, *then* u *is concave in* I.

PROOF: Let $x,y \in I$, $x \lesssim y$, let $\lambda \in (0,1)$, and set

$$h(t) = \lambda[u(x+t(1-\lambda)(y-x)) - u(x)] + (1-\lambda)[u(\lambda x+(1-\lambda)y) - u(\lambda x+(1-\lambda)y + t\lambda(y-x))]$$

for $t \in [0,1]$. Then $h(0) = 0$. Moreover, h possesses right derivatives $h'(t,1)$ for all $t \in [0,1]$ outside an at most countable set, and

$$h'(t,1) = \lambda(1-\lambda)[u'(x + t(1-\lambda)(y-x),y-x) - u'(\lambda x+(1-\lambda)y+t\lambda(y-x),y-x)].$$

But $x + t(1-\lambda)(y-x) \leq \lambda x + (1-\lambda)y + t\lambda(y-x)$ for all $t \in [0,1]$. Hence $h'(t,1) \geq 0$ for all $t \in [0,1]$ outside an at most countable set. By Proposition 2 in Chapter 1, §2 of Bourbaki (1949), the continuous function $h(t)$ is non-decreasing. In particular, $h(1) \geq h(0) = 0$. But

$$h(1) = \lambda[u(\lambda x+(1-\lambda)y) - u(x)] + (1-\lambda)[u(\lambda x+(1-\lambda)y) - u(y)]$$
$$= u(\lambda x+(1-\lambda)y) - [\lambda u(x) + (1-\lambda)u(y)].$$

Hence $u(x)$ is concave in I.

Let $\succsim$ denote, as in Section 2, a continuous, complete and convex preference ordering defined on the convex set $K \subset R^m$. In other words, we assume that a continuous, *quasi-concave* (utility) function v is given on K, such that $p \succsim q$ if and only if $v(p) \geq v(q)$. (We also assume that if $p \succ q$ and $0 < t < 1$, then $tp + (1-t)q \succ q$.) In order to motivate the point of view taken in the present section, let us assume, at first, that v is continuously differentiable on K ($v \in C^1(K)$). Suppose, in addition, that there exists a differentiable increasing real function w defined on the range $v(K)$ of v, such that the composite function $u = w(v)$ is actually *concave*. Then the chain rule implies that

$$\nabla u(p) = w'(v(p))\nabla v(p) \tag{17}$$

and Theorem 6 implies that the function

$$u(\lambda;p,y) = w'(v(p + \lambda y)) \langle\nabla v(p + \lambda y),y\rangle$$

is monotone non-decreasing in λ, even though $\langle\nabla v(p + \lambda y),y\rangle$ is not. Thus, the non-monotonicity of $\langle\nabla v(p + \lambda y),y\rangle = v(\lambda;p,y)$ is "corrected" by multiplying the gradient of v by a scalar factor $w'(v(p))$ which depends on p only through $v(p)$ (and so is constant on each indifference surface of $\succsim$). Our general strategy is to try and find such correcting functions $w'(v(p))$. However, it is not possible to represent a general preference ordering $\succsim$ by a C^1 quasi-concave utility function. This will force us to use some auxiliary (quasi-concave) utility functions (defined in general only on sublayers), and to replace (17) by the corresponding formula for directional derivatives

$$u'(p,y) = w'(v(p),1)v'(p,y). \tag{18}$$

The properties of the directional derivatives $v'(p,y)$ of those auxiliary functions will determine the concavifiability of $\succsim$ (i.e., whether or not there exists a *concave* u such that $u(p) \geq u(q)$ if and only if $p \succsim q$).

Thus let $\succsim$ be an arbitrary convex preference ordering defined on K, and let $a,b \in \text{int } K$, $a \prec b$. We may assume that the ordering $\succsim$ is strictly monotone on the interval $[a,b]$, i.e., that $0 \leq t_1 < t_2 \leq 1$ if and only if $a + t_1(b-a) \prec a + t_2(b-a)$. Otherwise, let b' be the closest point to a in $[a,b] \cap \{x : x \succsim b\}$. Then the convexity of $\succsim$ implies that for no $c \in [a,b']$, $c \neq b'$, can it be that $c \succsim b'$. Hence the ordering $\succsim$ is strictly monotone on $[a,b']$. For, if for some $0 \leq t_1 < t_2 \leq 1$ we have $t_1 < t_2$ but $a + t_1(b'-a) \succ a + t_2(b'-a)$, then $t_2 < 1$ and the set $\{x : x \succ a + t_2(b'-a)\}$ contains $a + t_1(b'-a)$ and b', hence is not convex.

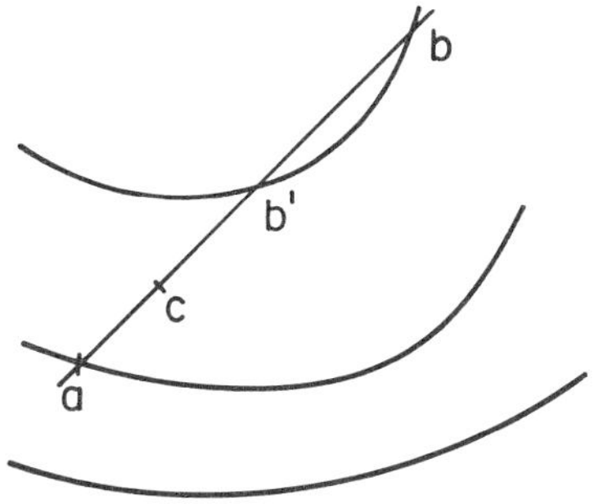

FIGURE 8. Replacing [a,b] by [a,b']

If $a + t_1(b'-a) \sim a + t_2(b'-a)$ then all the points $a + t(b'-a)$ are indifferent to each other for $t_1 \leq t \leq t_2$ and are strictly preferred by b'. Such a situation cannot occur for preference orderings satisfying the assumption made at the beginning of Section 2. Thus we may replace b by b', if necessary, to obtain the desired monotonicity. We claim that if $\succsim$ is also concavifiable, then the one-sided directional derivatives $u'(a + t(b-a), b-a)$ and $u'(a + t(b-a), a-b)$ are strictly positive and strictly negative, respectively, in the open interval $(0,1)$ if u is any concave utility function for $\succsim$. It suffices to prove that $u'(a + t(b-a), b-a) > 0$ for $0 < t < 1$. The non-negativity follows from the monotonicity of $\succsim$ in $[a,b]$. If we had $u'(a+t_0(b-a), b-a) = 0$ for some $0 < t_0 < 1$, then $u'(a+t(b-a), b-a) = 0$ for all $t_0 \leq t \leq 1$ by the monotonicity properties of $u'(p + \lambda y, y)$ and the non-negativity. But then $u(a + t(b-a))$ is constant on $t_0 \leq t \leq 1$ by Theorem 1 of Chapter 1, §2 in Bourbaki (1949), contradicting the strict monotonicity of $\succsim$ on $[a,b]$. In a similar fashion the negativity of $u'(a + t(b-a), a-b)$ is proved.

For any arbitrary interval $[a,b] \subset \text{int } K$ such that the preference ordering $\succsim$ is strictly monotone on $[a,b]$, we define a (usually non-concave) utility function $v_{a,b}(p)$ for $\succsim$ in the layer $\{p : a \precsim p \precsim b\}$ as follows: Let $a + t(b-a)$ be (the unique) point in $[a,b]$ which is indifferent to p, $0 \leq t \leq 1$. We set $v_{a,b}(p) = t$ (Note the similarity between this construction and that of the standard utility in Kannai, (1970)). We will sometimes supress the dependence on a,b and write $v(p)(= v_{a,b}(p))$. The function $v_{a,b}$ is well defined and continuous for all the preference orderings considered by us. (The functions $v_{a,b}$ were introduced independently by Crouzeix (1977).) The preceding discussion enables us to state the following necessary conditions for concavifiability.

PROPOSITION 8: Let $\succsim$ *be a convex preference ordering defined on an open convex set* K *in* R^m. *Let* $[a,b] \subset K$ *be an interval on which the ordering* $\succsim$ *is strictly monotone, and define* $v_{a,b}(p)$ *by*

$$a + v_{a,b}(p)(b-a) \sim p \quad \text{for} \quad a \precsim p \precsim b \tag{19}$$

Then a *<u>necessary</u>* *condition for concavifiability of* $\succsim$ *is that for all* $a',b' \in K$ *such that* $a \precsim a' \prec b' \precsim b$ *and* $\succsim$ *is strictly monotone on* $[a',b']$, *and for all* $0 < t < 1$, *the one-sided directional derivative* $v'_{a,b}(a'+t(b'-a'),b'-a')$ *exists and is different from zero and from* $+\infty$.

PROOF. Set $w_{a,b}(t) = u(a + t(b-a))$ for $0 < t < 1$ and some concave u representing $\succsim$. Then the directional derivatives of $w_{a,b}$ exist and are finite, $w'_{a,b}(t,+1) > 0$, $w'_{a,b}(t,-1) < 0$. Hence the function $w_{a,b}$ is invertible on $u((a,b))$ and the inverse function $w_{a,b}^{-1}$ possesses finite non-zero one-sided directional derivatives. Hence the function $v_{a,b}$, which can be written in the form $v_{a,b}(p) = w_{a,b}^{-1}(u(p))$, possesses finite one-sided directional derivatives everywhere. Moreover, if $\succsim$ is strictly monotone on $[a',b'] \subset K$, then

$$
\begin{aligned}
&v'_{a,b}(a'+t(b'-a'),b'-a')) \\
&= (w_{a,b}^{-1})'(u(a'+t(b'-a')),1)u'(a'+t(b'-a'),b'-a')
\end{aligned}
\tag{20}
$$

by the chain rule. But each of the terms on the right-hand side of (20) is positive. Hence $0 < v'_{a,b}(a' + t(b'-a'),b'-a')) < \infty$.

Note that Aumann (1975, Sec.9) speaks of the "de Finetti phenomenon" whenever $v'_{a,b}(a' + t(b'-a'),b'-a') = 0$ or $+\infty$.

The initiated reader will note the relationship between our Proposition 8, and the ideas involved with Proposition 3.2 in Kannai (1977). Using the auxiliary utility functions $v_{a,b}$ we can avoid the introduction of the (set-valued) functions adapted to the ordering $\succsim$ (in the sense of Kannai (1977, p.20)). It was remarked in Kannai (1977, pp. 28-29) that having a C^1 (quasi-concave) utility function v representing $\succsim$ eliminates the difficulties associated with the construction of an adapted function. In the approach presented here the functions $v_{a,b}$ and their directional derivatives play some of the roles which the C^1 function v and its gradient would have played, had they existed.

As a further step toward the goal of characterizing concavifiable preference orderings, we note that using the function $w_{a,b}$ defined in the proof of Proposition 8, we can write $u(p) = w_{a,b}(v_{a,b}(p))$ for all $a \precsim p \precsim b$. Let $a \prec p \prec q \prec b$. Then by an argument used earlier, the ordering $\succsim$ can be assumed to be strictly monotone at least in some subinterval $[p,r]$ of $[p,q]$, where $r \in (p,q)$. Hence $v'_{a,b}(p,q-p)$ is positive. Assume first that $v'_{a,b}(q,q-p)$ is also positive. Then the monotonicity of the directional derivatives of the concave function u implies the inequality

$$\frac{w'_{a,b}(v_{a,b}(p),1)}{w'_{a,b}(v_{a,b}(q),1)} \geq \frac{v'_{a,b}(q,q-p)}{v'_{a,b}(p,q-p)} \tag{21}$$

If $v'_{a,b}(q,q-p) \lesssim 0$ then (21) holds trivially. But the left-hand side of (21) depends only on the indifference classes of p and of q (besides its obvious dependence on a and b). Hence the following additional necessary condition for concavifiability (compare Proposition 3.3 in Kannai (1977)).

PROPOSITION 9: Let $\succsim$, a,b *and* $v_{a,b}$ *have the same meaning as in Proposition 8, and let all the assumptions and the conclusions of Proposition 8 hold for them. Let* $p,q \in K$, $a \prec p \prec q \prec b$. *Then a necessary condition for concavifiability is that*

$$\lambda(p,q) = \sup_{\substack{\bar{p} \sim p \\ \bar{q} \sim q}} \frac{v'_{a,b}(\bar{q},\bar{q}-\bar{p})}{v'_{a,b}(\bar{p},\bar{q}-\bar{p})} < \infty \tag{22}$$

The numbers $\lambda(p,q)$ serve as building blocks, from which the desired utility function will (if at all possible) be constructed. Thus, their role is not unlike that of the quantities $\alpha(p,q,r)$ introduced in Section 2 - without them being finite we cannot even try to construct a concave utility function representing the ordering $\succsim$, and expressions formed from them would do the required job.

We can now formulate the main result of this section.

THEOREM 10: Let $\succsim$ *be a convex preference ordering defined on an open convex set* $K \subset R^m$. *Suppose that for every* $a,b \in K$ *such that* $a \prec b$ *and such that the ordering* $\succsim$ *is strictly monotone in the interval* $[a,b]$, *the function* $v_{a,b}$ *satisfies the conditions of Propositions 8 and 9. Then* $\succsim$ *is concavifiable if and only if for every* a,b *as above and for every pair of points* $p,q \in K$, $a \prec p \prec q \prec b$,

$$\Lambda(p,q) = \sup \prod_{i=1}^{n-1} \lambda(p_i,p_{i+1}) < \infty, \tag{23}$$

where the supremum is taken over all finite sequences

$p = p_1 \prec p_2 \prec \cdots \prec p_n = q.$

PROOF. Using the same notation as in the proofs of Propositions 8 and 9, and given any sequence $a \prec p = p_1 \prec p_2 \prec \cdots \prec p_n = q \prec b$, we rewrite (21) as

$$\frac{w'_{a,b}(v_{a,b}(p_i),1)}{w'_{a,b}(v_{a,b}(p_{i+1}),1)} = \frac{w'_{a,b}(v_{a,b}(\bar{p}_i),1)}{w'_{a,b}(v_{a,b}(\bar{p}_{i+1}),1)} \tag{24}$$

$$\geqslant \frac{v'_{a,b}(\bar{p}_{i+1},\bar{p}_{i+1}-\bar{p}_i)}{v'_{a,b}(\bar{p}_i,\bar{p}_{i+1}-\bar{p}_i)}$$

for every $1 \leqslant i \leqslant n-1$ and $\bar{p}_i \sim p_i$, $\bar{p}_{i+1} \sim p_{i+1}$. Hence

$$\frac{w'_{a,b}(v_{a,b}(p_i),1)}{w'_{a,b}(v_{a,b}(p_{i+1}),1)} \geqslant \lambda(p_i,p_{i+1}) \tag{25}$$

for all $1 \leqslant i \leqslant n-1$. Multiplying (25) for $i = 1,\ldots,n-1$, we obtain the result

$$\frac{w'_{a,b}(v_{a,b}(p),1)}{w'_{a,b}(v_{a,b}(q),1)} = \frac{w'_{a,b}(v_{a,b}(p_1),1)}{w'_{a,b}(v_{a,b}(p_n),1)} \tag{26}$$

$$= \prod_{i=1}^{n-1} \frac{w'_{a,b}(v_{a,b}(p_i),1)}{w'_{a,b}(v_{a,b}(p_{i+1}),1)} \geqslant \prod_{i=1}^{n-1} \lambda(p_i,p_{i+1}),$$

where the left-hand side of (26) depends only on p and q and is independent of the particular sequence $p_2,\ldots,p_{n-1}$. The necessity of (23) follows, as well as the estimate

$$\frac{w'_{a,b}(v_{a,b}(p),1)}{w'_{a,b}(v_{a,b}(q),1)} \geq \Lambda(p,q). \qquad (27)$$

The (somewhat involved) proof of the sufficiency of (23) for the concavifiability of $\succsim$ will be given in Appendix C. Here we will only outline the main ideas. The first step consists of proving that Λ is multiplicative, i.e., that

$$\Lambda(p,r) = \Lambda(p,q)\Lambda(q,r) \qquad (28)$$

for all $a \prec p \prec q \prec r \precsim b$.

We then define a real-valued function $g_{a,b}$ on the open interval (0,1) by

$$g_{a,b}(t) = \begin{cases} \Lambda(a + t(b-a), \frac{a+b}{2}) & , \quad 0 < t < 1/2 \\ 1 & , \quad t = 1/2 \\ 1/\Lambda(\frac{a+b}{2}, a + t(b-a)) & , \quad 1/2 < t < 1 \end{cases} \qquad (29)$$

One then shows (among other things) that $g_{a,b}(t)$ is *monotone non-increasing* for $0 < t < 1$. Hence the function F(t), defined by

$$F(t) = \int_{\frac{1}{2}}^{t} g_{a,b}(s)ds \qquad (30)$$

is continuous, strictly increasing, and *concave* function of $t \in (0,1)$. Set

$$u_{a,b}(p) = F(v_{a,b}(p)) \qquad (31)$$

in the layer $\{p : a \prec p \prec b\}$. Then $u_{a,b}$ is shown to be a *least concave* utility function representing $\succsim$ in the layer. Using the uniqueness properties of least concave utility functions, one gets, as in Debreu (1976) and applying (30), a concave utility defined over all of K.

REMARK 2: The only modification one has to make in order to generalize Theorem 10 to a case where K is closed, is in adding conditions on the behavior of $\succsim$ near minimal points of $\succsim$. For, if $\succsim$ satisfies the conditions of Theorem 10 in int K, so that a (least) concave utility function u representing $\succsim$ does exist in int K, we may extend the definition of u to $\{p : p \in K$, p not minimal with respect to $\succsim\}$ by setting $u(p) = u(p')$, where $p' \sim p$, $p' \in$ int K. Then u is concave in its domain of definition, and we have to worry only over the behavior of u near a minimal point. Thus let $a \in K$ be a minimal point. (Then obviously $a \in \partial K$.) Let $b \in$ int K be arbitrary. We may choose b in such a way that $\succsim$ is strictly monotone on [a,b]. Then the function $v_{a,b}$ is well-defined in the layer $\{p : a \prec p \prec b\}$ and we may construct $g_{a,b}(t)$ for $t \in (0,1)$. We claim that $\succsim$ is concavifiable in K if and only if $g_{a,b}(t)$ is integrable in (0,1). We leave the details of the proof to the reader (compare also Theorem 3.5 in Kannai (1977)).

EXAMPLE 1: Let $\succsim$ be induced by the utility function $v = x_2/(1+x_1)$ in the unit square (this example has been treated in the earlier literature and also in Section 2, and is included here only for illustrative purposes). The closedness of K is not material here and in the next example - one could easily choose K open. Set $p_i = (0,i/n)$ for $0 \leqslant i \leqslant n/2$. (Note that $v = v_{a,b}$, where a = (0,0), b = (0,1).) The differentiability of v enables us to compute directional derivatives easily. Using (22) where (for each i separately) we choose $\bar{p} = (1,2i/n)$, $\bar{q} = p_{i+1}$, we get the result $\lambda(p_i,p_{i+1}) \geqslant 4$. Thus $\Lambda((0,0),(0,1/2)) = \infty$. Hence $\succsim$ is not concavifiable.

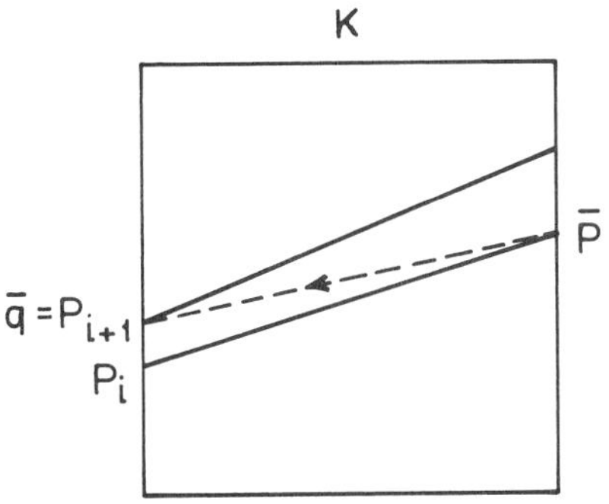

FIGURE 9. Example 1

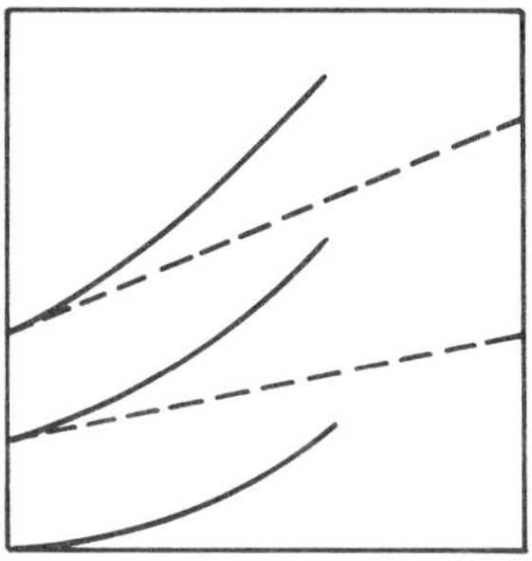

FIGURE 10. Example 2

EXAMPLE 2: Let $\succsim$ be induced by the utility function $(x_2 - x_1^4)/(1+x_1)$ in the unit square. (Note once more that $v = v_{a,b}$.) This preference ordering was introduced in Fenchel (1956), where it was proved that there exists no twice differentiable real function F of a real variable such that $F(v(x))$ is concave. We will show here, using two-points methods, that $\succsim$ is not concavifiable. Set $p_i^{(n)} = (0,(1+1/n)^{-i})$ for $n \leq i < \infty$ (here $p_{i+1}^{(n)} \prec p_i^{(n)}$).

For the computation of $\lambda(p_i^{(n)}, p_{i-1}^{(n)})$ set $\bar{q} = p_{i-1}^{(n)}$, $\bar{p} = (1/n,(1+1/n)^{-i+1}+1/n^4)$. Then $\bar{p} \sim p_i^{(n)}$, and it follows that

$$\lambda(p_i^{(n)}, p_{i-1}^{(n)}) \geq (1+\frac{1}{n}) \frac{(1+1/n)^{-i+1}n^3 - 1}{(1+1/n)^{-i}n^3 + 3} \tag{32}$$

if $i \leq n \log n$ then $(1 + 1/n)^{-i}n^2 \geq n$ and the right-hand side of (32) is greater than or equal to

$$(1 + \frac{1}{n}) \frac{(1 + 1/n)^{-i}n^3 + n - 1}{(1 + 1/n)^{-i}n^3 + 3} \geq 1 + \frac{1}{n} .$$

Setting now $p = (0,0)$, $q = (0,1/2)$ and noting that $\lambda(p_n^{(n)}, q) \geq 1$, $\lambda(p, p_{[n \log n]}^{(n)}) \geq 1$, we obtain

$$\Lambda(p,q) \geq \limsup \prod_{i=n+1}^{[n\log n]} \lambda(p_i^{(n)}, p_{i-1}^{(n)})$$

$$= \limsup \prod_{i=n+1}^{[n\log n]} (1+\frac{1}{n}) = \limsup n/e = \infty.$$

By Theorem 10, $\succsim$ is not concavifiable.

REMARK 3: The estimate

$$\lambda(p,q) \geq 1 \tag{33}$$

follows from (22) by choosing $\bar{p}$, $\bar{q}$ in (a,b). The condition (23) may be viewed as specifying the rate at which $\lambda(p,q)$ must tend to 1 as p and q get closer, so that the "infinite product" should converge. In Example 1, $\lambda(p,q)$ stays bounded away from 1, whereas in Example 2, $\lambda(p,q)$ does not approach 1 fast enough. (Compare also Theorem 15, equation (68).)

4. ONE POINT CONDITIONS

One-point conditions on the second derivatives of a twice-differentiable utility function v representing a convex preference relation $\succsim$, conditions which are equivalent to the existence of a twice-differentiable concave utility function u representing $\succsim$, have been derived by Fenchel in his lecture notes (1953) and paper (1956). Kannai (1977) characterized those twice differentiable v for which $\succsim$ is concavifiable, using Perron's integral. Here we will recall and reformulate Fenchel's conditions and will clarify them by means of several examples. We will also quote Kannai's Theorem. (In the next section we will characterize concavifiability using the Lebesgue integral.)

Following Fenchel, we consider a twice-differentiable function v defined on an open convex subset K of R^m. We are looking for a strictly monotone increasing real function F defined on v(K) such that u = F(v) is concave (compare (30) and (31)). Arguments similar to those playing roles in the proof of Proposition 8 lead us to Fenchel's first condition:

(I) The function v(x) has either no stationary values or has a global maximum as its only stationary value.

We look for a twice differentiable F. Denote the partial derivatives by subscripts (e.g., $v_i = \partial v/\partial x_i$). As is well-known, a twice-differentiable u is concave if and only if for every x, the matrix formed from the second-order derivatives of u is negative semi-definite. Thus we want the inequality

$$0 \geq \sum_{i,j=1}^{m} u_{ij}(x)\xi_i\xi_j = F''(v(x))\left(\sum_{j=1}^{m} v_j(x)\xi_j\right)^2 + \\ + F'(v(x)) \sum_{i,j=1}^{m} v_{ij}(x)\xi_i\xi_j \tag{34}$$

to hold for all real vectors $\xi \in R^m$. Choosing to consider at first those ξ which are parallel to the indifference surface at x, we obtain the well-known condition for quasi-concavity:

(II) For every $x \in K$ it is true that the quadratic form $\sum_{i,j=1}^{m} v_{ij}(x)\xi_i\xi_j$, when restricted to those ξ for which $\sum_{i=1}^{m} \xi_i v_i(x) = 0$, is negative semi-definite. (Note the I and II are the obvious differential formulation of the requirement that $\succ$ is a convex preference ordering satisfying the assumptions made at the beginning of Section 2.)

Let us introduce, for brevity, the following notation (to be used in the present section) in Section 5, and in Appendix E:

$$M(v,x,\xi) = \sum_{i,j=1}^{m} v_{i,j}(x)\xi_i\xi_j \tag{35}$$

The monotonicity of F along with the non-vanishing of grad u(x) for x non-maximal imply that $F'(v) > 0$ (unless x is maximal), and (34) may be rewritten as

$$\frac{F''(v(x))}{F'(v(x))} \leq - \frac{M(v,x,\xi)}{(\Sigma v_j(x)\xi_j)^2} \tag{36}$$

for all $\xi \in R^m$ for which $\Sigma \xi_j v_j(x) \neq 0$. Set

$$g(x,\xi) = \frac{M(v,x,\xi)}{\sum_{j=1}^{m} (v_j(x)\xi_j)^2} \tag{37}$$

for $x \in K$ and $\xi \in R^m$ such that $\sum_{j=1}^{m} \xi_j v_j(x) \neq 0$. The requirement that the left-hand side of (36) be finite implies that

$$g_1(x) = \sup_{\{\xi:\Sigma\xi_j v_j(x)\neq 0\}} g(x,\xi) < \infty \tag{38}$$

Fenchel evaluated the supremum in (38) and derived the condition

(III) Let $r(x)-1$ denote the rank of the form $M(v,x,\xi)$, restricted to the hyperplane $\sum_{j=1}^{m} \xi_j v_j(x) = 0$, where x is not a maximal point for v. Then the rank of the unrestricted form $M(v,x,\xi)$ cannot exceed $r(x)$.

Let S_i $(= S_i(x))$ denote the i-th elementary symmetric function of the eigenvalues of the quadratic form $M(v,x,\xi)$ defined for all $\xi \in R^m$, and denote by S_i^* the i-th elementary symmetric function of the eigenvalues of the same form restricted to $\{\xi: \sum_{j=1}^{m} \xi_j v_j(x) = 0\}$. With $r = r(x)$ we have from III that $S_r^* = S_{r+1} = 0$. Set $k = (\sum_{i=1}^{m} v_i(x)^2)^{1/2}$. Fenchel computed $g_1(x)$ and found that $g_1(x)$ is equal to $-S_r/(k^2 S_{r-1}^*)$. The left hand side of (36) depends on x only through $v(x)$. This observation leads to the following condition:

(IV) For each $t \in \text{int}[v(K)]$, we have

$$G(t) = \inf_{\{x:v(x)=t\}} [-g_1(x)] = \inf_{\{x:v(x)=t\}} \left(-\frac{S_r}{k^2 S_{r-1}^*}\right) > -\infty \tag{39}$$

The last of Fenchel's conditions is:

(V) The function $G(t)$ majorizes the logarithmic derivative of a function $H(t)$, where $H(t) > 0$ for $t \in \text{int}[v(K)]$ and H is differentiable for $t \in v(K)$.

Fenchel's fundamental theorem is:

THEOREM 11: The conditions I-V *are necessary and sufficient for the existence of a twice-differentiable function* $F(t)$, *such that* $F(v)$ *is a concave utility function for* $\succ$. *If* I-V *are fulfilled, then* F *can be given by* $F' = H$.

In Kannai (1977) the following was proved:

THEOREM 12: The preference relation $\succ$ *is concavifiable in* K *if and only if the function* G(t), *defined by (39), is Perron integrable in every compact subinterval of* int[v(K)]. *A least concave utility for* $\succ$ *is given by* $u(p) = F(v(p))$, *where* F *is defined in* int[v(K)] *by means of*

$$\int_{v(p_0)}^{v} \exp[(P)\int_{v(p_0)}^{t} G(s)ds]\,dt \qquad (40)$$

where $p_0 \in K$ *is an arbitrary non-maximal point.*

The Perron integral is defined in Kannai (1977, p.39); a full discussion can be found in Natanson (1960, Ch.XVI). (In the next section we will eliminate the need for considering Perron's integral.) Note that III and IV have to be satisfied for almost all t, as a Perron integrable function has to be finite almost everywhere.

In the remainder of the present section we will reinterpret and elucidate the meaning of the conditions III, IV and V.

An alternative way of stating conditions III and IV at a fixed non-maximal point $x \in K$ consists of choosing coordinates in such a way as to make the tangent hyperplane to the indifference surface $\{p: p \sim x\} = \{p: v(x) = v(p)\}$ equal to the hyperplane $\{x: x_m = 0\}$ and moving x to the origin. Thus $v_i = 0$ for $1 \leq i \leq m-1$ and $v_m = k$. The hyperplane $\Sigma v_i\xi_i = 0$ becomes $\xi_m = 0$. Choose now the coordinates $x_1,\ldots,x_{m-1}$ in such a way that the form $M(v,0,\xi)$ restricted to $\xi_m = 0$ is diagonal. Hence $v_{ij} = v_{ij}(0) = 0$ for $i \neq j$, $1 \leq i \leq m-1$, $1 \leq j \leq m-1$. By II $v_{ii} \leq 0$ for $1 \leq i \leq m-1$, and assume without loss of generality that $v_{ii} \neq 0$ for $1 \leq i \leq r-1 = r(x) - 1$, $v_{ii} = 0$ for $r \leq i \leq m-1$. The unrestricted matrix $V = (v_{ij})$, $1 \leq i, j \leq m$, is therefore of the form shown in Fig.11:

$$V = \begin{pmatrix} v_{1,1} & & & & & v_{m,1} \\ & \ddots & & \text{O} & & \\ & & v_{r-1,r-1} & & & \\ & \text{O} & & 0 & & v_{m,r} \\ & & & & \ddots & \\ v_{m,1} & & & v_{m,r} & & v_{m,m} \end{pmatrix}$$

FIGURE 11. The Hessian in special coordinates

where we have used the equality of mixed derivatives. If $v_{j,m} \neq 0$ for at least one j with $r \lesssim j \lesssim m-1$, then the j-th row and the m-th row are clearly linearly independent and adding any linear combination of the first r-1 rows would not change this, so that the rank of V is at least equal to r+1. Hence III can be stated as

III' $\quad v_{j,m} = 0 \quad$ for all $r \lesssim j \lesssim m-1$.

Expressed differently

III" If for any vector $e \in R^m$ with $\sum_{i=1}^{m} v_i e_i = 0$ the second derivative $v_{ee} = 0$, then $v_{en} = 0$ where n is the normal to the indifference surface at x.

If III' is satisfied then clearly S_r is equal to the determinant of the r x r matrix

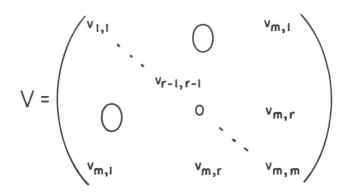

$$\begin{pmatrix} v_{1,1} & & 0 & v_{1,m} \\ & \ddots & & \\ 0 & & v_{r-1,r-1} & v_{r-1,m} \\ v_{1,m} & & v_{r-1,m} & v_{m,m} \end{pmatrix}$$

FIGURE 12. The non-zero portion of V

We can add $-v_{j,m}/v_{j,j}$ times the j-th column to the last column. Doing this for all $1 \leq j \leq r-1$ we find that

$$S_r = (v_{m,m} - \sum_{j=1}^{r-1} \frac{v_{j,m}^2}{v_{j,j}}) \prod_{i=1}^{r-1} v_{ii} = (v_{m,m} - \sum_{j=1}^{r-1} \frac{v_{j,m}^2}{v_{j,j}}) S^*_{r-1} \, . \tag{41}$$

Hence IV may be restated in the form

IV' For each $t \in (\alpha,\beta)$ we have

$$G(t) = \inf_{v(x)=t} \frac{1}{k^2} (-v_{m,m} + \sum_{j=1}^{r-1} \frac{v_{j,m}^2}{v_{j,j}}) > -\infty. \tag{42}$$

Let us note in passing that III' and IV' can also be deduced directly from (36) when the special coordinate system is used.

We can further reinterpret III by using locally the implicit function theorem so as to obtain a twice differentiable function

$$x_m = w(x_1,\ldots,x_{m-1},t) \tag{43}$$

such that

$$v(x_1,\ldots,x_{m-1},w(x_1,\ldots,x_{m-1},t)) \equiv t \tag{44}$$

For each fixed t, the equation (43) represents an indifference surface. (In some applications the ordering $\succsim$ is given by an equation (43) rather than by a utility.) From the implicit function theorem we obtain the formulas (also obtainable formally by differentiating (44) once)

$$v_m = \frac{1}{w_t} \, , \quad v_i = -\frac{w_i}{w_t} \tag{45}$$

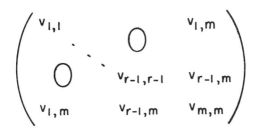

where here and in the following discussion $1 \leq i \leq m-1$ and $w_t = \frac{\partial w}{\partial t}$, etc. Differentiating (45) and using it repeatedly, we get the additional formulas

$$v_{m,m} = -\frac{w_{tt}}{w_t^3} \tag{46}$$

$$v_{i,m} = -\frac{w_{it}}{w_t^2} + \frac{w_{tt}w_i}{w_t^3} \tag{47}$$

$$v_{ii} = -\frac{w_{ii}}{w_t} + \frac{2\,w_{it}w_i}{w_t^2} - \frac{w_{tt}w_i^2}{w_t^3} \tag{48}$$

At the point $x_1 = \ldots = x_{n-1} = 0$, $t = v(0)$, we have $w_i = 0$ and III' can therefore be restated as

III''' If $w_{ii} = 0$ then $w_{it} = 0$.

Noting that w_{ii} is just the principal curvature of the indifference surface through the origin in the direction of the coordinate x_i and $w_{it} = \frac{\partial}{\partial t} w_i$ expresses the rate of change of the "direction" of the indifference surfaces as t changes, we see that the geometric meaning of III is that if the indifference surface is flat in a certain direction, then the component of its tangent at that direction must be stationary as a function of the utility. (It is this condition which is violated in examples 1 and 2, see Fig.9 and 10.) The preceding discussion also shows that IV' (or IV) and V concern the boundedness of the ratio of w_{ti}^2 to w_{ii}.

The conditions I-III are characterized by Fenchel (1956) as local conditions, whereas IV and V are listed as global ones. This means that those conditions may be violated globally even if locally everything is all right. But they may be violated locally too, as we will show presently. We note that if m=2 and $r(x) = 1$, then an easy computation shows that

$$\frac{-S_r}{k^2 S^*_{r-1}} = \frac{v_{12}^2 - v_{11}v_{22}}{v_{11}v_2^2 - 2v_{12}v_1v_2 + v_{22}v_1^2} \left(= \frac{1}{w_t}\left(w_{tt} - \frac{w_{1t}^2}{w_{11}}\right)\right) \tag{49}$$

the last equality holding at least locally.

EXAMPLE 3: Let $c(x)$ be the even piecewise linear function defined for $1 \geq x \geq 0$ by $c(x) = 0$ if $x = 0$ or $x = \frac{1}{n}$ and $c(x) = x$ for $x = \frac{1}{2}\left(\frac{1}{n} + \frac{1}{n+1}\right)$, $c(x)$ being linear in the intervals $\left[\frac{1}{n+1}, \frac{1}{2}\left(\frac{1}{n}+\frac{1}{n+1}\right)\right]$ and $\left[\frac{1}{2}\left(\frac{1}{n}+\frac{1}{n+1}\right), \frac{1}{n}\right]$. We define $c(x)$ by the equation $c(x) = c(-x)$ for $-1 \leq x \leq 0$. We will use the fact that there exists a positive ε such that

$$\int_0^x c(t)dt \geq \varepsilon x^2 \tag{50}$$

for $0 \leq x \leq 1$. We define the even functions $a(x)$ and $b(x)$ in $[-1,1]$ by setting $a(0) = a'(0) = b(0) = b'(0) = 0$, $b''(x) = c(x) + x^4$, $a''(x) = (c(x)+x^4)^2 = (b''(x))^2$. We set

$$x_2 = w(x_1,t) = a(x_1) + \frac{t^2x_1^2}{2} + tb(x_1)+t \tag{51}$$

Then $w_t(0,0) \neq 0$, so that by the implicit function theorem a C^2 utility $v(x_1,x_2)$ satisfying (44) can be constructed in a neighborhood of the origin. A straightforward computation shows that

$$w_{11} = a''(x_1) + t^2+tb''(x_1) \geq \frac{a''+t^2}{2} \geq 0$$

so that v induces a convex preference ordering, satisfying conditions I and II. On the indifference curve $t = 0$ we have $w_1 = a'(x_1)$, $w_t = 1+b(x_1)$, $w_{11} = a''(x_1)$, $w_{1t} = b'(x_1)$, and $w_{tt} = x_1^2$. Substitution in (49) shows that in this case

$$-\frac{S_r}{k^2S^*_{r-1}} = \frac{a''(x_1)x_1^2-b'^2}{(1+b)a''(x_1)} = \frac{x_1^2}{1+b} - \frac{b'^2}{(1+b)a''} \tag{52}$$

The first term on the right hand side of (52) tends to 0 as $x_1 \to 0$. It follows from (50) that $(b'(x_1))^2 \geq \varepsilon^2x_1^4$ for $0 < x_1 < 1$, but for $x_1 = \frac{1}{n}$ we have $a''(x_1) = x_1^8$, which makes (52) unbounded as $x_1 \to 0$ so that IV is *not satisfied locally* near the origin. Note that III is satisfied since the only point where an indifference curve has zero curvature is the origin and $v_{12}(0,0) = 0$.

EXAMPLE 4: Here we set $w(x_1,t) = t^4x_1^2+t^2x_1+t$. The implicit function theorem implies as above the existence of a C^2 function v, defined in a neighborhood of the origin in R^2, such that the equation (44) holds identically near the origin. The fact $w_{11} = 2t^4 \geqslant 0$ shows that the preference relation induced by $\succcurlyeq$ is convex. On the line $x_1 = 0$ we have the formulas $w_1 = t^2$, $w_t = 1$, $w_{11} = 2t^4$, $w_{1t} = 2t$, $w_{tt} = 0$. Substituting these values in (49) we see that

$$G(t) = \inf_{v(x)=t} \frac{-S_r}{k^2 S^*_{r-1}} \leqslant \frac{-S_r}{k^2 S^*_{r-1}} (x_1 = 0) = -\frac{2}{t^2}.$$

If condition V were satisfied for $t \to 0$ then a differentiable function $H(t)$ would exist with $H'(t)/H(t) \leqslant G(t)$ or $\ln H(t_1) - \ln H(t_2) < 2/t_1 - 2/t_2 \to -\infty$ as $t_2 \to 0$, a contradiction. Note that here condition III is satisfied (along the indifference line $x_2 = t = 0$) and condition IV is also satisfied for each t. The reader familiar with Perron integrals will be able to demonstrate (via Theorem 12) that the preference orderings defined in the examples are not concavifiable.

5. RELATIONS BETWEEN ONE, TWO AND THREE-POINT CONDITIONS

In the present section we shall prove that if $\succcurlyeq$ is derived from a twice-differentiable (quasi-concave) utility function v satisfying conditions I-V of Section 4, then the utility F(v) given by Theorem 11 is least concave if H is constructed in the most "natural" way, and we will characterize concavifiability. In particular we will show how the Lebesgue integral can replace Perron's integral in Theorem 12, thus improving upon the results of Kannai (1977). (Similar results were obtained by Crouzeix (1977).) We will also discuss the relations between the quantities $\alpha(p,q,r)$ (encountered in the three-points conditions), $\lambda(p,q)$ (encountered in the two-points conditions), and $G(t)$ (appearing in the discussion about one-point conditions), and interpret the right-hand sides of (14) and (23) as "Riemann sums" for integrals yielding least concave utility functions (compare (67)).

Thus let us assume, to begin with, that the preference ordering $\succsim$ is induced by a twice-differentiable utility function satisfying the Fenchel conditions I and II (Section 4), in an open convex set $K \subset R^m$. Let $a,b \in K$, $a \prec b$, and without loss of generality assume (as explained in Section 3) that v is strictly monotone increasing in the interval $[a,b]$. Then the real function mapping $t \in (0,1)$ into $v(a + t(b-a))$ is differentiable with a *positive* derivative (by I), hence is invertible with a twice-differentiable inverse $t(v)$, and the auxiliary function $v_{a,b}(p)$ defined by (19) in the layer $\{p : a \prec p \prec b\}$ is expressible as $v_{a,b}(p) = t(v(p))$. Hence $v_{a,b}$ is twice differentiable in the layer and satisfies there the conditions I and II. It is clear that the conditions III, IV and V hold for v in the layer if and only if they hold for $v_{a,b}$ there.

In Section 3 we have found out that $\succsim$ is concavifiable if and only if there exists a positive function defined on $(0,1)$ and denoted by $w'_{a,b}(t,1)$ such that for all $p,q \in K$, $a \prec p \prec q \prec b$ satisfying $v'_{a,b}(p+t(q-p),q-p) > 0$ for $0 \leq t \leq 1$, the inequality (21) holds. Taking the logarithms of (21) we obtain the inequality

$$\begin{aligned} \ell n[w'_{a,b}(v_{a,b}(p),1)] - \ell n[w'_{a,b}(v_{a,b}(q),1)] \geq \\ \ell n[v'_{a,b}(q,q-p)] - \ell n[v'_{a,b}(p,q-p)], \end{aligned} \tag{53}$$

and the definition (23) takes the form

$$\ell n\, \Lambda(p,q) = \sup \sum_{i=1}^{n-1} \ell n\, \lambda(p_i,p_{i+1}), \tag{54}$$

where the supremum is taken over all finite sequences $p = p_1 \prec p_2 \prec \cdots \prec p_n = q$. We know that the function $g_{a,b}(t)$, defined by (29) by means of Λ, satisfies (at least almost everywhere) the inequalities (21) (when we set $w'_{a,b}(t,1) = g_{a,b}(t)$). The expression (54) reminds us of an integral. Our main technical step in making the connection between the constructions of Section 3 and integration of expressions formed from $v_{a,b}$ is the following proposition.

PROPOSITION 13: Let I *be a real open interval, let* A *be an arbitrary set, and let* ϕ_α *be a differentiable function defined on a sub-interval* $I_\alpha \subset I$ *for every* $\alpha \in A$. *Assume that for a certain* $\alpha_0 \in A$, $I_{\alpha_0} = I$, *and* ϕ_{α_0} *is monotone non-decreasing. Then the following are equivalent:*

(i) The function

$$K(t) = \sup_{\alpha \in A} \phi'_\alpha(t) \tag{55}$$

is Lebesgue integrable in every compact subinterval of I.

(ii) There exists a real function $k(t)$ *defined on* I *such that for every* $\alpha \in A$ *and every* $t_1, t_2 \in I_\alpha$, $t_1 < t_2$, *the inequality*

$$k(t_2) - k(t_1) \geq \phi_\alpha(t_2) - \phi_\alpha(t_1) \tag{56}$$

holds.

(iii) For every $c,d \in I$, $c < d$,

$$L(c,d) = \sup \sum_{i=1}^{n-1} [\phi_{\alpha_i}(t_{i+1}) - \phi_{\alpha_i}(t_i)] < \infty, \tag{57}$$

where the supremum is taken over all finite sequences

$c = t_1 < t_2 < \ldots < t_n = d$, and $t_i, t_{i+1} \in I_{\alpha_i}$ for $1 \leq i \leq n-1$.

Moreover, if one (hence any) of the statements (i)-(iii) holds, then $k(t)$ can be chosen as an indefinite integral of $K(t)$, and

$$L(c,d) = k(d) - k(c) = \int_c^d K(t)\,dt \,. \tag{58}$$

The proof of Proposition 13 (which has nothing to do with convexity theory) is given in Appendix D.

We want to utilize the obvious similarities between (53), (54) and (56), (57), respectively. For this purpose we set $A = \{(x,\xi) : x \in K,\ \xi \in R^m, a \prec x \prec b,\ \langle \text{grad } v_{a,b}(x),\xi\rangle > 0\}$. Then the function $t = v_{a,b}(x + s\xi)$ is strictly increasing in some interval $s_1 < s < s_2$, where $s_1 < 0$, $s_2 > 0$, and has a positive derivative there. The convexity of $\succsim$ along with condition I ensure the existence of a maximal interval having this property (compare Proposition 5.2 in Kannai (1977)). Let (s_1,s_2) be this maximal interval, and set $I_{(x,\xi)} = (v_{a,b}(x + s_1\xi), v_{a,b}(x + s_2\xi)) \subset (0,1) = I$. Set

$$\phi_{(x,\xi)}(t) = \ell n[\, v'_{a,b}(x + s(t)\xi,\xi)] \tag{59}$$

where $s(t)$ is the inverse function of $v_{a,b}(x + s\xi)$, defined for $t \in I_{(x,\xi)}$. Choosing

$$k(t) = -\ell n[\, w'_{a,b}(t,1)]\,, \tag{60}$$

we assert that (53) is equivalent to (56). In fact, if $p,q \in K$, $a \prec p \prec q \prec b$ satisfy $v'_{a,b}(p + s(q-p),q-p) > 0$ for all $0 \le s \le 1$, then setting $x = p$, $\xi = q-p$ and using the continuity of $v'_{a,b}(p + t(q-p),q-p)$ (as a function of t) we find that $s_1 < 0$, $s_2 > 1$. Hence $v_{a,b}(p)$, $v_{a,b}(q) \in I_{p,q-p}$. Conversely, every inequality of the type (56) is clearly of the form (53) (with the choices (59) and (60)). The equivalence of (54) and (57) is clear, if one notes the definition (22) (compare also Theorem 5.3 in Kannai (1977)). To get the meaning of (55), let us differentiate $\phi_{x,\xi}(t)$ with respect to t:

$$\begin{aligned}(d/dt)\ell n[\, v'_{a,b}(x + s(t)\xi,\xi] &= (ds/dt)(d/ds)\ell n[\, v'_{a,b}(x + s\xi,\xi)] \\ &= (ds/dt)[\, v'_{a,b}(x + s\xi,\xi)]^{-1}(d/ds)v'_{a,b}(x + s\xi,\xi).\end{aligned} \tag{61}$$

But

$$dt/ds = dv_{a,b}(x + s\xi)/ds = \langle \text{grad } v_{a,b}(x + s\xi),\xi\rangle = v'_{a,b}(x + s\xi,\xi)$$

so that

$$ds/dt = [\, v'_{a,b}(x + s\xi,\xi)]^{-1} = \langle \text{grad } v_{a,b}(x + s\xi),\xi\rangle^{-1} \tag{62}$$

Also

$$(d/ds)v'_{a,b}(x + s\xi,\xi) = d/ds\langle \text{grad } v_{a,b}(x + s\xi),\xi\rangle = \\ = M(v_{a,b},x+s\xi,\xi) \tag{63}$$

(See (35)). Substituting (62) and (63) in (59) we obtain the important identity

$$\phi'_{x,\xi}(t) = \langle \text{grad } v_{a,b}(x + s\xi),\xi\rangle^{-2} M(v_{a,b},x+s\xi,\xi) \\ = g(x + s(t)\xi,\xi), \tag{64}$$

where we use the notation introduced in Section 4 (and set $v = v_{a,b}$). Noting that $g_1(x,\xi)$ is an even function of ξ, we have from (38) that $g_1(x) = \sup_{\{\xi:(x,\xi) \in A\}} g(x,\xi)$. It follows from (39) that

$$G(t) = -K(t) \tag{65}$$

Setting $\alpha_0 = ((a+b)/2,\ b-a)$ we find out that $\phi'_{\alpha_0}(t) = 0$ for all $0 < t < 1$. Hence all the conditions of Proposition 13 are satisfied, and we have proved

THEOREM 14: Let $a,b \in K$ *be such that the preference ordering* $\succsim$ *is strictly monotone on the interval* $[a,b]$, *and such that Fenchel's conditions* I *and* II *are satisfied for* $v_{a,b}$ *in the layer* $\{p : a \prec p \prec b\}$. *Then the preference ordering* $\succsim$ *is concavifiable in the layer if and only if the function* $G(t)$ *defined by* (39) *is Lebesgue integrable in every compact subinterval of* $(0,1)$. *A least concave utility function representing* $\succsim$ *in the layer is given by* $u(p) = F(v_{a,b}(p))$, *where* F *is represented by*

$$F(t) = \int_{1/2}^{t} \exp\,[-k(s)]ds \tag{66}$$

or, equivalently, by

$$u(p) = \int_{1/2}^{v(p)} \exp\,[\int_{1/2}^{t} G(s)ds]dt \tag{67}$$

Note that the representation (66) follows from (65), (54), (58) combined with (29) and (30). Note also that the constant 1/2 could be replaced in (66) and in (67) by arbitrary numbers contained in (0,1) (compare (40)). The crucial difference between Theorems 14 and 12 lies in the constant sign of $G(t)$ (or $(K(t))$, a fact which enables us to pass from Perron's to Lebesgue's integral (compare Theorem 2, §5, in Chapter XVI of Natanson (1961)). (If we drop - in Proposition 13 - the assumption about α_0, then the conclusions still hold, with the modifications that K is Perron integrable and the integral in (58) is Perron's integral.)

In Kannai (1977) the expression (14) was interpreted as a Riemann sum for integrals such as (67) (see the discussion on pp. 44-49 in Kannai (1977)). Here we will establish a link between the constructions of Section 2 (involving three points) and those involving $G(t)$ under the additional assumptions that $v \in C^2(K)$ and that all indifference surfaces have a non-vanishing Gaussian curvature and are compact. In this special case (where the concavifiability is well-known) many arguments can be simplified.

THEOREM 15: Let $\succ$ *be a preference ordering induced by a twice-continuously differentiable quasi-concave utility function* v *satisfying Fenchel's conditions* I *and* II. *Assume also that for all* $y \in K$, *the surface* $\{x : x \sim y\}$ *has a non-vanishing Gaussian curvature and is compact. Let* $a \prec b \in K$ *be such that the ordering* $\succ$ *is strictly increasing in the interval* $[a,b]$, *and let* $v_{a,b}$ *be given in the layer* $p : a \prec p \prec b$ *by* (19). *Then for every* $p \in (a,b)$,

$$\lim_{h \to 0_+} \frac{\lambda(p,p + h(b-a)) - 1}{h} = -G(v_{a,b}(p)) \tag{68}$$

and

$$\lim_{h \to 0_+} \frac{\alpha(p-h(b-a),p,p + h(b-a)) - 1}{h} = -G(v_{a,b}(p)). \tag{69}$$

The (somewhat computational) proof of Theorem 15 is given in Appendix E.

We can now interpret (14) as a Riemann sum for integrals like (40) (in the compact case, but this disagreement with the openness assumption of Section 4 does not really matter). Let a be minimal, b maximal with respect to $\succsim$ in K, such that $\succsim$ is monotone in [a,b]. Let $p \in K$ be neither maximal nor minimal, and such that $v_{a,b}(p)$ is rational (this last assumption is made only in order to shorten the computations). We will drop, in the sequel, the subindices a, b and will write $v = v_{a,b}$. For n divisible by the denominator of v(p), set

$$p_i = a + i(b-a)/n \qquad 0 \leq 1 \leq n . \tag{70}$$

Then $j = nv(p)$ and for n-k not too small, (69) implies that

$$\begin{aligned}\alpha_k\cdots\alpha_{n-1} &\approx [1 - G(k/n)/n][1 - G(k+1)/n)/n]\ldots[1 - G((n-1)/n)] \\ &= \exp\{-[G(k/n) + G((k+1)/n) + \ldots + G((n-1)/n)]/n\} \\ &\cdot[1-G(k/n)/n]\exp[G(k/n)/n]\ldots[1-G((n-1)/n)]\exp[G((n-1)/n)/n].\end{aligned}$$

If G is bounded, then

$$\begin{aligned}&[1-G(k/n)/n]\exp[G(k/n)/n]\ldots[1-G((n-1)/n)]\exp[G((n-1)/n)/n] = \\ &= [1 + 0(\tfrac{1}{n^2})][1 + 0(\tfrac{1}{n^2})]\ldots[1 + 0(\tfrac{1}{n^2})] = 1 + 0(\tfrac{1}{n}).\end{aligned}$$

Hence

$$\alpha_k\cdots\alpha_{n-1} \to \exp[-\int_{k/n}^{1} G(s)ds]. \tag{71}$$

It follows that the fraction appearing in (14) can be written (approximately) as

$$\frac{\exp[-\int_{j/n}^{1}G(s)ds]+\exp[-\int_{(j-1)/n}^{1}G(s)ds]+\ldots+\exp[-\int_{1/n}^{1}G(s)ds]}{1 + \exp[-\int_{(n-1)/n}^{1}G(s)ds]+\ldots+\exp[-\int_{1/n}^{1}G(s)ds]}.$$

Multiplying the numerator and the denominator by 1/n, we get a Riemann sum for the integral

$$[\int_0^{v(p)}\exp[-\int_t^1 G(s)ds]dt \ / \int_0^1\exp[-\int_t^1 G(s)ds]dt. \tag{72}$$

The integral in (72) is normalized so as to vanish if $v(p) = 0$ and to be equal to 1 if $v(p) = 1$, but otherwise it coincides with (67) (or (40)).

Recalling the fact that $\beta(p_1,p_2,p_3)$ as defined in (7) is equal to $\alpha(p_1,p_2,p_3)$, we have thus seen that the suggestions for constructing a concave utility, put forward by Fenchel (1953) for the general case (see pp.124-126), actually lead to the same functions as those constructed fully in the smooth case (pp.127-137). While we have outlined here the argument only for the regular (non-vanishing curvature) case, this equivalence was demonstrated in fuller generality in Kannai (1977).

6. IMPOSSIBILITY OF CONSTRUCTING CONCAVE UTILITY FUNCTIONS WHICH DEPEND CONTINUOUSLY ON THE PREFERENCES

There is a natural topology on the space P of the continuous preference orderings defined on a subset K of R^m. This is the closed convergence topology for the graphs, discussed at length by Hildenbrand (1974). [In many cases this topology coincides with the one introduced by Kannai (1970).] Let $P_{conc} \subset P$ be the subspace containing all concavifiable preference relations defined on K (K is now assumed to be convex). A natural problem is to find out whether there exists a continuous $F : K \times P_{conc} \to R$ (continuous in *both* variables!) such that $F(x,\succsim)$ is a concave utility function representing $\succsim$. In Kannai (1977) an example was exhibited showing that such an F cannot exist, even if we take K to be the unit square in R^2 and we want F to be defined only on $K \times P_1$ where P_1 is the set of monotone concavifiable preference orderings defined on K. Here we shall show that $F : K \times P_{conc} \to R$ such that $F(x,\succsim)$ is a concave utility function representing $\succsim$ does not exist even if K is a one-dimensional compact interval (compare the last example of Section 2). In fact, the following theorem is true:

THEOREM 16: Let K *be a closed interval in* R. *There exists a sequence* $\{\succsim_n\}$, $n = 1,2,\ldots$, *of concavifiable preference orderings converging to a concavifiable preference ordering* $\succsim$ *defined on* K, *such that if* $f_n(x)$ *is a concave utility function representing* $\succsim$ *on* K *for* $n = 1,2,\ldots,$ *then* $f_n(x)$ *does not converge to a utility function representing* $\succsim$ *on* K.

PROOF. Let $K = [-1,3]$, and set

$$v(x;\varepsilon) = \begin{cases} x & -1 \lesssim x \lesssim 0 \\ x^2 + 2\varepsilon x & 0 \lesssim x \lesssim \varepsilon \\ [(1-3\varepsilon^2)/(1-\varepsilon)](x-\varepsilon) + 3\varepsilon^2 & \varepsilon \lesssim x \lesssim 1 \\ 2 - x & 1 \lesssim x \lesssim 3 \end{cases} \tag{73}$$

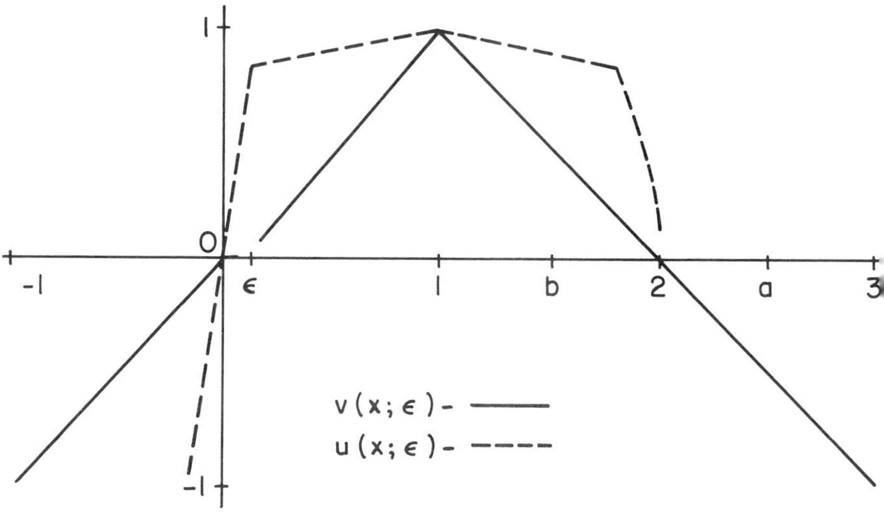

FIGURE 13. Concave and non-concave utility functions for $\succsim_{1/\epsilon}$

for $\varepsilon \geq 0$, sufficiently small. Setting $a = 2.5$, $b = 1.5$, we find out that $v(x;\varepsilon) = v_{a,b}(x;\varepsilon)$ (as defined in (19)). A short computation shows that for p,q in the layer $\{x : a \prec x \prec b\}$ such that $p < q < 1$, we have

$$\lambda_\varepsilon(p,q) = \begin{cases} (1 - 3\varepsilon^2)/(1-\varepsilon) & p < 0,\ \varepsilon < q \\ (q+\varepsilon)/(p+\varepsilon) & 0 < p < q < \varepsilon \\ (1 - 3\varepsilon^2)/[(1-\varepsilon)(2p + 2\varepsilon)] & 0 < p < \varepsilon < q, \end{cases} \tag{74}$$

and $\lambda_\varepsilon(p,q) = 1$ for other pairs $p \prec q$. Hence the corresponding multiplicative function $\Lambda_\varepsilon(p,q)$ is given as $\Lambda_\varepsilon(p,q) = M_\varepsilon(q)/M_\varepsilon(p)$, where

$$M_\varepsilon(p) = \begin{cases} (1 - 3\varepsilon^2)/(1-\varepsilon) & \varepsilon < p \\ 2(p+\varepsilon) & 0 < p < \varepsilon \\ 2\varepsilon & p < 0 \end{cases} \tag{75}$$

if $\varepsilon > 0$. If $u(x;\varepsilon)$ is an arbitrary concave utility function representing the preference ordering induced by $v(x;\varepsilon)$ then $u(x;\varepsilon)$ can be represented in the layer $\{x : a \prec x \prec b\}$ as $u = w(v_{a,b})$, where w satisfies the estimate (27). Application of (75) implies that if $t_1 < 1/2$, $1/2 + 3\varepsilon^2 < t_2$, $t_1, t_2 \in (0,1)$, then

$$\frac{w'(t_1,1)}{w'(t_2,1)} \geqslant \frac{(1 - 3\varepsilon^2)}{2\varepsilon(1-\varepsilon)} . \tag{76}$$

Proposition 2 in Chapter 1, §2 of Bourbaki (1949) (or integration) implies that

$$w(1) - w(3/4) \leqslant [2\varepsilon/(1 - 3\varepsilon^2)][w(1/2) - w(1/4)]$$

or, equivalently, that

$$u(1.5;\varepsilon) - u(1.75;\varepsilon) \leqslant [2\varepsilon/(1-3\varepsilon^2)][u(2;\varepsilon) - u(2.25;\varepsilon)] \tag{77}$$

for all concave utility functions representing the preference ordering induced by $v(x;\varepsilon)$. Let $\succcurlyeq_n$ be the ordering induced by $v(x;1/n)$ for n sufficiently large. If the sequence $f_n(x)$ of concave utility functions representing $\succcurlyeq_n$ converges to a utility function $f(x)$ representing the preference $\succcurlyeq$ induced by $v(x;0)$ (by construction $\succcurlyeq_n \to \succ$), then (77) implies that

$$0 \leqslant f_n(1.5) - f_n(1.75) \leqslant (3/n)[f_n(2) - f_n(2.25)] \to 0 \tag{78}$$

since $f_n(2) - f_n(2.25) \to f(2) - f(2.25)$. Hence $f(1.75) = f(1.5)$, so that f cannot serve as a utility function representing the ordering $\succcurlyeq$ for which the points 1.75 and 1.5 are not equivalent.

While the preference orderings $\succsim_n$ used in the proof of Theorem 16 cannot be represented by means of smooth utility functions, one could easily modify the example and have smooth concavifiable preferences $\succsim_n$ converge to a concavifiable $\succsim$, and the conclusion of Theorem 16 would still hold. Not even the choice of least concave utilities would work. If, however, we have $\Lambda_n(p,q) \to \Lambda(p,q)$, then the least concave utilities would converge. This suggests that a stronger topology than the usual closed convergence one is natural on P_{conc}. Note also that if we denote by $P_{conc,reg}$ the space of C^2 concavifiable preference orderings on the compact set K for which the Gaussian curvature of the indifference surfaces $\{x : x \sim y\}$ never vanishes, and we endow $P_{conc,reg}$ with the C^2 topology, then $F : K \times P_{conc,reg} \to R$, defined by letting $F(x,\succsim)$ be the least concave utility function representing $\succsim$ such that $F(x_0,\succsim) = 0$, $F(x_1,\succsim) = 1$, where x_0 and x_1 are minimal and maximal elements (respectively) with respect to $\succsim$, is continuous in both variables. In that case the functions G(t) are all smooth, as seen from III''' and IV' in Section 4. Proposition 13 and Theorem 14 imply then the continuity of F.

It is well-known [Kannai (1974) and Mas-Collel (1974); this was proved even earlier in a little-known paper by Mantel (1969)], that monotone concave preference orderings can be approximated by elements of P_1. It is also well-known that the limit of a sequence of concave functions is concave. Hence it is impossible to construct a continuous $F : K \times P \to K$ such that $F(x,\succsim)$ is a utility for each $\succsim \in P$ and such that $F(x,\succsim)$ is concave if $\succsim \in P_{conc}$. Here P is the space of all convex monotone orderings endowed with the closed graph topology. If we drop the concavity requirement, then a continuous F can be defined even on larger spaces P, as shown e.g. in Kannai (1970).

7. CARDINALITY OF LEAST CONCAVE UTILITY FUNCTIONS

Let $\succsim$ be a concavifiable preference ordering defined on $K \subset R^m$. The family F of least concave utility functions representing $\succsim$ in K has the following two properties: (i) If $u \in F$, then $ku \in F$ for all positive constants k; (ii) If $u \in F$, then $u+c \in F$ for all real constants c. Thus F satisfies the *formal* properties that a family of cardinal utility functions should satisfy.

As observed by Debreu (1976), the cardinal utility so obtained has an interpretation in the context of decision making under uncertainty. If we identify an element $p \in K$ with the probability having $\{p\}$ as support, and assume that a risk averse decision maker has a preference ordering on π - the set of probabilities on K, and if certain axioms are satisfied, then the decision maker has a von Neumann-Morgenstern utility v (defined on π) whose restriction to K is a concave function representing (the restriction of) $\succsim$ in K. Let u be a least concave utility function representing (the restriction of) $\succsim$ in K. Then $v = f(u)$ where f is concave: $v(K) \to R^1$, and we have separated the preferences of the decision maker for the commodity vectors in K represented by u, from his attitude towards risk as described by f (for further details, see Debreu (1976)).

A somewhat similar separation is manifest in the classical definition of complementarity of a pair of commodities (Kannai (1980)). The classical (ALEP) definition classifies the pair i,j as complementary (at x) if $u_{ij}(x) > 0$, and as substitutes if $u_{ij}(x) < 0$. This definition has fallen out of grace because the sign of $u_{ij}(x)$ is not an invariant of the ordering $\succsim$. If we assume, however, that $\succsim$ is monotone as well as concavifiable, then an elementary computation shows that the sign of $u_{ij}(x)$ is the same for all $u \in F$, and that by considering any other concave utility function $v = f(u)$ representing $\succsim$ the sign might change from positive to negative (but not vice-versa). Hence the suggested "rescued" classical ALEP definition: The commodities i and j, $i \neq j$, are said to be complementary at x if $u_{ij}(x) > 0$ for a least concave utility function representing the preference relation $\succsim$ of the consumer, and are said to be substitutes if $u_{ij}(x) < 0$. Thus, we can distinguish between a "built in the ordering" substitutes and between those peculiar to a person (the negative sign arising from the specific f - not from $\succsim$ itself).

We want also to argue that least concave utility functions are advantageous, in certain bargaining situations, over concave utility functions which are not least concave. It seems to us that this gives some additional intuitive justification for using least concave utility functions as cardinal utilities.

Let a be a vector of R^m with positive components. We think of a as the total resources available to two agents, who have to agree on an allocation. We assume that agent i has a continuous, monotone, and concavifiable preference ordering $\succsim_i$ on $K = \{x \in R^m : 0 \leq x \leq a\}$, $i = 1,2$. Let U_i ($i = 1,2$) denote the set of concave utility functions u representing $\succsim_i$ and normalized by $u(0) = 0$, $u(a) = 1$. For each pair $v_1, v_2 \in U_1 \times U_2$, consider the disposable hull $S(v_1,v_2)$ of the utility possibility set of K, i.e., the set $\{(s,t) : s \geq 0, t \geq 0$, and there exists an $x \in K$ such that $s \leq v_1(x)$, $t \leq v_2(a-x)\}$. The set $S(v_1,v_2)$ is convex (Chipman-Moore (1972)), and we consider the bargaining problem associated with this set (we can take the origin to be the threat point). If f is a strictly monotone increasing, continuous and concave function on [0,1] with $f(0) = 0$ and $f(1) = 1$, then $S(v_1,f(v_2)) \supset S(v_1,v_2)$, but (0,1) and (1,0) are extreme points for both sets. It is well known (as the author has learned from oral presentations by W. Thomson, A. Roth and U. Rothblum; compare also Kihlstrom, Roth and Schmeidler (1980), and Roth and Rothblum (1980)) that for many solutions of the bargaining problem [e.g., the Nash solution or the Kalai-Smorodinsky solution (Kalai-Smorodinsky, (1975)] this implies that agent 1 is better off in the selected outcome chosen from $S(v_1,f(v_2))$ than in the outcome chosen from $S(v_1,v_2)$ (this fact was proved implicitly in Kannai (1977) for the Nash solution of the bargaining problem). Consider now the following two-person non-zero sum game: The (pure) strategies spaces for the players are U_1 and U_2, and the payoff functions $M_i(v_1,v_2)$, $i = 1,2$, $v_1,v_2 \in U_1 \times U_2$, are given by the solution to the bargaining problem $S(v_1,v_2)$. It follows that the least concave utility functions form a Nash equilibrium point of this game. This Nash point is a strong equilibrium point (since the game is played essentially on the Pareto set, on which the interests of the players are essentially opposing).

Thus, it appears that if the preferences are fixed ("revealed") and the players cannot conceal them, then the best they can do is to pretend that their marginal utility decreases as little as possible.

APPENDIX A

PROOF OF THEOREM 2: The inequality (9) reads, for k = n-1,

$$f(p_{n-2}) \geqslant \frac{f(p_0)}{1+\alpha_{n-2} + \dots + \alpha_1 \dots \alpha_{n-2}} + \frac{(\alpha_{n-2} + \dots + \alpha_1 \dots \alpha_{n-2})}{1+\alpha_{n-2} + \dots + \alpha_1 \dots \alpha_{n-2}} f(p_{n-1}) \tag{79}$$

but (9) for k = n reads

$$f(p_{n-1}) \geqslant \frac{f(p_0)}{1+\alpha_{n-1} + \dots + \alpha_1 \dots \alpha_{n-1}} + \frac{(\alpha_{n-1} + \dots + \alpha_1 \dots \alpha_{n-1}) f(p_n)}{1+\alpha_{n-1} + \dots + \alpha_1 \dots \alpha_{n-1}} .$$

Insertion of this inequality in (79) yields

$$f(p_{n-2}) \geqslant \frac{f(p_0)}{1+\alpha_{n-2} + \dots + \alpha_1 \dots \alpha_{n-2}} + \frac{(\alpha_{n-2} + \dots + \alpha_1 \dots \alpha_{n-2})}{1+\alpha_{n-2} + \dots + \alpha_1 \dots \alpha_{n-2}}$$

$$\times \left[\frac{f(p_0)}{1+\alpha_{n-1} + \dots + \alpha_1 \dots \alpha_{n-1}} + \frac{(\alpha_{n-1} + \dots + \alpha_1 \dots \alpha_{n-1})}{1+\alpha_{n-1} + \dots + \alpha_1 \dots \alpha_{n-1}} f(p_n) \right]$$

$$= \frac{1+\alpha_{n-1} + \dots + \alpha_1 \dots \alpha_{n-1} + \alpha_{n-2} + \dots + \alpha_1 \dots \alpha_{n-2}}{(1+\alpha_{n-2} + \dots + \alpha_1 \dots \alpha_{n-2})(1+\alpha_{n-1} + \dots + \alpha_1 \dots \alpha_{n-1})} f(p_0)$$

$$+ \frac{(\alpha_{n-2} + \dots + \alpha_1 \dots \alpha_{n-2})\alpha_{n-1}(1+\alpha_{n-2} + \dots + \alpha_1 \dots \alpha_{n-2})}{(1+\alpha_{n-2} + \dots + \alpha_1 \dots \alpha_{n-2})(1+\alpha_{n-1} + \dots + \alpha_1 \dots \alpha_{n-1})} f(p_n)$$

$$= \frac{(1+\alpha_{n-1})(1+\alpha_{n-2} + \dots + \alpha_1 \dots \alpha_{n-2})}{(1+\alpha_{n-2} + \dots + \alpha_1 \dots \alpha_{n-2})(1+\alpha_{n-1} + \dots + \alpha_1 \dots \alpha_{n-1})} f(p_0)$$

$$+ \frac{\alpha_{n-2}\alpha_{n-1} + \dots + \alpha_1 \dots \alpha_{n-1}}{1+\alpha_{n-1} + \dots + \alpha_1 \dots \alpha_{n-1}} f(p_n)$$

$$= \frac{(1+\alpha_{n-1}) f(p_0) + (\alpha_{n-2}\alpha_{n-1} + \dots + \alpha_1 \dots \alpha_{n-1}) f(p_n)}{1+\alpha_{n-1} + \dots + \alpha_1 \dots \alpha_{n-1}}$$

and (10) is established for j = n-2. We now use backward induction on j to prove, in a similar fashion, (10) for all j. Assume, in fact, that (10) has been proved already for a certain j > 1. Setting j = k in (9) and applying (10), we obtain

$$f(p_{j-1}) \geq \frac{f(p_0)}{1+\alpha_{j-1}+\ldots+\alpha_1\cdots\alpha_{j-1}} + \frac{(\alpha_{j-1}+\ldots+\alpha_1\cdots\alpha_{j-1})f(p_j)}{1+\alpha_{j-1}+\ldots+\alpha_1\cdots\alpha_{j-1}}$$

$$\geq \left[\frac{1}{1+\alpha_{j-1}+\ldots+\alpha_1\cdots\alpha_{j-1}} + \frac{(\alpha_{j-1}+\ldots+\alpha_1\cdots\alpha_{j-1})(1+\alpha_{n-1}+\ldots+\alpha_{j+1}\cdots\alpha_{n-1})}{(1+\alpha_{j-1}+\ldots+\alpha_1\cdots\alpha_{j-1})(1+\alpha_{n-1}+\ldots+\alpha_1\cdots\alpha_{n-1})}\right] f(p_0)$$

$$+ \frac{(\alpha_{j-1}+\ldots+\alpha_1\cdots\alpha_{j-1})(\alpha_j\cdots\alpha_{n-1}+\ldots+\alpha_1\cdots\alpha_{n-1})}{(1+\alpha_{j-1}+\ldots+\alpha_1\cdots\alpha_{j-1})(1+\alpha_{n-1}+\ldots+\alpha_1\cdots\alpha_{n-1})} f(p_n) \qquad (80)$$

$$= \frac{[1+\alpha_{n-1}+\ldots+\alpha_1\cdots\alpha_{n-1}+(\alpha_{j-1}+\ldots+\alpha_1\cdots\alpha_{j-1})(1+\alpha_{n-1}+\ldots+\alpha_{j+1}\cdots\alpha_{n-1})]}{(1+\alpha_{j-1}+\ldots+\alpha_1\cdots\alpha_{j-1})(1+\alpha_{n-1}+\ldots+\alpha_1\cdots\alpha_{n-1})} f(p_0)$$

$$+ \frac{(\alpha_{j-1}+\ldots+\alpha_1\cdots\alpha_{j-1})\alpha_j\cdots\alpha_{n-1}(1+\alpha_{j-1}+\ldots+\alpha_1\cdots\alpha_{j-1})}{(1+\alpha_{j-1}+\ldots+\alpha_1\cdots\alpha_{j-1})(1+\alpha_{n-1}+\ldots+\alpha_1\cdots\alpha_{n-1})} f(p_n).$$

But

$$1+\alpha_{n-1}+\ldots+\alpha_1\cdots\alpha_{n-1}+(\alpha_{j-1}+\ldots+\alpha_1\cdots\alpha_{j-1})(1+\alpha_{n-1}+\ldots+\alpha_{j+1}\cdots\alpha_{n-1})$$

$$= 1+(\alpha_{j-1}+\ldots+\alpha_1\cdots\alpha_{j-1})+\alpha_{n-1}+(\alpha_{j-1}+\ldots+\alpha_1\cdots\alpha_{j-1})\alpha_{n-1}+\ldots+\alpha_{j+1}\cdots\alpha_{n-1}$$

$$+ (\alpha_{j-1}+\ldots+\alpha_1\cdots\alpha_{j-1})\alpha_{j+1}\cdots\alpha_{n-1}+\alpha_j\cdots\alpha_{n-1}+\ldots+\alpha_1\cdots\alpha_{n-1}$$

$$= (1+\alpha_{n-1}+\ldots+\alpha_{j+1}\cdots\alpha_{n-1}+\alpha_j\cdots\alpha_{n-1})(1+\alpha_{j-1}+\ldots+\alpha_1\cdots\alpha_{j-1}).$$

Substituting in the numerator of the fraction multiplying $f(p_0)$ in the extreme right hand side of (80), we see that

$$f(p_{j-1}) \geq \frac{(1+\alpha_{n-1}+\ldots+\alpha_j\cdots\alpha_{n-1})(1+\alpha_{j-1}+\ldots+\alpha_1\cdots\alpha_{j-1})}{(1+\alpha_{j-1}+\ldots+\alpha_1\cdots\alpha_{j-1})(1+\alpha_{n-1}+\ldots+\alpha_1\cdots\alpha_{n-1})} f(p_0)$$

$$+ \frac{(\alpha_{j-1}+\ldots+\alpha_1\cdots\alpha_{j-1})\alpha_j\cdots\alpha_{n-1}}{1+\alpha_{n-1}+\ldots+\alpha_1\cdots\alpha_{n-1}} f(p_n)$$

$$= \frac{(1+\alpha_{n-1}+\ldots+\alpha_j\cdots\alpha_{n-1})f(p_0)+(\alpha_{j-1}\cdots\alpha_{n-1}+\ldots+\alpha_1\cdots\alpha_{n-1})f(p_n)}{1+\alpha_{n-1}+\ldots+\alpha_1\cdots\alpha_{n-1}}$$

proving (10) for j-1.

APPENDIX B

Complete Proof of Sufficiency in Theorem 3:

Let us choose a sequence of finite sequences $\{p_{i,n}\}$, $0 \leq i \leq n < \infty$ of points in K with $p_{0,n} \prec p_{1,n} \prec \cdots \prec p_{n,n}$, where $p_{n,n}$ is maximal and $p_{0,n}$ is minimal w.r.t. $\succsim$, and such that the sequence is *dense* with respect to $\succsim$, in the sense that for any pair of points p, q $\in$ K with $p \prec q$, there exists a constant N such that for all $n \geq N$ there exists an index i with $p \prec p_{i,n} \prec q$. Obviously $\{p: p \succsim p_{0,n}\} = K$. Hence the concave function u_n constructed according to (13) is defined over all of K, for $n = 1,2,\ldots$, and $0 \leq u_n(x) \leq 1$ for all $x \in K$, $n = 1,2,\ldots$. By a classical selection theorem for concave functions [e.g., Rockafellar (1970, Theorem 10.9)], there exists a sequence u_{n_j} converging to a continuous concave function u defined on int K. If $p \prec q$ then by continuity there exist p' , q' $\in$ int K with $p \prec p' \prec q' \prec q$, and by hypothesis we have for all j large enough points p_{k,n_j} and p_{ℓ,n_j} with $p \prec p_{k,n_j} \prec p' \prec q' \prec p_{\ell,n_j} \prec q$. The construction (11) implies that $u_{n_j}(p_{\ell,n_j}) > u_{n_j}(p_{k,n_j})$ for j large enough. If $p'_{\ell,n_j} \sim p_{\ell,n_j}$ $(p'_{k,n_j} \sim p_{k,n_j})$ lies in the interval (p,p') ((q,q')) we obtain the inequality $u_{n_j}(p'_{\ell,n_j}) > u_{n_j}(p'_{k,n_j})$. Letting $p' \to p$ and $q' \to q$ we get the result $u(q) \geq u(p)$. Also u(p) = 0 if p is minimal and u(q) = 1 if q is maximal with respect to $\succsim$, since the same equations are true by construction for the functions u_n. Using the assumption that for all non-maximal p, $\{q: q \sim p\}$ does not contain a non-empty open set, we infer that $p \sim q$, p,q $\in$ int K implies that there exist sequences $p_n \to p$, $q_n \to q$ with $p_n \succ q$, $q_n \succ p$, so that $u(p_n) \geq u(q)$, $u(q_n) \geq u(p)$ by the preceding. But u is continuous in int K. Hence u(p) = u(q). If p is not maximal then (14) implies that u(p) < 1. To prove that u is a utility in int K we have thus to show only that if $q \prec p \prec b$ where b is a maximal point, then $u(q) < u(p)$. Let $p' \sim p$ be a point in the interval (q,b), i.e., there exists a $0 < t < 1$ such that $p' = tq + (1-t)b$. Then by concavity,

$$u(p) = u(p') \geq tu(q)+(1-t)u(b) = tu(q)+(1-t) > u(q).$$

We prove next that u is a utility in all of K. First we note that the equations $u(p) = 1$, if p is maximal, and $u(p) = 0$, if p is minimal, are valid also if $p \in \partial K$. If $p \in \partial K$ is neither maximal nor minimal there exists $p' \in \text{int } K$ with $p' \sim p$, and we set $u(p) = u(p')$. The function u is thus defined on all of K and clearly $p \succ q$ if and only if $u(p) > u(q)$. Let b be a maximal point and let $q \in \text{int } K$ be neither maximal nor minimal. For any $\varepsilon > 0$, set $r = tb+(1-t)q$ where $t = 0$ if $\varepsilon \geq 1-u(q)$ and $t = \dfrac{1-u(q)-\varepsilon}{1-u(q)}$ if $\varepsilon < 1-u(q)$. Then $u(r) \geq 1-\varepsilon$, so that for all p in a full neighborhood of b relative to K satisfying $p \succ r$ we have $|u(p)-u(b)| < \varepsilon$. If $p \in \partial K$ is neither maximal nor minimal, let $p' \in \text{int } K$ be such that $p' \sim p$. Our assumptions imply that if $p_n \to p$ then there exists a sequence $p'_n \to p'$ such that $p'_n \sim p_n$. Hence the continuity of u at $p' \in \text{int } K$ implies continuity at p. To prove continuity at the minimal points, set $\delta = \inf\limits_{p \in \text{int } K} u(p)$ (we want to prove $\delta = 0$). If $a \in \partial K$ is minimal, there exist no two points of K (different from a) such that a is a convex combination of those points, unless both points are also minimal. Hence the function v defined by $v(p) = u(p)$ if $p \succ a, v(p) = \delta$ if $p \sim a$ is (i) concave and (ii) $p \succ q$ if and only if $v(p) > v(q)$. The properties (i) and (ii) hold for $f(p) = \dfrac{v(p)-\delta}{1-\delta}$. Noting that no continuity assumption was used in the proof of Theorem 2, we conclude that (10) holds. Hence $f(p_{k,n}) \geq u_n(p_{k,n})$ for all k,n, and $f(p) \geq u(p)$ for all $p \in K$. For any $\varepsilon > 0$ there exists a non-minimal $q \in K$ with $v(q)-\delta < \varepsilon(1-\delta)$. Hence $u(p) \leq f(p) < \varepsilon$ for all $p \succ q$ and u is continuous *everywhere*. It is easy to see that u is indeed a concave utility in all of K.

The fact that u(p) is less than or equal to the sup in (14) follows from the construction. On the other hand, it follows from the proof of the necessity of (14), that for every concave utility u with u(K) = [0,1] u(p) must be greater than or equal to the sup. Hence the last statement of the Theorem, since the utility we have constructed satisfies u(K) = [0,1].

APPENDIX C

Complete Proof of Sufficiency in Theorem 10:

As a first step in proving sufficiency of (23), we construct a concave utility function $u_{a,b}$ in the (open) layer $\{p; a \prec p \prec b\}$. For this we prove first that Λ is multiplicative, i.e., that

$$\Lambda(p,r) = \Lambda(p,q)\,\Lambda(q,r) \tag{28}$$

for all $a \prec p \prec q \prec r \prec b$. It is clear that $\Lambda(p,r) \geq \Lambda(p,q)\,\Lambda(q,r)$. To prove the opposite inequality, we choose $\varepsilon > 0$ and let $p = p_1 \prec p_2 \prec \cdots \prec p_n = r$ be a sequence of elements of K such that

$$\Lambda(p,r) \leq \prod_{i=1}^{n-1} \lambda(p_i,p_{i+1}) + \varepsilon.$$

The only non-trivial case occurs when q is not one of the p_i's, so let $p_j \prec q \prec p_{j+1}$ $(1 \leq j < n)$. Let $\bar{p}_j \sim p_j$, $\bar{p}_{j+1} \sim p_{j+1}$, and let $\bar{q}$ be a point in the interval $(\bar{p}_j,\bar{p}_{j+1})$ such that $\bar{q} \sim q$. Then the proportionality of $\bar{p}_{j+1}-\bar{p}_j$ to $\bar{p}_{j+1}-\bar{q}$ and to $\bar{q}-\bar{p}_j$ implies the equation

$$\frac{v'_{a,b}(\bar{p}_{j+1},\bar{p}_{j+1}-\bar{p}_j)}{v'_{a,b}(\bar{p}_j,\bar{p}_{j+1}-\bar{p}_j)} = \frac{v'_{a,b}(\bar{q},\bar{q}-\bar{p}_j)}{v'_{a,b}(\bar{p}_j,\bar{q}-\bar{p}_j)} \cdot \frac{v'(\bar{p}_{j+1},\bar{p}_{j+1}-\bar{q})}{v'(\bar{q},\bar{p}_{j+1}-\bar{q})},$$

which in turn implies that $\lambda(p_j,p_{j+1}) \leq \lambda(p_j,q)\lambda(q,p_{j+1})$. Hence

$$\Lambda(p,r) \leq \prod_{i=1}^{n-1}\Lambda(p_i,p_{i+1}) + \varepsilon \leq \prod_{i=1}^{j-1}\lambda(p_i,p_{i+1})\lambda(p_j,q)\lambda(q,p_{j+1})\prod_{i=j+1}^{n-1}\lambda(p_i,p_{i+1}) + \varepsilon$$

$$\leq \Lambda(p,q)\,\Lambda(q,r) + \varepsilon.$$

The validity of the last inequality for every $\varepsilon > 0$ implies that $\Lambda(p,r) \leq \Lambda(p,q)\,\Lambda(q,r)$.

We now define a real-valued function $g_{a,b}$ on the open interval $(0,1)$ by

$$g_{a,b}(t) = \begin{cases} \Lambda(a + t(b-a), \frac{a+b}{2}) & 0 < t < 1/2 \\ 1 & t = 1/2 \\ 1/\Lambda(\frac{a+b}{2}, a + t(b-a)) & 1/2 < t < 1 \end{cases} \quad (29)$$

It follows from (28) and (29) that

$$\frac{g_{a,b}(t_1)}{g_{a,b}(t_2)} = \Lambda(a + t_1(b-a), a + t_2(b-a)) \quad (81)$$

for all $0 < t_1 < t_2 < 1$. It follows from (22), (23) and (81) that for all p,q such that $v_{a,b}(p) = t_1$, $v_{a,b}(q) = t_2$, we have

$$\frac{g_{a,b}(t_1)}{g_{a,b}(t_2)} \geq \frac{v'_{a,b}(q,q-p)}{v'_{a,b}(p,q-p)} .$$

For every $c,d \in K$ such that $a \prec c \prec d \prec b$ and such that the preference ordering $\succcurlyeq$ is strictly monotone on the interval $[c,d]$, we conclude that the function $\tilde{g}_{a,b}(t;c,d)$ defined for $t \in (0,1)$ by

$$\tilde{g}_{a,b}(t;c,d) = g_{a,b}(v_{a,b}(c+t(d-c))v'_{a,b}(c+t(d-c),d-c) \quad (82)$$

is *monotone non-increasing* for $t \in (0,1)$. In particular $\tilde{g}_{a,b}(t;a,b) = g_{a,b}(t)$. Hence $g_{a,b}(t)$ itself is monotone non-increasing. Define now

$$F(t) = \int_{1/2}^{t} g_{a,b}(s)\,ds. \quad (30)$$

Then F is a continuous, strictly increasing, concave function of $t \in (0,1)$. The monotone function $g_{a,b}(t)$ is continuous outside an at most countable set $A \subset (0,1)$. Hence $F(t)$ is (classically) differentiable with $F'(t) = g_{a,b}(t)$, for all $t \notin A$. Set

$$u_{a,b}(p) = F(v_{a,b}(p)). \quad (31)$$

Then $u_{a,b}(p)$ is a continuous utility function representing $\succsim$ in the layer $\{p : a \prec p \prec b\}$. The chain rule (applied for right derivatives) implies that

$$u'_{a,b}(c + t(d-c), d-c) = \tilde{g}_{a,b}(t;c,d) \qquad (83)$$

for $t \in (0,1)$, $v_{a,b}(c + t(d-c)) \notin A$, for every $c,d \in K$ such that $a \prec c \prec d \prec b$ and such that the preference ordering $\succsim$ is strictly monotone on the interval $[c,d]$. Thus the function $u_{a,b}(c + t(d-c))$, regarded as a function of the real variable t in the interval $I = (0,1)$ satisfies the assumptions of Proposition 7. Hence $u_{a,b}(c + t(d-c))$ is a concave function of t, and $u_{a,b}(p)$ is a *concave* function of p in (c,d) for every c,d as above.

The assumption made at the beginning of Section 2 can be reformulated as stating that if two distinct points p_1, q_1 which lie in an interval $[p,q]$ are indifferent ($p_1 \sim q_1$), then no point in $[p,q] \setminus [p_1,q_1]$ is strictly preferred to p_1. It follows that every interval $[p,q]$ can be decomposed as follows: $[p,q] = [p,p_1] \cup [p_1,p_2] \cup [p_2,q]$, where the preference ordering is strictly monotone increasing on $[p,p_1]$, the points in $[p_1,p_2]$ are indifferent to each other, and $\succsim$ is strictly decreasing on $[p_2,q]$ (one or two of the intervals might degenerate into a point). If $a \prec p$, $q \prec b$, then u is concave on each of the intervals. For it is concave on $[p,p_1]$ (set $p = c$, $p_1 = d$, and apply the result just proved), on $[p_1,p_2]$ (it is a constant there) and on $[p_2,q]$ (set $q = c$, $p_2 = d$). Moreover, $u_{a,b}$ is increasing from p to p_1 and decreasing from p_2 to q. Hence $u_{a,b}$ is concave in $[p,q]$ and thus is concave in the layer $\{p : a \prec p \prec b\}$.

Let u(p) be an arbitrary concave utility function representing $\succsim$ in the layer $\{p : a \prec p \prec b\}$. The positivity of $g_{a,b}(t)$ implies the invertibility of F(t). Hence u(p) can be written as

$$u(p) = w_{a,b}(F^{-1}(u_{a,b}(p)) = h(u_{a,b}(p)),$$

where $w_{a,b}$ is defined as in the proof of Proposition 8 and $h = w_{a,b} \circ F^{-1}$. Taking right derivatives, we find (outside an at most countable set) that

$$h'(t,1) = \frac{w'_{a,b}(F^{-1}(t),1)}{F'(t,1)} = \frac{w'_{a,b}(F^{-1}(t),1)}{g_{a,b}(t)} \tag{84}$$

Setting $p = a + F^{-1}(t_1)(b-a)$, $q = a + F^{-1}(t_2)(b-a)$, where $t_1 < t_2$, and applying (27) and (81) to (84), we conclude that $h'(t_1,1) \geq h'(t_2,1)$. It follows from Proposition 7 that $h(t)$ is a concave function of t in $F((0,1))$. Hence $u_{a,b}$ is a *least concave* utility function representing $\succsim$.

Using the uniqueness (up to multiplication by a positive scalar and translation) of least concave utility functions and Proposition 4, and letting a decrease, b increase (relative to $\succsim$), we obtain (imitating Debreu (1976), see also Theorem 2.6 in Kannai (1977) and its proof) the existence of a concave utility function representing $\succsim$ in all non-maximal points of K. If $\succsim$ has no maximal points in K, we are done. Otherwise let b be a maximal point of K, and let a be an arbitrary non-maximal point of K. As explained above, we may assume, without loss of generality, that $\succsim$ is strictly monotone increasing on [a,b]. Hence $u_{a,b}(p)$ is well-defined on the layer $\{p : a \prec p \prec b\}$ (by means of the formula (31)). The positivity of $g_{a,b}(s)$ for $s < 1$ as well as the monotonicity of $g_{a,b}(s)$ imply that F(1) can be computed by (30), and $u_{a,b}$ can be set equal to F(1) on the maximal points. It follows that the (extended) $u_{a,b}$ is continuous and concave in the layer $\{p : a \prec p \preceq b\}$. We can now "patch" together least concave utility functions defined in other sublayers of K, so as to obtain a utility function defined over all of K.

APPENDIX D

Proof of Proposition 13:

We first prove that (i) implies (ii). In fact, let k(t) be an indefinite Lebesgue integral of K(t) in I. Then k is differentiable almost everywhere in I and $k'(t) = K(t)$ a.e. Moreover, the non-negativity of $K(t)$ implies that for all $t \in I$,

$$\liminf_{h\to 0} \frac{k(t+h) - k(t)}{h} = \liminf_{h\to 0} \int_t^{t+h} \frac{K(t)}{h}\, dt \geq 0.$$

Let $\alpha \in A$ be arbitrary. Then the function $k(t) - \phi_\alpha(t)$ is differentiable almost everywhere in I_α with a non-negative derivative and

$$\liminf_{h\to 0} \frac{(k - \phi_\alpha)(t+h) - (k - \phi_\alpha)(t)}{h} \geq 0 - \phi'_\alpha(t) > -\infty$$

for all $t \in I_\alpha$. By Lemma 2 of §7, Chapter IX, in Natanson (1961), the function $k(t) - \phi_\alpha(t)$ is non-decreasing in I_α and (56) follows (along with the second equality in (58)).

Next we demonstrate the equivalence of (ii) and (iii). Let (ii) hold, and let $c = t_1 < t_2 < \ldots < t_n = d$ and α_i be chosen so that $t_i, t_{i+1} \in I_{\alpha_i}$. Applying (56) for each α_i separately ($1 \leq i \leq n-1$) we get the inequalities $k(t_{i+1}) - k(t_i) \geq \phi_{\alpha_i}(t_{i+1}) - \phi_{\alpha_i}(t_i)$. Summation of these inequalities yields the inequality

$$k(d) - k(c) = k(t_n) - k(t_1) \geq \sum_{i=1}^{n-1} [\phi_{\alpha_i}(t_{i+1}) - \phi_{\alpha_i}(t_i)].$$

The left hand side does not depend on the partition. Hence (57) follows (as well as the inequality $L(c,d) \leq k(d)-k(c)$). Assuming, conversely, that (iii) holds, we prove first that $L(c,d)$ is additive (compare (28)). Let $c,d,e \in I$, $c < d < e$. It is clear that $L(c,e) \geq L(c,d) + L(d,e)$. To prove the opposite inequality, let $\varepsilon > 0$ be arbitrary and let $c = t_1 < t_2 < \ldots < t_n = e$ be a partition of $[c,e]$, and let α_i be chosen for $1 \leq i \leq n-1$, so that $t_i, t_{i+1} \in I_{\alpha_i}$ and

$$L(c,e) \leq \sum_{i=1}^{n-1} [\phi_{\alpha_i}(t_{i+1}) - \phi_{\alpha_i}(t_i)] + \varepsilon.$$

The only non-trivial case occurs when d is not one of the t_i's, so let $t_j < d < t_{j+1}$ ($1 \leq j < n$). Then, trivially,

$$L(c,e) - \varepsilon \leq \sum_{i=1}^{j-1} [\phi_{\alpha_i}(t_{i+1}) - \phi_{\alpha_i}(t_i)] + \phi_{\alpha_j}(t_{j+1}) - \phi_{\alpha_j}(t_j) + \sum_{i=j+1}^{n-1} [\phi_{\alpha_i}(t_{i+1}) - \phi_{\alpha_i}(t_i)]$$

$$= \sum_{i=1}^{j-1} [\phi_{\alpha_i}(t_{i+1}) - \phi_{\alpha_i}(t_i)] + [\phi_{\alpha_j}(d) - \phi_{\alpha_j}(t_j)] + [\phi_{\alpha_j}(t_{j+1}) - \phi_{\alpha_j}(d)] + \sum_{i=j+1}^{n-1} [\phi_{\alpha_i}(t_{i+1}) - \phi_{\alpha_i}(t_i)]$$

$$\leq L(c,d) + L(d,e).$$

The validity of the inequality $L(c,e) - \varepsilon \leq L(c,d) + L(d,e)$ for every $\varepsilon > 0$ implies that $L(c,e) \leq L(c,d) + L(d,e)$.

Choosing now an arbitrary τ in I, we may set

$$k(t) = \begin{cases} L(\tau,t) & \text{for} \quad t > \tau \\ 0 & \text{for} \quad t = \tau \\ -L(t,\tau) & \text{for} \quad t < \tau \end{cases} \tag{85}$$

(compare (29)). Then for all $t_1, t_2 \in I$, $t_1 < t_2$, we have $k(t_2) - k(t_1) = L(t_1,t_2)$ and (56) follows from (57).

We finish the proof of Proposition *13* by showing that (iii) implies (i). In fact, let (iii) hold and let $k(t)$ be defined by (85), so that $k(t)$ satisfies (56) (as was proven already). In particular k is monotone non-decreasing, hence differentiable almost everywhere and all derived numbers of k are non-negative everywhere. Let $t \in I$ be a point where k is differentiable and let $t \in I_\alpha$ for some $\alpha \in A$. Then (56) implies that $k'(t) \geq \phi'_\alpha(t)$. Hence

$$k'(t) \geq \sup_{\alpha \in A} \phi'_\alpha(t) \quad \text{for almost all} \quad t \in I. \tag{86}$$

On the other hand, for all $c,d \in I$, $c < d$, and for every $\varepsilon > 0$, choose a partition $c = t_1 < t_2 < \ldots < t_n$ of $[c,d]$ and $\alpha_i \in A$, $1 \leq i < n-1$, such that $t_i, t_{i+1} \in I_{\alpha_i}$ and such that

$$L(c,d) - \varepsilon \leq \sum_{i=1}^{n-1} [\phi_{\alpha_i}(t_{i+1}) - \phi_{\alpha_i}(t_i)] \ . \tag{87}$$

Set now

$$k_\varepsilon(t) = \sum_{i=1}^{j-1} [\phi_{\alpha_i}(t_{i+1}) - \phi_{\alpha_i}(t_i)] + \phi_{\alpha_j}(t) - \phi_{\alpha_j}(t_j)$$

for $t \in [t_j, t_{j+1}]$, $1 \leq j \leq n$. Then $k_\varepsilon(t)$ is continuous in $[c,d]$ and differentiable everywhere except at the (finitely many) partition points (and even there the one-sided directional derivatives of k_ε exist and are finite), and $k'_\varepsilon(t) = \phi_{\alpha_i}(t)$ for $t \in (t_i, t_{i+1})$. Applying Lemma 2 of §7, Chapter IX, in Natanson (1961) once more we obtain the monotonicity of $k - k_\varepsilon$ in $[c,d]$. In particular, the non-negative function $k' - k'_\varepsilon$ is integrable in $[c,d]$. Let $h(t) = \sup_n k'_{1/n}(t)$. Then $h(t)$ is measurable and $h(t) \leq K(t)$ $(\leq k'(t))$ a.e. We claim that $h(t) = k'(t)$ almost everywhere. Otherwise, there exists a measurable set $B \subset [c,d]$ with $\mu(B) > 0$ and a positive δ such that $k'(t) - h(t) > \delta$ everywhere on B. Hence for all n,

$$\int_c^d [k'(t) - k'_{1/n}(t)]dt \geq \int_c^d [k'(t) - h(t)]dt > \delta\mu(B).$$

But by (87), $\int_c^d [k'(t) - k'_{1,n}(t)]dt < 1/n$, a contradiction. Hence $k'(t) = K(t)$ almost everywhere, and K is integrable in $[c,d]$ as it is the derivative of a non-decreasing function. We also obtain the first equality in (58).

APPENDIX E

Proof of Theorem 15:

We can rewrite (22) as

$$\lambda(p, p + h(b-a)) = \sup \frac{\langle \text{grad } v_{a,b}(x+k\xi), \xi \rangle}{\langle \text{grad } v_{a,b}(x), \xi \rangle} \tag{88}$$

where the sup is taken over

$\{(x,\xi,k) : x \in K, \xi \in R^m, k > 0, v_{a,b}(x)=v_{a,b}(p), v_{a,b}(x+k\xi)=v_{a,b}(x)+h\}$,

and the homogeneity of the directional derivatives is taken into account. Recall that every interval in K can be decomposed into at most 3 subintervals, in which $\succsim$ is (successively) strictly increasing, constant, and strictly decreasing. Without loss of generality we may assume that the denominator in the right hand side of (88) is positive. Then

$$\frac{\langle \operatorname{grad} v_{a,b}(x + k\xi),\xi\rangle}{\langle \operatorname{grad} v_{a,b}(x),\xi\rangle} - 1 = \frac{\langle \operatorname{grad} v_{a,b}(x + k\xi)\rangle - \langle \operatorname{grad} v_{a,b}(x),\xi\rangle}{\langle \operatorname{grad} v_{a,b}(x),\xi\rangle}$$

$$= \frac{\int_0^k M(v_{a,b}, x + s\xi,\xi)\,ds}{\langle \operatorname{grad} v_{a,b}(x),\xi\rangle} .$$

(See (35) for the meaning of the notation $M(v,x,\xi)$.) Hence the left hand side of (68) is equal to

$$\limsup_{h\to 0_+} \frac{\int_0^k M(v_{a,b}, x + s\xi,\xi)\,ds}{h\langle \operatorname{grad} v_{a,b}(x),\xi\rangle}$$

$$= \limsup_{h\to 0_+} \frac{k}{h} \cdot \frac{\int_0^k M(v_{a,b}, x + s\xi,\xi)\,ds}{k\langle \operatorname{grad} v_{a,b}(x),\xi\rangle} \tag{89}$$

$$= \limsup_{k\to 0_+} \frac{\frac{1}{k} \cdot \int_0^k M(v_{a,b}, x + s\xi,\xi)\,ds}{\frac{1}{k} \int_0^k \langle \operatorname{grad} v_{a,b}(x + s\xi),\xi\rangle ds \cdot \langle \operatorname{grad} v_{a,b}(x),\xi\rangle}$$

where the sup is taken over the set of (x,ξ,k) as above, and we have written $h = v_{a,b}(x + k\xi) - v_{a,b}(x)$ as an integral.

The assumptions of Theorem 15 imply that $r(x) = m$. Using the special coordinates employed in stating conditions III' and IV' of Section 4, we can write $g(x,\xi)$ in the form

$$g(x,\xi) = \sum_{i=1}^{m-1} v_{i,i}(x)(\xi_i/\xi_m)^2 + 2\sum_{i=1}^{m-1} v_{i,m}\xi_i/\xi_m + v_{m,m} ,$$

where $v_{i,i}(x) < 0$, $1 \leq i \leq m-1$. Thus the form $g(x,\xi)$ attains its maximum at the vector ξ satisfying

$$\frac{\xi_i}{\xi_m} = - \frac{v_{i,m}(x)}{v_{i,i}(x)}, \quad 1 \leq i \leq m-1. \tag{90}$$

The uniform continuity of the second derivatives of v in the indifference surfaces imply that the directions given by (90) are bounded away from the directions of tangent hyperplanes to the indifference surfaces (and that $g(x,\xi) \to -\infty$ as ξ tends to such a tangent hperplane). Using the uniform continuity as well as the homogeneity (in ξ) of $g(x,\xi)$ and of the expressions in (89), we find that the sup and the lim operations can be interchanged in (89). Hence the left hand side of (68) is equal, by (39), to the right hand side of (68).

As the proof of (69) is very similar to that of (68), we only sketch the argument. Let $x-k_1\xi, x, x+k_2\xi$ be collinear points in K with $v_{a,b}(x-k_1\xi) = v_{a,b}(p) - h$, $v_{a,b}(x) = v_{a,b}(p)$, and $v_{a,b}(x+k_2\xi) = v_{a,b}(p)+h$. Then $\succcurlyeq$ must be increasing in the interval $[x-k_1\xi,x]$. If $\succcurlyeq$ is not strictly increasing in $[x,x+k_2\xi]$, then there exists a minimal $k_3 \in (0,k_2)$ such that $x+k_3\xi \sim x+k_2\xi$ and by (4) only the triplet $x_1-k_1\xi,x,x+k_3\xi$ is relevant for the computation of α. Hence

$$\alpha(p - h(b-a),p,p + h(b-a)) = \sup \frac{k_1}{k_2}, \tag{91}$$

where the sup is taken over the set $\{(x,\xi,k_1,k_2) : x \in K,\ \xi \in R^m,$ $k_1,k_2 > 0,\ v_{a,b}(x) = v_{a,b}(p),\ v_{a,b}(x-k_1\xi) = v_{a,b}(p)-h,\ v_{a,b}(x+k_2\xi) =$ $= v_{a,b}(p)+h,\ \langle \text{grad } v_{a,b}(x+s\xi),\xi\rangle > 0 \text{ for } s \in (-k_1,k_2)\}$. Hence $v_{a,b}(x+s\xi) = t$ is invertible in $(v_{a,b}(p) - h, v_{a,b}(p) + h)$ and the derivative of the inverse function is equal to $\langle \text{grad } v_{a,b}(x+s\xi),\xi\rangle^{-1}$, and

$$k_2 = \int_{v(p)}^{v(p)+h} \langle \text{grad } v_{a,b}(x + s(t)\xi,\xi\rangle^{-1} dt ,$$

$$k_1 = \int_{v(p)-h}^{v(p)} \langle \text{grad } v_{a,b}(x + s(t)\xi,\xi\rangle^{-1} dt .$$

Substituting in (91), we see that the left hand side of (69) is equal to

$$\limsup_{h\to 0_+} \frac{\int_{v(p)-h}^{v(p)} \langle \operatorname{grad} v_{a,b}(x+s(t)\xi,\xi\rangle^{-1}dt - \int_{v(p)}^{v(p)+h} \langle \operatorname{grad} v_{a,b}(x+s(t)\xi,\xi\rangle^{-1}dt}{h \int_{v(p)}^{v(p)+h} \langle \operatorname{grad} v_{a,b}(x+s(t)\xi,\xi\rangle^{-1}dt}$$

$$= \limsup_{h\to 0_+} \frac{-\int_{v(p)-h}^{v(p)} \int_0^h (d/dt) \langle \operatorname{grad} v_{a,b}(x+s(t+\tau)\xi,\xi\rangle^{-1} \, dt \, d\tau}{h \int_{v(p)}^{v(p)+h} \langle \operatorname{grad} v_{a,b}(x+s(t)\xi,\xi\rangle^{-1} \, dt}$$

But

$$(d/dt) \langle \operatorname{grad} v_{a,b}(x+s(t+\tau)\xi,\xi\rangle^{-1} =$$

$$= -\langle \operatorname{grad} v_{a,b}(x+s(t+\tau)\xi,\xi\rangle^{-2} M(v_{a,b}, x + s(t+\tau)\xi,\xi)(ds/dt)$$

$$= -\langle \operatorname{grad} v_{a,b}(x+s(t+\tau)\xi,\xi\rangle^{-3} M(v_{a,b}, x + s(t+\tau)\xi,\xi).$$

Thus the left hand side of (69) is equal to

$$\limsup_{h\to 0_+} \frac{\left[\int_{v(p)-h}^{v(p)} \int_0^h \langle \operatorname{grad} v_{a,b}(x+s(t+\tau)\xi,\xi\rangle^{-3} M(v_{a,b}, x+s(t+\tau)\xi,\xi)\, d\tau\, dt\right]/h^2}{\left[\int_{v(p)-h}^{v(p)} \langle \operatorname{grad} v_{a,b}(x+s(t)\xi,\xi\rangle^{-1}dt\right]/h}.$$

Interchanging the sup and the lim, as in the proof of (68), we conclude that the left hand side of (69) is equal to

$$\sup_{\{x:v(x)=v(p)\}} \sup_{\{\xi:\langle \operatorname{grad} v_{a,b}(x),\xi\rangle>0\}} \langle \operatorname{grad} v_{a,b}(x),\xi\rangle^{-2} M(v_{a,b},x,\xi)$$

$$= -G(v(p)) .$$

APPENDIX F

The statement, made on page , that every $p \in \operatorname{conv}(A)$ can be written in the form $p = ta + (1-t)b$ for certain $a,b \in A$ and $t \in [0,1]$, is true in general only if for all $p \in K$, $\{q:q \succsim p\} = \operatorname{conv}(\{q;q\sim p\})$ (see Fig. 3 and Fig. 4). Theorem 3 can be proved in the general case by noting that for every compact, convex set $L \subset \text{int } K$, it is true that every p in $\text{conv}(A) \cap L$ can be written in the form $p = ta + (1-t)b$ for some $a, b \in A$ and $t \in [0,1]$ if n is large enough,

$$A = \bigcup_{i=0}^{n} \{ p \in K : p \sim p_{i,n} \}$$

and the sequence of finite sequences $\{p_{i,n}\}$, $0 \leqslant i \leqslant n < \infty$ is *dense* with respect to $\succsim$ (see Appendix B). Hence the extension of u_n from A to $\text{conv}(A) \cap L$ can be effected as on page (formula (13)), and a converging subsequence can be extracted from the sequence u_n. Choosing a sequence L_k of compact, convex subsets of int K with $\bigcup_k L_k = \text{int } K$ and applying a diagonal procedure, we find a subsequence u_{n_j} which converges to a contiguous concave function u defined on int K. The rest of the proof follows as in Appendix B.

REFERENCES

AUMANN, R.J. (1975). "Values of Markets with a Continuum of Traders", *Econometrica 43*, 611-646.

BOURBAKI, N. (1949). Fonctions d'une Variable Réelle, *Act. Sc. Ind. 1074*, Hermann, Paris.

CHIPMAN, J.S. and MOORE, J.C. (1972). "Social Utility and the Gains from Trade", *Journal of International Economics 2*, 157-172.

CROUZEIX, J.P. (1977). "Contributions à l'étude des Fonctions Quasi-convexes", Thèse, Université de Clermont-Ferrand.

DEBREU, G. (1959). *Theory of Value*, John Wiley and Sons, New York.

DEBREU, G. (1976). "Least Concave Utility Functions", *Journal of Mathematical Economics 3*, 121-129.

DE FINETTI, B. (1949). "Sulle Stratificazioni Convesse", *Annali di Matematica Pura ed Applicata (4) 30*, 173-183.

FENCHEL, W. (1953). "Convex Cones, Sets, and Functions", mimeo, Princeton University, Princeton, N.J.

FENCHEL, W. (1956). "Uber konvexe Funktionen mit vorgeschriebenen Niveaumannifgaltigkeiten", *Mathematische Zeitschrift 63*, 496-506.

HILDENBRAND, W. (1974). *Core and Equilibria of a Large Economy*, Princeton University Press, Princeton, N.J.

KALAI, E. and SMORODINSKY, M. (1975). "Other Solutions to Nash's Bargaining Problem", *Econometrica 43*, 513-518.

KANNAI, Y. (1970). "Continuity Properties of the Core of a Market", *Econometrica 38*, 791-815.

KANNAI, Y. (1974). "Approximation of Convex Preferences", *Journal of Mathematical Economics 1*, 101-106.

KANNAI, Y. (1977). "Concavifiability and Constructions of Concave Utility Functions", *Journal of Mathematical Economics 4*, 1-56.

KANNAI, Y. (1980). "The ALEP Definition of Complementarity and Least Concave Utility Functions", *Journal of Economic Theory 22*, 115-117.

KIHLSTROM, R.E., ROTH, A.E. and SCHMEIDLER, D. (1980). "Risk Aversion and Solutions to Nash's Bargaining Problem", to appear in *Proceedings of the Hagen-Bonn Meeting on Game Theory and Mathematical Economics*, (North Holland).

MANTEL, ROLF R. (1969). "On the Representation of Preferences by Concave Utility Functions", *Instituto Torcuato Di Tella Report.*

MAS-COLLEL, A. (1974). "Continuous and Smooth Consumers: Approximation Theorems", *Journal of Economic Theory 8*, 305-336.

MOULIN, H. (1973). "Concave Utility Functions Associated with Convex Preference Preorderings", *Cahiers de Mathématiques de la Décision, No. 7304.*

NATANSON, I.P. (1961). *Theory of Functions of a Real Variable, Vols. I and II*, F. Ungar, New York, English translation.

ROCKAFELLAR, R.T. (1970). *Convex Analysis*, Princeton University Press, Princeton, N.J.

ROTH, A.E. and ROTHBLUM, U.G. (1980). "Risk Aversion and Nash's Solution for Bargaining Games and Risky Outcomes", preprint.

ATTAINABLE SETS OF MARKETS: AN OVERVIEW*

Robert James Weber

The attainable set of a market is the set of all utility outcomes which can be achieved by the traders through a redistribution of goods. Attainable sets fall into two broad classes, according to whether the traders' utility functions are concave or are merely quasiconcave. Questions of major interest concern the characterization of these two classes, and the further classification of attainable sets according to their "complexity". The study of games which arise from markets is based on the analysis of attainable sets.

This paper surveys the current state of knowledge concerning these and related issues, and presents a set of open research problems.

1. DEFINITIONS AND BASIC RESULTS

Consider a market consisting of a set $N = \{1,2,\ldots,n\}$ of traders, and an m-dimensional commodity space $I^m = \{(y_1,\ldots,y_m): 0 \leq y_j \leq 1 \text{ for all } j\}$. For any collection $\{u_i\}_{i=1}^n$ of utility functions (real-valued functions on I^m) of the traders, the *attainable set* of the market is

$$A(u_1,\ldots,u_n) = \{x \in R^n: x \leq (u_1(y^1),\ldots,u_n(y^n)), \text{ where each } y^i \in I^m \text{ and } \Sigma y^i = (1,\ldots,1)\} .$$

We envision the traders arriving at a marketplace with initial commodity holdings which, when pooled, yield the commodity bundle $e^m \equiv (1,\ldots,1)$. Trader i has a preference relation over bundles in R^m_+, which is represented by the utility function u_i. The attainable set of the market is the set of all utility outcomes which can be achieved through some distribution of the available commodities among the traders.

Common assumptions in the study of markets are that the traders' preferences are complete, and that for every y in I^m, the "preference sets" $\{z: z \succsim_i y\}$ are closed and convex. A consequence of these

---

*The preparation of this paper was supported in part by the U.S. Office of Naval Research and the Center for Advanced Studies in Managerial Economics at Northwestern University.

ISBN 0-12-621120-5

assumptions is that the traders' utility functions are upper-semicontinuous and quasiconcave (see, for example, Debreu (1959)). A stronger assumption is that the traders' preferences are concavifiable; that is, that they can be represented by concave utility functions (Kannai (1981)).

Let U_1 be the collection of all upper-semicontinuous, quasiconcave utility functions, and let U_2 be the subcollection of continuous, concave utility functions. For $k = 1,2$, let $A_k(n)$ be the collection of all n-dimensional attainable sets arising from markets in which the traders' utility functions are in U_k.

Let V be an attainable set in $A_k(n)$. If $u_1,\ldots,u_n$ are functions in U_k defined on I^m (for some fixed m) such that $V = A(u_1,\ldots,u_n)$, then $\{u_i\}_{i=1}^n$ is a k-*representation* for V over I^m. The k-*complexity* of V is the least $m \geq 0$ such that there exists a k-representation for V over I^m.

A set X in R^n is the *comprehensive hull* of another set Y if $X = \{x \in R^n: x \leq y \text{ for some } y \in Y\}$; in this case, we say that X is *generated by* Y. Comprehensiveness is implicit in the definition of an attainable set. This embodies the assumption that any trader can unilaterally decrease his own utility. For later reference, we define a *corner* to be the comprehensive hull of a single point.

The following results are not difficult to prove:

THEOREM 1: Every set in $A_1(n)$ *is generated by a compact set.*

THEOREM 2: Every set in $A_2(n)$ *is generated by a compact, convex set.*

Theorem 1 can actually be extended slightly: if the functions $u_1,\ldots,u_n$ are upper-semicontinuous and bounded from below, then $A(u_1,\ldots,u_n)$ is compactly generated. The assumption of lower-boundedness cannot be eliminated (see Weber (1977)). Theorem 2 is proved by Billera and Bixby (1973a).

There are several useful operations which can be performed on attainable sets. If f and g are real-valued functions on I^m and I^ℓ respectively, define the functions $f \wedge g$ and $f \oplus g$ from $I^{m+\ell}$ to R as follows: if $(x,y) \in I^m \times I^\ell = I^{m+\ell}$, then $(f \wedge g)(x,y) = \min(f(x),g(y))$ and $(f \oplus g)(x,y) = f(x) + g(y)$. Both U_1 and U_2 are closed with respect to the operations $\wedge$ and $\oplus$.

PROPOSITION 1: Suppose V_1 *and* V_2 *are in* $A_k(n)$, *with* $V_1 = A(u_1,\ldots,u_n)$ *over* I^m *and* $V_2 = A(w_1,\ldots,w_n)$ *over* I^ℓ. *Take* $a > 0$ *and* $b \in R^n$. *Then*

(1) $aV_1 + b = A(au_1 + b_1,\ldots,au_n + b_n)$ over I^m.

(2) $V_1 \cap V_2 = A(u_1 \wedge w_1,\ldots,u_n \wedge w_n)$ over $I^{m+\ell}$.

(3) $V_1 + V_2 \equiv \{x_1 + x_2 : x_1 \in V_1 \text{ and } x_2 \in V_2\}$

$= A(u_1 \oplus w_1,\ldots,u_n \oplus w_n)$ over $I^{m+\ell}$.

Furthermore, these three derived sets are all in $A_k(n)$.

Part (1) of the proposition, in combination with Theorems 1 and 2, enables us, without loss of generality, to occasionally restrict our discussions to attainable sets generated by sets lying in the interior of the unit n-cube.

2. CHARACTERIZATION THEOREMS

Billera and Bixby (1973a) developed a full converse to Theorem 2.

THEOREM 3: Every convex, compactly-generated set in R^n *is in* $A_2(n)$.

To see the basic idea of their proof, fix a set V which is generated by a convex, compact subset of the unit n-cube. Let $D^i(h)$ be the corner in R^n generated by the point $(1,\ldots,-h,\ldots,1)$, and consider the set $V^i(h)$, defined as the convex hull of $V \cup D^i(h)$. This set is represented over I^{n-1} by the utility functions $u_j(y) = y_j$ for $j \neq i$, and $u_i(y) = \sup\{x_i : (e^{n-1}-y;x_i) \in V^i(h)\}$, where $e^{n-1} \equiv (1,\ldots,1) \in R^{n-1}$. If $h \geq n-1$, then every boundary point of V is a boundary point of $V^i(h)$ for at least one value of i. Therefore $\cap V^i(h) = V$, and by part (2) of Proposition 1, it follows that $V \in A_2(n)$. Incidentally, this construction demonstrates that every set in $A_2(n)$ has complexity no greater than n(n-1).

A precise characterization of $A_1(n)$ is not known. However, two large subsets of $A_1(n)$ have been determined. A compactly-generated set V in R^n is *convexifiable* if there are continuous, strictly increasing, real-valued functions $g_1,\ldots,g_n$ on R, such that

$$V(g_1,\ldots,g_n) \equiv \{x \in R^n\colon x \leq (g_1(z_1),\ldots,g_n(z_n)) \text{ for some } z \in V\}$$

is convex. For an attainable set to be convexifiable, there must be *some* utility representation of the traders' preferences which yields a convex attainable set.

THEOREM 4: Every convexifiable set in R^n *is in* $A_1(n)$.

The proof of this result is straightforward, and appears in Weber (1978). In essence, one convexifies V, adjoins the corner $D^i(0)$, obtains a Billera-Bixby representation of the new set, and then applies the inverses of the convexifying functions to the constructed utility functions. This procedure is carried out n times, with each trader i distinguished in turn; the resulting markets are then "intersected." In this manner, one obtains a representation of V involving at most n(n-1) commodities.

The following proposition, due to Mantel (1973) and Weber (1978), provides an inductive description of a family of convexifiable sets. A consequence of the proposition is that the convexifiable sets are dense (in the topology induced by the Hausdorff metric) in the collection of all compactly-generated sets. A set is *exponentially convexifiable* if for sufficiently large k the functions $g_1,\ldots,g_n$ defined by $g_i(x_i) = 1 - \exp(-kx_i)$ serve to convexify the set. Note that the only compactly-generated sets in R^1 are corners, and are exponentially convexifiable.

PROPOSITION 2: Let C *be a compact set in* R^n*, such that the comprehensive hull of* C *is exponentially convexifiable. Let* $f:R^n \to R$ *be twice continuously differentiable, and assume that the first-order partial derivatives of* f *are negative throughout* C*. Then the comprehensive hull of the compact set* $\{(z,f(z)) \in R^{n+1}: z \in C\}$ *is exponentially convexifiable.*

If the conditions of the proposition are not satisfied, then the set may in fact fail to be convexifiable. Two examples are illustrated on this page and the next.

The set in Figure 1 (which is shown to be non-convexifiable in Kannai and Mantel (1978) and Weber (1978)) *is* a member of $A_1(3)$. This can be seen through the following construction, which is based upon a construction given by Kannai and Mantel. Consider the comprehensive hull of the union of the three sets $\{x \in R^3_+: 2x_i + 2x_j + x_k \leq 1\}$ defined by permuting the indices $\{i,j,k\} = \{1,2,3\}$. This set has a representation over I^2, wherein all three traders' utility functions are defined by $u(x_1,x_2) = x_1/(2-x_2)$. Furthermore, the three sets generated by the corner on $(1,1,1/5)$, by $\{x \in R^3_+: x_1 + 2x_3 \leq 3/5, x_2 \leq 1\}$, and by $\{x \in R^3_+: x_2 + 2x_3 \leq 3/5, x_1 \leq 1\}$, are in $A_1(3)$ by virtue of their convexity. The intersection of these four attainable sets can be mapped into the set in Figure 1 through a positive affine transformation.

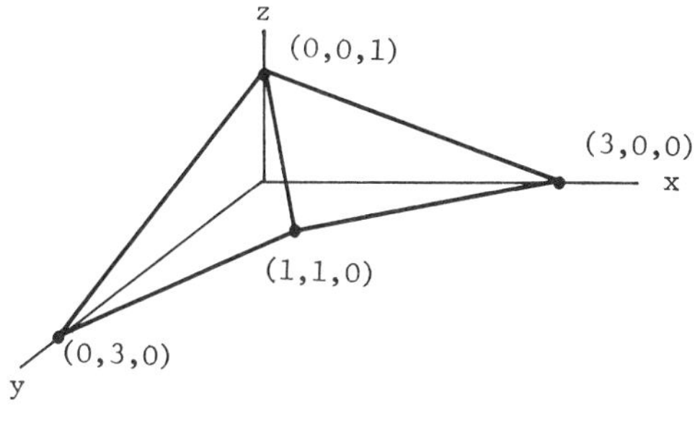

FIGURE 1: A nonconvexifiable set.

It is not known whether the set in Figure 2 is in $A_1(3)$.

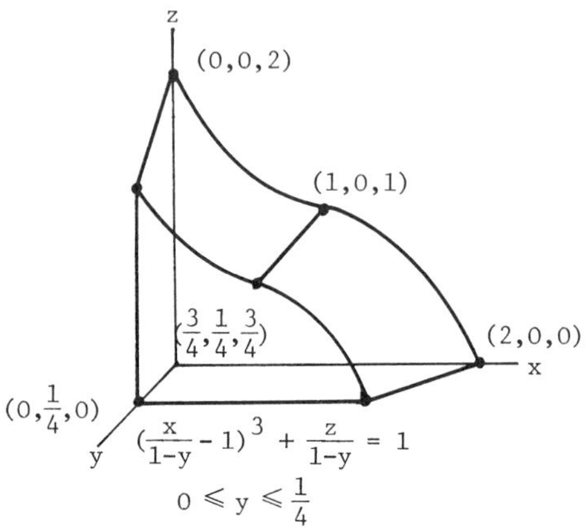

FIGURE 2: $\frac{\partial z}{\partial x} = 0$ for all points $(t,1-t,t)$.

A set in R^n is *finitely generated* if it is the union of finitely many corners. Although such sets contrast sharply with the convexifiable sets of Theorem 4, we have the following result.

THEOREM 5: Every finitely generated set in R^n *is in* $A_1(n)$.

This theorem is proved by Weber (1977) via a rather delicate construction which, in analogy with the construction for convexifiable sets, rescales the coordinate axes so as to bring the generating points to the boundary of a convex surface. A result from the same paper, which is necessary to this construction and is of some interest in its own right, is the following.

PROPOSITION 3: Take $0 < b < 1/n$, *and let* $p^0,\dots,p^n$ *be points in* R^n *of the form* $p^k = (b^{\ell_{k,1}},\dots,b^{\ell_{k,n}})$, *where all* $\ell_{k,j}$ *are non-negative integers. Assume that for some* $(t_1,\dots,t_n) \geq 0$ *with* $\Sigma t_k = 1$, *it is the case that* $p^0 \geq \Sigma t_k p^k$; *that is,* p^0 *lies on or above the convex hull of* $\{p^1,\dots,p^n\}$. *Then for some* k', $p^0 \geq p^{k'}$.

The construction used in the proof of Theorem 5 actually yields a representation in utility functions which are not only quasiconcave, but also monotone increasing and continuous. At one stage of the construction, a continuous function is quasiconcavified (that is, a new function is defined as the infimum of all quasiconcave functions greater than or equal to the original function). Because the domain of the function under consideration is polyhedral, this operation yields another continuous function. Interestingly, this operation of *quasiconcavification* does not necessarily preserve continuity when applied to a function on an arbitrary compact, convex domain. For example, consider the convex hull in R^3 of the circle $\{(x_1,x_2,0): (x_1-1)^2 + (x_2-1)^2 = 2\}$, and the points $(0,0,1)$ and $(0,0,-1)$. The function $f(x_1,x_2,x_3) = |x_3|$ is continuous on this set, but its quasiconcavification takes the value 1 at the origin, and is 0 at all other points on the circle. The *facial dimension* of a point x in a set K is the maximal dimension of any convex subset of K of which x is a relative interior point. The k-*skeleton* of K is the set of all points of K with facial dimension no greater than k; thus, for example, the 0-skeleton of a convex set is its set of extreme points. The crucial implication in the next proposition - that (a) implies (b) - is proved in Papadopoulou (1977).

PROPOSITION 4: Let K *be a compact, convex set in* R^n. *The following three assertions are equivalent:*

(a) *All* k-*skeletons of* K *are closed (for all* $0 \leq k \leq n$*).*

(b) *The convex hull of any relatively open subset of* K *is relatively open.*

(c) *The quasiconcavification of any continuous function on* K *is continuous.*

3. COMPLEXITY

The constructions used to prove Theorems 4 and 5 treat the n traders symmetrically (and in a sense, simultaneously), and require the use of n(n-1) commodities. With only slight modification, both constructions can be carried out inductively, by treating the traders in sequence and progressively using one commodity, then two more,..., then (n-1) more. This approach, which is developed in detail by Billera and Weber (1979), yields the following theorem.

THEOREM 6: If V *in* $A_1(n)$ *is convexifiable, or is finitely generated, then the* 1-*complexity of* V *is at most* n(n-1)/2.

A similar approach to the problem of 2-complexity is often, but not always, successful. Consider a set V in $A_2(n)$. Assume there is a set C contained in the boundary of V, which generates V and satisfies the following condition: there exists a closed set $Q \subset \{q \in R^n: \Sigma q_i = 1, q > 0\}$ such that for each $x \in C$, there is a $q \in Q$ for which $\Sigma q_i x_i \geq \Sigma q_i y_i$ for all $y \in V$. In this case, we say that V is (uniformly) *positively supported.* The collection of positively supported sets in $A_2(n)$ is dense in $A_2(n)$.

THEOREM 7: If V *in* $A_2(n)$ *is positively supported, then the 2-complexity of* V *is at most* n(n-1)/2.

The best currently-known universal upper bound on 2-complexity is due to Kalai (1975). Using a construction essentially different from those previously mentioned in this paper, he establishes the following result.

THEOREM 8: The 2-complexity of any set in $A_2(n)$ *is at most* $(n-1)^2 - (n-2) = n^2 - 3n + 3$.

In view of Theorem 7, it seems reasonable to conjecture that no set in $A_2(n)$ has 2-complexity greater than $n(n-1)/2$. Although supporting evidence is much weaker concerning lower bounds, it is commonly believed that there are sets in $A_2(n)$ requiring at least $n(n-1)/2$ commodities in their representations. The argument is that in complicated settings, it may be necessary to have a commodity *linking* each pair of traders. The construction used to prove Theorem 8 is of precisely this nature when specialized to the case n=3 (see Kalai and Smorodinsky (1975)).

The problem of establishing a lower bound for the complexity of an attainable set seems quite different from the upper-bounding problem. The only general result is due to Kalai and Smorodinsky.

THEOREM 9: For every $n \geq 3$, *there is a set in* $A_2(n)$ *of complexity* n.

The set referred to in Theorem 9 is generated by the convex hull of the (n+1) points $(1,0,\ldots,0),\ldots,(0,\ldots,0,1),(1/2,\ldots,1/2)$ in R^n. It can be shown that for any given trader, when the utility outcome is $(1/2,\ldots,1/2)$ there must be a commodity received only by him, as well as commodities shared exclusively by him with each of the other traders. Since each trader must receive a positive amount of *some* commodity at this utility outcome, there must be at least n commodities present in any representation of the set. A specific n-commodity representation is given by the utility functions $u_i(x_1,\ldots,x_n) = 1/2[x_i + \min_{1 \leq j \leq n} x_j]$.

A more detailed look at one aspect of the complexity issue is provided by Billera and Bixby (1976) in a characterization of sets in $A_2(n)$ which can be represented using a single commodity.

THEOREM 10: Let $V \in A_2(n)$ *be generated by a set in the unit n-cube* I^n, *and assume that* $\sup\{x_i : x \in V\} = 1$ *for each* $i \in N$. *(Every set in* $A_2(n)$ *can be affinely mapped into a set satisfying these conditions.) Then the complexity of* V *is at most* 1 *if and only if there are continuous, convex, nondecreasing functions* $h_i:[0,1] \to [0,1]$, *with* $h_i(0) = 0$, *such that*

$$V = \{x \in I^n: \Sigma h_i(x_i) = 1\} - R^n_+.$$

A geometric statement which follows from this result is that there is a convex rescaling of the coordinate axes which maps V into a set generated by a compact, convex subset of the unit simplex.

4. MARKET GAMES

Much of the work on attainable sets is an outgrowth of efforts to characterize those n-person games which can arise from markets. The first such effort was that of Shapley and Shubik (1969). They sought to show that certain games which are pathological with regard to the von Neumann-Morgenstern solution theory can occur in non-pathological economic settings.

A *game* (with transferable utility) is a real-valued function v on the subsets (coalitions) of a set N of players. Consider a market with trader set N, and assume that each trader i has a continuous, concave utility function u_i on R^m_+, and an initial endowment $\omega^i \in R^m_+$ of goods. The associated (transferable utility) *market game* is defined for each coalition S in N by

$$v(S) = \max\{\Sigma u_i(y^i): \text{all } y^i \geq 0, \text{ and } \sum_{i\in S} y^i = \sum_{i\in S} \omega^i\}.$$

For any coalition T in N, a T-*balanced collection* $\{\gamma_S\}_{S\subset T}$ is a set of non-negative numbers for which $\sum_{i\in S} \gamma_S = 1$ for every $i \in T$. A game is *totally balanced* if for all T, and all T-balanced collections $\{\gamma_S\}$,

$$\Sigma \gamma_S v(S) \leq v(T).$$

THEOREM 11: A game is a (transferable utility) market game if and only if it is totally balanced.

If utility is freely transferable among the traders, the attainable set of utility outcomes for any coalition is generated by a simplex. Such attainable sets are quite simple (cf. Theorem 10). Indeed, Shapley and Shubik showed that every market game arises from a market involving at most n commodities. Hart (1980) has recently refined this result by presenting a construction which employs only (n-1) commodities, and by showing that no lesser number will suffice for the *unanimity* game (defined by $v(N) = 1$, and $v(S) = 0$ for all $S \subsetneq N$).

In many (perhaps in most) markets, utility is not freely transferable. The concepts we have just defined can be extended to cover this more general case.

A *game* (without transferable utility) with player set N is a correspondence V which assigns to each coalition S a set $V(S) = C_S - R_+^S$, where $R^S \equiv \{x \in R^n: x_i = 0 \text{ for } i \notin S\}$ and $C_S \subset R^S$ is nonempty, compact, and convex. Corresponding to any market with trader set N there is a *market game*, defined for all $S \subset N$ by

$$V(S) = \{x \in R^S: x_i \leq u_i(y^i) \text{ for all } i \in S, \text{ where each } y^i \geq 0 \text{ and } \sum_{i \in S} y^i = \sum_{i \in S} \omega^i\} .$$

(After normalization of the commodity space, V(S) is the attainable set of a market with trader set S.)

A game is *totally balanced* if for all coalitions T in N, and all T-balanced collections $\{\gamma_S\}$,

$$\Sigma\, \gamma_S V(S) \subset V(T) .$$

Billera and Bixby (1973b) were able to prove the following partial analogue of Theorem 11.

THEOREM 12: Every market game is totally balanced. Let V *be totally balanced, and assume that each* V(S) *is a polyhedron. Then* V *is a market game.*

Subsequently, Mas-Colell (1975) provided another characterization result. A game V is *totally balanced with slack* if for every coalition T, and every T-balanced collection $\{\gamma_S\}$ in which $\gamma_T = 0$,

$$\Sigma\ \gamma_S V(S) \subset \text{Int } V(T) \ ,$$

where Int V(T) denotes the relative interior of V(T). The set of games which are totally balanced with slack is open and dense in the set of all totally balanced games.

THEOREM 13: Every game which is totally balanced with slack is a market game.

The Mas-Colell construction is distinctive, in that it uses the operation of market addition presented in part (3) of Proposition 1. Neither the Billera and Bixby nor the Mas-Colell construction bounds the number of commodities needed to represent a market game.

The definition of a game can be weakened further, by requiring only that each V(S) is compactly generated. One can then ask what games arise from markets in which the traders' utility functions are quasiconcave and upper-semicontinuous. There are no known results in this area.

5. RESEARCH PROBLEMS

1. Are all n-dimensional compactly-generated sets members of $A_1(n)$? (For example, is the set of Figure 2 attainable?)
2. What is the maximum complexity of sets in $A_1(n)$? in $A_2(n)$? (Although it is tempting to conjecture that both answers are $n(n-1)/2$, no n-dimensional attainable sets of complexity greater than n are currently known.)
3. Are all totally-balanced games market games? What types of games arise from quasiconcave markets? Is the number of commodities needed to represent any n-person market game bounded?

REFERENCES

BILLERA, L.J. (1974). "On Games Without Side Payments Arising from a General Class of Markets." *J. Math. Econom.* 1, 129-139.

_____ and BIXBY, R.E. (1973a). "A Characterization of Pareto Surfaces." *Proc. Amer. Math. Soc.* 41, 261-267.

_____ and _____ (1973b). "A Characterization of Polyhedral Market Games." *Internat. J. Game Theory* 2, 253-261.

_____ and _____ (1974). "Market Representations of n-Person Games." *Bull. Amer. Math. Soc.* 80, 522-526.

_____ and _____ (1976). "Pareto Surfaces of Complexity 1." *SIAM J. Appl. Math.* 30, 81-89.

_____ and WEBER, R.J. (1979). "Dense Families of Low-Complexity Attainable Sets of Markets." *J. Math. Econom.* 6, 67-73.

HART, S. (1980). "On the Number of Commodities Required to Represent a Market Game." Report No. 9/80, Institute for Advanced Studies, The Hebrew University of Jerusalem.

KALAI, E. (1975). "On Game-Type Subsets." *Internat. J. Game Theory* 4, 141-150.

_____ and SMORODINSKY, M. (1975). "On a Game-Theoretic Notion of Complexity." *Proc. Amer. Math. Soc.* 49, 416-420.

KANNAI, Y. and MANTEL, R. (1978). "Non-Convexifiable Pareto Sets." *Econometrica* 46, 571-575.

MANTEL, R. (1973). "Convexification of Pareto Sets." Unpublished manuscript.

MAS-COLELL, A. (1975). "A Further Result on the Representation of Games by Markets." *J. Econom. Theory* 10, 117-122.

SHAPLEY, L.S. and SHUBIK, M. (1969). "On Market Games." *J. Econom. Theory* 1, 9-25.

WEBER, R.J. (1977). "Attainable Sets of Quasiconcave Markets." *Proc. Amer. Math. Soc.* 64, 104-111.

_____ (1978). "Attainable Sets of Quasiconcave Markets, II: Convexifiable Sets." *Math. Oper. Res.* 3, 257-264.

SUPPLEMENTARY REFERENCES

DEBREU, G. (1959). *Theory of Value.* Cowles Foundation Monograph No. 17, Yale University Press.

KANNAI, Y. (1981). Concave Utility Functions - Existence, Constructions, and Cardinality. This volume, 543-611.

PAPADOPOULOU, S. (1977). "On the Geometry of Stable Compact Convex Sets." *Math. Ann.* 229, 193-200.

CONCAVITY AND QUASICONCAVITY IN THE THEORY OF PRODUCTION

Wolfgang Eichhorn

The notions of concavity and quasiconcavity are standard terms found in publications on production or utility theory which consider scalar-valued functions. In most cases these functions are assumed, additionally, to be homogeneous or homothetic and to satisfy the law of eventually diminishing marginal returns or Gossen's law, respectively. The purpose of this paper, which relies heavily on Eichhorn (1978), is to study connections between these notions. Particular emphasis is on sets of conditions that imply concavity.

1. PROPERTIES OF A SCALAR-VALUED PRODUCTION FUNCTION

A scalar-valued production function

$$F: \mathbb{R}_+^n \to \mathbb{R}_+ ,$$

where $\mathbb{R}_+ \equiv \{r | r \varepsilon \mathbb{R} , r \geqq 0\}$,

assigns to each input vector $x \varepsilon \mathbb{R}_+^n$ the maximum output obtainable with $x = (x_1, \ldots, x_n)$ per unit time, namely $F(x)$.

Shephard (1967, 212; 1970a, 22) assumes the following six properties for a production function F:

(1) $F(0) = 0$ (nothing comes from nothing);

(2) $\{u | u = F(x), x \varepsilon D \cap \mathbb{R}_+^n , D \text{ bounded}\}$ is bounded;

(3) if $x^* \geqq x$, then $F(x^*) \geqq F(x)$ (free disposal);

(4) for any $x \geq 0$ (that is, $x \geqq 0$, $x \neq 0$) such that $F(\lambda x) > 0$ for some $\lambda \varepsilon \mathbb{R}_{++} = \{r | r \varepsilon \mathbb{R} , r > 0\}$, $F(\lambda x) \to \infty$ as $\lambda \to \infty$ (attainability of output);

(5) F is *quasiconcave* on $\mathbb{R}_+^n$; and

(6) F is *upper semicontinuous* on $\mathbb{R}_+^n$, that is, for any $x \varepsilon \mathbb{R}_+^n$ and arbitrary $\alpha \varepsilon \mathbb{R}_{++}$ there exists a δ-neighborhood $N_\delta(x) \equiv \{y | y \varepsilon \mathbb{R}_+^n , |y-x| < \delta, \delta \varepsilon \mathbb{R}_{++}\}$ such that $F(x^*) < F(x) + \alpha$ for any $x^* \varepsilon N_\delta(x) \cap \mathbb{R}_+^n$.

Properties (1)-(6) may also be considered properties of a utility function $F: \mathbb{R}_+^n \to \mathbb{R}_+$. It is easy to prove that properties (1)-(6) are

ISBN 0-12-621120-5

independent, i.e., any five of them do not imply the remaining one, and that they neither imply any of the following four properties that play an important role in the theory of production (or utility):

(7) F is *concave* on $\mathbb{R}_{+}^{n}$;

(8) F *satisfies the law of diminishing marginal returns*, that is, all the product curves generated by partial factor variation or by proportional variation of less than n factors (variables) satisfy this law. A product curve *satisfies the law of diminishing marginal returns* if there exists a point such that the curve becomes (and remains) strictly concave to the right of this point.[1] If F is assumed to be a utility function, this concept is called *Gossen's law*;

(9) F is *homogeneous of degree* r ($r \in \mathbb{R}_{++}$), that is,

$$F(\lambda x) = \lambda^{r} F(x) \quad \text{for all } (\lambda, x) \in \mathbb{R}_{++} \times \mathbb{R}_{+}^{n}; \text{ and}$$

(10) F is *homothetic*, that is, for any $x \in \mathbb{R}_{+}^{n}$, $y \in \mathbb{R}_{+}^{n}$ and $\mu \in \mathbb{R}_{++}$, $F(x) = F(y)$ if and only if $F(\mu x) = F(\mu y)$.

2. HOMOGENEITY, CONVEXITY AND CONCAVITY OF PRODUCTION FUNCTIONS

When we speak of a *production (or utility) function*, from now on, we always mean a function F with $n \geqq 2$ satisfying (1), (2), and $F(x) \not\equiv 0$.

THEOREM 1: The product curves derived from a linearly homogeneous (i.e., r=1 *in (9)) production function by partial variation of any factor or by proportional variation of any complex of less than* n *factors satisfy:*

(i) the law of diminishing marginal returns; and

1 For detailed studies on the law of diminishing returns see Shephard (1970b) and Färe (1980).

(ii) the condition that intervals of strict concavity are always unbounded to the right

if and only if they are strictly concave everywhere.

REMARK. Theorem 1 does not hold for homogeneity of degree $r > 1$. For instance, the production function

$$F(x,y) = \frac{x^{1+\beta}y^{1+\gamma}}{x+y}, \quad F(0,0) = 0$$

for $\beta\varepsilon(0,1)$, $\gamma\varepsilon(0,1)$ satisfies conditions (i) and (ii), but the graphs of $x \to F(x,y)$ and $y \to F(x,y)$ are *not* strictly concave everywhere. For $r < 1$ see Theorem 2.

PROOF OF THEOREM 1: Let

$$F\colon \mathbb{R}_+^n \to \mathbb{R}_+, \; (x,y) \to F(x,y),$$

$$x = (x_1, \ldots, x_\ell), \; y = (y_1, \ldots, y_m), \; \ell+m=n, \; \ell\geqq 1, \; m\geqq 1,$$

be a linearly homogeneous production function. In view of (i) assume that $g(\xi) \equiv F(\xi x,y)$ is strictly concave for all $\xi \varepsilon (\alpha,\infty)$, where $\alpha \varepsilon \mathbb{R}_{++}$ is a constant depending on x and y. Because of (ii) we are interested in the case where g is convex for all $\xi \varepsilon (0,\alpha]$. By the linear homogeneity of F and the convexity of g and $\mu \varepsilon \mathbb{R}_{++}$, $\nu \varepsilon \mathbb{R}_{++}$, $\mu+\nu = 1$,

$$\begin{aligned} F[x, (\mu\xi + \nu\eta)y] &= (\mu\xi + \nu\eta)F\left(\frac{x}{\mu\xi + \nu\eta}, y\right) \\ &= (\mu\xi + \nu\eta)F\left[\left(\frac{\mu\xi}{\mu\xi + \nu\eta}\frac{1}{\xi} + \frac{\nu\eta}{\mu\xi + \nu\eta}\frac{1}{\eta}\right)x, y\right] \\ &\leqq \mu\xi F\left(\frac{x}{\xi}, y\right) + \nu\eta F\left(\frac{x}{\eta}, y\right) \\ &= \mu F(x,\xi y) + \nu F(x, \eta y) \end{aligned}$$

for all $1/\xi$, $1/\eta \in (0,\alpha)$. By the strict concavity of $g(\xi)$ for $\xi \varepsilon (\alpha,\infty)$,

$$F(x,(\mu\xi + \nu\eta)y) > \mu F(x,\xi y) + \nu F(x,\eta y) \quad (\mu\varepsilon\mathbb{R}_{++}, \nu\varepsilon\mathbb{R}_{++}, \mu+\nu=1)$$

for all $1/\xi, 1/\eta \in (\alpha,\infty)$, $\xi\neq\eta$. Hence, $\xi\to F(x,\xi y)$ is strictly concave for all $\xi \varepsilon (0,1/\alpha)$ and convex for all $\xi \varepsilon [1/\alpha,\infty)$, which contradicts (ii).

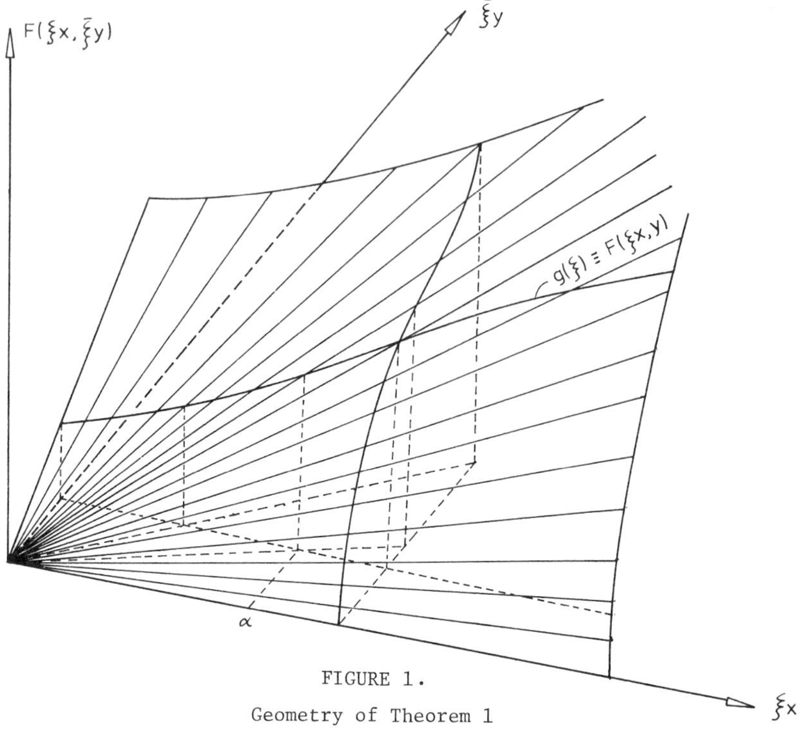

FIGURE 1.
Geometry of Theorem 1

Figure 1 illustrates Theorem 1 with an output surface of a linearly homogeneous production function that yields initially convex product curves when the factor complex x is increased proportionally.

COROLLARY 1: A linearly homogeneous production function does not satisfy the law of diminishing marginal returns if partial variation of at least one factor or proportional variation of at least one factor complex of less than n *factors yields a product curve that is initially convex.*

A similar nonexistence theorem when the homogeneity parameter $r < 1$ is:

THEOREM 2: There does not exist any twice continuously differentiable production function $F\colon \mathbb{R}_+^n \to \mathbb{R}_+$ *that is homogeneous of degree* $r \in (0,1]$, *and that satisfies the following three properties simultaneously:*

(11) *the total product curves generated by partial factor variation satisfy the law of diminishing marginal returns (8);*

(12) $F(x_1, \ldots, x_{n-1}, 0) = F(x_1, \ldots, x_{n-2}, 0, x_n) = \cdots$

$= F(0, x_2, \ldots, x_n) = 0$; *and*

(13) *there exist both an index* $p \in \{1, \ldots, n\}$ *and a constant factor quantity* $x_p^* \in \mathbb{R}_{++}$ *such that for all*

$(x_1, \ldots, x_{p-1}, x_{p+1}, \ldots, x_n) \in \prod_{i=1}^{n-1} (0,\delta)$

($\delta \in \mathbb{R}_{++}$ *a constant depending on* p *and* x_p^*) *the inequality*

$$\sum_{\substack{s,t=1 \\ s\neq p, t\neq p}}^{n} \frac{\partial^2 F(x_1,\ldots,x_{p-1},x_p^*,x_{p+1},\ldots,x_n)}{\partial x_s \partial x_t} \xi_s \xi_t \geqq 0 \tag{14}$$

holds with arbitrary $\xi_s \in \mathbb{R}$, $\xi_t \in \mathbb{R}$ $(s \neq p,\ t \neq p)$.

REMARKS. According to Property (12) it is necessary to apply positive amounts of all input quantities in order to obtain positive output quantities. Thus, all inputs are *essential* for production.

Property (13) implies that the function given by

$$\phi(x_1,\ldots,x_{p-1},x_{p+1},\ldots,x_n) \equiv F(x_1,\ldots,x_{p-1},x_p^*,x_{p+1},\ldots,x_n)$$

is convex in the (n-1)-dimensional cube $\prod_{i=1}^{n-1} (0,\delta)$. Note that (14) yields

$$\frac{\partial^2 F(x_1,\ldots,x_{p-1},x_p^*,x_{p+1},\ldots x_n)}{\partial x_q^2} \geqq 0 \text{ for } x_q \in (0,\delta),\ q\neq p,\ q=1,\ldots,n .$$

That is, the functions

$$h(x_q) = F(x_1,\ldots,x_{p-1},x_p^*,x_{p+1},\ldots,x_n)$$

are initially convex.

When $r = 1$, Theorem 2 remains true even without Property (12) (see Theorem 1). If $r \in (0,1)$, none of Properties (11)-(13) can be left out, as the following examples show.

The function $F: \mathbb{R}_+^2 \to \mathbb{R}_+$ given by

$$F(x,y) = \frac{x^2y^2}{x^{3+\gamma} + y^{3+\gamma}},\quad F(0,0) = 0 \qquad (\gamma \in (0,1))$$

satisfies (12) and (13), but not (11).

The function $F\colon \mathbb{R}_+^2 \to \mathbb{R}_+$ given by

$$F(x,y) = \frac{x^{2-\gamma}}{x+y} + y^{1-\gamma}, \quad F(0,0) = 0 \qquad (\gamma \in (0,1))$$

satisfies (11) and (13), but not (12).

The function $F\colon \mathbb{R}_+^2 \to \mathbb{R}_+$ given by

$$F(x,y) = (xy)^{1/3}, \quad F(0,0) = 0$$

satisfies (11) and (12), but not (13).

PROOF OF THEOREM 2: Let $F\colon \mathbb{R}_+^2 \to \mathbb{R}_+$ satisfy Property (13). Without loss of generality, choose $p = n$ and write x, c, and $G(x_1,\ldots,x_{n-1})$ for x_n, x_n^*, and $F(x_1,\ldots,x_{n-1},c)$, respectively. Then (14) becomes

$$\sum_{j,k=1}^{n-1} \frac{\partial^2 G(x_1,\ldots,x_{n-1})}{\partial x_j \partial x_k} \xi_j \xi_k \geqq 0, \tag{15}$$

where $(x_1,\ldots,x_{n-1}) \in \prod_{i=1}^{n-1} (0,\delta)$. Since F is homogeneous of degree r,

$$F(x_1,\ldots,x_{n-1},x) = \left(\frac{x}{c}\right)^r G\left(c\frac{x_1}{x},\ldots,c\frac{x_{n-1}}{x}\right), \tag{16}$$

where $x,c \in \mathbb{R}_{++}$. Differentiating (16) twice with respect to x and letting $z_i \equiv cx_i/x$, $i = 1,\ldots,n-1$,

$$\begin{aligned} \frac{\partial^2 F(x_1,\ldots,x_{n-1},x)}{\partial x^2} &= r(r-1)c^{-r}x^{r-2}G(z_1,\ldots,z_{n-1}) \\ &- 2(r-1)c^{1-r}x^{r-3}\sum_{i=1}^{n-1}\left(\frac{\partial G(z_1,\ldots,z_{n-1})}{\partial z_i} x_i\right) \\ &+ c^{2-r}x^{r-4}\sum_{j,k=1}^{n-1}\frac{\partial^2 G(z_1,\ldots,z_{n-1})}{\partial x_j \partial x_k} x_j x_k \,. \end{aligned} \tag{17}$$

We shall show that in the case where $r \in (0,1]$, because of (12), (13), expression (17) is greater than or equal to 0 for all $z_i \in (0,\delta)$, that is for all $x > cx_i/\delta$, which contradicts (11).

By (15), the last term in (17) is greater than or equal to 0. Thus Theorem 2 is proved by showing that (12), (13), and $r \in (0,1]$ yield the inequality

$$r(r-1)c^{-r}x^{r-2}G(z_1,\ldots,z_{n-1}) - 2(r-1)c^{1-r}x^{r-3}\sum_{i=1}^{n-1}\frac{\partial G(z_1,\ldots,z_{n-1})}{\partial z_i} \geqq 0 \qquad (18)$$

for all $(z_1,\ldots,z_{n-1}) \in \prod_{i=1}^{n-1}(0,\delta)$.

To prove this, consider the inequalities

$$\frac{r}{2}\frac{G(z_1,\ldots,z_{n-1})}{z_i} - \frac{\partial G(z_1,\ldots,z_{n-1})}{\partial z_i} \leqq 0, \; i = 1,\ldots,n-1, \qquad (19)$$

which hold true for all $(z_1,\ldots,z_{n-1}) \in \prod_{i=1}^{n-1}(0,\delta)$ because of both the convexity of G on $\prod_{i=1}^{n-1}[0,\delta)$ and (12); that is,

$$G(z_1,\ldots,z_{i-1},0,z_{i+1},\ldots,z_{n-1}) = 0, \; i = 1,\ldots,n-1.$$

Multiplication of inequality (19) by the nonpositive expression $2(r-1)c^{1-r}x^{r-3}x_i$ gives inequality (18).

3. HOMOTHETICITY, QUASICONCAVITY AND CONCAVITY OF PRODUCTION FUNCTIONS

We consider the class of upper semicontinuous production functions $F: \mathbb{R}_+^n \to \mathbb{R}_+$ that have the following properties:

(20) if $y \geqq x$, then $F(y) \geqq F(x)$ ("free disposal") and F has *nonincreasing returns to scale*, that is, for any $x \in \mathbb{R}_+^n$, $\lambda \to F(\lambda x)$ is nondecreasing and concave for $\lambda \in \mathbb{R}_+$;

(21) F is quasiconcave on $\mathbb{R}_+^n$; and

(22) F is homothetic.

Neither two of these properties are sufficient to yield concavity. For instance, the function $F: \mathbb{R}_+^2 \to \mathbb{R}_+$ given by Figure 1 satisfies (20) and (22) and is not concave; the function $G: \mathbb{R}_+^n \to \mathbb{R}_+$ given by

$$G(x_1,x_2) = \frac{x_1^{1.1}x_2^{1.4}}{x_1+x_2}, \quad G(0,0) = 0,$$

satisfies (21) and (22) and is not concave; and the function $H: \mathbb{R}_+^2 \to \mathbb{R}_+$ given by the graph in Figure 2 satisfies (20) and (21) and is not concave. The last example is due to Friedman (1973).

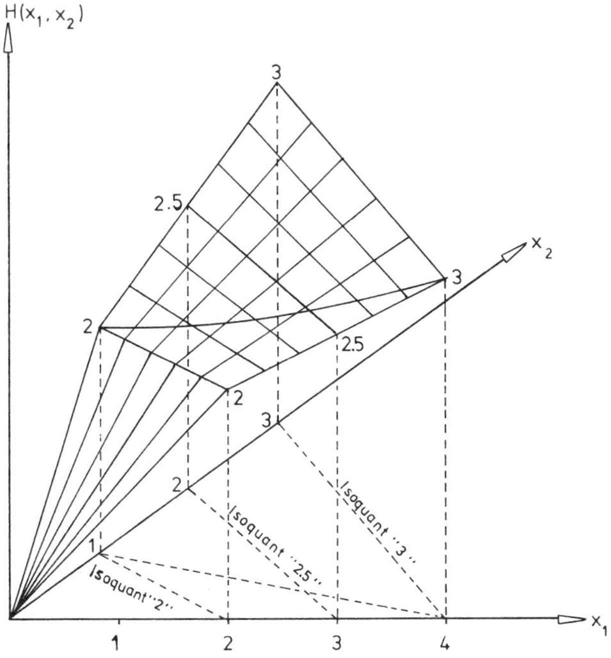

FIGURE 2.

A nonconcave output surface satisfying the assumptions of quasiconcavity and nonincreasing returns to scale.

THEOREM 3: An upper semicontinuous production function $F: \mathbb{R}_+^n \to \mathbb{R}_+$ *that satisfies assumptions (20)-(22) is concave on* $\mathbb{R}_+^n$.

PROOF (Friedman (1973)): Clearly F is concave on $\{z \mid z=\theta x+(1-\theta)y;\ \theta\in[0,1]\}$ when $F(x) = F(y)$, and when $x = \lambda y$, $\lambda \in \mathbb{R}_+$, from assumptions (21) and (20), respectively. Let $x \in \mathbb{R}_+^n$, $y \in \mathbb{R}_+^n$ not satisfy either of these conditions. Say $F(x) < F(y)$. Note that from assumptions (20), $F(x) \not\equiv 0$, and the upper semicontinuity of F it follows that $F(x) > 0$ whenever $x \neq 0$. Then by assumption (20) there is an unique $\mu \in \mathbb{R}_{++}$ such that $F(x) = F(\mu y)$. By (22), $F(x/\mu) = F(y)$. Now choose arbitrary $\theta \in [0,1]$, determining

$$z^{\theta} \equiv \theta x + (1-\theta)y,$$

and let

$$x^* \equiv \frac{1 - \theta + \theta\mu}{\mu} x, \quad y^* \equiv (1 - \theta + \theta\mu)y.$$

Then by (22), $F(x^*) = F(y^*)$ and

$$z^{\theta} = \frac{\theta\mu}{1 - \theta + \theta\mu} x^* + \frac{1 - \theta}{1 - \theta + \theta\mu} y^*$$

$$= \omega x + (1 - \omega)y,$$

where $\omega \equiv \theta\mu/(1-\theta+\theta\mu)$. By (21)

$$F(z^{\theta}) \geqq F(x^*) = F(y^*)$$

and by (20)

$$F(y^*) = F(\theta\mu y + (1-\theta)y)$$

$$\geqq \theta F(\mu y) + (1-\theta)F(y) = \theta F(x) + (1-\theta)F(y).$$

Therefore F is concave on $\mathbb{R}^n_+$.

REMARK. Strict concavity and, hence, the law of diminishing marginal returns obtains if (20) has strict concavity, (21) has strict quasiconcavity, and (22) holds.

Let $F: \mathbb{R}^n_+ \to \mathbb{R}_+$ be homogeneous of degree r, where $r \in (0,1]$. Then F satisfies assumption (22) and, moreover, the second assumption of (20). Thus we have the following well-known special case of Theorem 3 (see, e.g., Rader (1972)).

COROLLARY 2: The following properties of an upper semicontinuous function $F: \mathbb{R}^n_+ \to \mathbb{R}_+$ *(production or utility function, say) imply the concavity of* F:

i) homogeneity of degree r, *where* $r \in (0,1]$;

ii) quasiconcavity; and

iii) free disposal.

REFERENCES

EICHHORN, W. (1978). *Functional Equations in Economics.* Addison-Wesley, Reading, Massachusetts.

FÄRE, R. (1980). *Laws of Diminishing Returns.* Springer-Verlag, Berlin.

FRIEDMAN, J.W. (1973). "Concavity of Production Functions and Nonincreasing Returns to Scale." *Econometrica 41,* 981-984.

RADER, T. (1972). *Theory of Microeconomics.* Academic Press, New York.

SHEPHARD, R.W. (1967). "The Notion of a Production Function." *Unternehmensforschung 11,* 209-232.

SHEPHARD, R.W. (1970a). *Theory of Cost and Production Functions.* Princeton University Press, Princeton.

SHEPHARD, R.W. (1970b). "Proof of the Law of Diminishing Returns." *Zeitschrift für Nationalökonomie 30,* 7-34.

GENERALIZED CONVEXITY AND THE MEASUREMENT OF INEQUALITY

Wolfgang Eichhorn and Wilhelm Gehrig

The purpose of this note is to describe some connections between problems of the measurement of inequality in economics and problems of characterizing classes of well-known generalized convex functions. The results appear in both the economics and the mathematics literature.

We are interested in the measurement of the inequality of the income (or estate) distribution or of the concentration of industry.

Let $x = (x_1, \ldots, x_n) \in \mathbb{R}_+^n$, where $\mathbb{R}_+$ denotes the nonnegative reals, represent the income (or estate) distribution of n households or the turnover distribution of n firms.

We consider, as a particular set Σ of *inequality (concentration) measures*, the set of functions $I : \mathbb{R}_+^n \to \mathbb{R}$ that satisfy:

(i) the principle of progressive transfers:

$$I(x_\varepsilon^{ij}) < I(x) \text{ for all } x,\ x_\varepsilon^{ij} = (x_1, \ldots, x_i+\varepsilon, \ldots, x_j-\varepsilon, \ldots, x_n),$$

$$\text{whenever } x_i < x_j \text{ and } \varepsilon \in (0, (x_j - x_i)/2];\ \text{and}$$

(ii) symmetry:

$$I(Px) = I(x) \text{ for all } x \text{ and all permutation matrices } P.$$

The symmetry condition means that each household (firm) must be treated equally (the anonymity property). It is easy to defend as an attribute of a general measure of inequality. The principle of progressive transfers is harder to defend because it implies, for instance, that the inequality of (2,2,100) is *smaller* than that of (1,3,100). Examples of inequality measures exist in the economics literature for which $I(2,2,100) = I(1,3,100)$, for instance,

$$I(x) \equiv \frac{\sum_{i=1}^{n} \left| \frac{x_1+\ldots+x_n}{n} - x_i \right|}{x_1+\ldots+x_n}, \quad I(0) = 0$$

ISBN 0-12-621120-5

or

$$I(x) \equiv \frac{\max\{x_1,\dots,x_n\}-\min\{x_1,\dots,x_n\}}{x_1+\dots+x_n}, \quad I(0) = 0.$$

These and other examples of inequality measures, as in Sen (1973), satisfy

(i') *the weakened principle of progressive transfers*, that is, condition *(i)* with $\le$ instead of $<$, where $<$ remains valid whenever

$$x_i \le \frac{x_1+\dots+x_n}{n} < x_j = \max\{x_1,\dots,x_n\} \qquad \text{(Eichhorn (1980))}$$

and

(i") *the weak principle of progressive transfers*, that is, condition *(i)* with $\le$ instead of $<$.

We denote the set of functions satisfying *(i')*, *(ii)* or *(i")*, *(ii)* by Σ' and Σ'', respectively. Clearly, $\Sigma \subset \Sigma' \subset \Sigma''$.

In favor of conditions *(i)*, *(i')*, *(i")* it can be said that each of them implies the *mean value property* that every inequality (concentration) measure should have, namely,

$$I(\mu,\dots,\mu) \le I(x) \le \max_{P \in \mathcal{P}} I(P(\Sigma x_i, 0, \dots, 0)),$$

where

$\mu \equiv (x_1+\dots+ x_n)/2$, and $\mathcal{P}$ the set of all (n,n)-permutation matrices.

What can we say about the set Σ of inequality (concentration) measures? The following lemma gives a first indication.

LEMMA 1 (Hardy, Littlewood and Polya (1934), Berge (1963)), Dasgupta, Sen and Starrett (1973), Rothschild and Stiglitz (1973), Fields and Fei (1978), Marshall and Olkin (1979):

Let $x \in \mathbb{R}^n$ *(or* $\mathbb{R}^n_+$*) and* $y \in \mathbb{R}^n$ *(or* $\mathbb{R}^n_+$*) be ordered so that*

(0) $x_1 \le \dots \le x_n$, $y_1 \le \dots \le y_n$ *(w.l.o.g. because of (ii)).*

Then the following six conditions are equivalent:

(1) $x_1+\dots+x_k \ge y_1+\dots+y_k$ *for all* $k \in \{1,\dots,n-1\}$ *with strict inequality for at least one* k, *and* $x_1+\dots+x_n = y_1+\dots+y_n$;

(2) x *can be obtained from* y *by a finite sequence of operations of the form*

$$y_i^{\alpha+1} = y_i^\alpha + e^\alpha \leq y_j^\alpha \text{ for } i < j,\ e^\alpha > 0,$$

$$y_j^{\alpha+1} = y_j^\alpha - e^\alpha \geq y_i^\alpha,\ y_k^{\alpha+1} = y_k^\alpha \text{ if } k \neq i,j;$$

(3) x *can be obtained from* y *by a finite sequence of rank preserving operations of the form (2), that is,* $y_i^{\alpha+1}$ *has the same place, with respect to* $\leq$, *in vector* $y^{\alpha+1}$ *as* y_i^α *has in vector* y^α;

(4) *there exists a doubly stochastic*[1] *matrix* B *such that* x = By;

(5) *for any strictly Schur-convex (S-convex)*[2] *function* $S : \mathbb{R}^n \to \mathbb{R}$ *(or* $\mathbb{R}^n_+ \to \mathbb{R}$*) we have* $S(x) < S(y)$; *and*

(6) *for any strictly quasiconvex symmetric function* $Q : \mathbb{R}^n \to \mathbb{R}$ *(or* $\mathbb{R}^n_+ \to \mathbb{R}$*) we have* $Q(x) < Q(y)$.

This lemma does *not* imply that the set of all strictly quasiconvex symmetric functions is equal to our set Σ. This can be seen from the following proposition.

PROPOSITION 1: (a) If a function $F : \mathbb{R}^n \to \mathbb{R}$ *(or* $\mathbb{R}^n_+ \to \mathbb{R}$*) is strictly quasi-convex and symmetric then it is strictly* S*-convex.*

(b) The converse is not true.

PROOF: Let B be an arbitrary doubly stochastic (n,n)-matrix which is not a permutation matrix. As Berge (1963; 182) has shown, B can be written as:

---

[1] A square matrix is *doubly stochastic* if all its entries are nonnegative and each of its rows and columns sums to one. Note that in (4) x is a smoothing of y in the sense that each income (estate, turnover) level of x is a convex combination of income (estate, turnover) levels in y.

[2] A function $F:\mathbb{R}^n \to \mathbb{R}$ (or $\mathbb{R}^n_+ \to \mathbb{R}$) is said to be *strictly* S-*convex*, if it is S-*convex*, that is, if $F(By) \leq F(y)$ for all $y \in \mathbb{R}^n$ (or $\mathbb{R}^n_+$) and all doubly stochastic matrices B, and if $F(By) < F(y)$ whenever the vector By=x is *not* a permutation of the vector y. Note that the last condition is not equivalent to the condition that B be different from a permutation matrix.

$$B = \lambda_1 P_1 + \ldots + \lambda_m P_m,$$

where

$P_1, \ldots, P_m$ are permutation matrices and $m \geq 2$, $0 < \lambda_i < 1$, $\lambda_1 + \ldots + \lambda_m = 1$.

But then

$$\begin{aligned} F(By) &= F((\lambda_1 P_1 + \ldots + \lambda_m P_m)y) \\ &= F(\lambda_1 (P_1 y) + \ldots + \lambda_m (P_m y)) \\ &< \max_i F(P_i y) \quad \text{(by the strict quasiconvexity of } F \text{, if the vector } By =: x \text{ is not a permutation of the vector } y) \\ &= F(y) \quad \text{(by the symmetry of } F). \end{aligned}$$

(Another proof of (a) appears in Marshall and Olkin (1979; 69).)

The converse is not true since strict S-convexity does not imply strict quasiconvexity on lines (or line segments) that do not belong to hyperplanes (or subsets of these) of the form

$$\{y \in \mathbb{R}^n \,|\, y_1 + \ldots + y_n = \text{const}\} \quad (\text{or } \{y \in \mathbb{R}^n_+ \,|\, y_1 + \ldots + y_n = \text{const}\}). \qquad \square$$

The proposition implies that the set of all strictly quasiconvex and symmetric functions $F: \mathbb{R}^n_+ \to \mathbb{R}$ is a proper subset of Σ. Is this also true for the strictly S-convex functions $F: \mathbb{R}^n_+ \to \mathbb{R}$?

THEOREM 1 (Characterization of strictly S*-convex functions): The function* $I: \mathbb{R}^n_+ \to \mathbb{R}$ *belongs to the set* Σ*, and thus satisfies the principle of progressive transfers (i) and is symmetric (ii) if and only if it is strictly* S*-convex.*

Remark. Marshall and Olkin (1979; 54) *define* strict S-convexity by conditions *(i)* and *(ii)*.

PROOF: If I is strictly S-convex then I is symmetric (see Berge (1963; 220)) and satisfies the principle of progressive transfers (see the Lemma; equivalence of (2) and (4)).

It remains to prove that if I is symmetric and satisfies the principle of progressive transfers, it is strictly S-convex. Let us assume that I is *not* strictly S-convex and show that this yields a contradiction to assumption *(i)*. If I is not strictly S-convex then there exists an $y^o \in \mathbb{R}_+^n$ and a doubly stochastic matrix B^o satisfying

$$I(B^o y^o) \geq I(y^o),$$

where the vector $B^o y^o = x^o$ is not a permutation of the vector y^o. But $B^o y^o = x^o$ and y^o satisfy, after reordering according to *(0)*, condition (1) and thus condition (2). Hence we have a contradiction to the principle of progressive transfers, according to which

$$I(y^o) > I(y^1) > \ldots > I(y^N) = I(x^o) = I(B^o y^o). \qquad \square$$

Remark. It can also be shown that $I: \mathbb{R}_+^n \to \mathbb{R}$ belongs to the set Σ'', that is, it satisfies the weak principle of progressive transfers *(i″)* and is symmetric *(ii)* if and only if it is S-convex.

A characterization of the set Σ', whose functions are symmetric and satisfy the weakened principle of progressive transfers *(i′)*, has not yet been given. Such a characterization would be of interest for the following reason. The set Σ', which is a proper subset of the set Σ'' of the S-convex functions and contains the set Σ of the strictly S-convex functions properly, is well behaved from the measurement of inequality point of view. It is not too narrow as Σ is (see the examples), and not too wide as Σ'' is. For instance, the constant functions that are certainly not measures of inequality, belong to Σ''.

More generally, future research in this area should be directed to characterizations of particular measures of inequality by adding further conditions to "axioms" *(i)* or *(i′)* or *(i″)* and *(ii)*. See, in this connection, Atkinson (1970) and Blackorby and Donaldson (1978, 1980), who determine measures of inequality (or equality) in terms of social welfare (functions).

REFERENCES

ATKINSON, A.B. (1970). "On the Measurement of Inequality." *Journal of Economic Theory 2*, 244-263.

BERGE, C. (1963). *Topological Spaces*. Oliver and Boyd, Edinburgh and London.

BLACKORBY, C., and DONALDSON, D. (1978). "Measures of Relative Equality and Their Meaning in Terms of Social Welfare." *Journal of Economic Theory 18*, 59-80.

BLACKORBY, C., and DONALDSON, D. (1980). "A Theoretical Treatment of Indices of Absolute Inequality." *International Economic Review 21*, 107-136.

DASGUPTA, P., SEN, A., and STARRETT, D. (1973). "Notes on the Measurement of Inequality." *Journal of Economic Theory 6*, 180-187.

EICHHORN, W. (1980). "Wirtchaftliche Kennzahlen." *Quantitative Wirtchafts und Unternehmensforschung*. Ergebnisband des St. Galler Symposiums 1979, herausgeg. von R. Henn, B. Schips und P. Stähly. Springer-Verlag, Berlin.

FIELDS, G.S., and FEI, J.C.H. (1978). "On Inequality Comparisons." *Econometrica 46*, 303-316.

HARDY, G., LITTLEWOOD, J., and POLYA, G. (1934). *Inequalities*. Cambridge University Press, London.

MARSHALL, A.W., and OLKIN, J. (1979). *Inequalities: Theory of Majorization and Its Applications*. Academic Press, New York.

ROTHSCHILD, M., and STIGLITZ, J.E. (1973). "Some Further Results on the Measurement of Inequality." *Journal of Economic Theory 6*, 188-204.

SEN, A. (1973). *On Economic Inequality*. Clarendon Press, Oxford.

THE SIMPLICIAL DECOMPOSITION APPROACH IN OPTIMIZATION OVER POLYTOPES

Balder von Hohenbalken

This paper surveys simplicial decomposition algorithms which are due to von Hohenbalken (1975, 1977) and Wolfe (1976). Based on Carathéodory's theorem these methods can be applied to polyhedral feasible sets in external or internal representation and to objectives ranging from Euclidean distances to pseudoconcave functions. Exact and approximate versions of the algorithm and various applications are discussed.

1. INTRODUCTION

The term "simplicial decomposition" has here the following two meanings. One refers to the fact, stated in Carathéodory's theorem, that any compact, convex set X is the union of all simplices whose vertices belong to X. The other is that the type of algorithm discussed here also belongs to a class of methods descending from the decomposition principle for linear programs of Dantzig and Wolfe (1960). Typical representatives of this genre are Dantzig's columnar method (Dantzig (1963, Ch.24); Wagner (1969, Ch. 15)) and Huard's barycentric method (Broise, Huard, and Sentenac (1968, Ch.3)). Other approaches subsumable under this heading, but developed independently and originally based on different ideas are Wolfe's method to find nearest points in polytopes (Wolfe 1976) and Frank and Wolfe's (1956) algorithm. Geoffrion (1970) classified programming algorithms by strategies used. The (generalized) Frank-Wolfe method, our method and other decomposition algorithms involving barycentric representation use the general strategy of inner linearization followed by restriction.

Carathéodory's theorem retains its importance to our methods. However it is advantageous for comparison purposes to reformulate parts of the paper in terms of Dantzig-type decomposition. This mode of operation is to solve a given maximization problem by decomposing it into two parts called master and subprograms, or the master and satellite programs, or the minor and major cycles, which are solved repeatedly and

GENERALIZED CONCAVITY
IN OPTIMIZATION AND ECONOMICS

ISBN 0-12-621120-5

alternatingly. The master program usually represents a partial (often linearized) version of the original problem expressed in terms of barycentric coordinates corresponding to "generator" points fed to it by the subprogram. The primal solution point of the master program and/or associated Lagrangean multipliers are used to define the often linear (evaluated gradients are involved) objective function of the subprogram, which is maximized subject to all or some of the original constraints; the solution yields an additional generator point which is then utilized by the master program in the subsequent round, and so on. The routine is stopped by some exact or approximate optimality criterion applied to the solution of the subprogram.

The above procedure is applicable, *in principle*, to problems that are non-differentiable with general constraints or feasible sets given by finite point sets. Decomposition *per se* however, can only alleviate, not eliminate the algorithmic difficulties of nonlinearity. The satellite program always retains nonlinear constraints, and thus the burden of actual solution shifts back onto other algorithms to solve such problems. To be concrete and operational we shall confine our later discussion to methods for differentiable objective functions on polyhedral feasible sets. For these problems the subprograms of most decomposition methods become exact and finite procedures, and one can concentrate on optimal ways to solve the master program.

We now outline the special features of our simplicial decomposition method:

First, all constraints are assigned to the subprogram; thus the generator points it produces are always primally feasible and no dual variables are necessary to link the master to the subprogram.

Second, given the above, the criteria for selection of generator points and for optimality can be and are here formulated without using any Kuhn-Tucker-type results. Furthermore our algorithms remain well-defined and operational for almost all differentiable functions, for which they yield local maxima.

Third, the sets of generator points used by the master program are kept affinely independent, which implies that their convex hulls are simplices.

Fourth, the dimension of these simplices is kept minimal by dropping generators that do not carry (see section 2) relative maximizers. It is therefore possible to maintain strictly monotonic convergence to the maximum, despite irregular variations in the membership of the generator set.

Fifth, the simplices span linear varieties, into whose parallel subspaces the simplices are mapped. When redefined on these simplices, the master program becomes an essentially unconstrained maximization problem; this means that the best procedure to maximize any given function without restrictions can now be used to find constrained maximizers for it.

2. PRELIMINARIES

Let $x \in R^n$, with elements $x_i \in R$; $x_i > 0$ and $x_i \geqslant 0$ have the usual meaning; $x > 0$ means $x_i > 0$ for all i, $x \geqslant 0$ means $x_i \geqslant 0$ for all i; vectors with special properties (e.g., being a maximizer, an extreme point...) are superscripted: x^m, $\hat{x}^k$ etc. The gradient of $f : R^n \to R$ is denoted by $\frac{\partial f}{\partial x}$ or f_x; the Hessian by $\frac{\partial^2 f}{\partial x \partial x}$ or f_{xx}; evaluated at x^t, we write $\left.\frac{\partial f}{\partial x}\right|_{x=x^t}$ or f_{x^t} etc. Inner products are stated by mere juxtaposition; e.g. $f_{x^t}(x-x^t)$ is written for $\sum_{i=1}^{n} \left.\frac{\partial f}{\partial x_i}\right|_{x=x^t} (x_i - x_i^t)$; a transposition sign is used only to avoid ambiguities in matrix products, e.g. $D' f_{x^t x^t} D$. A vector $e \equiv (1,\ldots,1)$ with an appropriate number of components abbreviates summation: $ew = \sum_{i=1}^{k} w_i$.

Rockafellar (1970) distinguishes between two representations of a convex polyhedral set X, which are dual to each other: *externally* represented X as the intersection of a finite number of half-spaces and/or hyperplanes; and *internally* represented X as the convex hull of a finite point set.

The set of points $B \equiv \{\hat{x}^1,\ldots,\hat{x}^k\}$ is *affinely independent*, if the convex hull S of B is (k-1) - dimensional; B is then called an *affine basis*, and S is a *simplex*.

An *affine combination* is a linear combination whose weights sum to unity; the set M of all affine combinations of a given set B is a (linear) *manifold*; M is also the *affine hull* of B; if B is an affine basis, the unique weights expressing any $x \in M$ are called *barycentric coordinates*. L is the (unique) *subspace parallel* to the manifold M, if $L = M - \hat{x}$, for some $\hat{x} \in M$. L can also be expressed as the set of all linear combinations of the derived linear basis $D \equiv \{\hat{x}^1 - \hat{x}^k,\ldots,\hat{x}^{k-1} - \hat{x}^k\}$, whose members are linearly independent.

The *interior* of a convex set $S \subset R^n$ *relative* to its affine hull is denoted by *ri S*; clearly $S \neq \phi$ implies ri $S \neq \phi$. A vertex $\hat{x}^i$ of a simplex S is a *carrier* of $x \in S$ if the ith barycentric coordinate of x is positive; $x \in$ ri S if all vertices of S carry x. The vertex set of S is denoted by vert S.

The following results are fundamental to our approach to keep the set of generator points minimal and affinely independent:

THEOREM 1 (Carathéodory). Be $X \subset R^n$ *a nonempty compact and convex set; then every* $x \in X$ *lies in the relative interior of at least one simplex, whose vertices are extreme points of* X. *For proof see Rockafellar (1970, Theorem 17.1; 155; Corollary 18.5.1; 167).*

COROLLARY 1. If X *in addition is polyhedral, then the number of such simplices in which an* $x \in X$ *can lie is finite (Rockafellar (1970; Corollary 19.1.1)).*

COROLLARY 2. Assume the function $f: X \to R$ *is continuous and let* $S \subset X$ *be a nonempty simplex with vertex set* B*; then a maximizer of* f *on* S *exists and lies in the relative interior of a subsimplex of* S*, whose vertex set is a subset of* B.

PROOF. The corollary follows from Weierstrass theorem and Theorem 1.

For concreteness, the argument is carried on for differentiable functions, and gradients are used to define search directions and stopping criteria. Generalizations to not necessarily differentiable functions are possible (Wolfe,(1975)). We state now some well known results involving gradients, for reference and without proof.

LEMMA 1. Be $f: R \to R$ *differentiable,* $[x^t, x^b] \equiv \{(1-\alpha)x^t + \alpha x^b,\ 0 \leqslant \alpha \leqslant 1\}$ *a segment, whose end-points satisfy* $f_{x^t}(x^b - x^t) > 0$. *Then* $[x^t, x^b]$ *has points* y *that satisfy* $f(y) > f(x^t)$.

For proof see e.g. (van Moeseke (1965, Lemma 3.1.5)).

LEMMA 2. Be f *differentiable,* C *a compact convex polyhedral set, and* $x^t \in C$ *such that* $f(x^t) \geqslant f(x)$ *for all* $x \in X$. *Then*

a) $f_{x^t}(x - x^t) \leqslant 0$ *for all* $x \in C$

b) $f_{x^t}(\hat{x} - x^t) = 0$ *where* $\hat{x}$ *solves* $\max_{x \in C} f_{x^t} x$

c) $f_{x^t}(x - x^t) = 0$ *for all* $x \in C$ *when* $x^t \in \text{ri } C$;

A proof by contradiction is easy on the basis of Lemma 1.

LEMMA 3. Be f *a pseudoconcave function,* C *a convex set; then* $x^t \in \text{ri } C$ *is a global maximizer of* f *on* C *(i.e.* $f(x^t) \geqslant f(x)$ *for all* $x \in C$*) if and only if* $f_{x^t}(x - x^t) = 0$ *for all* $x \in C$.

The proof is immediate by Lemma 3 and the definition of pseudoconcavity.

3. THE STRUCTURE OF SIMPLICIAL DECOMPOSITION ALGORITHMS

Let the problem be

$$\max_{x \in X} f(x)$$

where $f : X \to R$ is differentiable, and X is a compact, convex, nonempty set. This problem is decomposed into master and subprogram which are solved alternately through successive rounds.

The *subprogram* is defined as

$$\max_{x \in X} f_{x^t} x$$

where x^t is an arbitrary point in the domain of f in the zeroth round, and the solution of the master program in all subsequent rounds. Each round adds a generator point $\hat{x}^k \in X$ which does *not* belong to the affine hull of $B^t = \{\hat{x}^1,\ldots,\hat{x}^{k-1}\}$. Since B^t is affinely independent itself it follows that the augmented set $B = \{\hat{x}^1,\ldots,\hat{x}^{k-1},\hat{x}^k\}$ is also affinely independent. B^t and B span simplices S^t and S resp., with S^t being facet of S.

The simplex $S \subset X$ in turn is the feasible set of the *master program*.

$$\max_{x \in X} f(x)$$

An exact or approximate solution of this master program yields $x^{t+1} \in \text{ri } S^{t+1} \subset S$, where S^{t+1} is either S itself or a proper face of S (excepting the facet S^t). Next, the generator set B is purged of points that do not carry x^{t+1}, which leaves $B^{t+1} = \text{vert } S^{t+1}$. The solution vector x^{t+1} thus lies in the relative interior of S^{t+1}. x^{t+1} also satisfies $f(x^{t+1}) > f(x^t)$ and is then used in the next round of the subprogram.

The procedure stops when the subprogram produces a point $\hat{x}^k$ that does not allow any further improvement, i.e., if $\hat{x}^k$ satisfies the optimality criterion $f_{x^t}(x^t - \hat{x}^k) = 0$.

With a polyhedral feasible set, the subprogram will be a *finite* and exact *procedure*, and remains essentially the same with any type of objective function. Only mild assumptions are needed to keep the *number* of iterations *finite*.

The procedure by which the master program is carried out has to be adapted for different classes of objective functions. For a narrow group of pseudoconcave functions (quadratics and the homogeneous $c'x - (x'Vx)^{1/2}$) a finite procedure has been constructed (von Hohenbalken (1975); Wolfe (1976)), whereby the whole algorithm becomes finite. Most other types of objective functions require infinite methods for their master programs, but because existing algorithms for unrestricted problems can be used virtually unaltered this poses few new problems.

If the constraints are also nonlinear, both sub-and master programs are infinite, and clearly both have to converge to insure convergence of the whole algorithm.

4. THE SUBPROGRAM

The subprogram of simplicial decomposition algorithms for polyhedral problems is a simple and finite procedure, that selects consecutive simplices in the feasible set X. From the initial step or the master program a simplex S^t with affine basis B^t is given, and an $x^t \in \text{ri } S^t$ such that $f_{x^t}(x-x^t) = 0$ for all $x \in S^t$. The subprogram finds an extreme point of X, say $\hat{x}^k$, that maximizes the linear form $f_{x^t}x$ on X, and then checks whether $f_{x^t}(\hat{x}^k-x^t) = 0$. If x^t is a global maximizer, then this is true (Lemma 2) and the algorithm terminates. If $f_{x^t}(\hat{x}^k-x^t) > 0$, an $x^{t+1} \in X$ can be found by the master program that satisfies $f(x^{t+1}) > f(x^t)$:

THEOREM 2. (von Hohenbalken (1977 ; 54)). Be S^t *the simplex spanned by the affine basis* $B^t \equiv \{\hat{x}^1,\ldots,\hat{x}^{k-1}\}$, $x^t \in \text{ri } S^t$ *such that* $f_{x^t}(x-x^t) = 0$ *for all* $x \in S^t$, *and let* $\hat{x}^k$ *satisfy* $f_{x^t}(\hat{x}^k-x^t) > 0$. *Then*

a) the augmented set $B \equiv \{\hat{x}^1,\ldots,\hat{x}^{k-1},\hat{x}^k\}$ *is an affine basis, that generates the new larger simplex* S;

b) S *contains at least one subsimplex* S^{t+1}, *that owns a relatively interior maximizer* x^{t+1} *that satisfies* $f(x^{t+1}) > f(x^t)$.

PROOF.

a) B^t also spans the manifold M^t; since $x^t \in \text{ri } S^t \subset M^t$, $f_{x^t}(x-x^t) = 0$ for all $x \in M^t$. $f_{x^t}(\hat{x}^k-x^t) > 0$ therefore implies $\hat{x}^k \notin M^t$, which makes the set $B \equiv \{\hat{x}^1,\ldots,\hat{x}^{k-1},\hat{x}^k\}$ affinely independent, since B^t was affinely independent.

b) Since both x^t, $\hat{x}^k \in S$ and S convex, the segment $[x^t,\hat{x}^k] \subset S$; by Lemma 1 there exist $y \in [x^t,\hat{x}^k]$ such that $f(y) > f(x^t)$. Therefore the relative maximizer x^{t+1} of f on S satisfies $f(x^{t+1}) \geq f(y) > f(x^t)$, and it can be found, by Corollary 2, in the relative interior of some subsimplex $S^{t+1} \subset S$.

The principal operation in the subprogram is to find an extreme point $\hat{x}^k \in X$, that maximizes the linear function $f_{x^t}x$; the procedure to be used obviously depends on the representation of the feasible set.

If X is externally given, $\hat{x}^k$ is located by the gradient-guided linear program

$$\max_{x \in X} f_{x^t}x, \quad X = \{x \in R^n | Ax \leq b, \ x \geq 0\} \ .$$

Its basic solution will be the desired extreme point $\hat{x}^k$.

If X is internally represented, one simply selects the largest component of a vector of inner products, i.e.

$$\max P f_{x^t} \text{ over all rows}$$

where P is a p x n matrix representing p points in R^n whose convex hull is X. The maximizing row of P, say $\hat{x}^k$, will be a satisfactory point of X.

On nonempty, compact and convex polyhedral feasible sets, the subprogram is a well defined and *finite procedure*. A condition on the objective function under which the *number* of rounds is *finite* is stated in the following proposition.

THEOREM 3: (von Hohenbalken, (1977, p. 55)). Let X *be a polyhedral feasible set,* f *a pseudoconcave objective function. Then the number of rounds is finite and the algorithm stops at the global maximum of* f *on* X.

PROOF. By Corollary 1, X contains a finite number of simplices (whose vertices are extreme points of X); by Lemma 3, the interior of each simplex contains at most one relative maximum (which is clearly relatively global); by Theorem 2, the objective function increases strictly between consecutive relative maximizers x^t and x^{t+1}, as long as $f_{x^t}(\hat{x}^k - x^t) > 0$; therefore no simplex can recur, and since each round proposes a new simplex, the number of rounds must be finite. The maximum on the last simplex to appear, say $f(x^T)$, is the maximum maximorum of all previous relative maxima, and x^T must satisfy $f_{x^T}(\hat{x}^k - x^T) = 0$ by Theorem 2. This condition is equivalent with x^T being a global maximizer of f on X, in view of Lemmas 2 and 3.

5. THE MASTER PROGRAM

What is true for most nonlinear programming algorithms holds also for simplicial decomposition methods: only for a small class of objective functions there is a finite and exact way to solve the master program. In addition to the linear case this class includes certain pseudoconcave functions, for instance the homogeneous function $cx-(xVx)^{1/2}$ and all pseudoconcave and semidefinite quadratics (von Hohenbalken (1975), Wolfe (1976) and Sacher (1980)). For most other pseudoconcave and for merely differentiable objective functions approximating procedures have to be used to solve the master program (von Hohenbalken (1977)).

Each round of the master program starts with the affine basis $B = \{\hat{x}^1, \ldots, \hat{x}^{k-1}, \hat{x}^k\}$ that spans the (k-1)-simplex S. By convention let $\hat{x}^k$ be the last accretion to B, which lies opposite the facet S^t of S that contains in its relative interior x^t, the solution of the previous round of the master program. We have $f(x^t) \geq f(x)$ for all $x \in S^t$ and $f_{x^t}\hat{x}^k > f_{x^t}x^t$.

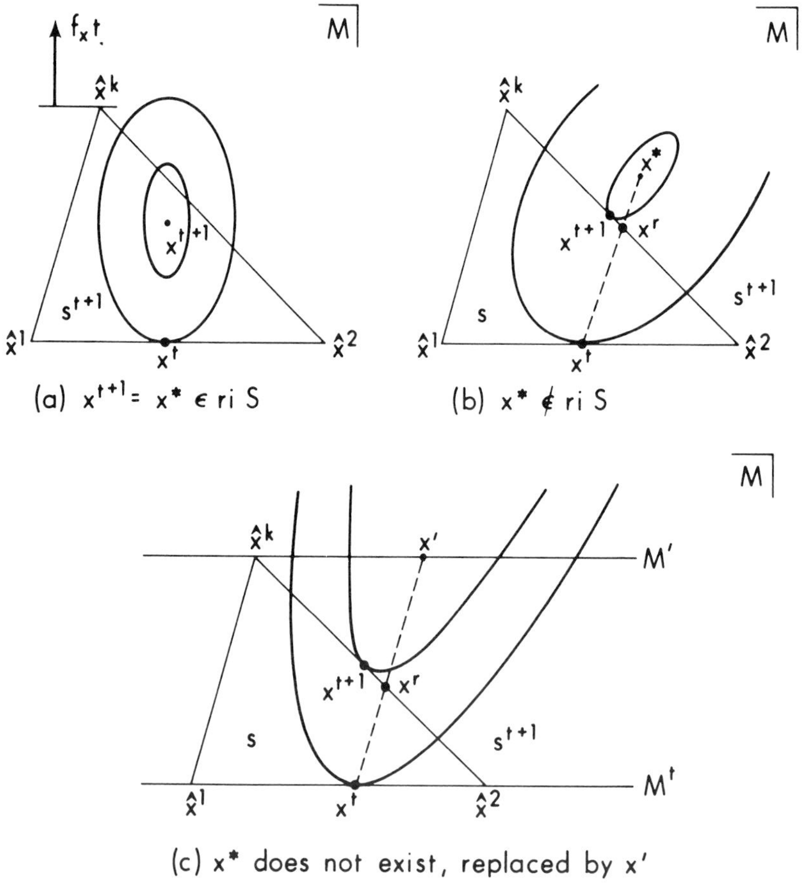

Figure 1. The finite master program.

From the second inequality above it follows by Corollary 2 and Lemma 1, that S must contain a maximizer x^{t+1} of f, such that $f(x^{t+1}) > f(x^t)$. The goal of the present round of the master program is to locate x^{t+1} and to determine the sub-simplex $S^{t+1} \subset S$ whose relative interior contains x^{t+1}. (By Carathéodory's theorem such an S^{t+1} always exists; it could be S itself or anyone of its proper faces, except, of course, the facet $S^t \subset S$).

Methods to find $x^{t+1} \in S^{t+1}$ have to diverge for different types of objective functions. For functions that allow, by a finite process, the exact determination of maximizers on manifolds, the master program is a finite procedure. Von Hohenbalken (1975) and Wolfe (1976) use calculus on manifolds to locate x^{t+1}. In the course of doing this one or more vertices of S might have to be dropped to find $S^{t+1} \subset S$ such that $x^{t+1} \in \text{ri } S^{t+1}$. Figure 1b shows the dropping procedure for the case when an unconstrained optimizer exists on the manifold M; Figure 1c shows the more sophisticated version that has to be adopted if no maximizer exists on M. For derivation and proofs see von Hohenbalken (1975)). To operate on generalized simplices (with vertices at infinity) Sacher (1980) employs Lemke's complementarity algorithm to find x^{t+1}.

For objective functions that do not admit a closed-form solution of the master program $\max_{x \in S} f(x)$, the problem is approximated by consecutive line searches in S. Because S is a simplex it is possible (and advantageous) to proceed in terms of barycentric coordinates; one uses the transformation $x = Bw$ to obtain $\max f(Bw)$, subject to $ew = 1$, $w \geq 0$. $B = \{\hat{x}^1, \ldots, \hat{x}^k\}$ is the affine basis of the simplex S. An even easier equivalent problem arises when the constraint $ew = 1$ is removed by eliminating w_k (the barycentric coordinate associated with $\hat{x}^k$); one obtains

$$\max_{y \geq 0} f(Dy + \hat{x}^k)$$

where $D = \{\hat{x}^1 - \hat{x}^k, \ldots, \hat{x}^{k-1} - \hat{x}^k\}$ and

$$y = (w_1, \ldots, w_{k-1}).$$

This last version of the master program, however, can be treated with methods of unconstrained optimization, because the remaining non-negativity constraints are never allowed to become binding: As soon as the line search (which proceeds in or into ri S) enters the boundary of S, non-carrying vertices are dropped from the basis B; the line search is then continued in the relative interior of some proper face of S; if *its* boundary is encountered, the algorithm reacts with further deletions from B, and so on. Reducing the dimensionality of the search thus *replaces* obeying constraints, which has the benefits indicated above.

The procedure stops when $x^{t+1} \in \text{ri } S^{t+1}$ is found, where S^{t+1} is a face of S (with possible dimensionality from 0 to k-1), and $f(x^{t+1}) \geq f(x)$ for all $x \in S^{t+1}$.

6. APPLICATIONS AND COMPUTATIONAL EXPERIENCE

The algorithm in its finite form was originally designed to solve the concave *homogeneous* problem

$$\max_{x \in X} \{f(x;m) \equiv c'x + m\,(x'Vx)^{\frac{1}{2}}\}$$

where V is positive definite, $m < 0$ and $X \subset R^n$ is compact, convex and polyhedral; the method provided the computational coping stone on a theory of decision-making under risk developed by van Moeseke (1965).

In the framework of *portfolio selection* one interprets c'x as expected net return, $(x'Vx)^{\frac{1}{2}}$ as the standard deviation of portfolio x, and m expresses the degree of risk aversion. As m percurs the real line, all efficient portfolios are revealed. For the special case where only one budget constraint and the short-selling prohibitions define the feasible set, i.e. when X is the simplex $\{x \in R^n | p'x = 1, x \geq 0\}$, the duality theory of homogeneous programming (van Moeseke (1965)) provides a nonsubjective criterion to select an *optimal* portfolio: portfolio $x^* \in X$ is optimal if x* is efficient and also allocates the budget with maximal caution, under the restriction that the marginal value of the budget dollar is not exceeded by its marginal cost.

The above theory and von Hohenbalken's algorithm have been applied to a 54 stock problem for the pension fund of the University of Alberta; see van Moeseke and von Hohenbalken (1974).

Despite being a by-product of the homogeneous method, it is the *quadratic* version of the author's algorithm that has reached prominence in professional portfolio optimization. For example, Rosenberg and Associates (BARRA Ltd. of Berkeley, California) sell a FORTRAN computer package together with forecasting and other services to investment houses for $30,000.-. The computational core of their portfolio selection procedure is the finite quadratic simplicial decomposition algorithm. The following "testimonial" is from the abstract of a paper by Rosenberg and Rudd (1979) : "We have programmed von Hohenbalken's algorithm to solve a realistic version of the optimal portfolio version problem, taking into account bounds and targets for portfolio yield and beta, upper limits on holdings, transactions costs, and a penalty for variance of total return implemented through a realistic model of the covariances among security returns. The algorithm seems to be ideally suited for this kind of problem; it approaches the optimum utility value with unprecendented rapidity and apparent robustness, and operates in a small workspace". Observe that Rosenberg and Rudd treat a *portfolio revision* problem, i.e., they compute an optimal vector of changes Δx of holdings from one given portfolio to another better one. Apart from being realistic this transformed problem allows otherwise nondifferentiable transactions costs to be incorporated in a smooth quadratic objective function. As an example, Rosenberg and Rudd give a realistic portfolio revision problem with 46 stocks in the portfolio and 106 acceptable assets in the universe. Twelve iterations of the algorithm, that took 2.5 seconds CPU time (on a CDC 7600 at Berkeley) and cost 50 cents, achieved 99% of the total potential gain in utility; (the exact optimum occurs after 30 further and more expensive iterations). This "early improvement" behaviour of the algorithm has been observed in most portfolio problems and can be exploited by early stopping rules. This feature and the ease of implicit upper bounding in the linear programming subroutine makes, according to Rosenberg and Rudd, the solution of problems with 1000 variables entirely feasible.

Recently J.S. Pang (1981) demonstrated "that the van de Panne-Whinston (1969) symmetric simplex method when applied to a certain implicit formulation of a quadratic program generates the same sequence of primal feasible vectors as does the von Hohenbalken simplicial decomposition algorithm specialized to the same program. Such an equivalence of the two algorithms extends earlier results for a least-distance program due to Cottle and Djang (1979).

Another important and widespread form of quadratic simplicial decomposition algorithms are the *least distance methods*; here the objective function is the (squared) Euclidean norm and the feasible set X is the convex hull of a finite point set. Let $y \in R^n, X \subset R^n$, and let Nr $y \in X$ be the point that minimizes $||x - y||$ over all $x \in X$; then

$y \in X$ iff $y = \text{Nr } y$; this amounts to a decision procedure for the non-trivial problem whether some point lies in an internally given polytope or not. Furthermore, as shown earlier, the simplicial decomposition procedure yields an affinely independent subset of vertices of X of which Nr y is a strictly convex combination (in other words, these vertices span a simplex S, and $\text{Nr } y \in \text{ri } S$).

Wolfe (1975, 1976) who independently developed least distance methods, uses them to find shortest subgradients in subdifferentials, an important routine in non-smooth optimization methods.

The above properties are employed in algorithms to derive the facial structure of polytopes, to find simplicial subdivisions, to determine the basis of a matrix and to find its Moore-Penrose inverse (von Hohenbalken (1978,1981),von Hohenbalken and Riddell (1979)). The polytope methods were actually developed for B. Clarke, a theoretical chemist at the University of Alberta. Clarke (1980) is advancing chemical stability research by converting the traditional reaction model to simplicial sets of parameters that make tractable former impasses of analysis.

The efficiency of least distance methods has been investigated and not found wanting by Wolfe (1975). Cottle and Djang (1979) show that Wolfe's nearest point algorithm generates the same sequence of feasible points as does the van de Panne-Whinston symmetric simplex algorithm applied to the associated quadratic programming problem.

Finally, some computational experience with infinite versions of simplicial decomposition algorithms. As shown above these methods have in fact exactly the same structure as their finite cousins, with the one difference that the master program cannot be solved exactly and finitely; in other words, the maximizer of f on the simplex S has to be approximated. It has been argued above that this can be done by direct implementation of any desired or suitable method for unconstrained optimization; since the subprogram remains a simple, finite and exact procedure, the convergence rate of the whole algorithm is the *same* as that of the procedure chosen to solve the master program.

For demonstration purposes, von Hohenbalken (1977) uses several quasi-Newton methods in conjunction with a nonmyopic line search in the master program to solve the three linearly constrained test problems in Colville (1968), contributed by Shell, Wood-Westinghouse and Gauthier-IBM France. The same procedures were employed to maximize, under linear constraints, posynomials of geometric programming and Fenchel's function, $(x-1) + [(1-x)^2 + 4(x+y)]^{1/2}$. Furthermore, because these line search procedures are impervious to multimodality and/or lack of pseudoconcavity, extrema for bilinear forms and Rosenbrock's curved ridge $[-100(y-x^2)^2 - (1-x)^2]$ were found.

REFERENCES

BROISE, P., HUARD, P. and SENTENAC, J. (1968). *Décomposition des Programmes Mathématiques*. Dunod, Paris.

CLARKE, B.L. (1980). "Stability of Complex Reaction Networks." *Advances in Chemical Physics*. (Prigogine and Rice, eds.). Wiley, New York.

COLVILLE, A.R. (1968). "A comparative study on nonlinear programming codes." IBM Technical Report No. 320-2949. New York Scientific Center.

COTTLE, R.W. and DJANG, A. (1979). "Algorithmic equivalence in quadratic programming, I: A least-distance programming problem." *J. or Optimization Theory and Applications 28*, 275-301.

DANTZIG, G.B. (1963). *Linear Programming and Extentions*. Princeton University Press, Princeton, N.J.

FRANK, M. and WOLFE, P. (1956). "An algorithm for quadratic programming." *Naval Research Quarterly 3*, 95-109.

GEOFFRION, A.M. (1970). "Elements of large-scale mathematical programming." *Management Science 16*, 652-691.

MANGASARIAN, O.L.(1969). *Nonlinear Programming*. McGraw-Hill, New York.

PANG, J.S. (1981). "An equivalence between two algorithms for quadratic programming." *Mathematical Programming 20*, 152-165.

ROCKAFELLAR, R.T. (1970). *Convex Analysis*. Princeton University Press, Princeton, N.J.

RUDD, A. and ROSENBERG, B. (1979). "Realistic portfolio optimization." *Studies in Management Sciences 11*, 21-46.

SACHER, R.S. (1980). "A decomposition algorithm for quadratic programming." *Mathematical Programming 18*, 16-30.
VAN DE PANNE, C. and WHINSTON, A. (1969). "The symmetric formulation of the simplex method for quadratic programming." *Econometrica 37*, 507-527.
VAN MOESEKE, P. (1965). "Stochastic linear programming: A study in resource allocation under risk." *Yale Economic Essays 5*, 196-254.
VAN MOESEKE, P. and VON HOHENBALKEN, B. (1974). "Efficient and optimal portfolios by homogeneous programming." *Zeitschrift Für Operations Research 18*, 205-214.
VON HOHENBALKEN, B. (1972). "Differentiable programming on polytopes." Research Paper Series, No. 14. Department of Economics, University of Alberta
VON HOHENBALKEN, B. (1975). "A finite algorithm to maximize certain pseudoconcave functions on polytopes." *Mathematical Programming 9*, 189-206.
VON HOHENBALKEN, B. (1977). "Simplical decomposition in nonlinear programming algorithms." *Mathematical Programming 13*, 49-68.
VON HOHENBALKEN, B. (1978). "A fast and simple method to find rank and basis for a matrix." *Econometrica 46* , 241-242.
VON HOHENBALKEN, B. (1978). "Least distance methods for the scheme of polytopes." *Mathematical Programming 15* , 1-11.
VON HOHENBALKEN. B. (1981). "Finding simplicial subdivisions of polytopes." Forthcoming in *Mathematical Programming.*
VON HOHENBALKEN, B. and RIDDELL, W.C. (1979). "A compact algorithm for the MOORE-PENROSE generalized inverse." *APL QUOTE QUAD 10*, 2.
WAGNER, H.M. (1969). *Principles of Operations Research.* Prentice-Hall, N.J.
WOLFE, P. (1975). "A method of conjugate subgradients for minimizing nondifferentiable functions." *Math. Programming Study 3, Non-differentiable Optimization*, (Balinski and Wolfe, eds.). North-Holland, Amsterdam, 145-173.
WOLFE, P. (1976). "Finding the nearest point in a polytope." *Mathematical Programming 11*, 128-149.

VECTOR-VALUED OPTIMIZATION

B.D. CRAVEN

This paper deals with constrained differentiable minimization problems in which the objective takes vector values. The concepts of strong minimum, vector minimum, and weak minimum are defined. Given a constraint qualification, a weak minimum implies the Kuhn-Tucker conditions. Conversely, if the Kuhn-Tucker conditions hold, then, subject to certain generalized convexity assumptions, conditions for a weak minimum are satisfied. There are analogous results for a strong minimum. Strong minimization also allows a vector duality theory in terms of a vector Lagrangian. Duality theorems for such problems are presented.

1. INTRODUCTION

Consider the constrained minimization problem

$$\text{Minimize } f(x) \text{ subject to } x \in H, \tag{1}$$

where x runs over some normed vector space X (say $X = \mathbb{R}^n$), the *feasible set* $H \subset X$ (with H often specified by inequalities such as $h(x) \leqslant 0$), and f(x) takes *vector values* in some space W (say $\mathbb{R}^r$). What is meant by a *minimum* for (1) is then a matter of definition. In some such problems, the components of f(x) fall into some natural order of priority; in some problems, the components of f(x) are deemed to be all measurable on some common scale (usually dollars), so that the problem reduces to minimizing some real function (often linear) of the components of f(x). In this paper, we shall discuss instead the case when the components of f(x) are *not* placed in a priority order (in some sense, they are equally important), and are not measurable on a common scale.

Several kinds of vector-valued minimum can be defined, in terms of a *convex cone* $P \subset W$. A subset $P \subset W$ is a *convex cone* if $p + q \in P$ and $\alpha p \in P$ whenever $p \in P$, $q \in P$ and $\alpha \in \mathbb{R}_+ \equiv (0,\infty) \subset \mathbb{R}$. In particular, the *nonnegative orthant* $\mathbb{R}^r_+ = \{w \in \mathbb{R}: \text{ all } w_j \geqslant 0\}$ is a convex cone. More generally, P is a *polyhedral cone* if $p \in P$ if and only if $Cp \in \mathbb{R}^s_+$, for some integer s and some real $s \times r$ matrix C. This means that $p \in P$ if and only if each of a certain set of linear combinations of elements of p (with coefficients

ISBN 0-12-621120-5

given by the rows of C) is nonnegative. From linear programming theory, it follows that P is a polyhedral cone if and only if $P = M(\mathbb{R}^k_+) \equiv \{Mz : z \in \mathbb{R}^k_+\}$ for some matrix M, and thus if P has a finite set of generators g_i (the columns of M), with each $p \in P$ of the form $p = \sum \alpha_i g_i$ with each $\alpha_i \in \mathbb{R}_+$. The *norm* (or length) of the vector x is denoted by $\| x \|$. The point $p \in \text{int } P$ (the *interior* of P) if $p + n \in P$ whenever $\|n\|$ is sufficiently small.

For the vector minimization problem (1), the point $\bar{x} \in H$ is

a *strong* local minimum if $f(x) - f(\bar{x}) \in P$ for each $x \in H \cap N$, (2)

(for some ball $N = \{x \in X : \| x - \bar{x} \| < \delta\}$, of sufficiently small radius δ);

a *vector* local minimum if $f(x) - f(\bar{x}) \in W \setminus (-P_o)$ for each (3)

$x \in H \cap N$, (where P_o means the cone P with the point 0 deleted, and $\setminus$ denotes set difference - thus $W \setminus (-P_o)$ means the complement of $-P_o$);

a *weak* local minimum if $f(x) - f(\bar{x}) \in W \setminus (-\text{ int } P)$ for

each $x \in H \cap N$, (where $-\text{ int } P \equiv \{-p : p \in \text{int } P\}$). (4)

If $H \cap N$ is replaced by H, then *global* minima are obtained, instead of local.

These concepts are illustrated in Figure 1, for which P is the indicated sector in $W = \mathbb{R}^2$. The values of $f(x) - f(\bar{x})$ must remain in the shaded region (when $x \in H \cap N$).

Clearly,

$$\text{strong minimum} \Rightarrow \text{vector minimum} \Rightarrow \text{weak minimum}. \tag{5}$$

If P is a polyhedral cone, then a *strong* minimum is described by

$$(\forall\, x \in H \cap N) \quad \sum_j c_{ij}[f_j(x) - f_j(\bar{x})] \geqslant 0 \tag{6}$$

for *all* indices i, where C has matrix elements C_{ij}. A *weak* minimum is described by (6) for *some* index i (since $[(\forall i) \sum_j C_{ij}[f_j(x) - f_j(\bar{x})] < 0]$ is contradicted.

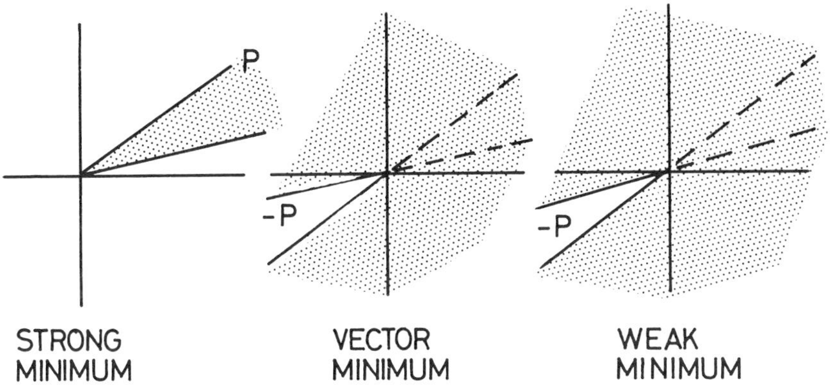

FIGURE 1. Three kinds of vector-valued minimum

A *maximum* is defined by minimizing -f.

One context where *weak minimum* arises is the planning of an industrial process, where it is sought to "minimize", subject to constraints, each of several incompatible variables (such as cost, wastage, and the negative of some quality factor). Often, some linear combination of these variables is taken as the criterion to be minimized, but usually there is no a priori reason for supposing that the *same* linear criterion remains valid, for all values of the variables. One may more reasonably choose a number of such combinations (each representing a different weighting of the decision variables, which describes some acceptable performance), and require at least one combination to be minimized. This gives a weak minimum.

In contrast, a strong minimum with respect to $R = \mathbb{R}_+^r$ requires the *simultaneous* minimization of each component $f_j(x)$, and this is rarely appropriate. Strong minimization with respect to a suitable polyhedral cone allows more flexibility. Consider, in particular (Craven, 1980a) a linear program with $x \in H \Leftrightarrow Ax - b \geqslant 0$, and set

$$f(x) = \begin{bmatrix} d_1^T x \\ d_2^T x \\ d_3^T x \end{bmatrix}, \quad C^T = \begin{bmatrix} 1 & 0 \\ \beta & 1 \\ \gamma & \delta \end{bmatrix}, \tag{7}$$

with $y \in P$ iff $Cy \in R_+^2$; β, γ, δ are real constants. Assume that $d_1^T s$ is minimized, subject to the constraints, at $x = \bar{x}$. Except in some limiting cases, a perturbation of $d_1^T x$ to $(d_1 + \beta^{-1} d_2)^T x$ or to $(d_1 + \gamma^{-1} \delta d_3)^T x$ leaves the optimal basis unchanged, while β and $\gamma^{-1}\delta$ remain in suitable intervals. For such β, γ, δ, the vector function $f(x)$ reaches a strong minimum with respect to the polyhedral cone $P = C^{-1}(R_+^2)$. Moreover, if β, γ, δ are fixed, in the interiors of their allowed intervals, a strong minimum of the problem, with respect to P is still reached, even when d_1, d_2, d_3, A, b are somewhat perturbed.

This paper will present results for *differentiable* minimization problems. Most of these results extend to problems with *convex* functions, not necessarily differentiable at all points, with derivatives replaced by *subdifferentials* (see e.g. Rockafellar, 1970, for definitions).

2. LAGRANGE MULTIPLIERS AND WEAK MINIMA

Consider the minimization problem

$$\text{Minimize } f(x) \text{ subject to } -k(x) \in J, \tag{8}$$

where f and k are differentiable functions, and J is a closed convex cone. (In particular, if $J = \mathbb{R}_+^p \times \{0\}$, where $\{0\}$ is the zero (polyhedral) cone in $\mathbb{R}^q$, then $-k(x) \in J$ represents p inequalities $k_j(x) \leqslant 0$ and q equalities $k_j(x) = 0$.) Suppose first that $f(x)$ is real valued; that (8) reaches a (local) minimum at $x = \bar{x}$, and that some smoothness requirement ("constraint qualification") holds. Then the Kuhn-Tucker conditions hold; they may be expressed as

$$\tau f'(\bar{x}) + \lambda k'(\bar{x}) = 0; \quad \lambda k(\bar{x}) = 0, \ 0 \neq \tau \in \mathbb{R}_+ \ ; \ \lambda \in J^*. \tag{9}$$

(So $\tau > 0$; usually τ is taken as 1.) Here J^* denotes the *dual cone* to J; if Z' is the *dual space* to Z (the set of continuous linear functions from Z into $\mathbb{R}$), then J^* is the set of $v \in Z'$ which map J into $\mathbb{R}_+$. (When $J = \mathbb{R}_+^p \times \{0\}$, then $J^* = \mathbb{R}_+^p \times \mathbb{R}^q$, and λ is a row vector of components λ_i; $\lambda_i \geqslant 0$ whenever λ_i corresponds to an *inequality* $k_i(x) \leqslant 0$.)

If, instead, $f(x)$ is vector valued, and a *weak minimum* of (8) is reached at $\bar{x}$, with respect to a cone P, then the Kuhn-Tucker necessary conditions are also obtained, but with $0 \neq \tau \in P^*$; thus

$$\tau f'(\bar{x}) + \lambda k'(\bar{x}) = 0; \quad \lambda k(\bar{x}) = 0 \; ; \; 0 \neq \tau \in P^*; \; \lambda \in J^*. \tag{10}$$

The proof (see Craven, 1977) is almost the same as when $f(x)$ is real. (The notion of *weak minimum* began here; see Craven (1972), Borwein (1974)).

Conversely, suppose that $\bar{x}$ satisfies $-k(\bar{x}) \in J$, together with (10). These will imply a minimum for (8), under additional assumptions of convexity. However, suitable generalized convexity will do. The relevant definitions are as follows. Consider a differentiable function $h : X \to Z$, and a cone $T \subset Z$. The function h is

T-convex if $h(x) - h(u) \geqslant h'(u)(x - u)$; (11)

T-quasiconvex if $h(x) - h(u) \in -T \Rightarrow h'(u)(x - u) \in -T$; (12)

T-pseudoconvex if $h'(u)(x - u) \in T \Rightarrow h(x) - h(u) \in T$; (13)

in each case, for all x and u in the domain of h; $h'(u)$ denotes the (Fréchet) derivative, thus $h(x) = h(u) + h'(u)(x - u) + o(\|x - u\|)$; $-T \equiv \{-t : t \in T\}$.

(h is *continuously differentiable* if also $h'(\cdot)$ is a continuous function.) If $Z = \mathbb{R}$ and $T = \mathbb{R}_+$, then (11), (12) and (13) reduce to the usual convexity of quasi- (pseudo-) convexity for real functions.

The "constraint qualification" assumed will be as follows. The contraint $-k(x) \in J$ is *locally solvable at* $\bar{x}$ if $-k(\bar{x}) \in J$, and whenever the direction d satisfies the linearized inequality $-k(\bar{x}) - k'(\bar{x})d \in J$ and $\|d\|$

is sufficiently small, there is for sufficiently small positive α a local solution $x = \bar{x} + \alpha d + \eta(\alpha)$ to $-k(x) \in J$, where $\eta(\alpha) = o(\alpha)$, meaning that $\|\eta(\alpha)\|/\alpha \to 0$ as $\alpha \downarrow 0$. In case $J = \mathbb{R}^p_+ \times \{0\}$, *locally solvable* is equivalent to the Kuhn-Tucker constraint qualification, and it holds, in particular, if the gradients of the active constraints at $\bar{x}$ are linearly independent. However, locally solvable extends to more complicated constraint systems (Craven, 1978). Unless the cone J is polyhedral, it must also be assumed that a certain cone K is closed. Define the linear mapping (or matrix) $M : \mathbb{R} \times X \to Z$ by $M(r,x) = k(\bar{x})r + k'(\bar{x})x$; denote its adjoint (or matrix transpose) by M^T. Then $C_o = M^T(J^*) \equiv \{M^T v : v \in J^*\}$ is a convex cone, but not always closed (containing all boundary points). (If the spaces are infinite dimensional, as for example in an optimal control problem, then *closed* must be taken in the "weak*" sense; see e.g. Craven (1978a;11).

Denote $b = -k(\bar{x})$; define the convex cone $J_b = \{\alpha(z - b) : \alpha \in \mathbb{R}_+,\ z \in J\}$. In case $Z = \mathbb{R}^m$ and $J = \mathbb{R}^m_+$, the *active constraints* are those $k_i(x) \leqslant 0$ for which $k_i(\bar{x}) = 0$. It follows readily that the active constraints are expressed by $-k(x) \in J_b$. Since the dual cone $(J_b)^* = \{u \in J^* : ub = 0\}$, the Kuhn-Tucker conditions (10) can be expressed in terms of active constraints only, by

$$\tau f'(\bar{x}) + \lambda k'(\bar{x}) = 0 \ ; \ \lambda \in (J_{-k(\bar{x})})^* \ . \tag{14}$$

The results for weak minimization are then as follows.

THEOREM 1: Let problem (8) reach a weak minimum at $x = \bar{x}$; *let* $-k(x) \in J$ *be locally solvable at* $\bar{x}$, *and let the cone* C_o *be (weak*) closed. Then the Kuhn-Tucker necessary conditions (10) hold, for some Lagrange multipliers* λ *and* τ.

PROOF: See Craven (1977). (The proof of Theorem 5, below, uses similar ideas.)

THEOREM 2: For problem (8) with constraint $-k(x) \in J$ *specified as* $[-g(x) \in S,\ h(x) = 0]$, *let the Kuhn-Tucker conditions hold at a feasible point* x; *thus let* $\tau f'(\bar{x}) + vg'(\bar{x}) + wh'(\bar{x}) = 0$; $vg(\bar{x}) = 0$; $0 \neq \tau \in P^*$, $v \in S^*$, $w \in Z'$; *where* S *is a convex cone. Let* g *be* S_b *-quasiconvex, where* $b = -g(\bar{x})$; *either let both* h *and* -h *be T-quasiconvex, with* $T \cap (-T) = \{0\}$, *or let* h *be T-quasiconvex with* $w \in T^*$, *where* T *is a closed convex cone. For each* $\theta \in P^*$, *let* θf *be pseudoconvex. Then* $\bar{x}$ *is a weak (local) minimum.*

PROOF: Since $v \in S^*$ and $vb = 0$, $v \in (S_b)^*$. Let x satisfy $-g(x) \in S$, $h(x)=0$. From $g(\bar{x}) - g(x) \in S_b$, and g assumed S_b -quasiconvex, it follows that $-g'(\bar{x})(x - \bar{x}) \in S_b$. Hence $vg'(\bar{x})(x - \bar{x}) \leqslant 0$. A similar argument with h and T shows, in one case, that $-h'(\bar{x})(x - \bar{x}) \in T \cap (-T) = \{0\}$, so that $wh'(\bar{x})(x - \bar{x}) \leqslant 0$, and in the other case that $-h'(\bar{x})(x - \bar{x}) \in T_{-h(\bar{x})}$; with $w \in T^*$ and $wh(\bar{x}) = 0$. $w \in (T_{-h(\bar{x})})^*$, so that again $wh'(\bar{x})(x - \bar{x}) \leqslant 0$. These inequalities combine with (15) to give $\tau f'(\bar{x})(x - \bar{x}) \geqslant 0$. By hypothesis, τf is pseudoconvex, hence $\tau f(x) \geqslant \tau f(\bar{x})$. Suppose, if possible, that $\bar{x}$ is not a weak minimum. Then there is a sequence $\{x_i\} \to \bar{x}$ of feasible points satisfying $(\forall i)\ f(x_i) - f(\bar{x}) \in -\operatorname{int} P$. Since $0 \neq \tau \in P^*$, $\tau f(x_i) - \tau f(\bar{x}) < 0$, contradicting $\tau f(x) \geqslant \tau f(\bar{x})$. //

Quasi-(pseudo-) convexity is only required here at the point $\bar{x}$, and with the function domain restricted to a neighbourhood of $\bar{x}$, to prove a *local* weak minimum. Of course, convexity implies both quasi- and pseudo-convexity; but we don't need all of convexity here. In particular, if $W = \mathbb{R}$ and $P = \mathbb{R}_+$, the requirement on f is just pseudoconvexity ($\equiv \mathbb{R}_+$ - pseudoconvexity) at $\bar{x}$. From Diewert, Avriel and Zang (1977), a function $\phi : \mathbb{R}^n \to \mathbb{R}$ is quasiconvex on an open convex set C if ϕ is positive semidefinite on the subspace orthogonal to C, and if $f'(x) \neq 0$ on C; this reference gives also various other criteria for pseudo- and quasi-convexity. Also (see Avriel and Schaible, 1978), ϕ is pseudoconvex on C (which implies quasiconvex) if ϕ is twice differentiable, and if $H_{i_1 \ldots i_k}(x)$ denotes the principal minor of the Hessian matrix

$\phi''(x)$ with rows and columns $i_1, \ldots, i_k$, and $B_{i_1 \ldots i_k}$ is the corresponding principal minor of the matrix

$$B(x) = \begin{bmatrix} 0 & \phi_1 & \phi_2 & \cdots & \phi_n \\ \phi_1 & \phi_{11} & \phi_{12} & \cdots & \phi_{1n} \\ \phi_2 & \phi_{21} & \phi_{22} & \cdots & \phi_{2n} \\ \cdot & \cdot & \cdot & & \cdot \\ \phi_n & \phi_{n1} & \phi_{n2} & & \phi_{nn} \end{bmatrix} \equiv \begin{bmatrix} 0 & \phi'(x) \\ \\ \phi'(x)^T & \phi''(x) \end{bmatrix} \tag{16}$$

then (for all $x \in C$) $B_{i_1 \ldots i_k} \leqslant 0$, and if $B_{i_1 \ldots i_k}(\bar{x}) = 0$ then $H_{i_1 \ldots i_k} \geqslant 0$ in a neighbourhood of $\bar{x} \in C$. (Here $\phi_i \equiv \partial\phi/\partial x_i$ and $\phi_{ij} \equiv \partial^2\phi/\partial x_i \partial x_j$.)

For other criteria for pseudo- (quasi-) convexity, see (inter alia) the book by Martos, and the cited papers by Ferland, Gupta and Bector, Mangasarian, and Schaible. Substantial open questions remain concerning criteria for P-quasiconvexity or P-pseudoconvexity, when P is a convex cone, not an orthant $\mathbb{R}^r_+$.

When $f(x)$ is real valued, there is (under suitable assumptions) a *dual problem* to (8), whose objective function is the *Langrangian* $f(x) + \lambda k(x)$. However, when $f(x)$ is vector valued, this approach does not succeed, because Theorem 1 does not give information about this Lagrangian, but only about $\tau f(x) + \lambda k(x)$, where now τ is *vector valued*, so cannot be replaced by 1. However, when a *strong* minimum is considered, a dual can be obtained, using a vector valued Lagrangian.

3. LAGRANGE MULTIPLIERS FOR STRONG MINIMA

Consider the pair of problems:

Minimize $f(x)$ subject to $-k(x) \in J$ (strongly with respect to P): (17)

$$\text{Maximize } f(u) + \Xi k(u) \text{ subject to } f'(u) + \Xi k'(u) = 0,$$

$$\Xi(J) \subset K \quad \text{(strongly with respect to U)}. \tag{18}$$

Here P, K, U are convex cones, to be specified later; the functions $f : X \to W$, $k : X \to Z$ are differentiable; $\Xi : Z \to W$ is a continuous linear mpa, taking the cone J into the cone K. The cones $P \subset W$ and $U \subset W$ define *partial orderings* in the space W, by $w \geqslant w' \Leftrightarrow w - w' \in P$ (and similarly for U). In strong (vector) minimization, these partial orderings replace the usual ordering (with respect to the cone $\mathbb{R}_+$) of the real line $\mathbb{R}$.

It will appear later that a duality theorem can be proved, under suitable assumptions, for the vector problems (17) and (18). This will involve, in part, a vector "weak duality" relation

$$f(x) \geqslant f(u) + \Xi k(u), \tag{19}$$

whenever x is feasible for (17) and (u,Ξ) is feasible for (18). The $\geqslant$ in (19) again represents a partial ordering, with respect to some cone. All of these partial orderings may, but need not, be taken with respect to the same convex cone P. Here, the *vector Lagrangian* is

$$L(u,\Xi) = f(u) + \Xi k(u). \tag{20}$$

In case $X = \mathbb{R}^n$, $W = \mathbb{R}^r$, and $Z = \mathbb{R}^m$, Ξ is (represented by) a $r \times m$ matrix. This contrasts with the Lagrange multiplier λ for weak minimization, which is (represented by) a row vector. Denote L_1 resp. L_2 the partial derivates of L with respect to its first and second (vector) arguments.

A Kuhn-Tucker theorem will be required for problem (17), involving the vector Lagrangian (20). Let $\underset{\sim}{F}$ denote the space W, with the ordering given by P. Results of Elster and Nehse (1976, 1978) show that the following statements are equivalent:

(a) $\underset{\sim}{F}$ is order complete;

(b) the Farkas theorem holds, with linear functional replaced by continuous linear mapping into $\underset{\sim}{F}$;

(c) the Hahn-Banach extension property holds, with *linear functional* replaced by linear mapping into $\underset{\sim}{F}$;

(d) a vector-valued Kuhn-Tucker theorem holds for a strong minimum of (17), assuming that (17) is a convex problem, satisfying a constraint qualification; the Lagrange multiplier Ξ is then a continuous linear mapping taking J into P.

Here, $\underset{\sim}{F}$ is *order complete* if every order-bounded set $A \subset \underset{\sim}{F}$ (for some $z \in \underset{\sim}{F}$, $a \leqslant z$ for every $a \in A$) has a least upper bound b ($a \leqslant b$ for each $a \in A$, and $[(\forall a \in A)\ a \leqslant z] \Rightarrow z \geqslant b$); note that $b \in F$, not necessarily $b \in A$. If the ordering $\leqslant$ in $\underset{\sim}{F}$ is given by a cone $\mathbb{R}^m_+$, then $\underset{\sim}{F}$ is order-complete (see e.g. Jameson, 1970), and the results of Elster and Nehse can be applied. However, there seems to be no general criterion for the order-completeness of $\underset{\sim}{F}$, directly in terms of the cone P. For example, it is not clear whether $\underset{\sim}{F}$ is order-complete when P is an arbitratry polyhedral cone. Moreover, the Kuhn-Tucker conditions may hold without order-completeness of $\underset{\sim}{F}$, if other parts of the problem, such as the function k. are suitably restricted.

The *Farkas theorem*, in its current generalized form, runs (for finite dimensions) as follows:

THEOREM 3: Let $A : \mathbb{R}^h \to \mathbb{R}^m$ *and* $B : \mathbb{R}^h \to \mathbb{R}^r$ *be (continuous) linear mappings; let* $J \subset \mathbb{R}^m$ *and* $K \subset \mathbb{R}^r$ *be closed convex cones. Either let (i)* $A(\mathbb{R}^h) = \mathbb{R}^m$, *or (ii) let the cone* $A^T(J^*)$ *be closed and the generators of* K *form a basis of* $\mathbb{R}^r$. *Then*

$$[Az \in J \Rightarrow Bz \in K] \Longleftrightarrow [(\exists V \in \underset{\sim}{M}(J,K))\quad B = VA], \tag{21}$$

denoting by $\underset{\sim}{M}(J,K)$ *the set of (continuous) linear mappings taking* J *into* K.

PROOF: The proof of case (i), assuming A *surjective*, is given in Craven (1972) Theorem 2.2, and Craven (1978a). Case (ii) is proved by Craven and Koliha (1977), Theorem 7. //

Note that *case (i) remains valid also when the spaces have infinite dimensions* (with application to control theory). It remains an open question whether (21) holds, merely assuming that $A^T(J^*)$ is closed (or weak * closed in infinite dimensions); if it is, then various results on strong minimization will be simplified. (A variant (and weaker) result of Ritter (1970) modifies (21) by replacing $\underset{\sim}{M}(J,K)$ by $\underset{\sim}{M}(J \cap \text{range}(A), K)$.)

The following Kuhn-Tucker theorem for strong (vector) minimization can now be proved (Craven, 1980a).

THEOREM 4: In problem (17), let f *and* k *be differentiable functions, and let* J *and* P *be closed convex cones; let (17) reach a strong minimum at* $x = \bar{x}$. *Assume either that* $k'(\bar{x})$ *is surjective (and then let* $K = P$*), or assume that the cone* $(k'(\bar{x}),k(\bar{x}))^T(J^*)$ *is (weak*) closed,* $K \supset P$, *the generators of* K *span the space* $\mathbb{R}^m$ *into which* k *maps, and let* $-k(x) \in J$ *be locally solvable at* $\bar{x}$. *Then, for some multiplier* Λ,

$$f'(\bar{x}) + \Lambda k'(\bar{x}) = 0; \ \Lambda k(\bar{x}) = 0; \ \Lambda \in \underset{\sim}{M}(J,K). \tag{22}$$

PROOF: If $k'(\bar{x})$ is surjective then $-k(x) \in J$ is locally solvable (Craven, 1978; 149), so that local solvability need not be assumed in this case.

Denote $A = [k'(\bar{x}),k(\bar{x})]$ and $B = [f'(\bar{x}),f(\bar{x})]$. Let $z = (d,\beta)$ satisfy $Az \in J$. For $\gamma > 0$, $-[k(\bar{x})\gamma + [k'(\bar{x})d + k(\bar{x})\beta]] \in J$. For γ large enough, $d' = (\beta + \gamma)^{-1}d$ satisfies $||d'|| < \delta$ and $-[k(\bar{x}) + k'(\bar{x})d'] \in J$. Since $-k(x) \in J$ is locally solvable, it has a solution $x = \bar{x} + \alpha d' + o(\alpha)$, for sufficiently small $\alpha > 0$. Since $\bar{x}$ strongly minimizes (17), $f(\bar{x} + \alpha d' + o(\alpha)) - f(\bar{x}) \in P$ for $\alpha \downarrow 0$. Since P is a convex cone, $f'(\bar{x})d' + o(\alpha)/\alpha \in P$; since P is closed, $f'(\bar{x})d = (\beta + \gamma)f'(\bar{x})d \in P$. Thus $Ax \in J \Rightarrow Bx = f'(x)d \in P \subset K$. From Theorem 3, under either hypothesis (i) or (ii), there exists $\Lambda \in \underset{\sim}{M}(J,K)$ such that $B = \Lambda A$, and this gives (22). //

The following theorem (Craven, 1980b) gives a vector version of *weak duality*. (Here, *weak* refers not to *weak minimization*, but to an inequality relating the two vector problems (17) and (18).) Let $Q \subset W$ be a convex cone.

THEOREM 5: Let x *be feasible for (17), and let* (u,Ξ) *be feasible for (18). For each continuous linear mapping* Γ *for which* $\Gamma(J) \subset K$, *assume that* $L(.,\Gamma)$ *is* Q-*pseudoconvex. Then*

$$f(x) - L(u,\Xi) \in Q + K \equiv \{\theta+\phi : \theta \in Q, \phi \in K\}. \tag{23}$$

PROOF: $L_1(u,\Xi) = 0 \Rightarrow L_1(u,\Xi)(x - u) = 0 \Rightarrow L(x,\Xi) - L(u,\Xi) \in Q$ by pseudoconvexity

$$\Rightarrow f(x) - L(u,\Xi) \in -\Xi\, k(x) + Q \subset K + Q. \quad //$$

This happens, in particular, when f is Q-convex and k is J-convex. Theorem 5 also generalizes a result of Gulati (1975, Theorem 5.3) to vector minimization. Compare also Mereau & Paquet (1973), and Mahajan & Vaȓtak (1977).

The Kuhn-Tucker conditions (22) were obtained under somewhat restrictive assumptions; these can, however, be weakened, and put in more manageable form. In order that $-k(x) \in J$ is locally solvable at $\bar{x}$, it suffices if

$$k(\bar{x}) + k'(\bar{x})(X) + J \subset N, \tag{24}$$

where k maps the normed vector space X into the normed vector space Z, and N denotes some ball $\{z \in Z : \|z\| < \varepsilon\}$. (This is proved in Craven (1978a), page 150, adapting a theorem of Robinson (1976).) Assume (temporarily) that int J is nonempty (thus restricting consideration to *inequality* constraints); this will lead both to a simpler approach to the criterion (24) (thus not needing Robinson's *convex processes*, which are multi-valued functions), and also to a more general Kuhn-Tucker theorem. The constraint $-k(x) \in J$ will be called *steady* at $\bar{x}$ if (24) holds, for some N.

When int J is nonempty, the dual cone J^* has a (weak *) compact convex *base* B, such that $J^* = \{\alpha b : \alpha \in \mathbb{R}_+, b \in B\}$ and $0 \notin B$. (See e.g., Craven (1977) Lemma 3, or Jameson (1970)). If $h \in$ int J, then B may be taken as $\{w \in J^* : wh = 1\}$. Denote by E the closure of the set of extreme points of B; then E defines a set of *generators* for J^*. The set E is finite exactly when J is a polyhedral cone. Then $-k(x) \in J \iff (\forall e \in E)$ $ek(x) \leqslant 0$.

Suppose that f(x) is strongly minimized at $\bar{x}$, subject to $-k(x) \in J$. For $\delta > 0$, denote $E_\delta = \{e \in E : -\delta \leqslant ek(\bar{x})\}$. Since $ek(\bar{x}) < -\delta$ for each $e \in E \setminus E_\delta$, ek(x) is bounded away from zero for $e \in E \setminus E_\delta$ and $\|x - \bar{x}\|$ sufficiently small. It follows that $\bar{x}$ remains a minimum when the *inactive constraints* $ek(x) \leqslant 0$ $(e \in E \setminus E_\delta)$ are omitted. (This does not hold when $\delta = 0$, except when E is finite, thus when J is polyhedral).

To handle the case when E is an infinite set, we may proceed as follows. Denote by C(E) the space of continuous real functions on E, with the uniform norm $\|\phi\| = \sup_{e \in E} |\phi(e)|$. Denote by $C_{lin}(E)$ the subspace of C(E), consisting of linear functions, and by $C^+_{lin}(E)$ the convex cone of nonnegative functions in $C_{lin}(E)$. Define the mapping $\psi : Z \to C_{lin}(E)$ by

$$(\forall e \in E)(\forall z \in Z) \quad \psi(z)(e) = e(z). \tag{25}$$

Then

$$-k(x) \in J \iff (\forall e \in E) \;\; ek(x) \leqslant 0 \iff -\psi k(z) \in C^+_{lin}(E). \tag{26}$$

In these terms, $E_\delta = \{e \in E : -\psi k(\bar{x})(e) \leqslant \delta\}$; and the less-constrained problem obtained by omitting inactive constraints (for $e \in E \setminus E_\delta$) takes the "cone form":

$$\text{Minimize } f(x) \text{ subject to } -\psi_\delta k(x) \in C^+_{lin}(E_\delta), \tag{27}$$

where $\psi_\delta : Z \to C_{lin}(E_\delta)$ is defined by $(\forall z \in Z, \forall e \in E_\delta)$ $\psi_\delta(z)(e) = ez$. The Kuhn-Tucker theorem can now be applied to (27), because of the following theorem (Craven (1980c)).

THEOREM 6: Let X *and* Y *be normed spaces (of any dimension); let* X_o *be a convex open subset of* X*; let* $k : X_o \to Y$ *be a continuously (Fréchet) differentiable function; let* J *be a closed convex cone in* Y*, with* $\text{int } J \neq \emptyset$ *and* $J \cap (-J) = \{0\}$*. Let* $-k(\bar{x}) \in J$*, and define the mappings* ψ *and* ψ_δ *as above, for* $\delta > 0$*. Then* ψ *is a continuous open mapping of* Z *onto* $C_{lin}(E)$*; and the constraint* $-k(x) \in J$ *is steady at* $\bar{x}$ *if and only if, for some* $\delta > 0$,

$$\psi_\delta\, k'(\bar{x})(X) = C_{lin}(E_\delta). \tag{28}$$

This result reduces the discussion of Kuhn-Tucker conditions for (17) to the "surjective" case (i) of Theorem 4. Note that, for this case, theorems 3 and 4 remain valid (with the same proofs) for infinite dimensional spaces, and so apply to the function space $C_{lin}(E_\delta)$ required by (27). Consider the problem

$$\text{Minimize } f(x) \text{ subject to } -g(x) \in S,\ h(x) = 0$$
$$\text{(strongly with respect to P).} \tag{29}$$

Here X,Y,Z,W are normed spaces (of any dimension); X_o is a convex open subset of X; $f : X_o \to W$ is (Fréchet) differentiable; $g : X_o \to Y$ and $h : X_o \to Z$ are continuously (Fréchet) differentiable; $P \subset W$ and $S \subset Y$ are closed convex cones with interiors. Redefine E_δ and ψ_δ to relate now to the cone S and the constraint $-g(x) \in S$ (instead of to $-k(x) \in J$ above). Assume $S \cap (-S) = \{0\}$.

THEOREM 7: Let problem (29) reach a strong minimum at $x = \bar{x}$*; let the constraint* $-g(x) \in S$ *be steady at* $\bar{x}$*, and let* $h'(\bar{x})$ *be surjective (onto* Z*). Then Lagrange multipliers* $\Theta : Y \to W$ *and* $\Pi : Z \to W$ *exist, satisfying*

$$f'(\bar{x}) + \Theta g'(\bar{x}) + \Pi h'(\bar{x}) = 0;\ \Theta g(\bar{x}) = 0;\ \Theta(S) \subset P. \tag{30}$$

PROOF: From Theorem 6, since $-g(x) \in S$ is steady at $\bar{x}$, the problem

$$\text{Minimize } f(x) \text{ subject to } -k(x) \equiv -\begin{bmatrix} \psi_\delta g(x) \\ h(x) \end{bmatrix} \in J \equiv \begin{bmatrix} C_{lin}(E_\delta) \end{bmatrix} \times \{0\}$$
$$\text{(strongly with respect to P),} \tag{31}$$

in which $[C_{lin}(E_\delta)] \times \{0\} \equiv \left\{\begin{bmatrix} u \\ 0 \end{bmatrix} : u \in C_{lin}(E_\delta)\right\}$, reaches a strong minimum also at $\bar{x}$. From Theorem 6, applied to g and S in place of k and J, $\psi_\delta g'(\bar{x})$ is surjective; and $h'(\bar{x})$ is surjective, by hypothesis. Hence, in (31), $k'(\bar{x})$ is surjective. By Theorem 4, case (i), there exists a Lagrange multiplier $\bar{V} = [\Delta,\Pi]$ for which

$$f'(\bar{x}) + [\Delta,\Pi]\begin{bmatrix} \psi_\delta g'(\bar{x}) \\ h'(x) \end{bmatrix} = 0 \ ; \ [\Delta,\Pi]\begin{bmatrix} \psi_\delta g(\bar{x}) \\ h(\bar{x}) \end{bmatrix} = 0 \ ;$$

$$\Delta(C^+_{lin}(E_\delta)) \subset P. \tag{32}$$

From Theorem 6, $\psi(Y) = C_{lin}(E)$, and hence $\psi_\delta(Y) = C_{lin}(E_\delta)$, $\psi_\delta(S) = C^+_{lin}(E_\delta)$. Define $\Theta = \Delta\psi_\delta$. Then, since $h(\bar{x}) = 0$, (32) gives $\Theta g(\bar{x}) = 0$, $\Theta(S) \subset P$, and $f'(\bar{x}) + \Theta g'(\bar{x}) + \Pi h'(\bar{x}) = 0$, which proves (30). //

THEOREM 8 (Converse Kuhn-Tucker): For problem (29), let the Kuhn-Tucker conditions (30) hold at a feasible point $\bar{x}$. *Assume either (a) that* $\tilde{L}(\cdot) = f(\cdot) + \Lambda g(\cdot) + \Pi h(\cdot)$ *is* P-*pseudoconvex, or (b) that* g *is* S_b-*quasiconvex, where* $b = -g(\bar{x})$, *and that both* h *and* -h *be* T-*quasiconvex with* $T \cap (-T) = \{0\}$, *or* h *is* T-*quasiconvex with* $\Pi(T) \subset P$, *where* T *is a closed convex cone. Let* f *be* P-*pseudoconvex. Then* $\bar{x}$ *is a strong minimum of (29).*

PROOF: Let x satisfy $-g(x) \in S$, $h(x) = 0$.

Case (a). Since $\tilde{L}'(\bar{x})(x - \bar{x}) = 0$ and $\tilde{L}$ is P-pseudoconvex, $\tilde{L}(x) - \tilde{L}(\bar{x}) \in P$. Then $f(x) - f(\bar{x}) \in \tilde{L}(x) - \tilde{L}(\bar{x}) + P \subset P + P \subset P$.

Case (b). As in the proof of Theorem 2, $-g'(\bar{x})(x - \bar{x}) \in S_b$ and $-h'(\bar{x})(x - \bar{x}) \in T$. Using $\Lambda(S) \subset P$ and $\Lambda g(\bar{x}) = 0$, $-\Lambda g'(\bar{x})(x - \bar{x}) \in P$; similarly $-\Pi h'(\bar{x})(x - \bar{x}) \in P$. From (30), $f'(\bar{x})(x - \bar{x}) \in P$. Since f is P-pseudoconvex, $f(x) - f(\bar{x}) \in P$. //

4. VECTOR DUALITY

Consider again the pair of strong optimization problems:

$$\text{Minimize } f(x) \text{ subject to } -k(x) \in J \text{ (strongly with respect to } P). \tag{17}$$

$$\text{Maximize } L(u,\Xi) \equiv f(u) + \Xi k(u) \text{ subject to } f'(u) + \Xi k'(u) = 0, \quad \Xi(J) \subset K \text{ (strongly with respect to } U). \tag{18}$$

Problem (18) will be called a (strong) *vector dual* of problem (17) if

(a) whenever x is feasible for (18) and (u,Ξ) is feasible for (18), $f(x) \geqslant L(u,\Xi)$, with $\geqslant$ defined in terms of some chosen convex cone; and

(b) if (17) reaches a strong minimum at $\bar{x}$, then (18) reaches a strong maximum at some $(\bar{u},\Lambda)$, and $f(\bar{x}) = L(\bar{u},\Lambda)$.

Similarly, (17) is a (strong) vector dual of (18) if (a) holds, and also

(c) if (18) reaches a strong vector maximum at $(\hat{u},\hat{\Lambda})$, then (17) reaches a strong vector minimum at some $\hat{x}$, with $f(\hat{x}) = L(\hat{u},\hat{\Lambda})$.

THEOREM 9 (Strong vector duality): Let problem (17) reach a strong minimum at $\bar{x}$, *and let the Kuhn-Tucker conditions (22) hold there. For each continuous linear mapping* Γ *for which* $\Gamma(J) \subset K$, *let* $L(.,\Gamma)$ *be* Q-*pseudoconvex, where* $Q + K \subset U$. *Then (18) is a strong vector dual to (17), with ordering* $\geqslant$ *defined by the cone* U.

PROOF: Property (a) follows from Theorem 5. From (22), $(\bar{x},\Lambda)$ is feasible for (18), and $f(\bar{x}) = L(\bar{x},\Lambda)$. From this and property (a), $(\bar{x},\Lambda)$ is a strong maximum for (18). //

The special case where $Q = P$ and $K + P = P = U$ is notable. There, all the orderings are given in terms of the single cone P. The Kuhn-Tucker conditions for (17) hold, in particular, when Theorem 7 is applicable.

To discuss *converse duality* (when is (17) a strong vector dual of

(18)?), the following definition is required. The constraint

$$L_1(u, \ ) = f'(u) + \Xi k'(u) = 0 \tag{33}$$

will be called *solvable near* $(\bar{u},\bar{\Lambda})$ (see Craven & Mond (1976)) if $L_1(\bar{u},\Lambda) = 0$, $\Lambda(J) \subset P$, and, whenever $(\Lambda+\Delta)(J) \subset P$ and $\alpha > 0$ is sufficiently small, the equation $L_1(u,\Lambda + \alpha\Delta) = 0$ has a solution $u = \bar{u} + \xi(\alpha,\Delta)$, for which $\xi(\alpha,\Delta) = 0(\alpha)$ as $\alpha \downarrow 0$ (thus $||\xi(\alpha,\Delta)|| \leqslant$ const. α as $\alpha \downarrow 0$). This holds, in particular, using the implicit function theorem, if L is twice continuously differentiable with respect to u, and this second derivative is invertible.

THEOREM 10 (Converse strong vector duality): In problems (17) and (18), let f *and* k *be continuously (Fréchet) differentiable; let* J *and* P *be closed convex cones, with* int $P \neq \emptyset$ *and* $P \cap (-P) = \{0\}$; *whenever* Γ *is a continuous linear mapping satisfying* $\Gamma(J) \subset P$, *let* $L(.,\Gamma)$ *be P-pseudoconvex; let* $U = P$. *Let (18) reach a strong maximum at* $(\bar{u},\Lambda)$, *and let* $L_1(u,\Xi) = 0$ *be solvable near this point. Then (17) is a strong vector dual to (18).*

PROOF: (Craven (1980b)) Let x be feasible for (17), and (u,Ξ) for (18). From Theorem 5, with $Q = K = P$, it follows that $f(x) - L(u,\Xi) \in P$.

Now assume that (18) reaches a strong maximum at $(\bar{u},\Lambda)$. It will be shown, in consequence, that $-k(\bar{u}) \in J$ and $\Lambda k(\bar{u}) = 0$. Thus $\bar{u}$ is feasible for (17), and $f(\bar{u}) = L(\bar{u},\Lambda)$. Suppose, if possible, that $\bar{u}$ is *not* a strong minimum for (17); then $f(\bar{u}) - f(\tilde{u}) \in P \setminus \{0\}$ for some $\tilde{u}$ feasible for (17). But also $f(\tilde{u}) - f(\bar{u}) = f(\tilde{u}) - L(\bar{u},\Lambda) \in P$, so that

$$f(u) - f(\bar{u}) \in (P \setminus \{0\}) \cap (-P) = \emptyset$$

since $P \cap (-P) = \{0\}$. Hence $\bar{u}$ is a strong minimum for (17), and strong vector duality follows.

To prove the above statements, let Δ satisfy $(\Lambda+\Delta)(J) \subset P$ and let $0 < \alpha < 1$. Since $(\bar{u},\Lambda)$ maximizes (18), and $(\Lambda+\alpha\Delta)(J) = [(1-\alpha)\Lambda + \alpha(\Lambda+\Delta)](J) \subset P$,

$$\begin{aligned} 0 &= L(\bar{u} + \xi(\alpha,\Delta), \Lambda+\alpha\Delta) - L(\bar{u},\Lambda) + p \quad \text{for some } p \in P, \\ &= L(\bar{u} + \xi(\alpha,\Delta), \Lambda+\alpha\Delta) - L(\bar{u} + \xi(\alpha,\Delta),\Lambda) + L(\bar{u} + \xi(\alpha,\Delta),\Lambda) \\ &\qquad - L(\bar{u},\Lambda) + p \\ &= L_2(\bar{u} + \xi(\alpha,\Delta),\Lambda)\alpha\Delta + L_1(\bar{u},\Lambda)\ \xi(\alpha,\Delta) + o(\alpha) + p \\ &\qquad \text{since } \xi(\alpha,\Delta) = O(\alpha) \text{ by hypothesis} \\ &= [L_2(\bar{u},\Lambda) + O(\alpha)]\alpha\Delta + 0 + o(\alpha) + p. \end{aligned} \tag{34}$$

Hence, letting $\alpha \downarrow 0$, $-L_2(\bar{u},\Lambda)\Delta \in P$. This holds, in particular, whenever $\Delta(J) \subset P$; hence $-\Delta k(\bar{u}) \in P$ whenever $\Delta(J) \subset P$. Suppose that $-k(\bar{u}) \notin J$. Since J is a closed convex cone, there exists $\rho \in J^*$ such that $\rho k(\bar{u}) > 0$. Define Δ by $(\forall z \in J)\ \Delta(z) = \rho(z)p_o$, where p_o is a fixed nonzero element of P; then a contradiction results, since $-\Delta k(\bar{u}) = -\rho k(\bar{u})p_o \notin P$, since $P \cap (-P) = \{0\}$. Now let $\Delta = -\frac{1}{2}\Lambda$; then $(\Lambda+\Delta)(J) = \frac{1}{2}\Lambda(J) \subset P$, hence $\frac{1}{2}\Lambda k(\bar{u}) \in P$. But also $-\Lambda k(\bar{u}) \in P$, hence $\Lambda k(\bar{u}) \in P \cap (-P) = \{0\}$. //

For another version of this result, see Craven (1980a), Theorem 7. It is there assumed that f is Q-convex, and k is P-convex, while Λ and the convex cone $Q \subset X$ satisfy $Q + \Lambda(J) \subset P$. On these hypotheses, the requirement that $\xi(\alpha,\Delta) = O(\alpha)$ can be weakened to $\xi(\alpha,\Delta) \to 0$ as $\alpha \downarrow 0$, since the convexity is used to obtain (34). Theorem 10 also generalizes Mahajan & Vertak (1977), Theorem 2.1.4, to cone constraints and non-differentiable $\xi(.,\Delta)$.

If the constraints of the "primal" problem (17) include an *equality* constraint $h(x) = 0$, then the (generalized) convexity required for a duality theorem severely restricts h. The usual assumption is that h is an affine function (linear + constant), to ensure convexity with respect to the "zero cone" $\{0\}$. However, such hypotheses can be relaxed, if the concept of *duality* is relaxed to *local duality* (see Craven (1975)). The definition of *local duality* differs from the preceding definition of duality by weakening the "weak duality" requirement (a) to hold only when $||x - \bar{x}||$, $||u - \bar{u}||$, and $||\Xi - \Lambda||$ are sufficiently small; thus "weak duality" is demanded only *locally*, in some neighbourhoods of the optima.

Consider now the pair of problems:

$$\text{Minimize } f(x) \text{ subject to } -g(x) \in S,\ -h(x) \in T \quad \text{(strongly with respect to P)}; \tag{35}$$

$$\text{Maximize } f(u) + \Theta g(u) \text{ subject to } \Theta(S) \subset P,\ \Pi(T) \subset P, \quad f'(u) + \Theta g'(u) + \Pi h'(u) = 0,\ -h(u) \in T \quad \text{(strongly with respect to U)}. \tag{36}$$

Here, S,T,P,U are closed convex cones; $S \cap (-S) = \{0\}$; Θ and Π denote continuous linear mappings. Observe that the primal constraint $-h(x) \in T$ is retained also in (36), which will turn out to be a local dual to (35), under suitable assumptions. Assume f and g (Fréchet) differentiable, g continuously differentiable.

Assume that (35) reaches a strong minimum at $\bar{x}$, and that the constraint $-h(x) \in T$ is *steady* at x. Then, as in (27), the minimum is still reached when some inactive constraints are omitted, and the modified constraint, say $-\hat{h}(x) \in \hat{T}$, has $\hat{h}'(x)$ surjective. Let the direction d satisfy $-\hat{h}(\bar{x}) + \hat{h}'(\bar{x})d \in T$; let $0 < \alpha < 1$. Local solvability applies, hence $-\hat{h}(x) \in \hat{T}$ has a solution $\tilde{x}(\alpha) = \bar{x} + \alpha d + o(\alpha)$ $(\alpha \downarrow 0)$. To show that *all* solutions near $\bar{x}$ have this form, assume also that $M = \hat{h}'(\bar{x})$ has *full rank* (its domain is the direct sum of its nullspace and a subspace $\underset{\sim}{V}$, where the restriction, A, of M to $\underset{\sim}{V}$ is bijective). Let $x_o = \bar{x} + \alpha d$ and $s = h(\bar{x}) + \alpha h'(\bar{x})d$. Then (see Craven (1978a;1980b)) the sequence $x_{n+1} = x_n + A^{-1}(h(x_n) - s)$ converges to a limit $\tilde{x}(\alpha)$ as above, and the contraction mapping theorem ensures uniqueness - there are no other solutions in a neighbourhood. Therefore there is a function ϕ, whose domain is a neighbourhood N of 0 in the cone $M^{-1}(0) \times \hat{T}_{-\hat{h}(\bar{x})}$, such that $\phi(N)$ is the set of all solutions to $-\hat{h}(x) \in \hat{T}$ in a neighbourhood of $\bar{x}$, and $\phi'(0)$ is the identity. For brevity, call $-h(x) \in T$ *fully steady* when this construction holds. Then

$$N = \{v : -\hat{h}(\bar{x}) - \hat{h}'(\bar{x})v \in \hat{T},\ ||v|| < \delta\}$$
$$= \{v : -\hat{h}'(\bar{x})v \in \hat{T}_{-\hat{h}(\bar{x})},\ ||v|| < \delta\}. \tag{37}$$

THEOREM 11 (Local duality): Let problem (35) reach a strong minimum at $\bar{x}$*; assume that the constraint* $-h(x) \in T$ *is* <u>*fully steady*</u> *at* $\bar{x}$*, and* $-g(x) \in S$ *is steady at* $\bar{x}$*. Let* $(f + \Gamma g)\circ\phi$ *be pseudoconvex whenever* $\Gamma(S) \subset P$*, or let* $f\circ\phi$ *be* P*-convex and* $g\circ\phi$ *be* S*-convex. Then (36) is a local dual to (35).*

PROOF: Since $-h(x) \in T$ is fully steady at $\bar{x}$, $f(x)$ is minimized strongly at $\bar{x}$, subject to $-g(x) \in S$, $x \in \phi(N)$. Hence $v = 0$ minimizes $(f\circ\phi)(v)$, subject to $-(g\circ\phi)(v) \in S$, $v \in \tilde{N}$, where $\tilde{N}$ is the cone generated by N. Since $-g(x) \in S$ is steady, so is $-(g\circ\phi)(v) \in S$, hence Theorem 6 shows that multiplier Θ exists with

$$(\forall\ d \in N)\ [(f\circ\phi)'(0) + \Theta(g\circ\phi)'(0)]\ d \in P,$$
$$\Theta(g\circ\phi)(0) = 0,\ \Theta(S) \subset P. \tag{38}$$

From (37), (38), and $\phi'(0)$ = identity, $-\hat{h}'(\bar{x})d \in \hat{T}_{-\hat{h}(\bar{x})}$ implies $[f'(\bar{x}) + \Xi g'(\bar{x})]d \subset P$. Since $\hat{h}'(\bar{x})$ is surjective, the Farkas theorem (Theorem 3) shows that $f'(\bar{x}) + \Theta g'(\bar{x}) = -\Pi h'(\bar{x})$, for some multiplier Π with $\Pi(\hat{T}_{-\hat{h}(\bar{x})}) \in P$. Hence, on putting back the inactive constraints,

$$f'(\bar{x}) + \Theta g'(\bar{x}) + \Pi h'(\bar{x}) = 0,\ \Theta g(\bar{x}) = 0,\ \Pi h(\bar{x}) = 0, \tag{39}$$

where $\Pi(T) \subset P$. Consequently $(\bar{x},\Theta,\Pi)$ is feasible for (36), and $f(\bar{x}) = f(\bar{x}) + \Theta g(\bar{x})$. Since also "weak duality" follows from Theorem 5, duality results. //

This theorem generalizes Theorem 3.2 of Craven (1975), which assumes f real and $T = \{0\}$. Some related results are found in Mahajan and Vartak (1977), section 2.3.

In order to apply this theorem, some criterion is required for convexity of functions $f\circ\phi$ and $g\circ\phi$. Assume now that f and g are continuously twice differentiable. Denote by q the canonical projection from

X onto $h'(\bar{x})^{-1}(0)$. Choose $s \in T$. For $0 < \alpha < 1$, substitute $x = \bar{x} + \alpha d + \alpha^2 r + o(\alpha^2)$ into $-h(x) = s \in T$, assuming that the initial direction d of the arc $x(\alpha)$ satisfies $h(\bar{x}) + h'(\bar{x})d = -s$. Then $-s + \alpha^2(h'(\bar{x})r + \frac{1}{2}h''(\bar{x})[d]^2) + o(\alpha^2) = s$, hence $h'(\bar{x})r + \frac{1}{2}h''(\bar{x})[d]^2 = 0$. This may be solved for r, constrained by $qr = 0$. Hence

$$\phi''(0) = -[h'(\bar{x})(1 - q)]^{-1}h''(\bar{x}), \tag{40}$$

(as in Craven (1975) for an equality constraint). Now let $m = \theta f$ for $0 \neq \theta \in P^*$. Since $\phi'(0)$ is the identity, (40) shows that

$$(m \circ \phi)''(v) = m''(\phi(v)) + m'(\phi(v)) \circ \phi(v) \quad (v \in N). \tag{41}$$

Then convexity of $m \circ \phi$ can be discussed in terms of eigenvalues of $(m \circ \phi)''(.)$. Also pseudoconvexity could be discussed in terms of the criterion (16).

5. DISCUSSION AND EXTENSIONS

Consider the problem:

$$\text{Minimize } f(x) \text{ subject to } h_i(x) \geqslant 0 \ (i=1,2,\ldots,m), \tag{42}$$

with $f : \mathbb{R}^n \to \mathbb{R}^r$, and $P \subset \mathbb{R}^r$ defined by $y \in P \iff Cy \in \mathbb{R}^s_+$, where C is a given matrix. A strong minimum of f(x) with respect to P is then equivalent to a strong minimum of Cf(x) with respect to $\mathbb{R}^s_+$. For the latter problem, the Kuhn-Tucker conditions give

$$Cf'(\bar{x}) + Wh'(\bar{x}),\ Wh(\bar{x}) = 0,\ W \geqslant 0,\ h(\bar{x}) \geqslant 0; \tag{43}$$

where $W \geqslant 0$ means that each component of the matrix W is $\geqslant 0$. If C happens to be invertible, then (43) leads to the following Kuhn-Tucker conditions for the original problem:

$$f'(\bar{x}) = \bar{V}h'(\bar{x}),\ \bar{V}h(\bar{x}) = 0,\ C\bar{V} \geqslant 0,\ h(\bar{x}) \geqslant 0, \tag{44}$$

for a suitable $r \times m$ matrix $\bar{V}$. In fact, (44) holds without requiring that C is invertible (from Theorem 4 applied to *polyhedral* P). Moreover, if the problem is also convex, then (44) is also sufficient for a strong minimum.

THEOREM 12 (Craven, 1980a): In problem (42), let f *satisfy either (a)* f *is linear, or (b)* f *is differentiable and each component of* Cf *is convex, or (c)* f *is twice differentiable and* $Cf''(x)$ *is positive semi-definite for each* x. *Let* $x = \bar{x}$ *satisfy the constraints of (42); let each* $-h_i(.)$ *be convex; let the gradients* $h_i'(\bar{x})$ *be linearly independent for those constraints active at* $\bar{x}$ *(thus when* $h_i(\bar{x}) = 0$*). Then (42) reaches a strong minimum at* $\bar{x}$*, with respect to* $P = C^{-1}(\mathbb{R}^s_+)$*, if and only if there exists an* $r \times m$ *matrix* $\bar{V}$ *satisfying (44).*

PROOF: Theorem 4 applies, since the matrix of gradients of active constraints is surjective. The matrix $\bar{V}$ maps $\mathbb{R}^n_+$ into P if and only if $C\bar{V}$ maps each basis vector of $\mathbb{R}^m_+$ into $\mathbb{R}^s_+$, thus if and only if $C\bar{V} \geqslant 0$. Hence a minimum implies (44). The converse follows from Theorem 5, because of the assumed convexity. //

If there are also equality constraints, then, as well as $C\bar{V} \geqslant 0$, each column $\bar{V}_j$ of $\bar{V}$ which corresponds to an equality constraint must satisfy $C\bar{V}_j = 0$.

The theorems of this paper allow generalizations in several directions. The hypotheses of differentiability assume more than is needed. The functions f and g need only be *Hadamard differentiable* instead of Fréchet); this means that, for each arc ω whose initial slope $\omega'(0)$ exists,

$$f(\omega(\alpha)) - f(\omega(0)) = \alpha f'(\omega(0))\omega'(0) + o(\alpha) \text{ as } \alpha \downarrow 0;$$

the derivative $f'(\omega(0))$ is again a linear mapping. If f is Lipschitz function, which has a linear Gâteaux derivative, then Hadamard differentiability follows. For the functions k and h, continuous (Fréchet) differentiability can be weakened somewhat. The property required is that

$$k(x + y) - k(x) = k'(a)y + \xi(x,y); \; ||\xi(x,y)|| \leqslant \varepsilon||y||$$

whenever both $||x - a|| < \delta(\varepsilon)$ and $||y|| < \delta(\varepsilon)$. The question remains open whether weaker derivatives will suffice, when local solvability is needed.

For weak minima, Lagrangian necessary conditions are known if either

(i) the functions are convex (then subdifferentials replace derivatives); or

(ii) each function is the sum of a convex function and a differentiable function (then subdifferentials, slightly generalized, are used - see Craven & Mond (1976); or

(iii) the functions are Lipschitz, and f is real valued (then Clarke's *generalized gradient* replaces derivative).

However, none of these nondifferentiable generalizations seems yet to apply to *strong* vector minimization. Conjecturally, they should.

Consider now a (weak or strong) vector minimization problem, assuming differentiability, but not any sort of convexity. If the minimization is unconstrained, then the *stationary points* (where the gradient $f'(x) = 0$) include the local minima, and also other critical points (local maxima and saddlepoints). Analogously, for a constrained minimization problem, a class of critical points called *quasimins* is defined, corresponding to stationary *Lagrangian*, and these include all minima (and other critical points as well). The precise definitions and results are as follows; they are given for *strong* minimization, but the concepts work equally well for *weak* minimization.

For the problem, minimize (strongly) f(x) subject to $x \in H$, the point $\bar{x}$ is a *strong quasimin* if, for some function θ satisfying $\theta(x) = o(||x - \bar{x}||)$,

$$f(x) - f(\bar{x}) - \theta(x) \in P \quad \text{as } x \to \bar{x} \text{ with } x \in H. \qquad (45)$$

Similarly, f(x) has a *strong quasimax* if $-f(x)$ has a strong quasimin.

Note that a strong quasimin is defined locally, not globally. (The corresponding definitions and results for *quasimin*, corresponding to *weak* minimization, were given in Craven (1977). The following analog of Theorem 5 then holds, for quasimin (Craven (1980a) Theorem 5).

THEOREM 13: In problem (17), let f *and* k *be (Fréchet) differentiable; let* J *and* P *be closed convex cones; let (17) reach a strong quasimin at* $\bar{x}$; *assume* <u>*either*</u> $k'(\bar{x})$ *is surjective (with* K=P*),* <u>*or*</u> $(k'(\bar{x}),k(\bar{x}))^T(J^*)$ *is (weak*) closed,* $K \supset P$*, the generators of* K *span the range space of* k*, and* $-k(x) \in J$ *is locally solvable at* $\bar{x}$. *Then the Kuhn-Tucker conditions (22) hold.*

PROOF: The proof is the same as for Theorem 4, except for the following portion. Since $\bar{x}$ is a strong quasimin, $f(\bar{x} + \alpha d' + o(\alpha)) - f(\bar{x}) - \theta(x) \in P$, for some $\theta(x) = o(||x - \bar{x}||)$. Since P is a convex cone,

$$[f'(\bar{x})\alpha d' + o(\alpha) + \theta(||x - \bar{x}||)]/\alpha \in P \text{ as } \alpha \downarrow 0,$$

so

$$f'(\bar{x})d' + o(\alpha)/\alpha \in P. \quad //$$

Likewise, in Theorem 7, the Kuhn-Tucker conditions (30) still hold, when $\bar{x}$ is only assumed to be a quasimin.

THEOREM 14 (Craven, 1980a): The Kuhn-Tucker conditions (22), at a feasible point $\bar{x}$, *imply that* $\bar{x}$ *is a quasimin of (17).*

PROOF: Suppose, if possible, that $\bar{x}$ is not a quasimin. Then, for some sequence $\{z_i\} \to 0$, $-k(\bar{x} + z_i) \in J$ and, whenever $\theta(z) = o(||z||)$, $f(\bar{x} + z_i) - f(\bar{x}) - \theta(z_i) \notin P$. Now, using (22),

$$\Lambda k'(\bar{x})z_i = \Lambda[k(\bar{x}+z_i) - k(\bar{x}) + o(||z_i||)] = \Lambda k(\bar{x}+z_i) - 0 + o(||z_i||);$$

and then

$$f(\bar{x}+z_i) - f(\bar{x}) = f'(\bar{x})z_i + o(||z_i||) = -\Lambda k'(\bar{x})z_i + o(||z_i||).$$

These combine, using $\Lambda k(\bar{x} + z_i) \in -P$, to give

$$f(\bar{x} + z_i) - f(\bar{x}) - \bar{\theta}(z_i) \in P, \text{ where } \bar{\theta}(z_i) = o(||z_i||).$$

A contradiction results when θ is chosen as $\bar{\theta}$. //

Consequently, problems (17) and (18) are related by the following *quasidual* property. If (17) reaches a strong quasimin at $\bar{x}$, and if the Kuhn-Tucker conditions (22) hold there, and $K \subset U$, then (18) reaches a strong quasimin at $(\bar{x},\Lambda)$, and $f(\bar{x}) = L(\bar{x},\Lambda)$ (the objective functions are equal there). To show this, let $(\bar{x} + p,\Lambda + \Gamma)$ be feasible for (18). Then

$$f(\bar{x}) - [f(\bar{x}+p) + (\Lambda+\Gamma)k(\bar{x}+p)] = -f'(\bar{x})p - \Gamma k(\bar{x}) - \Lambda k'(\bar{x})p$$
$$+ o(||p|| + ||\Gamma||) = -(\Lambda+\Gamma)k(\bar{x}) + o(||p|| + ||\Gamma||) \text{ since } \Lambda k(\bar{x}) = 0$$
$$\in K + o(||p|| + ||\Gamma||),$$

which proves the strong quasimax. (See Craven (1980a)).

In conclusion, note that the constraint qualification (local solvability, etc.) required for a duality theorem can be avoided, if the dual problem is somewhat modified. This has been done (for general cone-constraint problems) for weak minimization by Craven and Zlobec (1980). Conjecturally, there would be an analogous result also for strong minima.

REFERENCES

AVRIEL, M. and SCHAIBLE, S. (1978). "Second Order Characterizations of Pseudoconvex Functions." *Mathematical Programming 14,* 170-185.

BORWEIN, J.M. (1974). "Optimization with Respect to Partial Orderings." Ph.D. thesis, Oxford University.

CLARKE, Frank H. (1976). "A New Approach to Lagrange Multipliers." *Math. of Opns. Res. 1,* 165-174.

CRAVEN, B.D. (1972). "Nonlinear Programming in Locally Convex Spaces." *J. Optim. Theor. Appl. 10,* 197-210.

CRAVEN, B.D. (1975). "Converse Duality in Banach Space." *J. Optim. Theor. Appl. 16,* 229-236.

CRAVEN, B.D. (1977). "Lagrangian Conditions and Quasiduality." *Bull. Austral. Math. Soc. 16,* 325-339.

CRAVEN, B.D. (1978a). *Mathematical Programming and Control Theory.* Chapman & Hall, London.

CRAVEN, B.D. (1978b). "Optimization with Multiple Objectives." *Proc. 14th Annual Conf. of Opns. Res. Soc. of New Zealand (May 1978), Vol. 1,* Theory, 11-18.

CRAVEN, B.D. (1978c). "Multi-Criteria Optimization." *Math. Research Report 37,* University of Melbourne.

CRAVEN, B.D. (1980a). "Strong Vector Minimization and Duality." *Zeits. Angew. Math. Mech. 60,* 1-5.

CRAVEN, B.D. (1980b). "Lagrangian Conditions, Vector Minimization, and Local Duality." *Math. Research Report 37*, University of Melbourne.
CRAVEN, B.D. (1980c). "Implicit Function Theorems and Lagrange Multipliers." *Numerical functional Analysis and Optimization 2(6)*, 473-486.
CRAVEN, B.D. and KOLIHA, J.J. (1977). "Generalizations of Farkas' Theorem." *SIAM J. Math. Analysis 8*, 983-997.
CRAVEN, B.D. and MOND, B. (1976). "Lagrangian Conditions for Quasi-differentiable Optimization." *Survey of Mathematical Programming* (Proceedings of 9th International Symposium on Mathematical Programming, Budapest, 1976), (Vol. 1, A. Prékopa, ed.), 177-192, North-Holland, Amsterdam.
CRAVEN, B.D. and ZLOBEC, S. (1980). "Complete Characterization of Optimality for Convex Programming in Banach Spaces." *Applicable Analysis* (to appear).
DIEWERT, W.E., AVRIEL, M. and ZANG, I. (1981). "Nine Kinds of Quasi-concavity and Concavity." *J. of Economic Theory* (forthcoming).
ELSTER, K.-H. and NEHSE, R. (1976). "Nichtkonvexe Optimierungsprobleme." 21 Intern. wiss, Koll. TH Ilmenau, Sektion Mathematik, Rechentechnik und ökonomische Kybernetik, 69-72.
ELSTER, K.-H. and NEHSE, R. (1978). "Necessary and Sufficient Conditions for the Order-completeness of Partially Ordered Vector Spaces." *Math. Nachr. 81*, 301-311.
FERLAND, J. (1978). "Matrix Criteria for Pseudoconvex Functions in the Class C^2." *Linear Algebra Appl. 21*, 47-57.
GULATI, T.R. (1975). "Optimality Criteria and Duality Theory in Complex Mathematical Programming." Ph.D. thesis, Dept. of Math., Indian Institute of Technology, Delhi.
GUPTA, S.K. and BECTOR, C.R. (1968). "Nature of Quotients, Products and Rational Powers of Convex (Concave)-like Functions." *Math. Student 36*, 63-67.
JAMESON, G. (1970). *Ordered Linear Spaces*. Lecture Notes in Math. 141, Springer-Verlag, Berlin.
MAHAJAN, D.G. and VARTAK, M.N. (1977). "Generalizations of Some Duality Theorems in Nonlinear Programming." *Math. Prog. 12*, 293-317.
MANGASARIAN, O.L. (1965). "Pseudoconvex Functions." *SIAM J. Control 3*, 281-290.
MANGASARIAN, O.L. (1970). "Convexity, Pseudoconvexity and Quasiconvexity of Composite Functions." *Cahiers du Centre d'Étude de Recherche Opérationelle 12*, 114-122.
MARTOS, B. (1975). *Nonlinear Programming Theory and Methods*. North-Holland, Amsterdam.
MEREAU, P. and PAQUET, J.G. (1973). "The Use of Pseudoconvexity and Quasiconvexity in Sufficient Conditions for Global Constrained Minima." *Int. J. Control 18*, 831-838.
NEHSE, R. (1978). "The Hahn-Banach Property and Equivalent Conditions." *Comment. Math. Univ. Carolinas 19*, 165-177.
RITTER, K. (1970). "Optimization Theory in Linear Spaces, Part III: Mathematical Programming in Partially Ordered Banach Spaces." *Math. Ann. 184*, 133-154.
ROBINSON, S.M. (1976). "Stability Theory for Systems of Inequalities, Part II: Differentiable Nonlinear Systems." *SIAM J. Numerical Analysis 13*, 497-513.
ROCKAFELLAR, R.T. (1970). *Convex Analysis*. Princeton Univ. Press.
SCHAIBLE, S. (1972). "Quasiconvex Optimization in General Real Linear Spaces." *Zeits. Opns. Res. 16*, 205-213.
SCHAIBLE, S. (1973). "Quasiconcave, Strictly Quasiconcave and Pseudoconcave Functions." *Opns. Res. Verfahren 17*, 308-316.

SCHAIBLE, S. (1978). *Analyse und Anwendungen von Quotientenprogrammen.* Verlag Anton Hain, Meisenheim am Glan, Germany.

ACKNOWLEDGEMENT

My thanks are due to Ms. Anne Wennagel, for finding many references for me in Science Citation Index.

SOME INFLUENCES OF GENERALIZED AND ORDINARY CONVEXITY IN DISJUNCTIVE AND INTEGER PROGRAMMING

Robert Jeroslow[1]

Three extensions of linear programming are convex, disjunctive, and integer programming. Each generalization represents a different direction, and is attuned to specific distinctive features of the phenomena studied. While e.g. in convexity, and certain of its generalizations, line segment containment provides the crucial property of polyhedra which is retained while a curvature of the feasible region is then permitted, in integer programming the linearity of the region is retained while the discrete nature of the variables departs entirely from the continuum and is the primary complicating factor. Nevertheless, developments in convex programming and its generalizations have influenced disjunctive and integer programming.

Conversely, parts of the infinitary disjunctive programming may be useful in nonconvex nonlinear programming. Similarly "integer analogues" recently discovered in integer programming represent developments somewhat parallel to the generalized duality schemes which extend Lagrangean duality for convex programs to more general duality results for nonconvex programs. These integer analogues are in fact the primary focus of our present report.

Most of the influence of convexity in disjunctive and integer programming is in the theory of cutting-planes, which we earlier surveyed in detail in Jeroslow (1977, 1978). Here we make some additional remarks from the somewhat different perspective of developments in nonlinear programming.

1. INTRODUCTION

Convexity is a generalization of linearity. If each linear constraint of a linear program is replaced by a convex constraint, a *convex program* arises.

However, there are other ways of producing important extensions of the linear program, and these include: adding (generally non-convex) logical conditions, thereby producing what is called a *disjunctive program* in Balas (1979); or adding the requirement that the variables be integer, producing an *integer program*.

---

[1]The author's research reported in this paper has been supported by grant ENG-7900284 of the National Science Foundation.

ISBN 0-12-621120-5

These three extensions of the linear program are different in nature, but elementary results from convexity have been used in disjunctive programming and integer programming. We have surveyed these uses of convexity as part of the earlier papers Jeroslow (1977, 1978). Partially in order not to repeat ourselves, our emphasis here will be to reverse the perspective given earlier, and show how the ideas from disjunctive programming may be beneficial in nonconvex nonlinear programming.

In addition, recent results in integer programming (see e.g. Blair and Jeroslow (1980), Shrijver (1979), and Wolsey (1978, 1979)) have led to the concept of an *integer analogue* to a statement of linear duality. The recent work has been strongly influenced by Gomory (1963). To some extent, this development corresponds to the development of generalized duality schemes for nonconvex programming (see e.g. Tind and Wolsey (1978)).

Our focus throughout is on cutting-planes, i.e. valid linear inequalities implied by a given set of constraints. The primary influence of convexity in disjunctive and integer programming has been on the theory of cutting-planes. Also, cutting-planes relate to nonlinear programming in a more direct fashion than one might first intuit.

For example, an approach to the study of the general nonlinear program

$$\begin{array}{rl} \inf & f(x) \\ \text{subject to} & g(x) \leq 0,\ x \in K, \end{array} \tag{1}$$

where $f:R^n \to R$ and $g:R^n \to R^m$, is to develop all the valid linear inequalities for the set $S = \{(z,w) \mid \text{for some } x\in K,\ f(x) \leq z \text{ and } g(x) \leq w\}$ as was done in Duffin and Jeroslow (1979) in the convex case. A valid linear inequality $z + \sum_1 \lambda_i w_i \geq u$ is equivalent to the Lagrangean statement: $\inf_x \{f(x) + \sum_1 \lambda_i g_i(x)\} \geq u$.

For a second example, let $\{a^1, \ldots, a^t\}$ be any finite set of vectors in R^m, $S^k = \{(z,w) \mid (z,w) \in S \text{ and } a^k w = \max_i a^i w\}$. If, for each k,

$z+\sum_{1}^{k}\lambda_i w_i \geq u$ is a valid linear inequality for S^k it is not difficult to show that, for some $p \geq 0$, $z + p\|w\| \geq u$ is valid for S, where $\|w\|$ denotes the norm of w. The latter fact is, in turn, equivalent to the norm-penalty statement: $\inf_x \{f(x) + p\|g(x)\|\} \geq u$.

2. DISJUNCTIVE PROGRAMMING: CO-PROPOSITIONS

We consider a propositional logic, built up from finitary or infinitary uses of the logical connectives '$\wedge$' (for: 'and') and 'V' (for: 'or'), starting from linear inequality statements of the form $ax \geq b$ (see Tait (1968) for such propositional logic). The symbol 'V' for 'or' is also called the 'disjunction'.

Any system of convex constraints can be stated in this logic, as e.g. the requirement of being in the level set of a quasi-convex function. In fact, the constraint $z \geq x^2$ is equivalent to

$$z \geq 2x_o (x - x_o) + x_o^2, \text{ for all } x_o \varepsilon R. \tag{2}$$

To state it another way, $z \geq x^2$ is equivalent to this infinite 'and' statement of the propositional logic: $\bigwedge_{x_o \varepsilon R}(z - 2x_o x \geq - x_o^2)$. The infinite 'and' is simply a notational variant of (2). Note that convex constraints constitute that part of the logic in which only the 'and' connective '$\wedge$' is used. The logic contains propositions asserting many nonconvex statements, via the disjunction 'V', as for example $(x \leq 1) \vee (x \geq 2)$.

We now describe an inductive assignment of closed convex cones to propositions of this logic, which is called the 'co-proposition assignment'. We denote propositions by Greek letters α, β, γ, etc., and the 'co-proposition' assigned to a proposition α is denoted $CT(\alpha)$. On occasion, we write $\alpha(x)$ to emphasize the dependence of the proposition α on $x \varepsilon R^n$.

If α is a linear inequality statement $ax \geq b$, let

$$CT(\alpha) = \{\lambda(a, -b) + \theta(0,1) \mid \lambda, \theta \geq 0; \lambda, \theta \in R\}. \tag{3}$$

If $\alpha = \bigwedge_{h\in H} \alpha_h$, for H a (possibly infinite) index set, we define

$$CT(\alpha) = \text{clconv}\,(\bigcup_F\{\sum_{h\in F} CT(\alpha_h) \mid F \subseteq H, F \text{ finite}\}) \tag{4}$$

where $\sum_{h\in F} CT(\alpha_h) = \{\sum_{h\in F} a_h \mid a_h \in CT(h) \text{ for each } h\in F\}$. If $\alpha = \bigvee_{h\in H} \alpha_h$, we define

$$CT(\alpha) = \bigcap_{h\in H} CT(\alpha_h). \tag{5}$$

Quite possibly $CT(\alpha) = \{(0,b) \mid b \leq 0\}$, which indicates that no non-trivial linear inequalities are obtained from α. If H is finite and $CT(\alpha_h)$ is a polyhedral cone for $h\in H$, then (4) simplifies to:

$$CT(\alpha) = \sum_{h\in H} CT(\alpha_h) \tag{4'}$$

The co-proposition assignment has the property that:

$$\text{If } \alpha = \alpha(x) \text{ is true for } x\in R^n, \text{ and } (\pi, -\pi_o)\in CT(\alpha), \text{ then } \pi x \geq \pi_o \text{ is true.} \tag{6}$$

Indeed, (6) is correct for the ground step (3) of our inductive construction, and it is a property preserved by the inductive steps (4) and (5). Indeed, (4) in essence provides that the sum of valid linear inequalities, and their closure, yield valid inequalities. Similarly (5) provides that those inequalities common to all propositions α_h, $h\in H$, must be valid, provided only that at least one of these propositions holds.

As one application of the co-propositions, we obtain cutting-planes from the nonconvex condition:

$$x \notin C_1 \cup C_2 \cup \ldots \cup C_t \quad \text{and } x \geq 0 \tag{7}$$

where $C_k = \{x\in R^n \mid ax \leq b, (a,b)\in H_k\}$, $1 \leq k \leq t$, is a closed convex set, and where H_k is an arbitrary non-empty index set.

The co-propositions provide this family of inequalities:

$$\sum_{j=1}^{n} x_j \max\left\{ \sum_{k=1}^{t} \lambda_k(a^k, b^k)\, a_j^k \;\middle|\; (a^k, b^k) \varepsilon H_k \text{ for } k=1,\ldots,t \right\}$$
$$\geq \min\left\{ \sum_{k=1}^{t} \lambda_k(a^k, b^k)\, b^k \;\middle|\; (a^k, b^k) \varepsilon H_k \text{ for } k=1,\ldots,t \right\}. \tag{8}$$

In (8), $x = (x_1,\ldots,x_n)$; a_j^k is the j-th component of a^k; and we are permitted to arbitrarily select $\lambda_k(a^k, b^k) \geq 0$ as $(a^k, b^k) \varepsilon H_k$ varies.

An interesting special case occurs when $t = 1$, all $b^1 > 0$ and the choice $\lambda_1(a^1, b^1) = b^1$ is made for each $(a^k, b^k) = (a,b)\varepsilon H = H_1$. This gives the cutting-planes:

$$\sum_{j=1}^{n} x_j \max\{a_j/b \mid (a,b)\varepsilon H\} \geq 1 \tag{8$'$}$$

For $|H_1|$ finite, these are some of the cuts obtained in Glover (1973) and Balas (1975); see Jeroslow (1977) for a discussion of relationships between disjunctive constructions and the intersection cut constructions. The co-propositions were introduced in Jeroslow (1974) as a generalization of disjunctive constructions.

For t=1, the cut is as drawn in Figure 1. Specifically, a plane is passed through the intersection points of the convex set $C=C_1$ with the co-ordinate axes. Since $x \notin C$, we can restrict x to be in the half space that lies to the side of the hyperplane which does not contain the origin. For t=2, two cuts from the family of cuts are drawn in Figure 2. Both figures assume that the intercepts exists and that the origin lies in the interior of the convex regions, as depicted.

Our limited space has required us to sketch only a few fundamental points about the disjunctive methods, which were first introduced in Balas (1975, 1979); in particular, Balas (1979) contains an important result on "facial constraints", which we have not touched on here and which has a number of consequences.

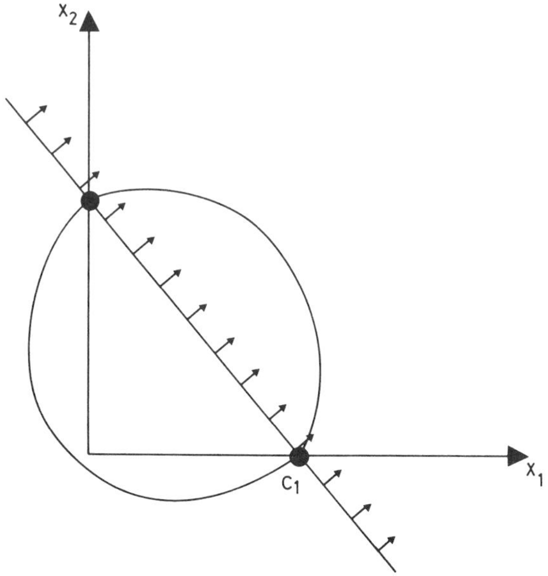

FIGURE 1. The Usual Intersection or Convexity Cut as a Disjunctive Cut (positive intercepts case).

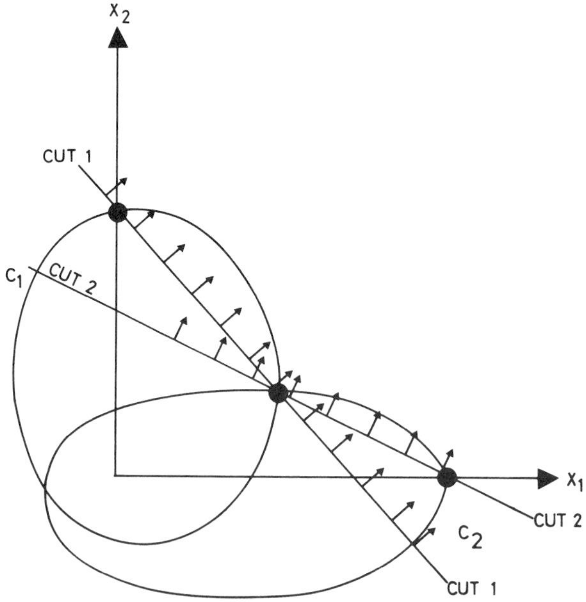

FIGURE 2. Example of a Family of Cuts from the Disjunctive Methods.

3. INTEGER PROGRAMMING: ANALOGUES

A *Chvatal function* is one which is built up from linear functions λb, with λ rational, by repeatedly rounding-up to the nearest integer, and then taking rational non-negative combinations of the functions thus obtained. In the original linear functions λb we can have λ of arbitrary signs, but thereafter the non-negativity of multipliers must be observed. For example, a Chvatal function of two variables b_1, b_2 is given by $f(b_1, b_2) = 3[-b_1 + 2b_2] + 2(3b_1 + [-b_2])$, where $[u]$ denotes the round-up of the real number u (i.e., the smallest integer not smaller than u).

Chvatal functions appear to play the role in integer programming, that linear functions play in the dual form of two equivalent linear statements. In other words, among the various equivalency theorems regarding linear inequalities, one tends to obtain true statements when the variables of the "primal" are required to be integer, and the linear functions of the "dual" are allowed to become Chvatal functions. We assume that all quantities of an integer program are rational, and that the proper choice of "primal" and "dual" statements has been made.

For example, it is well known that the linear program

$$\begin{aligned} &\min cx \\ &\text{subject } Ax = b \\ &\qquad x \geqslant 0 \end{aligned} \tag{9}$$

has as its dual the program

$$\begin{aligned} &\max \theta b \\ &\text{subject } \theta A \leqslant c. \end{aligned} \tag{10}$$

According to the heuristic principle annunciated in the last paragraph, the dual of the integer program in rationals

$$\begin{aligned} &\min cx \\ &\text{subject } Ax = b \\ &\qquad x \geqslant 0 \text{ and } x \text{ integer} \end{aligned} \tag{9)'$$

ought to be

$$\begin{aligned} &\max f(b) \\ &\text{subject to } f(a^j) \leqslant c_j \\ &\qquad f \text{ Chvatal} \end{aligned} \tag{10)'$$

where $A = [a^j]$ (cols) and $c = (c_j)$. Indeed, it is the case, that if (9)' is consistent, its value is that of (10)' (see Blair and Jeroslow (1980)).

As a second illustration of the heuristic principle, recall the linear theorem that a finitely generated polyhedral cone, i.e. a set of the form $\{b \mid \text{there is an } x \geq 0 \text{ with } Ax = b\}$, has a definition in terms of homogeneous linear inequalities. If the later definition is termed "dual", then we would conclude that a set of the form

$$\{b \mid \text{there is an integer } x \geq 0 \text{ with } Ax = b\} \tag{11}$$

will have a definition of the form

$$\{b \mid f_i(b) \leq 0,\ i = 1,\ldots, p\} \tag{12}$$

for certain Chvatal functions $f_1,\ldots, f_p$ (at least when A is rational). This is indeed the case, though we remark that in (12) the direction of the homogeneous inequalities cannot be reversed.

A third linear theorem is that the optimal value of a linear program, as a function of its right-hand-side (r.h.s.) b, i.e. the function given by

$$L(b) = \inf \{ cx \mid Ax = b,\ x \geq 0 \} \tag{13}$$

is the maximum of a finite number of linear functions, where it is defined. ($L(b)$ is defined precisely if there is $x \geq 0$ with $Ax = b$). And it is indeed the case that the value function of an integer program, given by

$$G(b) = \inf \{ cx \mid Ax = b,\ x \geq 0 \text{ and integer}\} \tag{14}$$

is the maximum of finitely many Chvatal functions, where it is defined. A fourth linear theorem states that $L(b)$ is defined exactly where a certain finite set of linear functions are all nonpositive; and, as one would expect, $G(b)$ is defined where a certian finite set of Chvatal functions are all nonpositive.

The role of Chvatal functions as an *integer analogue* of linear functions is even more pronounced. To each Chvatal function f is associated a linear function $\bar{f}$ called its *carrier*. The carrier is, intuitively, obtained by erasing all round-up operations and collecting terms. The carrier of the Chvatal function in the first paragraph of this section is therefore $\bar{f}(b_1, b_2) = 3(-b_1 + 2b_2) + 2(3b_1 - b_2) = 3b_1 + 4b_2$. Now the role that the Chvatal function f plays in the discrete version of a linear theorem, appears to be the role its carrier $\bar{f}$ plays in the original theorem. For example, the carrier of G(b) in (14) is the linear optimal value L(b) of (13).

A fifth linear theorem states that, if $Ax = b$, $x \geqslant 0$ is inconsistent, there is a linear form θw such that $\theta Ax \leqslant 0$ for all $x \geqslant 0$ and $\theta b > 0$. This linear theorem is a version of the Farkas Lemma; for other linear theorems of the alternative, see Mangasarian (1969, Table 2.4.1). As one would expect, if there is no solution to $Ax = b$, $x \geqslant 0$ and integer, then there is a Chvatal function f with $f(Ax) \leqslant 0$ for all integer $x \geqslant 0$, and $f(b) > 0$.

A sixth linear theorem, the Finite Basis Theorem for Cones, states that the solution set to a finite set of homogeneous linear inequalities $\{b \mid Eb \leqslant 0\}$ for some rational matrix E, has a finite basis, i.e. there is a matrix A such that $Eb \leqslant 0$ if and only if $Ax = b$ for some $x \geqslant 0$. Viewing the statement "$Eb \leqslant 0$" as the "dual" statement it turns out *not* to be the case that for all finite sets of Chvatal functions $f_1, \ldots, f_t$ there exists a rational matrix A with:

$$\begin{array}{c}\text{b rational and } f_i(b) \leqslant 0 \text{ for } i = 1, \ldots, t \text{ if and only if} \\ \text{there is an } x \geqslant 0 \text{ integer with } Ax = b.\end{array} \tag{15x}$$

However, this conjecture was almost correct, as one need only change "rational" to "integer". Indeed, there is an integer matrix A with:

$$\begin{array}{c}\text{b integer and } f_i(b) \leqslant 0 \text{ for } i = 1, \ldots, t \text{ if and only if} \\ \text{there is an } x \geqslant 0 \text{ integer with } Ax = b\end{array} \tag{15}$$

The discovery of integer analogues is recent, and we do not know the extent and complete nature of the phenomenon. The mixed-integer program can be indirectly treated by the ideas presented here, but a direct treatment is not possible since optimal value functions of a mixed-integer program are not closed under the inductive operations which construct Chvatal functions. On the other hand, constraint sets of the form

$$\begin{aligned} Ax + By &= Cb \\ x, y &\geq 0 \\ x &\text{ integer} \end{aligned}$$

in which a general rational matrix C pre-multiplies the right-hand-side, do allow much of the treatment of integer programming to go over to the mixed case. We shall report our recent joint results in the near future.[2]

4. CONCLUSIONS

We have shown how ideas from convexity and generalized convexity have influenced disjunctive programming, and we have indicated that even the concept of a linear function can be generalized and adapted to the discrete setting.

The generalizations that one studies depend on which aspect of linearity or convexity is retained, and which new feature of some non-convexity one chooses to underscore. We can expect further fruitful generalizations in the future.

---

[2] We have recently discovered that the value functions of these pre-multiplied constraint sets are precisely the closure of the class of Chvatal functions under the operations of maximum and infimal convolution. Once again, the influence of ideas from convexity is present.

REFERENCES

BALAS, E. (1975). "Disjunctive Programming: Cutting-Planes from Logical Conditions." In *Nonlinear Programming 2*, (O.L. Mangasarian, R.R. Meyer, and S.M. Robinson, eds.). Academic Press, New York, 279-312.

BALAS, E. (1979). "Disjunctive Programming." *Annals of Discrete Mathematics 5*, 3-51.

BLAIR, C.E., and JEROSLOW, R.G. (1980). "The Value Function of an Integer Program." *Mathematical Programming*, to appear.

BLAIR, C.E., and JEROSLOW, R.G. (1978). "A Converse for Disjunctive Constraints." *Journal of Optimization Theory and Its Applications.*

DUFFIN, R.J., and JEROSLOW, R.G. (1979). "The Limiting Lagrangean." College of Management, Georgia Institute of Technology and Carnegie-Mellon University.

GLOVER, F. (1973). "Convexity Cuts and Cut Search." *Operations Research 21*, 123-124.

GLOVER, F. (1975). "Polyhedral Annexation in Mixed Integer and Combinatorial Programming." *Mathematical Programming 9*, 161-188.

GOMORY, R.E. (1963). "An Algorithm for Integer Solutions to Linear Programs." In *Recent Advances in Mathematical Programming*, (Graves and Wolfe, eds.). 269-302.

JEROSLOW, R. (1977). "Cutting-Plane Theory: Disjunctive Methods." *Annals of Discrete Mathematics 1*, 293-330.

JEROSLOW, R. (1978). "Cutting-Plane Theory: Algebraic Methods." *Discrete Mathematics 23*, 121-150.

JEROSLOW, R. (1974). "Cutting-Planes for Relaxations of Integer Programs." MSRR No. 347, GSIA, Carnegie-Mellon University.

JOHNSON, E.L. (1979). "The Group Problem for Mixed Integer Programming." *Mathematical Programming Studies 2*, 137-179.

MANGASARIAN, O.L. (1969). *Nonlinear Programming.* McGraw-Hill, New York.

ROCKAFELLAR, R.T. (1970). *Convex Analysis.* Princeton University Press, Princeton, New Jersey.

SHRIJVER, A. (1979). "On Cutting Planes." Department of Mathematics, Eindhoven University.

STOER, J., and WITZGALL, C. (1970). *Convexity and Optimization in Finite Dimensions I.* Springer-Verlag, Berlin.

TAIT, W.W. (1968). "Normal Derivability in Classical Logic." In *The Syntax and Semantics of Infinitary Languages*, (Jon Barwise, ed.). Lecture Notes in Mathematics No.72, 204-236. Springer-Verlag, Berlin-Heidelberg.

TIND, J., and WOLSEY, L. (1978). "A Unifying Framework for Duality Theory." In *Mathematical Programming.* D.P. 7834, CORE, Université Catholique de Louvain.

WOLSEY, L. (1978). "Integer Programming Duality: Price Functions and Sensitivity Analysis." London School of Economics and CORE.

WOLSEY, L. (1979). "The b-hull of an Integer Program." London School of Economics and CORE.

YOUNG, R.D. (1971). "Hypercylindrically-deduced Cuts in Zero-One Integer Programs." *Operations Research 19*, 1393-1405.

PART VII

APPLICATION TO STOCHASTIC SYSTEMS

UNIVARIATE AND MULTIVARIATE TOTAL POSITIVITY, GENERALIZED CONVEXITY AND RELATED INEQUALITIES

Samuel Karlin   Yosef Rinott

Various univariate and multivariate concepts of *total positivity* are set forth. Multivariate total positivity and generalized convexity play an underlying role in determining measures of association and orderings among vector random variables and classes of probability inequalities on coverage of regions. In the analysis of reliability systems multidimensional analogs of the concept of increasing failure rate and bounds on system reliability are related to multivariate total positivity. A host of statistical applications in areas such as the analysis of symmetric sampling schemes from finite populations, multivariate ranking and selection procedures, multivariate matching problems, slippage problems, and various properties of power functions fall in this domain. In the analysis of sampling plans comparisons of various functionals such as variances of estimates under different sampling schemes lead to inequalities which are related to total positivity. Probabilistic inequalities arise in ranking and selection problems in comparisons of probabilities of correct selection (or ranking) for different procedures, or in computing bounds for such probabilities. Concepts of positive dependence related to total positivity are relevant to problems of matching of observations, and are widely used in reliability theory where system-components are assumed to have positively dependent life-times. This paper surveys some basic ideas and examples, in order to provide the background and motivation to more recent results and problems related to multivariate concepts of total positivity.

1. INTRODUCTION

Recent work in areas such as multivariate statistical analysis, correlation inequalities for particle systems in statistical mechanics, and certain optimization models in operations research indicates the natural potential and relevance of multivariate formulations of total positivity. The purpose of this paper is to describe a number of multivariate results and applications (Section 5). The univariate total positivity theory (Sections 2-4) provides an essential antecedant for the multivariate theory.

ISBN 0-12-621120-5

A systematic treatment of univariate forms of total positivity and its applications, and bibliography appear in Karlin (1968). Multivariate concepts of total positivity and applications will be treated in detail in a forthcoming monograph, Karlin and Rinott (1981).

In Section 2 we delineate some notions and results concerning totally positive kernels. Section 3 provides applications of total positivity to optimization problems in statistical decision theory. We briefly discuss in Section 4 the notions of Tchebycheff systems and generalized convex functions. The formulation of multivariate total positivity and some applications are described in Section 5.

2. TOTAL POSITIVITY AND VARIATION DIMINISHING TRANSFORMATIONS

In the following survey (Sections 2-4) some of the definitions and statements will not be absolutely rigorous in order to avoid technical details and simplify the exposition. Further elaborations of the concepts and proofs are covered in the references. The results of this section are presented in detail in Karlin (1968). Let $S(x_1,\ldots,x_m)$ denote the number of sign changes with respect to zero, i.e., the number of times the sequence $x_1,\ldots,x_m$ changes from positive to negative, or from negative to positive, zero terms being discarded. For example, $S(-1,1,2,1) = 1 = S(-1,0,1,0,2,0,1)$, $S(-1,0,1,0,1,-2) = 2$.

DEFINITION 1: Let $f(t)$ *be defined on* I*, an ordered set. Let* $S(f) = \sup S(f(t_1),\ldots,f(t_m))$ *where the supremum is extended over all* $t_1 < \ldots < t_m$ *in* I*,* m *arbitrary and* $S(x_1,\ldots,x_m)$ *is the number of sign changes of the sequence* $x_1,\ldots,x_m$*, zero terms being discarded.*

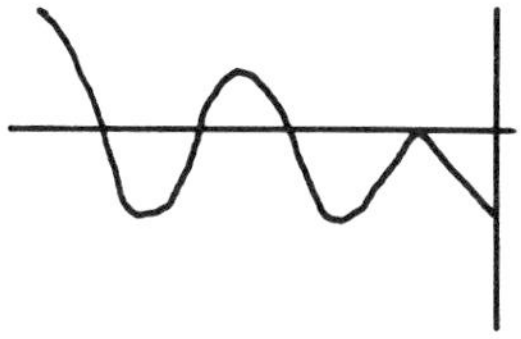

FIGURE 1: A schematic function with $S(f) = 3$

Before presenting a formal definition of totally positive kernels we describe its principal utilitarian characterization (compare with Theorem 1). *A kernel* $K(x,y)$ *defined on a product of linearly ordered sets* $X \times Y$ *is totally positive if for all suitably integrable functions* f *and* σ*-finite measures* μ *on* Y

$$g(x) = \int K(x,y)f(y)d\mu(y) \tag{1}$$

entails $S(g) \leq S(f)$.

An alternative formulation is the following:

DEFINITION 2: A real valued function (kernel) $K(x,y)$ *of two variables ranging over linearly ordered sets* X *and* Y *respectively is said to be totally positive of order* r (TP_r) *if for all*

$$x_1 < \dots < x_m,\ y_1 < \dots < y_m,\ x_i \in X,\ y_j \in Y,\ 1 \leq m \leq r \tag{2}$$

we have

$$\begin{vmatrix} K(x_1,y_1) & \dots & K(x_1,y_m) \\ \vdots & & \\ K(x_m,y_1) & \dots & K(x_m,y_m) \end{vmatrix} \geq 0 \tag{3}$$

In the case of strict inequalities in (3) K is said to be strictly TP_r (STP_r).

For example, the property TP_2 is equivalent to $K(x,y) \geq 0$ and

$$\begin{vmatrix} K(x_1,y_1) & K(x_1,y_2) \\ K(x_2,y_1) & K(x_2,y_2) \end{vmatrix} \geq 0 \tag{4}$$

If $K(x,y) > 0$, then (4) can be expressed as $\frac{K(x,y_1)}{K(x,y_2)}$ is increasing in x for any $y_1 < y_2$.

Related concepts are obtained when the determinants in (3) maintain a constant sign, not necessarily positive; e.g., when $K \geq 0$ satisfies (4) with the direction of the inequality reversed, K is said to be reverse rule of order 2 (RR_2).

Examples: The exponential family of densities $K(x,y) = \beta(y)\gamma(x)\, e^{\phi(x)\psi(y)}$, with ϕ and ψ increasing is TP, i.e., it is TP_r for all r. This includes for example $K(x,y) = e^{xy}$ and $K(x,y) = (2\pi)^{-\frac{1}{2}} e^{-(x-y)^2/2}$ which are TP on $X,Y = (-\infty,\infty)$. Also, $K(x,p) = \binom{n}{x} p^x(1-p)^{n-x}$ is TP in the variables $x \in \{0,1,\ldots,n\}$ and $p \in [0,1]$, and $K(x,\lambda) = e^{-\lambda}\frac{\lambda^x}{x!}$, $x = 0,1,\ldots,\lambda \in (0,\infty)$ is TP.

The total positivity nature of more elaborate examples, including the noncentral t and F distributions, can be deduced with the aid of the following composition formula.

LEMMA 1: If $K(x,z)$ *is* TP_r *and* $L(z,y)$ *is* TP_s *then the kernel defined by*

$$M(x,y) = \int K(x,z)L(z,y)d\mu(z) \tag{5}$$

(where $d\mu$ *is a* σ*-finite measure) is* $TP_{\min(r,s)}$.

Under certain mild technical assumptions on the functions and measures involved one can prove for $g(x) = \int K(x,y)f(y)d\mu(y)$:

THEOREM 1: If $K(x,y)$ *is* TP_r *then* $S(g) \leq S(f)$ *provided* $S(f) \leq r-1$. *If* $S(g) = S(f)$ *then* f *and* g *exhibit the same sequence of sign changes in their respective domains.*

Theorem 1 readily implies:

COROLLARY 1: Suppose $K(x,y)d\mu(y) \equiv 1$ *for all* $x \in X$ *and* $K(x,y)$ *is* TP_2. *If* f *is monotone on* Y *then* $g(x) = \int K(x,y)f(y)d\mu(y)$ *is monotone on* X.

If in addition K *transforms a linear function to a linear function, i.e., for* $f(y) = y$, $g(x) = \int K(x,y)f(y)d\mu(y)$ *is of the form* $g(x) = ax+b$, *and* K *is* TP_3, *then for any convex function* f, *its transform* g *will also retain convexity.*

If we regard $K(x,y)d\mu(y)$ as a probability density with parameter x, then the first part of Corollary 1 can be expressed as follows: if U and V are random variables with densities $K(x_1,y)d\mu(y)$ and $K(x_2,y)d\mu(y)$ respectively, where $x_1 < x_2$, then $Ef(U) \leq Ef(V)$ for all increasing f (V is stochastically larger than U).

The concepts of convex cones of functions and their dual cones (see definition below) are germane in understanding the structure behind many classes of inequalities.

DEFINITION 3: Given a cone of functions C *define the dual cone* C* *to be the collection of signed measures* $d\mu$ *satisfying*

$$\int \phi d\mu \geq 0 \quad \text{for all } \phi \in C. \tag{6}$$

Corollary 1 shows that for the cone of increasing functions, a TP_2 kernel $K(x,y)$ defines for any $x_1 < x_2$ the signed measure $[K(x_2,y) - K(x_1,y)]d\mu(y)$ in C*.

More details on cones of generalized convex functions, generalized convexity preserving transformations and dual cones are given in Section 4. Multivariate generalizations are described in Section 5.

3. OPTIMIZATION IN STATISTICAL DECISION THEORY

Consider the choice between two alternative decisions d_0, d_1 on the basis of an observed random variables X (possibly some function of a sample consisting of several observations). Let X have the density function $f(x,\omega)d\mu(x)$ where ω is a parameter taking values in an ordered set Ω.

A (randomized) decision rule can be described as a function $\phi(x) \in [0,1]$, where $\phi(x) = 1$ means that when $X = x$ is observed decision d_1 is taken, $\phi(x) = 0$ means that decision d_0 is taken, and $0 < \phi(x) < 1$ means that d_1 is taken with probability $\phi(x)$ and d_0 with probability $1 - \phi(x)$. Note that the probability of decision d_1 is $\int \phi(x)f(x,\omega)d\mu(x)$.

In the context of testing hypotheses concerning the parameter ω we would like to determine whether $\omega \in \Omega_0$ or $\omega \in \Omega_1$ where $\Omega = \Omega_0 \cup \Omega_1$ and d_i is interpreted as the determination $\omega \in \Omega_i$.

The classical Neyman-Pearson approach for unbiased tests (see Lehmann (1959)) can be described as the following *optimization* problem: Find a decision function ϕ which will

$$\textit{maximize} \int \phi(x)f(x,\omega)d\mu(x) \quad \text{for} \quad \omega \in \Omega_1 \quad \textit{subject to} \tag{7}$$

$$\int \phi(x)f(x,\omega)d\mu(x) \leq \alpha \quad \text{for} \quad \omega \in \Omega_0 \ (\alpha \text{ fixed}), \tag{8a}$$

and

$$\int \phi(x)f(x,\omega)d\mu(x) \geq \alpha \quad \text{for} \quad \omega \in \Omega_1. \tag{8b}$$

As as example suppose $\Omega = R$ (the real line), $\Omega_0 = (-\infty,\omega_0]$, $\Omega_1 = (\omega_0,\infty)$, for some $\omega_0 \in R$, and $f(x,\omega)$ is TP_2.

Let ϕ^* satisfying

$$\phi^*(x) = \begin{cases} 1 & x > \lambda \\ 0 & x < \lambda \end{cases} \tag{9}$$

be a decision function for which (8) holds. Suppose ϕ is another decision function satisfying (8). Under obvious continuity conditions we can assume by (8) that

$$\int\phi^*(x)f(x,\omega_0)d\mu(x) = \int\phi(x)f(x,\omega_0)d\mu(x) = \alpha. \tag{10}$$

Note that $\phi^*(x) - \phi(x)$ exhibits one sign change from - to +. Theorem 1 implies that $\int[\phi^*(x) - \phi(x)]f(x,\omega)d\mu(x)$ has one sign change which by (10) occurs at $\omega = \omega_0$. Thus

$$\int \phi^*(x)f(x,\omega)d\mu(x) \leq \int \phi(x)f(x,\omega)d\mu(x) \quad \text{for } \omega \in \Omega_0$$

and

$$\int \phi^*(x)f(x,\omega)d\mu(x) \geq \int \phi(x)f(x,\omega)d\mu(x) \quad \text{for } \omega \in \Omega_1.$$

It follows that ϕ^* is essentially the unique solution of the optimization problem (7)-(8).

Similar results hold for $\Omega_0 = [\omega_0, \omega_0']$, $\Omega_1 = R \setminus \Omega_0$ where now $f(x,\omega)$ is assumed to be TP_3 and

$$\phi^*(x) = \begin{cases} 0 & \lambda_1 < x < \lambda_2 \\ 1 & \text{otherwise} \end{cases} \tag{11}$$

Consider now $L_i(\omega)$ to be the loss functions when the decision d_i, $i = 0,1$, is taken. Typically, in our first example one may take $L_0(\omega) - L_1(\omega)$ to have one sign change (from - to +) occurring at ω_0, while in the second example it is natural to assume that $L_0(\omega) - L_1(\omega)$ has two sign changes. More generally, suppose that $L_0(\omega) - L_1(\omega)$ has n sign changes. The expected loss is expressed as $\int [\phi(x)L_1(\omega) + (1 - \phi(x))L_0(\omega)]f(x,\omega)d\mu(x)$. A *Bayesian* approach consists of defining a "prior" distribution on Ω, say $d\sigma$, and solving the following optimization problem: Find a decision function ϕ to:

$$\textit{minimize} \int \{\int [\phi(x)L_1(\omega) + (1 - \phi(x))L_0(\omega)]f(x,\omega)d\mu(x)\}d\sigma(\omega). \tag{12}$$

For any x consider

$$\Lambda(x) = \int (L_1(\omega) - L_0(\omega))f(x,\omega)d\sigma(\omega). \tag{13}$$

Observations of (12)-(13) indicate that the solution of (12) has $\phi(x) = 1$ if $\Lambda(x) < 0$ and $\phi(x) = 0$ if $\Lambda(x) > 0$. By Theorem 1 $\Lambda(x)$ has at most n sign changes provided that $f(x,\omega)$ is TP_{n+1}, and in this case the optimal decision function has the form

$$\phi^*(x) = \begin{cases} 1 & \lambda_{2i} < x < \lambda_{2i+1} \\ 0 & \text{otherwise} \end{cases} \qquad i = 0,\ldots,\frac{n-1}{2}.$$

Decision functions of the form (9) and (10) will again be obtained if $L_0 - L_1$ has one and two sign changes, respectively.

The same approach can be applied to obtain minimax and admissibility properties of similar decision functions, and to suitable multidecision problems (see Karlin (1958) and references therein).

4. GENERALIZED CONVEXITY AND TOTAL POSITIVITY

Generalized convex functions are briefly discussed in this section. For details and bibliography the reader is referred to Karlin and Studden (1966).

DEFINITION 4: Let $u_1(t),\ldots,u_n(t)$ *be continuous functions defined on the interval* (a,b). *For* $x \in (a,b)$, $y = 1,\ldots,n$ *define* $K(x,y) = u_y(x)$. *The functions* $u_1,\ldots,u_n$ *are said to constitute a Tchebycheff system* (T-system) *on* (a,b) *if the kernel* $K(x,y)$ *is* STP_n *(see Section 2).*

Explicitly a T-system is defined by the property

$$\begin{vmatrix} u_1(t_1) & u_1(t_2) & \cdots & u_1(t_n) \\ u_2(t_1) & u_2(t_2) & \cdots & u_2(t_n) \\ \vdots & & & \\ u_n(t_1) & u_n(t_2) & \cdots & u_n(t_n) \end{vmatrix} > 0 \tag{14}$$

for all $a < t_1 < \ldots < t_n < b$. Note that (14) implies that for any set of real constants $\{c_k\}$ not all zero, the function $\sum_{k=1}^{n} c_k u_k(t)$ does not vanish more than $n - 1$ times on (a,b).

DEFINITION 5: A function ϕ *on* (a,b) *is said to belong to the cone* $C(u_1,\ldots,u_n)$ *of generalized convex functions with respect to the T-system* $u_1,\ldots,u_n$ *if*

$$\begin{vmatrix} u_1(t_1) & u_1(t_2) & \cdots & u_1(t_{n+1}) \\ u_2(t_1) & u_2(t_2) & \cdots & u_2(t_{n+1}) \\ \vdots & & & \\ u_n(t_1) & u_n(t_2) & \cdots & u_n(t_{n+1}) \\ \phi(t_1) & \phi(t_2) & \cdots & \phi(t_{n+1}) \end{vmatrix} \geq 0 \tag{15}$$

for all $a < t_1 < \ldots < t_{n+1} < b$.

Examples: For $u_1 \equiv 1$ the cone $C(u_1)$ consists of all non-decreasing functions on (a,b). If $u_2(t) = ct + d$, $c > 0$, then $C(u_1,u_2)$ is the cone of convex functions on (a,b). For $u_i(t) = t^{i-1}$, $i = 1,\ldots,n$, $C(u_1,\ldots,u_n)$ consists of all functions ϕ having $\phi^{(n)} \geqslant 0$ and limits of such functions.

Cone-preserving transformations as in Corollary 1 are subsumed by the following result:

Suppose $K(x,y)$ *is* STP_n *and set* $v_i(x) = \int K(x,y)u_i(y)d\mu(y)$ *(where* μ *is a measure on* (a,b)*). Then the kernel* $K(x,y)$ *transforms a function* $f \in C(u_1,\ldots,u_n)$ *to a function* $g \in C(v_1,\ldots,v_n)$ *defined by* $g(x) = \int K(x,y)f(y)d\mu(y)$.

Properties of the cones $C(u_1,\ldots,u_n)$ and their dual cones (see definition 3) and their structure, characterizations of extreme rays and related differential operators are discussed in Karlin and Studden (1966).

We conclude this section with a simple theorem and examples indicating the applicability of the theory to derivation of inequalities.

DEFINITION 6: A signed measure $d\mu$ *is said to have* k *sign changes on* (a,b) *if there exists a division of* (a,b) *into disjoint consecutive intervals* $J_0,\ldots,J_k$ *such that* $d\mu$ *is of alternating sign and non-null on* $J_0,\ldots,J_k$. *(In the case* $d\mu(t) = f(t)dt$ *for some continuous* f*, the number of sign changes of* $d\mu$ *is equivalent to the number of ordinary sign changes of* f*).*

THEOREM 2: Let $d\mu$ *satisfy*

$$\int_a^b u_j(t)d\mu(t) = 0, \qquad j = 1,\ldots,n. \tag{16}$$

Assume that $d\mu$ *has exactly* n *sign changes on* (a,b) *and that it is non-negative and non-null on some interval extending to the endpoint* b. *Then* $d\mu \in C^*(u_1,\ldots,u_n)$ *(the dual cone of* $C(u_1,\ldots,u_n)$*, see Definition 3).*

As a simple illustration we prove the following rearrangement inequality: *Let* f *be a probability density in* R *and* ϕ, *and* ψ *two non-decreasing (or two non-increasing) functions. Then*

$$\int \phi(x)\psi(x)f(x)dx \geq \left(\int \phi(x)f(x)dx\right)\left(\int \psi(x)f(x)dx\right). \tag{17}$$

PROOF: Assume without loss of generality that $\psi \geq 0$. Set $\gamma = \int \psi(x)f(x)dx$. The signed measure $d\mu(t) = \left[\frac{\psi(t)f(t)}{\gamma} - f(t)\right]dt$ has one sign change and the conditions of Theorem 2 hold for the cone $C(u_1)$ with $u_1 \equiv 1$. Since $\phi \in C(u_1)$, Theorem 2 implies $d\mu \in C^*(u_1)$, i.e., $\int \phi(t)d\mu(t) \geq 0$ which reduces to (17).

For a second application let X_i, $i = 1,\ldots,n$, be independent binomial variables with $P(X_i = 1) = 1 - P(X_i = 0) = p_i$, and set

$$(p_1',\ldots,p_n') = (p_1,\ldots,p_n)T \tag{18}$$

where T is an $n \times n$ doubly stochastic matrix. (When (18) holds for some T, $\underline{p} = (p_1,\ldots,p_n)$ is said to *majorize* $\underline{p}' = (p_1',\ldots,p_n')$.) Let $X = \sum_{i=1}^{n} X_i$ and let $d\mu_p$ and $d\mu_{p'}$ be the probability measures associated with the random variables X when the X_i are governed by $\underline{p} = (p_1,\ldots,p_n)$ and $\underline{p}' = (p_1',\ldots,p_n')$, respectively. It can be shown that the signed measure $d\mu = d\mu_{p'} - d\mu_p$ satisfies the assumptions of Theorem 2 with respect to the cone of convex functions, $C(1,ct+d)$, $c > 0$. Theorem 2 implies $E_{\underline{p}}\phi(X) \leq E_{\underline{p}'}\phi(X)$ for all convex ϕ. Equivalently the function $\psi(\underline{p}) = \psi(p_1,\ldots,p_n) = E_{\underline{p}}\phi(x)$ is monotone with respect to the partial ordering of majorization.

A detailed survey of majorization and related concepts was given recently by Marshall and Olkin (1979). Multivariate extensions and applications to entropy comparisons are given in Karlin and Rinott (1980d,e).

5. MULTIVARIATE CONCEPTS OF TOTAL POSITIVITY

Multivariate concepts related to total positivity have recently been studies by many authors in various branches of mathematics. We now survey some recent results on multivariate total positivity. For further details, proofs, references, and applications see, e.g., Barlow and Proschan (1975), Kemperman (1977), Karlin and Rinott (1980a,b,c).

A natural extension of TP_2 kernels to multivariate functions defined on a lattice X formed as a produce of ordered sets X_i, where $X = X_1 \times \ldots \times X_n$, is to require that the function k on X be non-negative and

$$\begin{vmatrix} k(\underline{x} \wedge \underline{y}) & k(\underline{y}) \\ k(\underline{x}) & k(\underline{x} \vee \underline{y}) \end{vmatrix} \geq 0 \tag{19}$$

where v and $\wedge$ are the lattice operations defined for $\underline{x} = (x_1,\ldots,x_n)$, $\underline{y} = (y_1,\ldots,y_n) \in X$ by $\underline{x} \vee \underline{y} = (\max(x_1,y_1),\ldots,\max(x_n,y_n))$, $\underline{x} \wedge \underline{y} = (\min(x_1,y_1),\ldots,\min(x_n,y_n))$. (Compare with (4).) Equivalently, $k \geq 0$ satisfies

$$k(\underline{x} \vee \underline{y})k(\underline{x} \wedge \underline{y}) \geq k(\underline{x})k(\underline{y}). \tag{20}$$

A kernel k *satisfying (20) is called Multivariate* $TP_2(MTP_2)$.

Examples of MTP_2 functions include multivariate probability densities associated with the following distributions:

(i) the multivariate normal under a suitable covariance structure (Barlow and Proschan (1975))

(ii) absolute values of multivariate normal variables under a certain covariance structure (see Karlin and Rinott (1980c))

(iii) the distribution of characteristic roots of certain Wishart matrices appearing in multivariate hypotheses testing problems (Dykstra and Hewett (1978), Perlman and Olkin (1978))

(iv) the multivariate logistic distribution

(v) multivariate versions of the Gamma, F and t distributions

(vi) the joint distribution of components of any one dimensional diffusion process (Barlow and Proschan (1975))

(vii) the joint distribution of order statistics from a sample of independent random variables

(viii) the negative multinomial distribution (see Karlin and Rinott (1980a)).

We next discuss some results concerning MTP_2 functions and some extensions.

Let $d\sigma(\underline{x})$ be the product of the measures $d\sigma(x_i)$ on X_i. We abbreviate $d\sigma(\underline{x}) = d\underline{x}$.

THEOREM 3: (Ahlswede and Daykin (1978).) Let f_1, f_2, f_3, f_4 *be non-negative functions on* X *satisfying for all* $\underline{x}, \underline{y} \in X$

$$f_1(\underline{x}) f_2(\underline{y}) \leq f_3(\underline{x} \vee \underline{y}) f_4(\underline{x} \wedge \underline{y}). \tag{21}$$

Then

$$\left(\int f_1(\underline{x}) d\underline{x}\right) \left(\int f_2(\underline{x}) d\underline{x}\right) \leq \left(\int f_3(\underline{x}) d\underline{x}\right) \left(\int f_4(\underline{x}) d\underline{x}\right). \tag{22}$$

(The proof proceeds by induction on the dimension n.)

An appropriate choice of the functions f_i, i = 1,...,4, yields:

THEOREM 4: (Holley (1974), Preston (1974), Kemperman (1977).) Let $f_1(\underline{x})$ *and* $f_2(\underline{x})$ *be probability densities on* X *satisfying*

$$f_1(\underline{x}) f_2(\underline{y}) \leq f_2(\underline{x} \vee \underline{y}) f_1(\underline{x} \wedge \underline{y}). \tag{23}$$

Then for any increasing function ϕ *on* X

$$\int \phi(\underline{x}) f_1(\underline{x}) d\underline{x} \leq \int \phi(\underline{x}) f_2(\underline{x}) d\underline{x}. \tag{24}$$

Thus, f_2 is stochastically larger than f_1 or, equivalently, $(f_2(\underline{x}) - f_1(\underline{x}))d\underline{x}$ belongs to the dual of the cone of increasing functions on X. (Compare with Section 2.) Another consequence is the following result which extends Corollary 1 concerning the preservation of monotonicity.

COROLLARY 2: Let $f(\underline{x},\underline{\lambda})$ *be* MTP_2 *in* $(\underline{x},\underline{\lambda}) = (x_1,\ldots,x_n, \lambda_1,\ldots,\lambda_p)$ *and assume* $\int f(\underline{x},\underline{\lambda})d\underline{x} \equiv 1$ *for all* $\underline{\lambda}$ *in a given parameter space* $\Lambda = \Lambda_1 \times \ldots \times \Lambda_p$ *where* Λ_i *are ordered sets. Then* $\int \phi(\underline{x})f(\underline{x},\underline{\lambda})d\underline{x}$ *is increasing in* $\underline{\lambda}$ *for any increasing function* ϕ *on* X.

Further examples and results concerning MTP_2 functions are obtained by the following lemma which extends Lemma 1:

LEMMA 2: Let $X = X_1 \times \ldots \times X_n$ *be a product lattice and similarly consider* $Y = Y_1 \times \ldots \times Y_m$ *and* $Z = Z_1 \times \ldots \times Z_p$. *If* $K(\underline{x},\underline{z})$ *is* MTP_2 *on* $X \times Z$ *and* $L(\underline{z},\underline{y})$ *is* MTP_2 *on* $Z \times Y$ *then the kernel* M *defined on* $X \times Y$ *by*

$$M(\underline{x},\underline{y}) = \int K(\underline{x},\underline{z})L(\underline{z},\underline{y})d\sigma(\underline{z})$$

(where $d\sigma$ is a product measure on Z) *is* MTP_2 *in all its variables.*

The following theorem is the basis of many applications.

THEOREM 5: (Sarkar (1969), Fortuin, Ginibre and Kasteleyn (1971).) Let $K(\underline{x})$ *be a* MTP_2 *probability density on* X. *Then for any pair of increasing (or decreasing) functions* ϕ *and* ψ *on* X

$$\int \phi(\underline{x})\psi(\underline{x})k(\underline{x})d\underline{x} > \left(\int \phi(\underline{x})k(\underline{x})d\underline{x}\right)\left(\int \psi(\underline{x})k(\underline{x})d\underline{x}\right). \tag{25}$$

(Compare with (17).)

If the random vector $\underline{X}$ possesses the density $k(\underline{x})$, then (25) can be expressed by

$$E\{\phi(\underline{X})\psi(\underline{X})\} \geq \{E\phi(\underline{X})\}\{E\psi(\underline{X})\}. \tag{26}$$

When (26) holds the components of $\underline{X}$ are said to be *associated.* Association of random variables was defined in the context of reliability theory where the random lifetimes of components of an operating system are expected to be positively dependent. Using inequalities derived from (26) bounds on system reliability are obtained (see Barlow and Proschan (1975)). Other applications appear in statistical mechanics (see e.g. Fortuin et al. (1971)), limit theorems for associated random variables (Newman and Wright (1980)), and probability inequalities.

A random vector $\underline{X} = (X_1,\ldots,X_n)$ having a joint MTP_2 density satisfies the following inequalities as special cases of the above theorems.

(i) $\mathrm{Cov}(X_i,X_j) \geq 0$;

(ii) $E\{\phi(X_1,\ldots,X_k) \mid X_{k+1} = x_{k+1},\ldots,X_n = x_n\}$ is increasing in $x_{k+1},\ldots,x_n$ for any increasing function ϕ;

(iii) $E\{\prod_{i=1}^{n} \phi_i(X_i)\} \geq \prod_{i=1}^{n} E\{\phi_i(X_i)\}$ for ϕ_i increasing (or decreasing) and in particular

$$P(X_1 \leq c_1,\ldots,X_n \leq c_n) \geq \prod_{i=1}^{n} P(X_i \leq c_i). \tag{27}$$

Inequalities of the type (27) (applied also to absolute values of random variables) provide bounds for probabilities associated with multivariate confidence sets related to statistical estimation of parameters.

When the inequality (20) is reversed, namely

$$k(\underline{x} \vee \underline{y})d(\underline{x} \wedge \underline{y}) \leq k(\underline{x})k(\underline{y}) \tag{28}$$

k is called a Multivariate reverse rule of order 2 (MRR_2; compare with the definition of RR_2 kernels in Section 2). The multinomial distribution, multivariate hypergeometric, the Dirichlet family of distributions and the multivariate normal distribution for special covariance structures all have MRR_2 densities.

Under some additional conditions requiring that the MRR_2 property is preserved for certain marginals of the distributions (these conditions will not be specified here; they are valid for all the above examples, see Karlin and Rinott (1980b)), the following result obtains.

Let $\underline{X} = (X_1,\ldots,X_n)$ *have an* MRR_2 *joint density, and let* ϕ_i, $i = 1,\ldots,n$ *be increasing (or decreasing) logconcave functions. (Equivalently,* $\phi_i(x-y)$ *is* TP_2*). Then*

$$\begin{aligned} &E\{\phi_1(X_1)\ \cdots\ \phi_k(X_k)\phi_{k+1}(X_{k+1})\ \cdots\ \phi_n(X_n)\} \\ &\leq E\{\phi_1(X_1)\ \cdots\ \phi_k(X_k)\}E\{\phi_{k+1}(X_{k+1})\ \cdots\ \phi_n(X_n)\}. \end{aligned} \tag{29}$$

As special cases of (29) we obtain

$$\mathrm{Cov}(X_i, X_j) \leq 0$$

$$P(X_1 \leq c_1, \ldots, X_n \leq c_n) \leq \prod_{i=1}^{n} P(X_i \leq c_i)$$

and moment inequalities of the type

$$E\{\prod_{i=1}^{n} X_i^{\alpha_i}\} \leq \prod_{i=1}^{n} E(X_i^{\alpha_i}) \qquad \alpha_i \leq 0.$$

For details, examples, and further results and applications see Karlin and Rinott (1980b). Further higher order concepts of multivariate total positivity, convexity preserving transformations, related convexity notions such as multivariate log-convexity and Schur convexity, and multivariate sign changes will be explored in Karlin and Rinott (1981).

REFERENCES

AHLSWEDE, R. and DAYKIN, D.E. (1978). An inequality for weights of two families of sets, their unions and intersections. *Z. Wahrscheinlichkeitstheorie verw. Geb. 93:* 183-185.

BARLOW, R.E. and PROSCHAN F. (1975). *Statistical Theory of Reliability and Life Testing.* Holt, Rinehart and Winston, New York.

DYKSTRA, R.L. and HEWETT, J.E. (1978). Positive dependence of the roots of a Wishart matrix. *Ann. Statist. 6:* 235-238.

FORTUIN, C.M., GINIBRE, J. and KASTELEYN, P.W. (1971). Correlation inequalities on some partially ordered sets. *Commun. Math. Phys. 22:* 89-103.

HOLLEY, R. (1974). Remarks on the FKG inequalities. *Commun. Math. Phys. 36:* 227-231.

KARLIN, S. (1958). Polya-type distributions, IV: Some principles of selecting a single procedure from a complete class. *Ann. Math. Statist. 29:* 1-21.

KARLIN, S. (1968). *Total Positivity.* Stanford University Press.

KARLIN, S. and STUDDEN, W.J. (1966). *Tchebycheff Systems: With Applications in Analysis and Statistics.* Interscience Publ., New York.

KARLIN, S. and RINOTT, Y. (1980a). Classes of orderings of measures and related correlation inequalities: I. Multivariate totally positive distributions. To appear in *J. of Multivariate Analysis.*

KARLIN, S. and RINOTT, Y. (1980b). Classes of orderings of measures and related correlation inequalities: II. Multivariate reverse rule distributions. To appear in *J. of Multivariate Analysis.*

KARLIN, S. and RINOTT, Y. (1980c). Total positivity properties of absolute value multinormal variables with applications to confidence interval estimates and related probabilistic inequalities. Submitted for publication.

KARLIN, S. and RINOTT, Y. (1980d). Classes of entropy inequalities: I. The univariate case. To appear in *Advances of Applied Probability.*

KARLIN, S. and RINOTT, Y. (1980e). Classes of entropy inequalities: II. The multivariate case. To appear in *Advances of Applied Probability*.
KARLIN, S. and RINOTT, Y. (1981). *Multivariate Monotonicity, Generalized Convexity and Ordering Relations, with Applications in Analysis and Statistics*. Monograph in preparation.
LEHMANN, E.L. (1959). *Testing Statistical Hypotheses*. Wiley, New York.
MARSHALL, A.W. and OLKIN, I. (1979). *Inequalities: Theory of Majorization and its Applications*. Academic Press.
KEMPERMAN, J.H.B. (1977). On the FKG inequalities for measures on a partially ordered space. *Indag. Math. 39:* 313-331.
NEWMAN, C.M. and WRIGHT, A.L. (1980). An invariance principle for certain dependent sequences. To appear in *Ann. Probability*.
PERLMAN, M.D. and OLKIN, I. (1978). Unbiasedness of invariant tests for MANOVA and other multivariate problems. To appear in *Ann. Statist.*
PRESTON, C.J. (1974). A generalization of the FKG inequalities. *Commun. Math. Phys. 36:* 233-241.
SARKAR, T.K. (1969). Some lower bounds fo reliability. Tech. Report No.124, Dept. of Operations Research and Statistics, Stanford University.

GENERALIZED CONCAVE FUNCTIONS IN STOCHASTIC PROGRAMMING AND PORTFOLIO THEORY*

J.G. Kallberg and W.T. Ziemba

In this paper we describe some generalized concave functions that arise in a natural way in the analysis of particular stochastic programming problems. The logarithmic concave probability measure theory of Prékopa and the properties of logconcave and logconvex functions primarily developed by Klinger and Mangasarian are reviewed and illustrated by problems arising in flood control reservoir system design, and aspiration, fractile and two stage stochastic programming. Portfolio analysis problems are generally concave stochastic programming problems with a large number of decision and random variables. Generalized concave functions arise in at least three circumstances: one may wish to decompose the portfolio problem into subproblems that are easier to analyze and solve, it may not be economically feasible to determine exact solutions hence good approximate solutions are required, and one may have an imperfect security market in which the level of investment in various assets changes the distribution of security returns. These three cases are illustrated, respectively, by the two stage normal or stable return distribution model with a risk free asset, the joint lognormal distribution case when the utility function is a power or logarithmic function, and by parimutuel horserace betting models.

1. LOGCONCAVE FUNCTIONS AND MEASURES

Logconcave (and logconvex) *functions* are important types of generalized concave functions in statistics and optimization. A non-negative function defined on the convex set $X \subset R^n$, $f: X \to [0,\infty)$, is logconcave if $\forall x_1, x_2 \varepsilon X$ and $\forall \lambda \varepsilon (0,1)$

$$f(\lambda x_1+(1-\lambda)x_2) \geqslant (f(x_1))^{\lambda}(f(x_2))^{1-\lambda}. \qquad (1)$$

These functions are called logconcave (abbreviated as L-concave) because f is logconcave if and only if *log f* is *concave*. Similarly f is logconvex (L-convex) if and only if log f is convex. In this case the inequality in (1) is reversed. Thus L-concave and L-convex functions are transformable into concave and convex functions, respectively, with log as the transform.

*This research was supported by the National Sciences and Engineering Research Council of Canada and the Graduate School of Business, New York University. Without implicating them we would like to thank A. Ben-Tal, B.D. Craven, D. B. Hausch, A. Prékopa, Y. Rinott, S. Schaible, R. Wets, and I. Zang for helpful comments on an earlier draft of this paper.

ISBN 0-12-621120-5

Avriel and Zang (1974), see also Avriel (1976) and Zang (1981), have shown that such transformable functions must be at least semistrict quasiconcave (semistrict quasiconvex) and are pseudoconcave (pseudoconvex) if they are differentiable. L-convex and L-concave functions are related by their reciprocals (when defined), viz., 1/L-concave is L-convex and 1/L-convex is L-concave. The function f is strictly L-concave if (1) holds strictly for $x_1 \neq x_2$ and strictly L-convex if the reverse inequality is strict for $x_1 \neq x_2$.

L-concave and L-convex functions have a long history dating at least to Fetke (1912) who was concerned with convolutions of logconcave functions. Later Artin (1931) used such concepts in the study of the gamma function. Barlow, Marshall and Proschan (1963) in their study of reliability problems noted that a component had an increasing (decreasing) failure rate if the complement of the failure probability distribution, namely $1-F(x)$, is L-concave (L-convex). Miller and Wagner (1965) showed that the feasible region for linear joint chance constraints is described by concave inequalities if the uncertainty occurs only in the right hand side coefficients and these random variables have independent probability distributions $F_i(b_i)$, where each $1-F_i(b_i)$ is L-concave. Related concepts have been used by a number of authors in the statistics and optimization literature. See in particular Marshall and Olkin (1979) and Karlin and Rinott (1981 and forthcoming). A comprehensive development of the properties of L-concave and L-convex functions appears in Klinger and Mangasarian (1968).

Logconcave *measures* are more recent and were introduced by Prékopa (1971). A recent survey of their theoretical properties appears in Prékopa (1980); see also the papers in Prékopa (1978). A measure P, defined on the Borel subsets of R^n is logconcave if for each pair A,B of convex subsets of R^n, and $\forall\lambda\varepsilon(0,1)$,

$$P(\lambda A+(1-\lambda)B) \geqslant (P(A))^{\lambda}(P(B))^{1-\lambda}. \tag{2}$$

Here + and scalar multiplication are the usual set operations, i.e., $C + D = \{c+d|c\in C, d\in D\}$ and $\gamma C = \{\gamma c|c\in C\}$ for $\gamma\in R$. If P is a logconcave measure in R^n then $g(x) \equiv P(A+x)$ is a logconcave function of $x\in R^n$. For $A \equiv \{y|y\leqslant 0\}$ we have $g(x) = Pr\{t \leqslant x\}$ so the cumulative distribution function of a random vector generated from a logconcave probability measure is a logconcave function.

In addition to considerable work by Prékopa and his coworkers (some of which is referenced below), Borell (1975) and Rinott (1974,1976) have developed additional theory and applications of logconcave measures.

In sections 1-3 of this paper we survey some of the important results of logconcave functions and measures. The basic references utilized here and for additional study are Klinger and Mangasarian (1968), Rinott (1976) and Prékopa (1980). Refer to these papers for proofs of results. Results appearing here are stated without proof unless they appear to be new or add to the clarity of the exposition.

A resumé of properties of L-concave and L-convex *functions* appears in Table 1. The results and counterexamples in this table are for the most part adapted from Klinger and Mangasarian (1968), and Prékopa (1980). Some of these results are special cases of results for G-concave (see Zang (1981) and Avriel, Diewert, Schaible and Zang (forthcoming) or $(h,\emptyset)$ concave functions (see Avriel and Zang (1974), Avriel (1976) and Ben-Tal (1977)).

TABLE 1: RESUMÉ OF PROPERTIES OF L-CONCAVE AND L-CONVEX FUNCTIONS

1. f(x) L-convex (L-concave) $\Leftrightarrow$ log f(x) convex (concave)

2a. f(x) L-convex $\Rightarrow$ f(x) convex
(This result follows directly from (1) since $f = e^{\log f}$.)

2b. f(x) L-convex $\not\Leftarrow$ f(x) nonnegative convex
(e.g. let f(x) = x on $X_+ \equiv \{x \mid x \geq 0\}$)

3a. f(x) L-concave $\Leftarrow$ f(x), nonnegative, concave

3b. f(x) L-concave $\not\Rightarrow$ f(x) concave
(e.g. $f(x) = x^2$ on X_+)

4a. f(x) L-concave $\Leftarrow$ f(x), nonnegative, linear

4b. f(x) L-convex $\not\Leftarrow$ f(x), nonnegative, linear, nonconstant
(e.g. f(x) = x on X_+)

5. $f_i(x)$ L-convex (L-concave) $\Rightarrow$ $\prod_{i=1}^{m} (f_i(x))^{\alpha_i}$, $\alpha_i > 0$ L-convex (L-concave)
$\not\Leftarrow$
(e.g. for L-convexity, $f_1(x) = x, f_2(x) = x^{-2}$, $\alpha_1 = \alpha_2 = \frac{1}{2}$, m = 2 on $X_{++} \equiv \{x \mid x > 0\}$)

6a. L-convex + L-convex +...+ L-convex $\Rightarrow$ L-convex (strictly if at least two are not proportional)

6b. L-concave + L-concave $\not\Rightarrow$ L-concave
(e.g. $f_1(x) = x^3, f_2(x) = x + 3$, X = [1,2])

For results 7-9 assume as appropriate that the functions are positive.

7a. $\dfrac{\text{L-concave}}{\text{L-convex}} \Rightarrow$ L-concave

7b. $\dfrac{\text{L-concave} \cdot \text{L-concave} \cdots \text{L-concave}}{\text{L-convex} \cdot \text{L-convex} \cdots \text{L-convex}} \Rightarrow$ L-concave
(combining 5 and 7a)

8. $\dfrac{1}{\text{L-concave}} \Leftrightarrow$ L-convex

$\dfrac{1}{\text{L-convex}} \Leftrightarrow$ L-concave

9a. $\dfrac{1}{\text{L-concave}} \Rightarrow$ convex

9b. L-concave $\not\Leftarrow$ $\dfrac{1}{\text{nonnegative convex}}$
(e.g. $f(x) = 1 + x^2$)

10. g(y) L-convex (L-concave), nondecreasing, f(x) convex (concave) $\Rightarrow$ g(f(x)) L-convex (L-concave)
$\not\Leftarrow$
(e.g. for L-convexity, g(x) = y, $f(x) = x^{-2}$, on X_{++})

11.

$$\alpha^{g(x)} \text{ L-convex } \Leftrightarrow \begin{cases} \alpha > 1 & g(x) \text{ convex} \\ & \text{or} \\ 0 < \alpha < 1 & g(x) \text{ concave} \\ & \text{or} \\ \alpha = 1 & g(x) \text{ L-convex} \end{cases}$$

$$\alpha^{g(x)} \text{ L-concave } \Leftrightarrow \begin{cases} \alpha > 1 & g(x) \text{ concave} \\ & \text{or} \\ 0 < \alpha < 1 & g(x) \text{ convex} \\ & \text{or} \\ \alpha = 1 & g(x) \text{ L-concave} \end{cases}$$

12. $\alpha > 0,\ c \in R^n,\ \beta \in R \Rightarrow \alpha^{cx+\beta}$ L-convex and L-concave

13a. $f(x)$ L-concave $\Rightarrow$ $f(x)$ semistrictly quasiconcave

13b. $f(x)$ L-concave and upper semicontinuous (usc) $\Rightarrow$ $f(x)$ quasiconcave

13c. $f(x)$ L-concave and differentiable $\Rightarrow$ $f(x)$ pseudoconcave

14a. $f(x)$ L-convex $\Rightarrow$ $f(x)$ semistrictly quasiconvex

14b. $f(x)$ L-convex and lower semicontinuous (lsc) $\Rightarrow$ $f(x)$ quasiconvex

14c. $f(x)$ L-convex and differentiable $\Rightarrow$ $f(x)$ pseudoconvex
(these results follow from 2a; the converse does not hold for 13a-c and 14a-c, e.g. $f(x) = x_1/x_2$, $X = \{x \mid x_i > 0,\ i = 1,2\}$, example due to S. Schaible)

15. $f(x)$ L-concave $\underset{\not\Leftarrow}{\Rightarrow}$ $\{x \mid f(x) \geqslant 0\}$ is a convex set
(e.g. $f(x) = x_1/x_2$ on $X = \{x \mid x_i > 0,\ i=1,2\}$)

16. $f(x)$ positive, differentiable, L-convex (L-concave) $\Leftrightarrow$

$$\frac{\nabla f(x^1)^T (x^2 - x^1)}{f(x^1)} \underset{(\geqslant)}{\leqslant} \log\left[\frac{f(x^2)}{f(x^1)}\right]$$

$$\Leftrightarrow \left[\frac{\nabla f(x^2)}{f(x^2)} - \frac{\nabla f(x^1)}{f(x^1)}\right]^T (x^2 - x^1) \underset{(\leqslant)}{\geqslant} 0$$

17. $f(x), g(x)$ L-concave $\Rightarrow$ $f(x-y)g(y)$ L-concave in (x,y)

18. $f(x,y)$ L-concave, $(x,y) \in R^n x R^m$, A convex, $A \subset R^m$ $\Rightarrow$ $g(x) \equiv \int_A f(x,y)dy$ L-concave
(hence marginals of L-concave densitites are L-concave)

19. Nondegenerate normal, Gumbel, uniform, beta, Dirichlet, gamma, Weibull, and Wishart distributions are logconcave (the last five only on a subset of their possible parameter values, see e.g. Prékopa (1971), Borell (1975) and Marshall and Olkin (1979)).

The most important *sufficient* condition that yields logconcave measures is

THEOREM 1 (Prékopa (1971, 1980)): Suppose P *is a probability measure in* R^n *generated by a density of the form* $f(x) = e^{-Q(x)}$ $x \in R^n$. *Then if* Q *is a convex function (i.e.* f *is an L-concave function) it follows that* P *is a logarithmic concave probability measure.*

This result can be used to immediately establish the logconcavity of common distributions such as the normal, uniform and beta distributions. See Property 19 in Table 1 for other logconcave distributions.

This theorem is generalized by the following result of Borell (1975) and Rinott (1976).

THEOREM 2: Let P *be a probability measure on* R^n *generated by a density* f, *i.e.,* $P(A) = \int_A f(x)dx$ *for any Borel set* $A \subset R^n$. *Then* P *satisfies*

$$P(\lambda A+(1-\lambda)B) \geqslant (\lambda(P(A))^s + (1-\lambda)(P(B))^s)^{1/s} \tag{3}$$

$\forall \lambda \varepsilon (0,1)$, *for any measurable Borel subsets* A *and* B *of* R^n, *iff there exists a function* g:*range* f $\rightarrow$ R, *such that*

$g^{s/(1-sn)}$ *is convex* $(-\infty \leqslant s < 0)$

$\log g$ *is concave* $(s = 0)$

$g^{s/(1-sn)}$ *is concave* $(0 < s < 1/n)$.

If P satisfies (3) it is said to be convex of order s (Borell (1975); see also Avriel (1973), and Lindberg (1981)). Since $s = -\infty$ implies that the right side of (3) is min(P(A),P(B)), convexity of order $-\infty$ is quasiconvexity. Borell (1975) proved that quasi-concave measures $P(A)(= \int_A f(x)dx)$ arise when they are generated from continuous probability densities f in R^n (such as the multivariate t and F densities) that have the property that $f^{-1/n}$ is convex on R^n. If s = 0, P is logconcave and one has a generalization of Theorem 1. The case s = 1/n holds for Lebesgue measure and is known as the Brunn-Minkowski inequality (Brunn (1887)). A proof of Theorem 2 along with a number of examples illustrating its usefulness in statistics appears in Rinott (1976).

A useful result in stochastic programming is the following theorem of Prékopa (1972, 1980).

THEOREM 3: a) If $g(x,y) = R^n \times R^m \to R^\ell$ *is concave and* $\xi \in R^m$ *has a logconcave probability distribution, then the function*

$$h(x) \equiv P[g_1(x,\xi) \geqslant 0,\ldots,g_\ell(x,\xi) \geqslant 0] \qquad (4)$$

is logconcave on R^n.

b) If g *is concave and* ξ *has a quasiconcave probability distribution then* h *is quasiconcave on* R^n.

A consequence of Theorem 3 (see property 13b in Table 1) is that the feasible region for *joint* chance constrained programming problems with logconcave or quasiconcave probability distributions involving concave inequalities is a convex set. Hence problems such as the one described in section 3 can be solved by standard convex programming algorithms.

As an example of Theorem 3 suppose, following Miller and Wagner (1965), that each $g_i(x,\xi) = \xi_i - a_i^T x$, where the ξ_i have *independent* logconcave distributions F_i. Then $\log H(x) = \sum_i \log G_i(a_i^T x)$, where the $G_i \equiv 1 - F_i$ are the tail distributions of the ξ_i. Thus a chance constraint of the form $P[a_1^T x \leqslant \xi_1,\ldots,a_n^T x \leqslant \xi_n] \geqslant \beta$ is equivalent to $g(x) \geqslant \hat{\beta}$, where g is the *explicit concave* function $\sum_i \log G_i(a_i^T x)$ and $\hat{\beta} \equiv \log \beta$.

Theorem 3 also enables us to generalize some results relating to stochastic programming. For example, Jagannathan and Rao (1973) show that if ξ has a uniform or normal distribution then

$$g(x_1,x_2) = P(x_1 \leqslant \xi \leqslant x_2)$$

is logconcave in R^2. Theorem 3a implies that this result is valid if ξ has any logconcave density since $P(x_1 \leqslant \xi \leqslant x_2) = P(g_1(x,\xi) \geqslant 0,\ g_2(x,\xi) \geqslant 0)$ using the identifications $g_1(x,\xi) \equiv \xi - x_1$ and $g_2(x,\xi) \equiv x_2 - \xi$. Prékopa (1980) also demonstrates the following related result for quasiconcave densities: the function $h(x) \equiv P(\xi - c \leqslant a^T x \leqslant \xi + d)$ is quasiconcave on R^n if $\xi \in R$ has a quasiconcave density and $c \geqslant 0$, $c \in R$, $d \geqslant 0$, $d \in R$ and $a \in R^n$ are constants. This result follows directly from the fact that $g(x) \equiv \int_{I+x} f(t)dt$ is quasiconcave if f is a nonnegative quasiconcave function in R and $I \subset R$ is an interval.

The results in Theorem 3 are also useful for distributions, such as the lognormal, which are not logconcave. We say that $\xi \equiv (\xi_1,\ldots,\xi_n)^T \in R^n$ has a multivariate lognormal distribution if $y \equiv (\log \xi_1,\ldots,\log \xi_n)^T$ has a multivariate normal distribution. By Property 1 of Table 1 it follows that ξ is *not* logconcave. However we have

COROLLARY 1: If $\xi \in R^n$ is a positive lognormally distributed random vector and $\xi_0 \in R$ has a logconcave distribution where ξ_0 and ξ are independent. Then the function

$$h(x) \equiv P(\xi^T x \leqslant \xi_0) \tag{5}$$

is logconcave on the positive orthant of R^n.

PROOF: Let $z_i \equiv \log x_i$ and $y_i \equiv \log \xi_i$, $i = 1,\ldots,n$. Now

$$h(x) = P(\xi^T x \leqslant \xi_0) = P(e^{y_1+z_1}+\ldots+e^{y_n+z_n}-\xi_0 \leqslant 0)$$

$$= P(-e^{y_1+z_1}-\ldots-e^{y_n+z_n} + \xi_0 \geqslant 0) \equiv P(g(z,y) \geqslant 0).$$

The function $g = R^n \times R^{n+1} \to R$ is concave and $y \equiv (\xi_0,\ldots,\xi_n)^T$ has a log-concave density thus the result follows from Theorem 3a. ||

The model (5) may be interpreted as a chance constraint guaranteeing that sufficient capital (ξ_0) be available for investment in securities $i = 1,\ldots,n$ in amounts $x_1,\ldots,x_n$ with return distribution $F(\xi_1,\ldots,\xi_n)$.

The transformation used in the proof of Corollary 1 utilizes the fact that lognormal distributions are positive. Positive numbers can be represented in the form e^y hence one has:

COROLLARY 2: (Prékopa (1980)): Let the matrix ξ *have elements* ξ_{ij}, $i=1,\ldots,m$, $j=1,\ldots,n$ *and suppose only the first* $r \leqslant n$ *columns contain random variables. The set of ordered pairs* (i,j) *for which* ξ_{ij} *is random,* $1 \leqslant i \leqslant m$, $1 \leqslant j \leqslant r$ *is denoted by* A. *Let* B *denote the set of subscripts where the right hand side vector* ξ_0 *is random in the expression* $P(\xi^T x \leqslant \xi_0)$. *If the* ξ_{ij}, $(i,j) \in A$ *are positive with probability one and the constants* $\xi_{ij}, (i,j) \notin A$ *are nonnegative and the joint distribution of log* $\xi_{ij}, (i,j) \in A$ *and log* ξ_{0i} $i \in B$ *is logconcave then*

$$h(e^{x_1},\ldots,e^{x_r},x_{r+1},\ldots,x_n) = P(\xi^T x \leqslant \xi_0)$$

is logconcave on R^n.

We may prove the same results assuming (ξ,ξ_0) is joint normally distributed. The transformation of the inner product used in Corollary 1 is the generalized z-vector addition due to Ben-Tal (1977) with $z(x) = (\log x_1,\ldots,\log x_n)$. Hence this result can be proven succintly by observing that logconcave (and more generally, $(h,\emptyset)$-concave) functions preserve this generalized algebraic operation. Some results concerning the convexity of the feasible regions in chance constrained programming problems involving normal distributions due to Van de Panne and Popp (1963) and Prékopa (1974) are summarized in:

THEOREM 4: a) Suppose $\xi \equiv (-\xi_0, \xi_1, \ldots, \xi_n)^T \sim N(\bar{\xi}, \Sigma)$ $x \equiv (x_1, \ldots, x_n)^T$ *and* $\beta \geqslant 1/2$. *Then the set* $X \equiv \{x | P(\xi_1 x_1 + \ldots + \xi_n x_n \geqslant \xi_0) \geqslant \beta\}$ *is convex and* $X = \{x | \bar{\xi}^T \binom{x}{1} - K_\beta [(x,1)^T \Sigma \binom{x}{1}]^{1/2} \geqslant 0\}$ *where* K_β *is the* β *fractile of the* cdf *of the* N(0,1) *variate,* ψ, *i.e.* $\psi(K_\beta) = \beta$.

b) Suppose

$$\xi \equiv \begin{vmatrix} -\xi_{01}, \xi_{11}, \ldots, \xi_{im} \\ \vdots \quad \vdots \quad\quad \vdots \\ -\xi_{0m}, \xi_{n1}, \ldots, \xi_{nm} \end{vmatrix} \sim N(\bar{\xi}, \Sigma),$$

where the cross-covariance matrices of the columns (ξ^i) *(or the rows) of* ξ *are constant multiples of a fixed covariance matrix* Σ_0, *i.e.,*

$E\left[(\xi^i - \bar{\xi}^i)^T (\xi^j - \bar{\xi}^j)\right] = s_{ij} \Sigma_0$, $i,j = 0, \ldots, n$ (*or* $0, \ldots, m$). *Then the set* $X \equiv \{x | P(\xi^T x \geqslant \xi_0) \geqslant \beta\}$ *is convex for* $\beta \geqslant 1/2$.

Theorem 4 can be generalized to apply for multivariate stable distributions whose dispersion functions are convex; see section 5 below for technique and examples.

The Kuhn-Tucker conditions for the program

$$\min f(x) \qquad \text{s.t. } g(x) \geqslant 0,\ x \in R^n \tag{6}$$

for $f: R^n \to R$ and $g: R^n \to R^m$ are necessary and sufficient (assuming a constraint qualification) if f is pseudoconvex and g is quasiconcave. Thus the logconcavity of h, see (4), has application in stochastic programming viewing h as the objective or as a chance constraint.

2. APPLICATION TO ASPIRATION AND FRACTILE PROGRAMMING

An example of a two-stage stochastic program is the aspiration model

$$\min_{x\in X} z(x) = P(\phi(x) + \inf_{y\in Y} \psi(y) > 0) \tag{7}$$

$$\text{s.t.} \quad h(x,y,\xi) \geqslant 0 \quad \text{a.s.} \tag{8}$$

where ϕ: $R^n \to R$ corresponds to the first-stage cost,

ψ: $R^m \to R$ corresponds to the second-stage cost, $X \subset R^n$ and $Y \subset R^m$ are convex sets,

h: $R^n \times R^m \times R^q \to R^p$ defines the linkage constraint set,

$\xi \in R^q$ is a random vector, and the tolerance level is set to zero without loss of generality.

We can interpret (7) as minimizing the probability of an unfavorable event. In stage one we choose x and obtain cost $\phi(x)$ and then observe the realization of ξ. We then take the second stage (recourse) action y to minimize $\psi(y)$. An example of this type is the portfolio revision model (see, e.g., Kallberg and Ziemba (1981) who discuss an expected value model similar to (7)-(8)). Here x is the allocation of investment into q risky securities, ξ represents the end of period prices and y represents the portfolio adjustments taken after observing ξ;$-\phi$ and $-\psi$ are the investor's utility functions over returns in the two periods.

Bereanu (1980) proved that (7)-(8) has a logconcave objective if ξ has a logconcave distribution and ϕ,ψ and h are linear. This result also holds for concave functions.

THEOREM 5: Suppose ϕ: $R^n \to R$, ψ: $R^m \to R$, *and* g: $R^n \times R^m \times R^q \to R^p$ *are concave functions. If* ξ *has a logconcave probability distribution on a convex set* $\Xi \subset R^q$, *then* z *is logconcave on the convex set*

$$T \equiv \{x | z(x) > 0\}.$$

PROOF: Let $v(\xi,x) \equiv \inf_{y \in Y}[\phi(x)+\psi(y)]$ for (ξ,x,y) satisfying (8).

Clearly, v is concave. For $x \in T$, define

$$K(x) \equiv \{\xi \mid v(\xi,x) \leqslant 0\}.$$

If $\xi_i \in K(x_i)$, $i = 1,2$, then for $\lambda \in (0,1)$

$$v(\lambda\xi_1+(1-\lambda)\xi_2,\ \lambda x_1+(1-\lambda)x_2) \leqslant 0.$$

Thus

$$\lambda K(x_1) + (1-\lambda)K(x_2) \subset K(\lambda x_1+(1-\lambda)x_2).$$

By definition

$$z(\lambda x_1+(1-\lambda)x_2) \geqslant P(\xi \mid \xi \in \lambda K(x_1)+(1-\lambda)K(x_2)).$$

Since ξ has a logconcave distribution

$$z(\lambda x_1+(1-\lambda)x_2) \geqslant (z(x_1))^{\lambda}(z(x_2))^{1-\lambda}.$$

Hence z is a logconcave which implies T is convex. ||

In the static case with normal distributions ($\xi \sim N(\bar{\xi},\not{\Sigma})$, $\not{\Sigma}$ positive definite) the aspiration model

$$\begin{aligned} p \equiv \max\ & P(\xi^T x \geqslant \alpha) \\ \text{s.t. } & Ax \leqslant b,\ x \geqslant 0 \end{aligned} \tag{9}$$

for a given aspiration level α is equivalent to, see Charnes and Cooper (1963) or Parikh (1968),

$$\begin{aligned} \max\ & f(x) \equiv \frac{\bar{\xi}^T x-\alpha}{(x^T \not{\Sigma} x)^{1/2}} \\ \text{s.t. } & Ax \leqslant b,\ x \geqslant 0 \end{aligned} \tag{10}$$

assuming $x = 0$ is not optimal (e.g. $\bar{\xi} \geqslant 0$, $\alpha \leqslant 0$). Since the denominator of f is positive and convex, f is pseudoconcave (Mangasarian (1971)) on $X \equiv \{x \mid Ax \leqslant b,\ x \geqslant 0\} \cap \{x \mid \bar{\xi}^T x \geqslant \alpha\}$, and $p \geqslant 1/2$. Problems of the form (10) are frequently equivalent to convex quadratic programs.

For example, Schaible (1978) and Bereanu (1980) have shown that

$$\begin{aligned} \max\ f(x) &= \frac{\bar{\xi}^T x-\alpha}{(x^T \Sigma x)^{1/2}} \\ \text{s.t.}\quad & e^T x \leqslant 1,\ x \geqslant 0 \end{aligned} \tag{11}$$

is equivalent to

$$\begin{aligned} m \equiv \min\ & y^T \Sigma y \\ \text{s.t.}\quad & \sum_{i=1}^{n} (\bar{\xi}_i-\alpha) y_i \geqslant 1,\ y \geqslant 0 \end{aligned} \tag{12}$$

assuming that $\alpha < \max_i \bar{\xi}_i$. The corresponding maximum probability is $F(m^{-1/2}) \geqslant 1/2$ and the unique solution of (11) is $x^* = \frac{\alpha y^*}{\bar{\xi}^T y^*-1}$ where F is the cumulative standard normal distribution. General analyses of transformations of this type appear in Bradley and Frey (1974) and Schaible (1974, 1978).

The related fractile model

$$\begin{aligned} &\max\ d \\ &\quad \text{s.t.}\ P(\xi^T x \geqslant d) \geqslant \beta \\ &\qquad \text{s.t.}\ Ax \leqslant b,\ x \geqslant 0 \end{aligned} \tag{13}$$

for a given fractile $\beta \in (1/2,1)$ with $\xi \sim N(\bar{\xi},\Sigma)$, Σ positive semidefinite is equivalent (see Kataoka (1963), Geoffrion (1967), or Parikh (1968)) to the concave program

$$\begin{aligned} \max\ & \bar{\xi}^T x - K_\beta (x^T \Sigma x)^{1/2} \\ \text{s.t.}\quad & Ax \leqslant \beta,\ x \geqslant 0, \end{aligned}$$

where $K_\beta = F^{-1}(\beta)$ and F is the cumulative standard normal distribution.

3. APPLICATIONS

A typical application is the following model due to Prékopa and Szantai (1978). The general goal is to design a flood control system.

In particular one must determine the location and size of floodwater reservoirs within a given water system.

They model the flood control reservoir system as a rooted directed tree where the root represents the "downstream end" of the water system and the edges represent the connecting waterways. The direction is determined by water flow and edges correspond to potential reservoirs. The objective is to determine the size of these reservoirs to minimize the sum of construction costs and penalties for terminal water flow in excess of system capacity (viz. flood). Constraints include a chance constraint ensuring that the probability of containment is sufficiently large. The system may be viewed as follows:

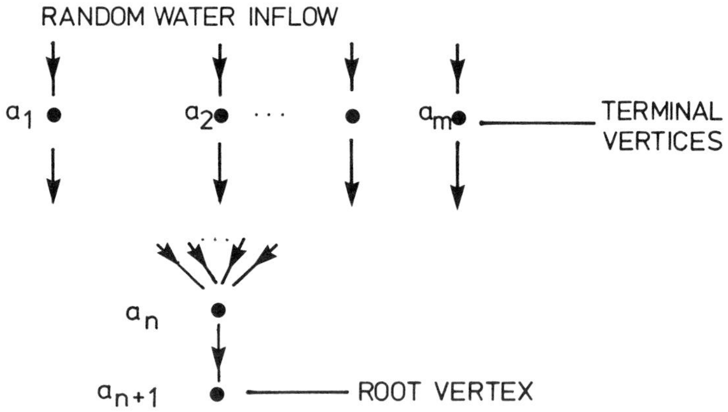

It is assumed that there is a random water inflow $(x_1,\ldots,x_m)$ at each of the *terminal vertices*. The flows at the other vertices

$(x_{m+1},\ldots,x_{n+1})$ are then determined by the recursion,

$$x_j = \sum_{i\in A_j} (x_i - \min(x_i,K_i)), \quad j = m+1,\ldots,n+1. \tag{14}$$

Here A_j is the set of first-order antecedents of a_j and K_i is the reservoir capacity on the edge from a_i (potentially zero). The event $x_n > K_n$ (water flow exceeding "final" reservoir capacity) corresponds to a flood. The function

$$g(x,K) = K_n - x_n$$

is concave in (x,K).

The program is

$$\min_{\{K_i\}} \left[\sum_i |c_i(K_i)| + E(\mu(x_n,K_n))\right] \tag{15}$$

$$\text{s.t.} \quad P(x_n \leqslant K_n) \geqslant p$$

$$0 \leqslant K_i \leqslant V_i, \quad i=1,\ldots,n+1$$

where for vertex i,

$c_i(K_i)$ is the building cost function,

μ is the "penalty" for the event $x_n > K_n$,

p is a tolerance probability, and

V_i is an upper bound on reservoir capacity.

Some structure is required to make (15) amenable to standard nonlinear programming techniques. Prékopa and Szantai assume that

c_i is linear,

$$\mu(x_n,K_n) = \begin{cases} q(x_n-K_n) & \text{if } x_n > K_n \\ 0 & \text{otherwise,} \end{cases}$$

q is linear, and

$x \equiv (x_1,\ldots,x_n)$ has a nondegenerate normal distribution.

It then follows that $E(\mu)$ is convex in (x,K), Theorem 3 demonstrates that the chance constraint in (15) is logconcave and hence (15) is a convex program and can be solved by standard algorithms. Prékopa and Szantai suggest use of an adaptation of Veinott's (1967) supporting hyperplane method and illustrate its use with numerical examples.

Similar approaches have been used in inventory theory (Prékopa and Kelle (1978)) and in a stochastic linear programming model used for the electrical energy sector of Hungary (Prékopa, Ganczer, Deak and Patyi (1980)). For a survey of applications in Hungary see Prékopa (1978). Applications to statistics and other areas are discussed in Marshall and Olkin (1979), and Karlin and Rinott (1981 and forthcoming). Klinger and Mangasarian (1978) describe applications to geometric programming, reliability theory (see also Rinott (1976)), Laplace transforms (e.g., Schoenberg (1951) uses Polya frequency functions of order two which are logconcave) and information theory.

4. PORTFOLIO SELECTION PROBLEMS

The classic static portfolio selection problem is the *concave* program

$$\begin{aligned} \text{maximize } Z(x) &= E_{\xi} u(w_0 \xi^T x), \\ \text{s.t. } e^T x &= 1, \; x \geqslant 0, \end{aligned} \tag{16}$$

where x_i is the fraction of the investor's initial wealth w_0 invested in asset i, $x \equiv (x_1,\ldots,x_n)^T$, $e \equiv (1,\ldots,1)^T$, ξ_i is the random return received per dollar invested in asset i, $\xi \equiv (\xi_1,\ldots,\xi_n)^T$ has the joint distribution function G which is independent of the choice of x, u is the investor's concave utility function defined on final wealth and E_ξ denotes mathematical expectation with respect to ξ. The concavity assumption is equivalent to the supposition that the investor is *risk averse* and thus would never prefer a fair gamble to the status quo. That is: if the status quo is $w_0(=\lambda w_1+(1-\lambda)w_2)$ and a gamble results in final wealth of w_1 and w_2 with probabilities λ and $(1-\lambda)$, respectively, then $u(w_0) \geq \lambda u(w_1) + (1-\lambda)u(w_2)$. In practice, it is generally advisable to add additional features to the formulation (16). For example if the investment choices must satisfy constraints related to the minimum and maximum investment in particular assets, then one has the additional constraints, $\alpha_i \leq x_i \leq \beta_i$. If the investor wishes to include the possibility of short sales, in which he sells x_i dollars of investment i and must buy (repay) at ξ_i, then $\alpha_i < 0$. The investor may also wish to borrow y$ in margin funds at cost c(y). The problem is then

$$\begin{aligned} &\text{maximize } z(x,y) \equiv E_\xi u[\xi^T x - c(y)], \\ &\qquad\qquad \text{s.t. } \alpha \leq x \leq \beta,\ 0 \leq y \leq \delta,\ e^T x = w_0 + y, \end{aligned} \tag{17}$$

where $\alpha \equiv (\alpha_1,\ldots,\alpha_n)^T$, $\beta = (\beta_1,\ldots,\beta_n)^T$ and δ is the maximum amount that can be borrowed. Under the plausible assumptions that u is nondecreasing (i.e. more wealth is always preferable) and concave and c is convex (i.e. the marginal borrowing rate is nondecreasing), (17) is also a concave program involving only *concave* functions. In this formulation to preserve concavity we interpret the x_i as absolute investment amounts and utilize a budget constraint instead of a share constraint. General algorithms to solve problems (16) and (17) are discussed, e.g. in Ziemba (1978), and utilize numerical quadrature schemes to compute function and gradient evaluations with standard nonlinear programming algorithms. See Kallberg and Ziemba (1980) for a specialized algorithm of this type. The efficiency of the algorithms depends primarily upon n, the number of investment alternatives, G the cumulative distribution function of returns, and, u, the utility function. In practice n is generally quite large ($n \geq 200$ is common). Except for special cases, such as discrete distributions, quadratic utility, or exponential utility combined with normal distributions, univariate numerical quadratures or approximations are required unless G can be expressed in closed form. An attractive situation arises when ξ has a multivariate normal distribution in which case $\xi^T x$ has a univariate normal distribution and the computations are manageable even for very large n. The normal distribution case also simplifies because, as the following theorem indicates, it is always possible to utilize a *convenient* utility function to model investor preferences.

THEOREM 6: Suppose $\xi \sim N(\mu, \Sigma)$, *where all* μ_i *and* σ_{ij} *are finite,* $u_i \in C^2$, $u_i' > 0$, $u_i'' < 0$, $i = 1,2$. *Consider problems*

$$\text{(P1)} \quad \{\max E_\xi u_1(w_1 \xi^T x) \mid e^T x = 1,\ x \geqslant 0\} \quad \textit{and}$$

$$\text{(P2)} \quad \{\max E_\xi u_2(w_2 \xi^T x) \mid e^T x = 1,\ x \geqslant 0\}.$$

Assume that P1 *and* P2 *have optimal solutions and that* x* *solves* P1. *If*

$$w_1 \frac{E_\xi u_1''(w_1 \xi^T x^*)}{E_\xi u_1'(w_1 \xi^T x^*)} = w_2 \frac{E_\xi u_2''(w_2 \xi^T x^*)}{E_\xi u_2'(w_2 \xi^T x^*)} \tag{18}$$

then x* *solves* P2.

This result is discussed and proven in Kallberg and Ziemba (1978, 1979) and was developed in collaboration with Mark Rubinstein. It is an *exact global* result *independent* of the *variance* of ξ and indicates that one may safely replace utility function u_2 for u_1 where u_2 is chosen to satisfy (18). In practice for a given parameter value β_1 for u_1 a parameter value β_2 for u_2 is *chosen* to satisfy (18). The measure in (18) has no obvious interpretation in terms of investor risk aversion and it is not clear how such a measure might be estimated. Hence Kallberg and Ziemba considered the approximation to (18) where

$$w_i \frac{E_\xi u_i''(w_i \xi^T x^*)}{E_\xi u_i'(w_i \xi^T x^*)} \quad \text{is replaced by}$$

$$w_i E_\xi \left| \frac{u_i''(w_i \xi^T x^*)}{u_i'(w_i \xi^T x^*)} \right| . \tag{19}$$

Now (19) is the average Arrow-Pratt relative risk aversion (see Pratt (1964)) of portfolio x^* (the minus sign omitted). Extensive numerical analysis reported in Kallberg and Ziemba (1978) indicate that

$$w_1 \int \frac{u_1''(\xi^T x)}{u_1'(\xi^T x)} \, dG(\xi) = w_2 \int \frac{u_2''(\xi^T x)}{u_2'(\xi^T x)} \, dG(\xi)$$

implies that $x^1 \cong x^2$. By $x^1 \cong x^2$ we mean that the approximate portfolio x^2 has basically the same *cash equivalent* (i.e. the *fixed* amount $u^{-1}[E_\xi u(\xi^T x)]$ which is equivalent to the *random* amount $\xi^T x$) as the portfolio x^1. For example, with quarterly data the percentage cash equivalent error is less than 10^{-2}%. The portfolio weights x^1 and x^2 are virtually identical as well. The accuracy of the approximation depends upon the variance of $\xi^T x$; for data of a year's duration or less the accuracy is no worse than data estimation errors. Therefore, for normal distributions one can safely assume that the investor has a simple utility function such as a quadratic or exponential and can relate investor preferences to the parameter of this function. The accuracy of this approximation has not been investigated for non-normal data.

The discussion so far has concerned concave functions. Generalized concave functions arise in at least three situations:

1) one wishes to decompose the portfolio problem into subproblems that are easier to solve and to analyze,
2) it is not economically feasible to solve the exact portfolio problem and "good" approximate solutions are

required, and

3) the security market is not perfect and the level of investment changes the distribution of the security returns.

Particular models that illustrate these situations are described in the next three sections.

5. THE TWO STAGE APPROACH

A common strategy when faced with a difficult optimization problem is to attempt to *decompose* the problem into more tractable sub-problems whose solutions can be combined to determine an optimal solution. In this section we describe such an instance for the case of a risk averse investor faced with the choice among several assets which have either multivariate normal or symmetric stable distributions. The key assumption is that there exists a risk free asset that can be bought or sold in any amount at a fixed price. The portfolio problem involves n assets and hence has n random variables and n decision variables. In the two stage approach[1] one utilizes stochastic dominance constructs to find one subproblem that yields optimal *relative* portfolio weights for the risky assets. A second subproblem determines an optimal *mix* of the risk free asset and a composite asset formed by weighting the risky assets by their optimal relative weights determined in stage one. The first stage problem only utilizes information regarding the probability distribution of the random investments and determines relative portfolio weights that are optimal for any convergent (see footnote 2 below)

[1]The two-stage procedure was suggested by Lintner (1965). Ziemba et al. (1974) showed how to perform the second stage calculations. The extension to the stable distribution case is due to Ziemba (1974). The discussion here is adapted from Ziemba (1978).

concave utility function. This problem is *deterministic* with n decision variables. It involves the maximization of a pseudoconcave fractional function over the unit simplex. The second stage problem is a concave, stochastic search problem to determine a single variable, the amount invested in the risk free asset to maximize expected utility for some given concave utility function. The search problem involves only a *single* random variable, the normal or stable risky composite asset which may be considered to be a *mutual fund*. Investors with different risk attitudes choose *different* combinations of the (*same*) mutual fund and the risk free asset.

Let $\hat{x} \equiv (x_1,\ldots,x_n)^T$ be the vector of portfolio allocations for the risky assets and x_0 be the allocation made to the safe asset. The vector of risky asset returns is $\hat{\xi} \equiv (\xi_1,\ldots,\xi_n)^T$ and the safe asset has the fixed return $\bar{\xi}_0$. Let $x^T \equiv (x_0,\hat{x})^T$ and $\xi^T \equiv (\bar{\xi}_0,\hat{\xi})^T$. Suppose that the only constraint on x_0 is the budget constraint. The constraints on $\hat{x}$ are assumed to be described by the homogeneous concave inequalities, $g_j(\hat{x}) \geqslant 0$, $j=1,\ldots,m$, which include the constraint $\hat{x} \geqslant 0$ as well as the budget constraint. The investor's problem is

$$\begin{aligned} &\text{maximize} \quad E_\xi u(\xi^T x) \\ &\qquad \text{s.t. } x_0 + e^T\hat{x} = 1,\ g_j(\hat{x}) \geqslant 0,\ j = 1,\ldots,m, \end{aligned} \tag{20}$$

where u, the investor's utility function is assumed to be monotone non-decreasing, and concave, and $e \equiv (1,\ldots,1)^T$.

In this section we develop the results for the joint normally distributed case ($\xi \sim N(\mu,\Sigma)$) and briefly describe several generalized concave functions that arise for particular types of multivariate stable distributions.

The following well known result indicates that it is sufficient to limit consideration to points that lie on a mean-standard deviation efficient surface. It is the basis for Markowitz's (1952) mean-variance approach to portfolio selection.

THEOREM 7: Suppose $\xi_i \sim N(\mu_i, \sigma_i^2)$, $i=1,2$, *are distinct normal variates with distributions* $G(X)$ *and* $F(X)$, *respectively, where* $|\mu_i| < \infty$, $0 < \sigma_i < \infty$, $\mu_1 \leqslant \mu_2$ *and* $G(z) > H(z)$ *for some* z. *Then* $\int u(w)dG(w) \geqslant \int u(w)dH(w)$ *for all convergent*[2] *concave nondecreasing* u *if and only if* $\sigma_1 \leqslant \sigma_2$.

PROOF: Sufficiency: (a) Suppose $\mu_1 = \mu_2 = 0$. Let $k = \sigma_2/\sigma_1 \geqslant 1$. Then

$$\int u(w)dH(w) - \int u(w)dG(w)$$

$$= \int_{-\infty}^{\infty} [u(kw) - u(w)]dG(w) \qquad \text{(by symmetry)}$$

$$= \int_0^{\infty} \underbrace{\{[u(kw) - u(w)] - [u(-w) - u(-kw)]\}}_{\leqslant 0 \text{ (by concavity)}} dG(w)$$

(b) Suppose $\mu_1 \geqslant \mu_2$. Let $\tilde{G}$ and $\tilde{H}$ be the cumulative distributions for x and y, respectively, when their means are translated to zero. Let $\varepsilon \equiv \mu_1 - \mu_2 \geqslant 0$. Then

$$\int u(w)dH(w) = \int u(w+\mu_2)d\tilde{H}(w) \leqslant \int u(w+\mu_2)d\tilde{G}(w)$$

$$\leqslant \int u(w+\mu_2+\varepsilon)dG(w) \leqslant \int u(w)dG(w)$$

where the first inequality follows from (a). This simple proof of sufficiency is due to Hillier (1963).

[2]A sufficient assumption is $\lim_{w\to-\infty} \frac{|u(w)|}{Ae^{Bw^2}} = L$ for some $0 \leqslant L < \infty$ where $A > 0$, and $B > 0$ satisfies $x^T \not{\Sigma} x < 1/\sqrt{2B}$; see Chipman (1973) and Ziemba (1974).

Necessity: For any $\mu_1 \geqslant \mu_2$ having $\sigma_1 > \sigma_2$ a concave nondecreasing u can be found such that $\int u(w)\,dG(w) < \int u(w)\,dH(w)$, since G and H cross only once; see Theorem 3 in Hanoch and Levy (1969).

The efficient surface may be obtained by solving the parametric concave program

$$\phi(\beta) \equiv \text{maximize } \bar{\xi}^T x - 1, \tag{21}$$
$$\text{subject to } x_0 + e^T\hat{x} = 1,$$
$$g_j(\hat{x}) \geqslant 0, \quad j = 1,\ldots,m,$$
$$f(\hat{x}) \leqslant \beta,$$

for all $\beta > 0$, where $f(\hat{x}) \equiv (\hat{x}^T \not{\Sigma} \hat{x})^{\frac{1}{2}}$. If $x_0 = 0$, the optimal objective value, say, $\phi_0(\beta)$, is a concave function of β (see Figure 1). For general x_0, $\phi(\beta)$ consists of all points on the ray L that emanates from the point $(\bar{\xi}_0, 0)$ and supports the function $\phi_0(\beta)$ at the point M having mean $\bar{\xi}_c$ and standard deviation σ_c (where the subscript c denotes the composite asset, see below). Since $\bar{\xi}^T x$ and $f(\hat{x})$ are positively homogeneous, points on L are linear combinations of $x_0 = 1$ (total investment in the safe asset) and the point M. Since L is a ray, the optimal ratios of the risky assets, namely,

$$x_i^* \Big/ \sum_{j=1}^{n} x_j^*, \quad i \neq 0$$

are independent of u. Hence all investors having concave, nondecreasing utility functions will invest their wealth in a combination of the risky composite asset, i.e., mutual fund M and the safe asset i = 0. This is Tobin's (1958) famous separation theorem. Breen (1968), Stone (1970), Bradley (1975) and Ziemba (1978) have proven a series of theorems indicating that the result holds for homogeneous dispersion measures f and constraints g. The statement given here is taken from Ziemba (1978).

THEOREM 8: Let the efficiency problem be given by (21), where it is assumed that the g_j *are homogeneous concave functions and contain the constraint* $\hat{x} \geqslant 0$, f *is positive, convex, and positively homogeneous,* $\beta > 0$, u *is nondecreasing and concave and (20) has an optimal solution* x^*. *Then the relative proportions invested in the risky assets*

$$x_i^* \Big/ \sum_{j=1}^{n} x_j^*, \quad i \neq 0$$

are independent of u.

PROOF. Suppose $x_0^* \neq 0$. Since x^* solves (20) it must solve (21) for some β. Hence it must satisfy the Kuhn-Tucker conditions for (21) (which are necessary (homogeneity of f and g guarantee satisfaction of Slater's constraint qualification) and sufficient under the assumptions):

(a) $f(\hat{x}) \leqslant \beta$, $g_j(\hat{x}) \geqslant 0$, $\quad j = 1,\ldots,m$

$$x_0 + e^T\hat{x} = 1,$$

(b) (i) $\bar{\xi}_i - \lambda \dfrac{\partial f(\hat{x})}{\partial x_i} - \mu + \displaystyle\sum_{j=1}^{m} \alpha_j \dfrac{\partial g_j(\hat{x})}{\partial x_i} = 0$, $i = 1,\ldots,n$.

(ii) $\bar{\xi}_0 - \mu = 0$,

and

(c) $\lambda \geqslant 0$, $\alpha_j \geqslant 0$, $\quad j = 1,\ldots,m$.

It suffices to show that for all $\sigma > 0$ there exists an x_0^* such that $x^{**} = (\gamma x_0^*, \sigma x_1^*, \ldots, \sigma x_n^*)$ also solves the Kuhn-Tucker conditions for all $\beta^{**} \geqslant \beta\sigma$. Conditions (a) are satisfied because

$$f(x^*) \leqslant \beta \Rightarrow \quad f(\sigma\hat{x}^*) = \sigma f(\hat{x}^*) \leqslant \sigma\beta$$

$$g_j(\hat{x}^*) \geqslant 0 \Rightarrow \quad g_j(\sigma\hat{x}^*) = \quad g_j(\hat{x}^*) \geqslant 0,$$

$$x_0^* + e^T\hat{x}^* = 1 \Rightarrow \gamma x_0^* + \sigma e^T\hat{x}^* = 1$$

for

$$\gamma \equiv \frac{1-\sigma e^T \hat{x}^*}{x_0^*} \neq 0.$$

Suppose $x_0^* = 0$, then $\hat{x}^*$ corresponds to a portfolio associated with the point M having mean $\bar{\xi}_c$ and standard deviation σ_c, which is independent of u even if M is not unique.

One may find the slope of L and the point M by solving[3]

$$\begin{aligned} &\text{maximize} \quad && g(\hat{x}) \equiv \frac{\bar{\hat{\xi}}^T \hat{x} - \xi_0}{f(\hat{x})} \\ &\text{s.t.} && e^T \hat{x} = 1, \\ & && g_j(\hat{x}) \geqslant 0, \quad j = 1,\ldots,m. \end{aligned} \tag{22}$$

---

[3] That (22) and (23) are equivalent to (21) at $x_0 = 0$ is usually taken for granted in the finance literature. For completeness a proof is included here. For $\beta > 0$ (21) can be written as

(A) $\max n(x)$
s.t. $d(x) \leqslant \beta,\ x \in C,$

where $n(x) \equiv \bar{\xi}^T x, d(x) = f(x)$ and $C \equiv \{x | e^T x = 1,\ g(x) \geqslant 0\}$, where the ^'s are deleted from the x's for convenience. Suppose x* solves A and $d(x^*) = \beta$. Using the substitution $x = \alpha^{-1} y$ yields the equivalent problem

(B) $\max_{y,\alpha} n(\alpha^{-1} y) = \alpha^{-1} n(y)$
s.t. $d(y) \leqslant \alpha\beta,\ \alpha > 0,\ y \in \{y | e^T y = \alpha,\ g(y) \geqslant 0\}.$

Hence (B) attains a maximum at $y^* = \alpha x^*$, in particular when $\alpha = 1/\beta$. Then the more restrictive problem

(C) $\max_{y,\alpha} \beta n(y)$
s.t. $d(y) = 1 \quad y \in \{y | e^T y = \alpha,\ g(y) \geqslant 0\},$
$\alpha \geqslant 0,$

is also minimized at (y,α), $(\beta^{-1} x^*, \beta^{-1})$. Using the transformation $x = \frac{y}{\alpha}$, $\alpha > 0$ yields the equivalent fractional program (i.e. (23))

(D) $\max_{x \in C} n(x)/d(x)$

since $d(y) = 1 \Rightarrow 1 = d(\alpha x) = \alpha d(x)$ or $\alpha = 1/d(x) = n(y) = \alpha n(x) = n(x)/d(x)$. Hence x* solves (A) if and only if it solves (D). This proof was supplied by B.D. Craven. It is an application of the general results in Bradley and Frey (1974) to problem (21).

Letting $\bar{\tilde{\xi}}_i = \bar{\xi}_i - \bar{\xi}_0$, i = 1,...,n, (22) becomes

$$\begin{aligned} &\text{maximize} \quad && g(\hat{x}) = \frac{\bar{\tilde{\xi}}^T \hat{x}}{f(\hat{x})} \\ &\text{s.t.} && e^T \hat{x} = 1, \\ & && g_j(\hat{x}) \geqslant 0, \quad j = 1,\ldots,m. \end{aligned} \tag{23}$$

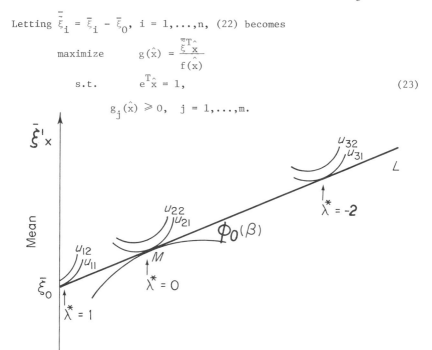

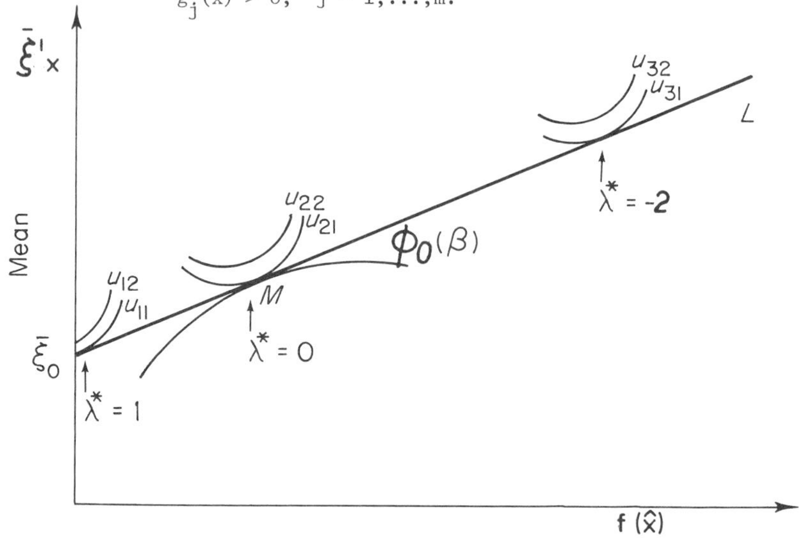

Figure 1 Geometry of the Generalized Tobin Separation Theorem[4]

We assume that $\bar{\tilde{\xi}}^T \hat{x} > 0$ for all *interesting* feasible $\hat{x}$. Since the $\bar{\tilde{\xi}}$ are the risk premiums, and $\hat{x} \geqslant 0$, this assumption is extremely mild because an optimal solution to (20) is $\hat{x}^* = 0$, $\hat{x}_0^* = 1$ if $\bar{\tilde{\xi}}^T \hat{x} \leqslant 0$ for all feasible $\hat{x}$.

Since $g(\hat{x})$ is the ratio of a positive linear function to a strictly convex positive function it is strictly pseudoconcave and hence has a unique maximum on the compact set of feasible portfolio allocations.

Since $g(\hat{x})$ is positively homogeneous of degree zero, i.e. $g(\hat{x}) = g(\lambda\hat{x})$ for all $\lambda > 0$, one may delete the constraints $e^T\hat{x} = 1$ from (23) and solve the simpler problem

---

[4] Note that $\lambda^* = 1, 0$ and -2 mean that for their respective utility functions it is optimal to invest only in the safe asset, only in the risky composite asset and, to borrow twice one's initial wealth at the risk free rate to invest in the risky composite asset.

$$\text{maximize} \quad g(\hat{x}) = \bar{\xi}^T\hat{x}/f(\hat{x}) \tag{24}$$
$$\text{subject to} \quad g_j(\hat{x}) \geqslant 0, \quad j = 1,\ldots,m,$$

and then rescale the optimal solution to (24), say $\hat{x}^0$, so that its components sum to 1 and $x_i^* = x_i^0/e^T\hat{x}^0$. The problem (24) simplifies further for specific g_j. For example, suppose only the constraint $\hat{x} \geqslant 0$ is present then since $f(\hat{x}) = (1/\sqrt{2})(\hat{x}^T \not{Z}\hat{x})^{1/2}$, the Kuhn-Tucker conditions of (24) (which are necessary and sufficient) indicate after substitution an equivalent problem is the quadratic program

$$\text{maximize} \quad \bar{\xi}^T\hat{x} - \hat{x}^T\not{Z}\hat{x}/2 \tag{25}$$
$$\text{subject to} \quad \hat{x} \geqslant 0.$$

This in turn is equivalent to the linear complementary problem: find $w,z \geqslant 0$, such that $w = \not{Z}z - \bar{\xi}$, $w^Tz = 0$ where $x^* = z^*/e^Tz^*$. See Ziemba et al. (1974) for details and other special cases when the returns have a joint normal distribution.

The analysis presented in this section is also valid for some *particular* types of *symmetric* stable distributions having the same characteristic exponent $\alpha \in (1,2]$, i.e. power to which they are convergent; $\alpha = 2$ is normal. They are developed in detail in Ziemba (1974) and utilize generalizations of Theorems 7 and 8. For the purpose of this volume we indicate some of the functional forms of f that arise. In each case f is convex or strictly convex and positively homogeneous *on* $\{\hat{x} | e^T\hat{x} = 1,\ \hat{x} \geqslant 0\}$.

a) If the investments have independent stable distributions then

$$f(\hat{x}) = \left\{ \sum_{i=1}^{n} S_i \hat{x}_i^{\alpha} \right\}^{1/\alpha} \tag{26}$$

which is strictly convex when the dispersions S_i are positive.

b) For Press's class of stable distributions (see Press (1972) and Ziemba (1974)) one may have, for example,

$$f(\hat{x}) = \left(\frac{1}{2}\right)^{1/\alpha} (\hat{x}^T \Omega_1 \hat{x})^{1/2} \tag{27}$$

which is strictly convex assuming Ω_1 is positive definite, or

$$f(\hat{x}) = \frac{1}{2} \left\{ \sum_{i=1}^{m} (\hat{x}^T \Omega_j \hat{x})^{\alpha/2} \right\}^{1/\alpha} \tag{28}$$

which is strictly convex assuming all the Ω_j are positive semi-definite and some Ω_j is positive definite.

c) For Samuelson's class of stable distributions

$$f(\hat{x}) = \left\{ \sum_{j=1}^{k} \left| \sum_{i=1}^{n} a_{ij} \hat{x}_i \right|^{\alpha} S_j^{\alpha} \right\}^{1/\alpha} \tag{29}$$

which is convex assuming the dispersion coefficients S_j are positive.

For (26)-(28) the function g is strictly pseudoconcave and (24) has a unique maximum. For (29) g is pseudoconcave and (24) has a (possibly non-unique) maximum.

Bradley and Frey (1974) have studied the solution and equivalence properties of fractional programming problems involving homogeneous functions. Their general results indicate that (24) is equivalent (has the same set of optimal solutions) to the convex program

$$\begin{aligned} &\text{minimize} && f(\hat{x}), \\ &\text{subject to} && g_j(\hat{x}) \geqslant 0, \quad j = 1,\ldots,m, \quad \bar{\bar{\xi}}^T \hat{x} = 1. \end{aligned} \tag{30}$$

Problem (30) is generally easier to solve than (24) for functions (26)-(29). When ξ is joint normal and only the $\hat{x} \geqslant 0$ constraint is present, (30) yields

$$\begin{aligned} &\text{minimize} && \hat{x}^T \not{\Sigma} \hat{x}, \\ &\text{subject to} && \hat{x} \geqslant 0, \quad \bar{\bar{\xi}}^T \hat{x} = 1 \end{aligned} \tag{31}$$

a problem equivalent to (25).

Once the first stage problem has been solved to obtain $\hat{x}^*$, the second-stage problem is then to determine the optimal ratios of risky to nonrisky assets. The risky composite asset is

$$R = \hat{\xi}^T \hat{x}^* \sim N(\bar{\bar{\xi}}^T \hat{x}^*, \hat{x}^{*T} \not{\Sigma} \hat{x}^*).$$

The best combination of R and the risk free asset may be found by solving (see Figure 1)

$$\begin{aligned} &\text{maximize} \quad \phi(\lambda) \equiv E_R u[\lambda\bar{\xi}_0 + (1-\lambda)R] \\ &\text{subject to} \quad -\infty < \lambda \leqslant 1. \end{aligned} \tag{32}$$

Problem (32) is a concave stochastic program with one random variable R and one decision variable λ. It is easy to solve by combining a search method such as golden sections or the Bolzano bisecting search method with a univariate numerical quadrature scheme. Numerical results with portfolio data appear in Ziemba et al. (1974). For example, with the logarithmic utility function $u(w) = \log(\beta+\gamma w)$, for $\gamma = 10$, with quarterly and yearly data of Canadian Pension Funds, one obtains Figure 2. The curves shows the optimal proportions to be invested in the risky composite and the safe asset for various investor risk attitudes described by the parameter β.

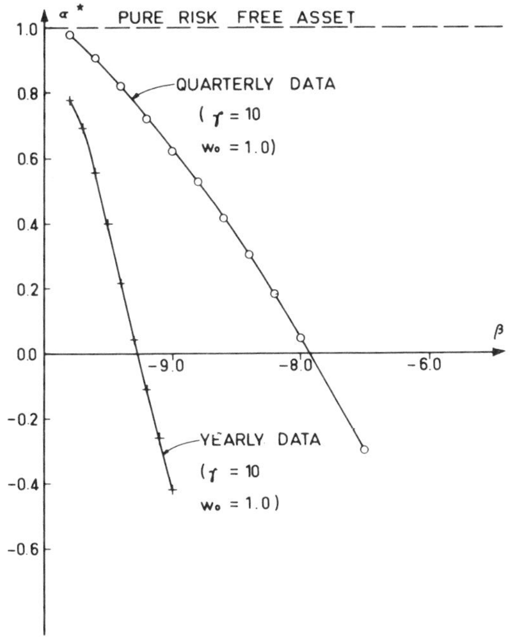

Figure 2 Logarithmic Utility Function-
Quarterly and Yearly Data: α^* vs. β.

Note that α^* is a decreasing function of β. This is not surprising since for w and γ fixed $\beta + \gamma w$ represents perceived wealth and

$$-\frac{\partial^2 u(\beta+\gamma w)/\partial\beta^2}{\partial u(\beta+\gamma w)/\partial\beta} = \frac{\gamma}{\beta+\gamma w}$$

which is decreasing in β, see Pratt (1964). Hence for increased β one would expect that there would be increased investment in the risky asset, meaning that α^* would decrease. The investor is less risk averse when faced with yearly data. The reasons for this behavior is that the risk premium per unit of variance, namely $(\bar{R} - \bar{\xi}_0)/\sigma_R^2$ is much higher for the yearly data. Hence the expected utility maximizing α^* is higher for the quarterly data.

The final optimal portfolio is to invest α^* in the risk free asset and $(1-\alpha^*)\hat{x}_i^*$ in risky asset i, i = 1,...,n.

In the case when the returns have stable distributions one must use discrete or other approximations since the density is not known in closed form except for the Normal and Cauchy distributions. One gets a similar expression to (32) with such an approximation; see Ziemba (1974, 1978) for details and numerical results.

6. THE LOGNORMAL DISTRIBUTION CASE

We consider the specific case of (16) when u(w) is the power function $u(w) = s(\gamma)w^\gamma$, $\gamma\neq 0$, $s(\gamma) = \pm 1$ if $\gamma \gtrless 0$ and ξ has a multivariate lognormal distribution. This is an important case to consider because the power utility function and lognormal return distribution have a number of desirable properties. For example assuming $\gamma < 1$, u(w) is strictly increasing, strictly concave and has strictly decreasing absolute risk aversion (i.e. as wealth increases the investor is more and more willing to invest in a given favourable gamble, see Pratt (1964)). It also induces optimal myopic behaviour in some dynamic portfolio revision models, see Hakansson (1971).

Since $\xi = (\xi_1,\ldots,\xi_n)^T$ has a multivariate lognormal distribution, $y \equiv (\log\xi_1,\ldots,\log\xi_n)^T \sim N(\mu,\Sigma)$, where the covariance matrix Σ has typical element $\sigma_{ij} = \text{cov}(\log\xi_i, \log\xi_j)$. It is assumed that all $|\mu_i| < \infty$, $|\sigma_{ij}| < \infty$ and Σ is positive definite. Several authors, e.g. Lintner (1972) and Rosenberg (1973), have found that actual security price data are often well approximated by the lognormal distribution. In addition the lognormal assumption indicates that the investment returns have limited liability since $P\{\xi^T x \geqslant 0\} = 1$ for all $x \geqslant 0$.

The problem is

$$\begin{aligned}\text{maximize } Z(x) &\equiv \int_0^\infty \cdots \int_0^\infty s(\gamma)(\xi^T x)^\gamma dG(\xi) \\ &= \int_0^\infty \cdots \int_0^\infty s(\gamma)(\Sigma x_i e^{y_i})^\gamma dN(y_1,\ldots,y_n),\end{aligned} \tag{33}$$

which is a stochastic program with n random variable and n decision variables. The computation of an optimal solution x* is extremely difficult even for moderate n, because an exact solution of (33) requires that the objective or its gradient be numerically evaluated for many values of x^k. Thus good approximations are desirable. Ohlson and Ziemba (1976) have considered the approximation that obtains when it is assumed that $\xi^T x$ is lognormal for any feasible x. Under this assumption (33) reduces to the explicit deterministic nonlinear program[5]

$$\begin{aligned}&\text{maximize } A(x) \equiv \log \textstyle\sum a_i x_i - \phi \log \sum\sum c_{ij} x_i x_j \\ &\qquad \text{s.t.} \qquad e^T x = 1, \; x \geqslant 0\end{aligned} \tag{34}$$

where

$$\phi \equiv \frac{1-\gamma}{2(2-\gamma)} \in (0,1/2), \qquad a_i \equiv \exp(\mu_i + 1/2\sigma_{ii})$$

---

5 A similar analysis to what follows can be developed for the case when $u(w) = \log w$, see Ohlson (1972). The results follow basically from the fact that

$$\lim_{\alpha\to 0} \frac{w^\alpha - 1}{\alpha} = \log w .$$

and $\quad c_{ij} \equiv \exp(\mu_i+\mu_j+1/2(\sigma_{ii}+\sigma_{jj}+2\sigma_{ij}))$.

The expression A(x) is derived as follows. If W is lognormal then so is W^γ and $s(\gamma)EW^\gamma = s(\gamma)\exp\{\gamma E(\log W) + 1/2\gamma^2 V(\log W)\}$, see Aitcheson and Brown (1966). Maximizing this latter expression is equivalent to maximizing $\{E(\log W) + 1/2\gamma V(\log W)\}$. Now

$$\exp\{E(\log W) + 1/2V(\log W)\} = EW = \sum \exp\{\mu_i+1/2\sigma_{ii}\}x_i, \tag{35}$$

and

$$\begin{aligned}\exp\{2E(\log W) &+ 2V(\log W)\} \\ &= EW^2 = \sum\sum \exp\{\mu_i+\mu_j+1/2(\sigma_{ii}+\sigma_{jj}+2\sigma_{ij})\}x_i x_j.\end{aligned} \tag{36}$$

Taking the logs of both sides of the first equality sign in (35) and (36) and solving gives

$$E(\log W) = 2 \log EW - 1/2 \log EW^2 \tag{37}$$

and

$$V(\log W) = -2 \log EW + \log EW^2 . \tag{38}$$

Combining the expressions on the right side of (35) and (36) with (37) and (38) and simplifying yields A(x).

Ohlson and Ziemba (1976) have shown that (34) has a unique optimal solution $\hat{x}$. Moreover $\hat{x}$ is mean-variance efficient, i.e. it maximizes the mean for any given variance. They also consider some of the qualitative properties of x^* and $\hat{x}$, and compare $\hat{x}$ with solutions x^α obtained by taking Taylor series approximations to u of order two, around $w = \alpha$ and solving the resulting quadratic program. On balance the solution $\hat{x}$ has much to commend itself in the sense that its qualitative properties are very similar to those of x^* and much better than those of x^α.

To provide a test of how close $\hat{x}$ and x^*, and $Z(\hat{x})$ and $Z(x^*)$ are to each other, Dexter, Yu and Ziemba (1980) computed these quantities for n = 2,3, and 4 for various values of γ, using monthly data from the CRSP tape (1970) for the 200 months between February 1926 and July 1942.

(Above n=4 the computations of the exact solution became prohibitively expensive.) The estimates of the μ_i and σ_{ij} were taken to be the sample means, variances and covariances of the logarithms of the relative prices. The solution of (34) is complicated because of the log functions. Hence it is more advantageous to solve

$$\text{maximize} \qquad B(x) \equiv \frac{\sum a_i x_i}{\{\sum\sum c_{ij} x_i x_j\}^{\phi}} \tag{39}$$

$$\text{s.t.} \qquad e^T x = 1, \quad x \geqslant 0.$$

Ohlson and Ziemba (1976) proved that $C = \{c_{ij}\}$ is positive definite and presented a proof that $\log x^T Cx$ is convex which would imply that $A(x)$ was concave and that (34) is a concave program. Unfortunately this proof is in error, e.g. in the n=1 case $\log cx^2 = 2 \log x + \log c$ which is concave not convex. Schaible and Ziemba (1981) have investigated the concavity properties of $A(x)$ and $B(x)$. A brief summary of their results concerning $B(x)$ is contained in the following theorem. Let L be the matrix whose typical element is $\exp(\sigma_{ij})$. L is related to C by $C = \text{diag}\ a\ L\ \text{diag}\ a$, where each $a_i \equiv \exp(\mu_i + \sigma_{ii}) > 0$, and L as well as C is positive definite. Let $\alpha \equiv (e^T L^{-1} e)\ell_{ii}^{max}$, where $e \equiv (1,\ldots,1)^T$ and ℓ_{ii}^{max} is the largest diagonal element of L. We say that L is a class I matrix if $1 \leqslant \alpha < 2$ otherwise it is of class II.

THEOREM 9: For class I matrices L*:*

a) there exists $\phi^* \in (0, 1/2)$, *such that* $B(x)$ *is concave on* $X \equiv \{x | x \geqslant 0\}$ *for* $\phi \in (0,\phi^*]$,

b) for $\phi \in (\phi^*,1/2)$ $B(x)$ *is not quasiconcave on* X, *and*

c) for $\phi \in (\phi^*,1/2)$ *there exists a convex cone* $K \subset X$ *such that* $B(x)$ *is concave on* K, *where* K *is a shrinking function of* ϕ, *i.e.* $K(\phi_1) \subset K(\phi_2)$ *if* $\phi_1 > \phi_2$.

For class II matrices :

d) B(x) *is never quasiconcave on* X *for any* $\phi \in (0,\frac{1}{2})$.

Proofs as well as additional results appear in Schaible and Ziemba (1981).

The numerical results described by Dexter, Yu and Ziemba (1980) indicate that the approximate solution is virtually identical to the exact[6] solution and the x values are very similar. For example for n=3 the percentage cash equivalent error $(\frac{u^{-1}[Z(x^*)]-u^{-1}[Z(\hat{x})]}{u^{-1}[Z(x^*)]} \times 100)$ for varying γ is always less than 10^{-3}% and the portfolio weights are as shown in Figure 3. For all the data in Dexter, Yu and Ziemba (1980), $\phi^* \cong 0.495$, L is a class I matrix and B(x) is concave. For examination of various ϕ^* resulting from other data see Schaible and Ziemba (1981). The ratio of computing times between the exact and approximate solutions is essentially exponential in n; see Figure 4. Hence the approximation is a very useful one if one trades off accuracy for computing costs.

---

[6]By *exact* we mean optimal to a tolerance of 10^{-12} using the following version of (33):

$$\text{maximize} \quad \int_{\mu_1-3\sigma_1}^{\mu_1+3\sigma_1} \cdots \int_{\mu_n-3\sigma_n}^{\mu_n+3\sigma_n} (\sum x_i e^{y_i})^{\gamma} dN(y_1,\ldots,y_n)$$

$$\text{s.t.} \quad e^T x = 1, \quad x \geqslant 0,$$

where $\sigma_j \equiv \sigma_{jj}^{1/2}$ is the standard deviation of log ξ_j and the constants have been eliminated.

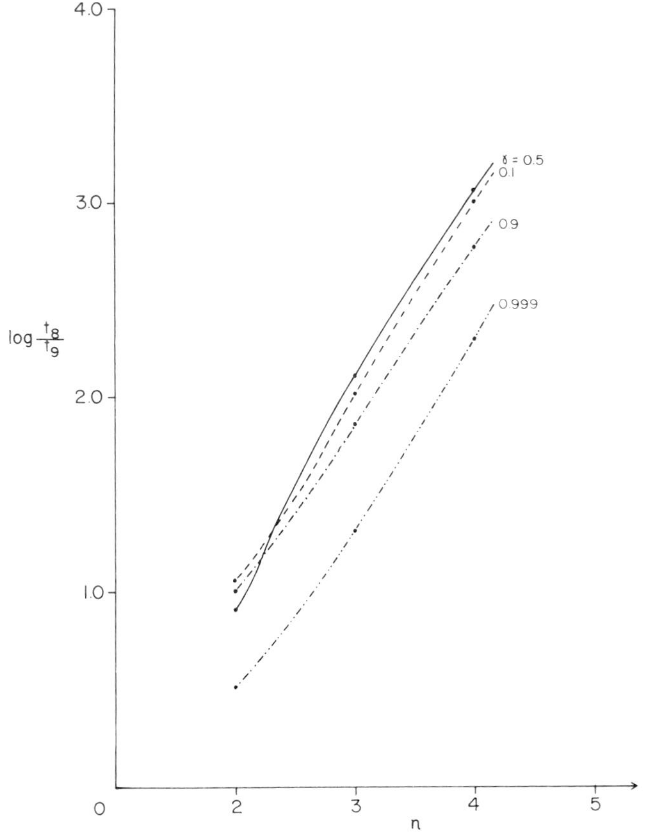

Figure 4: Ratio of original versus approximating problem computational times

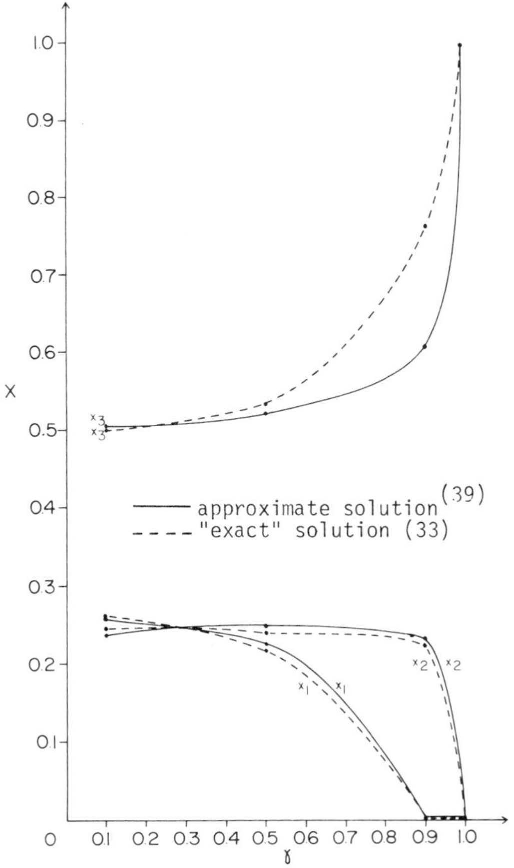

Figure 3: Optimal portfolio weights versus γ when n = 3

7. PARIMUTUEL BETTING MODELS[7]

Suppose in a given horserace there are n horses and the ith horse's chances of winning is q_i. Let W_i be the amount bet on i by the crowd, $W \equiv \Sigma W_i$ and 1-Q be the percentage track take. Then an investor/gambler with initial wealth w_0 will maximize his long run rate of growth of assets[8] by choosing bets w_i to

$$\text{maximize } \phi(w_1,\ldots,w_n) = \sum_{i=1}^{n} q_i \log\left\{\left[\frac{Q(W+\Sigma w_j)-w_i-W_i}{w_i+W_i}\right] w_i + w_0 - \sum_{j\neq i} w_j\right\} \tag{40}$$

$$\text{s.t.} \quad \sum_{i=1}^{n} w_i \leqslant w_0, \quad w_i \geqslant 0, \quad i = 1,\ldots,n.$$

The payoff is determined as follows. Suppose horse i wins then all bettors on horse i are returned their bets $(w_i + W_i)$ plus their share of the profits $\frac{Q(W+\Sigma w_i)-w_i-W_i}{w_i + W_i}$ per dollar bet on i. Hence the expected utility of final wealth is the objective function in (40).

The portfolio model (40), like the other related model in this section, is distinguished by the fact that the return distributions are affected by the investment choices. This is natural since the amounts bet influence the odds. It is this feature that destroys the concavity of (40) even with linear utility. We see also that if one neglects this feature then one has concavity. Some concavity/generalized concavity properties of ϕ are presented in the following theorem. For convenience let

$$\phi(w) \equiv \sum_{i=1}^{n} q_i \log(f_i(w)) \quad \text{and} \quad h(w) \equiv \sum_{i=1}^{n} q_i f_i(w).$$

---

7 In this section we present results, obtained in collaboration with D.B. Hausch, concerning the concavity and generalized concavity of two objective functions arising in parimutuel betting models. Some open questions remain and they are described in the text.

8 See e.g. the papers by Breiman and Thorpe in Ziemba-Vickson (1975) for discussion and proof that maximizing the expected logarithm of final wealth indeed maximizes (asymptotically) the rate of asset growth in repeated independent investment trials.

THEOREM 10:

a) In general ϕ is not concave even if log utility is replaced by linear utility unless the bets by the investor are assumed not to influence the odds.

b) Each $f_i(w)$ is pseudoconcave, and

c) ϕ is concave transformable hence local maxima are global.

PROOF:

(a) In the linear utility case suppose $n = 2$, $Q = .8$, $W_1 = 1$, $W_2 = 10$, $w_0 = 300$, $q_1 = 10/11$, $q_2 = 1/11$, $w^1 = (1,10)$ and $w^2 = (19,192)$. Then $\frac{1}{2}h(w^1) + \frac{1}{2}h(w^2) = 277.764 > 277.735 = h(\frac{1}{2}w^1+\frac{1}{2}w^2)$ hence $h(w)$ is not concave. In the log utility case with the same data except $w_0 = 10^6$, ϕ is not concave because $\frac{1}{2}\phi(w^1) + \frac{1}{2}\phi(w^2) = 5.999990342 > 5.99990329 = \phi(\frac{1}{2}w^1+\frac{1}{2}w^2)$. If the bets do not influence the odds then $\phi(w) = \sum q_i \log [g_i(w)]$ where $g_i(w) \equiv (QW-W_i)w_i/W_i - \sum_{j\neq i} w_j + w_0$. Hence ϕ is the sum of concave functions of linear functions which is concave. When log utility is replaced by linear utility ϕ is linear hence concave.

(b) $$f_i(w) = \{Q(W+\Sigma w_j) - (w_i+W_i)\} \frac{w_i}{w_i+W_i} - \sum_{j\neq i} w_j + w_0$$

$$= Qw_i \frac{W+\Sigma w_j}{w_i+W_i} - w_i - \sum w_j + w_i + w_0$$

$$= Qw_i \frac{W+\Sigma w_j}{w_i+W_i} + (Q-1)\sum w_j - Q\sum w_j + w_0$$

$$= Q(W+\Sigma w_j) - QW_i \frac{W+\Sigma w_j}{w_i+W_i} + (Q-1)\sum w_j - Q\sum w_j + w_0$$

$$= (Q-1) \sum w_j + QW + w_0 - QW_i \frac{W+\Sigma w_j}{w_i+W_i}$$

$$= (Q-1)\sum(w_j+W_j) + (1-Q)\sum W_j + Q\sum W_j + w_0 - \frac{QW_i\sum(w_j+W_j)}{w_i + W_i}$$

$$= \underbrace{(Q-1)\sum(w_j+W_j)}_{\text{linear}} + \underbrace{\sum W_j + w_0}_{\text{constant}} - \underbrace{\frac{QW_i\sum(w_j+W_j)}{w_i + W_i}}_{\text{fractional}} . \qquad (41)$$

Thus (41) is of the form

$$f(x) = a_0 + a^T x + b^T x / c^T x \tag{42}$$

where $a_0 \equiv W + w_0$, $x \equiv (w_1+W_1,\ldots,w_n+W_n)$,

$a \equiv (Q-1,\ldots,Q-1)$, $b \equiv (-QW_i,\ldots,-QW_i)$,

$c \equiv (0,\ldots,0,1,0,\ldots,0)$ and $\downarrow$ indicates the ith column. In (41) $a = \left[\frac{1-Q}{QW_i}\right] b \equiv \mu b$. Since $Q \in (0,1)$ and $W_i > 0$ it follows that $\mu > 0$.

Thus by Proposition II of Schaible (1977), (42) is quasiconcave and pseudoconcave (on an open convex set) of $a^T x \leqslant 0$ (i.e. $x \geqslant 0$ or $w_i+W_i \geqslant 0$).

(c) If $x_j \equiv w_j+W_j$ then (41) becomes

$$f_i(x) = -\alpha \sum x_i + \beta_i - \gamma_i \sum x_j / x_i$$

where $\alpha \equiv (1-Q) > 0$, $\beta_i = \sum W_i + w_0 > 0$

and $\gamma_i = QW_i > 0$.

Letting $z_i \equiv \log x_i$, i.e., $x_i = e^{z_i}$ gives

$$f_i(z) = -\alpha \sum e^{z_j} + \beta_i - \gamma_i \sum e^{z_j - z_i}$$

which is concave. Hence log f_i and ϕ are also concave. That the transformed constraint set is still convex is verified as follows. The constraint $w_i \geqslant 0$ becomes $x_i \geqslant W_i$ or $e^{z_i} \geqslant W_i$ or $z_i \geqslant \log W_i$. Now the budget constraint $\sum w_i \leqslant w_0$ becomes $\sum x_i \leqslant w_0+W$ or $\sum e^{z_i} \leqslant w_0+W$, which describes a convex set. Thus problem (40) maps into an equivalent concave maximization problem via a 1-1 transformation. This implies that problem (40) possesses the property that local maxima are global.

Remarks: Thus to obtain concavity one must either use the approximation that the investor's bets do not influence the odds or have rather severe conditions arise, e.g. $Q\sum w_j = w_i$, $w_i > 0$ for all i. The latter condition is unlikely to be satisfied in practice. The result in (b) indicates that the objective function in (40) is the sum of concave functions of pseudoconcave functions. It is not known if ϕ is generally pseudoconcave even in the linear utility function case. However, since ϕ is concave-transformable local maxima are global maxima. The linear utility function case is known as Isaac's model (1953). In this case one has an explicit solution as follows. Assume that there is an *underbet* great enough to overcome the track take,

$$\text{i.e. } \max_j q_j/(W_j/W) > 1/Q \tag{43}$$

then at least one $w_i^* > 0$. Renumber the horses so $p_1 \leqslant \ldots \leqslant p_n$ where $p_i \equiv q_i/W_i$. Then $w_i^* = \lambda_i\sqrt{q_i W_i} - W_i$, $i = t,\ldots,n$, where $\lambda_i^2 \equiv Q\sum_{j=1}^{i-1} W_j/(1-Q\sum_{j=i}^{n} q_j)$ and the optimal solution has the form $w_1^* = \ldots w_{t-1}^* = 0$, $w_t^* > 0,\ldots,w_n^* > 0$.

The difficulty with the model (40) is that efficiency studies, see e.g. Snyder (1978), indicate that the probability that horse i wins, q_i, is accurately estimated by the relative betting frequency of the betting public, W_i/W, except for stable tail biases.[9] Hence it is unlikely that condition (43) will be satisfied except in rare instances and even then there is no obvious simple way to estimate the q_i. However, the model (40) may be useful in cases where expert opinion or factor analytic methods yield estimates of the q_i that satisfy (43).

Hausch, Ziemba and Rubinstein (1980) suggest instead that there are more likely to be inefficiencies in the place and show[10] markets than the win market. There are a number of reasons for this, in particular because

---

[9] Specifically longshots (high odds horses) are overbet and favorites (low odds horses) are underbet. However, even the latter bias is only about nine percent compared to the about eighteen percent track take.

[10] A horse is said to *place* if it finishes first or second, it *shows* if it finishes first, second or third. The order of finish is immaterial, however, the payoff depends on the amount bet on the other horse(s) that "finish in the money", see (44).

it is much more complicated to calculate the odds for place and show bets. The basic assumptions of their model are: 1) there is an *efficient* market to win so $q_i = (W_i/W)$, 2) following Harville (1973) assume that the probability that i is first and j is second is $q_iq_j/(1-q_i)$ and of an ijk finish is $q_iq_jq_k/(1-q_i)(1-q_i-q_j) \equiv q_{ijk}$, and 3) the same assumptions as (40) with the place and show expressions replacing the win expressions. This yields the model

$$\underset{\{p_\ell\}\{s_\ell\}}{\text{Maximize}}\ \psi(p,s) \equiv \sum_{i=1}^{n} \sum_{\substack{j=1\\ j\neq i}}^{n} \sum_{\substack{k=1\\ k\neq i,j}}^{n} q_{ijk} \log \left[\begin{array}{l} \dfrac{Q(P+\sum_{\ell=1}^{n} p_\ell)-(p_i+p_j+P_i+P_j)}{2} \\ \times \left[\dfrac{p_i}{p_i+P_i}+\dfrac{p_j}{p_j+P_j}\right] \\ +\dfrac{Q(S+\sum_{\ell=1}^{n} s_\ell)-(s_i+s_j+s_k+S_i+S_j+S_k)}{3} \\ \times \left[\dfrac{s_i}{s_i+S_i}+\dfrac{s_j}{s_j+S_j}+\dfrac{s_k}{s_k+S_k}\right] \\ +w_0 - \sum_{\substack{\ell=1\\ \ell\neq i,j,k}}^{n} s_\ell - \sum_{\substack{\ell=1\\ \ell\neq i,j}}^{n} p_\ell \end{array} \right] \tag{44}$$

$$\text{s.t.} \sum_{\ell=1}^{n} (p_\ell+s_\ell) \leqslant w_0, \quad p_\ell \geqslant 0, \quad s_\ell \geqslant 0, \qquad \ell = 1,\ldots,n,$$

where Q = 1-the track take, W_i, P_j and S_k are the total dollar amounts bet to win, place and show on the indicated horse by the crowd, respectively, $q_i \equiv W_i/W$ is the theoretical probability that horse i wins, w_0 is initial wealth and p_ℓ and s_ℓ are the investor's bets to place and show on horse ℓ, respectively.

The argument of the log represents final wealth if there is an ijk finish so the expectation yields expected utility of final wealth. The payoffs are calculated[11] so that for horses in the money the betting amounts are returned and the resulting portion of the pool in excess of the track take is divided equally among the winning horses. We have the following partial results concerning the concavity, generalized concavity and solution properties of (44).

THEOREM 11:

a) If each p_ℓ and s_ℓ is small in relation to P_ℓ and S_ℓ, (i.e. the investor's bets make an insignificant effect on the odds) and these terms are neglected accordingly when they appear with such a P_ℓ or P and S_ℓ or S, respectively then ψ is concave for logarithmic as well as linear utility.

b) In general ψ is not concave.

c) If there is linear utility and an a priori decision has been made, e.g. on the basis of expected returns, to bet on horse ℓ to place and horse m to show then ψ is concave and there is an exact closed form solution.

PROOF:

(a) Under the assumptions

$$\psi(p,s) = \sum_i \sum_{j\neq i} \sum_{k\neq i,j} q_{ijk} \log(g_{ij}(p)+h_{ijk}(s)+w_0)$$

where

$$g_{ij}(p) \equiv \left(\frac{QP - (P_i+P_j)}{2}\right)\left(\frac{p_i}{P_i} + \frac{p_j}{P_j}\right) - \sum_{\ell\neq i,j} p_\ell,$$

$$h_{ijk}(s) \equiv \left(\frac{QS - (S_i+S_j+S_k)}{3}\right)\left(\frac{s_i}{S_i} + \frac{s_j}{S_j} + \frac{s_k}{S_k}\right) - \sum_{\ell\neq i,j,k} s_\ell,$$

---

[11]This formulation omits the features of minus pools and breakage that slightly alter the exact model of the betting situation. See Hausch, Ziemba, and Rubinstein (1980) for discussion.

which is the sum of concave functions of linear functions hence is concave. When log utility is replaced by linear utility ψ is linear in p,s hence concave.

(b) Let $f(p,s) = \sum_i \sum_{j\neq i} \sum_{k\neq i,j} q_{ijk}(g_{ij}(p) + h_{ijk}(s) + w_0)$,

$$g_{ij}(p) \equiv \left(\frac{Q(P+\Sigma p_\ell)-p_i-p_j-P_i-P_j}{2}\right)\left(\frac{p_i}{p_i+P_i} + \frac{p_j}{p_j+P_j}\right) - \sum_{\ell\neq i,j} p_\ell$$

$$h_{ijk}(s) \equiv \left(\frac{Q(S+\Sigma s_\ell)-s_i-s_j-s_k-S_i-S_j-S_k}{3}\right)\left(\frac{s_i}{s_i+S_i} + \frac{s_j}{s_j+S_j} + \frac{s_k}{s_k+S_k}\right) - \sum_{\ell\neq i,j,k} s_\ell .$$

In the linear case suppose $n = 4$, $Q = .8$, $P_1 = S_1 = 97$, $P_2 = P_3 = P_4 = S_2 = S_3 = S_4 = 1$, $w_0 = 5000$, $q_1 = .01$, $q_2 = q_3 = q_4 = .33$, $p^1 = s^1 = (97,1,1,1)$ and $p^2 = s^2 = (1999,19,19,19)$. Then $\frac{1}{2}f(p^1,s^1) + \frac{1}{2}f(p^2,s^2) = 4562.64 > 4557.97 = f(\frac{1}{2}p^1+\frac{1}{2}p^2,\ \frac{1}{2}s^1+\frac{1}{2}s^2)$, hence f is not concave. In the log utility case with the same data except $w_0 = 10^6$, ψ is not concave because $\frac{1}{2}\psi(p^1,s^1) + \frac{1}{2}\psi(p^2,s^2) = 5.9998107 > 5.9998080 = \psi(\frac{1}{2}p^1+\frac{1}{2}p^2,\ \frac{1}{2}s^1+\frac{1}{2}s^2)$.

(c) When there is linear utility and one bet on horse ℓ to place and one bet on horse m to show the objective function becomes

$$\sum_{\substack{i=1\\ i\neq\ell}}^{n}\left[\hat{q}_{i\ell}\left(\frac{Q(P+p_\ell)-(p_\ell+P_\ell+P_i)}{2}\right)\left(\frac{p_\ell}{p_\ell+P_\ell}\right) + p_\ell\right]$$

$$+ \sum_{\substack{i=1\\ i\neq\ell}}^{n}\sum_{\substack{j=1\\ j\neq i,m}}^{n}\left[\tilde{q}_{ijm}\left(\frac{Q(S+s_m)-(s_m+S_m+S_i+S_j)}{3}\right)\left(\frac{s_m}{s_m+S_m}\right) + s_m\right]$$

$$- p_\ell - s_m + w_0$$

$$= \left(\frac{Q+1}{2}\, p_\ell\right) \sum_{\substack{i=1\\ i\neq \ell}}^{n} \hat{q}_{i\ell} + \left(\frac{p_\ell}{p_\ell + P_\ell}\right) \sum_{\substack{i=1\\ i\neq \ell}}^{n} \frac{\hat{q}_{i\ell}\left(Q \sum\limits_{\substack{k=1\\ k\neq \ell}}^{n} P_k - P_i\right)}{2}$$

$$+ \left(\frac{Q+2}{3}\, s_m\right) \sum_{\substack{i=1\\ i\neq m}}^{n} \sum_{\substack{j=1\\ j\neq i,m}}^{n} \tilde{q}_{ijm} + \frac{s_m}{s_m + S_m} \sum_{\substack{i=1\\ i\neq m}}^{n} \sum_{\substack{j=1\\ j\neq i,m}}^{n} \tilde{q}_{ijm} \left(\frac{Q \sum\limits_{\substack{k=1\\ k\neq m}}^{n} S_k - S_i - S_j}{3}\right)$$

$$- p_\ell - s_m + w_0 , \tag{45}$$

where

$$\hat{q}_{i\ell} \equiv \frac{q_i\, q_\ell}{1-q_i} + \frac{q_i\, q_\ell}{1-q_\ell} , \quad \text{and}$$

$$\tilde{q}_{ijm} \equiv \frac{q_i\, q_j\, q_m}{(1-q_i)(1-q_i-q_j)} + \frac{q_i\, q_j\, q_m}{(1-q_i)(1-q_i-q_m)} + \frac{q_i\, q_j\, q_m}{(1-q_m)(1-q_m-q_j)} .$$

Since (45) consists of sums of linear terms plus terms of the form $x/(a+x)$ for $a \geqslant 0$ it is concave. Hence the Kuhn-Tucker conditions are necessary and sufficient and they yield the following solution, where $\ell = m$ is allowed:

$$p^* \equiv -P_\ell + \sqrt{\frac{P_\ell(Q\hat{q}_\ell(P-P_\ell)-\hat{R}_\ell)}{2-\hat{q}_\ell(Q+1)}}$$

$$s^*_m \equiv -S_m + \sqrt{\frac{S_m(Q\tilde{q}_m(S-S_m)-\tilde{R}_m)}{3-\tilde{q}_m(Q+2)}} , \quad \text{where}$$

$$\hat{q} \equiv \sum_{i\neq \ell} \left(\frac{q_i q_\ell}{1-q_i}\right) + \sum_{j\neq \ell} \left(\frac{q_\ell q_j}{1-q_\ell}\right) \quad \text{and}$$

$$\tilde{q}_m \equiv \sum_{i\neq m} \sum_{j\neq i,m} \left(\frac{q_i q_j q_m}{(1-q_i)(1-q_i-q_j)}\right) + \sum_{i\neq m} \sum_{k\neq i,m} \left(\frac{q_i q_m q_k}{(1-q_i)(1-q_i-q_m)}\right)$$

$$+ \sum_{j\neq m} \sum_{k\neq j,m} \left(\frac{q_m q_j q_k}{(1-q_m)(1-q_m-q_j)}\right) ,$$

$$\hat{R}_\ell \equiv \sum_{i\neq\ell}\left(\frac{q_i q_\ell}{1-q_i}\right)P_i + \sum_{j\neq\ell}\left(\frac{q_\ell q_j}{1-q_\ell}\right)P_j, \quad \text{and}$$

$$\tilde{R}_m \equiv \sum_{i\neq m}\sum_{j\neq i,m}\left(\frac{q_i q_j q_m}{(1-q_i)(1-q_i-q_j)}\right)(S_i+S_j)$$

$$+\sum_{i\neq m}\sum_{k\neq i,m}\left(\frac{q_i q_m q_k}{(1-q_i)(1-q_i-q_m)}\right)(S_i+S_k)$$

$$+\sum_{j\neq m}\sum_{k\neq j,m}\left(\frac{q_m q_j q_k}{(1-q_m)(1-q_m-q_j)}\right)(S_j+S_k)\,.$$

(i) If $p_\ell^* + s_m^* \leqslant w_0$, $p_\ell^* \geqslant 0$, and $s_m^* \geqslant 0$, then it is optimal to bet p_ℓ^* on horse ℓ to place and s_m^* on horse m to show.

(ii) If $p_\ell^* \leqslant 0$ and $s_m^* \leqslant 0$ bet nothing.

(iii) If $p_\ell^* \leqslant 0$ and $s_m^* > 0$ bet $\min(s_m^*, w_0)$ on horse m to show.

(iv) If $p_\ell^* > 0$ and $s_m^* \leqslant 0$ bet $\min(p_\ell^*, w_0)$ on horse ℓ to place.

(v) If $p_\ell^* + s_m^* > w_0$ and $p_\ell^* > 0$, $s_m^* > 0$ then the optimal amounts to bet are s_m^{**} and p_ℓ^{**}, from the Kuhn-Tucker conditions

$$w_0 - p_\ell^{**} - s_m^{**} \geqslant 0,$$
$$p_\ell^{**} \geqslant 0,$$
$$s_m^{**} \geqslant 0,$$

$$\lambda_1(w_0 - p_\ell^{**} - s_m^{**}) = 0, \qquad \lambda_1, \lambda_2, \lambda_3 \geqslant 0$$
$$\lambda_2 p_\ell^{**} = 0,$$
$$\lambda_3 s_m^{**} = 0,$$
$$\frac{\partial f(p_\ell^{**}, s_m^{**})}{\partial p_\ell} - \lambda_1 + \lambda_2 = 0, \quad \text{and}$$
$$\frac{\partial f(p_\ell^{**}, s_m^{**})}{\partial s_m} - \lambda_1 + \lambda_3 = 0.$$

Remarks. It is not known whether or not the function $\psi(p,s)$ is pseudoconcave in general, even for linear utility, or in suitable instances that are likely to occur in practice. In fact results anologous to those in Theorem 10 b and c that the *argument* is pseudoconcave and the objective function is concave-transformable are *not* obtainable using the methods used in the proof of that theorem. Hausch, Ziemba and Rubinstein (1980) solved several *thousand* models similar to (44) using GRG. In about twenty cases solutions were obtained using multiple starting values and in all instances the final solutions were identical. Hence it appears that ψ has a type of possibly local generalized concavity or connectivity, see Martin (1981), that yields optimal solutions using gradient algorithms. The linear exact solution result (c) is useful because generally it is optimal even with the exact logarithmic solution to bet on at most one horse. However one tends to bet too much with linear utility since one essentially bets until the odds are driven to be unprofitable in an expected value sense or the budget constraint is binding. The logarithmic formulation yields a more reasonable betting scheme that reflects the risks as well as the possible returns. Hausch, Ziemba and Rubinstein (1980) present results of the use of the model (44) and approximations to make it fully implementable in practice. Figure 5 illustrates typical behaviour. See also Ziemba and Hausch (1982) for a layman's account of the final system with the results of hypothetical and actual betting, etc.

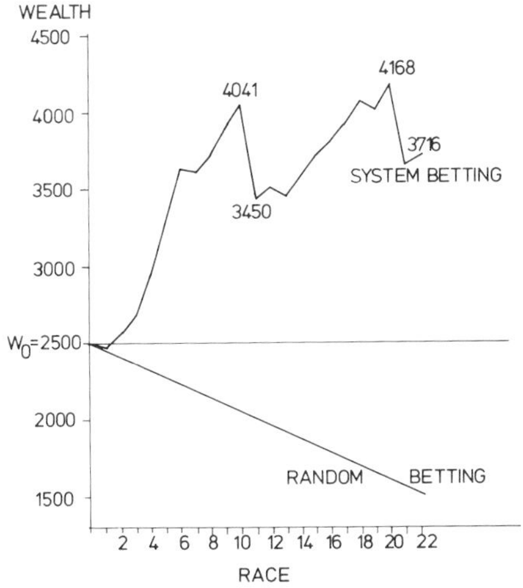

FIGURE 5 Results from Summer 1980 Exhibition Park Betting: 9 Days 22 Races

REFERENCES

AITCHISON, J. and BROWN, J.A.C. (1966). *The Lognormal Distribution.* Cambridge University Press, Cambridge.

ARTIN, E. (1931). "Einführung in die Theorie der Gammafunktion." *Hamburger Mathematische Einzelschriften, 11,* 1-35.

AVRIEL, M. (1973). "Solution of Certain Nonlinear Programs Involving r-Convex Functions." *Journal of Optimization Theory and Applications, 11,* 159-174.

AVRIEL, M. and ZANG, I. (1974). "Generalized Convex Functions With Applications to Nonlinear Programming." Chapter 2 in *Mathematical Programs for Activity Analysis,* (P. Van Moeseke, ed.), North-Holland Publishing Co., Amsterdam.

AVRIEL, M. (1976). *Nonlinear Programming Analysis and Methods.* Prentice-Hall, Englewood Cliffs, New Jersey.

AVRIEL, M., DIEWERT, W.E., SCHAIBLE, S. and ZANG, I. *Generalized Concavity.* (Forthcoming.)

BARLOW, R.E., MARSHALL, A.W. and PROSCHAN, F. (1963). "Properties of Probability Distributions With Monotone Hazard Rate." *Annals of Mathematical Statistics, 34,* 375-389.

BEN-TAL, A. (1977). "On Generalized Means and Generalized Convex Functions." *Journal of Optimization Theory and Applications, 21,* 1-13.

BEREANU, B. (1980). "Some Numerical Methods in Stochastic Linear Programming Under Risk and Uncertainty." In *Stochastic Programming,* (M.A.H. Dempster, ed.), Academic Press Inc., New York, 169-205.

BORELL, C. (1975). "Convex Set Functions in D-Space." *Periodica Mathematica Hungarica, 6,* 111-136.

BRADLEY, S.P. (1975). "Computing Optimal Portfolios Using Risk-Expected Return Efficient Points." Mimeo, Graduate School of Business, Harvard University, Boston, Massachusetts.

BRADLEY, S.P. and FREY, JR., S.C. (1974). "Fractional Programming With Homogeneous Functions." *Operations Research, 22,* 350-357.

BREEN, W. (1968). "Homogeneous Risk Measures and the Construction of Composite Assets." *Journal of Financial and Quantitative Analysis, 3,* 405-413.

BRUNN, H. (1887). "Über Ovale und Eiflächen." Inaugural Dissertation, München.

CENTER FOR RESEARCH IN SECURITY PRICES MONTHLY SECURITY PRICE INFORMATION TAPE, University of Chicago, 1970.

CHARNES, A. and COOPER, W.W. (1963). "Deterministic Equivalents for Optimizing and Satisficing Under Chance Constraints." *Operations Research, 11,* 18-39.

CHIPMAN, J. (1973). "On the Ordering of Portfolios in Terms of Mean and Variance." *Review of Economic Studies, 40,* 167-190.

DEXTER, A.S., YU, J.N.W. and ZIEMBA, W.T. (1980). "Portfolio Selection in a Lognormal Market When the Investor has a Power Utility Function: Computational Results." In *Stochastic Programming,* (M.A.H. Dempster, ed.), Academic Press Inc., New York, 507-523.

FETKE, M. (1912). "Über ein Problem von Laguerre." *Rend. Circ. Mat. Palermo, 34,* 89-100, 110-120.

GEOFFRION, A.M. (1967). "Stochastic Programming With Aspiration or Fractile Criteria." *Management Science, 13,* 672-679.

HAKANSSON, N.H. (1971). "On Optimal Myopic Portfolio Policies, With and Without Serial Correlation of Yields." *J. Business, 44,* 324-334.

HARDY, G.H., LITTLEWOOD, J.E. and POYLA, G. (1934). *Inequalities.* Cambridge University Press, Cambridge.

HARVILLE, D.A. (1973). "Assigning Probabilities to the Outcomes of Multi-Entry Competitions." *Journal of the American Statistical Association, 68*, 312-316.

HANOCH, G. and LEVY, H. (1969). "The Efficiency Analysis of Choices Involving Risk." *Review of Economic Studies, 36*, 335-346.

HAUSCH, D.B., ZIEMBA, W.T. and RUBINSTEIN, M.E. (1980). "Efficiency of the Market for Racetrack Betting." UBC Faculty of Commerce W.P. No.712, forthcoming in *Management Science*.

HILLIER, F.S. (1963). *The Evaluation of Risky Interrelated Investments.* North-Holland, Amsterdam.

ISAACS, R. (1953). "Optimal Horse Race Bets." *American Mathematical Monthly*, 310-315.

JAGANNATHAN, R. and RAO, M.R. (1973). "A Class of Nonlinear Chance-Constrained Programming Models With Joint Constraints." *Operations Research, 21*, 360-364.

KALLBERG, J.G. and ZIEMBA, W.T. (1978). "Comparison of Alternative Utility Functions in Portfolio Selection Problems." Faculty of Commerce Working Paper No.609, University of British Columbia.

KALLBERG, J.G. and ZIEMBA, W.T. (1979). "On the Robustness of the Arrow-Pratt Risk Aversion Measure." *Economics Letters, 2*, 21-26.

KALLBERG, J.G. and ZIEMBA, W.T. (1980). "A Modified Frank-Wolfe Algorithm Using Function and Gradient Approximations with Application to Portfolio Selection." In *Recent Results in Stochastic Programming*, (P. Kall and A. Prékopa, eds.), Springer-Verlag, N.Y., 139-159.

KALLBERG, J.G. and ZIEMBA, W.T. (1981). "An Algorithm for Portfolio Revision: Theory, Computational Algorithm, and Empirical Results." In *Applications of Management Science, Vol.1*, (R.L. Schultz, ed.), JAI Press Inc., Greenwich, 267-292.

KATAOKA, S. (1963). "A Stochastic Programming Model." *Econometrica, 31*, 181-196.

KARLIN, S. and RINOTT, Y. *Multivariate Monotonicity, Generalized Convexity and Ordering Relations, With Applications in Analysis and Statistics.* (Forthcoming.)

KARLIN, S. and RINOTT, Y. (1981). "Univariate and Multivariate Total Positivity, Generalized Convexity and Related Inequalities." This volume, 703-718.

KLINGER, A. and MANGASARIAN, O.L. (1968). "Logarithmic Convexity and Geometric Programming." *Journal of Math. Analysis and Applications, 24*, 388-408.

LINTNER, J. (1965). "The Valuation of Risk Assets and the Selection of Risky Investments in Stock Portfolios and Capital Budgets." *Review of Economics and Statistics, 47*, 13-37.

LINTNER, J. (1972). "Equilibrium in a Random Walk and Lognormal Securities Market." Discussion Paper Number 235. Harvard Institute of Economics Research, Cambridge, Mass.

MANGASARIAN, O.L. (1971). "Convexity, Pseudo-Convexity and Quasi-Convexity of Composite Functions." *Cahiers du Centre d'Études de Recherche Operationelle, 12*, 114-122.

MARKOWITZ, H.M. (1952). "Portfolio Selection." *Journal of Finance, 6*, 77-91.

MARSHALL, A.W. and OLKIN, I. (1979). *Inequalities: Theory of Majorization and Its Applications.* Academic Press, New York.

MARTIN, D.H. (1981). "Connectedness of Level Sets as a Generalization of Concavity." This volume, 95-107.

MILLER, B.L. and WAGNER, H.M. (1965). "Chance-Constrained Programming With Joint Constraints." *Operations Research, 13*, 930-945.

OHLSON, J.A. (1972). "Optimal Portfolio Selection in a Lognormal Market When the Investor's Utility Function is Logarithmic." Research Paper No.117, Graduate School of Business, Stanford University.

OHLSON, J.A. and ZIEMBA, W.T. (1976). "Portfolio Selection in a Lognormal Market When the Investor has a Power Utility Function." *J. Financial Quant. Analysis, 11,* 57-71.

PARIKH, S.C. (1968). "Notes on Stochastic Programming." Mimeo, University of California, Berkeley.

PRATT, J. (1964). "Risk Aversion in the Small and in the Large." *Econometrica, 32,* 122-136.

PRÉKOPA, A. (1971). "Logarithmic Concave Measures with Application to Stochastic Programming." *Actu. Sci. Math., 32,* 301-316.

PRÉKOPA, A. (1972). "A Class of Stochastic Programming Decision Problems." *Math. Operationsforsch. Statist., 3,* 349-354.

PRÉKOPA, A. (1974). "Programming Under Probabilistic Constraints with a Random Technology Matrix." *Math. Operationsforsch. Statist., 5,* 109-116.

PRÉKOPA, A. ed. (1978). *Studies in Applied Stochastic Programming I.* Tanulmanyok 80, Budapest.

PRÉKOPA, A. (1980). "Logarithmic Concave Measures and Related Topics." In *Stochastic Programming,* (M.A.H. Dempster, ed.), Academic Press, New York, 63-82.

PRÉKOPA, A. and KELLE, P. (1978). "Reliability Type Inventory Models Based on Stochastic Programming." *Math. Prog. Study, 9,* 43-58.

PRÉKOPA, A. and SZANTAI, T. (1978). "Flood Control Reservoir System Design Using Stochastic Programming." *Math. Prog. Study, 9,* 138-151.

PRÉKOPA, A., GANCZER, S., DEAK, I. and PATYI, K. (1980). "The Stabil Stochastic Programming Model and Its Experimental Application to the Electrical Energy Sector of the Hungarian Economy." In *Stochastic Programming,* (M.A.H. Dempster, ed.), Academic Press, New York, 369-385.

PRESS, S.J. (1972). "Multivariate Stable Distributions." *Journal of Multivariate Analysis, 2,* 444-463.

RINOTT, Y. (1974). "Some Multivariate Concepts of Monotonicity, Convexity and Total Positivity with Applications to Probability Inequalities." Ph.D. thesis, Weizmann Inst. of Science, Israel.

RINOTT, Y. (1976). "On Convexity of Measures." *The Annals of Probability, Vol.4, No.6,* 1020-1026.

ROSENBERG, B. (1973). "The Behaviour of Random Variables with Non-Stationary Variance and the Distribution of Security Prices." Unpublished paper, University of California, Berkeley.

SCHAIBLE, S. (1974). "Parameter-Free Convex Equivalent and Dual Programs of Fractional Programs." *Zeitschrift für Operations Research, 18,* 187-196.

SCHAIBLE, S. (1977). "A Note on the Sum of a Linear and a Linear-Fractional Function." *Naval Research Logistics Quarterly,* 691-693.

SCHAIBLE, S. (1978). *Analyse und Anwendungen von Quotientenprogrammen.* Hain-Verlag, Meisenheim.

SCHAIBLE, S. and ZIEMBA, W.T. (1981). "On the Concavity Properties of a Fractional Program Arising in Portfolio Theory." UBC Faculty of Commerce Working Paper, July.

SCHOENBERG, I.J. (1951). "On Polya Frequency Functions." *J. Anal. Math., 1,* 331-374.

STONE, B.K. (1970). *Risk, Return and Equilibrium.* MIT Press, Cambridge, Massachusetts.

SNYDER, W.W. (1978). "Horse Racing: Testing the Efficient Markets Model." *Journal of Finance, 33,* 1109-1118.

VEINOTT, A.F. (1967). "The Supporting Hyperplane Method for Unimodal Programming." *Operations Research,* 147-152.

ZANG, I. (1981). "Concavifiability of C^2 Functions: A Unified Exposition." This volume, 131-152.

ZIEMBA, W.T. (1974). "Choosing Investment Portfolios When the Returns Have Stable Distributions." In *Mathematical Programming in Theory and Practice,* (P.L. Hammer and G. Zoutendijk, eds.), North-Holland Publishing Co., Amsterdam.

ZIEMBA, W.T. (1978). "Portfolio Applications: Computational Aspects." In *Stochastic Dominance,* (G.A. Whitmore and M.C. Findlay, eds.), Lexington Books, D.C. Heath, Lexington, Mass., 197-260.

ZIEMBA, W.T., PARKAN, C. and BROOKS-HILL, F.J. (1974). "Calculating Investment Portfolios With Risk Free Borrowing and Lending." *Management Science, 27,* 209-222.

ZIEMBA, W.T. and VICKSON, R.G., eds. (1975). *Stochastic Optimization Models in Finance.* Academic Press Inc., New York.

ZIEMBA, W.T. and HAUSCH, D.B. (1982). *Beat the Races.* (Forthcoming.)